Environmental Studies and Climate Change

Currently, anthropogenic activities have caused unprecedented destruction of the environment at alarming rates, leading to undesirable alterations in air, land, and water. The process of environmental degradation has been accelerated by industrial processes, which result in waste as well as over-consumption of natural resources. The ecological balance has been disturbed, and resources have shrunk. All this has resulted in climate change, which has emerged as a major concern in the 21st century. Changes in the environment are driven by demand for energy, water, and food to raise the standard of living. These are also responsible for climate change, with contributions from deforestation and CO_2 emissions from fossil fuels such as coal and petroleum.

The present volume discusses some of the main issues regarding environmental degradation and the causes as well as the impact of climate change, which is impacting the ecosystem. The effects of various pollutants, causes of climate change with case studies on geochemistry and glaciers, etc., and measures to reduce the impact on biodiversity, health, etc. are discussed in detail in its chapters.

In a nutshell, this volume discusses in detail the following issues:

- Anthropogenic and natural factors in environmental degradation
- Climate change history, causes, and threats to abiotic and biotic systems
- Case studies on the impact of climate change and living systems
- Mitigation and preparedness for the future

Translating Animal Science Research
Series Editor: RC Sobti, Department of Biotechnology, Panjab University, Chandigarh, India

Biodiversity: Threats and Conservation
RC Sobti

Environmental Studies and Climate Change
RC Sobti, S.K. Malhotra, Kamal Jaiswal, and Sanjeev Puri

Genomics, Proteomics and Biotechnology
RC Sobti, Manishi Mukesh, and Aastha Sobti

Biomedical Research, Medicine and Disease
RC Sobti and Aastha Sobti

For more information, visit webpage: www.routledge.com/Translating-Animal-Science-Research/book-series/TASR?pd=published,forthcoming&pg=1&pp=12&so=pub&view=list?pd=published,forthcoming&pg=1&pp=12&so=pub&view=list

Environmental Studies and Climate Change

Edited by
Prof R.C. Sobti, Prof S.K. Malhotra, Prof Kamal Jaiswal, and Prof Sanjeev Puri

CRC Press
Taylor & Francis Group
Boca Raton London New York

CRC Press is an imprint of the
Taylor & Francis Group, an informa business

First edition published 2023
by CRC Press
6000 Broken Sound Parkway NW, Suite 300, Boca Raton, FL 33487-2742

and by CRC Press
4 Park Square, Milton Park, Abingdon, Oxon, OX14 4RN

CRC Press is an imprint of Taylor & Francis Group, LLC

Library of Congress Cataloging-in-Publication Data
Names: Sobti, R. C., editor.
Title: Environmental studies and climate change / edited by RC Sobti,
S.K. Malhotra, Kamal Jaiswal, and Sanjeev Puri.
Description: First edition. | Boca Raton, FL : CRC Press, 2023. |
Series: Translating animal science research | Includes bibliographical references and index.
Identifiers: LCCN 2022023848 (print) | LCCN 2022023849 (ebook) |
ISBN 9781032115535 (hbk) | ISBN 9781032116327 (pbk) |
ISBN 9781003220824 (ebk)
Subjects: LCSH: Climatic changes. |
Environmental degradation. | Environmental risk assessment.
Classification: LCC QC903 .E59 2023 (print) |
LCC QC903 (ebook) | DDC 577.27–dc23/eng20221006
LC record available at https://lccn.loc.gov/2022023848
LC ebook record available at https://lccn.loc.gov/2022023849

ISBN: 9781032115535 (hbk)
ISBN: 9781032116327 (pbk)
ISBN: 9781003220824 (ebk)

DOI: 10.1201/9781003220824

Typeset in Times
by Newgen Publishing UK

Contents

PART II Palaeoclimate Studies and Computational Analysis of Climate Change

PART III Climate Change: Challenges and Management Strategies

Acknowledgements

R.C. Sobti acknowledges the Indian National Science Academy, New Delhi for providing a platform to continue academic pursuits under the Senior Scientists Programme.

R.C. Sobti is also thankful to his wife Dr. Vipin Sobti and daughters Er. Aditi and Dr. Aastha (and their spouses Er. Vineet and Er. Ankit, respectively) for their support in preparing the manuscript of the book. Special thanks to granddaughter Irene for her constant support.

Editor Biographies

Ranbir Chander Sobti, former Vice-Chancellor Panjab University, Chandigarh & Babasaheb Bhimrao Ambedkar University (Central University), Lucknow, is a scientist, an able administrator, and a dynamic institution builder.

He has published more than 300 papers in journals of national and international repute, such as *Mutation Research*, *Carcinogenesis*, *Archives of Toxicology*, *Cancer Genetics and Cytogenetics*, *Molecular Cell Biochemistry*, and *PLOS ONE*, to name a few. He has also had more than 50 books published by international publishers such as Springer, Academic Press, CRC Press, Elsevier, and others.

He is a Fellow of the Third World Academy of Sciences, National Academy of Sciences India, Indian National Science Academy, National Academy of Medical Sciences, National Academy of Agricultural Sciences, Canadian Academy of Cardiovascular Diseases, and a few others. He was the General President of the Indian Science Congress for the 102nd session held at University of Jammu in 2013. Dr. Sobti is the recipient of many prestigious awards, such as the Indian National Science Academy Young Scientist Medal, the University Grants Commission Career Award, Punjab Rattan Award, JC Bose Oration and Sriram Oration Awards, Life Time Achievement Awards of the Punjab Academy of Sciences, Zoological Society of India, and the Environment Academy of India, besides many other medals and awards at national and international levels.

Professor Sandeep K. Malhotra has been associated with teaching and research at the University of Allahabad for the past 46 years, during which he worked in the Department of Biotechnology and successfully handled a large number of research projects sanctioned by national agencies. He has the distinction of being decorated with 15 Gold Medals for pursuing research interests, the silver lining being the University Grants Commission Career Award. He has one patent, no. 282490, and more than 250 research publications, including 9 books and 13 reviews, to his credit. He is a Fellow of various associations.

Professor Kamal Jaiswal, Department of Applied Animal Science, Babasaheb Bhimrao Ambedkar University, Lucknow, has over 23 years' teaching and research experience and is an expert in parasitology and sericulture. He has published over 75 research papers in various reputed national/international research journals. He is the recipient of fellowships of the Zoological Society of India and the Society of Life Sciences and a number of awards/medals. He has successfully completed a World Bank-aided project on identification of "silk worm raised for UP's climate condition."

Professor Sanjeev Puri, Professor of Biotechnology, University Institute of Engineering & Technology and founder of Coordinator Centre for Stem Cell & Tissue Engineering, Panjab University, Chandigarh, India, has published more than 90 peer-reviewed research publications in national/international journals, as well as many book chapters, and edited more than five books. He has published two national patents and identified polycyclic aromatic hydrocarbon-hydrolyzing microorganisms for soils. Dr. Puri has active collaborations with both national and international institutes. He has ***been*** editor and also on the editorial board/reviewers' panel of reputed journals.

Contributors

Sachin Agarwal
Environmental Studies, SOITS
Indira Gandhi National Open University
New Delhi, India

Monika Aggarwal
University Institute of Applied Management Sciences
Panjab University
Chandigarh, India

Shikha Ahalavat
Department of Zoology & Aquaculture
Haryana Agriculture University
Hisar, Haryana, India

Fayaz Ahmad
Department of Zoology
University of Kashmir
Srinagar, Kashmir, India

Imtiaz Ahmed
Department of Zoology
University of Kashmir
Srinagar, Kashmir, India

Ahmad Ali
Department of Zoology
Panjab University
Chandigarh, India

Tufail Banday
Sher-e-Kashmir, University of Agricultural Science and Technology
Srinagar, Jammu and Kashmir, India

Chandra Bhan
Laboratory of Environmental Nanotechnology and Bioremediation
Department of Environmental Science
Babasaheb Bhimrao Ambedkar University
India

Sushil Kumar Bharti
Department of Geology
Babasaheb Bhimrao Ambedkar University (A Central University)
Lucknow, Uttar Pradesh, India
Swami Vivekanand Subharti University
Meerut, India

N.S. Bisht
Department of Botany and Microbiology
Hemvati Nandan Bahuguna Garhwal (Central) University Srinagar (G)
Uttarakhand, India

Nibedita Borgohin
Department of Nematology
Assam Agricultural University
Jorhat, Assam, India

Sunanda Burman
Mosquito, Microbiology and Nanotechnology Research Units
Parasitology Laboratory
Department of Zoology
The University of Burdwan
Purba Bardhaman, West Bengal, India

Latchoumycandane Calivarathan
Molecular Pharmacology and Toxicology Laboratory
Department of Life Sciences
School of Life Sciences, Central University of Tamil Nadu
Neelakudi Campus
Thiruvarur, India

Goutam Chandra
Mosquito, Microbiology and Nanotechnology Research Units
Parasitology Laboratory
Department of Zoology, The University of Burdwan
Purba Bardhaman, West Bengal, India

Nisha Changotra
GPS Nihalpur
Department of School Education
Government of Jammu and Kashmi
Kathua, Jammu and Kashmir, India

Ashok Kumar Chaubey
Department of Zoology
Chaudhary Charan Singh University
Meerut, India

Reshu Chaudhary
Department of Biotechnology
Keral Verma Subharti College of Science
Swami Vivekanand Subharti University
Meerut, Uttar Pradesh, India

Archana Chauhan
Department of Zoology
Panjab University
Chandigarh, India

Indu Chhabra
Department of Computer Applications
Panjab University
Chandigarh, India

Rasy Fayaz Choh
Department of Zoology
University of Kashmir
Srinagar, India

Amit Krishna De
The Indian Science Congress Association
Kolkata, India

S. Dhar
Department of Environmental Sciences
Central University of Jammu
Samba, Jammu and Kashmir, India

Phuntsog Dolma
Department of Zoology
Panjab University
Chandigarh, India

Divya Dubey
Department of Environmental Science
School of Earth and Environmental Sciences
Babasaheb Bhimrao Ambedkar University
Lucknow, India

Vineet Kumar Dubey
Wildlife Institute of India
Dehradun, Uttarakhand, India

Venkatesh Dutta
Department of Environmental Science
School of Earth and Environmental Sciences
Babasaheb Bhimrao Ambedkar University
Lucknow, India

Francesco Fazio
Department of Veterinary Sciences
University of Messina
Messina, Italy

Alok Sagar Gautam
Department of Physics
Hemvati Nandan Bahuguna Garhwal University
Uttarakhand, India

Sneha Gautam
Karunya Institute of Technology and Sciences Coimbatore
Tamil Nadu, India

Anjani R.K. Gollakota
National Yunlin University of Science and Technology
Douliou
Yunlin County, Taiwan, ROC

Anu Gupta
Department of Computer Applications
Panjab University, Chandigarh, India

Aadil Hussain
College of Temperate Sericulture
Sher-e-Kashmir University of Agricultural Sciences and Technology
Jammu and Kashmir, India

Aamina Hussan
Department of Zoology
University of Kashmir
Kashmir, India

Kashifa Iqbal
Department of Environmental Science
Babasaheb Bhimrao Ambedkar University (Central)
Lucknow, India

Kousar Jain
Department of Zoology
University of Kashmir
Srinagar, Kashmir, India

Neeshma Jaiswal
Department of Zoology
Babasaheb Bhimrao Ambedkar University
Lucknow, India

Pranitha Janardhanan
Department of Biochemistry and Molecular Biology
School of Life Sciences
Pondicherry University
Pondicherry, India

Jeyaraj Antony Johnson
Wildlife Institute of India
Dehradun, Uttarakhand, India

Anuragini Kadwalia
Department of Zoology
Panjab University
Chandigarh, India

Neerja Kapoor
Department of Zoology
CMP PG Constituent College (A Constituent College of the University of Allahabad)
Prayagraj, Uttar Pradesh, India

Manpreet Kaur
Energy Research Centre
Panjab University
Chandigarh, India

Anupama Kaushik
Dr. SSB University Institutes of Chemical Engineering and Technology
Panjab University
Chandigarh, India

Irfan Latief Khan
Shere-e-Kashmir
College of Temperate Sericulture
Keral Verma Subharti College of Science
Jammu and Kashmir, India

Ashu Khosla
Department of Geology
Panjab University, Chandigarh, India

Ibraq Khurshid
Department of Zoology
Central University of Kashmir
Ganderbal, Kashmir, India

Richa Kothari
Department of Environmental Sciences
Central University of Jammu
Samba, Jammu and Kashmir, India

Gagandeep Kour
Department of Environmental Sciences
Central University of Jammu
Samba, Jammu and Kashmir, India

Amit Kumar
Department of Biotechnology
Keral Verma Faculty of Science
Swami Vivekanand Subharti University
Meerut, India

Narendra Kumar
Department of Environmental Science
Babasaheb Bhimrao Ambedkar University (A Central University)
Lucknow, Uttar Pradesh, India

Puneet Kumar
Department of Zoology
Aligarh Muslim University
Aligarh, Uttar Pradesh, India

Rajeev Kumar
Department of Agricultural Botany (Genetics and Plant Breeding)
C.S.S.S. (P.G.) College
Meerut, Uttar Pradesh, India

Santosh Kumar
Department of Geology
Centre of Advanced Study
Kumaun University
Nainital, India

Saroj Kumar
Department of Environmental Science
School of Earth and Environmental Sciences
Babasaheb Bhimrao Ambedkar University
Lucknow, India

Keshav Malhotra
Department of Evening Studies – Multi Disciplinary Research Centre
Panjab University
Chandigarh, India

Meenu Maniradhan
Molecular Pharmacology and Toxicology Laboratory
Department of Life Sciences
School of Life Sciences
Central University of Tamil Nadu
Neelakudi Campus
Thiruvarur, India

M.R. Mir Mehreen Manzoor
College of Temperate Sericulture
Sher-e-Kashmir University of Agricultural Sciences and Technology
Jammu and Kashmir, India

Sameena Maqbool
College of Temperate Sericulture
Sher-e-Kashmir University of Agricultural Sciences and Technology
Jammu and Kashmir, India

Premendu Prakash Mathur
Department of Biochemistry and Molecular Biology
School of Life Sciences
Pondicherry University
Pondicherry, India

Shaista Mehraj
College of Temperate Sericulture
Sher-e-Kashmir University of Agricultural Sciences and Technology
Jammu and Kashmir, India

Umer Aziz Mir
Department of Zoology
University of Kashmir
Srinagar, India

Minakshi Mishra
Department of Statistics
Babasaheb Bhimrao Ambedkar University
Lucknow, Uttar Pradesh, India

Anshuman Misra
Department of Geology
Kumaun University
Nainital, India

Archna Narula
M S Ramaiah Institute of Technology
Bangalore, India

Shaista S. Nissa
Sher-e-Kashmir University of Agriculture Sciences and Technology
Srinagar, Jammu and Kashmir, India

Veena Pande
Department of Biotechnology
Kumaun University
Sir J.C. Bose Technical Campus
Nainital, India

Abhishek Pandey
Department of Geology
Centre of Advanced Study
Kumaun University
Nainital, India

Pooja Pant
Department of Biotechnology
Kumaun University
Sir J.C. Bose Technical Campus
Nainital, India

D. Pathania
Department of Environmental Sciences
Central University of Jammu
Samba, Jammu and Kashmir, India

Rajeev Patnaik
Department of Geology
Panjab University
Chandigarh, India

M. Prashanth
Geology Discipline Group
School of Sciences
Indira Gandhi National Open University
New Delhi, India

Bhanu Pratap
Department of Environmental Sciences
Babasaheb Bhimrao Ambedkar University
Lucknow, India

Pushpraj
Department of Statistics
Babasaheb Bhimrao Ambedkar University
Lucknow, Uttar Pradesh, India

Shubham Raina
Department of Environmental Sciences
Central University of Jammu
Samba, Jammu and Kashmir, India

M.I. Rather
Department of Environmental Sciences
University of Kashmir
Srinagar, Jammu and Kashmir, India

Gaurav Rattan
Dr. SSB University Institutes of Chemical Engineering and Technology
Panjab University
Chandigarh, India

Yogesh Kumar Rawal
Department of Zoology
Panjab University
Chandigarh, India

Jyoti Rawat
Department of Biotechnology
Kumaun University
Sir J.C. Bose Technical Campus
Nainital, India

Reshu
Department of Zoology
Babasaheb Bhimrao Ambedkar University
Lucknow, India

Boyina Rupini
Environmental Studies, SOITS
Indira Gandhi National Open University
New Delhi, India
and
C.S.S.S. (P.G.) College
Meerut, Uttar Pradesh, India

Rana Samad
Department of Zoology
Kirori Mal College
University of Delhi
Delhi, India

Dinesh K. Saxena
Department of Botany
Bareilly College
Bareilly, India

Ruchi Saxena
Department of Chemistry
Narishiksha Niketan (PG) College
Lucknow, India

R.S. Sengar
Department of Agriculture Biotechnology
College of Agriculture
Sardar Vallabhbhai Patel University of Agriculture and Technology
Meerut, India

Syed Shanaz
Sher-e-Kashmir University of Agriculture Sciences and Technology
Srinagar, Jammu and Kashmir, India

Aashna Sharma
Wildlife Institute of India
Dehradun, India

Isha Sharma
Department of Botany
School of Life Science
Cluster University Jammu
Jammu, India

R.K. Sharma
College of Temperate Sericulture
Sher-e-Kashmir University of Agricultural Sciences and Technology
Jammu and Kashmir, India

Zubair Ahmad Sheikh
Department of Zoology
University of Kashmir
Srinagar, Jammu and Kashmir, India

Chi-Min Shu
National Yunlin University of Science and Technology
Douliou
Yunlin County, Taiwan, ROC

Anita Singh
Department of Environmental Sciences
Central University of Jammu
Samba, Jammu and Kashmir, India

Har Mohan Singh
School of Energy Management
Shri Mata Vaishno Devi University
Jammu, Jammu and Kashmir, India

Jiwan Singh
Laboratory of Environmental Nanotechnology and Bioremediation
Department of Environmental Science
Babasaheb Bhimrao Ambedkar University
Lucknow, India

Seema Singh
Department of Zoology
Nari Shiksha Niketan (PG) College
Lucknow, India

Sonal Singhal
Department of Chemistry
Panjab University
Chandigarh, India

Kuppusamy Sivakumar
Wildlife Institute of India
Dehradun, Uttarakhand, India

Rashmi Srivastava
Department of Zoology
Babasaheb Bhimrao Ambedkar University
Lucknow, India

Shahana Riyaz Tramboo
Sher-e-Kashmir University of Agriculture Sciences and Technology
Srinagar, Jammu and Kashmir, India

Amit Tripathi
Department of Zoology
University of Lucknow
Uttar Pradesh, India

Arjun Tyagi
Department of Electrical Engineering
Netaji Subhas University of Technology
New Delhi, India

V.V. Tyagi
School of Energy Management
Shri Mata Vaishno Devi University
Jammu, Jammu and Kashmir, India

Akshat Uniyal
Department of Botany and Microbiology
Hemvati Nandan Bahuguna Garhwal (Central) University Srinagar (G)
Uttarakhand, India

Priyanka Uniyal
School of Pharmaceutical Sciences and Technology
Sardar Bhagwan Singh University
Uttarakhand, India

Sravanthi Veluturla
M S Ramaiah Institute of Technology
Bangalore, India

Omkar Verma
Geology Discipline Group
School of Sciences
Indira Gandhi National Open University
New Delhi, India

Anita Yadav
Department of Zoology
CMP PG Constituent College (A Constituent College of the University of Allahabad)
Prayagraj, Uttar Pradesh, India

Awadesh Kumar Yadav
Department of Geology
Babasaheb Bhimrao Ambedkar University
Lucknow, India

U.R. Zargar
Department of Zoology
Government Degree College
Anantnag, India

Part I

Environment Management and Monitoring

1 Environment Studies and Climate Change

A Bibliometric Analysis

Monika Aggarwal
University Institute of Applied Management Sciences, Panjab University, Chandigarh

Keshav Malhotra
Department of Evening Studies – Multi Disciplinary Research Centre, Panjab University, Chandigarh

R.C. Sobti
Department of Biotechnology, Panjab University, Chandigarh

CONTENTS

1.1 INTRODUCTION: ENVIRONMENT AND CLIMATE CHANGE

Earth has undergone significant environmental changes, but the past 10,000 years have been unusual (Rockstrom et al., 2009). The literature is quite rich in the context of providing evidence of the ecological impact of climate change on earth (Walther et al., 2002), whether it is marine (Jackson et al., 2001), freshwater (Oki and Kanae, 2006), air, soil, forests, food systems (Godfray et al., 2010), etc. There are emissions of greenhouse gasses as never before. According to the United Nations Environment Programme (UNEP) website, the temperature on earth will rise by a grave 3°C in this century. The curve can be reduced to 1.5°C if the emissions are cut by 30 gigatonnes annually within the timespan of this decade. Rising temperature is a threat that results in melting of glaciers and increase in sea level, further leading to disturbances in coastal regions. Carbon emissions pollute water and damage marine ecosystems. The significance of water as a resource for socio-economic development and health cannot be ignored (Xiang et al., 2021). Not only damage to marine life but the wildfires in Australia, California, the Congo Basin and the Amazon are also the result of climate change. This implies that there is a dire need to be efficient with resources, curbing deforestation, due to which agriproducts are becoming contaminated (Rajão et al., 2020), and fuel waste, along with significantly reducing emissions from all sectors. There is a need to make changes in lifestyle

DOI: 10.1201/9781003220824-2

and find nature-based solutions that encourage reforestation and a clean environment. The gravity of human-induced climate change depends not only on the magnitude of the change but also on the potential for irreversibility (Solomon et al., 2009). There is a ray of hope as well. Khan et al. (2020) opined that that the role of a reliable carbon emissions measurement is important for devising a relevant climate policy to deal with environmental problems. However, it has been stated that the emissions of the wealthiest 1% of the world are more than twice those of the poorest 50%; wealth can also play a key role in creating as well as solving the problem of environmental degradation. The results show that the successful implementation of technology to harness renewable energy in the energy sector of developing countries has been able to cut emission by 12.5 gigatonnes annually. Technology gives hope, because there are greener alternatives. Further, the Covid-19 crisis offers a good opportunity to formulate and implement low-carbon, green solutions and policies. Evidently, there is no time to lose. More and more countries are realizing that while making the net zero emission pledge under the 2015 Paris Agreement. The growing concern has boosted academic interest in the domain of environment studies and climate change, and this is evident from the research documents that have been published so far. The present study is an endeavour to conduct a bibliometric analysis on published research in this domain.

1.2 RESEARCH METHODOLOGY

The published documents were searched from the SCOPUS database using the keywords "Environment Studies" OR "Environment" OR "Environmental Studies" AND "Climate Change", and 21595 documents were found.

Table 1.1 shows the acceptance and rejection criteria for further processing of these documents. After following these criteria, 15767 articles were finally processed for analysis as stage-2 using VOSviewer software. The descriptive data of these 15767 articles are stated in the next section.

1.3 DESCRIPTIVE ANALYSIS

Table 1.2 shows the number of documents published from the year 1938 till 2022, including ahead of publication documents. The table shows that till the year 1990, only 15 documents, i.e. 0.10%, were published out of the total documents under study. From 1991 to 2000, 334 documents, i.e. 2%, were published. From the year 2001 to 2010, 17.64% (2781 documents) were published, and most papers (12637 documents) were published in this decade, i.e. from the year 2011 till August 2021. This shows that research in the domain of environment and climate change has only gained momentum in the last decade.

Considering the total time period, i.e. from the year 1938 till August 2021, it was found that the United States has published the maximum 6167 documents, followed by the United Kingdom with 2950 documents. China has produced 2627 documents, followed by Australia with 1859 documents.

Table 1.3 shows the list of countries producing more than 100 documents. India too has contributed 537 documents in SCOPUS journals. The domain also attracts a lot of funding support, as these publications are funded by around 159 organizations. The National Natural Science Foundation of China, the National Science Foundation, the European Commission, the Natural Environment Research Council and UK Research and Innovation are the top five funding agencies.

Table 1.4 shows the list of funding agencies that have funded more than 100 publications out of the total publications considered for the purpose of this chapter.

1.4 BIBLIOMETRIC ANALYSIS

1.4.1 Author's Keywords Co-occurrence Analysis

Co-occurrence analysis of author's keywords (full counting method) shows that out of 7068 keywords, 180 meet the threshold of minimum occurrence of 5 times.

TABLE 1.1
Acceptance and Rejection Criteria for Selection of Documents

Base	Acceptance	Rejection
Type of Document	Articles (16464)	Conference papers (2102) Review (1691) Book Chapter (383) Editorial (290) Note (256) Short Survey (236) Letter (111) Book (50) Retracted (5)
Source	Journal (19068)	Book Series (428) Book (366) Conference Proceedings (1646) Trade Journal (86) Undefined (1)
Language	English (20993)	Chinese (329) French (127) Spanish (63) German (62) Russian (27) Japanese and Portuguese (11 each) Italian (9) Korean (6)

Source: Author's compilation.
Note: Numbers given in brackets indicate the total documents.

TABLE 1.2
Documents Published by Year

Year	Number of Documents	Year	Number of Documents	Year	Number of Documents	Year	Number of Documents
1938	1	1993	17	2003	108	2013	952
1959	1	1994	30	2004	141	2014	877
1965	1	1995	29	2005	156	2015	1040
1968	1	1996	32	2006	233	2016	1213
1970	1	1997	31	2007	343	2017	1254
1974	1	1998	30	2008	423	2018	1373
1989	4	1999	60	2009	505	2019	1478
1990	5	2000	80	2010	664	2020	1568
1991	11	2001	87	2011	826	2021	1145
1992	14	2002	121	2012	910	2022	1

Source: Authors' compilation.

TABLE 1.3
Publications by Country

Country	No.	Country	No.	Country	No.	Country	No.
United States	6167	Sweden	736	South Africa	350	Argentina	154
United Kingdom	2950	Norway	652	Belgium	330	Taiwan	153
China	2627	Japan	572	New Zealand	322	Turkey	151
Australia	1859	India	537	Poland	295	Israel	146
Germany	1785	Denmark	508	Greece	260	Indonesia	144
Canada	1679	Brazil	438	South Korea	247	Romania	138
France	1330	Portugal	381	Chile	195	Ireland	134
Spain	1306	Finland	378	Czech Republic	191	Hong Kong	130
Italy	1243	Austria	372	Mexico	182	Iran	129
Switzerland	860	Russian Federation	366	Malaysia	162	Singapore	109
Netherlands	742					Pakistan	101

Source: Authors' compilation.

TABLE 1.4
Funding Agencies Sponsoring Research in Environment and Climate Change

National Natural Science Foundation of China	1227
National Science Foundation	1148
European Commission	787
Natural Environment Research Council	513
UK Research and Innovation	452
Seventh Framework Programme	432
Chinese Academy of Sciences	395
Ministry of Science and Technology of the People's Republic of China	286
Government of Canada	269
Natural Sciences and Engineering Research Council of Canada	249
National Key Research and Development Program of China	234
Australian Research Council	214
Deutsche Forschungsgemeinschaft	208
National Aeronautics and Space Administration	200
European Regional Development Fund	185
Horizon 2020 Framework Programme	166
National Oceanic and Atmospheric Administration	165
Ministry of Education, Culture, Sports, Science and Technology	154
U.S. Department of Energy	154
Department of Education and Training	151
Japan Society for the Promotion of Science	150
Conselho Nacional de Desenvolvimento Científico e Tecnológico	148
Ministerio de Economía y Competitividad	143
U.S. Department of Commerce	133
U.S. Department of the Interior	128
Fundação para a Ciência e a Tecnologia	120
Ministry of Education of the People's Republic of China	120
Bundesministerium für Bildung und Forschung	115
European Research Council	114

TABLE 1.4 (Continued)
Funding Agencies Sponsoring Research in Environment and Climate Change

Schweizerischer Nationalfonds zur Förderung der Wissenschaftlichen Forschung	114
Ministario da Ciancia, Tecnologiae Inovaao	112
National Basic Research Program of China (973 Program)	110
U.S. Geological Survey	110
Fundamental Research Funds for the Central Universities	108
U.S. Department of Agriculture	108
Norges Forskningsrad	102

Source: Authors' compilation.

TABLE 1.5
Author Keywords Co-occurrence Analysis

Keyword	Occurrences	Total Link Strength	Keyword	Occurrences	Total Link Strength
Climate Change	676	669	Thermal Comfort	17	22
Environment	108	160	Precipitation	16	31
Temperature	48	64	Sustainable Development	16	15
Global Warming	38	42	China	15	12
Remote Sensing	35	32	Phenology	15	17
Sustainability	32	58	Climate Variability	14	16
Drought	27	42	Renewable Energy	14	14
Resilience	26	42	Warming	13	15
Built Environment	24	38	Water Quality	13	16
Biodiversity	23	35	Climate Change Adaptation	12	12
Tibetan Plateau	22	31	Eutrophication	12	21
Arctic	21	28	Salinity	12	21
Climate	21	26	Conservation	11	16
Human Activities	19	24	Covid-19	11	7
Climate Warming	18	16	Ecosystem Services	11	11
Adaptation	17	31	Mediterranean	11	11
Holocene	17	15	Or In Environment And Climate Change	11	2
Life Cycle Assessment	17	11	Pollution	11	17

Source: Authors' compilation.

Table 1.5 shows the list of keywords that occurred more than 10 times. Other than climate change and environment, which were the key concern of the study, temperature, global warming remote sensing, sustainability, drought and resilience were the most frequently co-occurring keywords. Particularly in the Indian context, the studies have been related to keywords like climate change, environment, life cycle assessment, sustainability and Modis (NASA's MODIS instrument). The grey areas that need to be further explored are local adaptation, risk assessment, traditional knowledge, small-scale fisheries, water chemistry and social determinants of health.

Figure 1.1 shows the network of author's keywords co-occurrence analysis. Keywords in the same colour fall in the same cluster.

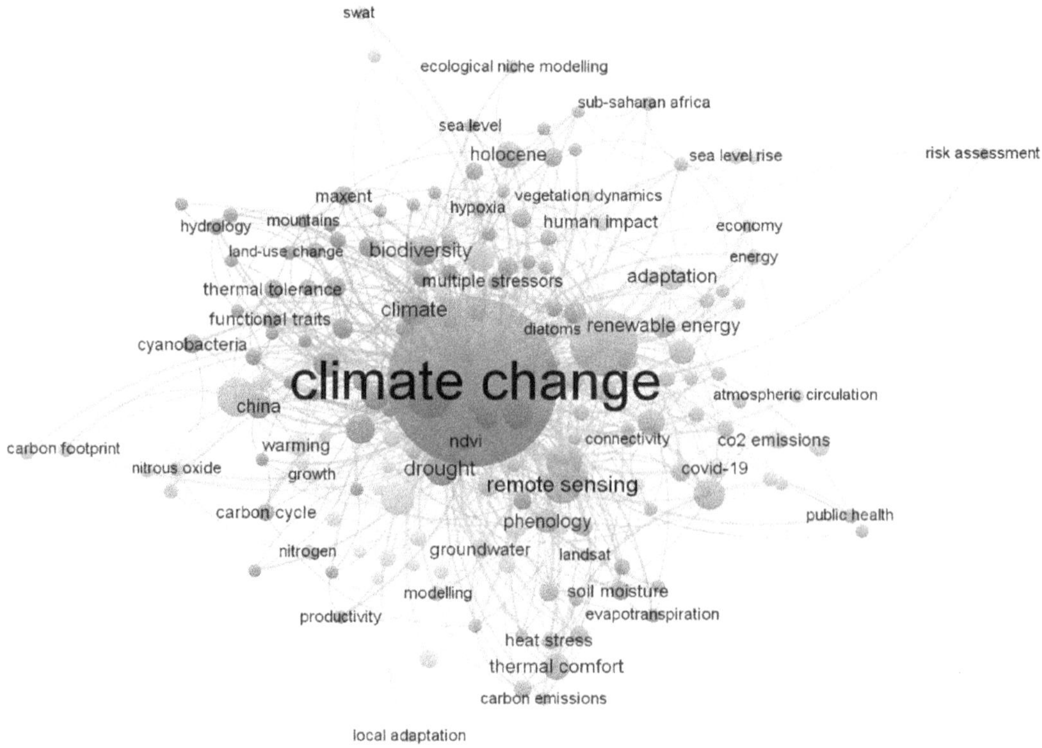

FIGURE 1.1 Author's keywords co-occurrence analysis.

(Authors' compilation using VOSviewer.)

1.4.2 Citation Analysis by Country (on the Basis of First 2000 Documents)

On the basis of recent data of the first 2000 documents, out of 180 countries, 66 meet the threshold of a minimum 5 citations. However, the largest set of connected items consists of 39 items, as shown in Figure 1.2

Figure 1.2 shows the citation analysis by country. These countries have been divided into seven clusters on the basis of documents cited as presented in Table 1.6. Cluster 1 comprises Germany, Sweden, Norway, Greece, the Netherlands, Poland, Belgium, Israel, Hungary, Latvia and Slovakia. Countries including China, Australia, Canada, Spain, Finland, Chile, Argentina, the Russian Federation, Nepal and New Zealand are in cluster 2. Cluster 3 is formed by the United States, Switzerland, Denmark, the Czech Republic, Iran, Austria and Bulgaria. Cluster 4 comprises the United Kingdom, France, South Africa, Thailand, Kenya and Tanzania. Cluster 5 comprises Italy, Brazil, Portugal, South Korea and Vietnam. China has produced the most documents (486) in recent years, with most citations as well (1279). The United States is at number two in terms of both documents produced and citations. Germany is at number three in terms of citations, followed by the United Kingdom. Germany is ahead in citations as compared with the United Kingdom, but the United Kingdom has produced more documents as compared with Germany. India has produced 70 papers in the last 2 years with 118 citations. But it is not in any cluster, as the total strength index is zero in the case of India.

Table 1.6 shows the list of countries, documents produced, number of citations and the cluster in which they fall. The following section provides a deep dive into the top 20 papers having more than 1300 citations. A summary of the same is presented in Table 1.7.

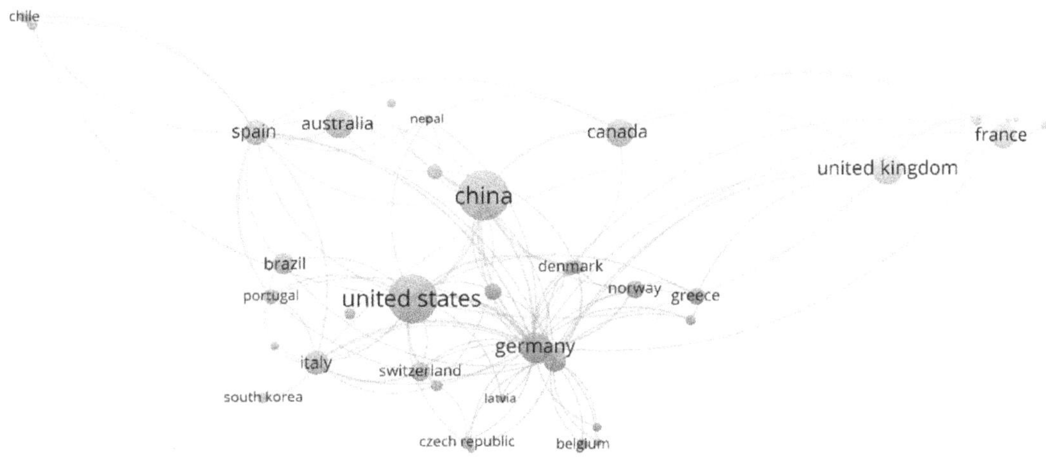

FIGURE 1.2 Citation analysis by country.

(Authors' compilation using VOSviewer.)

TABLE 1.6
Citation Analysis by Country

Country	Documents	Citations	Cluster	Country	Documents	Citations	Cluster
China	486	1279	2	Iran	32	80	3
United States	464	1195	3	Austria	33	73	3
Germany	193	494	1	Belgium	28	71	1
United Kingdom	217	476	4	Chile	30	70	2
Australia	133	448	2	South Korea	30	69	5
Canada	144	406	2	South Africa	37	68	4
Spain	140	349	2	Israel	14	57	1
France	117	341	4	Argentina	20	53	2
Italy	127	298	5	Thailand	13	48	4
Sweden	78	282	1	Russian Federation	28	43	2
Brazil	71	236	5	Nepal	13	42	2
Switzerland	78	195	3	New Zealand	32	42	2
Norway	54	170	1	Hungary	10	41	1
Greece	21	146	1	Latvia	5	41	1
Netherlands	48	143	1	Viet Nam	13	41	5
Denmark	47	133	3	Bulgaria	5	37	3
Finland	40	125	2	Slovakia	10	34	1
Portugal	45	117	5	Kenya	10	12	4
Poland	44	104	1	Tanzania	5	6	4
Czech Republic	32	84	3	India	70	118	

Source: Authors' compilation.

Table 1.7 shows the summary of the top 20 documents in terms of their citations. It also highlights the main focus of each document, name of journal and number of citations. In the context of climate change and humanity, Walther et al. (2002) found a consistent pattern of the impact of climate change across ecological systems, from the polar terrestrial to the tropical marine environment.

TABLE 1.7
Summary of Top 20 Documents

Sr. No.	Title	Journal	Citations	Focus
1	Ecological responses to recent climate change	*Nature*	6586	Ecology change
2	Food security: The challenge of feeding 9 billion people	*Science*	5576	Food security
3	A safe operating space for humanity	*Nature*	5510	Boundaries for human development
4	Historical overfishing and the recent collapse of coastal ecosystems	*Science*	4348	Impact of overfishing on coastal disturbance
5	Solutions for a cultivated planet	*Nature*	3742	Agriculture environment
6	Transformation of the nitrogen cycle: Recent trends, questions, and potential solutions	*Science*	3697	Global nitrogen cycle
7	Climate change, human impacts, and the resilience of coral reefs	*Science*	2537	Coral reefs
8	Rapid range shifts of species associated with high levels of climate warming	*Science*	2511	Shift in species
9	Genetic consequences of climatic oscillations in the Quaternary	*Philosophical Transactions of the Royal Society B: Biological Sciences*	2349	DNA
10	Climate-driven increases in global terrestrial net primary production from 1982 to 1999	*Science*	2291	Plant growth
11	Ethanol can contribute to energy and environmental goals	*Science*	2068	Biofuel
12	Global hydrological cycles and world water resources	*Science*	2036	Water cycles
13	Ecology: Climate change and distribution shifts in marine fishes	*Science*	1913	Marine
14	Recent developments in Life Cycle Assessment	*Journal of Environmental Management*	1755	Life cycle assessment tools
15	Climate warming and disease risks for terrestrial and marine biota	*Science*	1726	Infectious disease
16	Irreversible climate change due to carbon dioxide emissions	*Proceedings of the National Academy of Sciences of the United States of America*	1616	Carbon dioxide concentration
17	The impact of climate change on the world's marine ecosystems	*Science*	1604	Ocean
18	Climate: Blooms like it hot	*Science*	1554	Cyanobacterial blooms
19	Impact of climate change on marine pelagic phenology and trophic mismatch	*Nature*	1404	Plankton taxa
20	Climate-driven trends in contemporary ocean productivity	*Nature*	1376	Marine food web

Source: Authors' compilation.

Further, considering trends in global warming, the authors conclude that the ecological changes are clearly visible. Hewitt (2004) found that the associated population dynamics varied with life history and geography, and the present genetic constitution of populations and species carries attenuated signals of these past dynamics. Rockstrom et al. (2009) identified and quantified planetary boundaries that should not be crossed, because crossing certain biophysical thresholds could have disastrous consequences for humanity. In the context of water and ocean studies, Hoegh-Guldberg and Bruno (2010) conclude that climate change is clearly altering the ocean ecosystem. Godfray et al. (2010) observed that continuous growth of population leads to competition for land, water and energy, and overexploitation of fisheries, impacting the global ability to produce food. Further, the demand for food will increase for at least another 40 years. The authors recommended that a multidimensional linked global strategy is required to ensure sustainable and equity food security and to mitigate the threat of climate change. Similarly, Jackson et al. (2001) opined that palaeoecological, archaeological and historical data suggested that overfishing caused ecological extinction, and this has led to all other human disturbances to the coastal ecosystem, including pollution, water quality degradation and anthropogenic climate change. Hughes et al. (2003) concluded that the coral reefs are threatened globally due to the diversity and scale of human impact. The adoption of strong policy measures is required in order to reduce the rate of global warming. Oki and Kanae (2006) expected that climate change will accelerate water cycles. Due to this, there will be a slowing of the increase in people living under water stress. Considering data of 66 plankton taxa for the period from 1958 to 2002, Edwards and Richardson (2004) investigated the emergence of climate warning signals across trophic levels. They concluded that not only is the marine pelagic community responding to climate changes, but the level of response differs throughout the community and the seasonal cycle, leading to a mismatch between trophic levels and functional groups. Behrenfeld et al. (2006) found that the link between the physical environment and ocean biology functions through changes in upper-ocean temperature and stratification, which influence the availability of nutrients for phytoplankton growth. In the case of soil, Foley et al. (2011) found that our agricultural systems are continuously degrading land, water, biodiversity and climate on a global level. On the other hand, the increasing population is increasing demand for agriculture and natural resources. It was recommended that halting agricultural expansion, terminating harvesting on underperforming land, increasing crop efficiency, reducing waste and changing diets can help reduce agricultural environment footprints and enhance quality food production. Galloway et al. (2008) studied the nitrogen cycle and found that due to increased burning of fossil fuels, the rising demand for nitrogen in agriculture and industry and its inefficient use, the global nitrogen cycle is rapidly transforming. This is causing environmental and human health problems. Nemani et al. (2003) explored the responses of vegetation to climatic changes by analysing data for 18 years (1982 to 1999) pertaining to climatic and satellite observations of vegetation activity. The authors found that global climate changes have eased several critical climatic constraints on plant growth. The reasons identified included decreased cloud cover and increased solar radiation. Chen et al. (2011) used meta-analysis to estimate the distribution of species. The authors found that distribution is shifting to higher elevation at a rate two to three times faster as compared with previous reporting. This is due to external drivers of change, including global warming, and multiple internal traits of individual species themselves. Farrell et al. (2006) evaluated six representative analyses of fuel ethanol to study the potential effect of increased use of biofuel. The authors concluded that there is a lack of understanding of the environmental effects of biofuel production. Studies reporting negative net energy used obsolete data. It was concluded that cellulosic technology is required for large-scale use of ethanol for fuel. Finnveden et al. (2009) studied the strengths and weaknesses and recent developments in life cycle assessment tools used to access the environmental impact and resources used in a product's life cycle. Harvell et al. (2002) opined that there may be population decline and species extinction due to infectious diseases. Climate warming may increase pathogen development and disease transmission. The ability to predict epidemics in the world population can be improved if the independent

TABLE 1.8
Top Cited Recent Publications

Document	Citations
Fadare O.O. (2020)	129
Kanniah K.D. (2020)	101
Khan Z. (2020)	89
Rajão R. (2020)	57
Xiang X. (2021)	40

Source: Authors' compilation.

and interactive effects of multiple climate drivers on disease impact can be separated. Solomon et al. (2009) concluded that increased carbon dioxide concentration results in climate change, which is largely irreversible for 100 years after emission stops. Paerl and Huisman (2008) established that a link exists between global warming and the worldwide proliferation of harmful cyanobacterial blooms.

A shift has been observed in recent studies published in the year 2020 onwards. Table 1.8 shows the top five most cited recent publications after the year 2020.

Table 1.8 shows the list of the top five authors producing the top cited papers published in the years 2020 and 2021. Studies related to Covid-19, environment and climate change include Fadare and Okoffo (2020) with the most citations. The researchers found that the presence of plastics in the environment has contributed significantly to climate change, specifically because of carbon emission and its risk associated with the global food chain. The uncategorized disposal of face masks used during the Covid-19 pandemic added a further threat to the environment, as plastic particles propagate microbes such as invasive pathogens. There is no unified international regulation on plastic pollution management. The authors also opined that plastic pollution may be the next world pandemic. Similarly, Kanniah et al. (2020) also focused on air pollution, climate change and Covid-19. The researchers quantified the changes in aerosol and air pollutants associated with the general shutdown in Malaysia and Southeast Asia during the Covid-19 pandemic using aerosol optical depth (AOD) observations from the Himawari-8 satellite, along with tropospheric NO_2 column density from Aura-OMI. A substantial decrease in AOD over the sea and in pollution over the oceanic region was found. As compared with 2018 and 2019, during lockdown, a notable decrease in PM_{10}, $PM_{2.5}$, NO_2, SO_2 and CO concentrations was observed in Malaysia. On carbon emissions and environment, Khan et al. (2020) used second-generation panel co-integration methodologies to explore the unidentified determinants of CO_2 emissions in G7 countries from the year1990 to 2017. It was found that a stable long-run relationship exists between CO_2 emissions, trade, income, environmental innovation and renewable energy consumption. Further, policy with a focus on renewable energy sources significantly affects consumption-based CO_2 emissions and vice versa. On environmental crises due to deforestation, Rajão et al. (2020) examined the interlinkage between illegal deforestation in Amazon and Cerrado. It was found that 2% of the property in these areas is responsible for 62% of all potentially illegal deforestation. Further, 20% of soy and 17% of beef are contaminated. Xiang et al.(2021) conclude that climate change will intensify the specific challenges in water resource management.

1.6 CONCLUSION

There is ample literature support for the harmful impact of climate change on the environment. This human-induced change can only be solved by human intervention. The present research focuses on the documents published in this domain in SCOPUS journals. The descriptive data shows that the research in environment studies and climate change gained momentum in recent decades, particularly after the year 2011. In terms of number of publications, the United States has published the greatest number of documents, followed by the United Kingdom, China, Australia, Germany, Sweden, Norway, Japan and India. Moving on to the analysis of the 2000 most recent documents, it was found that China is taking the lead in publication, followed by the United States, the United Kingdom, Germany and Australia. The research has attracted numerous funding agencies. The National Natural Science Foundation of China, the National Science Foundation and the European Commission are the top funding agencies. Bibliometric analysis shows that climate change, environment, temperature, global warming and remote sensing are the most frequently used keywords in the literature in the domain under study. Further, there are few environmental studies in the domain of environment and climate change related to local adoption, risk assessment, traditional knowledge, small-scale fisheries, water chemistry and social determinants of health. It is recommended that these areas can be explored further. The focus of the top 20 most cited papers is on ecology change, food security, boundaries for human development, impact of overfishing on coastal disturbance, agriculture environment, global nitrogen cycle, coral reefs, shift in species, DNA, plant growth, biofuel, water cycles, marine, life cycle assessment tools, infectious disease, carbon dioxide concentration, ocean, cyanobacterial blooms and plankton taxa. Recent studies are concentrating on CO_2 emissions, Covid-19-related issues such as plastic, face masks and the impact of lockdown on the environment, and climate change. The literature recommends that further studies can focus on eco-friendly alternatives to find sustainable solutions to plastic waste (Fadare and Okoffo, 2020) and renewable energy sources (Khan et al., 2020), quantifying the reduction in harmful air pollutants (Kanniah et al., 2020), the impact of raising awareness, the development of strategies to decrease nitrogen-containing waste (Galloway et al., 2008), use of biofuel (Farrell et al., 2006), and marine ecosystems (Hoegh-Guldberg and Bruno, 2010). The literature supports the fact that there are ecosystem disturbances, ecological change and harmful impacts on the environment due to global warming, population dynamics, pollution, crossing biophysical thresholds and other human interventions. The quality of soil, water, marine environments, agriculture and air has been degraded, but this can be prevented from further depletion. If climate change continues, then there will be enormous challenges and costs for society and all humankind.

REFERENCES

Behrenfeld, M.J., R.T. O'Malley, D.A. Siegel, C.R. McClain, J.L. Sarmiento, G.C. Feldman, A.J. Milligan, P.G. Falkowski, R.M. Letelier, and E.S. Boss. 2006. 'Climate-Driven Trends in Contemporary Ocean Productivity', *Nature, 444*(7120): 752–755. doi:10.1038/nature0531

Chen, I.C., J.K. Hill, R. Ohlemuller, D.B. Roy, and C.D. Thomas. 2011. 'Rapid Range Shifts of Species Associated with High Levels of Climate Warming', *Science, 333*(6045): 1024–1026. doi:10.1126/science.1206432

Edwards, M., and A.J. Richardson. 2004. 'Impact of Climate Change on Marine Pelagic Phenology and Trophic Mismatch', *Nature, 430*(7002): 881–884. doi:10.1038/nature02808

Fadare, O.O., and E.D. Okoffo. 2020. 'Covid-19 Face Masks: A Potential Source of Microplastic Fibres in the Environment', *Science of the Total Environment, 737*: 140279.

Farrell, A.E., R.J. Plevin, B.T. Turner, A.D. Jones, M. O'Hare, and D.M. Kammen. 2006. 'Ethanol Can Contribute to Energy and Environmental Goals', *Science, 311*(5760): 506–508. doi:10.1126/science.1121416

Finnveden, G., M.Z. Hauschild, T. Ekvall, J. Guinée, R. Heijungs, S. Hellweg, A. Koehler, D. Pennington, and S. Suh. 2009. 'Recent Developments in Life Cycle Assessment', *Journal of Environmental Management, 91*(1): 1–21. doi:10.1016/j.jenvman.2009.06.018

Foley, J.A., N. Ramankutty, K.A. Brauman, E.S. Cassidy, J.S. Gerber, M. Johnston, M.D. Muller, C. O'Connell, D.K. Ray, P.C. West, C. Balzer, and E.M. Bannett. 2011. 'Solutions for a Cultivated Planet', *Nature, 478*(7369): 337–342. doi:10.1038/nature10452

Galloway, J.N., A.R. Townsend, J.W. Erisman, M. Bekunda, Z. Cai, J.R. Freney, L.A. Martinelli, S.P. Seitzinger, and M.A. Sutton. 2008. 'Transformation of the Nitrogen Cycle: Recent Trends, Questions and Potential Solutions', *Science, 320*(5878): 889–892. doi:10.1126/science.1136674

Godfray, H.C.J., J.R. Beddington, I.R. Crute, L. Haddad, D. Lawrence, J.F. Muir, J. Pretty, S. Robinson, S.M. Thomas, and C. Toulmin. 2010. 'Food Security: The Challenge of Feeding 9 Billion People', *Science, 327*(5967): 812–818. doi:10.1126/science.1185383

Harvell, C.D., C.E. Mitchell, J.R. Ward, S. Altizer, A.P. Dobson, R.S. Ostfeld, and M.D. Samuel. 2002. 'Climate Warming and Disease Risks for Terrestrial and Marine Biota', *Science, 296*(5576): 2158–2162. doi:10.1126/science.1063699

Hewitt, G.M. 2004. 'Genetic Consequences of Climatic Oscillations in the Quaternary', *Philosophical Transactions of the Royal Society B: Biological Sciences, 359*(1442): 183–195. doi:10.1098/rstb.2003.1388

Hoegh-Guldberg, O., and J.F. Bruno. 2010. 'The Impact of Climate Change on the World's Marine Ecosystems', *Science, 328*(5985): 1523–1528. doi:10.1126/science.1189930

Hughes, T.P., A.H. Baird, D.R. Bellwood, M. Card, S.R. Connolly, C. Folke, R. Grosberg, O. Hoegh-Guldberg, J.B.C. Jackson, J. Kleypas, J.M. Lough, P. Marshall, and J. Roughgarden. 2003. 'Climate Change, Human Impacts, and the Resilience of Coral Reefs', *Science, 301*(5635): 929–933. doi:10.1126/science.1085046

Jackson, J.B.C., M.X. Kirby, W.H. Berger, K.A. Bjorndal, L.W. Botsford, B.J. Bourque, R.H. Bradbury, R. Cooke, J. Erlandson, J.A. Estes, T.P. Hughes, S. Kidwell, C.B. Lange, H.S. Lenihan, J.M. Pandolfi, C.H. Peterson, R.S. Steneck, M.J. Tegner, and R.R. Warner. 2001. 'Historical Overfishing and the Recent Collapse of Coastal Ecosystems', *Science, 293*(5530): 629–637. doi:10.1126/science.1059199

Kanniah, K.D., N.A.F. Kamarul Zaman, D.G. Kaskaoutis, and M.T. Latif. 2020. 'Covid-19's Impact on Atmospheric Environment in Southeast Asia Region', *Science of the Total Environment, 736*: 1–11. doi: 10.1016/j.scitotenv.2020.139658

Khan, Z., S. Ali, M. Umar, D. Kirikkaledi, and Z. Jiao. 2020. 'Consumption-based Carbon Emissions and International Trade in G7 Countries: The Role of Environmental Innovation and Renewable Energy', *Science of the Total Environment, 730*: 138945. doi: 10.1016/j.scitotenv. 2020.138945

Nemani, R.R., C.D. Keeling, H. Hashimoto, W.M. Jolly, S.C. Piper, C.J. Tucker, R.B. Myneni, and S.W. Running. 2003. 'Climate-driven Increases in Global Terrestrial Net Primary Production from 1982 to 1999', *Science, 300*(5625): 1560–1563. doi:10.1126/science.1082750

Oki, T., and S. Kanae. 2006. 'Global Hydrological Cycles and World Water Resources', *Science, 313*(5790): 1068–1072. doi:10.1126/science.1128845

Paerl, H.W., and J. Huisman. 2008. 'Climate: Blooms Like It Hot', *Science, 320*(5872): 57–58. doi:10.1126/science.1155398

Rajão, R., B. Soares-filho, F. Nunes, J. Borner, I. Machado, D. Assis, A. Oliveira, I. Pinto, V. Ribeiro, I. Rausch, H. Gibbs, and D. Figueira. 2020. 'The Rotten Apples of Brazil's Agribusiness', *Science, 369*(6501): 246–248. doi: 10.1126/science.aba6646

Rockstrom, J., W. Steffen, K. Noone, A. Persson, F.S. Chapin, E.F. Lambin, T.M. Lenton, M. Scheffer, C. Folke, H.J. Schellnhuber, T. Hughes, S.V.D. Leeuw, H. Rodhe, S. Sorlin, P.K. Snyder, R. Costanza, M. Falkenmark, L. Karlberg, R.W. Corell, V.J. Fabry, J. Hansen, B. Walker, D. Liverman, K. Richardson, P. Crutzen, and J.A. Foley. 2009. 'A Safe Operating Space for Humanity', *Nature, 461*(7263): 472–475. doi:10.1038/461472a

Solomon, S., G. Plattner, R. Knutti, and P. Friedlingstein. 2009. 'Irreversible Climate Change Due to Carbon Dioxide Emissions', *Proceedings of the National Academy of Sciences of the United States of America, 106*(6): 1704–1709. doi:10.1073/pnas.0812721106

Walther, G., E. Post, P. Convey, A. Menzel, C. Parmesan, T.J.C. Beebee, and F. Bairlei. 2002. 'Ecological Responses to Recent Climate Change', *Nature, 416*(6879): 389–395. doi:10.1038/416389a

Xiang, X., Q. Li, S. Khan, and O.I. Khalaf. 2021. 'Urban Water Resource Management for Sustainable Environment Planning Using Artificial Intelligence Techniques', *Environmental Impact Assessment Review, 86*: 106515. doi: 10.1016/j.eiar.2020.106515

2 Environmental Implications of Persistent Organic Pollutants (POPs)

Sushil Kumar Bharti
Department of Geology, Babasaheb Bhimrao Ambedkar University (A Central University), Lucknow, Uttar Pradesh, India

Awadhesh Kumar Yadav
Department of Botany, K.P.Singh College, Hardoi, Uttar Pradesh, India

Narendra Kumar
Department of Environmental Science, Babasaheb Bhimrao Ambedkar University (A Central University), Lucknow, Uttar Pradesh, India

CONTENTS

2.1 INTRODUCTION

During the 18th century, IPOPs were commercially produced to control pests and ensure food security, whereas UPOPs such as polyaromatic hydrocarbons (PAHs), pentachlorobenzene (PCB), dioxin, furan, and perfluorooctanesulfonic acid (PFOS) are by-products of industrial processes (Godfray et al., 2010; Herzig et al., 2019; Yang et al., 2019; Liu et al., 2018; Pariatamby & Kee, 2016). In the mid-19th century, the use of IPOPs increased significantly due to the benefits to modern society (Anderson, 2010). POPs are long-lived organic hydrophobic or lipophilic chemicals capable of migrating to higher latitudes through long-range transport, termed grasshopper effects (Suami et al., 2020; Alharbi et al., 2018; Wania & Mackay, 1996). Further, as they move through the food chain,

DOI: 10.1201/9781003220824-3

they accumulate in the fatty tissues of living organisms and impose serious health effects (Bacaloni et al., 2011; Whylie et al., 2003; Wania & Mackay, 1996). IPOPs such as pentachlorobenzene (PCB), dichlorodiphenyltrichloroethane (DDT), Aldrin, Dieldrin, Endrin, Heptachlor, hexachlorocyclohexane (HCH), Mirex, Toxaphene, Chlordecone, pentachlorophenol, α-Endosulfan, β-Endosulfan, and Dicofol are used to control vectors of diseases such as plague, malaria, and typhus (Nascimento et al., 2017; Chakraborty et al., 2010; Srimurali et al., 2015; Devi et al., 2011).

IPOPs are also used in various manufacturing processes and agricultural practices as insecticides, herbicides, and rodenticides (Kumar & Pannu, 2018; Mishra et al., 2007). In 2003, POPs were regulated by the Aarhus Protocol on POPs in the United Nations Economic Commission for Europe (UNECE, 1947) Convention on Long-Range Transboundary Air Pollution (CLRTAP, 1979) (Table 2.1) (Kim et al., 2013). However, in 2004 the Stockholm Convention, an international treaty, entered into force to protect human health and the environment from POP contamination (Herzig et al., 2019). According to the United Nations Environment Programme, POPs have a unique physicochemical characteristic: once they are released into the environment, they will remain intact for several years (Sharkey et al., 2020). Further, the semi-volatile nature of POPs favours their long-range transport, and they are distributed easily among soil, water, and ambient air. Moreover, they can accumulate in fatty tissue due to their low water and high lipid solubility (Bacaloni et al., 2011), entering the food chain at different levels and being considered toxic to flora and fauna.

A significant fraction of POPs (15–40%) is dispersed in the atmosphere during agriculture/industrial/commercial applications through volatilization or spray drift processes (Nascimento et al., 2017; Degrendele et al., 2016; Socorro et al., 2016; Coscolla et al., 2014). In recent decades, advances in analytical techniques, i.e. gas chromatography with electron capture detector (ECD), gas chromatography with mass spectroscopy (MS), and high-performance liquid chromatography (HPLC), have enabled the detection of residual DDT, dichlorodiphenyldichloroethylene (DDE), Dieldrin, and PCBs in environmental and biological samples at low levels (Muir & Howard, 2006). Several vital publications, scientific discoveries, and human exposure incidents set the stage for discussion on POP control (Carson & Edwards, 2009; Jensen et al., 1969; Yoshimura et al., 2003). A detailed description of POPs has been provided in Table 2.1.

2.2 DISCOVERY AND PROHIBITION OF PERSISTENT ORGANIC POLLUTANTS

On 24 June 1998, UNECE Convention on Long-Range Transboundary Air Pollution adopted the Protocol on POPs in Aarhus, Denmark (Wettestad, 2002). The objective of the UNECE Protocol and the Stockholm Convention is to verify the manufacture, releases, emissions, and technical grade of POPs and their metabolites, with the aim of restricting and promptly prohibiting those POPs that have adverse effects on flora and fauna. Details regarding discovery/synthesis and prohibition are given in Table 2.2.

2.3 TYPES AND CLASSIFICATION OF INTENTIONAL AND UNINTENTIONAL PERSISTENT ORGANIC POLLUTANTS

Organic compounds with a half-life or residence time greater than 48 hours in the atmosphere are referred to as POPs as per the Stockholm Convention, a treaty to protect human health and the environment from POPs (World Health Organization, 2003). Based on their physicochemical characteristics, POPs were classified as dirty dozen or legacy POPs after the approval of the UNEP Stockholm Convention, 2001 (United Nations Environment Programme, 2001). However, additional POPs were identified and later classified as intentionally and unintentionally produced POPs. The classification of POPs has been summarized in Figures 2.1, 2.2, and 2.3 and Table 2.3.

TABLE 2.1
POPs Listed in Stockholm Convention and Aarhus Protocol

	Stockholm Convention				The Aarhus Protocol on Persistent Organic Pollutants		
Persistent Organic Pollutants (POPs)	**Annex A**	**Annex B**	**Annex C**	**Industrial Chemical**	**Annex I**	**Annex II**	**Annex III**
Aldrin	✓				✓		
Dieldrin	✓				✓		
Chlordane	✓				✓		
Endrin	✓				✓		
Heptachlor	✓				✓		
Hexachlorobenzene	✓				✓		
Mirex	✓				✓		
Toxaphene	✓				✓		
Polychlorinated biphenyls (PCBs)				✓	✓	✓	
Dichlorodiphenyltrichloroethane (DDT)		✓		.		✓	
Polychlorinated dibenzodioxins (PCDD)			✓	✓			✓
Polychlorinated dibenzofurans (PCDF)			✓	✓			✓
α-Hexachlorocyclohexane (α-HCH)	✓					✓	
β-Hexachlorocyclohexane (β-HCH)	✓					✓	
γ-Hexachlorocyclohexane (γ- HCH)	✓			✓		✓	
Chlordecone	✓				✓		
Decabromodiphenyl ether				✓			✓
Dicofol	✓						✓
α-Endosulfan	✓						✓
β- Endosulfan	✓						✓
Hexabromobiphenyl				✓	✓		
Hexabromocyclododecane				✓			✓
Hexabromodiphenyl ether			✓	✓		✓	
Hexachlorobutadiene			✓			✓	
Perfluorooctanoic sulfonic acid (PFOS)			✓			✓	
Pentachlorobenzene (PCBs)			✓		✓		
Polychlorinated naphthalene (PCN)				✓			✓
Polycyclic aromatic hydrocarbon (PAH)			✓				✓

Note: This table is modified from Teran et al. (2012). The dirty dozen are represented in italics. in addition, Annex A and I are aimed at elimination, Annex B and II aimed at restriction, and Annex C and III aimed at unintentional production and industrial chemicals.

2.4 PROPERTIES AND SOURCES OF INTENTIONALLY PRODUCED POPS

Intentionally produced POPs are generally recalcitrant and may be found in different environmental matrices (Okeke et al., 2002; De Cruz et al., 1996). Direct inhalation, dermal absorption, consumption of leafy vegetables, meat, fish, and dairy products, and volatilization after application of POPs are the primary sources of human exposure to intentional and unintentional POPs (Kumar et al., 2018; Nascimento et al., 2017; Degrendele et al., 2016; Socorro et al., 2016; Coscolla et al., 2014; Chakraborty et al., 2010; Srimurali et al., 2015; Devi et al., 2011). Lindane is one of the most common intentionally produced POPs that enter the food chain and bioaccumulate at different trophic levels (Zuloaga et al., 2000; Okeke et al., 2002; De Cruz et al., 1996; Johri et al., 1996). However, fugitive dust is the ultimate source of lindane. Lindane can be transferred to an uncontaminated site from

TABLE 2.2
Comparative Account of Discovery/Synthesis and Prohibition of POPs

Compound	Discovered/Synthesized	Prohibition
Dichloro diphenyl trichloroethane (DDT)	DDT was first synthesized in 1874 by Othmar Zeidler (Zeidler, 1874) and later described as an insecticide in 1939 and finally commercially synthesized in 1945 (World Health Organization, 1979).	Until 1973, DDT was banned in the United States, but some developing countries have continuously used DDT to combat malaria (Agency for Toxic Substances and Disease Registry Office of Innovation and Analytics, Toxicology Section, 2002).
Pentachlorinated biphenyls (PCBs)	First discovered in 1881, their commercial production began in 1929 synthesized in 1929 (Schmidt and Schultz, 1881).	The use of PCBs in any form is prohibited after 2025, and disposal must strictly follow the provisions of the Hazardous Waste (Management. Handling and Transboundary Movement) Rules, 2008 (Robertson et al., 2018)
Polybrominated diphenyl ethers (PBDEs)	PBDEs have been in use since the 1960s. They are synthetic compounds used as fire and flame retardant additives in commercial and household products (Horton et al., 2013).	In the late 1990s, Penta and Octa BDE restrictions were applied when their historical production was less than 10% (Abbasi et al., 2019).
Polycyclic aromatic hydrocarbons (PAHs)	Cook, Hewett, and Hieger, in 1929, working on coal tar and its carcinogenicity, accidentally discovered benzo (α) pyrene, a PAH (Cook et al., 1933).	The European Union, National Institute for Occupational Safety and Health (NIOSH), and US Environmental Protection Agency (USEPA) regulate PAH concentrations and prohibit the use of specific PAHs in consumer goods as of 2015 (Kim et al., 2013).
Perfluorooctane sulfonate (PFOS)	PFOS has recently been identified as a POP and included in Annex B of the Stockholm Convention on Persistent Organic Pollutants (Kelly et al., 2009)	The European Union bans the use of PFOS due to its persistence in the environment, bioaccumulation, and toxicity to humans (Tang et al., 2006).
Dioxins and furans	Dioxins and furans are not produced intentionally, except for research use and as standards, but they are released unintentionally as by-products during the combustion of organic matter (Kulkarni et al., 2008; Davy, 2004).	In 2004, dioxins and furans were globally banned under the Stockholm Convention on Persistent Organic Pollutants POPs (Schiavon et al., 2016)

agricultural fields, forestry warehouses, and hazardous waste sites through long-range transport (Kumar & Pannu, 2018; Mishra et al., 2007). Besides lindane, DDT was used extensively during World War II to combat malaria, typhus, and several vector-borne diseases (Tao et al., 2007; Ricking & Schwarzbauer, 2012; Tesfahunegny, 2016).

The use of DDT has been prohibited in developed countries due to its harmful effects. In contrast, some developing countries, such as China, India, Indonesia, and Mexico, continued to use DDT against malaria and vector-borne diseases (Iwata et al., 1994; Fiedler, 2008; van den Berg, 2009). According to Chakraborty et al. (2010), DDT was banned in 1989 for agricultural practice in India, with a mandate to use 10,000 t/year for vector-borne disease control. To combat malaria and visceral leishmaniasis, the National Malaria Control Program (Tren, 2002) was launched, in which 20% of global consumption of DDT was used for residual spraying in rural areas/peri-urban

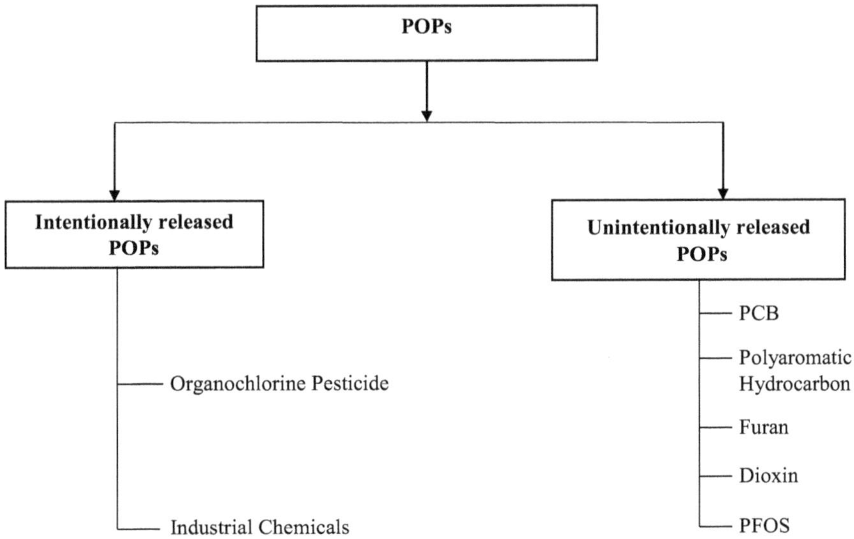

FIGURE 2.1 Types and classification of POPs produced intentionally and unintentionally. (From Gaur et al., 2018 and Ashraf, 2017.)

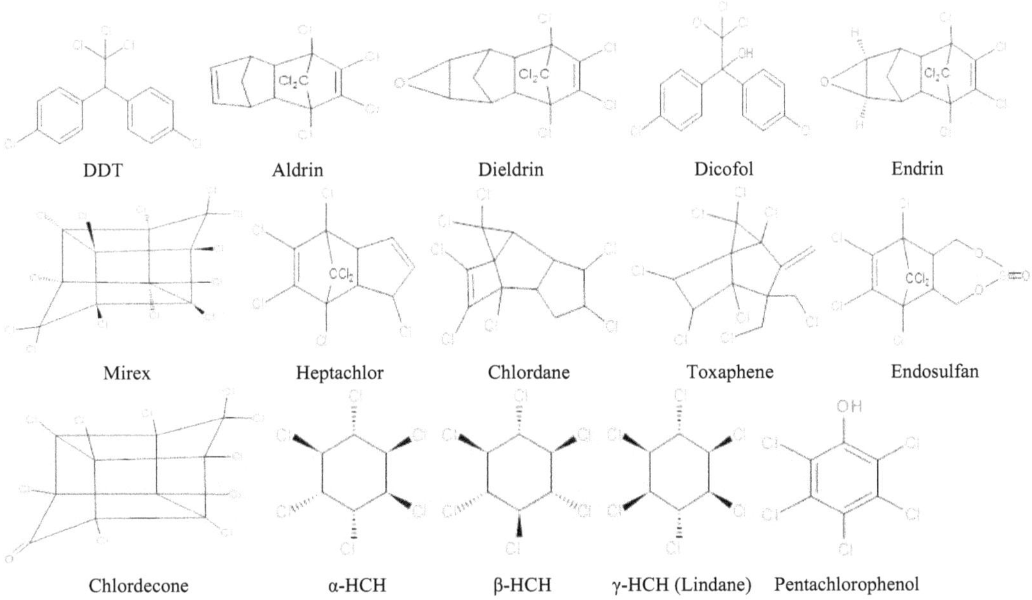

FIGURE 2.2 Intentionally produced persistent organic pollutants and their chemical structures.

areas of India (Gupta, 2004). Dichlorodiphenyldichloroethylene (DDE), a metabolite formed by dehydrohalogenation of DDT, is transferred to offspring more readily than some of its counterparts (Faroon et al., 2019; Blaustein et al., 2010). Unlike DDT, Dicofol is confined to soil and sediments rather than the hydrosphere and atmosphere, indicating inadequate volatility (Elmore & La Merrill, 2019). During 1950–70, Aldrin and Dieldrin were widely used as broad-spectrum soil insecticides to protect various food crops and were effective against invertebrates such as worms, beetles, and

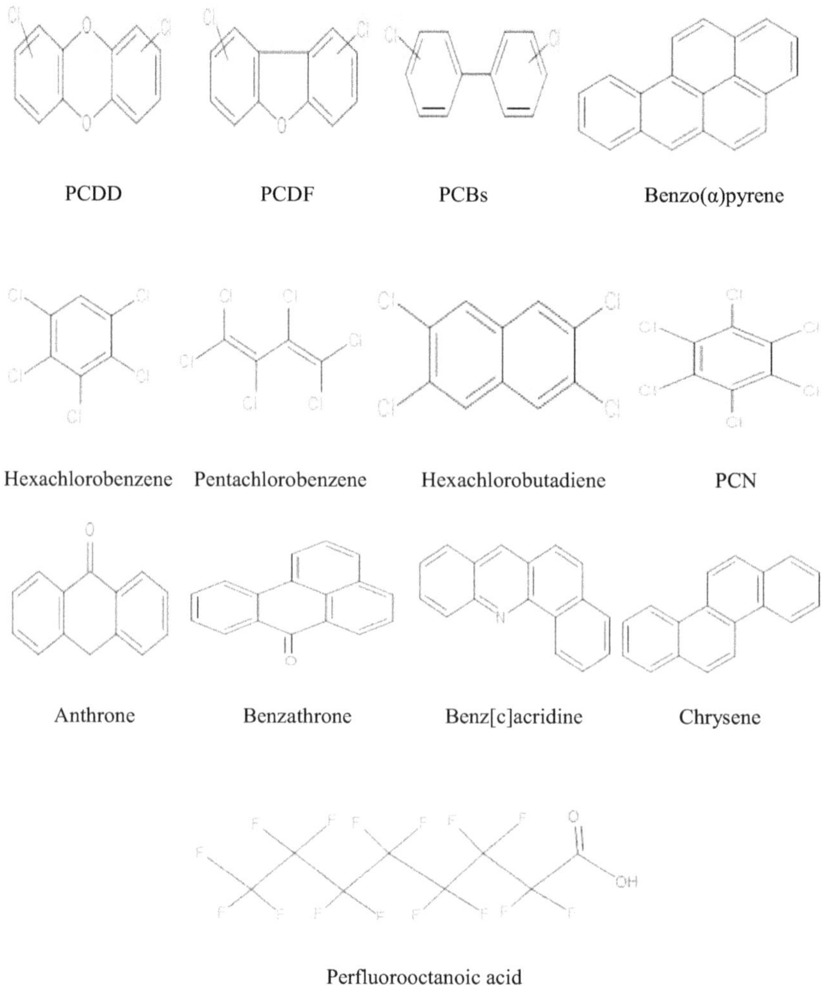

FIGURE 2.3 Unintentionally produced persistent organic pollutants and their chemical structures.

termites (Jayaraj et al., 2016). During the application, traces of Aldrin and Dieldrin were added to the atmospheric environment and, by volatilization, entered the food chain (Jayaraj et al., 2016; Chakraborty et al., 2010; Srimurali et al., 2015; Devi et al., 2011). Chlordane and Heptachlor undergo enantiomeric enrichment in the environment when used as insecticides against termites (Dearth & Hites, 1991).

Later, the production and use of chlordane and heptachlor were prohibited due to their environmental persistence, bioaccumulation along the food chain, and associated health risks (Kurt-Karakus et al., 2005; Purudue et al., 2007). Endrin belongs to the Aldrin and Dieldrin family but is effective against cotton pests and even toxic to non-target species (Zitko, 2003). Additionally, acute exposure to Endrin is considered more toxic than its counterparts due to its rapid degradation in biota (Matsumoto et al., 2009; Mawussi et al., 2009). Mirex is an intentionally produced POP used to control the populations of imported fire ants during 1961 (Faroon et al., 1995; Kaiser, 1978). The Mirex traces in aquatic organisms indicate a higher bioconcentration factor in water followed by sediment (Carlson et al., 1976).

TABLE 2.3
Comprehensive Statement on Acute and Chronic Exposure to Intentionally and Unintentionally Produced POPs

POPs		Molecular Weight	Residence Time	Health Hazards	References
Intentionally produced POPs	Lindane	290.81	15 months	Acute exposure to lindane can cause nausea, headache, and loss of consciousness. In contrast, chronic exposure reduced the number of lymphocytes, which may promote oncogenesis.	Madaj et al., 2018; Rani et al., 2017; Jayaraj et al., 2016; Croom et al., 2015.
	DDT	354.48	2–15 years	Mainly produces neurological and immunodeficiency disorders on acute to chronic exposure.	Wurl & Obbard, 2005; Tren & Bate, 2004; Bian et al., 2009; Tao et al., 2007.
	Dicofol	370.5	60 days	It is considered toxic to both flora and fauna; however, it can reduce the quality of the eggshell, but it is not carcinogenic.	Sánchez et al., 2010; Rasenberg & van de Plassche, 2003.
	Aldrin Dieldrin	364.9 380.9	4–7 years 9 months	Inhaling Aldrin/Dieldrin produces adverse neurological and reproductive disorders, but it is not carcinogenic.	Stevenson et al., 1999; Sielken & Valdez-Flores, 1999; Sarcinelli et al., 2003
	Chlordane Heptachlor	409.8 373.3	10 years 2 years	Acute exposure to the metabolites of chlordane and heptachlor in humans causes disorders of the nervous system, digestive system, and liver; while chronic exposure leads to the development of liver cancer, non-Hodgkin's lymphoma, and acute myeloid leukaemia.	Engel et al., 2019; Park et al., 2020
	Endrin	380.9	1 day to 12 years	Acute and chronic exposure to Endrin produces severe neurotoxic effects, nausea, vomiting, dizziness, stomach pain, headache, unconsciousness, seizures, and central nervous system depression.	Jinxian et al., 2005
	Mirex	545.55	10 years	Direct inhalation of Mirex can disrupt the endocrine system, which directly or indirectly affects women's ovulation, pregnancy, and endometrial growth.	Yu et al., 2013; Thakur & Pathania, 2020.
Unintentionally produced POPs	Dioxin	285.16	10–22 years	Acute exposures of humans to dioxin are associated with chloracne, hyperpigmentation, impaired liver function, and lipid metabolism; however, chronic exposure leads to depression of the immune, endocrine, and nervous systems. Furthermore, WHO classified it as carcinogenic by nature on prolonged exposure.	Hays & Aylward, 2003; Solomon-Wisdom & Ndana, 2012; Guo et al., 2018
	Furan	269.16	10–22 years	Furan is a potent hepatotoxin and hepatocarcinogen in rodents and has been classified by the International Agency for Research on Cancer (IARC) as possibly carcinogenic to humans (group 2B).	Moro et al., 2012; IARC Working Group on the Evaluation of Carcinogenic Risks to Humans, International Agency for Research on Cancer, & World Health Organization, 1995
	PCB	255.18	10–22 years	In humans, acute exposure to PCB produces acne-like eruption, pigmentation of the skin, nails, and conjunctivas, and increased ocular secretions. Chronic exposure leads to dysfunction of the neurological and endocrine systems.	Safe, 1990; Robertson et al., 2018
	PAH	252.32	1–5 weeks	Chronic exposure to PAHs such as benzo [a] pyrene (BaP) can produce oncogenic, mutagenic and teratogenic effects in humans.	Juhasz & Naidu, 2000; Ravindra et al., 2008; Roslund et al., 2018
	PFOS	499.12	5–6 years	According to the Organisation for Economic Co-operation and Development (OECD), chronic exposure to PFOS can result in reduced body weight, hepatotoxicity, reduced cholesterol, and a steep dose–response curve for mortality.	Wang et al., 2013; US Environmental Protection Agency, 2009

2.5 PROPERTIES AND SOURCES OF UNINTENTIONALLY PRODUCED POPS

There are several POPs that were produced during combustion and other chemical processes as a by-product, and they were listed as unintentionally produced POPs in Annex C of the Stockholm Convention, 2009 (Herzig et al., 2019; Yang et al., 2019; Liu et al., 2018). However, pentachlorobenzene (PeCBz), polychlorobenzene (PCN), and hexachlorobutadiene (HcBD) were also added to Annex C in 2009, 2015, and 2017, respectively (Sharkey et al., 2020; United Nations Environment Programme, 2017). PCB is one of the unintentionally produced POPs produced worldwide from 1930 to the 1980s (Jing et al., 2018). It consists of two phenyl rings with chlorine atoms, which can be attached to any of ten different positions to produce 209 different congeners and ten chiral compounds (McFarland & Clarke, 1989; Alder et al., 1993). Generally, PCBs are used extensively as plasticizers, coolants, lubricants, and heat exchange fluids in transformers, condensers, and capacitors (Wong et al., 2007). During improper storage and disposal of PCBs, some fractions of PCB can be added to the ambient air via evaporation (Jing et al., 2018).

After being added to the environment, it can travel long distances with the prevailing wind (Klanova et al., 2008). A higher level of PCB contamination has been observed in commercial and industrial areas compared with rural areas due to their proximity to the sources of pollution (Wilcke et al., 2006). For PCB, soil can be considered a better sink than the hydrosphere and the atmosphere due to its long retention time. Polybrominated diphenyl ethers (PBDEs) are structurally similar to PCBs and resistant to physical, chemical, and biological degradation, leading to their excellent flame-retardant properties (Loganathan & Lam, 2011). Apart from PCBs, unintentionally produced POPs, i.e. HCB, PeCBz, and HcBD, are emitted from waste incineration, cement production, and smelting processes of ferrous and non-ferrous metals, accompanied by the release of PCDD/PCDF (Hedman et al., 2006; United Nations Environmental Programme, 2017). HCB and pentachlorobenzene (PeCBz) are formed as chlorinated impurities of chlorinated solvents, pigments, and pesticides (Weber et al., 2011; Takasuga, 2012).

PAHs are produced unintentionally from incomplete combustion of organic matter, while open and diffuse biomass burning is the primary source of PAHs in the ambient environment (Bartrons et al., 2016). In addition to biomass combustion, residential heating, vehicle exhaust, asphalt production, and coal gasification are familiar anthropogenic PAH sources (Baklanov et al., 2007). Further, PFOS (an unintentionally produced POP) is a fluorinated anion used as a surfactant in carpets, leather, paper, packaging, fabric, and stuffing and surface protectors (European Food Safety Authority, 2008). Dioxins and furans are produced unintentionally during combustion processes, such as forest fires, industrial operations, the metallurgical industry, and municipal, industrial, and hospital wastes (Kulkarni et al., 2008; Anderson & Fisher, 2002).

2.6 RISK ASSOCIATED WITH INTENTIONALLY AND UNINTENTIONALLY PRODUCED POPS

A comprehensive statement of acute and chronic exposure to intentionally and unintentionally produced POPs is summarized in Table 2.3.

2.7 TRANSPORT AND FATE OF POPS

Long-range distribution of POPs among different environmental matrices is governed by physicochemical characteristics such as vapour pressure, Henry's constant, and partition coefficients (Bharti et al., 2020; Robinson et al., 2019; Buccini, 2003). Further, greater surface area and longer retention time of POPs enhance their levels in different environmental settings and produce uncertainty in their distribution, degradation, and circulation (Hanedar et al., 2019; Salihoglu et al., 2011; Harrad et al., 1994; Sarkar et al., 2008). Longer residence time and persistency in the atmosphere favour accumulation in water, sediments, vegetation, snow, ice, and permafrost (Bailey et al., 2000; Grannas et al., 2013). During recent decades, in remote places like the Arctic, colder temperature favours the distribution/retention of POPs (Pereira et al., 2017). On the other hand, crust, hydrosphere, and atmosphere can also act as sinks or sources of POPs. However, the complexity of POPs and their fate can be studied with the help of receptor modelling and emission inventories (Breivik et al., 2006).

2.8 IMPACTS OF INTENTIONALLY AND UNINTENTIONALLY PRODUCED POPS ON CLIMATE CHANGE

Climate change has pronounced effects on the environment. But, there is uncertainty in predicting their effects on human beings (Nadal et al., 2015). To resolve the uncertainty of climate change, the World Climate Research Program (WCRP) was begun with the joint efforts of the International Scientific Council (ISC) and the World Meteorological Organization (WMO) in 1980 (Teran et al., 2012). Further, the fourth assessment of the United Nations Intergovernmental Panel on Climate Change (IPCC) also addressed the scientific progress on climate change during 2007 (Solomon et al., 2007; Rummukainen et al., 2014). POPs have contributed significantly to altering regional meteorology, while anthropogenic activities synergize the negative impacts on the environment (Hagemen et al., 2015).

In addition, variation in temperature, wind speed, and wind direction can influence the precipitation pattern, resulting in uneven distribution of solar radiation (Budyko, 1969; Ricke et al., 2010). The uneven distribution of solar radiation on the landmass and ocean surfaces ultimately affects the heat budget of the Earth's surface, which leads to a significant contribution to global warming (Lewis et al., 1990; Ming et al., 2014). There are three aspects to understanding the effect of climate change on POPs. Firstly, it is likely to influence the environmental behaviour of POPs by enhancing the volatilization from the principal and subordinate sources (Diffenbaugh et al., 2012; Fussel, 2009). Air–surface exchange, wet and dry deposition, and reaction rates favour the partitioning of POPs in different environmental settings (Teran et al., 2012). Secondly, POPs can alter the chemical solubility of other pollutants, changing the physicochemical characteristics and decomposition rate of other pollutants in different environmental settings (Zhou & Ma, 2013). Interestingly, qualitative information is not enough to understand the role of POPs in climate change (Woehrnschimmel et al., 2013). The shortage of adequate data, complex mechanisms of transport, toxicological and biological processes, and specific knowledge gaps make it challenging to predict the role of POPs in climate change (Whitmarsh, 2011). Nevertheless, with the help of receptor modelling, several scientists predicted the overall impact of climate change on precipitation, temperature, sea-level rise, carbon sequestration, wind direction, and atmospheric radiative forcing (Asseng et al., 2015; Ma et al., 2012).

2.8.1 Consequences of Elevated Temperature for the Dispersal of POPs

Temperature is the degree of hotness or coldness of the Earth's surface (Ahrens, 2007). The temperature anomaly is illustrated by the blue and red lines in Figure 2.4. The temperature can be influenced naturally by the inclination angle of solar radiation, distribution of land and water, and rotation and revolution of the Earth (Ricke et al., 2010; Rummukainen et al., 2014; Socorro et al., 2016; Solomon et al., 2007). Additionally, some other conditions, such as inversion of temperature, low mixing height, low wind speed, wind direction, and high relative humidity favour the accumulation of various atmospheric pollutants, including POPs, in lower strata of the atmosphere (Hou & Wu, 2016). Variation in temperature and relative humidity can alter physicochemical characteristics such as vapour pressure and octanol–air partition coefficient K_{oa}, which subsequently influence the partitioning of POPs between the gas and particle phases (Cousins & Jones, 1998; Cousins & Mackay, 2001; Lohmann & Lammel, 2004; Pankow, 1987).

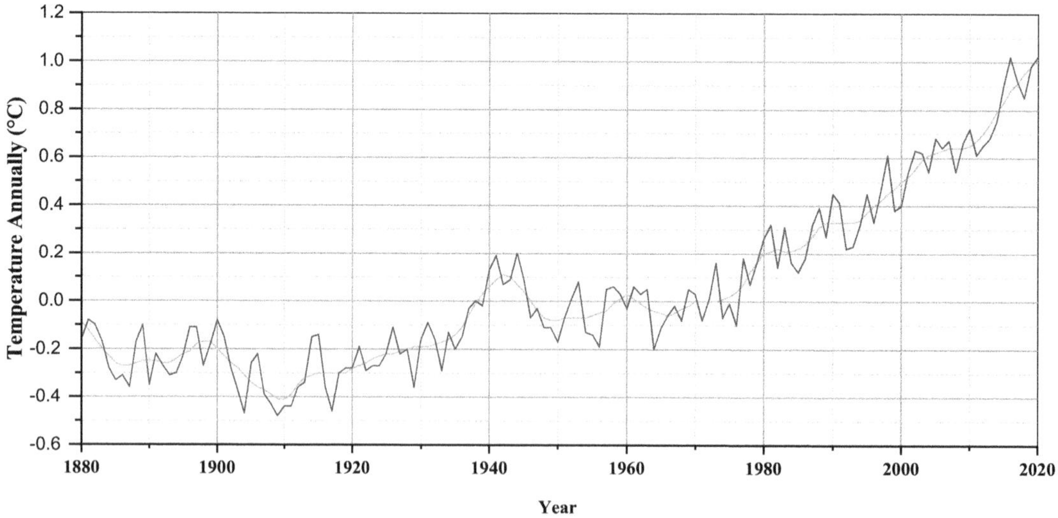

FIGURE 2.4 Temperature anomaly is modified and restructured from the Goddard Institute for Space Studies (GISS), New York dataset, 2021. The solid blue line is the Earth's annual mean temperature, and the solid red line is the five-year lowess smooth.

Continuous temperature monitoring by scientists at Goddard Institute for Space Studies (GISS), New York, revealed that Earth's average surface temperature has increased by 1.02°C since 1880 (Hansen et al., 2019). However, from 1975 onwards, the temperature increased significantly at a rate of 0.15–0.20°C per decade, increasing the levels of water vapour in the atmosphere (van Diedenhoven, 2021). An increase of 1.02°C in global annual temperature can favourably influence the Henry's law constants and promote higher partitioning from water into the atmosphere for POPs (Hansen et al., 2019; van Diedenhoven, 2021).Subsequently, it can alter the atmospheric and oceanic transport, global distribution, partitioning between different environmental matrices, secondary emission, and bioaccumulation in the food chain (Lamon et al., 2009; Kallenborn et al., 2012; Bustnes et al., 2010). This also demonstrates the widely cited hypothesis that traces of the POPs were also present in remote areas like the Arctic (Kallenborn et al., 2012; van Diedenhoven, 2021).

2.8.2 Role of Rainfall in the Dispersal of POPs

According to the IPCC 6th Assessment Report, global average annual precipitation will increase by the end of this century (Wehner, 2020). The dataset of precipitation obtained from National Oceanic and Atmospheric Administration (NOAA), 2021, Washington, DC, shows that precipitation patterns changed drastically in the last five decades (Figure 2.5).As a result, wet areas have become wetter and dry regions have become drier worldwide. Enhanced precipitation reduces atmospheric pollution by washout effects or acting as a natural scrubber (Mitchell & Jones, 2005). At the same time, humid conditions enhance the rate of adsorption of POPs, which intensifies pesticide runoff by northward-flowing rivers. However, reduced precipitation increases the hydrophilicity of aerosols and the particle radius, which ultimately enhance the concentration of POPs in the environment (Bharti et al., 2017; Stemmler & Lammel, 2012; Meijer & Dijkstra, 2009).

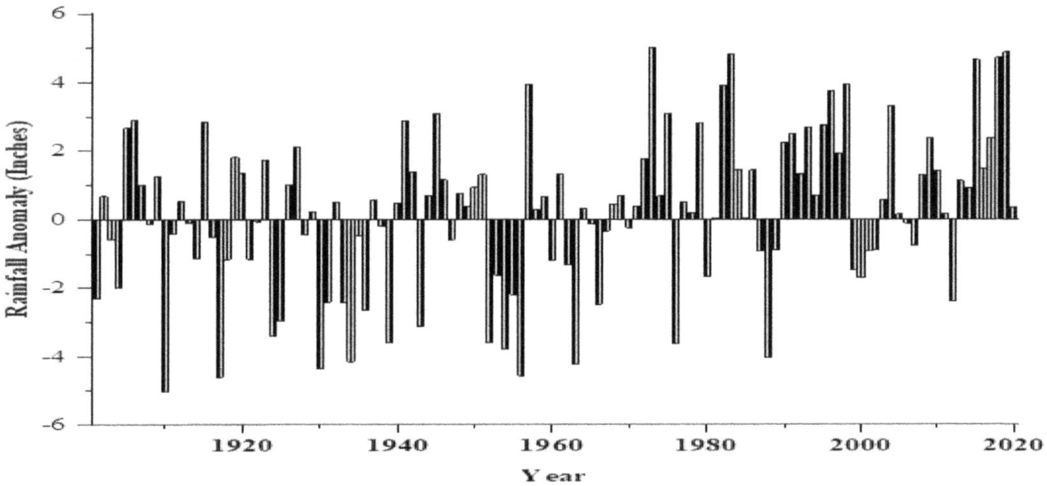

FIGURE 2.5 Rainfall anomalies.

(Modified and restructured from the dataset (1901–2019) obtained from National Oceanic and Atmospheric Administration (NOAA), 2021 & New York.)

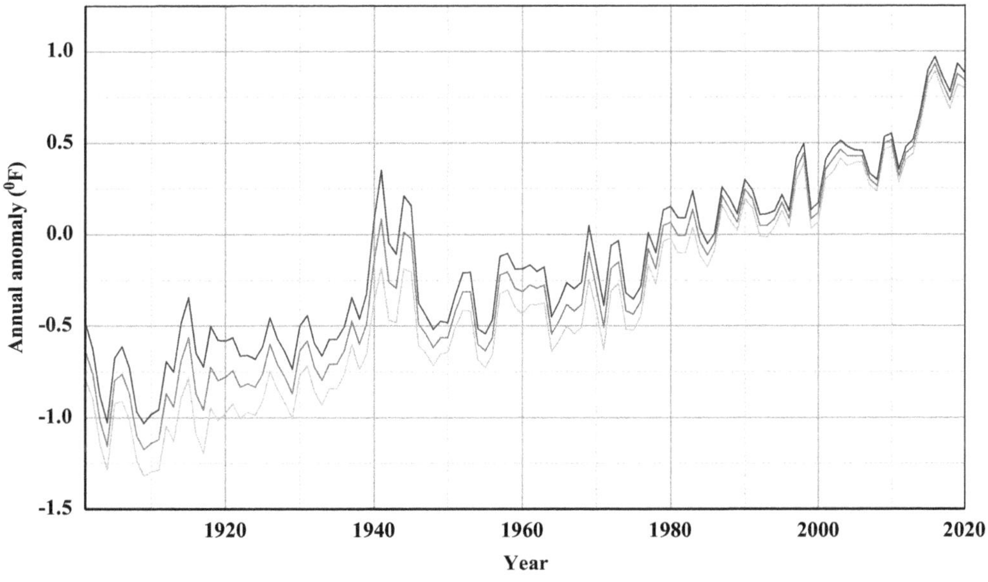

FIGURE 2.6 Sea-surface temperature (SST).

(Modified and restructured from the dataset (1900–2020) of NOAA (National Oceanic and Atmospheric Administration), accessed August 2021. www.ncdc.noaa.gov/data-access/marineocean-data/extended-reconstructed-sea-surface-temperature-ersst.)

2.8.3 Sea-Level and Dispersal of POPs

2.8.4 Role of CO2 Rise in the Dispersal of POPs

Since the pre-industrial era, the global average concentration of carbon dioxide (CO_2) in the atmosphere has increased from 278 to 411 parts per million, as shown in Figures 2.6 and 2.7 (Intergovernmental Panel on Climate Change (IPCC), 2014). High levels of CO_2 intensify the

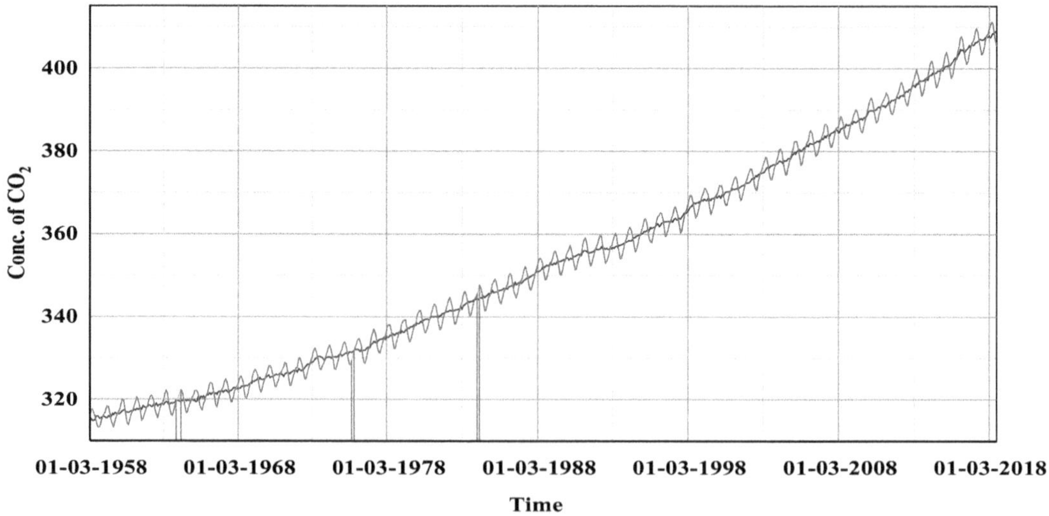

FIGURE 2.7 Average concentration of CO_2.

(Dataset retrieved from the National Center for Atmospheric Research Staff (Eds) and last modified 27 August 2021.)

absorption of infrared radiation, which eventually increases the heating capacity of water vapour. As the level of atmospheric CO_2 increases, acidification increases, which can alter the pathway of POPs in the food web (Alkordi et al., 2016; Ma et al., 2016). In addition, elevated levels of CO_2 can alter plant growth, biomass, microbial functional diversity, and the bioavailability of POPs in the soil (Amundsen et al., 2013). According to IPCC report 2015, global warming will promote the proliferation of insects and other phytopathogens; as a result, subsequent use of POPs is likely to increase under changing climatic conditions, leading to an increased load of pesticide residues in the soil (Nadal et al., 2015). However, without pesticides, it is impossible to meet an expected 50% increase in food demand by 2030 and possibly double by 2050 (Banwart, 2011).

2.9 CONCLUSION

There has been tremendous pressure on the agricultural sector to ensure food security for the growing world population. Several POPs have been found intentionally and unintentionally in the agriculture sector, particularly for growing hybrid crops. Unintentionally released POPs such as PFOS, PeCBz, HcBD, and PCN were discovered recently and added to annexures of the Stockholm Convention as per their characteristics. Since POPs can migrate to remote areas due to their unique physicochemical characteristics, regional meteorology plays a significant role in transporting them to remote areas far from the place of origin and use. The presence of POPs in Antarctica and the Arctic region confirms their long-range transport. Further, POPs can accumulate and amplify in the food chain, producing adverse effects on human health. POPs will play a crucial role in the era of climate change by influencing human health, crop quality, and environmental dynamism.

REFERENCES

Abbasi, G., Li, L., & Breivik, K. (2019). Global historical stocks and emissions of PBDEs. *Environmental Science & Technology*, *53*(11), 6330–6340.

Agency for Toxic Substances and Disease Registry Office of Innovation and Analytics, Toxicology Section. (2002). Profile for DDT, DDE and DDD, US Department of Health and Human Services. *Public Health Service, 497*, 17–26.

Ahrens, C. D. (2007). *Meteorology Today*, Eighth Edition. Thomson, USA

Alder, A. C., Haggblom, M. M., Oppenheimer, S. R., & Young, L. Y. (1993). Reductive dechlorination of polychlorinated biphenyls in anaerobic sediments. *Environmental Science & Technology*, *27*(3), 530–538.

Alharbi, O. M., Khattab, R. A., & Ali, I. (2018). Health and environmental effects of persistent organic pollutants. *Journal of Molecular Liquids, 263*, 442–453.

Alkordi, M. H., Weseliński, Ł. J., D'Elia, V., Barman, S., Cadiau, A., Hedhili, M. N., ... & Eddaoudi, M. (2016). CO_2 conversion: the potential of porous-organic polymers (POPs) for catalytic CO_2–epoxide insertion. *Journal of Materials Chemistry A*, *4*(19), 7453–7460.

Amundsen, H., Anderson, L., Andersson, A., Azetsu-Scott, K., Bellerby, R., Beman, M., ... & Yamamoto-Kauai, M. (2013). AMAP Assessment 2013: Arctic Ocean Acidification. Arctic Monitoring and Assessment Programme (AMAP).

Anderson, D. R., & Fisher, R. (2002). Sources of dioxins in the United Kingdom: the steel industry and other sources. *Chemosphere*, *46*(3), 371–381.

Anderson, K. (2010). Globalization's effects on world agricultural trade, 1960–2050. *Philosophical Transactions of the Royal Society B: Biological Sciences, 365*(1554), 3007–3021.

Ashraf, M. A. (2017). Persistent organic pollutants (POPs): a global issue, a global challenge. *Environmental Science Pollution Research*, 24, 4223–4227. doi: 10.1007/s11356-015-5225-9.

Asseng, S., Ewert, F., Martre, P., Rötter, R. P., Lobell, D. B., Cammarano, D., ... & Zhu, Y. (2015). Rising temperatures reduce global wheat production. *Nature Climate Change*, *5*(2), 143–147.

Bacaloni, A., Insogna, S., & Zoccolillo, L. (2011). Remote zones air quality-persistent organic pollutants: sources, sampling and analysis. *Air Quality Monitoring, Assessment and Management*, 223–240. DOI:10.5772/17024.

Bailey, R., Barrie, L. A., Halsall, C. J., Fellin, P., & Muir, D. C. G. (2000). Atmospheric organochlorine pesticides in the western Canadian Arctic: evidence of transpacific transport. *Journal of Geophysical Research: Atmospheres, 105*(D9), 11805–11811.

Baklanov, A., Hänninen, O., Slørdal, L. H., Kukkonen, J., Bjergene, N., Fay, B., ... & Odegaard, V. (2007). Integrated systems for forecasting urban meteorology, air pollution and population exposure. *Atmospheric Chemistry and Physics, 7*(3), 855–874.

Banwart, S. (2011). Save our soils. *Nature, 474*(7350), 151–152.

Bartrons, M., Catalan, J., & Penuelas, J. (2016). Spatial and temporal trends of organic pollutants in vegetation from remote and rural areas. *Scientific Reports, 6*(1), 1–10.

Bharti, S. K., Kumar, D., Anand, S., Barman, S. C., & Kumar, N. (2017). Characterization and morphological analysis of individual aerosol of PM10 in urban area of Lucknow, India. *Micron,* 103, 90–98.

Bharti, S. K., Barman, S. C., & Kumar, N. (2020). Organochlorine pesticides (OCPs) in atmospheric particulate matter: sources and effects. In: Shukla, V., Kumar, N. (eds) *Environmental Concerns and Sustainable Development* (pp. 97–111). Springer, Singapore.

Bian, X., Liu, H., Gan, J., Li, R., & Yang, J. (2009). HCH and DDT residues in bivalves *Anodonta woodiana* from the Taihu Lake, China. *Archives of Environmental Contamination and Toxicology, 56*(1), 67–76.

Blaustein, A. R., Walls, S. C., Bancroft, B. A., Lawler, J. J., Searle, C. L., & Gervasi, S. S. (2010). Direct and indirect effects of climate change on amphibian populations. *Diversity, 2*(2), 281–313.

Breivik, K., Vestreng, V., Rozovskaya, O., & Pacyna, J. M. (2006). Atmospheric emissions of some POPs in Europe: a discussion of existing inventories and data needs. *Environmental Science & Policy, 9*(7–8), 663–674.

Buccini, J. (2003). The development of a global treaty on persistent organic pollutants (POPs). In *Persistent Organic Pollutants* (pp. 13–30). Springer, Berlin, Heidelberg.

Budyko, M. I. (1969). The effect of solar radiation variations on the climate of the Earth. *Tellus, 21*(5), 611–619.

Bustnes, J. O., Gabrielsen, G. W., & Verreault, J. (2010). Climate variability and temporal trends of persistent organic pollutants in the Arctic: a study of glaucous gulls. *Environmental Science & Technology, 44*(8), 3155–3161.

Carlson, D. A., Konyha, K. D., Wheeler, W. B., Marshall, G. P., & Zaylskie, R. G. (1976). Mirex in the environment: its degradation to Kepone and related compounds. *Science, 194*(4268), 939–941.

Carson, G.R., Edwards, N.M. (2009). Criteria of wheat and flour quality. In: Khan, K., Shewry, P. (eds) *Wheat Chemistry and Technology*, 4th ed (pp. 97–118.). American Association of Cereal Chemists, St Paul, MN.

Chakraborty, P., Zhang, G., Li, J., Xu, Y., Liu, X., Tanabe, S., & Jones, K. C. (2010). Selected organochlorine pesticides in the atmosphere of major Indian cities: levels, regional versus local variations, and sources. *Environmental Science & Technology, 44*(21), 8038–8043.

Cook, J. W., Hewett, C. L., & Hieger, I. (1933). The isolation of a cancer-producing hydrocarbon from coal tar-Parts I, II and III. *Journal of the Chemical Society (Resumed),* 395–405.

Coscolla, C., Munoz, A., Borras, E., Vera, T., Rodenas, M., & Yusa, V. (2014). Particle size distributions of currently used pesticides in ambient air of an agricultural Mediterranean area. *Atmospheric Environment, 95*, 29–35.

Cousins, I. T., & Jones, K. C. (1998). Air–soil exchange of semi-volatile organic compounds (SOCs) in the UK. *Environmental Pollution, 102*(1), 105–118.

Cousins, I. T., & Mackay, D. (2001). Gas–particle partitioning of organic compounds and its interpretation using relative solubilities. *Environmental Science & Technology, 35*(4), 643–647.

Croom, E. L., Shafer, T. J., Evans, M. V., Mundy, W. R., Eklund, C. R., Johnstone, A. F., ... & Pegram, R. A. (2015). Improving in vitro to in vivo extrapolation by incorporating toxicokinetic measurements: a case study of lindane-induced neurotoxicity. *Toxicology and Applied Pharmacology, 283*(1), 9–19.

Davy, C. W. (2004). Legislation with respect to dioxins in the workplace. *Environment International, 30*(2), 219–233.

De Cruz, I., Lacroix, G., Mougin, C., & Grolleau, G. (1996). Residues of chlorinated pesticides in eggs of the gray heron (*Ardea cinerea* L.): contribution of capillary gas chromatography ion-trap mass detection. *Journal of High Resolution Chromatography, 19*(1), 62–64.

Dearth, M. A., & Hites, R. A. (1991). Chlordane accumulation in people. *Environmental Science & Technology, 25*(7), 1279–1285.

Degrendele, C., Okonski, K., Melymuk, L., Landlová, L., Kukučka, P., Audy, O., ... & Klánová, J. (2016). Pesticides in the atmosphere: a comparison of gas-particle partitioning and particle size distribution of legacy and current-use pesticides. *Atmospheric Chemistry and Physics, 16*(3), 1531–1544.

Devi, N. L., Qi, S., Chakraborty, P., Zhang, G., & Yadav, I. C. (2011). Passive air sampling of organochlorine pesticides in a northeastern state of India, Manipur. *Journal of Environmental Sciences, 23*(5), 808–815.

Diffenbaugh, N. S., Hertel, T. W., Scherer, M., & Verma, M. (2012). Response of corn markets to climate volatility under alternative energy futures. *Nature Climate Change, 2*(7), 514–518.

Elmore, S. E., & La Merrill, M. A. (2019). Oxidative phosphorylation impairment by DDT and DDE. *Frontiers in Endocrinology, 10*, 122.

Engel, L. S., Zabor, E. C., Satagopan, J., Widell, A., Rothman, N., O'Brien, T. R., ... & Grimsrud, T. K. (2019). Prediagnostic serum organochlorine insecticide concentrations and primary liver cancer: a case–control study nested within two prospective cohorts. *International Journal of Cancer, 145*(9), 2360–2371.

European Food Safety Authority (2008). Perfluorooctane sulfonate (PFOS), perfluorooctanoic acid (PFOA) and their salts: scientific opinion of the panel on contaminants in the food chain. *EFSA Journal, 6*(7), 653.

Faroon, O., Kueberuwa, S., Smith, L., & Derosa, C. (1995). ATSDR evaluation of health effects of chemicals: II. Mirex and chlordecone: health effects, toxicokinetics, human exposure, and environmental fate. *Toxicology and Industrial Health, 11*(6), 1–195.

Faroon, O., Ruiz, P., Jones, D., Keith, S., Scinicariello, F., McClure, P., ... & Stevens, S. (2019). Toxicological profile for DDT, DDE, and DDD: draft for public comment.

Fiedler, H. (2008). Stockholm convention on POPs: obligations and implementation. In: Mehmetli, E., Koumanova, B. (eds) *The Fate of Persistent Organic Pollutants in the Environment* (pp. 3–12). Springer, Dordrecht.

Fussel, H. M. (2010). Review and Quantitative Analysis of Indices of Climate Change Exposure, Adaptive Capacity, Sensitivity, and Impacts. Background note to the World Development Report 2010, The International Bank for Reconstruction and Development / The World Bank, Washington, DC, USA, 34 pp.

Gaur, N., Narasimhulu, K., & PydiSetty, Y. (2018). Recent advances in the bio-remediation of persistent organic pollutants and its effect on environment. *Journal of Cleaner Production, 198*, 1602–1631.

Godfray, H. C. J., Beddington, J. R., Crute, I. R., Haddad, L., Lawrence, D., Muir, J. F., ... & Toulmin, C. (2010). Food security: the challenge of feeding 9 billion people. *Science, 327*(5967), 812–818.

Grannas, A. M., Bogdal, C., Hageman, K. J., Halsall, C., Harner, T., Hung, H., ... & Wania, F. (2013). The role of the global cryosphere in the fate of organic contaminants. *Atmospheric Chemistry and Physics, 13*(6), 3271–3305.

Guo, Z., Xie, H. Q., Zhang, P., Luo, Y., Xu, T., Liu, Y., ... & Zhao, B. (2018). Dioxins as potential risk factors for autism spectrum disorder. *Environment International, 121*, 906–915.

Gupta, P. K. (2004). Pesticide exposure—Indian scene. *Toxicology, 198*(1–3), 83–90.

Hagemen, K. J., Bogdal, C. and Scheringer, M. (2015).Long-range and regional atmospheric transport of POPs and implications for global cycling. In: Barcelo, D. (ed.) *Comprehensive Analytical Chemistry* (pp. 363–387). Elsevier, Amsterdam, Netherlands.

Hanedar, A., Güneş, E., Kaykioğlu, G., Çelik, S. Ö., & Cabi, E. (2019). Presence and distributions of POPS in soil, atmospheric deposition, and bioindicator samples in an industrial-agricultural area in Turkey. *Environmental Monitoring and Assessment, 191*(1), 1–15.

Hansen, J., Sato, M., Ruedy, R., Schmidt, G. A., Lo, K., & Persin, A. (2019). *Global Temperature in 2018 and Beyond.* Earth Institute, Columbia University, New York.

Harrad, S. J., Sewart, A. P., Alcock, R., Boumphrey, R., Burnett, V., Duarte-Davidson, R., ... & Jones, K. C. (1994). Polychlorinated biphenyls (PCBs) in the British environment: sinks, sources and temporal trends. *Environmental Pollution, 85*(2), 131–146.

Hays, S. M., & Aylward, L. L. (2003). Dioxin risks in perspective: past, present, and future. *Regulatory Toxicology and Pharmacology, 37*(2), 202–217.

Hedman, B., Näslund, M., & Marklund, S. (2006). Emission of PCDD/F, PCB, and HCB from combustion of firewood and pellets in residential stoves and boilers. *Environmental Science & Technology, 40*(16), 4968–4975.

Herzig, R., Lohmann, N., & Meier, R. (2019). Temporal change of the accumulation of persistent organic pollutants (POPs) and polycyclic aromatic hydrocarbons (PAHs) in lichens in Switzerland between 1995 and 2014. *Environmental Science and Pollution Research, 26*(11), 10562–10575.

Horton, M. K., Bousleiman, S., Jones, R., Sjodin, A., Liu, X., Whyatt, R., ... & Factor-Litvak, P. (2013). Predictors of serum concentrations of polybrominated flame retardants among healthy pregnant women in an urban environment: a cross-sectional study. *Environmental Health, 12*(1), 1–13.

Hou, P., & Wu, S. (2016). Long-term changes in extreme air pollution meteorology and the implications for air quality. *Scientific Reports, 6*(1), 1–9.

IARC Working Group on the Evaluation of Carcinogenic Risks to Humans, International Agency for Research on Cancer, & World Health Organization (1995). *Dry Cleaning, Some Chlorinated Solvents and Other Industrial Chemicals* (Vol. 63). World Health Organization, Geneva, Switzerland.

Intergovernmental Panel on Climate Change (IPCC) (2014). *Climate Change 2013: The Physical Science Basis. Contribution of Working Group I to the Fifth Assessment Report of the Intergovernmental Panel on Climate Change*, Stocker, T. F. et al. (eds) Cambridge University Press, Cambridge, UK.

Iwata, H., Tanabe, S., Sakai, N., Nishimura, A., & Tatsukawa, R. (1994). Geographical distribution of persistent organochlorines in air, water and sediments from Asia and Oceania, and their implications for global redistribution from lower latitudes. *Environmental Pollution, 85*(1), 15–33.

Jayaraj, R., Megha, P., & Sreedev, P. (2016). Organochlorine pesticides, their toxic effects on living organisms and their fate in the environment. *Interdisciplinary Toxicology, 9*(3–4), 90.

Jensen, S., Johnels, A. G., Olsson, M., & Otterlind, G. (1969). DDT and PCB in marine animals from Swedish waters. *Nature, 224*(5216), 247–250.

Jing, R., Fusi, S., & Kjellerup, B. V. (2018). Remediation of polychlorinated biphenyls (PCBs) in contaminated soils and sediment: state of knowledge and perspectives. *Frontiers in Environmental Science, 6*, 79.

Jinxian, H., Huijuan, L., Jiuhui, Q., Jia, R., Haining, L., & Guoting, L. (2005). Dieldrin and endrin removal from water by triolein-embedded adsorbent. *Chinese Science Bulletin, 50*(23), 2696–2700.

Johri, A. K., Dua, M., Tuteja, D., Saxena, R., Saxena, D. M., & Lal, R. (1996). Genetic manipulations of microorganisms for the degradation of hexachlorocyclohexane. *FEMS Microbiology Reviews, 19*(2), 69–84.

Juhasz, A. L., & Naidu, R. (2000). Bioremediation of high molecular weight polycyclic aromatic hydrocarbons: a review of the microbial degradation of benzo [a] pyrene. *International Biodeterioration & Biodegradation, 45*(1–2), 57–88.

Kaiser, K. L. (1978). Pesticide report: the rise and fall of Mirex. *Environmental Science & Technology, 12*(5), 520–528.

Kallenborn, R., Halsall, C., Dellong, M., & Carlsson, P. (2012). The influence of climate change on the global distribution and fate processes of anthropogenic persistent organic pollutants. *Journal of Environmental Monitoring, 14*(11), 2854–2869.

Kelly, B. C., Ikonomou, M. G., Blair, J. D., Surridge, B., Hoover, D., Grace, R., & Gobas, F. A. (2009). Perfluoroalkyl contaminants in an Arctic marine food web: trophic magnification and wildlife exposure. *Environmental Science & Technology, 43*(11), 4037–4043.

Kim, K. H., Jahan, S. A., Kabir, E., & Brown, R. J. (2013). A review of airborne polycyclic aromatic hydrocarbons (PAHs) and their human health effects. *Environment International, 60*, 71–80.

Klanova, J., Eupr, P., Kohoutek, J., & Harner, T. (2008). Assessing the influence of meteorological parameters on the performance of polyurethane foam-based passive air samplers. *Environmental Science & Technology, 42*(2), 550–555.

Kulkarni, P. S., Crespo, J. G., & Afonso, C. A. (2008). Dioxins sources and current remediation technologies—a review. *Environment International, 34*(1), 139–153.

Kumar, D., & Pannu, R. (2018). Perspectives of lindane (γ-hexachlorocyclohexane) biodegradation from the environment: a review. *Bioresources and Bioprocessing, 5*(1), 1–18.

Kumar, N., Kulsoom, M., Shukla, V., Kumar, D., Kumar, S., Tiwari, J., & Dwivedi, N. (2018). Profiling of heavy metal and pesticide residues in medicinal plants. *Environmental Science and Pollution Research, 25*(29), 29505–29510.

Kurt-Karakus, P. B., Bidleman, T. F., & Jones, K. C. (2005). Chiral organochlorine pesticide signatures in global background soils. *Environmental Science & Technology, 39*(22), 8671–8677.

Lamon, L., Von Waldow, H., MacLeod, M., Scheringer, M., Marcomini, A., & Hungerbühler, K. (2009). Modeling the global levels and distribution of polychlorinated biphenyls in air under a climate change scenario. *Environmental Science & Technology, 43*(15), 5818–5824.

Lewis, M. R., Carr, M. E., Feldman, G. C., Esaias, W., & McClain, C. (1990). Influence of penetrating solar radiation on the heat budget of the equatorial Pacific Ocean. *Nature, 347*(6293), 543–545.

Liu, X., Fiedler, H., Gong, W., Wang, B., & Yu, G. (2018). Potential sources of unintentional produced PCB, HCB, and PeCBz in China: a preliminary overview. *Frontiers of Environmental Science & Engineering, 12*(6), 1.

Loganathan, B. G., & Lam, P. K. S. (Eds.) (2011). *Global Contamination Trends of Persistent Organic Chemicals*. CRC Press, Boca Raton.

Lohmann, R., & Lammel, G. (2004). Adsorptive and absorptive contributions to the gas-particle partitioning of polycyclic aromatic hydrocarbons: state of knowledge and recommended parametrization for modeling. *Environmental Science & Technology, 38*(14), 3793–3803.

Ma, J., Hung, H., & Macdonald, R. W. (2016). The influence of global climate change on the environmental fate of persistent organic pollutants: a review with emphasis on the Northern Hemisphere and the Arctic as a receptor. *Global and Planetary Change, 146*, 89–108.

Macdonald, R. W. (2005). Climate change, risks and contaminants: a perspective from studying the Arctic. *Human and Ecological Risk Assessment, 11*(6), 1099–1104.

Madaj, R., Sobiecka, E., & Kalinowska, H. (2018). Lindane, kepone and pentachlorobenzene: chloropesticides banned by Stockholm convention. *International Journal of Environmental Science and Technology, 15*(2), 471–480.

Matsumoto, E., Kawanaka, Y., Yun, S. J., & Oyaizu, H. (2009). Bioremediation of the organochlorine pesticides, dieldrin and endrin, and their occurrence in the environment. *Applied Microbiology and Biotechnology, 84*(2), 205–216.

Mawussi, G., Sanda, K., Merlina, G., & Pinelli, E. (2009). Assessment of average exposure to organochlorine pesticides in southern Togo from water, maize (*Zea mays*) and cowpea (*Vigna unguiculata*). *Food Additives & Contaminants: Part A, 26*(3), 348–354.

McFarland, V. A., & Clarke, J. U. (1989). Environmental occurrence, abundance, and potential toxicity of polychlorinated biphenyl congeners: considerations for a congener-specific analysis. *Environmental Health Perspectives, 81*, 225–239.

Meijer, P. T., & Dijkstra, H. A. (2009). The response of Mediterranean thermohaline circulation to climate change: a minimal model. *Climate of the Past, 5*(4), 713–720.

Ming, T., Liu, W., & Caillol, S. (2014). Fighting global warming by climate engineering: is the Earth radiation management and the solar radiation management any option for fighting climate change? *Renewable and Sustainable Energy Reviews, 31*, 792–834.

Mishra, C., Sharma, S., & Kakkar, P. (2007). A study to evaluate heavy metals and organochlorine pesticide residue in *Zingiber officinale* Rosc. collected from different ecological zones of India. *Bulletin of Environmental Contamination and Toxicology, 79*(1), 95–98.

Mitchell, T. D., & Jones, P. D. (2005). An improved method of constructing a database of monthly climate observations and associated high-resolution grids. *International Journal of Climatology: A Journal of the Royal Meteorological Society, 25*(6), 693–712.

Moro, S., Chipman, J. K., Wegener, J. W., Hamberger, C., Dekant, W., & Mally, A. (2012). Furan in heat-treated foods: formation, exposure, toxicity, and aspects of risk assessment. *Molecular Nutrition & Food Research, 56*(8), 1197–1211.

Muir, D. C., & Howard, P. H. (2006). Are there other persistent organic pollutants? A challenge for environmental chemists. *Environmental Science & Technology, 40*(23), 7157–7166.

Nadal, M., Marquès, M., Mari, M., & Domingo, J. L. (2015). Climate change and environmental concentrations of POPs: a review. *Environmental Research, 143*, 177–185.

Nascimento, M. M., da Rocha, G. O., & de Andrade, J. B. (2017). Pesticides in fine airborne particles: from a green analysis method to atmospheric characterization and risk assessment. *Scientific Reports, 7*(1), 1–11.

Okeke, B. C., Siddique, T., Arbestain, M. C., & Frankenberger, W. T. (2002). Biodegradation of γ-hexachlorocyclohexane (lindane) and α-hexachlorocyclohexane in water and a soil slurry by a Pandoraea species. *Journal of Agricultural and Food Chemistry, 50*(9), 2548–2555.

Pankow, J. F. (1987). Review and comparative analysis of the theories on partitioning between the gas and aerosol particulate phases in the atmosphere. *Atmospheric Environment (1967), 21*(11), 2275–2283.

Pariatamby, A., & Kee, Y. L. (2016). Persistent organic pollutants management and remediation. *Procedia Environmental Sciences, 31*, 842–848.

Park, E. Y., Park, E., Kim, J., Oh, J. K., Kim, B., Hong, Y. C., & Lim, M. K. (2020). Impact of environmental exposure to persistent organic pollutants on lung cancer risk. *Environment International, 143*, 105925.

Pereira, É. R., Merib, J., Cadorim, H. R., Schneider, M., Carvalho, G. S., Duarte, F. A., ... & Feldmann, J. (2017). Development of a fast screening method for the direct determination of chlorinated persistent organic pollutants in fish oil by high-resolution continuum source graphite furnace molecular absorption spectrometry. *Food Control, 78*, 456–462.

Purudue, M. P., Hoppin, J. A., Blair, A., Dosemeci, M., & Alavanja, M. C. (2007). Occupational exposure to organochlorine insecticides and cancer incidence in the Agricultural Health Study. *International Journal of Cancer, 120*(3), 642–649.

Rani, Z. T., Hugo, A., Hugo, C. J., Vimiso, P., & Muchenje, V. (2017). Effect of post-slaughter handling during distribution on microbiological quality and safety of meat in the formal and informal sectors of South Africa: a review. *South African Journal of Animal Science, 47*(3), 255–267.

Rasenberg, M. H. C., & van de Plassche, E. J. (2003). "Risk Profile and Summary Report for Dicofol," Dossier prepared for the UNECE Convention on Long-range Transboundary Air Pollution's Expert Group on POPs. Royal Haskoning, Breda, The Netherlands.

Ravindra, K., Sokhi, R., & Van Grieken, R. (2008). Atmospheric polycyclic aromatic hydrocarbons: source attribution, emission factors and regulation. *Atmospheric Environment, 42*(13), 2895–2921.

Ricke, K. L., Morgan, M. G., & Allen, M. R. (2010). Regional climate response to solar-radiation management. *Nature Geoscience, 3*(8), 537–541.

Ricking, M., & Schwarzbauer, J. (2012). DDT isomers and metabolites in the environment: an overview. *Environmental Chemistry Letters, 10*(4), 317–323.

Robertson, L. W., Weber, R., Nakano, T., & Johansson, N. (2018). PCBs risk evaluation, environmental protection, and management: 50-year research and counting for elimination by 2028. *Environmental Science and Pollution Research, 25*(17), 16269–16276.

Robinson, H., Buccini, G., Curry, L., & Perez-Escamilla, R. (2019). The World Health Organization Code and exclusive breastfeeding in China, India, and Vietnam. *Maternal & Child Nutrition, 15*(1), e12685.

Roslund, M. I., Grönroos, M., Rantalainen, A. L., Jumpponen, A., Romantschuk, M., Parajuli, A., ... & Sinkkonen, A. (2018). Half-lives of PAHs and temporal microbiota changes in commonly used urban landscaping materials. *PeerJ, 6*, e4508.

Rummukainen, M., Fuhrer, J., & Gregory, P. (2014). Climate projections for 2050. In: Fuhrer, J., Gregory, P. (eds) *Climate Change Impact and Adaptation in Agricultural Systems* (pp. 7–16). CABI, Wallingford.

Safe, S. (1990). Polychlorinated biphenyls (PCBs), dibenzo-p-dioxins (PCDDs), dibenzofurans (PCDFs), and related compounds: environmental and mechanistic considerations which support the development of toxic equivalency factors (TEFs). *Critical Reviews in Toxicology, 21*(1), 51–88.

Salihoglu, G., Salihoglu, N. K., Aksoy, E., & Tasdemir, Y. (2011). Spatial and temporal distribution of polychlorinated biphenyl (PCB) concentrations in soils of an industrialized city in Turkey. *Journal of Environmental Management, 92*(3), 724–732.

Sarcinelli, P. N., Pereira, A. C. S., Mesquita, S. A., Oliveira-Silva, J. J., Meyer, A., Menezes, M. A., ... & Wolff, M. (2003). Dietary and reproductive determinants of plasma organochlorine levels in pregnant women in Rio de Janeiro. *Environmental Research, 91*(3), 143–150.

Sarkar, S. K., Bhattacharya, B. D., Bhattacharya, A., Chatterjee, M., Alam, A., Satpathy, K. K., & Jonathan, M. P. (2008). Occurrence, distribution and possible sources of organochlorine pesticide residues in tropical coastal environment of India: an overview. *Environment International, 34*(7), 1062–1071.

Schiavon, M., Torretta, V., Rada, E. C., & Ragazzi, M. (2016). State of the art and advances in the impact assessment of dioxins and dioxin-like compounds. *Environmental Monitoring and Assessment, 188*(1), 57.

Schmidt, H., & Schultz, G. (1881). Ueber Benzidin (α-Diamidodiphenyl). *Justus Liebigs Annalen der Chimie, 207*(3), 320–347.

Sharkey, M., Harrad, S., Abdallah, M. A. E., Drage, D. S., & Berresheim, H. (2020). Phasing-out of legacy brominated flame retardants: the UNEP Stockholm Convention and other legislative action worldwide. *Environment International, 144*, 106041.

Sielken Jr, R. L., & Valdez-Flores, C. (1999). Probabilistic risk assessment's use of trees and distributions to reflect uncertainty and variability and to overcome the limitations of default assumptions. *Environment International, 25*(6–7), 755–772.

Socorro, J., Durand, A., Temime-Roussel, B., Gligorovski, S., Wortham, H., & Quivet, E. (2016). The persistence of pesticides in atmospheric particulate phase: an emerging air quality issue. *Scientific Reports, 6*(1), 1–7.

Solomon, S., Manning, M., Marquis, M., & Qin, D. (2007). *Climate Change 2007-the Physical Science Basis: Working Group I Contribution to the Fourth Assessment Report of the IPCC (Vol.* 4). Cambridge University Press, Cambridge, UK.

Solomon-Wisdom, G. O., & Ndana, R. W. (2012). Dioxin effect on human health-a review. *Journal of Global Biosciences, 1*, 10–16.

Srimurali, S., Govindaraj, S., Kumar, S. K., & Rajendran, R. B. (2015). Distribution of organochlorine pesticides in atmospheric air of Tamilnadu, southern India. *International Journal of Environmental Science and Technology, 12*(6), 1957–1964.

Stemmler, I., & Lammel, G. (2012). Long-term trends of continental-scale PCB patterns studied using a global atmosphere–ocean general circulation model. *Environmental Science and Pollution Research, 19*(6), 1971–1980.

Stevenson, D. E., Walborg Jr, E. F., North, D. W., Sielken Jr, R. L., Ross, C. E., Wright, A. S., & Klaunig, J. E. (1999). Monograph: reassessment of human cancer risk of aldrin/dieldrin. *Toxicology Letters, 109*(3), 123–186.

Suami, R. B., Sivalingam, P., Al Salah, D. M., Grandjean, D., Mulaji, C. K., Mpiana, P. T., & Poté, J. (2020). Heavy metals and persistent organic pollutants contamination in river, estuary, and marine sediments from Atlantic Coast of Democratic Republic of the Congo. *Environmental Science and Pollution Research, 27*(16), 20000–20013.

Takasuga, T., Nakano, T., & Shibata, Y. (2012). Unintentional POPs (PCBs, PCBz, PCNs) contamination in articles containing chlorinated paraffins and related impacted chlorinated paraffin products. *Organohalogen Compounds, 74*, 1437–1440.

Tang, C. Y., Fu, Q. S., Robertson, A. P., Criddle, C. S., & Leckie, J. O. (2006). Use of reverse osmosis membranes to remove perfluorooctane sulfonate (PFOS) from semiconductor wastewater. *Environmental Science & Technology, 40*(23), 7343–7349.

Tao, S., Li, B. G., He, X. C., Liu, W. X., & Shi, Z. (2007). Spatial and temporal variations and possible sources of dichlorodiphenyltrichloroethane (DDT) and its metabolites in rivers in Tianjin, China. *Chemosphere, 68*(1), 10–16.

Teran, T., Lamon, L., & Marcomini, A. (2012). Climate change effects on POPs' environmental behaviour: a scientific perspective for future regulatory actions. *Atmospheric Pollution Research, 3*(4), 466–476.

Tesfahunegny, W. (2016). Impact of pesticides on birds from DDT to current fatality: a literature review. *Journal of Zoology Studies, 3*(2), 44–55.

Thakur, M., & Pathania, D. (2020). Environmental fate of organic pollutants and effect on human health. In: Singh, P., Kumar, A., & Borthakur, A., *Abatement of Environmental Pollutants* (pp. 245–262). Elsevier, Amsterdam, Netherlands.
Tren, R. (2002). Malaria and Climate Change. Working papers series, Julian Simon Centre for Policy Research.
Tren, R., & Bate, R. (2004). South Africa's War against Malaria. Lessons from the Developing World. Policy Analysis.
United Nations Environment Programme (2001). Stockholm convention on persistent organic pollutants (POPs).
United Nations Environment Programme (2017). *Guidance on Best Available Techniques and Best Environmental Practices for the Recycling and Disposal of Articles Containing Polybrominated Diphenyl Ethers (PBDEs) Listed under the Stockholm Convention on Persistent Organic Pollutants.* United Nations Institute for Training and Research (UNITAR), Geneva, Switzerland.
US Environmental Protection Agency (2009). Long-chain perfluorinated chemicals (PFCs) action plan. "Health Effects Support Document for Aldrin/Dieldrin."
van den Berg, H. (2009). Global status of DDT and its alternatives for use in vector control to prevent disease. *Environmental Health Perspectives, 117*(11), 1656–1663.
van Diedenhoven, B. (2021). Variation of ice microphysical properties with temperature and humidity at tops of convective clouds. *Geophysical Research Letters*, *48*(16), e2021GL093673.
Wang, S., Huang, J., Yang, Y., Hui, Y., Ge, Y., Larssen, T., ... & Harman, C. (2013). First report of a Chinese PFOS alternative overlooked for 30 years: its toxicity, persistence, and presence in the environment. *Environmental Science & Technology*, *47*(18), 10163–10170.
Wania, F., & Mackay, D. (1996). Peer reviewed: tracking the distribution of persistent organic pollutants. *Environmental Science & Technology, 30*(9), 390A–396A.
Weber, R., Watson, A., Malkov, M., Costner, P., & Vijgen, J. (2011). Unintentionally produced hexachlorobenzene and pentachlorobenzene POPs waste from solvent production–the need to establish emission factors and inventories. *Organohalogen Compounds, 73*, 2205–2208.
Wehner, M. F. (2020). Characterization of long period return values of extreme daily temperature and precipitation in the CMIP6 models: part 2, projections of future change. *Weather and Climate Extremes, 30*, 100284.
Wettestad, J. (2002). The convention on long-range transboundary air pollution (CLRTAP). In: Miles, E.L., Underdal, A., Andresen, S., Wettestad, J., Skjaerseth, J.B., & Carlin, E.M. (eds). *Environmental Regime Effectiveness: Confronting Theory with Evidence* (pp. 197–221). MIT Press, Cambridge, MA.
Whitmarsh, L. (2011). Scepticism and uncertainty about climate change: dimensions, determinants and change over time. *Global Environmental Change, 21*(2), 690–700.
Whylie, P., Albaiges, J., Barra, R., Bouwman, H., Dyke, D., & Wania, F. (2003). *Regionally Based Assessment of Persistent Toxic Substances. Global Report.* United Nations Environment Programme (UNEP), Geneva, Switzerland.
Wilcke, W., Krauss, M., Safronov, G., Fokin, A. D., & Kaupenjohann, M. (2006). Polychlorinated biphenyls (PCBs) in soils of the Moscow region: concentrations and small-scale distribution along an urban–rural transect. *Environmental Pollution, 141*(2), 327–335.
Woehrnschimmel, R., Kral, C., Muellner, F., Wild, S., Neudorfer, H., & Dangl, F. (2013). A combined hysteresis and eddy-current model developed for a wide frequency range in electric machine applications. In *2013 IEEE Energy Conversion Congress and Exposition* (pp. 3180–3185).
Wong, M. H., Wu, S. C., Deng, W. J., Yu, X. Z., Luo, Q., Leung, A. O. W., ... & Wong, A. S. (2007). Export of toxic chemicals–a review of the case of uncontrolled electronic-waste recycling. *Environmental Pollution, 149*(2), 131–140.
World Health Organization (1979). *DDT and Its Derivatives*. World Health Organization, Geneva, Switzerland.
World Health Organization (2003). Health risks of persistent organic pollutants from long-range transboundary air pollution. In: *Hexachlorocyclohexane* (pp. 61–85). Joint WHO/Convention Task Force on the Health Aspects of Air Pollution, Geneva, Switzerland.
Wurl, O., & Obbard, J. P. (2005). Organochlorine pesticides, polychlorinated biphenyls and polybrominated diphenyl ethers in Singapore's coastal marine sediments. *Chemosphere, 58*(7), 925–933.
Yang, L., Zheng, M., Zhao, Y., Yang, Y., Li, C., & Liu, G. (2019). Unintentional persistent organic pollutants in cement kiln co-processing solid wastes. *Ecotoxicology and Environmental Safety, 182*, 109373.

Yoshimura, T., Yasuo, S., Watanabe, M., Iigo, M., Yamamura, T., Hirunagi, K., & Ebihara, S. (2003). Light-induced hormone conversion of T 4 to T 3 regulates photoperiodic response of gonads in birds. *Nature, 426*(6963), 178–181.

Yu, Y., Huang, J., Zhang, W., Zhang, K., Deng, S., & Yu, G. (2013). Mechanochemical destruction of mirex co-ground with iron and quartz in a planetary ball mill. *Chemosphere, 90*(5), 1729–1735.

Zeidler, O. (1874). I. Verbindungen von Chloral mit Brom-und Chlorbenzol. *Berichte der Deutschen Chemischen Gesellschaft*, *7*(2), 1180–1181.

Zhao, D., & Li, Y. R. (2015). Climate change and sugarcane production: potential impact and mitigation strategies. *International Journal of Agronomy*, *5*(1), 1–11.

Zhou, Z., & Ma, L. (2013). Research advance of climate change influence on environmental behavior of POPs. *Environmental Pollution Control, 35*, 84–87.

Zitko, V. (2003). Chlorinated pesticides: Aldrin, DDT, endrin, dieldrin, mirex. In: Fiedler, H. (Ed.), *The Handbook of Environmental Chemistry: Part O, Persistent Organic Pollutants* (Vol. 3) (pp. 47–90). Springer-Verlag, Berlin.

Zuloaga, O., Etxebarria, N., Fernández, L. A., & Madariaga, J. M. (2000). Optimization and comparison of MAE, ASE and Soxhlet extraction for the determination of HCH isomers in soil samples. *Fresenius' Journal of Analytical Chemistry, 367*(8), 733–737.

3 DNA Barcoding and Metabarcoding

A Potential Tool for Environmental Biodiversity Monitoring

Archana Chauhan, Phuntsog Dolma, Anuragini Kadwalia, and Ahmad Ali
Department of Zoology, Panjab University, Chandigarh

R.C. Sobti
Department of Biotechnology, Panjab University, Chandigarh

CONTENTS

3.1 INTRODUCTION

Biodiversity is defined as the diversity of living organisms on planet Earth at genetic, species, and ecosystem level. Biodiversity analysis is one of the important areas that require extensive research, as it improves our understanding of life on our planet. The devastating effects of anthropogenic activities such as pollution, increasing population, animal trafficking, deforestation, use of pesticides, and plastic waste, etc., on various species. Due to such human activities, ecosystems are put under immense pressure, which adversely affects the flora and fauna. The loss of habitat due to deforestation and other human activities such as water pollution poses a higher level of threat to endemic and narrowly distributed taxonomic groups. For example, amphibians mostly occur in the tropical areas and have small geographical ranges. Recent investigations have revealed that one-third or more of the 6,300 amphibian species are threatened with extinction (Wake and Vredenburg, 2008). Over a few centuries and millennia, there has been a considerable loss in the number of species, which has led scientists to argue that a sixth mass extinction may be under way. So, biodiversity surveys have become critical for monitoring environmental health and for the management of natural resources. They should be done in all ecosystems, such as aquatic (ocean, lakes, ponds, rivers), land (agricultural land, desert, mountain), forest, grassland, etc., as early detection

DOI: 10.1201/9781003220824-4

of vulnerable, threatened species or invasive species can result in framing management and conservation strategies on time.

3.2 METHODS FOR BIODIVERSITY MONITORING

Traditionally, different methods are employed for assessing biodiversity, depending upon the habitat and behaviour of organisms. For example, great crested newts normally live on land and breed in ponds and pools. So, conventional survey methods for their detection include trapping, torchlight surveys, and egg counts (Rees et al., 2014). Similarly, for detection of fishes, methods such as electrofishing, gill nets, and trap nets are traditionally used for studying biodiversity. These survey methods can be time consuming and costly and need manpower. Also, morphological keys are sometimes only effective for a particular life stage or gender. Recently, advancements in DNA-based methods for identification of species and the reduced cost of sequencing techniques have provided an opportunity for employing DNA barcoding and metabarcoding for environment biodiversity analysis. While DNA barcoding focuses mainly on the detection of single species, the metagenomics approach is mainly concerned with revealing the overall diversity of a habitat using environmental DNA (eDNA). Combined with next-generation sequencing (NGS), eDNA analysis has become a powerful tool for biodiversity studies. eDNA refers to the genetic material that can be obtained from environmental samples such as snow, water, soil, or air. An organism continuously sheds skin, scales, faeces, gametes, wounded tissues, etc. into its environment (soil, water, air), and these become the source of eDNA. With the help of NGS, these smaller fragments of eDNA can be sequenced for identification up to the level of species or even subspecies with the help of bioinformatics tools.

3.3 DNA BARCODING

DNA barcoding was first defined by Hebert et al. (2003). Just as a barcode is printed on an item and used in the supermarket for identifying a specific item, a DNA barcode is a short fragment of DNA (500–1000 bp approx.) that is used to identify a particular known species. The ideal concept of DNA barcoding emphasizes finding one gene sequence that could be used for species detection across all taxonomic groups, from viruses to plants and animals. However, that gene has not yet been found, so different barcode DNA sequences are used for animals, plants, microbes, and viruses (Purty and Chatterjee, 2016). Usually, those segments of DNA that have low intraspecific and high interspecific variation are selected as barcodes, so that they can be used to identify and distinguish between known species. Here, the meaning of known species is those whose genomes are sequenced and the sequences submitted to available reference databases. Generally, mitochondrial DNA is preferred as a barcode due to its high copy number, less recombination, and maternal inheritance. For animals, cytochrome c oxidase I (COI) is commonly used as a barcode (Hebert et al., 2003), while 12S rRNA and 18S rRNA genes are also used. 16S rRNA genes are used for bacteria (Hiorns et al., 1997; Großkopf et al., 1998), 18S rRNA genes for eukaryotes (cytoplasmic and mitochondrial) (Hadziavdic et al., 2014; Yang et al., 2014), rbcL and matK for plants (Kress and Erickson, 2007), and ITS region for fungi (Schoch et al., 2012). For animals, a 650 bp region of mitochondrial cytochrome c oxidase I (COI) gene is considered as a "barcode of life" and provides identification up to species level of approximately 98% for large taxonomic groups of animals such as birds, mammals, fishes, and various arthropods. While full-length barcodes are used for barcode library construction, minibarcodes are applicable in the detection of museum-preserved, processed food and degraded samples. In the case of these samples with degraded DNA, such a large fragment of DNA is difficult to amplify due to degradation of DNA. For preservation of specimens in museums, formalin is required, which results in degradation of DNA, making the retrieval of a full-length barcode difficult. However, some minibarcodes can be amplified from such samples (Baird et al.,

2011). Therefore, for such samples with degraded DNA, minibarcodes are used, which are approximately 100 to 300 bp long (Hajibabaei et al., 2007).In the case of DNA barcoding using eDNA, minibarcodes of approximately 200 bp are preferred. This is because eDNA that is obtained from the environment is degraded as compared with DNA extracted from an organism's tissues, such as muscle, scales, hair, feathers, etc. In order to amplify the DNA barcode of the sample in question for identification purposes, the selection of the primer plays an important role. Several primers have been designed to amplify DNA barcodes and minibarcodes (Meusnier et al., 2008; Dubey et al., 2010; Sønstebø et al., 2010), or anyone can design their own primers with the help of software like PRIMER3. After the amplification of target barcode by polymerase chain reaction (PCR) using primers, sequencing is done on the appropriate sequencing platforms, such as Sanger sequencing or NGS, depending upon the type of sample being used (Figure 3.1). If the sample is taken from the organism itself and is in good condition, such as muscle tissue, scales, hair, etc., Sanger sequencing can detect the organism up to species level. But in the case of degraded DNA, NGS can provide better results, as it is more sensitive than traditional Sanger sequencing. After the sequencing, the sequence is compared with the reference sequences available in databases offered by platforms such as BOLD systems, GenBank, etc. for species detection. Over the past few decades, eDNA has emerged as a potential source for species detection using DNA barcoding. DNA barcoding using eDNA focuses on the detection of single species from the environment. The entire biotic community of the particular environment can also be analysed using barcoding of eDNA, but this is popularly known as metabarcoding and is explained later in this chapter. For DNA barcoding using eDNA, specific primers have been designed to amplify the barcode region. These primers are mainly species-specific primers, and the DNA can be amplified using conventional PCR or quantitative PCR (qPCR). Since eDNA is fragmented and can be present in lower quantities in the case of rare or endangered species, qPCR is preferred over conventional PCR, which is followed by NGS for authentic species identification. Using eDNA, barcoding can be used for the detection of invasive or rare species (Ficetola et al., 2008; Takahara et al., 2013; Wilcox et al., 2013).

3.4 METABARCODING

DNA metabarcoding is a special case of barcoding, which is applied to identify different taxa in samples that contain more than one organism (Dormontt et al., 2018). Such samples include soil, water, faeces, permafrost, etc. In DNA barcoding, DNA is extracted from the organism's tissue, such as muscles, feathers, etc., for its identification, and the focus is on the identification of single species, while in metabarcoding, DNA is extracted from the environment (eDNA) in which the organism is living and tends to identify the biodiversity of the sample. eDNA refers to the genetic material that can be obtained from environmental samples such as snow, water, soil, or air. An organism continuously sheds skin, scales, faeces, gametes, wounded tissues, etc. into its environment (soil, water, air), and these become the source of eDNA. eDNA metabarcoding was earlier used for the detection of microbial and fungal communities, but over the past decade, it has also been widely used for the detection of plant and animal species (Deiner et al., 2017). Since eDNA in the environment is more prone to degradation or fragmentation, metabarcoding is mainly aimed at amplifying short DNA fragments of fewer than 200 base pairs. With the help of NGS, these smaller fragments of eDNA can be sequenced for the taxon identification up to the level of species (Figure 3.2).

3.5 ESTIMATING BIODIVERSITY BY METABARCODING

i) Sample collection: Sample collection is a crucial step, as it determines the quality of data that will be generated after sequencing. Sterilization of all the sample collection apparatus is necessary, as any negligence in this may lead to contamination or detection of false positives. Consider, for

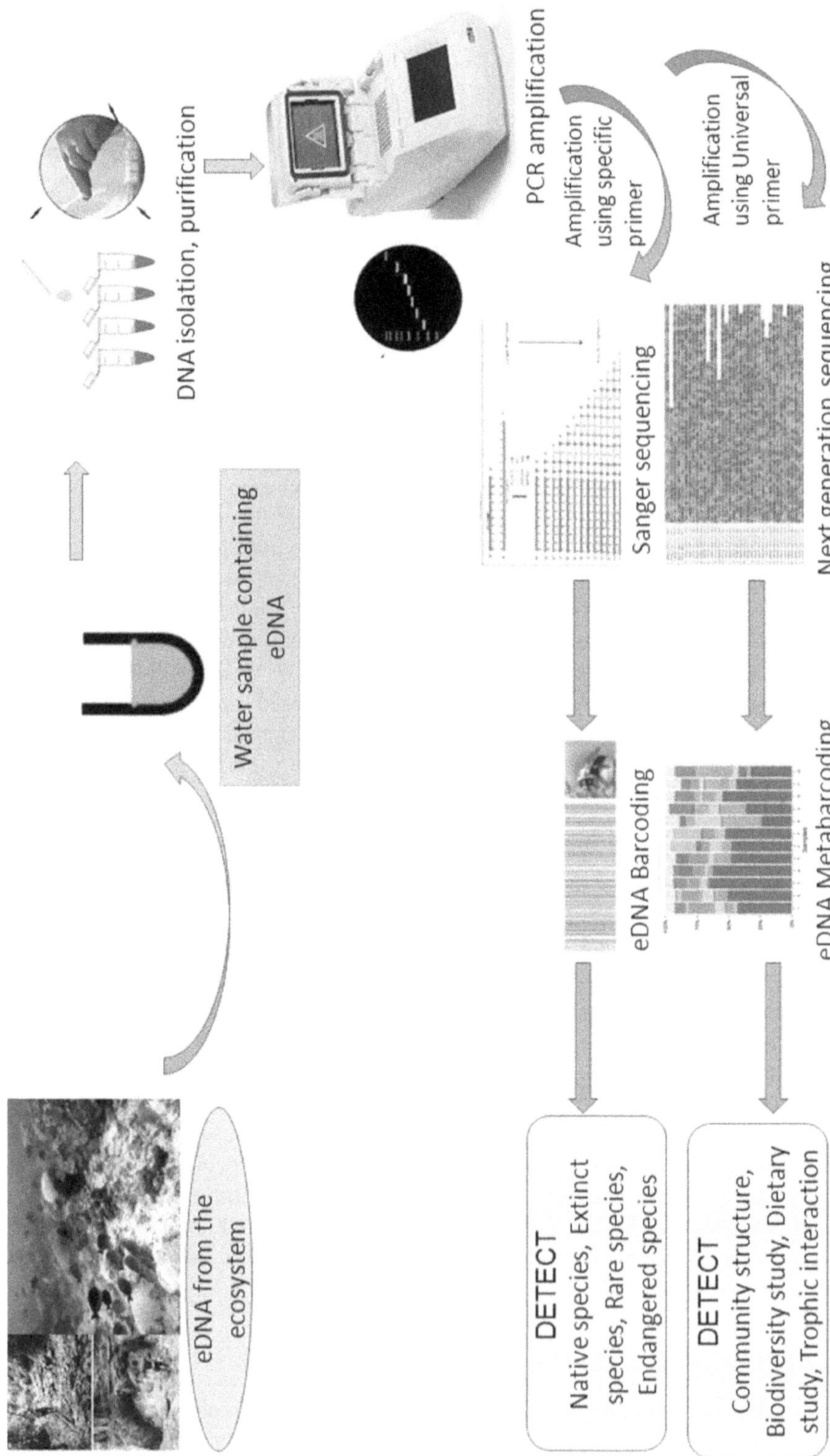

FIGURE 3.1 DNA barcoding and metabarcoding workflow.

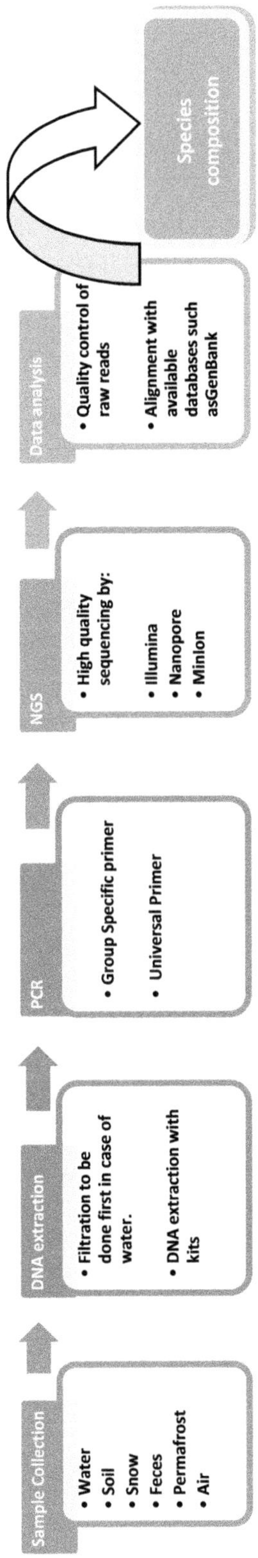

FIGURE 3.2 Metabarcoding workflow for analysing biodiversity.

FIGURE 3.3 Sampling of water at different points in a lake. Different sampling points ensure more biodiversity to be identified as more eDNA is collected. Dots of various colours indicate eDNA from variety of organisms.

example, that you are trying to identify the bacterial diversity of a high-altitude lake. If there is any bacterial contamination in the sample apparatus, that species will be shown to be present in that lake, although it is actually absent, leading to the detection of a false positive. Another important aspect of sample collection is the volume of sample taken during sampling and the range of sampling. Sampling a greater range increases the chances of identification of more species. So, the volume and range of sampling should be regarded as important (Figure 3.3).

ii) DNA extraction: Sample collection is followed by DNA extraction from the sample. Several commercial kits are available for this purpose, which are solely dedicated to the extraction of DNA from environmental samples and give good-quality DNA for further analysis (Shu et al., 2020). DNA extraction should be done in a separate room dedicated to DNA extraction to prevent contamination.

iii) PCR: In the eDNA technique, PCR is done to amplify a specific region (minibarcodes) of the DNA, which can be compared with the reference database available (such as NCBI, MitoFish, etc.), leading to the identification of the desired taxa (genus, species etc.). For biodiversity analysis using eDNA, two basic approaches are used; one is to detect the presence of a single species in a particular habitat (DNA barcoding), and the other is to estimate the overall biodiversity of a particular group, such as bacteria, mammals, fishes, amphibians, etc., by metagenomics in the habitat of interest. For both approaches, different kinds of primers are employed. For metabarcoding using eDNA, specific primers have been designed to amplify the minibarcode region, since eDNA is fragmented. These primers can be specific to a group, such as fishes (Miya et al., 2015; Valentini et al., 2016), Batrachia (Valentini et al., 2016), decapod crustaceans (Komai et al., 2019), cephalopod molluscs (de Jonge et al., 2021), etc., or broad-range "universal" primers, such as vertebrate-specific primers (Kelly et al., 2014) or metazoan primers (Lim et al., 2016). In some cases, species-specific primers are also used. The ultimate goal of all these primers is to amplify the target barcode for species identification.

iv) NGS and data analysis: After PCR, amplicons are subjected to sequencing. Metabarcoding requires high-throughput sequencing, which can be achieved by NGS platforms in which thousands of amplicons are sequenced in parallel. NGS platforms like Illumina Miseq and MinION are being used for metabarcoding. The use of eDNA combined with NGS is a powerful tool for monitoring species composition in aquatic bodies and hence will help with more accurate biodiversity estimation (Rees et al., 2014). The raw data thus obtained from NGS is subjected to data filtration, quality control, etc. with the help of various software available for bioinformatics analysis. The filtered sequences are then compared with the reference database available. This reference database is similar to the one used in DNA barcoding. The alignment of sequences obtained from NGS with the reference database leads to the identification of desired taxa (genus, species).

3.5.1 Applications of DNA Barcoding and Metabarcoding

Identification of species at any life stage: DNA barcoding can be used for the identification of species at any stage of life, such as egg, larva, adult, etc., or damaged specimens. There is no need for prior knowledge of species, such as morphological characters, for the identification of the organism. In this way, a large number of species can be recognized, belonging to a broad range of taxonomic groups, from bacteria to insects to mammals, and a specialized taxonomist is not required for this purpose.

Identification of cryptic species: The term "species" is defined as a group of closely related individuals that are morphologically identical and can interbreed, whereas cryptic species include groups of organisms that are morphologically identical but cannot interbreed and hence should be considered as different species. Since these organisms are morphologically similar, they cannot be distinguished by traditional taxonomy. These taxonomic ambiguities can be resolved by using DNA barcoding methods, which can distinguish between species on the basis of sequence divergence within and between species. Since the species are morphologically identical, reproductive isolation along with molecular data is considered as the confirmation of cryptic species. Although for some cases of cryptic species reproductive incompatibility has been confirmed by crossbreeding experiments, for some organisms, it is not possible to conduct crossbreeding experiments, so morphological, ecological, and physiological data can be combined with genetic data to resolve such taxonomic ambiguities (Paterson et al., 2016).

Resolving taxonomic problems due to phenotypic plasticity: When a species is exposed to different environments, the genotype produces different phenotypes. This is known as phenotypic plasticity. In some cases, due to phenotypic plasticity, members of the same species occupying different habitats acquire distinct morphological features and hence are sometimes assigned as different species while their genotype is still similar. DNA barcoding can be employed to resolve such taxonomic issues arising due to plasticity of characters on the basis of sequence divergence and genetic distances.

Minimizing food piracy: DNA barcoding can also play an important role in identifying mislabelling of food at the cooked, raw, or frozen stage. As this technique is able to detect degraded DNA, frozen or cooked food can be checked for any mislabelling or adulteration. Minibarcodes are generally employed for this purpose. Also, identification of food by barcoding becomes very important in cases when endangered species are sold as food, which is prohibited for their conservation (Hobbs et al., 2019).

Biodiversity monitoring: eDNA metabarcoding is non-invasive, as it does not need the organism's tissue or the whole organism itself. Also, it is more sensitive than the traditional survey methods. It could serve as an alternative (or complementary) tool for biodiversity monitoring. As it is able to analyse the entire range of biodiversity, from prokaryotes to amphibians to mammals, from environmental samples only, it could lead to the detection of the entire flora and fauna of a particular habitat in a more cost-effective manner than if the whole biodiversity analysis of a single habitat were done by traditional methods.

Detection of invasive species: The eDNA metabarcoding technique has been successfully employed for the detection of invasive species such as fishes (Lim et al., 2016), amphibians (Dejean et al., 2012), crustaceans (Trέguier et al., 2014), etc. in aquatic bodies. In such habitats, early detection of invasive species is difficult by traditional methods such as electrofishing, hydroacoustic surveys, etc. Metabarcoding can detect the presence of invasive species even in low numbers, which can provide a chance to frame strategies for management purposes.

Detection of endangered species: Metabarcoding is such a sensitive technique that it is able to detect the presence of a very few copies of eDNA released by the organism, so it can be used to detect the presence of rare or endangered species, which are very low in number. eDNA is even able to detect the presence of organisms in their habitats when they are believed to have disappeared (Lopes et al., 2020).

Analysis of diet: The metabarcoding of faecal matter and stomach contents has provided new insights into the diet analysis of various organisms, as it provides a more detailed picture of the diet consumed by the organism in question. The traditional method of diet analysis involves examination of fully and partially digested food material, which requires a lot of expertise and labour, and it is particularly difficult for endangered species or those that are aquatic or small in size (Ruppert et al., 2019).This is based on extraction of DNA from faecal and stomach contents, which contains the DNA of the diet (seeds, plants, insects, animals, etc.). Faeces are generally considered as a good source of DNA, as they are easy to obtain even in the case of elusive species, but the quality of the DNA obtained is often much lower due to the digestion process before and the decomposition processes after defecation (Hawlitschek et al., 2018). Such DNA can be amplified using minibarcodes and then sequenced with NGS.

3.5.2 Limitations of DNA Barcoding and Metabarcoding

Though DNA barcoding and metabarcoding are revolutionary for species detection and biotic community analysis, there are some limitations to these techniques. First of all, species identification by these methods cannot give information about the life stage of the organism. In the case of barcoding using eDNA, we cannot predict the life stage of the organism from which the eDNA comes, i.e. whether the organism is in the egg stage, juvenile, adult, etc. Secondly, this technique is entirely based on the reference database, as the sequenced fragment of DNA needs to be aligned to the reference database. So, the correct detection of the species in question depends on the availability and authenticity of a reference database. If there is incorrect identification of the species in the reference database, it will result in incorrect species identification by the researcher. Detection of false negatives and false positives is also a challenge that needs to be taken care of. Also, whereas traditional methods are based on the live or real-time identification of the target organism, eDNA-based identification cannot give information about the state of the organism, whether it is alive or dead, or the current presence of the species.

3.6 CONCLUSION

With the advancements of DNA-based approaches and the introduction of affordable NGS techniques, in-depth analysis of various ecosystems is being done, leading to an entirely new perspective on biodiversity assessment (Figure 3.4). eDNA complemented with NGS has completely revolutionized biodiversity analysis methods, enabling the understanding of species composition without harming and disturbing the organism itself and also saving a lot of manpower and time. However, these methods are still very new, and standardization is needed for certain things such as sample collection, sufficient volume of sample collected, choice of primers, improvement of reference database, etc. Also, there are limitations to DNA-based biodiversity analysis, as discussed earlier. So, biodiversity analysis using DNA barcoding and metabarcoding has great potential for

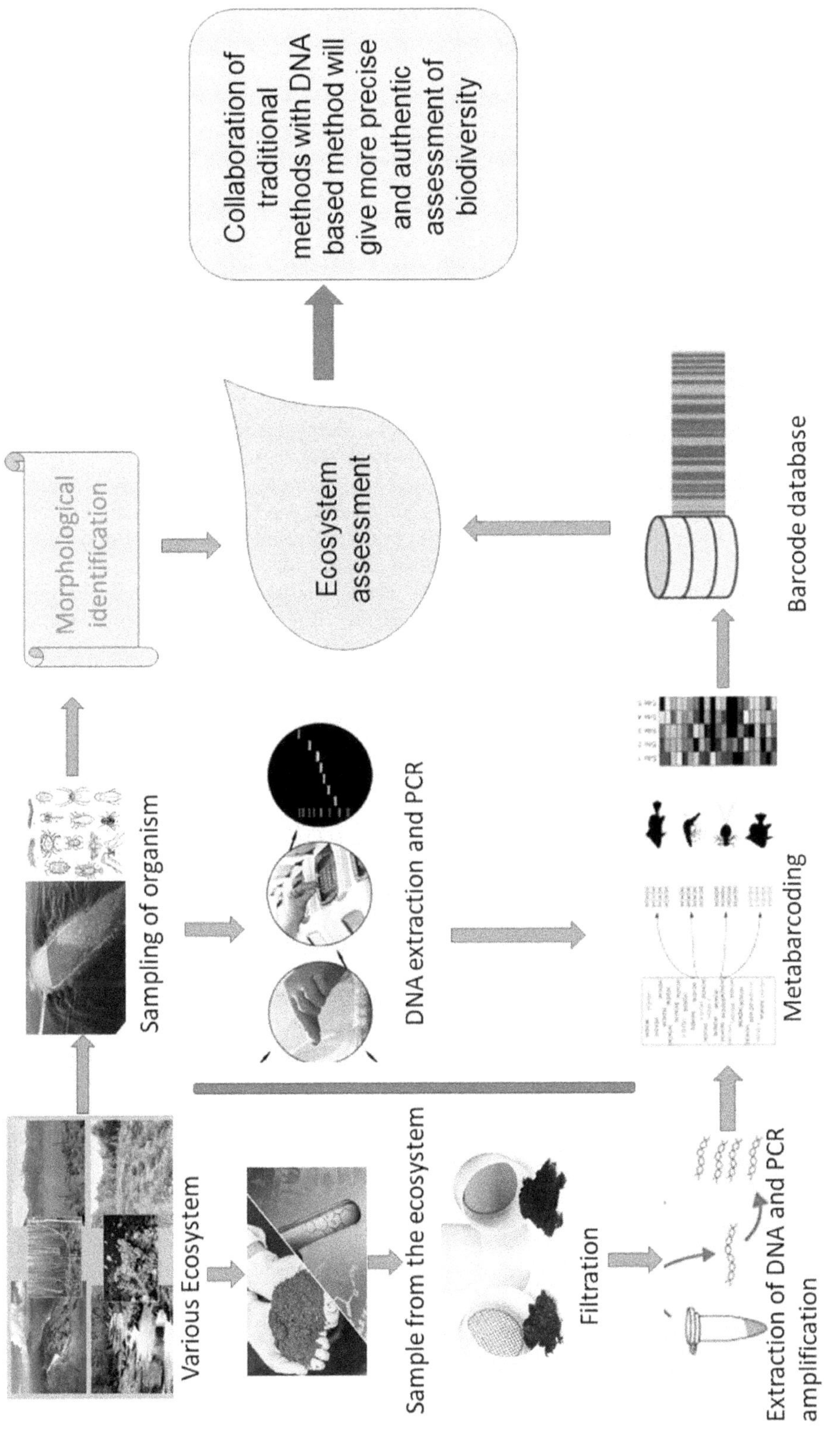

FIGURE 3.4 Holistic approach combining traditional methods with DNA-based barcoding and metabarcoding for biodiversity monitoring.

identifying species and biotic community analysis, yet it must be complemented with traditional methods for better resolution regarding taxonomic ambiguities.

REFERENCES

Baird, D.J., Pascoe, T.J., Zhou, X., & Hajibabaei, M. (2011). Building freshwater macroinvertebrate DNA-barcode libraries from reference collection material: formalin preservation vs specimen age. *Journal of the North American Benthological Society* 30: 125–130.

de Jonge, D.S.W., Merten, V., Bayer, T., Puebla, O., Reusch, T.B.H., & Hoving, H.J.T. (2021). A novel metabarcoding primer pair for environmental DNA analysis of Cephalopoda (Mollusca) targeting the nuclear 18S rRNA region. *Royal Society Open Science* 8: 201388.

Deiner, K., Bik, H.M., Mächler, E., Seymour, M., Lacoursière-Roussel, A., Altermatt, F., Creer, S., Bista, I., Lodge, D.M., de Vere, N., Pfrender, M.E., & Bernatchez, L. (2017). Environmental DNA metabarcoding: transforming how we survey animal and plant communities. *Molecular Ecology* 26: 5872–5895.

Dejean, T., Valentini, A., Miquel, C., Taberlet, P., Bellemain, E., & Miaud, C. (2012). Improved detection of an alien invasive species through environmental DNA barcoding: the example of the American bullfrog *Lithobates catesbeianus*. *Journal of Applied Ecology* 49(4): 953–959.

Dormontt, E.E., van Dijk, K.-j., Bell, K.L., Biffin, E., Breed, M.F., Byrne, M., Caddy-Retalic, S., Encinas-Viso, F., Nevill, P.G., Shapcott, A., Young, J.M., Waycott, M., & Lowe, A.J. (2018). Advancing DNA barcoding and metabarcoding applications for plants requires systematic analysis of herbarium collections – an Australian perspective. *Frontiers in Ecology and Evolution* 6: 134.

Dubey, B., Poorlin Ramakodi, M., & Haque, I. (2010). DNA mini-barcoding: an approach for forensic identification of some endangered Indian snake species. *Forensic Science International: Genetics* 5: 181–184.

Ficetola, G.F., Miaud, C., Pompanon, F., & Taberlet, P. (2008). Species detection using environmental DNA from water samples. *Biology Letters* 4: 423–425.

Großkopf, R., Janssen, P.H., & Liesack, W. (1998). Diversity and structure of the methanogenic community in anoxic rice paddy soil microcosms as examined by cultivation and direct 16S rRNA gene sequence retrieval. *Applied and Environmental Microbiology* 64 (3): 960–969.

Hadziavdic, K., Lekang, K., Lanzen, A., Jonassen, I., Thompson, E.M., & Troedsson, C. (2014). Characterization of the 18S rRNA gene for designing universal eukaryote specific primers. *PLoS One* 9: e87624.

Hajibabaei, M., Singer, G.A.C., Clare, E.L., & Hebert, P.D.N. (2007). Design and applicability of DNA arrays and DNA barcodes in biodiversity monitoring. *BMC Biology* 5: 24.

Hawlitschek, O., Fernández-González, A., Balmori-de la Puente, A., & Castresana, J. (2018). A pipeline for metabarcoding and diet analysis from fecal samples developed for a small semi-aquatic mammal. *PLoS One* 13: e0201763.

Hebert, P.D.N., Cywinska, A., Ball, S.L., deWaard, J.R. (2003). Biological identifications through DNA barcodes. *Proceedings of the Royal Society of London. Series B: Biological Sciences* 270: 313–321.

Hiorns, W.D., Methé, B.A., Nierzwicki-Bauer, S.A., & Zehr, J.P. (1997). Bacterial diversity in Adirondack Mountain lakes as revealed by 16S rRNA gene sequences. *Applied and Environmental Microbiology* 63: 2957–2960.

Hobbs, C.A.D., Potts, R.W.A., Bjerregaard Walsh, M., Usher, J., & Griffiths, A.M. (2019). Using DNA barcoding to investigate patterns of species utilisation in UK shark products reveals threatened species on sale. *Scientific Reports* 9: 1028.

Kelly, R.P., Port, J.A., Yamahara, K.M., & Crowder, L.B. (2014). Using environmental DNA to census marine fishes in a large mesocosm. *PLoS One* 9: e86175.

Komai, T., Gotoh, R.O., Sado, T., & Miya, M. (2019). Development of a new set of PCR primers for eDNA metabarcoding decapod crustaceans. *Metabarcoding and Metagenomics* 3: e33835.

Kress, W.J., & Erickson, D.L. (2007). A two-locus global DNA barcode for land plants: the coding rbcL gene complements the non-coding trnH-psbA spacer region. *PLoS One* 2: e508.

Lim, N.K., Tay, Y.C., Srivathsan, A., Tan, J.W., Kwik, J.T., Baloğlu, B., ... & Yeo, D.C. (2016). Next-generation freshwater bioassessment: eDNA metabarcoding with a conserved metazoan primer reveals species-rich and reservoir-specific communities. *Royal Society Open Science* 3(11): 160635.

Lopes, C.M., Baêta, D., Valentini, A., Lyra, M.L., Sabbag, A.F., Gasparini, J.L., ... & Zamudio, K.R. (2020). Lost and found: frogs in a biodiversity hotspot rediscovered with environmental DNA. *Molecular Ecology* 30(13), 3289–3298.

Meusnier, I., Singer, G.A.C., Landry, J.-F., Hickey, D.A., Hebert, P.D.N., & Hajibabaei, M. (2008). A universal DNA mini-barcode for biodiversity analysis. *BMC Genomics* 9: 214.

Miya, M., Sato, Y., Fukunaga, T., Sado, T., Poulsen, J.Y., Sato, K., ... & Iwasaki, W. (2015). MiFish, a set of universal PCR primers for metabarcoding environmental DNA from fishes: detection of more than 230 subtropical marine species. *Royal Society Open Science* 2(7): 150088.

Paterson, I.D., Mangan, R., Downie, D.A., Coetzee, J.A., Hill, M.P., Burke, A.M., Downey, P.O., Henry, T.J., & Compton, S.G. (2016). Two in one: cryptic species discovered in biological control agent populations using molecular data and crossbreeding experiments. *Ecology and Evolution* 6: 6139–6150.

Purty, R.S., & Chatterjee, S. (2016). DNA barcoding: an effective technique in molecular taxonomy. *Austin Journal of Biotechnology & Bioengineering* 3(1): 1059.

Rees, H.C., Maddison, B.C., Middleditch, D.J., Patmore, J.R.M., & Gough, K.C., (2014). REVIEW: The detection of aquatic animal species using environmental DNA – a review of eDNA as a survey tool in ecology. *Journal of Applied Ecology* 51: 1450–1459.

Ruppert, K.M., Kline, R.J., & Rahman, M.S. (2019). Past, present, and future perspectives of environmental DNA (eDNA) metabarcoding: a systematic review in methods, monitoring, and applications of global eDNA. *Global Ecology and Conservation* 17: e00547.

Schoch, C.L., Seifert, K.A., Huhndorf, S., Robert, V., Spouge, J.L., Levesque, C.A., & Chen, W. (2012). Nuclear ribosomal internal transcribed spacer (ITS) region as a universal DNA barcode marker for *Fungi. Proceedings of the National Academy of Sciences* 109: 6241.

Shu, L., Ludwig, A., & Peng, Z. (2020). Standards for methods utilizing environmental DNA for detection of fish species. *Genes* 11: 296.

Sønstebø, J.H., Gielly, L., Brysting, A.K., Elven, R., Edwards, M., Haile, J., Willerslev, E., Coissac, É., Rioux, D., Sannier, J., Taberlet, P., & Brochmann, C. (2010). Using next-generation sequencing for molecular reconstruction of past Arctic vegetation and climate. *Molecular Ecology Resources* 10: 1009–1018.

Takahara, T., Minamoto, T., & Doi, H. (2013). Using environmental DNA to estimate the distribution of an invasive fish species in ponds. *PLoS One* 8: e56584.

Tréguier, A., Paillisson, J.-M., Dejean, T., Valentini, A., Schlaepfer, M.A., & Roussel, J.M. (2014). Environmental DNA surveillance for invertebrate species: advantages and technical limitations to detect invasive crayfish *Procambarus clarkii* in freshwater ponds. *Journal of Applied Ecology* 51: 871–879.

Valentini, A., Taberlet, P., Miaud, C., Civade, R., Herder, J., Thomsen, P. F., ... & Dejean, T. (2016). Next-generation monitoring of aquatic biodiversity using environmental DNA metabarcoding. *Molecular Ecology* 25(4): 929–942.

Wake, D.B., & Vredenburg, V.T. (2008). Are we in the midst of the sixth mass extinction? A view from the world of amphibians. *Proceedings of the National Academy of Sciences* 105: 11466–11473.

Wilcox, T.M., McKelvey, K.S., Young, M.K., Jane, S.F., Lowe, W.H., Whiteley, A.R., & Schwartz, M.K. (2013). Robust detection of rare species using environmental DNA: the importance of primer specificity. *PLoS One* 8: e59520.

Yang, L., Tan, Z., Wang, D., Xue, L., Guan, M. X., Huang, T., & Li, R. (2014). Species identification through mitochondrial rRNA genetic analysis. *Scientific Reports* 4(1): 1–11.

4 Effect of Flow Alteration on River Ecology

State of the Art

Kashifa Iqbal and Venkatesh Dutta
Department of Environmental Science, School of Earth and Environmental Sciences (SEES); Babasaheb Bhimrao Ambedkar University, Lucknow

CONTENTS

4.1 INTRODUCTION

Natural environmental flows are required for the healthy functioning as well as physical and biological habitat formation of an aquatic ecosystem. The biota living in the stream is well adapted for various natural alterations. Since time immemorial, human beings have been altering the natural hydrology of free-flowing streams for their existence and development, exploiting the natural

DOI: 10.1201/9781003220824-5

flows of rivers for irrigation in agriculture, transportation, trade, water supply for the industrial and domestic needs, flood control for security and survival purposes, and hydropower generation for meeting the ever-increasing energy demands of the burgeoning population. Currently, humans use over half of the water available globally, which is expected to further increase to 70% by the year 2025. About 70% of all water usage worldwide is currently for agriculture. Irrigation is the primary usage for most of the freshwater. By 2050, when there will be between 9.4 and 10.2 billion people on the planet, there will be a greater burden on the water system. For the prosperity and progression of different sectors, like agriculture, hydropower, and industry as well as the domestic supply, various efforts are being made, such as impoundments, diversion weirs, water abstraction, and exploitation of aquifers for groundwater. These anthropogenic influences on the natural regime of a river have caused massive, uncontrolled effects on riverine ecosystems, homogenizing the hydrologic, geomorphic, and ecological dynamics that govern the riverine ecosystem structure and functions (Dutta, 2021; Wohl et al., 2017; Poff et al., 2007).Around 60% of global rivers have been subjected to hydrologic alteration. The riverine ecosystem is suffering radical and worsening modifications as a consequence of human-induced changes in streamflow, landscape, and climate. Over half of the rivers globally are controlled by dams (Li et al., 2013; Nilsson et al., 2005). Furthermore, these anthropogenically induced problems exacerbate the situation when biological difficulties occur, such as exotic/invasive species that remodel the food web, thereby disturbing the flexibility of the ecosystem. In this way, the biodiversity of the riverine ecosystem is affected (Vorosmarty et al., 2010).This is further accelerated by the negative effects of climate change, which lead to additional impacts on the ecosystem (Harley et al., 2012).

Large rivers in particular, as well as their lowland alluvial floodplain environments, experience the greatest degradation because of development and the upstream influences (Poff et al., 2010). Thus, these freshwater ecosystems, being the most varied and fruitful ecosystems, are also the most damaged globally (Reid et al., 2019). From small headwaters to large rivers, streams reveal a gradient of environmental factors, which comprises flow regime alteration (Mohammed et al., 2018), temperature (Pyne & Poff, 2017), and channel morphology and solar input (Ferencz & Cardenas, 2017). The concept of a stream continuum (Wilhelm et al., 2015) associates the fluctuation in physical factors occurring from small streams to large rivers. A river basin is formed from the catchments of the various tributaries of a river (e.g., Ganga River Basin). The water balance of a particular river basin is calculated from the water gains (precipitation) and losses (evapotranspiration and runoff), which also comprise storage phases (soil water, groundwater, ice, snow). At different locations within the catchment, the observed discharge (m^3/s) is created based on weather-related and bio-geophysical factors (Table 4.1).

Various processes involved in organizing and shaping the physical habitat as well as the allied biotic communities are governed by the river flow. Variability in the flow is an intrinsic feature of

TABLE 4.1
Elements (Weather-Related and Bio-Geophysical) Governing River Flow

Weather-related elements	Bio-geophysical elements
– Precipitation kind (rain, hail, sleet or snowfall) – Quantity, concentration, extent, and expansion of precipitation above the watershed – Soil dampness as an outcome of former precipitation – Atmospheric situations influencing permeation as well as evapotranspiration (evaporation from land surface and vegetation)	– Catchment area – Altitude – Landscape, slope of the land – Catchment profile and drainage grid designs – Kind of soil, land cover, and pattern of land use and florae of the area – Surface water bodies in the catchment area that inhibit or interrupt runoff in the river downstream

river systems around the world, which is equally responsible for its ecological functioning (Altermatt et al., 2013). The river size, inclusive of the geographic disparity in the climate, geology, topography, and vegetative cover, helps shape the regional flow regimes. The magnitude, frequency, duration, timing, and rate of change of a stream are vital components that are essential for nourishing biodiversity and enhancing ecosystem integrity (Rice et al., 2012). The habitat is greatly influenced by the movement of water and sediments in the channel itself and between the channel and the floodplain. Various habitat structures are formed and sustained through an extensive range of flows in a river body (Stromberg et al., 2007). The flow regime of a river also helps in directing special varied habitat types that are required by some riverine species to complete their life cycle (Silva et al., 2018). Some adaptive features have been developed by the riverine biodiversity to withstand the changes in habitat that occur as a consequence of the natural flow variation. Certain species are dependent on regular and periodic vicissitudes in the stream flows for finishing their life cycle (Lytle & Poff, 2004; Kiernan et al., 2012). Aquatic species have evolved special adaptability due to the natural flow regimes (Arthington et al., 2010). The flow fluctuation, considered as the master variable, determines the physical and chemical features coupled with the biological features. The biological diversity is augmented when the level of hydrological variability is only moderate (Rolls et al., 2018).

Probably, the association between the biota and the physical nature of the aquatic environment is determined by large events inducing the form and shape of the channel (Bunn & Arthington, 2002). Land-use changes occurring in the catchment area beside the water resource due to development projects bring about certain aspects of the flow regime that subsequently result in the decline of aquatic biota through these mechanisms.

4.2 SALIENT FEATURES OF NATURAL FLOW REGIME IN RIVERS

Flow, in particular, is responsible for the various aspects that influence the physical environment in rivers and also define the biotic composition. The persistence and multiplicative potential of aquatic biota are greatly influenced by the flow instabilities, because their life cycle stratagems are directly affected by natural flow regimes (Olden & Naiman, 2010; Bunn & Arthington, 2002). Even though the biota are naturally adapted to the discharge dynamics, nevertheless, these might pose positive or negative consequences (e.g., stranding, drift, diminished productivity) when the force of flows is remarkably high or the timing of such flows is unfamiliar to the organisms (Nagrodski et al., 2012; Unfer et al., 2011). Besides the natural discharge dynamics, the anthropogenic flow variations induced through human activities are harmful to the aquatic ecology.

4.3 NATURAL FLOWS: PROVISIONING ASPECT FOR A HEALTHIER LIFE

A wide variety of provisioning services, such as clean water and plants, including food, are provided by the environmental flows or e-flows. Amongst these, the most common that come to mind when considering rivers are fish and its products, which contain good value for human nutrition (the primary animal protein source of rural people in developing countries). High flow events are greatly responsible for the migration and spawning of fishes, as some stages of their life cycle are reliant on natural flow dynamics. Flow regime profoundly determines the growth of floodplain agriculture. The modification of the natural river dynamics interrupts the basic flood cycles and therefore disturbs influxes of nutrient-laden sediments to the agricultural fields in floodplains.

4.4 NATURAL FLOWS: REGULATING AND SECURITY ASPECT

The regulating services of the ecosystem, such as erosion, pollution, pest control, and floods, are controlled by the environmental flows. The banks of the river are covered by the riparian vegetation; nevertheless, these are also reliant on the various flow regimes: for example, high flow events like floods for dispersal of seeds and fruits. Several estuaries and coastal wetlands are jeopardized by the building of dams, which steadies the flow and traps sediment. As a consequence of the deterioration and impoverishment of the freshwater ecosystem, its intrinsic ability to mitigate the harmful impacts of very high flows is hampered, thereby threatening the safety of individuals and communities. The devastation caused by Hurricane Katrina and the tsunami in Southern Asia due to the loss of mangroves and natural barriers is one classic example.

4.5 NATURAL FLOWS: SPIRITUAL ASPECT

Rivers are an epitome of religious, aesthetic, historical, and archaeological principles that are principal pillars of a country's heritage, as shown in Table 4.2. They fulfil our spiritual desires or needs.

TABLE 4.2
Relations Existing amid Ecosystem Services and the Ecological Processes Maintained by Environmental Flows and Human Welfare

Ecosystem Services	Human Welfare	Environmental Flows Component and Ecological Processes
Provisioning Aspect		
Encourage the distribution of a wide variety of diverse fundamental provisioning amenities like clean water, plants, and food as well.	Basic requirements for a normal life	**Fisheries:** The life cycle of various fish species profoundly relies on the natural water flow dynamics, i.e. high flow events (floods) are significant for migrating as well as the spawning of the fishes. **Medicinal florae and fruits:** Some particular plants blossom in the riverine floodplains during the extreme low events (drought). High flow events aid in dispersing the seeds plus fruits of riparian vegetation.
Regulating Aspect		
Assists in the control of contamination, pests, and overflows.	Refuge, wellbeing	**Checking of floods**: River banks check floods by the riparian vegetation. The soil-moisture level of the banks is controlled by the flows. The nutrients and seeds as well as fruits are deposited by the high flow events. **Pollution control**: Extreme high flow events rejuvenate the river water quality, cleaning out waste and contaminants. **Pest control**: A stream having flow variability is resilient to the introduction of some invasive species, whereas dammed and altered streams favour the establishment of exotic species.
Cultural Aspect		
Religious, leisure, and aesthetic amenities.	Benefits social relations	Ample amount of flow to amplify aesthetics as well as cultural worth.
Supporting Aspect		
Nutrient as well as sediment cycling, biodiversity maintenance.	Fundamental for normal life, refuge, wellbeing, respectable social connections	A healthy species balance is maintained by high flow events (floods) between aquatic and riparian communities. Organism heterogeneity is sustained in a floodplain forest.
Spiritual Aspect		
	Religious and spiritual fulfilment	Substantial flow required.

Innumerable instances are found globally of people living in close association with rivers. For example, various earlier civilizations – the Indus Valley civilization (Indus River), Chinese civilization (Yellow River), Ancient Egypt (Nile River), Mesopotamia (Euphrates and Tigris Rivers) – were located around rivers. Each river is considered as a unique personality, due to its very nature of meandering and flowing, and a character that is influenced both by its environment and location as well as by the culture it supports. In many societies in South Asia, rivers are also considered as companions, spiritual guides, or friends, just like individuals. In India, the River Ganga is believed to be the *Mokshdayini*; i.e., bathing in Ganga releases the soul from death and rebirth cycles. Also, several earlier towns, which are now important pilgrimage centres, are situated on the banks of Ganga, such as Rishikesh, Haridwar, Allahabad, Mirzapur, Gangasagar, Varanasi, and many more (Dwivedi et al., 2018).

4.6 CONSEQUENCES OF AN ALTERED FLOW REGIME OF A RIVER

Though rivers were earlier divided by natural agents like waterfalls, cascades, and beaver dams, nonetheless, anthropogenic instabilities have put an extra burden on the habitats (Boggie et al., 2018; Fuller et al., 2015).These human-created disruptions work locally as well as at the landscape level to modify the riverine habitats. The alterations that occur on the river channel scale are localized disturbances, including structures like embankments, groynes, etc. used for subjugating the river flow, while those occurring on the vast regional scale are landscape disturbances like urbanization, deforestation, etc., functioning all through the catchment (Cooper et al., 2017). These disturbances are less harmful individually, but when they act in combination, they have manifold, accumulating effects on the aquatic habitat (Carnie et al., 2016; Schinegger et al., 2012). Dams can have great impacts at the localized as well as the landscape extent (Nilsson et al., 2005) and can even disturb the smooth and healthy processes of freshwater ecosystems, whereas the rivers' response to such disturbances may be diverse and multifaceted (Van Cappellen & Maavara, 2016; Poff, 2014). The physical framework of the river-course is seriously affected by damming, resulting in the loss of heterogeneous habitations like pools and riffles. Dams also cause severe fluxes in the thermal regime and variation in the rivers' flow regime (Carlisle et al., 2014). Damming interrupts the longitudinal integrity of the river due to the obstacles, causing disruption in the course of its flow, which in turn results in reduced water flow, amplified sedimentation, increased water transparency, etc. (Birk et al., 2012).The dynamic equilibrium prevailing between the water and sediment drifting in a natural unrestricted free-flowing river is beneficial for all the aquatic biota living in it: vertebrates, invertebrates, and also the floral species (Figure 4.1). This dynamicity is disturbed by the anthropogenic involvements (dams, dykes) in their natural courses (Figure 4.2) (Besacier-Monbertrand et al., 2014).The question that at once comes to mind – why is the river flow being disrupted? – has a plausible explanation in the societal context of meeting demands like water supply, irrigation, hydropower, flood control, and transportation as well.

Over half of the greatest river networks are incarcerated globally (Nilsson et al., 2005; Ripl, 2003). Yet, the demand for new dams continues to pose a threat to the biodiverse aquatic systems in the world (Winemiller et al., 2016). The ever-increasing excesses of the hydroclimate are also contributing to the deteriorating ecosystems all over the geographical extent (Kirkman et al., 2013; Ruhi et al., 2016). Various lotic water systems switch their flow regimes from perennial to intermittent as a result of rainfall fluctuations and escalated evapotranspiration (Rawat et al., 2016; Jaeger et al., 2014), thereby modifying mean runoff annually as well as affecting high and low flows of the river.

The biodiversity is greatly affected across diverse levels of biological organization because of the tamed flow regimes, which in turn reduces the tenacity of the endemic species of the region, which are well habituated to the prior surroundings, while generating favourable conditions for

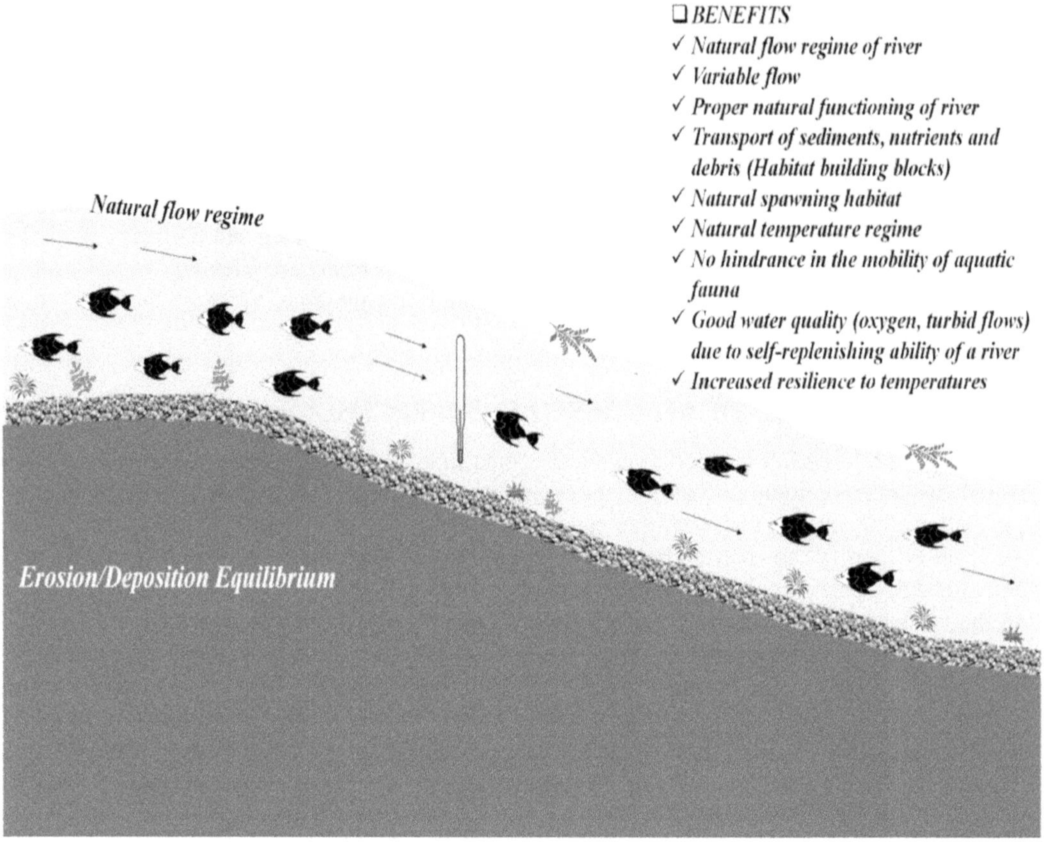

FIGURE 4.1 Natural free-flowing river at dynamic equilibrium.

exotic species to establish themselves and proliferate (Macel et al., 2014; Warren et al., 2013). For example, the native Amazon basin macrophyte *Eichhornia crassipes* (water hyacinth), being highly invasive, has entered the freshwater aquatic ecosystems of North America, Europe, Africa, Asia, and Australia, causing devastating effects like reduced light penetration and dissolved oxygen depletion, thereby causing fish mortality (Tellez et al., 2008) attributed to its high growth rates and potential for dispersal, concealing the large open surfaces of water at the cost of the native macrophytes. In Spain's Guadiana River and Australia's floodplain wetlands, an invasive helophyte, *Typha* spp., has flourished because of the modified wet-dry cycles and the amplified flow steadiness as a consequence of dam construction (Téllez et al., 2008; Ruhi et al., 2019).

In the same way as the havoc created by non-native macrophytes in the flow-modified regions, there are also examples of non-native ichthyofaunal invasion as well as establishment by replacing native aquatic organisms in the water-scarce regions of the Mediterranean, the southwestern United States, Australia, and Asia, where there is an extra burden on freshwater resources for use in agriculture as well as for the healthy functioning of the ecosystem (Whitney et al., 2016; Propst et al., 2008). Alteration and invasion are often considered as coupled stressors that also interact (Table 4.3). Flow events enable aquatic organisms to enter the floodplain linking riverine ecosystems by organic matter appropriations and drifting organisms (Robinson et al., 2004). There is a coupling of various natural disturbances, like magnitude, frequency, and duration, as well as the timing of the different flow forms with the life cycle events of the biota, such as spawning, recruitment, and seed dispersal (Sabater et al., 2019). The conservation of the distinguishing aspects of a natural flow regime

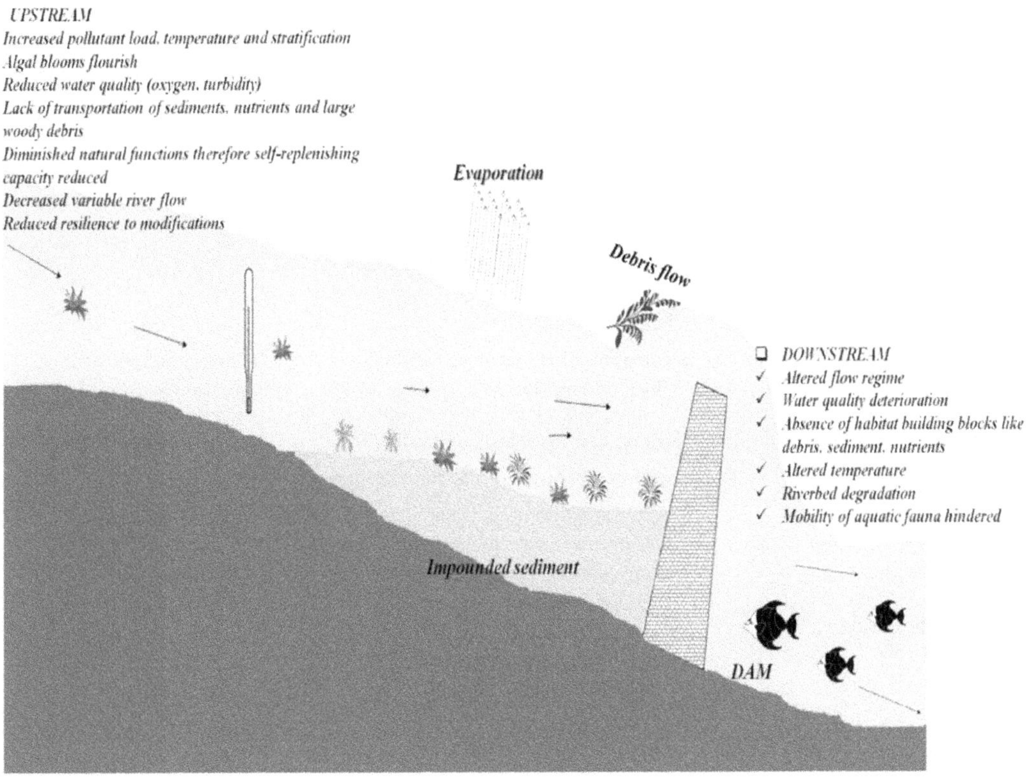

FIGURE 4.2 Anthropogenic influence on a river (dammed river).

amplifies the stability of biodiversity, possessing adaptive morphological, behavioural, and life cycle qualities (Lytle & Poff, 2004).

4.7 STATE OF THE ART IN ENVIRONMENTAL FLOW ASSESSMENT

Environmental flow is the water left in our rivers to ensure downstream environmental, social, and economic benefits. Across the globe, it is also termed ecological flow and instream flow. For evaluating the impacts of various projects, various environmental flow techniques have been applied. A good amount of geomorphological and ecological data has been used after collection and intensive analysis (Hughes & Louw, 2010). In many countries, the major drawback is the lack of data availability that links ecological conditions to specific flows. So, to overcome this, methods have been developed that depend on hydrological indices obtained from historical data (Kashaigili et al., 2007; Liu et al., 2016). Figure 4.3 gives an overview of environmental flow assessment methods required for healthy ecological functioning in order of their priorities.

The four aspects that should be considered when selecting a method for estimating the environmental flow are hydrological aspects, hydraulic aspects, biological aspects, and cultural/recreation/religious/social aspects (ecosystem services) (Baghel et al., 2019). For the evaluation of environmental flow assessments, basically two models are implemented (Acreman et al., 2014b): (1) the natural flow model for abating the flow regime alteration to revive the natural conditions for conserving biodiversity, and (2) the management-based model, for which the environmental flows are planned for accomplishing certain explicit goals like ecosystem services. The International Water Management Institute (IWMI) classifies environmental flows based on their line of action or principles as per the categorization depicted in Figure 4.4.

TABLE 4.3
Relations Existing Amid Ecosystem Services and the Ecological Processes Maintained by Environmental Flows and Human Welfare

Cause(s)	Consequence(s)	Mechanism	Food Web Theory
Reduced perturbation regimes (downstream)	Augmentation of food chain	Additional trophic levels in the food web are sustained by the flow steadiness induced as a result of damming, which encourages the institution of non-native predators (e.g. piscivore fishes).	Dynamic solidity or "disturbance" hypothesis of food-chain length (Pimm & Lawton, 1977)
	Revamping of food web	Exotic primary and secondary consumers might find supplies more quickly using dietary changes. In the absence of potent predators, the invasive species experience increased productivity.	Omnivory and insertion mechanisms (Post & Takimoto, 2007)
	Food web simplification	The exotic taxa complete undeveloped fundamental niches in the presence of new thermal and flow regimes, and once these are filled up, the invasive species inhibit the realized niches of the native population, employing hostile connections, resulting in loss of nodes and links.	Food web niche model (Thompson et al., 2012; Nordström et al., 2015; Layman et al., 2015)
Lentic conditions (e.g. reservoir)	Food chain lengthening	Maximum storage and organic matter processing rates are due to the increased residence time, which augments production, permitting additional (exotic) trophic levels.	"Productivity" or resource accessibility hypothesis of food-chain length (Albouy et al., 2014) R* rule or resource-ratio hypothesis (Tilman, 2004; Wilson et al., 2007)
	Top-down effects	Lentic conditions and keystone herbivores may favour strong effects of exotic predators residing at the top trophic level, putting pressure on those at the lower trophic levels.	Trophic cascades (Knight et al., 2005)
Increased habitat capacity (in the reservoir, tail waters)	Food chain lengthening	The productive aquatic ecosystem size increases due to the vast water-logged regions and extended hydroperiods giving additional (invasive) trophic levels.	Productive space and ecosystem size hypotheses of food chain length (Post et al., 2000; Takimoto & Post, 2013)

4.7.1 Hydrobiological or Holistic Technique

The hydrobiological or holistic technique uses biological, hydrological, hydraulic data coupled with habitat simulation models that apparently take a holistic approach to environmental flow estimation, comprising certain ecosystem components like geomorphology, hydraulic habitat, water quality, riparian and aquatic vegetation, macroinvertebrates, fish, and other vertebrates that rely on the river ecosystem (Durance & Ormerod, 2007; Kumara & Srikantaswamy, 2011).

4.7.2 Building Block Technique (BB)

The BB technique is based on the principle that riverine organisms depend upon the fundamental elements or building blocks of the flow regime. Different flow regimes (high, medium, low)

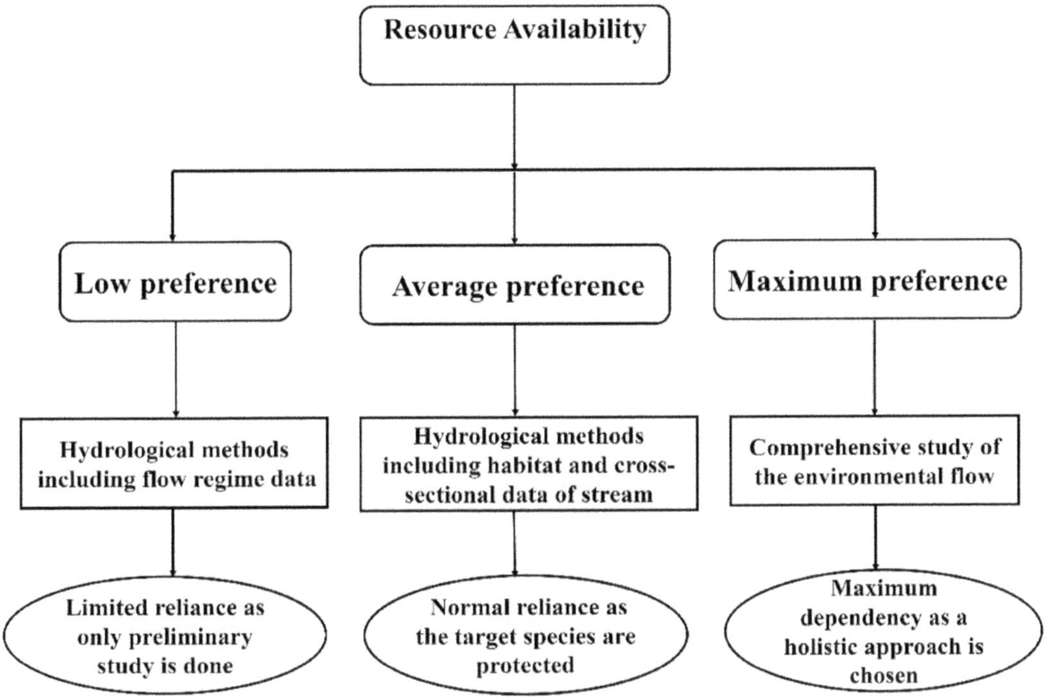

FIGURE 4.3 Environmental flow assessment methods on a priority basis.

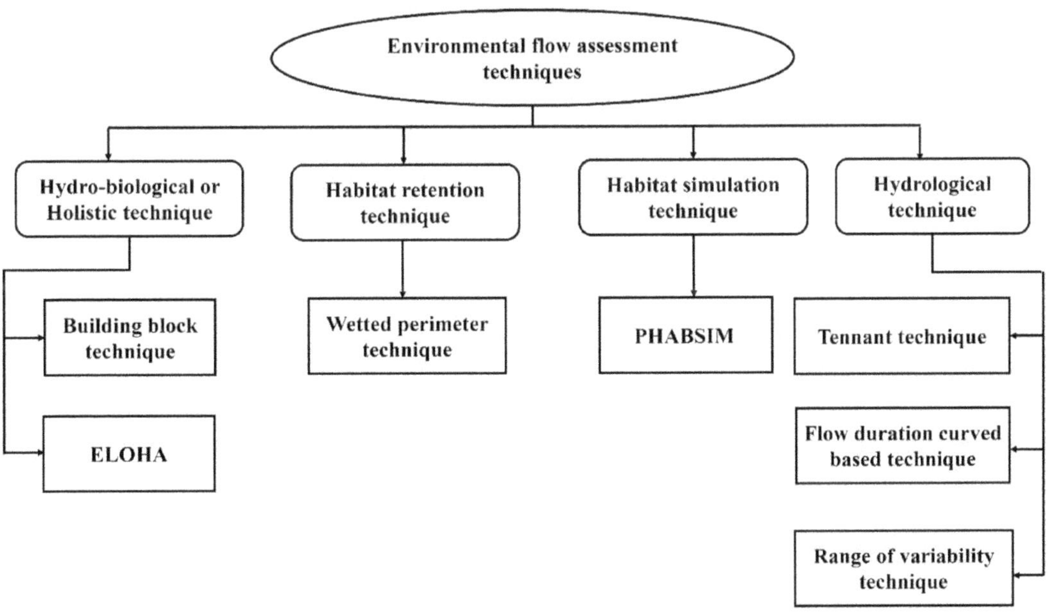

FIGURE 4.4 International Water Management Institute (IWMI) classification of environmental flows.

TABLE 4.4
Ecological Roles Sustained by Natural Environmental Flows

Base flows	• Give suitable habitation for aquatic biota • Maintain adequate water temperature, water chemistry, and dissolved oxygen • Level of the water table and soil moistness in plants should be adequate in floodplain areas • Supply sufficient water for consumption to land-dwelling faunal species • Sustain the eggs of ichthyofauna and amphibians in stream water • Facilitate the fishes to migrate to suitable feeding and spawning zones • Sustain hypothetical creatures that prefer to dwell in drenched sediments
High flows	• Build a physical identity of streams as well as pools and riffles • Control the proportions of river bed substrates, which comprise sand, gravel, cobble • Inhibit the riparian vegetation from invading the streams • Re-establish the standard water quality state after extended low flows by washing away the contaminants and waste products • Avoid siltation and ventilate eggs living in the spawning gravels
Extreme flows	• Signal to the fishes that these are suitable migration and spawning grounds • Activate new life cycle stages (in smaller and larger organisms) • Allocate suitable nursery zones for young ones and facilitate the ichthyofauna to spawn on floodplain areas • Provide new feeding prospects for fishes and river birds like duck, swan, screamer • Replenish the water table of the floodplain areas • Sustain the floodplain forest diversity through extended inundation (dissimilar species of plant have diverse tolerance levels) • Encourage the plants' richness and dispersal on floodplains • Nutrient deposition on an alluvial plain • Gravel and cobbles accumulation in spawning grounds • Wash away organic matter and large woody debris (habitat-forming structures) into the mainstream • Evict exotic and invading species from aquatic riparian zones • Dispersal of the seeds and fruits of riparian flora • Trigger stream lateral movement, hence establishing new habitations, like some secondary channels and oxbow lake) • Plant seedlings are exposed to soil moisture for an extended time duration

responsible for healthy ecological functioning are based upon the time, duration, frequency, and magnitude of the flows in a river (Figure 4.5). Therefore, these flow regime factors should be properly maintained for a pristine and healthy river. The blocks consist of low flows, medium flows, and high flows essential for the channel maintenance, as given in Table 4.4, which generally depends upon specialists from two distinct fields, i.e., physical scientists like hydrologists, hydrogeologists, and geomorphologists, and biological scientists such as aquatic ecologists. Based on data modelling and their professional skill/decision, the experts agree to the building blocks of the flow regime.

4.7.3 Ecological Limits of Hydrological Abstractions (ELOHA)

The ecological limits of hydrologic alteration (ELOHA) technique is a fusion of various hydrologic methods and environmental flow paradigms supporting wide-ranging flow management at a particular location. There are four steps in the ELOHA technique: 1. fabricate a hydrologic base of flow data; 2. natural river form categorization, 3. establish flow–ecology relationships with each river form; 4. implement a policy for accomplishing river condition targets. ELOHA works on the concept that when there is an increase in the extent of flow alteration from the present baseline

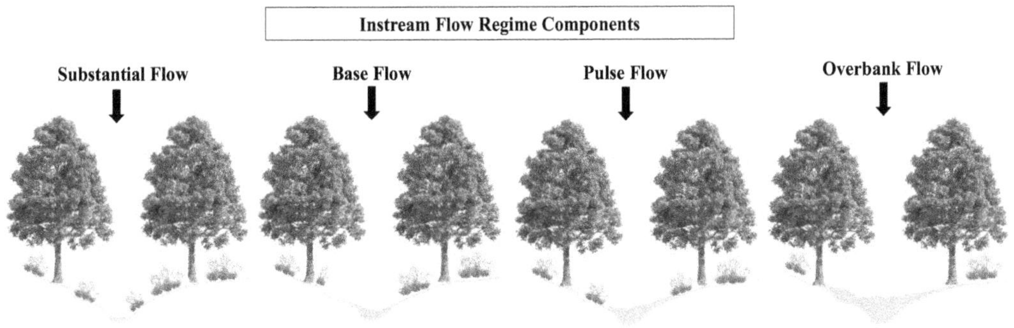

FIGURE 4.5 Various river flow regimes.

conditions, it leads to increased ecological changes. The hydrological data collected from all over the region gives a foundation to compare the contemporary flow regimes with reference situations. ELOHA can be well achieved by testing the associations between the altered flow and ecological features with the prevailing and freshly gathered data from the field.

4.7.4 Hydraulic Geometry or Wetted Perimeter (WP) Techniques

The WP technique is a quite simple and reliable method that finds the association between the river flow and simple hydraulic features like river depth, velocity, or wetted perimeter to estimate an adequate flow. It generally makes use of the hydrological and hydraulic modelling data, with minimal use of ecological data. This technique is preferred over empirical or hydrological methods, as it integrates the flow data with hydraulic parameters over the river cross-section. If there is a variation in a WP at a cross-section, especially in the case of a riffle, which is considered the most productive benthic habitat, together with discharge, this establishes the basis for an environmental flow approval. This is also termed the habitat retention method.

4.7.5 Habitat Simulation Techniques

This technique augments its efficacy by integrating the hydraulics of a river with the favourable habitat features of the target biota. The hydraulic models are generated by combining flow data with hydraulic parameters like river depth, velocity or WP, substratum composition and cover, and complex hydraulic indices at various cross-sections of a stream. Thereafter, the biological surroundings inhabited by the riverine biota are coupled with the hydraulic and discharge relationship data to lay out a more comprehensive and modelled study of the preferential physical as well as ecological habitat of the riverine biodiversity at specific flows. Accordingly, the basis for the flow recommendation is formed by incorporating the hydrological, hydraulic, and biological response data. PHABSIM (Physical Habitat Simulation Model) is a technique used for habitat simulation that incorporates hydrology, stream morphology, and microhabitat preferences to generate relationships between river flow and habitat availability by target fish or invertebrate species to provide the environmental flows (Sedighkia et al., 2017; Ayll6n et al., 2012).

4.7.6 Hydrological Methods

Although the riverine ecology is affected by multiple factors like temperature, water quality, and turbidity, streamflow is still the major factor or master variable that holds distinct importance in aquatic ecosystems (Zeiringer et al., 2018), as it influences aquatic habitat, river morphology, biotic life, river connectivity, and water quality. Flow also regulates the abiotic components of the aquatic system, like dissolved oxygen, water temperature, suspended and bedload sediment size, and channel bed stability. Stream flows, principally high flows, also influence the morphological formation of channel and flood plains. A substantial quantity of water in a stream is considered great for the health of the aquatic organisms, and flow is also considered an indicator of the temporal integrity of a river ecosystem (Poff et al., 2017).

4.7.7 Tennant Method

The Tennant technique was developed by Tennant in 1976 for studying the mean annual flows in the 11 rivers of Montana, Nebraska, and Wyoming states for the protection of the health of a riverine ecosystem. The average annual flows of a river are needed for the ecological integrity of an aquatic ecosystem. The aquatic fauna require a certain percentage of flow for themselves. For survival, 10% of the flow is required; for a satisfactory habitat, the requirement is 30% of flow; and for excellent wellbeing, 60% of the flow is required (Poff et al., 2017).

4.7.8 Flow Duration Curve Based Methods

A flow duration curve (FDC) is a graph plot between variation in the flow and percentage time equalled or exceeded. These can be made by using either the complete time-series flow data or flow data relating to a definite period (like months) in different years. These curves can be developed for a specific area or by integrating the data obtained from various areas from a province that is homogeneous in the context of its hydrology and meteorology as well. The environmental flow categories, together with their management plans and illustrations, are given in Table 4.5 (Ahn et al., 2018).

4.7.9 Range of Variability Technique

An extensive statistical analysis of the various ecological parameters is carried out to evaluate the flow regime modifications in the Range of Variability Technique (RVT) (Baghel et al., 2019; Poff et al., 2017). This technique gives 32 grouped hydrological indicators that rely upon the long-term daily flow regime features, specifically magnitude, timing, duration, frequency, and rate of discharge. These indicators depict various flow unpredictability aspects. There exists a close association between hydrological and ecological variables, and these indicators are also checked for their modification to show the flow regime variation. Table 4.6 provides an outline of the status of environmental flow techniques globally, which gives an overview of the various methods followed for environmental flow evaluation worldwide.

4.7.10 Shortcomings of the Existing E-flows Assessment Methods

Every ecosystem is unique in its own way, and so are the fluvial ecosystems. Environmental flow techniques are significant for the safety of fishery resources and for environmental impact assessments as well. Flow naturalization that affects the natural hydrological regime alteration of the analysed channel should be taken into account when determining the flow characteristics (Książek et al., 2019). There are various methods of e-flow estimation, including the hydrological, hydraulic, habitat simulation, and flow duration curves. However, the most widely used method is the hydrological method, which is used around 33% of the time, as it is the easiest and simplest, but at the same time, it has some drawbacks when it comes to environmental flows in respect of faunal habitat requirements (Książek et al., 2019).The chief problem with the hydrological methods utilizing flow characteristics is flow naturalization. Enormous pressure on groundwater and surface water bodies disrupts flows, particularly low flows and base flows. The natural hydrological regime of a water

TABLE 4.5
Environmental Management Categories (EMC) and Management Outlook

Categories	Illustration	Management Outlook
A	Pristine streams having trivial alteration within the stream and the riparian habitat.	Rivers and basins secured. New hydro-projects (dams, diversions, etc.) forbidden.
B	In spite of the hydro resource advancement and basin alteration in minimally altered rivers, the stream biodiversity and habitations are unimpaired.	Allowance of agricultural irrigation and water supply.
C	The riverine biodiversity habitations are disturbed, yet the fundamental functioning is maintained. Fewer susceptible species are left. Presence of invasive species.	Due to socio-economic advancement involving the construction of dams and weirs, etc., there arise numerous instabilities in the environment, resulting in poor water quality and habitat destruction.
D	The natural territories of the biodiversity and the fundamental operations are disturbed significantly. Comparatively lower species richness. Reduced sensitive species. Prevalence of invasive species.	A considerable amount of deterioration in habitat and water quality due to the growth of hydro resources (dams, diversions, transfers).
E	Dwindling of habitations and also their accessibility. Species richness diminished significantly. Survival of tolerant species. Lack of reproduction in the endemic species. Invasion of exotic species.	Unchecked use of water resources due to overpopulation and urbanization. Intensive involvement by management is required to reinstate the flow pattern of a river to a higher management class.
F	Human alteration has caused havoc due to a complete loss of natural biodiversity and habitations. The worst scenario is that the fundamental functioning of the ecosystem is irreversibly devastated.	Intolerable status for any stream. If the slightest possibility exists, the flow regime of the river should be re-established to a higher class.

body is disrupted by surface and underground water consumption, sewage releases, reservoirs, lakes, and other human-induced morphological alterations. Secondly, in hydrological methods, low flows cannot be accurately measured from a stage–discharge curve, as low flow measurements can be affected by errors because of the extrapolation of the flow rate curve due to lack of measurements. Thirdly, flow assessment is a major hindrance in ungauged catchments. The hydraulic geometry method, which is used around 11% of the time among the e-flow methods, offers comparatively lower values of e-flows than hydrological methods, but the choice of criteria is the main issue when using hydraulic methods. The habitat necessities of the aquatic organisms and the nature of the stream or the waterbody should be properly defined before making use of the hydraulic method.

While estimating e-flows through various methods, certain aspects of a water body should not be trivialized, like the water quality, geochemical aspects of the river-bed, or the climate of the area. Many components, like discharge, flow rate, etc., are considered when developing models for e-flow estimation, but important aspects, like water quality, depicting the physiochemical status, and climatic environment, are not included in such models. Good water quality is a vital characteristic for the smooth and healthy functioning of the aquatic ecosystem. The e-flows calculated on the water quality cannot compute the flow requirements efficiently and accurately, because the e-flow allocations will exceed the natural hydrological river flow in severely polluted fluvial ecosystems (Sharma & Dutta, 2020). It is essential to develop methods for e-flow estimation that are not influenced by flow errors.

TABLE 4.6
Environmental Flow (EF) Methods in Practice/Suggested for Use in the Future in Several Nations

Nation	Environmental Flow Methods	Rivers, Length (l), and EF (m^3/s)	Extensively Used/Favoured Methods
Alaska, United States	Instream flow incremental methodology (IFIM); Tennant technique, which comprises amendments on grounds of fish information and expert decisions; and several other hydrological indexes.	Wulik river L 130 km EF 98.6 m^3/s	Regular use of the Tennant method and its reformed form. In special situations, IFIM is employed.
Australia	Several methods, such as river hydraulics and habitat simulation; instream flow incremental methodology; holistic method; Tennant Technique; flow restoration methodology, and habitat analysis technique.	Murray river L 2,508 km EF 66% of mean annual flow (MAF)	Instream flow incremental methodology and holistic methods.
Austria	Various methodologies and habitat modelling.	–	Undetermined.
Britain	Several methods like instream flow incremental methodology; hydrological tools and techniques; alternate methods comprising Scott Wilson Kirkpatrick method and holistic techniques like River Babingley methodology.	Don river L 110 km EF 36–44% MAF	Undetermined.
Canada	Methods like instream flow incremental methodology, comprising the Fish Rule Curve Method; Tennant technique, comprising average annual flow estimation; Wetted perimeter technique and several water quality techniques.	–	Application of environmental flow techniques, and modified forms of Tennant and some other techniques.
Denmark	Hydrological methods and some modelling approaches.	–	Use of median minimum system.
Finland	Methods reliant on the fish habitat modelling, Habitat imitation technique by making use of the geographic information system (GIS) method.	Kutinjoki river EF 2.4–4.8	Undetermined.
France	Techniques for habitat simulation; hydrological methods.	–	Evaluation of Habitat Method (EVHA): applied in about 70 cases.
Germany	Hydrological models, proficient judgement related to certain special cases, habitat simulation method.	–	Annual estimation of an average of minimum daily flows; skilled judgement for estimating 100 flows.
India	Holistic methods; hydrological modelling approaches.	Tungabhadra River; L 531 km EF NA Punarbhaba River; L 160 km EF 0.39	
Indonesia	Instream flow incremental method and various techniques.	Sekampung River EF 3.5	
Italy	Daily and annual mean flows; hydrological methods, Tennant technique; WP approach; other techniques.	Vomano River L 76 km	Hydrological techniques.

TABLE 4.6 (Continued)
Environmental Flow (EF) Methods in Practice/Suggested for Use in the Future in Several Nations

Nation	Environmental Flow Methods	Rivers, Length (l), and EF (m^3/s)	Extensively Used/Favoured Methods
Japan	Multivariate habitat suitability approach and multi-dimensional hydraulic modelling.	–	Undetermined.
Netherlands	Approach related to eco-tope sorting, habitat model, habitat suitability, policy, and substitutes study methods, habitat simulation modelling using GIS approach, hydrological model, some other techniques.	–	Undetermined.
New Zealand	Numerous hydraulic, hydrological, and habitat imitation techniques, Orth & Leonard and various other approaches for regionalization, instream flow incremental method, and some other related techniques.	Wairau River L 10 km EF 8.4	IFIM.
Norway	Habitat modelling approaches making use of the hybrid technique.	–	River System Simulator (RSS) and microhabitat modelling
South Africa	Hydrological methods like instream flow incremental methodology, riparian vegetation flow techniques, geomorphological alteration flow models, multivariate statistical methods to carry out hydrological and ecological regionalization.	Buzi River L 250 km EF 57% mean annual runoff (MAR)	Building Block Methodology (BBM) and associated suite of methods for ecological reserve determination.
Spain	Instream flow incremental methodology, hybrid techniques, holistic approach and past flow series methods, multivariate biomass techniques, hydrological method like basic flow technique.	–	Undetermined.
United States	Water quality techniques, hydrological models based on historical flow accounts, habitat imitation technique, hydraulic rating method, several hybrid methods; instream flow incremental methodology; Tennant technique, Tessman and Bayha amendment; wetted perimeter technique; skilled opinion; Texas system; range of variability approach, Singh model.	Tennessee River L 1049 km	Instream flow incremental methodology has been used at many locations and mentioned in several cases; Tennant technique, wetted perimeter method; average base flow are frequently exploited in several places.

Source: Revised from Tharme, 2003.

Some studies have been performed using physiochemical aspects (Sharma & Dutta, 2020; Acreman et al., 2014a; Dhanya & Kumar, 2015), but modelling that emphasizes these aspects has not yet been developed. Methods for e-flow estimation cannot be the same at a global level, as we have diverse river types, which could be temperate or tropical rivers; snow-fed, ground-fed, or spring-fed rivers; perennial or intermittent rivers. So, e-flow estimation methods should be sufficiently dynamic to give the output more precisely with a change in the river type. These e-flow modelling methods do not pay attention to what flow-sensitive communities require for their sustenance in the aquatic environment (Sharma & Dutta, 2020; Bogan et al., 2017; Theodoropoulos et al., 2019).

E-flow modelling methods sometimes do not find application in the context of data-deficient rivers, as they require long-term flow data, and also, a dearth of evaluation of ecosystem services, pollution problems, and stream–aquifer exchange makes it an onerous task (Datry et al., 2017).The hydrological methods do not report the extensive ranges of the variable flow regimes and geological and morphological discrepancy of the fluvial systems (Skoulikidis et al., 2017). The contemporary hydrological methods do not address the problems of variable flow requirement of intermittent streams in tropical countries (Theodoropoulos et al., 2019).The e-flow methods do not combine the requirements as well as the contribution of the riparian vegetation, the aquatic ecosystem, which in itself is quite wide, fluvial geomorphology, and the role of the groundwater in flow weightage estimations, which are the drawbacks of these methods. The environmental flow categories should be designed based on the river type and keeping in mind the ecological water required by the aquatic diversity living in those rivers.

4.7.11 Development and Management Strategies to Improve Environmental Flows

The water flow from the existent hydro structures is reliant on the dam type, the water discharge provision from the dams, and the main passages for the release of the water. Certain measures are swiftly instigated and yield comparatively good results for environmental flows. There are various structures used for water management, along with various approaches and measures that advance the environmental flows, as depicted in Table 4.7. The various structures must be as per distinct

TABLE 4.7
Paradigmatic Infrastructural Construction in Addition to the Management Approaches Chosen for Advancing Environmental Flows (EFs)

Water Management		Illustrative Approaches and Initiatives Adopted to Advance EFs	
Operations	**Infrastructure**	**Approaches**	**Initiatives**
Water withdrawal and storage, taming of stream flows	Weirs, dams (small and big), embankments, and stream channelization.	• The timing, quantity, as well as the quality of the downstream water availability would be upgraded, and the amount of water withdrawal and flow alteration reduced by handling water demand.	• To modify the layout quality for incorporating new amenities and alter the prevailing functionality tactics of the reservoir • Install passages of standing dams • Neutralize/shut down the dams for re-establishing flows.
Withdrawal and replenishment of the groundwater	Hand-pumps (most basic type), bore-wells, tube-wells, groundwater replenishment basins, rainwater harvesting (surface runoff, rooftop), individual or community based, etc.	• Abate groundwater withdrawals that lower the water table • To advance the flow accessibility to ecosystems that are reliant on groundwater • To enhance flood- and stormwater intrusion into groundwater and upgrade its quality.	• To alter the rate of withdrawals by keeping a check via imposing monetary values • Pioneer structures that preserve the storm/flood water at several scales for recharging the water table • For instigating sustainable management of groundwater table and aquifer • Develop/launch infrastructures that serve several water purposes.

TABLE 4.7 (Continued)
Paradigmatic Infrastructural Construction in Addition to the Management Approaches Chosen for Advancing Environmental Flows (EFs)

Water Management		Illustrative Approaches and Initiatives Adopted to Advance EFs	
Operations	**Infrastructure**	**Approaches**	**Initiatives**
Transference and circulation to consumers	Circulation canals (primary and tertiary), channels, pipelines, aqueducts, etc.	• Enhance the transport network efficacy and lessen the pressure on the supply system by avoiding accidental losses in the transportation network.	• Fix the problem of outflows/ leakage in the water circulation network and infrastructure.
Maintenance of water quality	Water treatment services, drainage networks.	• Upgrade the water treatment technologies • Regulate and restrict the agricultural, municipal, and industrial contaminants from inflowing into the water channels. • For natural cleansing, replenish the marshes and wetlands in addition to the e-flows.	• Launch/instigate novel and innovative water treatment amenities and structures for achieving higher water quality standards and at the same time, eradicate/amend activities and structures that pollute the groundwater.
Management of water demand by consumers	Mechanism specially designed for the preservation and sustainable management of water systems.	• Reduction in the amount of groundwater and surface water withdrawals, at the same time reusing and recycling water in viable places.	• Amount of water being abstracted from groundwater and supplied to end-users should be controlled and measured. • Proper initiatives should be taken for endorsing water conservation, and water-efficient technologies should be invented for reusing and recycling of water.
Non-conventional source	Salt removal from seawater/brackish water, water recycling, multipurpose water management by adapting accustomed water gathering methods, etc.	• The non-conventional set-up should be instituted to consolidate the water supply grids and advance water resources management in a unified manner.	• Establish viable and advantageous infrastructure such as desalination equipment. • Implementation of integrated surface and groundwater management by promoting rainwater harvesting (runoff and rooftop) practices in cities and villages.
Management of the drainage basin and watershed	Sustainable agricultural activities that check erosion for effective terrestrial management, including forest and foliage management.	• Enhancing the water withholding capacities of the drainage basins/watershed by lessening the unchecked terrestrial/water erosion, thereby maintaining the stability of soil and avoiding sediment erosion in streams.	• Adapt certain watershed management strategies like enhancing the forest foliage cover and sustainable land use and farming activities by integrating the water collection technologies available locally.

conditions. All infrastructure options and measures should be seen as context specific, complementary, and effective over different timeframes.

4.7.12 Effect of the Altered Flow Regime on Macrophytes Living in a Riverine Environment

Recently, an extensive amount of invasive species, including the macrophytes (vascular flora, stoneworts, bryophytes), has occupied the aquatic ecosystem in the form of interwinding belts and clusters, which is also a significant biotic feature of the littoral zone (Thomaz et al., 2015; Stuber et al., 2016). The adjoining region to the shore is colonized by various plant-like helophytes (e.g. *Verbascum thapsus*, *Thymus vulgaris*, etc.), floating leaves vegetation like nympheids (*Nymphaea elegans*, *Nymphaea candida*, *Euryale ferox*, etc.) and submerged leaves vegetation like elodeids (*Potamogeton crispus*, *Myriophyllum aquaticum*), which also determines the presence of rich fauna associated with them in distinct ways (trophic and paratrophic), invertebrates (epiphytic fauna) as well as vertebrates (amphibians, reptiles, fish, and birds). In shallow water bodies where the majority of the bottom surface is insulated, the role of macrophytes in the functioning of the entire ecosystem can be sufficiently significant to provide the basis for the designation of one of the alternative regimes; a macrophyte-dominated state (Dokulil & Donabaum, 2018; Capon et al., 2015).

Macrophytes exhibit their dominance in distinct ecosystems like lakes, wetlands, rivers, reservoirs, streams, and marine environments, including rapids and falls. Macrophytic primary produce exceeds the production of any other primary producers (Cremona et al., 2016; Francoeur et al., 2014). Macrophytes liberate some nutrients, such as phosphorus and nitrogen, that are consumed by micro-algae and bacteria, which specifically use organic carbon, which in turn plays a major role in nutrient cycling. These flora also give a significant supply of food to the grazing as well as the detritivore food webs, which in turn influences the community structure as well as the composition and diversity of the aquatic biota. There is a clear distinction amongst the micro- and macroinvertebrates (Nguyen et al., 2015), water birds (Guadagnin et al., 2012), and the assemblage of fishes among the various types of plants. The planktivorous fishes prefer submersed macrophytes, while the carnivorous fishes prefer the roots of free-floating macrophytes. The free-floating macrophytes act as refugia of prey, and when investigated for fish clusters, their approach from bottom-up and top-down is different from that of submersed plants (Thomaz and Cunha, 2010).

In a study on the Brazilian southeastern coast, the fish and the crustaceans that were used as prey by fishes used the macrophytes for their carbon intake, which has been revealed by analysing the carbon isotopes in fish biomass. There exists some contradiction, in that the colonization by exotic macrophytes of the aquatic environment exhibits negative (Stiers et al., 2011) as well as positive (Henninger et al., 2009) impact on the concentration of invertebrate taxa. Human-induced alterations like water level regulation and increment in the intensity of underwater light (Carniatto et al., 2014) enable *Hydrilla verticillate* (a submerged macrophyte) to flourish, which has created chaos in the Parana river and its associated channels in Brazil (Carniatto et al., 2014). These compete for light with the indigenous aquatic plants of that region. Due to flow obstruction, the water flow becomes low and is stabilized, which results in the appearance of submersed aquatic macrophytes (SAM) in reservoirs (Grzybkowska et al., 2017). Microhabitats having a good amount of food resources for both invertebrates and vertebrates (Bakker et al., 2016; Grzybkowska et al., 2017) can be created with a dense growth of macrophytes. Organisms like zooplankton and young fishes (Hilt et al., 2018) also get shelter from SAM after being washed out from reservoirs. The existence of juvenile and highly prevalent ichthyofaunal species is possible due to the prevalence of SAM, which increases the habitat complexity. While the young ones live on the benthic invertebrates, periphyton and zooplankton, the grown-ups rely on distinct feeding groups.

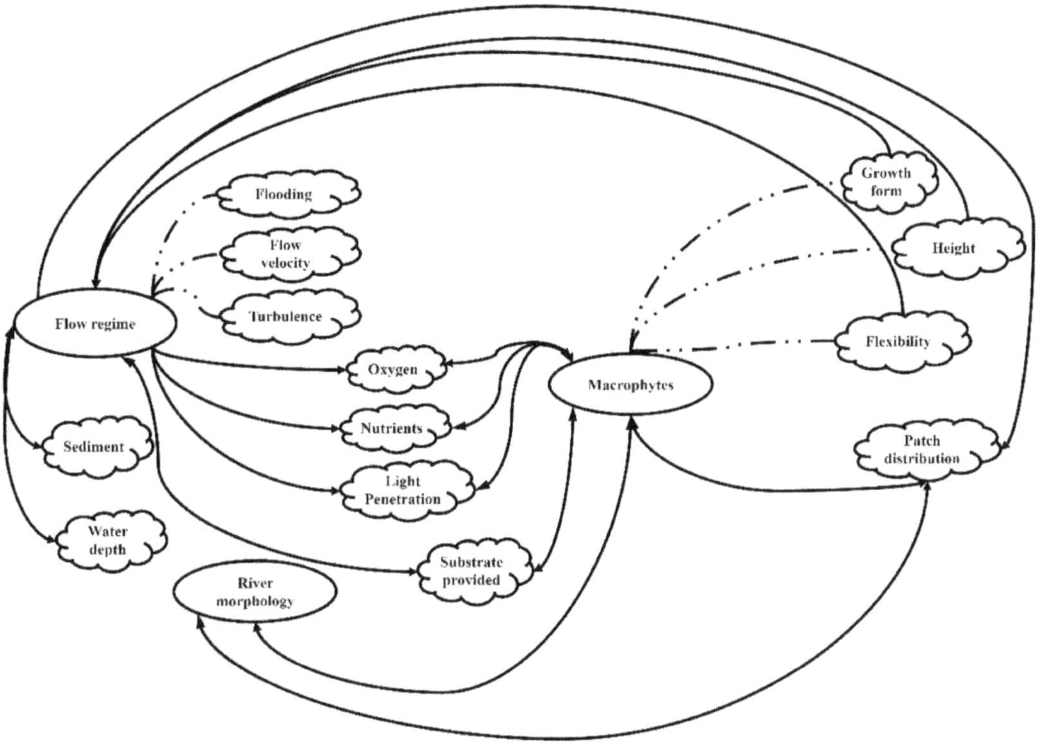

FIGURE 4.6 The inter-connection among physical features and aquatic macrophytes in a river.

Physical aspects like the water current, availability of light, and bottom substrate play a major role in determining the distribution pattern of freshwater macrophytes (Figure 4.6). These aspects, coupled with the water and sediment chemistry, play a significant role in aquatic foliage regionalization downstream (Steffen et al., 2014).Macrophytes dwelling in temperate lowland rivers are mainly controlled by some important features like velocity and discharge rate, light, substrate and siltation, competition, nutrient status, and management (Vukov et al., 2018). Nevertheless, in such temperate lowland areas, the discharge and velocity play a critical role in colonization as well as the establishment of the macrophytic vegetation (Janauer et al., 2010). For a healthy and proper functioning of the primary producer dynamics in a river, the anthropogenic alterations, in addition to the hydromorphological modifications, play a significant role in the diversity and distribution of the macrophytes (Bertrin et al., 2018). The macrophytic processes are greatly impacted by factors such as the shelter and habitation to the periphyton, zooplankton, invertebrates as well as vertebrates, affecting the functioning of the biochemical cycles in the aquatic ecosystem (Liu et al., 2018). In the large lowland rivers, macrophytes become worthwhile indicator flora, as they respond well to most of the influential factors (Camargo, 2018).

4.7.13 Effect of the Altered Flow Regime on the Ichthyofaunal and Other Aquatic Diversity

Flow regime modification ultimately results in the annexation and establishment of invasive and alien ichthyofaunal species in a riverine environment. In a riverine ecosystem, the influence of flow change is clearly apparent across the various assemblages, like plants, invertebrates, and fish, as described in Table 4.8. While reflecting the cohesive impact of the ecological alterations, fish groupings inclusive of widespread species signify an assortment of distinct trophic states, like herbivores, insectivores, piscivores, omnivores, etc., at the same time. Hence, the presence of these faunal species could be used to surmise the existence of additional biota in the riverine environment, as the adults are mostly found at the top trophic level. From the larval to the juvenile stage and then as adults, these species move through various trophic levels, including primary producer levels. The unified ecological wellbeing of a river is reflected in the ichthyofaunal community structure.

The information regarding ecological circumstances required by fish assemblages for accomplishing their life cycle plays a significant role in the comprehensive understanding of river functioning. If the species richness in a river is sufficiently high and the structure of the fish assemblage is sufficiently diverse, then knowledge of the environmental conditions needed by fish to complete their life cycles can contribute greatly to an understanding of the functioning of that river. For the monitoring and management of some useful flows required for healthy functioning of a river, this information gives special guidelines for such flows to sustain their requirements. It is inferred that when the environmental flows required for fish sustenance are effective and appropriate, the flow regime required for invertebrates will be achieved too, as these are found in prodigious numbers. The ichthyofaunal diversity is an important indicator of the ecological significance and understanding of a river, as sentinels of ecological integrity can be used for the endorsement of specific environmental flows. Because of their comparatively long lives and agility, fishes are regarded as valuable indicators for estimating effects over many years in a riverine ecosystem. The extreme low flows (baseflow) conditions determine the hydraulic habitation and also the life cycle phases needed by various species of fish, whereas the high flows determine the life cycle indications or habitat necessities. Throughout the phases of the life cycle in a distinct habitat, certain fishes require permanent base flows, whereas others require high flows for migration and spawning. Such flows in high quantity are also required for starting the development of the gonads and cleaning spawning beds and nursery areas (Figure 4.7).Various benthic fauna, including the ichthyofauna, obtain a habitable environment due to diverse bedload materials that are responsible for the formation of pools, riffles, and bars. The spawning habitats and promising nursery grounds determine the species richness and fish community structure. Comparatively greater densities, higher biomass, and diversity as well as abundance of species are present in the natural, unaffected pristine channels (Dutta et al., 2018).

Throughout the gradient of a river, various distinct types of habitats are formed from the segment of the headstream to the downstream segment because of the spatial organization. Many different habitat forms exist in a river channel, as given in Table 4.8 (Dutta et al., 2018), but three major habitats are assumed to exist while studying the ecological aspect of a river, as shown in Figure 4.8:

(1) Pools: zones of deep, slow-flowing water current located outside the bends of the meandering river. These have stagnant water with somewhat negligible water flow and sustain some species like *Garra* sp., *Puntius* sp., and *Pethia* sp.
(2) Runs: zones of a few metres in a river having no turbulence, with medium to high water current velocity. Certain species like *Salmophasia* sp. and *Dawkinsia* sp. are present in these run segments.
(3) Riffle/Rapids: zones of fast-flowing turbulent water current with shallow depth over boulders, stones, and cobbles. Species like *Barilius* sp., *Bhavania*, and *Homaloptera* sp. are found in these habitats.

The ichthyofaunal species diversity exhibits an inclination towards distinct habitats of pools, riffles, cascades, runs, and backwaters (Dobrev et al., 2018). The channel where the organisms live and the environmental pressures present there determine a lot of the morphological structure of the body of an aquatic animal. Such morphological distinctions help stimulate diversity in communities, because every species chooses and dwells in a habitation befitting its morphology. From a study

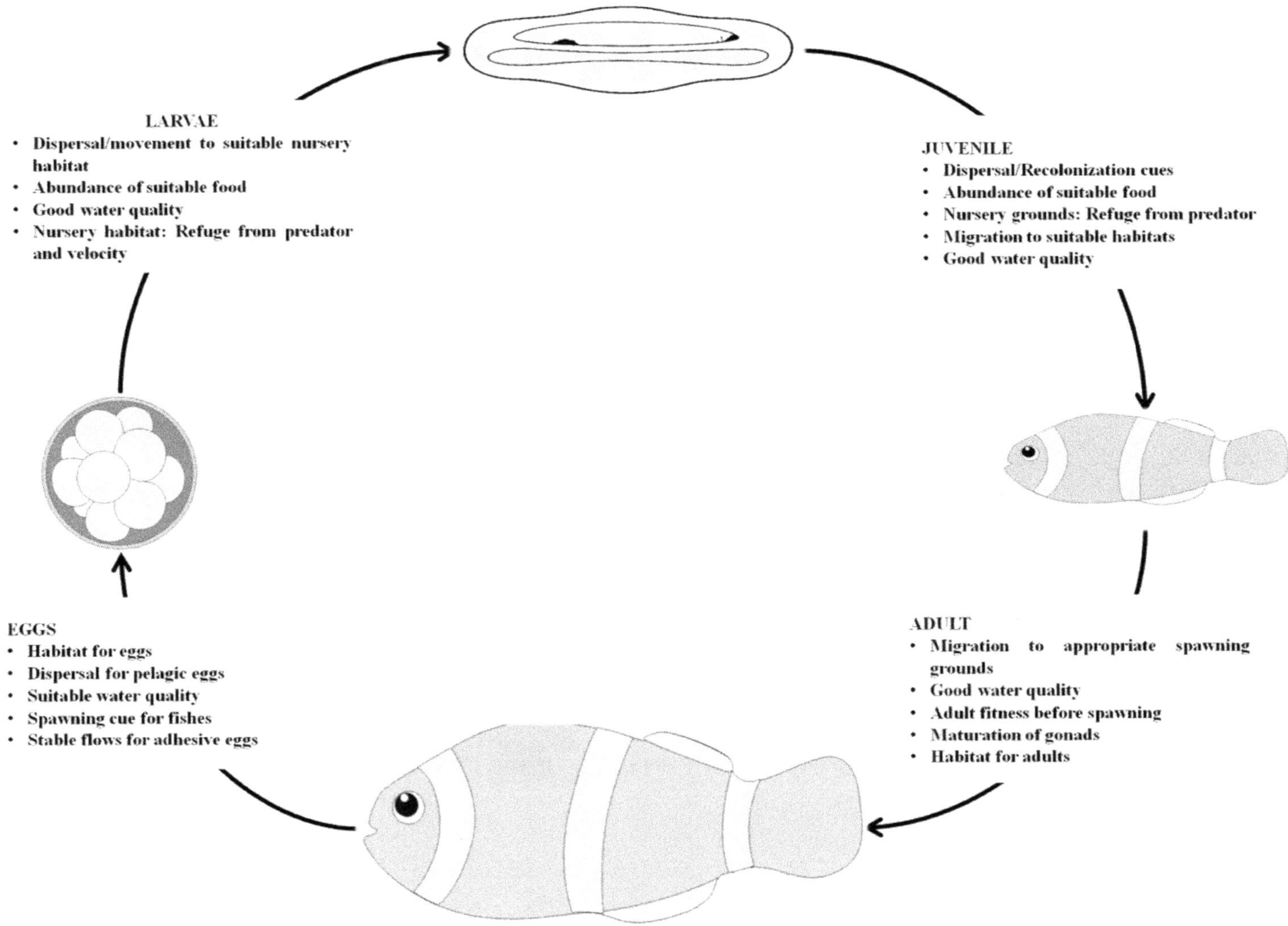

FIGURE 4.7 Natural flow required for the ichthyofaunal life cycle stages.

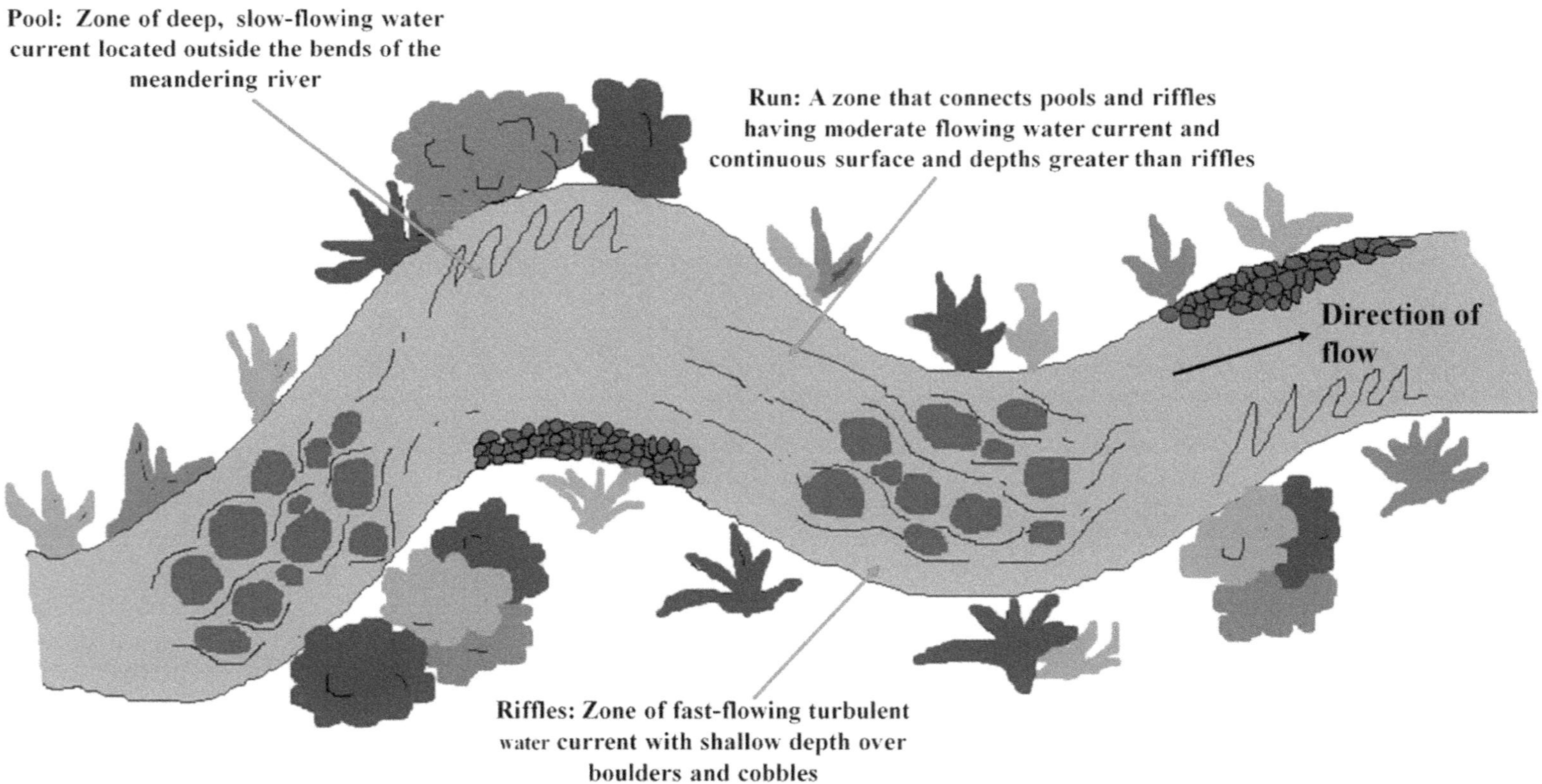

FIGURE 4.8 Major fish habitats found within a river.

TABLE 4.8
Diverse Habitat Forms Existing within a River Channel along with the Species Preferring such Habitats

Habitat Forms	Habitat Characteristic	Species Dwelling in Such Distinct Habitats
Open river	River without channel confluences and pools	A mixed assemblage of fishes like *Leuciscus leucius*, *L. cephalus*, *Cottus gobio*, *Barbatula barbatula*, *Phoxinus phoxinus*, *Leuciscus idus*, *Tor khudree* (Deccan mahseer), *Devario malabaricus* (Malabar danio), *Hypselobarbus* sp., *Salmophasia boopis* (razor-belly minnow), species of Balitoridae, Loricariidae, Gyrinocheilidae family
Deep pools	Deep pools with high turbidity and coarse substrate	Species such as *Chitala chitala*, *Wallago attu*, *Ompok pabda*, *Clarias*, *Channa punctatus*, and *C. striatatus* inhabit this habitat, as do many catfish species of genus *Mystus*, like *M. singhala*, *M. aor*, and *M. tengra*
Backwater and shallow pools	Medium-sized pools with low water flow	Species like *Notopterus*, *Cirrhinus*, *Rita*, *Labeo*, *Channa striatus*, *Heteropneustes fossilis*, *Tilapia*, *Mastacembelus armatus*, *Notemigonus crysoleucas*, and *Gonialosa manmina* prefer this habitat
Channel confluence	Junction of two rivers	Ideal place for some freshwater fishes like *Labeo rohita*, *Catla catla*, *Tor tor*, *Tenualosa ilisha*, *Gonialosa manmina*, *Chela laubuca*, *Amblypharyngodon mola*, etc.
Wetlands-connected ditches	Medium-sized to small ditches connected to riverside wetlands	Minnows like *Pimephales promelas*, common carps, some catfishes of *Siluriformes*, *Anabas testudineus*, *P. fluviatilis*, *Systomus sarana*, etc.
Grassed banks	Stabilized stream banks	*Punctius* sp., *Cyprinus carpio*, *Myxus (Trachystoma) petardi*, *Lepomis macrochirus*, *Notemigonus crysoleucas*, *Gonialosa modesta*, *Amblypharyngodon mola*
Well-vegetated banks	Well-vegetated banks maintain the stream temperature	*Gambusia holbrook*, *Ctenopharyngodon idella*, *Gonialosa manmina*, *Gonialosa modesta*, *Amblypharyngodon mola*

Source: Modified from Dutta et al., 2018.

done in Indiana on the White, Salamonie, Wabash, and Mississinewa rivers, it is apparent that flow velocity determines preference of fish habitat, where the pool habitats are characterized by greater numbers of fish with a low depth ratio, whereas riffle habitats are characterized by greater numbers of fishes with a high depth ratio, which are more typically found in riffle areas.

Due to stream depth, flow velocity, and cover, which are considered imperative factors, the habitat heterogeneity increases, which in turn increases the heterogeneity of faunal species (Cruz et al., 2018). Habitat structure also determines the ichthyofaunal community structure, such as abundance pattern and taxonomic composition (Ferreira et al., 2018). In a study carried out in the northern midwestern United States with streams showing a variable flow regime, a deviation in fish community structure and functional association has been reported due to provincial variation (Epstein et al., 2018). Flow regime alteration also influences the fish community structure at small regional scales in the drainage network. Relatively small, generalist, and physiologically tolerant species are found in the streams that have low flows for a prolonged period (Lennox et al., 2019). Because of the altering flows, the ichthyofaunal population becomes vulnerable to stranding as a result of the different species' behavioural response, which is the resultant of many factors, such as substrate characteristics, body size, species type, temperature, time, substrate characteristics, and the rate of flow reductions (Table 4.9) (Bradford, 1997).

TABLE 4.9
Description of Several Biotic Responses to Different Flow Regimes

	Altered Flow Variables	Biotic Responses	References
		Biotic Responses to Altered Flow Regimes with Flow-induced Changes in Habitat	
1.	Base flow reliability amplified; flow changeability reduced	Aquatic macrophytes grow disproportionately	Preston et al., 2018
		Reduction in fish populations	Almodóvar et al., 2019
		Increased standing crop and reduced diversity of macroinvertebrates	de la Fuente et al., 2018
2.	Erratic (diurnal) patterns of inflow below hydroelectric dams	Reduction in species richness of benthic macroinvertebrates	Bertoncin et al., 2019
		Fish stranding	Heggenes et al., 2018
3.	Conversion of lotic habitat to lentic	Elimination of salmonids and pelagic spawning fishes and the dominance of generalist fish species	Tamario et al., 2018
		Loss of fishes adapted to turbid river habitats	Mori et al., 2018
		Loss of fishes due to inundation of spawning grounds	Dahlke et al., 2018
		Life-history Responses to Altered Flow Regimes	
1.	Spates timing	Constant low flows are needed for spawning and recruitment of riverine ichthyofaunal diversity	de Magalhaes Lopes et al., 2018
2.	Rising flows timing	Loss of indications for fish spawning and movement	Bunn & Arthington, 2002
3.	Modified temperature regimes below dams	Delayed spawning in fish	Jumani et al., 2018
		Eradication of temperature-specific species of fishes	Luoma et al., 2018
4.	Reduced seasonality	Lessened synchrony of breeding in gammarid shrimps	Bunn & Arthington, 2002
		Biotic Responses to the Loss of Longitudinal or Lateral Connectivity	
1.	Presence of instream barriers	Loss of migratory fish species	Lin et al., 2018
2.	Reduced frequency, duration, and area of inundation of flood plain wetlands	Reduced spawning areas and/or recruitment success of lowland river fish	Pal & Saha, 2018; Sandi et al., 2018
		Biotic Responses to Disturbed Flow Regimes with Invasion and Success of Exotic and Introduced Species	
1.	Reduced flow variability and increased seasonal stability	Favour populations of exotic fish species (carp, mosquito fish)	Thompson et al., 2018
2.	Conversion of lotic to lentic habitat	The proliferation of exotic fish species	Raghavan et al., 2008
3.	Inter-basin transfers of water	Translocation of fish species	Ready et al., 2018; Arthington et al., 2018

4.8 CONCLUSION

The fundamental aspect for managing a healthy riverine ecology is to maintain a natural flow regime, thereby smoothing its biogeochemical cycling and sediment fluxes in the river, which in turn is responsible for good aquatic biodiversity. Due to human-induced hydrological impoundments like dams, weirs, etc., which reduce the downstream connectivity of a naturally flowing unregulated river, the ichthyofaunal and other aquatic populations (invertebrates, waterfowl, etc.) decline, as these faunal species are sensitive to variable flow regimes. The lotic habitats (rivers, streams) occupied by the aquatic fauna have been homogenized into stagnant lentic habitats (i.e., pools). Consequently, a good quantity of healthy riverine aquatic biota is found in an unaltered and free-flowing natural river when compared with altered and regulated ones. Certain species sensitive to natural flow regimes have long, cylindrical morphology, preferring locations having a higher concentration of dissolved oxygen. When impounded, the rivers heat up rapidly, which in turn depletes the oxygen concentration, thereby influencing the species composition of the river. Fishes are faunal species that can find new habitats if their habitat is lost, provided there is interconnectivity between the mainstream and its associated tributaries. It is a fact that dams pose immediate effects in addition to the long-term impacts on the ichthyofaunal as well as other aquatic populations (vertebrates, invertebrates, etc.) living in a riverine environment. So, keeping in mind future projects that are yet to be undertaken, it becomes very important to formulate and design projects such that they reduce the harmful and negative effects of impoundments on the aquatic ecosystem and concomitantly improve current projects by establishing ecological connectivity within a stream. Multidisciplinary knowledge and understanding should be applied for the protection of some of the pristine rivers currently existing in their natural regime. The hydrological models and methods developed so far are doing well in their respective terms, but a comprehensive understanding of the ongoing activities within an aquatic ecosystem can only be achieved when the models also consider the habitat preferences of the aquatic biota during their various life cycle stages and when the distinct aspects of the theory of food web connectivity are well studied and incorporated within these models. Also, the models should be region specific and sufficiently dynamic to at once consider and incorporate the changing climatic conditions (temperature, precipitation) of any region while incorporating the archaic history of the region as well. Refining current techniques of e-flow estimation is necessary for a thorough and comprehensive evaluation of variable flow regimes. Future research on e-flow estimation should take into account the following factors: (i) multiple requirements for aquatic-riparian habitat during both high-flow floods and low-flow droughts; (ii) interactions between surface and groundwater to account for baseflows in maintaining lean flows; and (iii) water quality to assess the impact of pollution loading and water abstraction on riverine ecosystems. This would be a comprehensive strategy in practice; otherwise, we would simply study freshwater ecosystems and the biota that goes along with them without ever being able to replicate them accurately on the ground.

ACKNOWLEDGEMENTS

Kashif Iqbal gets the DST-INSPIRE JRF Fellowship (IF 180622) from the Department of Science and Technology, New Delhi. The authors also wish to extend their heartful thanks to the Department of Environmental Science, Babasaheb Bhimrao Ambedkar University, Lucknow for providing the required facilities for carrying out this study.

REFERENCES

Acreman, M., Arthington, A.H., Colloff, M.J., Couch, C., Crossman, N.D., Dyer, F., Overton, I., Pollino, C.A., Stewardson, M.J., & Young, W. (2014a). Environmental flows for natural, hybrid, and novel riverine ecosystems in a changing world. *Frontiers in Ecology and the Environment,* 12(8), 466–473. https://doi.org/10.1890/130134

Acreman, M.C., Overton, I.C., King, J., Wood, P.J., Cowx, I.G., Dunbar, M.J., Kendy, E., & Young, W.J. (2014b). The changing role of ecohydrological science in guiding environmental flows. *Hydrological Sciences Journal*, 59(3–4), 433–450. https://doi.org/10.1080/02626667.2014.886019

Ahn, J.M., Kwon, H.G., Yang, D.S., & Kim, Y.S. (2018). Assessing environmental flows of coordinated operation of dams and weirs in the Geum River basin under climate change scenarios. *Science of the Total Environment,* 643, 912–925. https://doi.org/10.1016/j.scitotenv.2018.06.225

Albouy, C., Velez, L., Coll, M., Colloca, F., Le Loc'h, F., Mouillot, D., & Gravel, D. (2014). From projected species distribution to food-web structure under climate change. *Global Change Biology*, 20(3), 730–741. https://doi.org/10.1111/gcb.12467

Almodóvar, A., Ayllón, D., Nicola, G.G., Jonsson, B., & Elvira, B. (2019). Climate-driven biophysical changes in feeding and breeding environments explain the decline of southernmost European Atlantic salmon populations. *Canadian Journal of Fisheries and Aquatic Sciences*, 76(9), 1581–1595. https://doi.org/10.1139/cjfas-2018-0297

Altermatt, F., Seymour, M., & Martinez, N. (2013). River network properties shape α-diversity and community similarity patterns of aquatic insect communities across major drainage basins. *Journal of Biogeography*, 40(12), 2249–2260. https://doi.org/10.1111/jbi.12178

Arthington, Á.H., Naiman, R.J., Mcclain, M.E., & Nilsson, C. (2010). Preserving the biodiversity and ecological services of rivers: new challenges and research opportunities. *Freshwater Biology*, 55(1), 1–16. https://doi.org/10.1111/j.1365-2427.2009.02340.x

Arthington, A.H., Finlayson, C.M., & Pittock, J. (2018). Freshwater ecological principles. In Finlayson, C.M., Arthington, A.H., & Pittock, J. (eds) *Freshwater Ecosystems in Protected Areas*. Academic Press, London, 34–53. http://doi.org:10.4324/9781315226385-3

Ayllón, D., Almodóvar, A., Nicola, G.G., & Elvira, B. (2012). The influence of variable habitat suitability criteria on PHABSIM habitat index results. *River Research and Applications*, 28(8), 1179–1188. https://doi.org/10.1002/rra.1496

Baghel, D.S., Gaur, A., Karthik, M., & Dohare, D. (2019). Global trends in environmental flow assessment: an overview. *Journal of the Institution of Engineers (India): Series A*, 100(1), 191–197. https://doi.org/10.1007/s40030-018-0332-5

Bakker, E.S., Wood, K.A., Pagès, J.F., Veen, G.C., Christianen, M.J., Santamaría, L., Nolet, B.A., & Hilt, S. (2016). Herbivory on freshwater and marine macrophytes: a review and perspective. *Aquatic Botany*, 135, 18–36. https://doi.org/10.1016/j.aquabot.2016.04.008

Bertoncin, A.P., Pinha, G.D., Baumgartner, M.T., & Mormul, R.P. (2019). Extreme drought events can promote homogenization of benthic macroinvertebrate assemblages in a floodplain pond in Brazil. *Hydrobiologia*, 826(1), 379–393. https://doi.org/10.1007/s10750-018-3756-z

Bertrin, V., Boutry, S., Alard, D., Haury, J., Jan, G., Moreira, S., & Ribaudo, C. (2018). Prediction of macrophyte distribution: the role of natural versus anthropogenic physical disturbances. *Applied Vegetation Science*, 21(3), 395–410. https://doi.org/10.1111/avsc.12378

Besacier-Monbertrand, A. L., Paillex, A., & Castella, E. (2014). Short-term impacts of lateral hydrological connectivity restoration on aquatic macroinvertebrates. *River Research and Applications*, 30(5), 557–570. https://doi.org/10.1002/rra.2597

Birk, S., Bonne, W., Borja, A., Brucet, S., Courrat, A., Poikane, S., Solimini, A., Van De Bund, W., Zampoukas, N., & Hering, D. (2012). Three hundred ways to assess Europe's surface waters: an almost complete overview of biological methods to implement the Water Framework Directive. *Ecological Indicators*, 18, 31–41. https://doi.org/10.1016/j.ecolind.2011.10.009

Bogan, M.T., Chester, E.T., Datry, T., Murphy, A.L., Robson, B.J., Ruhi, A., Stubbington, R., & Whitney, J.E. (2017). Resistance, resilience, and community recovery in intermittent rivers and ephemeral streams. In Datry, T., Bonada, N., & Boulton, A. (eds) *Intermittent Rivers and Ephemeral Streams*. Academic Press, London, 349–376.

Boggie, M.A., Collins, D.P., Donnelly, J.P., & Carleton, S.A. (2018). Land use, anthropogenic disturbance, and riverine features drive patterns of habitat selection by a wintering waterbird in a semi-arid environment. *PloS One*, 13(11), e0206222. https://doi.org/10.1371/journal.pone.0206222

Bradford, M.J. (1997). An experimental study of stranding of juvenile salmonids on gravel bars and in sidechannels during rapid flow decreases. *Regulated Rivers: Research & Management: An International Journal Devoted to River Research and Management*, 13(5), 395–401.https://doi.org/10.1002/(sici)1099-1646(199709/10)13:5%3C395::aid-rrr464%3E3.0.co;2-l

Bunn, S.E., & Arthington, A.H. (2002). Basic principles and ecological consequences of altered flow regimes for aquatic biodiversity. *Environmental Management*, 30(4), 492–507. https://doi.org/10.1007/s00267-002-2737-0

Camargo, J.A. (2018). Responses of aquatic macrophytes to anthropogenic pressures: comparison between macrophyte metrics and indices. *Environmental Monitoring and Assessment*, 190(3), 1–11. https://doi.org/10.1007/s10661-018-6549-y

Capon, S.J., Lynch, A.J.J., Bond, N., Chessman, B.C., Davis, J., Davidson, N., Finlayson, M., Gell, P.A., Hohnberg, D., Humphrey, C., & Kingsford, R.T. (2015). Regime shifts, thresholds and multiple stable states in freshwater ecosystems; a critical appraisal of the evidence. *Science of the Total Environment*, 534,122–130. https://doi.org/10.1016/j.scitotenv.2015.02.045

Carlisle, D.M., Nelson, S.M., & Eng, K. (2014). Macroinvertebrate community condition associated with the severity of streamflow alteration. *River Research and Applications*, 30(1), 29–39. https://doi.org/10.1002/rra.2626

Carniatto, N., Fugi, R., Thomaz, S.M., & Cunha, E.R. (2014). The invasive submerged macrophyte *Hydrilla verticillata* as a foraging habitat for small-sized fish. *Natureza & Conservação*, 12(1), 30–35. https://doi.org/10.4322/natcon.2014.006

Carnie, R., Tonina, D., McKean, J.A., & Isaak, D. (2016). Habitat connectivity as a metric for aquatic microhabitat quality: application to Chinook salmon spawning habitat. *Ecohydrology,* 9(6), 982–994. https://doi.org/10.1002/eco.1696

Cooper, A.R., Infante, D.M., Daniel, W.M., Wehrly, K.E., Wang, L., & Brenden, T.O. (2017). Assessment of dam effects on streams and fish assemblages of the conterminous USA. *Science of the Total Environment,* 586, 879–889. https://doi.org/10.1016/j.scitotenv.2017.02.067

Cremona, F., Laas, A., Arvola, L., Pierson, D., Noges, P., & Noges, T. (2016). Numerical exploration of the planktonic to benthic primary production ratios in lakes of the Baltic Sea catchment. *Ecosystems*, 19(8), 1386–1400. https://doi.org/10.1007/s10021-016-0006-y

Cruz, P.R., Baumgartner, M.T., & Gomes, L.C. (2018). Connection promoted by annual floods shaping the response community attributes of shallow lakes fish inhabit the upper Paraná River floodplain. *Environmental Biology of Fishes*, 101(12), 1723–1738. https://doi.org/10.1007/s10641-018-0824-1

Dahlke, F.T., Butzin, M., Nahrgang, J., Puvanendran, V., Mortensen, A., Pörtner, H. O., & Storch, D. (2018). Northern cod species face spawning habitat losses if global warming exceeds 1.5°C. *Science Advances*, 4(11), eaas8821. https://doi.org/10.1126/sciadv.aas8821

Datry, T., Singer, G., Sauquet, E., Jorda-Capdevila, D., Von Schiller, D., Subbington, R., Magand, C., Pařil, P., Miliša, M., Acuña, V., Alves, M., Augeard, B., Brunke, M., Cid, N., Csabai, Z., England, J., Froebrich, J., Koundouri, P., Lamouroux, N., Martí, E., Morais, M., Munné, A., Mutz, M., Pesic, V., Previšić, A., Reynaud, A., Robinson, C., Sadler, J., Skoulikidis, N., Terrier, B., Tockner, K., Vesely, D., & Zoppini, A. (2017). Science and management of intermittent rivers and ephemeral streams (SMIRES). *Research Ideas and Outcomes,* 3, e21774. https://doi.org/10.3897/rio.3.e21774

de la Fuente, M., Bonada, N., Bêche, L., Dahm, C.N., Mendez, P.K., Tockner, K., Uehlinger, U., & Acuña, V. (2018). Evolutionary responses of aquatic macroinvertebrates to two contrasting flow regimes. *Hydrobiologia*, 808(1), 353–370. https://doi.org/10.1007/s10750-017-3437-3

de Magalhaes Lopes, J., Alves, C.B.M., Peressin, A., & Pompeu, P.S. (2018). Influence of rainfall, hydrological fluctuations, and lunar phase on spawning migration timing of the Neotropical fish *Prochilodus costatus*. *Hydrobiologia*, 818(1), 145–161. https://doi.org/10.1007/s10750-018-3601-4

Dhanya, C.T., & Kumar, A. (2015). Making a case for estimating environmental flow under climate change. *Current Science*, 109(6), 1019–1020.

Dobrev, Y., Trajković, S., Gocić, M., Milićević, D., & Lissev, N. (2018). Rock ramp impact on the riverine hydraulic. *Facta Universitatis, Series: Architecture and Civil Engineering,* 16(1), 113–123. https://doi.org/10.2298/fuace170425009d

Dokulil, M.T., & Donabaum, K. (2018). Concepts for restoration: nutrient loading, thresholds and alternative stable states. *The Alte Donau: Successful Restoration and Sustainable Management*, 10, 3–14. http://doi.org:10.1007/978-3-319-93270-5_1

Durance, I., & Ormerod, S.J. (2007). Climate change effects on upland stream macroinvertebrates over a 25-year period. *Global Change Biology*, 13(5), 942–957. https://doi.org/10.1111/j.1365-2486.2007.01340.x

Dutta, V., Sharma, U., Iqbal, K., Kumar, R., & Pathak, A.K. (2018). Impact of river channelization and riverfront development on fluvial habitat: evidence from Gomti River, a tributary of Ganges, India. *Environmental Sustainability*, 1(2), 167–184. https://doi.org/10.1007/s42398-018-0016-0

Dwivedi, S., Mishra, S., & Tripathi, R.D. (2018). Ganga water pollution: a potential health threat to inhabitants of Ganga basin. *Environment International*, 117, 327–338. https://doi.org/10.1016/j.envint.2018.05.015

Epstein, J.M., Pine III, W.E., Romagosa, C.M., Scott, M.C., Phillips, C.T., Marion, C.A., & Baiser, B. (2018). State- and regional-scale patterns and drivers of freshwater fish functional diversity in the southeastern USA. *Transactions of the American Fisheries Society*, 147(6), 1179–1198. https://doi.org/10.1002/tafs.10110

Ferencz, S.B., & Cardenas, M.B. (2017). Diel stream temperature regimes of Bukovsky regions of the conterminous United States. *Geophysical Research Letters*, 44(5), 2264–2271. https://doi.org/10.1002/2017gl072641

Ferreira, M.C., Begot, T.O., da Silveira Prudente, B., Juen, L., & de Assis Montag, L.F. (2018). Effects of oil palm plantations on habitat structure and fish assemblages in Amazon streams. *Environmental Biology of Fishes*, 101(4), 547–562. https://doi.org/10.1007/s10641-018-0716-4

Francoeur, S.N., Winslow, K.P., Miller, D., Stow, C.A., Cha, Y., & Peacor, S.D. (2014). Spatial and temporal patterns of macroscopic benthic primary producers in Saginaw Bay, Lake Huron. *Journal of Great Lakes Research*, 40, 53–63. https://doi.org/10.1016/j.jglr.2014.01.011

Fuller, M.R., Doyle, M.W., & Strayer, D.L. (2015). Causes and consequences of habitat fragmentation in river networks. *Annals of the New York Academy of Sciences*, 1355(1), 31–51. https://doi.org/10.1111/nyas.12853

Gomes, M.P., de Albuquerque, C.Q., Andrades, R., Martins, A.S., Robinson, L.A., & Spencer, M. (2018). Influence of detached macroalgae on fish size and condition in nearshore habitats. *Estuarine, Coastal and Shelf Science*, 211, 227–237. https://doi.org/10.1016/j.ecss.2017.08.023

Grzybkowska, M., Kucharski, L., Dukowska, M., Takeda, A.M., Lik, J., & Leszczyńska, J. (2017). Submersed aquatic macrophytes and associated fauna as an effect of dam operation on a large lowland river. *Ecological Engineering*, 99, 256–264. https://doi.org/10.1016/j.ecoleng.2016.11.023

Guadagnin, D.L., Peter, A.S., Rolon, A.S., Stenert, C., & Maltchik, L. (2012). Does non-intentional flooding of rice fields after cultivation contribute to waterbird conservation in southern Brazil? *Waterbirds*, 35(3), 371–380. https://doi.org/10.1675/063.035.0302

Harley, C.D., Anderson, K.M., Demes, K.W., Jorve, J.P., Kordas, R.L., Coyle, T.A., & Graham, M.H. (2012). Effects of climate change on global seaweed communities. *Journal of Phycology*, 48(5), 1064–1078. https://doi.org/10.1111/j.1529-8817.2012.01224.x

Heggenes, J., Alfredsen, K., Bustos, A.A., Huusko, A., & Stickler, M. (2018). Be cool: a review of hydrophysical changes and fish responses in winter in hydropower-regulated northern streams. *Environmental Biology of Fishes*, 101(1), 1–21. https://doi.org/10.1007/s10641-017-0677-z

Henninger, T.O., Froneman, P.W., Richoux, N.B., & Hodgson, A.N. (2009). The role of macrophytes as a refuge and food source for the estuarine isopod *Exosphaeroma hylocoetes*. *Estuarine, Coastal and Shelf Science*, 82(2), 285–293. https://doi.org/10.1016/j.ecss.2009.01.017

Hilt, S., Alirangues Nuñez, M.M., Bakker, E.S., Blindow, I., Davidson, T.A., Gillefalk, M., Hansson, L.A., Janse, J.H., Janssen, A.B., Jeppesen, E., & Kabus, T. (2018). Response of submerged macrophyte communities to external and internal restoration measures in north temperate shallow lakes. *Frontiers in Plant Science*, 9, 194. https://doi.org/10.3389/fpls.2018.00194

Hughes, D.A., & Louw, D. (2010). Integrating hydrology, hydraulics and ecological response into a flexible approach to the determination of environmental water requirements for rivers. *Environmental Modelling & Software*, 25(8), 910–918. https://doi.org/10.1016/j.envsoft.2010.03.004

Jaeger, K.L., Olden, J.D., & Pelland, N.A. (2014). Climate change poised to threaten hydrologic connectivity and endemic fishes in dryland streams. *Proceedings of the National Academy of Sciences*, 111(38), 13894–13899. https://doi.org/10.1073/pnas.1320890111

Janauer, G.A., Schmidt-Mumm, U., & Schmidt, B. (2010). Aquatic macrophytes and water current velocity in the Danube River. *Ecological Engineering*, 36(9), 1138–1145. https://doi.org/10.1016/j.ecoleng.2010.05.002

Jumani, S., Rao, S., Kelkar, N., Machado, S., Krishnaswamy, J., & Vaidyanathan, S. (2018). Fish community responses to stream flow alterations and habitat modifications by small hydropower projects in the Western Ghats biodiversity hotspot, India. *Aquatic Conservation: Marine and Freshwater Ecosystems*, 28(4), 979–993. https://doi.org/10.1002/aqc.2904

Kashaigili, J.J., Mccartney, M., & Mahoo, H.F. (2007). Estimation of environmental flows in the Great Ruaha River catchment, Tanzania. *Physics and Chemistry of the Earth, Parts A/B/C*, 32(15–18), 1007–1014. https://doi.org/10.1016/j.pce.2007.07.005

Kiernan, J.D., Moyle, P.B., & Crain, P.K. (2012). Restoring native fish assemblages to a regulated California stream using the natural flow regime concept. *Ecological Applications*, 22(5), 1472–1482. https://doi.org/10.1890/1051-0761-22.5.1472

Kirkman, S.P., Yemane, D., Kathena, J., Mafwila, S.K., Nsiangango, S.E., Samaai, T., Axelsen, B., & Singh, L. (2013). Identifying and characterizing demersal fish biodiversity hotspots in the Benguela Current Large

Marine Ecosystem: relevance in the light of global changes. *ICES Journal of Marine Science*, 70(5), 943–954. https://doi.org/10.1093/icesjms/fst040

Knight, T.M., McCoy, M.W., Chase, J.M., McCoy, K.A., & Holt, R.D. (2005). Trophic cascades across ecosystems. *Nature*, 437(7060), 880–883. http://doi:10.1038/nature03962

Książek, L., Woś, A., Florek, J., Wyrębek, M., Młyński, D., & Wałęga, A. (2019). Combined use of the hydraulic and hydrological methods to calculate the environmental flow: Wisloka river, Poland: case study. *Environmental Monitoring and Assessment*, 191(4), 1–17. https://doi.org/10.1007/s10661-019-7402-7

Kumara, B.H., & Srikantaswamy, S. (2011). Environmental flow requirements in Tungabhadra River, Karnataka, India. *Natural Resources Research*, 20(3), 193–205. http://doi.org:10.1007/s11053-011-9143-3

Layman, C.A., Giery, S.T., Buhler, S., Rossi, R., Penland, T., Henson, M.N., Bogdanoff, A.K., Cove, M.V., Irizarry, A.D., Schalk, C.M., & Archer, S.K. (2015). A primer on the history of food web ecology: fundamental contributions of fourteen researchers. *Food Webs*, 4, 14–24. https://doi.org/10.1016/j.fooweb.2015.07.001

Lennox, R.J., Crook, D.A., Moyle, P.B., Struthers, D.P., & Cooke, S.J. (2019). Toward a better understanding of freshwater fish responses to an increasingly drought-stricken world. *Reviews in Fish Biology and Fisheries,* 29(1), 71–92. https://doi.org/10.1007/s11160-018-09545-9

Li, K., Zhu, C., Wu, L., & Huang, L. (2013). Problems caused by the Three Gorges Dam construction in the Yangtze River basin: a review. *Environmental Reviews*, 21(3), 127–135. https://doi.org/10.1139/er-2012-0051

Lin, H.Y., Brown, C.J., Dwyer, R.G., Harding, D.J., Roberts, D.T., Fuller, R.A., Linke, S., & Possingham, H.P. (2018). Impacts of fishing, river flow and connectivity loss on the conservation of a migratory fish population. *Aquatic Conservation: Marine and Freshwater Ecosystems*, 28(1), 45–54. https://doi.org/10.1002/aqc.2831

Liu, S., Zhao, T., Zhu, Y., Qu, X., He, Z., Giesy, J. P., & Meng, W. (2018). Molecular characterization of macrophyte-derived dissolved organic matters and their implications for lakes. *Science of the Total Environment*, 616, 602–613. https://doi.org/10.1016/j.scitotenv.2017.10.289

Liu, Z., Törnros, T., & Menzel, L. (2016). A probabilistic prediction network for hydrological drought identification and environmental flow assessment. *Water Resources Research*, 52(8), 6243–6262. https://doi.org/10.1002/2016wr019106

Luoma, J.A., Severson, T.J., Barbour, M.T., & Wise, J.K. (2018). Effects of temperature and exposure duration on four potential rapid-response tools for zebra mussel (*Dreissena polymorpha*) eradication. *Management of Biological Invasions*, 9(4), 425–438. https://doi.org/10.3391/mbi.2018.9.4.06

Lytle, D.A., & Poff, N.L. (2004). Adaptation to natural flow regimes. *Trends in Ecology & Evolution*, 19(2), 94–100. https://doi.org/10.1016/j.tree.2003.10.002

Macel, M., de Vos, R.C., Jansen, J.J., van der Putten, W.H., & van Dam, N.M. (2014). Novel chemistry of invasive plants: exotic species have more unique metabolomic profiles than native congeners. *Ecology and Evolution*, 4(13), 2777–2786. https://doi.org/10.1002/ece3.1132

Mohammed, I.N., Bolten, J.D., Srinivasan, R., & Lakshmi, V. (2018). Satellite observations and modeling to understand the Lower Mekong River Basin streamflow variability. *Journal of Hydrology,* 564, 559–573. https://doi.org/10.1016/j.jhydrol.2018.07.030

Mori, T., Kato, Y., Takagi, T., Onoda, Y., & Kayaba, Y. (2018). Turbid water induces refuge behaviour of a commercially important ayu: a field experiment for interstream movement using multiple artificial streams. *Ecology of Freshwater Fish*, 27(4), 1015–1022. https://doi.org/10.1111/eff.12411

Nagrodski, A., Raby, G.D., Hasler, C.T., Taylor, M.K., & Cooke, S.J. (2012). Fish stranding in freshwater systems: sources, consequences, and mitigation. *Journal of Environmental Management*, 103, 133–141. https://doi.org/10.1016/j.jenvman.2012.03.007

Nguyen, T.H.T., Boets, P., Lock, K., Ambarita, M.N.D., Forio, M.A.E., Sasha, P., ... & Goethals, P.L. (2015). Habitat suitability of the invasive water hyacinth and its relation to water quality and macroinvertebrate diversity in a tropical reservoir. *Limnologica*, 52, 67–74.

Nilsson, C., Reidy, C.A., Dynesius, M., & Revenga, C. (2005). Fragmentation and flow regulation of the world's large river systems. *Science*, 308(5720), 405–408. https://doi.org/10.1126/science.1107887

Nordström, M.C., Aarnio, K., Törnroos, A., & Bonsdorff, E. (2015). Nestedness of trophic links and biological traits in a marine food web. *Ecosphere*, 6(9), 1–14. https://doi.org/10.1890/es14-00515.1

Olden, J.D., & Naiman, R. J. (2010). Incorporating thermal regimes into environmental flows assessments: modifying dam operations to restore freshwater ecosystem integrity. *Freshwater Biology*, 55(1), 86–107. https://doi.org/10.1111/j.1365-2427.2009.02179.x

Pal, S., & Saha, T.K. (2018). Identifying dam-induced wetland changes using an inundation frequency approach: the case of the Atreyee River basin of Indo-Bangladesh. *Ecohydrology & Hydrobiology*, 18(1), 66–81. https://doi.org/10.1016/j.ecohyd.2017.11.001

Pimm, S.L., & Lawton, J.H. (1977). Number of trophic levels in ecological communities. *Nature*, 268(5618), 329–331. https://doi.org/10.1038/268329a0

Poff, N.L. (2014). Rivers of the Anthropocene?. *Frontiers in Ecology and the Environment*, 12(8), 427–427. https://doi.org/10.1890/1540-9295-12.8.427

Poff, N.L., Olden, J.D., Merritt, D.M., & Pepin, D.M. (2007). Homogenization of regional river dynamics by dams and global biodiversity implications. *Proceedings of the National Academy of Sciences*, 104(14), 5732–5737. https://doi.org/10.1073/pnas.0609812104

Poff, N.L., Richter, B.D., Arthington, A.H., Bunn, S.E., Naiman, R.J., Kendy, E., Acreman, M., Apse, C., Bledsoe, B.P., Freeman, M.C., & Henriksen, J. (2010). The ecological limits of hydrologic alteration (ELOHA): a new framework for developing regional environmental flow standards. *Freshwater Biology*, 55(1), 147–170. https://doi.org/10.1111/j.1365-2427.2009.02204.x

Poff, N.L., Tharme, R.E., & Arthington, A.H. (2017). Evolution of environmental flows assessment science, principles, and methodologies. In Horne, A.C., Webb, J.A., Stewardson, M.J., Richter, B., & Acreman, M. (eds) *Water for the Environment*. Elsevier, United Kingdom, 203–236. https://doi.org/10.1016/b978-0-12-803907-6.00011-5

Post, D.M., & Takimoto, G. (2007). Proximate structural mechanisms for variation in food-chain length. *Oikos*, 116(5), 775–782. https://doi.org/10.1111/j.2007.0030-1299.15552.x

Post, D.M., Pace, M.L., & Hairston, N.G. (2000). Ecosystem size determines food-chain length in lakes. *Nature*, 405(6790), 1047–1049. https://doi.org/10.1038/35016565

Preston, D.L., Hedman, H.D., & Johnson, P.T. (2018). Nutrient availability and invasive fish jointly drive community dynamics in an experimental aquatic system. *Ecosphere*, 9(3), e02153. https://doi.org/10.1002/ecs2.2153

Propst, D.L., Gido, K.B., & Stefferud, J.A. (2008). Natural flow regimes, nonnative fishes, and native fish persistence in arid-land river systems. *Ecological Applications*, 18(5), 1236–1252. https://doi.org/10.1890/07-1489.1

Pyne, M.I., & Poff, N.L. (2017). Vulnerability of stream community composition and function to projected thermal warming and hydrologic change across ecoregions in the western United States. *Global Change Biology*, 23(1), 77–93. https://doi.org/10.1111/gcb.13437

Raghavan, R., Prasad, G., Anvar-Ali, P.H., & Pereira, B. (2008). Exotic fish species in a global biodiversity hotspot: observations from River Chalakudy, part of Western Ghats, Kerala, India. *Biological Invasions*, 10(1), 37–40. https://doi.org/10.1007/s10530-007-9104-2

Rawat, J.S., Govind, A., Rawat, G., Joshi, M., Rai, S.P., & Gahlot, N. (2016). Perennial to ephemeral transformation of a Lesser Himalayan watershed. *Current Science*, 111(4), 686–693. https://doi.org/10.18520/cs/v111/i4/686-693

Ready, R.C., Poe, G.L., Lauber, T.B., Connelly, N.A., Stedman, R.C., & Rudstam, L.G. (2018). The potential impact of aquatic nuisance species on recreational fishing in the Great Lakes and Upper Mississippi and Ohio River Basins. *Journal of Environmental Management*, 206, 304–318. https://doi.org/10.1016/j.jenvman.2017.10.025

Reid, A.J., Carlson, A.K., Creed, I.F., Eliason, E.J., Gell, P.A., Johnson, P.T., Kidd, K.A., MacCormack, T.J., Olden, J.D., Ormerod, S.J., & Smol, J.P. (2019). Emerging threats and persistent conservation challenges for freshwater biodiversity. *Biological Reviews*, 94(3), 849–873. https://doi.org/10.1111/brv.12480

Rice, J., Arvanitidis, C., Borja, A., Frid, C., Hiddink, J.G., Krause, J., Lorance, P., Ragnarsson, S.Á., Sköld, M., Trabucco, B., & Enserink, L. (2012). Indicators for sea-floor integrity under the European Marine Strategy Framework Directive. *Ecological Indicators*, 12(1), 174–184. https://doi.org/10.1016/j.ecolind.2011.03.021

Ripl, W. (2003). Water: the bloodstream of the biosphere. *Philosophical Transactions of the Royal Society of London. Series B: Biological Sciences*, 358(1440), 1921–1934. https://doi.org/10.1098/rstb.2003.1378

Robinson, C.T., Uehlinger, U.R.S., & Monaghan, M.T. (2004). Stream ecosystem response to multiple experimental floods from a reservoir. *River Research and Applications*, 20(4), 359–377. https://doi.org/10.1002/rra.743

Rolls, R.J., Heino, J., Ryder, D.S., Chessman, B.C., Growns, I.O., Thompson, R.M., & Gido, K.B. (2018). Scaling biodiversity responses to hydrological regimes. *Biological Reviews*, 93(2), 971–995. https://doi.org/10.1111/brv.12381

Ruhi, A., Catford, J.A., Cross, W.F., Escoriza, D., & Olden, J.D. (2019). Understanding the nexus between hydrological alteration and biological invasions. In Sabater, S., Elosegi, A., & Ludwig, R. (eds) *Multiple Stressors in River Ecosystems: Status, Impacts and Prospects for the Future*. Elsevier, Netherlands, 45–64. https://doi.org/10.1016/b978-0-12-811713-2.00003-0

Ruhí, A., Olden, J.D., & Sabo, J.L. (2016). Declining streamflow induces collapse and replacement of native fish in the American Southwest. *Frontiers in Ecology and the Environment,* 14(9), 465–472. https://doi.org/10.1002/fee.1424

Sabater, S., Elosegi, A., & Ludwig, R. (2019). Defining multiple stressor implications. In Sabater, S., Elosegi, A., & Ludwig, R. (eds) *Multiple Stressors in River Ecosystems: Status, Impacts and Prospects for the Future*. Elsevier, Netherlands, 1–22. https://doi.org/10.1016/b978-0-12-811713-2.00001-7

Sandi, S.G., Saco, P.M., Kuczera, G., Wen, L., Saintilan, N., & Rodriguez, J.F. (2018). Predicting floodplain inundation and vegetation dynamics in arid wetlands. *E3S Web of Conferences,* 40, 02019. EDP Sciences. https://doi.org/10.1051/e3sconf/20184002019

Schinegger, R., Trautwein, C., Melcher, A., & Schmutz, S. (2012). Multiple human pressures and their spatial patterns in European running waters. *Water and Environment Journal*, 26(2), 261–273. https://doi.org/10.1111/j.1747-6593.2011.00285.x

Sedighkia, M., Ayyoubzadeh, S.A., & Haji Esmaeili, M. (2017). Habitat simulation technique as a powerful tool for instream flow needs assessment and river ecosystem management. *Environmental Energy and Economic Research*, 1(2), 171–182. http://doi.org/10.22097/EEER.2017.47244

Sharma, U., & Dutta, V. (2020). Establishing environmental flows for intermittent tropical rivers: why hydrological methods are not adequate? *International Journal of Environmental Science & Technology (IJEST)*, 17(5), 21. https://doi.org/10.1007/s13762-020-02680-6

Silva, A.T., Lucas, M.C., Castro-Santos, T., Katopodis, C., Baumgartner, L.J., Thiem, J.D., Aarestrup, K., Pompeu, P.S., O'Brien, G.C., Braun, D.C., & Burnett, N.J. (2018). The future of fish passage science, engineering, and practice. *Fish and Fisheries*, 19(2), 340–362. https://doi.org/10.1111/faf.12258

Skoulikidis, N.T., Vardakas, L., Amaxidis, Y., & Michalopoulos, P. (2017). Biogeochemical processes controlling aquatic quality during drying and rewetting events in a Mediterranean non-perennial river reach. *Science of the Total Environment*, 575, 378–389. https://doi.org/10.1016/j.scitotenv.2016.10.015

Steffen, K., Leuschner, C., Müller, U., Wiegleb, G., & Becker, T. (2014). Relationships between macrophyte vegetation and physical and chemical conditions in northwest German running waters. *Aquatic Botany*, 113, 46–55. https://doi.org/10.1016/j.aquabot.2013.10.006

Stiers, I., Crohain, N., Josens, G., & Triest, L. (2011). Impact of three aquatic invasive species on native plants and macroinvertebrates in temperate ponds. *Biological Invasions*, 13(12), 2715–2726. https://doi.org/10.1007/s10530-011-9942-9

Stromberg, J.C., Lite, S.J., Marler, R., Paradzick, C., Shafroth, P.B., Shorrock, D., White, J.M., & White, M.S. (2007). Altered stream-flow regimes and invasive plant species: the Tamarix case. *Global Ecology and Biogeography*, 16(3), 381–393. https://doi.org/10.1111/j.1466-8238.2007.00297.x

Stuber, O.S., Kirkman, L.K., Hepinstall-Cymerman, J., & Martin, G.I. (2016). The ecological condition of geographically isolated wetlands in the south-eastern United States: the relationship between landscape level assessments and macrophyte assemblages. *Ecological Indicators*, 62, 191–200. https://doi.org/10.1016/j.ecolind.2015.11.037

Takimoto, G., & Post, D.M. (2013). Environmental determinants of food-chain length: a meta-analysis. *Ecological Research*, 28(5), 675–681. https://doi.org/10.1007/s11284-012-0943-7

Tamario, C., Degerman, E., Donadi, S., Spjut, D., & Sandin, L. (2018). Nature-like fishways as compensatory lotic habitats. *River Research and Applications*, 34(3), 253–261. https://doi.org/10.1002/rra.3246

Téllez, T.R., López, E.M.D.R., Granado, G.L., Pérez, E.A., López, R.M., & Guzmán, J.M.S. (2008). The water hyacinth, *Eichhornia crassipes*: an invasive plant in the Guadiana River Basin (Spain). *Aquatic Invasions,* 3(1), 42–53. https://doi.org/10.3391/ai.2008.3.1.8

Tharme, R.E. (2003). A global perspective on environmental flow assessment: emerging trends in the development and application of environmental flow methodologies for rivers. *River Research and Applications*, 19(5–6), 397–441. http://doi:10.1002/rra.736

Theodoropoulos, C., Papadaki, C., Vardakas, L., Dimitriou, E., Kalogianni, E., & Skoulikidis, N. (2019). Conceptualization and pilot application of a model-based environmental flow assessment adapted for intermittent rivers. *Aquatic Sciences*, 81(1), 1–17. https://doi.org/10.1007/s00027-018-0605-0

Thomaz, S.M., & Cunha, E.R.D. (2010). The role of macrophytes in habitat structuring in aquatic ecosystems: methods of measurement, causes and consequences on animal assemblages' composition and biodiversity. *Acta Limnologica Brasiliensia*, 22, 218–236. https://doi.org/10.4322/actalb.02202011

Thomaz, S.M., Mormul, R.P., & Michelan, T.S. (2015). Propagule pressure, invasibility of freshwater ecosystems by macrophytes and their ecological impacts: a review of tropical freshwater ecosystems. *Hydrobiologia*, 746(1), 39–59. https://doi.org/10.1007/s10750-014-2044-9

Thompson, R.M., Brose, U., Dunne, J.A., Hall Jr, R.O., Hladyz, S., Kitching, R.L., Martinez, N.D., Rantala, H., Romanuk, T.N., Stouffer, D.B., & Tylianakis, J.M. (2012). Food webs: reconciling the structure and function of biodiversity. *Trends in Ecology & Evolution,* 27(12), 689–697. https://doi.org/10.1016/j.tree.2012.08.005

Thompson, R.M., King, A.J., Kingsford, R.M., Mac Nally, R., & Poff, N.L. (2018). Legacies, lags and long-term trends: effective flow restoration in a changed and changing world. *Freshwater Biology*, 63(8), 986–995. https://doi.org/10.1111/fwb.13029

Tilman, D. (2004). Niche tradeoffs, neutrality, and community structure: a stochastic theory of resource competition, invasion, and community assembly. *Proceedings of the National Academy of Sciences*, 101(30), 10854–10861. https://doi.org/10.1073/pnas.0403458101

Unfer, G., Hauer, C., & Lautsch, E. (2011). The influence of hydrology on the recruitment of brown trout in an Alpine river, the Ybbs River, Austria. *Ecology of Freshwater Fish*, 20(3), 438–448. https://doi.org/10.1111/j.1600-0633.2010.00456.x

Van Cappellen, P., & Maavara, T. (2016). Rivers in the Anthropocene: global scale modifications of riverine nutrient fluxes by damming. *Ecohydrology & Hydrobiology*, 16(2), 106–111. https://doi.org/10.1016/j.ecohyd.2016.04.001

Vorosmarty, C.J., McIntyre, P.B., Gessner, M.O., Dudgeon, D., Prusevich, A., Green, P., Glidden, S., Bunn, S.E., Sullivan, C.A., Liermann, C.R., & Davies, P.M. (2010). Global threats to human water security and river biodiversity. *Nature*, 467(7315), 555–561. https://doi.org/10.1038/nature09440

Vukov, D., Ilić, M., Ćuk, M., Radulović, S., Igić, R., & Janauer, G.A. (2018). Combined effects of physical environmental conditions and anthropogenic alterations are associated with macrophyte habitat fragmentation in rivers-study of the Danube in Serbia. *Science of the Total Environment,* 634, 780–790. https://doi.org/10.1016/j.scitotenv.2018.03.367

Warren, R.J., Ursell, T., Keiser, A.D., & Bradford, M.A. (2013). Habitat, dispersal and propagule pressure control exotic plant infilling within an invaded range. *Ecosphere*, 4(2), 1–12. https://doi.org/10.1890/es12-00393.1

Whitney, J.E., Gido, K.B., Pilger, T.J., Propst, D.L., & Turner, T.F. (2016). Metapopulation analysis indicates native and non-native fishes respond differently to effects of wildfire on desert streams. *Ecology of Freshwater Fish*, 25(3), 376–392. https://doi.org/10.1111/eff.12217

Wilhelm, L., Besemer, K., Fragner, L., Peter, H., Weckwerth, W., & Battin, T.J. (2015). Altitudinal patterns of diversity and functional traits of metabolically active microorganisms in stream biofilms. *The ISME Journal*, 9(11), 2454–2464. https://doi.org/10.1038/ismej.2015.56

Wilson, J.B., Spijkerman, E., & Huisman, J. (2007). Is there really insufficient support for Tilman's R* concept? A comment on Miller et al. *The American Naturalist*, 169(5), 700–706. https://doi.org/10.1086/513113

Winemiller, K.O., McIntyre, P.B., Castello, L., Fluet-Chouinard, E., Giarrizzo, T., Nam, S., Baird, I.G., Darwall, W., Lujan, N.K., Harrison, I., & Stiassny, M.L.J. (2016). Balancing hydropower and biodiversity in the Amazon, Congo, and Mekong. *Science*, 351(6269), 128–129. https://doi.org/10.1126/science.aac7082

Wohl, E., Hall Jr, R.O., Lininger, K.B., Sutfin, N.A., & Walters, D.M. (2017). Carbon dynamics of river corridors and the effects of human alterations. *Ecological Monographs*, 87(3), 379–409. https://doi.org/10.1002/ecm.1261

Zeiringer, B., Seliger, C., Greimel, F., & Schmutz, S. (2018). River hydrology, flow alteration, and environmental flow. In Schmutz, S., & Sendzimir, J. (eds) *Riverine Ecosystem Management.* Springer, Cham, 67–89. https://doi.org/10.1007/978-3-319-73250-3_4

Zemke-White, W.L., & Clements, K.D. (1999). Chlorophyte and rhodophyte starches as factors in diet choice by marine herbivorous fish. *Journal of Experimental Marine Biology and Ecology*, 240(1), 137–149. https://doi.org/10.1016/s0022-0981(99)00056-8

5 Elemental Mobility in the Near-Surface Environment

A Study from Bhowali-Bhimtal and Berinag Regions of Kumaun Lesser Himalaya, India

Abhishek Pandey and Santosh Kumar
Department of Geology, Centre of Advanced Study, Kumaun University, Nainital

CONTENTS

5.1 INTRODUCTION

During chemical weathering, the earth's primary materials (hard bedrock) break down and are altered into secondary products that are in equilibrium with the newly imposed physico-chemical conditions. In this process, the elemental constituents of the materials are rearranged and mobilized under the influence of processes associated with the hydrosphere, atmosphere and biosphere (White and Brantley, 1995; Duzgoren-Aydin *et al.*, 2002). Although the weathering of earth's material is a common global phenomenon, studies on elemental mobility during the alteration process have not been given much attention (Dearman, 1995; Hencher and McNicoll, 1995; Price, 1995; Moon and Jayavardane, 2004). Although there are some studies that have been focused on elemental mobility and water–rock interactions during weathering processes (Goldich, 1938; Harris and Adams, 1966; Ruxton, 1968;

DOI: 10.1201/9781003220824-6

Nesbitt, 1979; Nesbitt *et al.*, 1980; Fritz and Regland, 1980; Chesworth *et al.*, 1981; Cramer and Nesbitt, 1983; Fritz and Mohr, 1984; Gouveia *et al.*, 1993; Condie *et al.*, 1995; Ohlender *et al.*, 1996; Minarik *et al.*, 1998; Tapia *et al.*, 1998; Tripathi and Rajamani, 1999; Sharma and Rajamani, 2000; Rajamani *et al.*, 2009; Babechuk *et al.*, 2014; Vyshnavi *et al.*, 2014 and references therein), the phenomenon of chemical weathering of lithounits in an orogenic belt is not yet fully understood.

Understanding the weathering process becomes complex because of the variety of controlling factors involved in it, for instance climate, biological activity, topography and composition of parent rock, which exhibit a complex relationship with each other. During chemical weathering, the mobilization and reorganization of certain elements is particularly difficult to decipher because of the involvement of various processes, like dissolution of primary minerals, formation of secondary phases, transport of materials, co-precipitation and ion-exchange of various minerals (Harris and Adams, 1966; Nesbitt, 1979; Nesbitt *et al.*, 1980; Fritz and Ragland, 1980; Chesworth *et al.*, 1981; Cramer and Nesbitt, 1983; Fritz and Mohr, 1984). Many researchers assume that elemental mobility is correlated with breakdown of mineral constituents (Colman, 1982; Eggleton *et al.*, 1987; Nesbitt and Wilson, 1992; Gavshin *et al.*, 1997; Hill *et al.*, 2000; Patino *et al.*, 2003) along with other sets of processes, such as solubility of parent mineralogy, pressure, temperature, redox potential and type of leaching agent, that generally control the mobility of elements during weathering.

The present study is an attempt to understand the elemental mobility during weathering of metavolcanics and associated rock units located in Bhowali-Bhimtal and Berinag regions of Kumaun Lesser Himalaya by using magnetic susceptibility and the geochemistry of major, trace and rare earth elements (REE). Most of the *in situ* weathered crust profiles develop when the intensity of weathering with time precedes erosion. Such profiles develop a typical layered structure, which is divided into different horizons. Each horizon is differentiated according to physical and chemical properties like colour, grain size, texture, and mineralogical composition, etc. The degree of weathering is usually highest at the outer top oxidizing layer and gradually acquires a reducing trend towards the base or centre of the parent rock. In order to understand the relationship of fresh as well as weathered counterparts of the parent rock, samples have been collected from the various soil horizons (wherever possible), including the fresh parent rock, to study the elemental mobility in weathered or soil-forming processes *in situ*.

5.2 GEOLOGICAL FRAMEWORK

The present study is conducted for two different locations, namely Bhowali-Bhimtal and surrounding regions, which falls in the Nainital district, and the Berinag area, which is located in the Pithoragarh district. Both of these areas are part of Kumaun Lesser Himalaya (KLH), representing the outer and inner parts of the KLH, respectively.

5.2.1 Geological Characteristics of Bhowali-Bhimtal and Adjoining Region

The lithological constituents of the Bhowali-Bhimtal region are principally composed of Bhowali volcanics (BV) and Nagthat quartzites, which are considered to be penecontemporaneous based on field observations. Valdiya (1980a) has described them together as Nagthat Formation, belonging to the Jaunsar Group, although at a later stage, some other workers considered the BV and associated quartzite to be of Mesoproterozoic age and placed them below the Jaunsar Group (Shankar *et al.*, 1989, 2002). Palaeoproterozoic Amritpur granitoids (ca. 1850 Ma) probably form the oldest exposed lithounits of the area, which is separated tectonically from the Middle Miocene Siwalik Group by the Main Boundary Thrust (MBT) towards the southwest. Siwalik Group is composed of the youngest rocks exposed in the adjoining localities close to the study area. Towards southeast, Amritpur granitoids (AG) come into contact with the Ramgarh Group, consisting essentially of metamorphics (phyllites), while in its northwestern extremity AG are associated with the quartzite of Nagthat

Formation, which tectonically overlies the AG along the Salari Thrust (Valdiya, 1980a). The assemblage of purple, fawn, white and green quartz-arenites characterizes the Nagthat Formation (Auden, 1934), and green tuffaceous phyllite and quartzite along with penecontemporaneous BV are essential components of Nagthat lithology.

The exposures of BV cover an area of about 1000 square kilometres, and most prominent exposures are observed along the Bhowali-Bhimtal-Naukuchiatal transect. These are observed to be variably weathered in places. The BV comprise amygdaloidal, vesicular and massive basalts converted partially or wholly into amphibolite, green schist and chlorite schist in the vicinity of the thrust zone.

The BV are overlain successively by Jantwalgaon limestone, Betalghat Formation, Bhumiadhar Formation and Lariakanta quartzite, which collectively form part of the Jaunsar Group. This has been further overlain by slates and quartz-arenites of the Blaini Formation, which is overridden by carbonate and dolomites of the Krol Formation, most prominently exposed in the Nainital and adjoining localities. Dykes and sills of relatively fresh and metamorphosed dolerite and microgabbro intrude the BV; these are obviously younger in age. These younger intrusives are also injected into the Blaini and Krol Formations. The older basalt volcanisms, such as BV, are penecontemporaneous with the Nagthat sediments, whereas younger mafic magmatism, such as sills and dykes, could be the post-Krol. A detailed geological map of the Bhowali-Bhimtal region where BV are prominently exposed is shown in Figure 5.1, which is a compiled version based on the authors' own observations and existing geological maps (Bartarya, 1988; Valdiya, 1988; Shankar *et al.*, 1989, 2002).

5.2.2 Geological Characteristics of Berinag Region

The Berinag volcanics (BNV) have been considered equivalent to BV because of their more-or-less similar geological setup, as discussed by earlier workers (e.g. Heim and Gansser, 1939; Valdiya, 1980a,b; Pant, 1985). The geological map (after Pant, 1985) of Berinag-Gangolihat localities is shown in Figure 5.2. The Rautgara Formation of eastern Kumaun Himalaya is mainly constituted by slates and is successively overlain by the limestones, dolomites and magnesites belonging to the Gangolihat Formation, which has a thrusted contact with the quartzites and metavolcanics, both belonging to Berinag Formation along the Berinag thrust. This is further overlain by the Askot Formation, mainly constituted by schist and mylonite gneiss. Most of the BNV are moderately to highly weathered and metamorphosed to slates, phyllites and chlorite schist. The schistosity is rather more developed in this area, which is characterized by flaky minerals and soapy feel. At some places, huge bodies of quartzite have shown sharp contact with BNV, while at other places, it occurs interbedded with volcanics. The BNV is tholeiitic and formed by partial melting of a garnet peridotite (upper mantle) source (Pandey and Kumar, 2006).

5.3 METHODS OF STUDY

5.3.1 Sampling Method

Sets of samples were collected from various soil horizons, including weathered and fresh (wherever possible) variants of the soil profile. It may be noted that the concentration of trace elements varies with depth, soil type and size fraction. Standard scientific methods have been adopted for sampling the soil profile, where colour and grain size have been chosen as principal criteria to differentiate between various horizons. In the steep terrains, samples were taken from mountain top ridges and at the base. Various types of field equipment, i.e. shovels or hoes and different kinds of auger, were used for collection of soil samples in hard and soft terrenes. Most of the surface samples are taken from a depth of 15–20 cm, though the profiles may vary from 0.40 to 1 metre or even more. The quantity of samples may vary from 5 to 500 g, depending upon the size fraction.

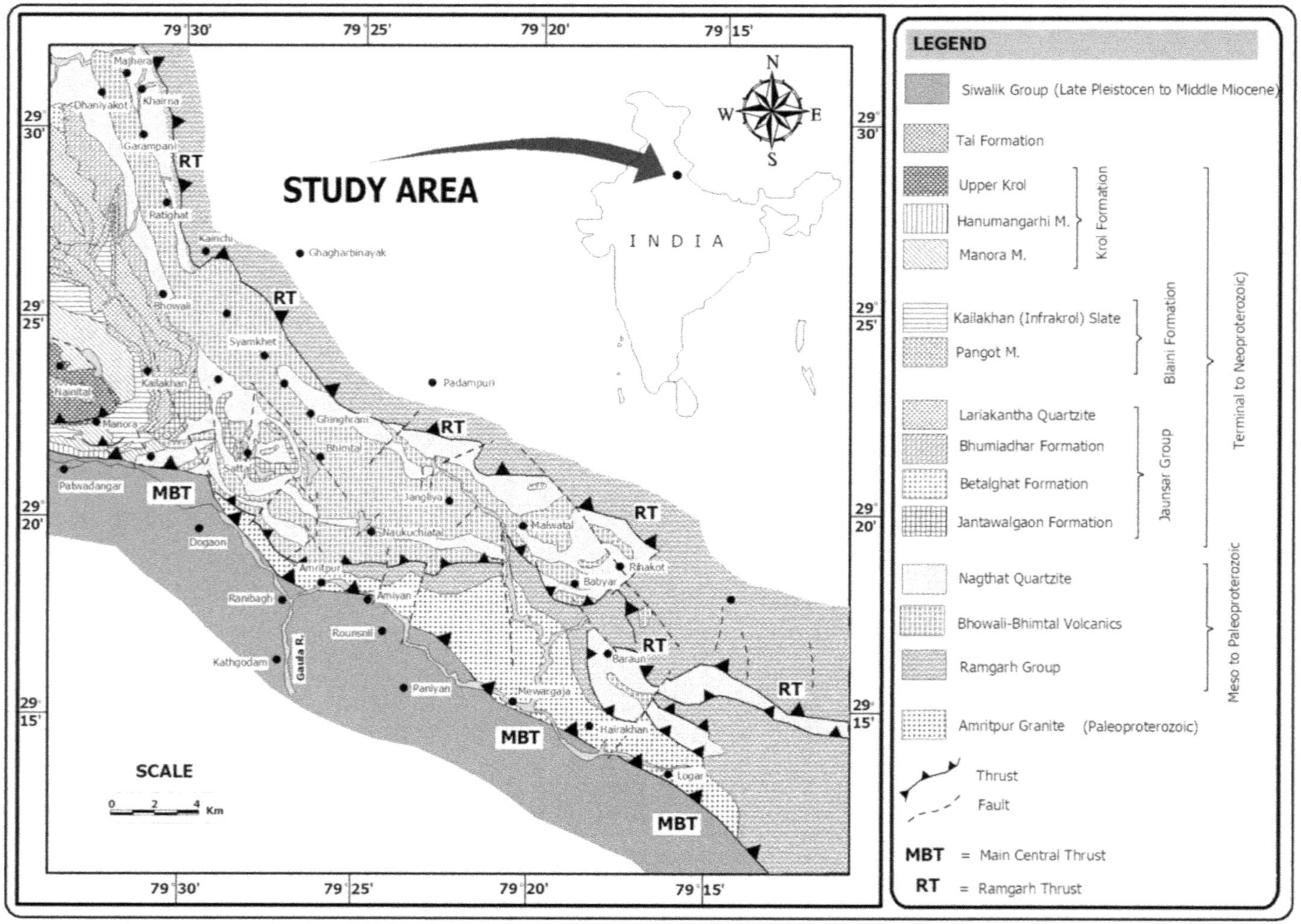

FIGURE 5.1 Geological map of Kumaun Lesser Himalaya showing the different lithological units.
(Compiled from Bartarya, 1988; Valdiya, 1988; Shankar *et al.*, 1989, 2002.)

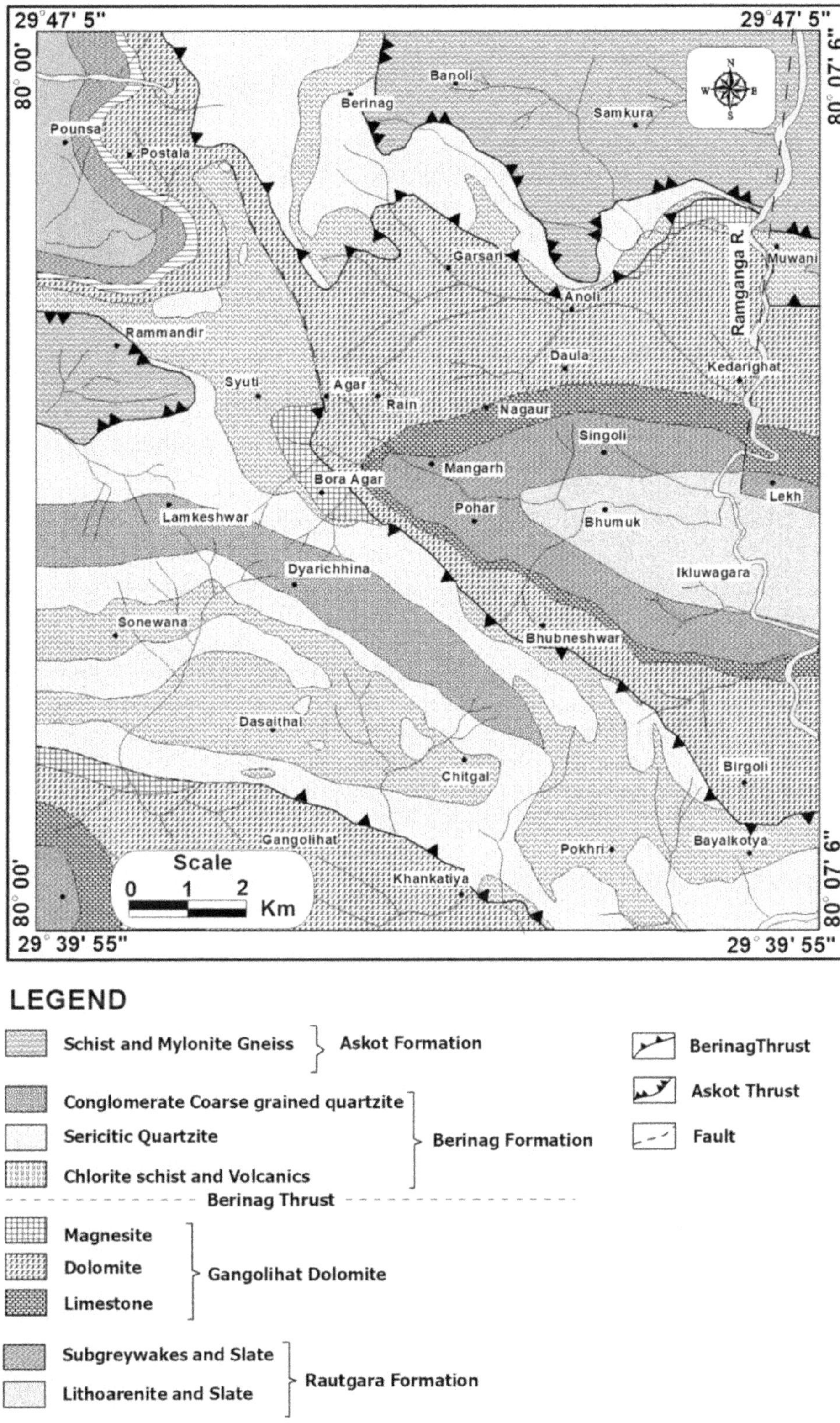

FIGURE 5.2 Geological map of the Berinag region.

(After Pant, 1985.)

5.3.2 Sample Preparation

The preparation of the samples was carried out in the field and in the laboratory. Wherever soil and weathered materials were stony, the dried samples were sieved down to a 2-mm sieve in the field. Normally, the soil samples are dried, crushed and sieved (0.5–1-mm sieves). About 15–20 g of the samples is fine-grain powdered up to −80 to −200 mesh for analytical purposes. Homogeneous and representative portions of soil samples are processed from a large proportion of samples by the cone and quartering method. The loss on ignition (LOI) was determined using the following procedure:

1) 1 g of the powdered sample is weighed on the electronic balance at room temperature.
2) The sample is transferred to a Teflon crucible and then placed in a muffle furnace, and the temperature is gradually increased to 900°C during 45 minutes. This allows evaporation of all volatile materials from the sample.
3) The crucible is taken out of the furnace, cooled to room temperature and weighed again.
4) The difference in weight before and after heating the sample is the LOI.

5.3.3 Analytical Methods

Whole-rock major oxide (SiO_2, TiO_2, Al_2O_3, $Fe_2O_3^t$, MnO, MgO, CaO, Na_2O, K_2O and P_2O_5) and trace (Rb, Ba, Sr, Y, Zr, Nb) element analyses were carried out for representative BV and BNV samples using X-ray fluorescence (XRF) at Wadia Institute of Himalayan Geology (WIHG), Dehra Dun. The XRF machine is a SIEMENS SRS 3000 fitted with a sequential X-ray spectrometer with end window Rh X-ray tube. The elemental analysis was performed using pressed powder pellets following an intensity-based model (Lucas-Tooth and Pyne, 1964) and using international reference materials. The operating conditions for major oxides were: no filter, vacuum path, 20/40 kV, and for trace elements: no filter, vacuum path, 55/60 kV. The overall accuracy (% residual standard deviation (RSD)) is lower than 5% for major oxides and lower than 12% for trace elements, the average precision being better than 1.5% in all cases (Saini et al., 1998). The minimum detection limit for most of the trace elements is ~5 ppm.

Trace elements, including the rare earth elements (Sc, V, Cr, Co, Ni, Cu, Zn, Ga, Cs, Hf, Ta, Pb, Th, U, La, Ce, Pr, Nd, Sm, Eu, Gd, Tb, Dy, Ho, Er, Tm, Yb and Lu), were analysed from representative BV and BNV samples at National Geophysical Research Institute (NGRI), Hyderabad, using inductively coupled plasma-mass spectrometry (ICP-MS). The model of ICP machine was ELAN DRC II (Perkin-Elmer Sciex Instrument, US). The system was optimized for maximum intensity (~ 40,000 counts/s) across the mass range using a 1 mg/ml solution of Mg, Rh, In, Ba, Ce, Pb and U. A dual detector was used. The formation levels of oxide (CeO) and doubly charged ions (Ba^{++}) were kept at 22% by optimizing nebular gas flow, lens voltage and other parameters before the analysis was performed. Appropriate rock reference standards were used to minimize the matrix effects. To make the solution for ICP-MS analysis, 50 mg of the sample as well as the rock standards was dissolved in a matrix of 7 ml HF, 3 ml HNO_3 and 1 ml $HClO_4$ and covered with a lid after adding 1 ml of 10 mg/ml Rh to act as an internal standard, and the solution kept overnight for digestion. The solution was then heated on a hotplate at 200°C for about 1 hour, after which the lid was removed and the mixture evaporated to almost dryness. The evaporation process was repeated thrice after adding a mixture of 3 ml HF and 1 ml $HClO_4$. Then the contents were dissolved using 10 ml of 1:1 HNO_3 and the volume made to 100 ml. The solutions thus prepared were finally used to determine the trace and rare earth elements using ICP-MS.

Six elements (Rb, Sr, Ba, Nb, Zr and Y) were analysed by ICP-MS as well as by XRF. The results (ppm) for these elements obtained by both techniques were compared. Good correlations ($R^2 = 0.98$, 0.99 and 0.88) were obtained for Rb, Sr and Ba, respectively, whereas Nb, Zr and Y do not provide good and acceptable correlations. The XRF results for these six elements were preferred to those from ICP-MS, as recommended elsewhere (Norman *et al.*, 1989).

5.4 RESULTS AND INTERPRETATION

5.4.1 Bedrock and Status of Weathering

The Bhowali-Bhimtal region of KLH was initially selected for the study, and for comparison, some of the weathered crust profiles of Berinag region have also been selected. Based on geology, topography, altitude and bedrock exposures, a few sites were selected for sampling the materials. The selected sites are mostly constituted by Paleoproterozoic granitoids (ca. 1850 Ma) in the Amritpur region, metavolcanics in Amritpur, and the penecontemporaneous Nagthat sedimentaries and quartzites in the Bhowali-Bhimtal and Ratighat regions. The locations of bedrock and weathered crust profile sample sites are shown in Figure 5.3.

The studied region is tectonically active and lacks idealized soil profiles. Mostly, the bedrocks were found to be asymmetrically weathered, mostly controlled by the degree of metamorphism, deformation (accretion), lithology (mostly basalt and metabasalt as chlorite and green schist) and availability of water springs as seepage from local or hanging aquifers. *In situ* profiles were adjudged based on partial preservation of original foliations, variation in grain size constituents and their interrelationships with bedrock. In the shear zones, rocks are found to be highly crushed, pulverized and strongly weathered. The water seepage through this zone and retention of water in this zone promote the development of weathered profiles locally and in patches. This process is induced by alternately dry and humid climates. The weathered products show variegated yellow brown, pale brown, deep brown, and greenish and bluish colours. Particularly in the shear zone, the weathered products have been converted to clay (Figure 5.4a,b).

In the Berinag region similar features can also be observed, where the weathered crust profiles are commonly developed due to weathering of volcanics. The locations of bedrock and weathered crust profile sample sites are shown in Figure 5.3. The shearing and shattering of the volcanics are common (Figure 5.4c), and the highly weathered volcanics have commonly been turned into chlorite schist (Figure 5.4d).

Colour and grain size variation of the profiles were the major criteria to sample the materials. However, an attempt was made to understand the degree of weathering using the field measurement of magnetic susceptibility (MS, i.e. c = $\times10^{-3}$ SI unit), which is thought to diminish from bedrock (parent rocks) to highly weathered profiles, provided the sample surface is chosen with care and is free from iron leaching for MS measurements. The MS is an important physical property of the rocks. It has been observed that the weathered and fresh bedrock samples have different MS values. For determining the MS values in the field, the SM-30 model MS meter, Geofyzica, Brno was used. The observed MS values for various weathered crust profile samples along various transects are presented in Table 5.1. The MS values show a wide range, though most of them measure less than 1. The elevated MS values (9.02–14.7$\times10^{-3}$ SI) of sample Gh-6a are due to iron leaching on the surface, and become notably lower (2.5–2.6$\times10^{-3}$ SI) in the core of the same exposure. The samples Gh-4a (MS value 15.8–19.1$\times10^{-3}$ SI), Gh-10a (MS value 4.49–4.85$\times10^{-3}$ SI) and Gh10b (MS value 43.1–46.7$\times10^{-3}$ SI) also show elevated values due to iron leaching on the surface. The observed variations of MS values are heterogeneous, which is suggestive of different degrees of leaching of ferromagnesian minerals, which are relatively easily mobile during chemical weathering.

5.4.2 Microscopic Evidence of Mineral Alteration

Parent rock (basalt) has been studied in rock thin sections. The evidence of mineralogical changes/weathering/alteration has been recorded from the volcanics of the Bhowali-Bhimtal and Berinag

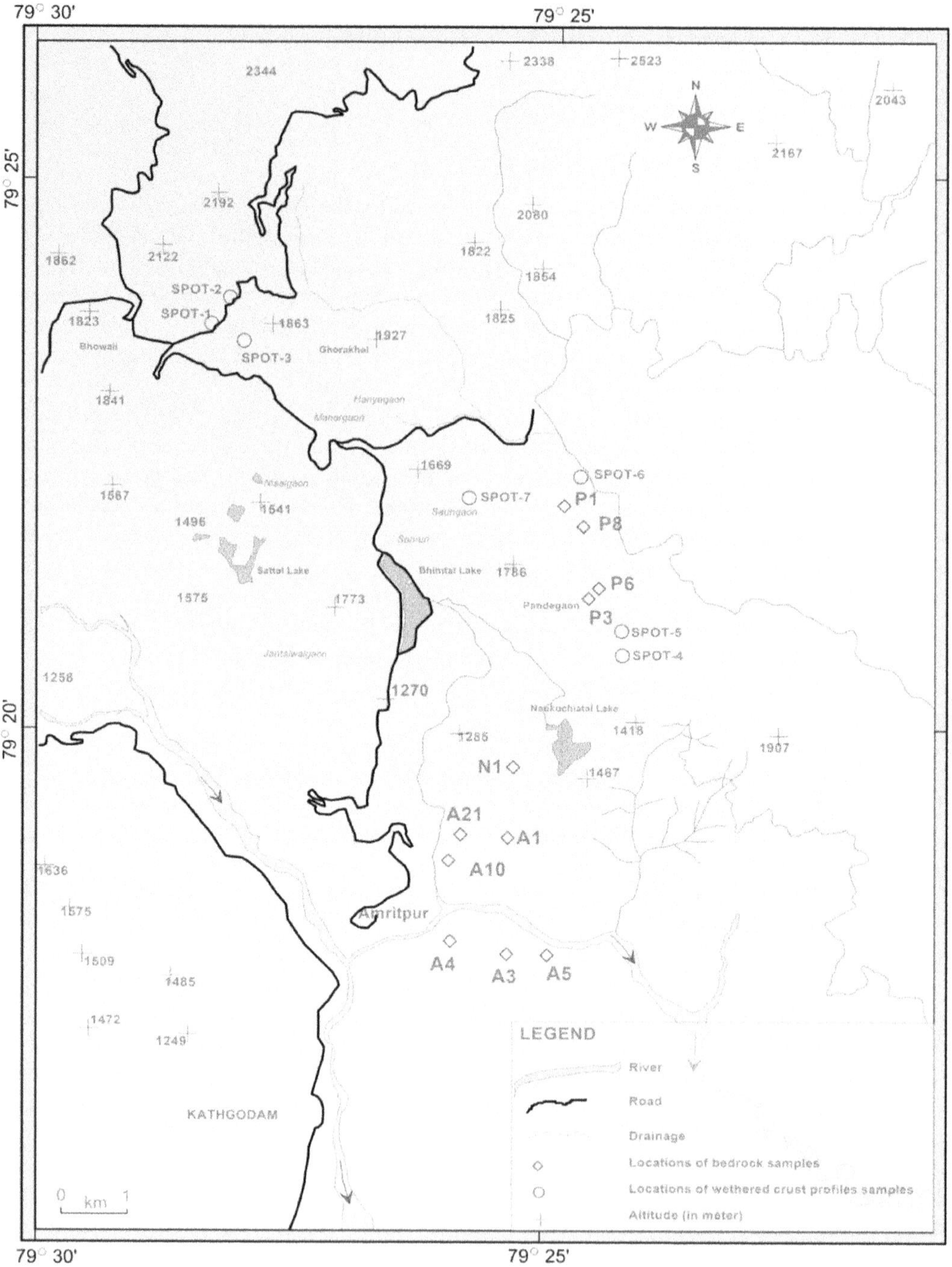

FIGURE 5.3 Location map of the studied samples. Altitude (m above msl) is also shown on the map.

regions. The plagioclase has been mostly found as albitic, and the original clinopyroxene composition has been mostly converted into amphiboles and chlorite. Most of the clinopyroxenes have changed to amphiboles (actinolite, tremolite), uralite and chlorite all along the crystal margins and cleavage planes but also exhibit radiating growth. Uralitization of clinopyroxene appears deuteric in nature and occurs as sheaf-like pseudomorph aggregates of tiny needles occupying the

TABLE 5.1
Magnetic Susceptibility (MS) Values (× 10^{-3} SI unit) of Various Samples of Weathered Crust Profiles

Sr. No.	Gh-1	Gh-2	Gh-3	Gh-4a	Gh-4b	Gh-4c	Gh-5	Gh-6a	Gh-6b	Gh-7	Gh-8	Gh-9	Gh-10	Gh-10	Gh-11	J-1	J-2	J-3
1	0.412	0.331	0.204	15.8	1.08	5.43	6.6	13.1[1]	2.5	0.105	0.001	0.512	4.49	43.1[1]	0.594	0.255	0.21	0.419
2	0.332	0.23	0.232	19.1	0.801	5.77	7.79	11.0[1]	2.6	0.131	0	0.523	5.54	46.7[1]	0.678	0.26	0.174	0.506
3	0.342	0.413	0.296	18.3	0.803	3.97	5.93	11.5[1]	–	0.157	–	–	4.85	44.8[1]	0.705	0.196	0.209	0.384
4	–	0.22	0.284	–	–	5.96	5.97	12.2[1]	–	0.129	–	–	–	–	–	0.189	–	–
5	–	–	–	–	–	6.45	5.64	9.02[1]	–	–	–	–	–	–	–	–	–	–
6	–	–	–	–	–	–	–	15.6[1]	–	–	–	–	–	–	–	–	–	–
7	–	–	–	–	–	–	–	14.7[1]	–	–	–	–	–	–	–	–	–	–

[1] The elevated MS values are recorded on the surface (sample Gh-6a), whereas the low MS values of Gh-6b are recorded from the core of the exposure.

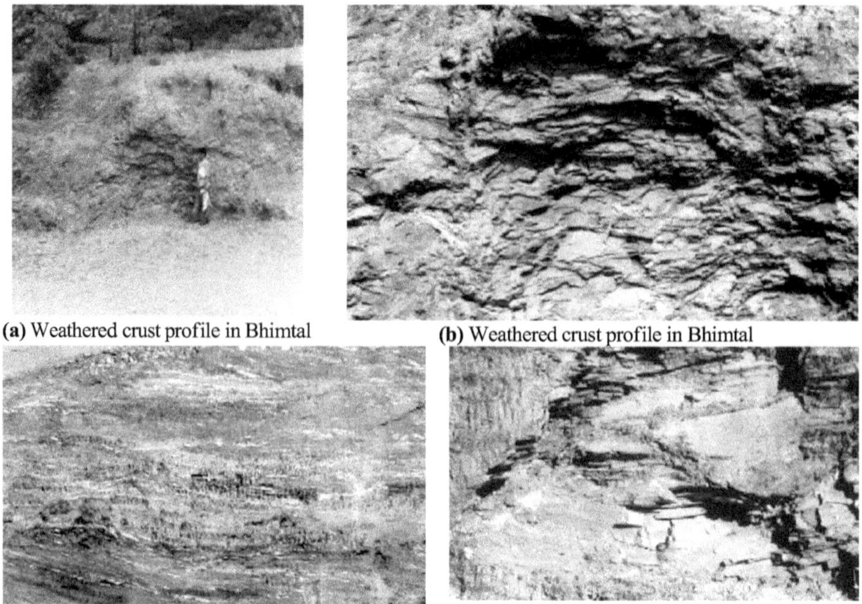

(a) Weathered crust profile in Bhimtal (b) Weathered crust profile in Bhimtal

(c) Sheared rocks in Berinag (d) chloritised phyllitic rocks in Berinag

FIGURE 5.4 Weathered and sheared rocks in the Bhowali-Bhimtal and Berinag regions of KLH.

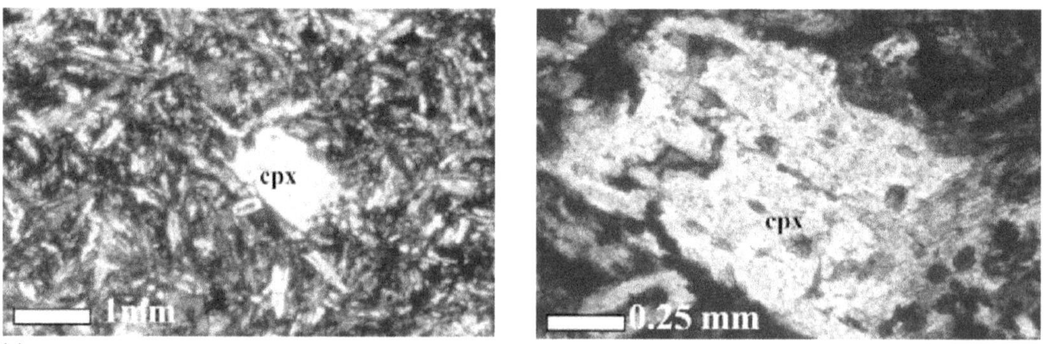

(a) (b)

FIGURE 5.5 (a) Clinopyroxene phenocryst embedded in fine-grained groundmass dominating in plagioclase. Under Cross Nichols. (b) Closer view of clinopyroxene shown in Figure 5.5a. Under Cross Nichols. Note that boundaries of clinopyroxene are showing alteration into amphiboles.

interstitial spaces between the plagioclase laths (Figure 5.5a,b). In the fine-grained varieties, the groundmass comprises the randomly oriented acicular (needle-shaped) plagioclase and clinopyroxene, within which lie the variolitic aggregates of plagioclase laths. Plagioclase is the dominant mineral species and is easily recognizable by its tabular shape and Carlsbad and albite twinning. In many cases, the outer boundaries of the plagioclase are deformed and mixed with the groundmass. Amphiboles are present as alteration products of clinopyroxenes and can be recognized by their greenish yellow to deep green colour, commonly developed at the crystal boundaries. The majority of the crystals, however, exhibit second-order bright blue, pink and yellowish-brown interference colours. Amphiboles exhibit strong pleochroism, varying from greenish yellow to yellowish green and sometimes to deep bluish green. The extinction angle varies from 20° to 30°, which is typical of

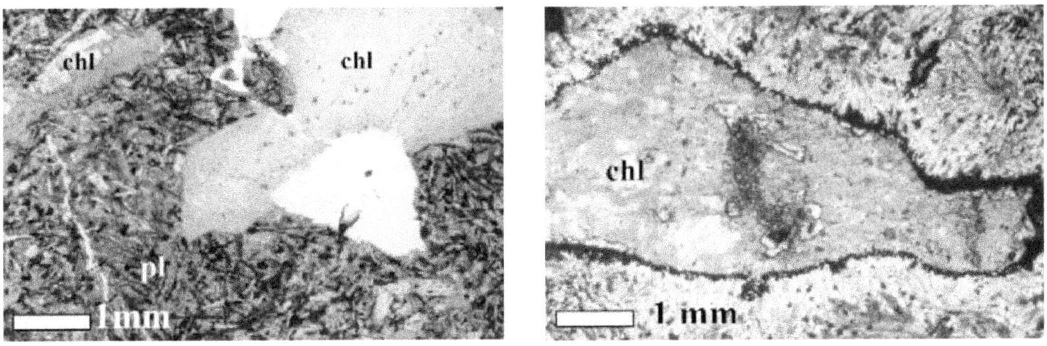

FIGURE 5.6 Chloritized clinopyroxenes embedded in fine-grained groundmass dominantly formed by tiny grains of plagioclase. Under Cross Nichols. Note that the plagioclase is compositionally albitized during alteration. The bar scale on the right-hand side of the photomicrograph equals 0.25 mm.

secondary (metamorphic) amphiboles. It is observed under microscope that constituting minerals of parent rocks (protolith) were altered to varying degrees; however, megascopically they appear fresh and massive.

Chlorite is the most widespread secondary mineral in the volcanics of the Bhowali-Bhimtal region, and occurs commonly as a massive and radiating needle-shaped mineral with deep green colour showing nematoblastic texture (Figure 5.6). The pleochroism varies from pale to medium green colour. Chlorite has been formed in BV as a metamorphic product under the conditions of greenschist facies. The dark green colour of the BV is mostly due to the presence of chlorite and to a certain extent to the presence of amphiboles.

Electron probe micro-analysis of fresh clinopyroxene relics (N = 37) of weathered and fresh basalts classifies them as diopside augite and their evolution in quartz-tholeiitic ($0.5 < Ti/Mg < 0.8$) basaltic magma in a temperature range of 1050–1100°C at pressure <2 kbar, empirically estimated based on eigenvectors (X_{PT}, Y_{PT}, X_m, Y_m) derived from the major element of clinopyroxenes (Pandey and Kumar, 2007). Electron probe micro-analysis of biotite (N = 6) from Palaeoproterozoic ilmenite series (c = 0.006 to 0.098×10^{-3} SI unit) leucogranites of the Amritpur region can be classified as ferribiotite enriched in siderophyllite (crustal-derived) components, which crystallized in a peraluminous (S-type) granite melt (Kumar *et al.*, 2006).

5.4.3 Geochemistry of Bedrock and Weathered Crust Profiles

The geochemistry of bedrocks and their weathered equivalents, if compared, can provide an important clue to the mobility of elements prevailing in the near-surface environment. The major, trace and rare earth element (REE) geochemistry of the weathered crust and its fresh counterparts was studied. The major element analysis for 20 weathered and 20 unweathered samples of the Bhowali-Bhimtal area and 3 weathered and 2 unweathered samples of the Berinag area was performed using WD-XRF. The result of major element analysis of weathered samples of the Bhowali-Bhimtal and Berinag regions is shown in Tables 5.2 and 5.3, respectively. The trace and REE chemical analysis for 20 weathered and 20 unweathered samples of the Bhowali-Bhimtal area and 3 weathered and 2 unweathered samples of the Berinag area was carried out using ICP-MS. The results of trace element analysis of weathered samples of the Bhowali-Bhimtal and Berinag regions are shown in Tables 5.4 and 5.5, respectively. The REE analytical results of weathered samples of the Bhowali-Bhimtal and Berinag regions are summarized in Tables 5.6 and 5.7, respectively. The present discussion is based on a comprehensive account of comparative geochemistry of weathered crust and its unweathered parent rocks.

TABLE 5.2
Major Oxide (wt. %) Analysis and CIPW Norm (wt. %) of Weathered Crust Samples of Bhowali-Bhimtal Region, Kumaun Lesser Himalaya

	1	2	3	4	5	6	7	8	9	10	11	12	13	14	15	16	17	18	19	20
	Gh1	Gh2	Gh3	Gh4a	Gh4b	Gh4c	Gh5	Gh6	Gh7	Gh9	Gh10	Gh11	J1	J2	AP-1	P3	JL2	R3	ST1	M2
SiO_2	50.87	58.38	59.00	48.17	48.10	45.45	46.28	44.07	61.21	46.91	49.25	33.92	48.43	47.54	36.93	47.94	44.57	41.03	44.53	52.98
TiO_2	1.24	0.00	0.91	2.14	2.01	2.12	1.80	2.17	0.93	2.64	2.12	2.40	1.12	1.25	2.95	1.05	1.11	1.98	1.89	1.79
Al_2O_3	17.47	17.00	15.27	14.85	13.64	14.43	13.47	13.19	17.45	12.45	13.07	11.52	15.15	15.71	14.26	14.46	12.97	17.05	13.06	12.42
$Fe_2O_3^t$	13.82	6.50	9.40	18.95	16.83	20.40	20.70	20.60	6.40	19.17	18.96	33.81	12.89	15.04	22.35	9.14	17.22	22.27	18.01	17.45
MnO	0.11	0.07	0.09	0.24	0.19	0.25	0.16	0.21	0.07	0.13	0.21	0.26	0.18	0.24	0.07	0.07	0.17	0.07	0.08	0.09
MgO	4.86	3.90	4.89	6.06	7.76	5.87	7.11	9.49	2.96	10.90	6.48	10.96	8.92	7.30	12.52	19.82	10.53	9.52	15.40	7.24
CaO	0.40	0.74	0.35	0.59	0.77	0.45	0.82	0.95	0.38	0.42	3.37	1.70	6.10	4.63	0.75	1.06	2.91	0.58	0.53	2.01
Na_2O	0.05	2.73	0.49	1.00	1.60	0.62	0.90	1.23	0.42	0.15	2.96	0.01	2.67	1.93	0.01	0.01	0.10	0.01	1.10	2.11
K_2O	4.06	5.41	6.25	0.82	1.09	1.14	1.42	0.49	5.79	1.93	2.26	0.13	0.65	1.75	2.91	1.75	3.55	3.49	2.08	2.76
P_2O_5	0.07	0.25	0.07	0.09	0.10	0.09	0.14	0.11	0.07	0.13	0.20	0.07	0.08	0.07	0.31	0.12	0.12	0.22	0.12	0.26
LOI	7.38	3.22	3.05	10.37	9.35	11.66	9.11	9.71	4.64	7.77	3.23	7.80	6.28	7.03	6.60	7.52	6.72	5.99	5.20	2.23
Total	**100.33**	**99.77**	**99.76**	**103.27**	**101.45**	**102.48**	**101.91**	**102.21**	**100.33**	**102.59**	**102.10**	**102.58**	**102.46**	**102.50**	**99.64**	**102.95**	**99.97**	**102.19**	**102.00**	**101.35**
qtz	22.71	13.84	21.60	22.81	15.87	19.71	14.08	9.55	30.78	13.57	–	–	–	1.05	–	5.39	–	1.40	–	8.26
or	26.13	33.27	38.47	5.33	7.09	7.54	9.24	3.20	35.92	12.21	13.73	0.85	4.03	11.01	18.88	10.94	22.87	21.86	12.88	16.69
ab	0.47	24.05	4.30	9.24	14.94	5.92	8.39	11.47	3.76	1.32	25.72	0.09	23.72	17.34	0.09	0.09	0.91	0.09	9.77	18.29
an	1.71	2.13	1.34	2.59	3.46	1.88	3.47	4.41	1.51	1.33	15.85	8.67	28.88	23.91	1.87	4.77	14.84	1.52	1.94	8.48
crn	13.51	6.14	7.54	12.52	9.56	12.98	10.20	10.10	10.46	10.38	0.10	9.20	–	2.53	11.49	11.51	4.33	13.49	8.74	2.99
di	–	–	–	–	–	–	–	–	–	–	–	–	1.95	–	–	–	–	–	–	–
hy	28.11	14.24	21.05	36.46	38.90	42.18	45.47	51.44	12.96	50.94	30.39	51.52	30.34	37.94	41.94	62.77	48.86	51.90	51.04	35.70
ol	–	–	–	–	–	–	–	–	–	–	3.62	18.17	5.66	–	13.39	–	1.43	–	7.40	–
mag	4.64	2.60	3.76	6.39	5.71	5.06	5.03	5.02	2.58	4.55	6.00	6.35	3.00	3.53	5.42	2.14	4.15	5.22	4.18	5.50
ilm	2.56	3.14	1.80	4.44	4.21	4.51	3.77	4.55	1.86	5.37	4.13	4.97	2.24	2.52	6.14	2.11	2.29	3.98	3.76	3.48
ap	0.16	0.60	0.16	0.22	0.27	0.23	0.35	0.27	0.18	0.32	0.47	0.18	0.19	0.17	0.78	0.29	0.31	0.53	0.29	0.61
$Mg^{\#}$	46.92	61.79	58.35	44.58	53.70	40.21	44.54	51.85	55.49	57.06	46.25	42.15	61.80	53.16	56.69	83.52	58.83	49.97	66.64	51.08
FeO^t/MgO	2.56	1.50	1.73	2.81	1.95	3.13	2.62	1.95	1.94	1.58	2.63	2.78	1.30	1.85	1.61	0.42	1.47	2.11	1.05	2.17
Salic	51.02	73.29	65.70	39.97	41.36	35.03	35.19	28.62	71.96	28.44	55.30	9.61	56.63	53.31	20.84	21.19	38.62	24.88	24.58	51.72
Femic	35.31	19.98	26.60	47.29	48.82	51.76	54.27	61.00	17.40	60.86	44.14	81.01	42.46	43.99	66.89	67.02	56.74	61.10	66.38	44.68
CI	10.94	9.21	10.23	14.16	18.38	13.37	17.10	22.66	6.94	21.75	27.47	38.70	47.51	37.44	37.33	41.32	34.85	32.63	30.15	21.41
DI	49.31	71.16	64.36	37.38	37.90	33.16	31.72	24.22	70.45	27.10	39.44	0.94	27.75	29.40	18.97	16.42	23.78	23.35	22.65	43.24
SI	22.38	21.59	24.05	23.91	29.90	22.32	25.06	31.57	19.61	35.72	22.23	26.14	37.12	29.50	34.88	66.18	35.16	28.50	43.92	25.69
AR	1.60	1.89	2.52	1.27	1.46	1.18	1.29	1.28	2.07	1.38	1.93	1.02	1.37	1.44	1.48	1.26	1.60	1.50	1.61	1.83

Note: Mg# = Mg number; t = total; CIPW normative minerals were calculated on anhydrous basis and normalizing the Fe2O3/FeO = 0.2 as recommended for basalt by Middlemost (1989) using the computer program SINCLAS (Verma et al., 2002). The standard mineral symbols are used after Kretz (1983) and are abbreviated in small letters as recommended for CIPW wt. % norm (McBirney, 1984). BA-basaltic andesite; B-basalt; subal-subalkaline; FOI-foidite; TEP-tephrite; LOI-loss on ignition; CI-crystallization index; SI-solidification index; AR-alkalinity ratio; CIA-chemical index of alteration.

TABLE 5.3
Trace Elements (ppm) Analysis of Representative Weathered Crust Samples from Bhowali-Bhimtal Region, Kumaun Lesser Himalaya

	1	2	3	4	5	6	7	8	9	10	11	12	13	14	15	16	17	18	19	20
	Gh1	Gh2	Gh3	Gh-4a	Gh4b	Gh4c	Gh5	Gh6	Gh7	Gh9	Gh10	Gh11	J1	J2	P3	AP-1	JL2	R3	ST1	M2
Cr	90.69	129.53	73.04	102.32	99.23	92.88	106.58	105.12	100.58	52.39	53.13	54.31	145.88	145.85	155.29	86.56	54.69	52.96	124.01	27.01
Ni	50.28	38.03	38.37	58.44	51.78	56.49	53.15	60.66	44.75	41.09	43.85	50.27	112.19	115.75	132.51	25.35	35.96	88.57	35.99	29.81
Co	38.28	18.98	36.56	56.87	50.43	58.48	52.75	64.90	38.00	55.33	66.78	71.82	57.67	56.55	57.86	60.91	44.29	64.62	46.99	43.61
Sc	27.43	39.59	21.31	41.79	39.45	43.17	36.09	39.34	20.82	38.90	40.10	40.13	36.27	38.25	42.19	27.15	30.67	40.38	27.96	31.66
V	162.68	197.04	127.23	308.39	283.95	313.79	335.14	283.50	123.02	331.82	363.37	400.20	212.27	234.60	295.87	384.31	256.37	351.14	306.82	315.32
Cu	190.74	62.13	113.59	203.90	43.59	190.15	151.98	89.32	85.74	58.95	228.34	18.02	80.41	103.53	15.92	21.96	113.11	15.90	19.34	138.04
Pb	13.72	7.85	14.24	11.32	7.99	16.55	9.43	11.20	45.12	4.94	10.68	2.09	4.52	5.61	4.00	5.73	5.67	5.32	4.30	5.96
Zn	101.28	109.78	98.95	140.65	167.71	161.02	144.30	218.92	94.45	140.43	205.10	174.72	107.86	117.04	76.26	129.90	104.84	71.57	120.35	135.60
Rb	65.40	77.70	97.60	25.90	52.50	38.10	71.30	16.40	99.09	32.50	47.50	2.30	19.10	54.40	35.10	29.80	62.20	71.60	53.70	65.10
Cs	4.26	2.13	2.45	4.39	7.15	5.71	7.71	1.15	5.42	2.18	5.10	0.18	0.95	1.20	0.71	2.25	1.88	1.21	3.12	3.46
Ba	956	1974	977	350	276	345	266	173	1377	398	547	167	231	315	223	135	332	101	121	398
Sr	21	68	35	53	43	16	30	26	31	8	67	5	128	128	5	8	49	5	5	106
Ga	25.12	18.23	21.15	23.02	23.75	24.19	21.12	22.80	20.53	23.08	26.33	25.54	18.01	20.68	22.26	29.37	26.40	20.26	16.52	23.63
Ta	1.35	1.80	1.20	0.70	0.76	0.76	0.57	0.71	1.19	1.09	1.08	1.13	0.47	0.44	0.48	1.09	1.61	1.08	1.68	2.36
Nb	11.30	16.10	10.40	10.20	7.40	8.20	6.20	8.90	9.60	11.20	12.10	13.20	5.00	5.20	5.50	15.80	7.20	10.10	8.60	11.20
Hf	4.15	4.79	4.26	2.00	1.35	2.07	1.30	1.35	8.29	1.43	6.06	1.27	1.85	1.84	0.41	1.44	0.75	0.36	0.50	1.30
Zr	215	203	204	154	143	154	137	169	195	210	227	215	104	109	98	197	124	177	109	213
Y	52.20	65.10	56.50	51.70	56.80	68.00	58.70	46.50	57.50	66.70	71.50	47.90	28.00	40.80	31.80	50.50	50.40	54.20	38.00	70.10
Th	9.63	12.05	8.09	2.30	1.73	2.50	1.97	1.46	7.84	3.21	3.08	2.09	1.85	1.73	1.14	4.34	1.70	2.37	1.53	3.31
U	1.85	1.80	1.10	1.30	1.20	1.78	1.87	0.62	1.31	0.58	0.90	0.32	0.40	0.38	0.14	0.50	0.30	0.17	0.25	0.48

The elements Rb, Ba, Sr, Nb, Zr and Y are analysed by both XRF and ICP-MS techniques, and the values shown here are XRF analysed; all other elements are analysed by ICP-MS only.

TABLE 5.4
Rare Earth Elements (ppm) Analysis and Important Ratios of Representative Weathered Crust Samples from Bhowali-Bhimtal Region, Kumaun Lesser Himalaya

	1	2	3	4	5	6	7	8	9	10	11	12	13	14	15	16	17	18	19	20
	Gh1	Gh2	Gh3	Gh-4a	Gh4b	Gh4c	Gh5	Gh6	Gh7	Gh9	Gh10	Gh11	J1	J2	P3	AP-1	JL2	R3	ST1	M2
La	40.156	45.75	42.523	23.041	15.425	29.22	21.086	22.359	38.4	39.546	36.99	34.167	13.78	15.88	10.564	28.301	13.62	6.889	11.101	24.178
Ce	80.253	89.043	75.114	48.46	29.276	46.959	44.75	46.692	81.387	75.316	77.348	70.079	29.613	35.236	25.01	59.384	36.825	19.398	26.378	60.576
Pr	7.503	9.065	7.44	5.332	3.712	6.84	4.99	4.694	7.888	8.75	7.926	6.859	3.195	3.268	2.686	5.997	3.957	2.209	2.924	6.371
Nd	36.511	44.017	35.835	28.932	20.195	36.328	26.757	24.492	37.406	46.406	14.749	34.866	16.561	17.233	15.335	32.223	20.752	12.448	16.701	33.714
Sm	6.719	8.567	6.282	6.397	4.697	7.641	5.592	4.681	6.314	9.698	8.298	6.544	3.542	3.606	3.411	6.377	4.542	2.758	3.591	6.699
Eu	1.604	2.111	1.508	1.916	1.326	2.137	1.528	1.338	1.612	2.487	2.272	1.465	1.032	1.325	0.775	1.765	1.477	0.682	1.106	2.065
Gd	6.996	8.633	6.045	7.554	5.847	8.759	5.887	4.679	6.302	11.068	9.609	7.386	4.173	4.191	3.983	7.035	5.577	3.027	4.28	8.298
Tb	0.93	1.256	0.723	1.139	0.922	1.283	0.814	0.564	0.737	1.603	1.463	1.034	0.643	0.655	0.613	0.902	0.833	0.359	0.606	1.316
Dy	4.374	6.294	3.072	5.792	4.822	6.131	3.921	2.369	3.239	7.609	7.513	5.403	3.432	3.456	3.199	4.005	4.237	1.43	2.74	6.223
Ho	0.854	1.26	0.545	1.129	0.943	1.214	0.725	0.424	0.565	1.34	1.45	1.083	0.701	0.691	0.691	0.685	0.776	0.221	0.477	1.25
Er	2.57	3.977	1.722	3.292	2.846	3.427	2.061	1.228	1.738	3.623	4.181	3.217	2.098	2.089	2.207	1.921	2.388	0.64	1.363	3.742
Tm	0.405	0.619	0.264	0.512	0.429	0.526	0.31	0.193	0.272	0.508	0.626	0.494	0.323	0.314	0.326	0.248	0.35	0.083	0.177	0.573
Yb	2.067	3.313	1.412	2.544	2.113	2.771	1.519	1.025	1.459	2.371	2.924	2.371	1.64	1.636	1.601	1.39	1.684	0.467	0.942	2.787
Lu	0.29	0.454	0.195	0.312	0.259	0.339	0.198	0.143	0.211	0.255	0.342	0.267	0.206	0.211	0.17	0.178	0.204	0.054	0.127	0.344
ΣREE	151.076	178.609	140.157	113.311	77.387	124.355	99.052	92.522	149.13	171.034	138.701	141.068	67.159	73.911	60.007	122.11	83.602	43.776	61.412	133.958
La_N	109.42	124.66	115.87	62.78	42.03	79.62	57.46	60.92	104.63	107.75	100.79	93.10	37.55	43.27	28.78	77.11	37.11	18.77	30.25	65.88
Ce_N	83.86	93.04	78.49	50.64	30.59	49.07	46.76	48.79	85.04	78.70	80.82	73.23	30.94	36.82	26.13	62.05	38.48	20.27	27.56	63.30
Pr_N	54.77	66.17	54.31	38.92	27.09	49.93	36.42	34.26	57.58	63.87	57.85	50.07	23.32	23.85	19.61	43.77	28.88	16.12	21.34	46.50
Nd_N	51.35	61.91	50.40	40.69	28.40	51.09	37.63	34.45	52.61	65.27	20.74	49.04	23.29	24.24	21.57	45.32	29.19	17.51	23.49	47.42
Sm_N	29.09	37.09	27.19	27.69	20.33	33.08	24.21	20.26	27.33	41.98	35.92	28.33	15.33	15.61	14.77	27.61	19.66	11.94	15.55	29.00
Eu_N	18.44	24.26	17.33	22.02	15.24	24.56	17.56	15.38	18.53	28.59	26.11	16.84	11.86	15.23	8.91	20.29	16.98	7.84	12.71	23.74
Gd_N	22.86	28.21	19.75	24.69	19.11	28.62	19.24	15.29	20.59	36.17	31.40	24.14	13.64	13.70	13.02	22.99	18.23	9.89	13.99	27.12
Tb_N	16.03	21.66	12.47	19.64	15.90	22.12	14.03	9.72	12.71	27.64	25.22	17.83	11.09	11.29	10.57	15.55	14.36	6.19	10.45	22.69
Dy_N	11.48	16.52	8.06	15.20	12.66	16.09	10.29	6.22	8.50	19.97	19.72	14.18	9.01	9.07	8.40	10.51	11.12	3.75	7.19	16.33
Ho_N	10.04	14.81	6.40	13.27	11.08	14.27	8.52	4.98	6.64	15.75	17.04	12.73	8.24	8.12	8.12	8.05	9.12	2.60	5.61	14.69
Er_N	10.32	15.97	6.92	13.22	11.43	13.76	8.28	4.93	6.98	14.55	16.79	12.92	8.43	8.39	8.86	7.71	9.59	2.57	5.47	15.03
Tm_N	11.38	17.39	7.42	14.38	12.05	14.78	8.71	5.42	7.64	14.27	17.58	13.88	9.07	8.82	9.16	6.97	9.83	2.33	4.97	16.10
Yb_N	8.33	13.36	5.69	10.26	8.52	11.17	6.13	4.13	5.88	9.56	11.79	9.56	6.61	6.60	6.46	5.60	6.79	1.88	3.80	11.24
Lu_N	7.61	11.92	5.12	8.19	6.80	8.90	5.20	3.75	5.54	6.69	8.98	7.01	5.41	5.54	4.46	4.67	5.35	1.42	3.33	9.03
Eu*	25.97	32.65	23.47	26.19	19.72	30.85	21.72	17.78	23.96	39.08	33.66	26.23	14.49	14.65	13.89	25.30	18.94	10.92	14.77	28.06
$(La/Lu)_N$	14.38	10.46	22.64	7.67	6.18	8.95	11.06	16.23	18.89	16.10	11.23	13.28	6.94	7.81	6.45	16.51	6.93	13.24	9.07	7.30
$(Ce/Yb)_N$	10.06	6.96	13.79	4.94	3.59	4.39	7.63	11.80	14.46	8.23	6.86	7.66	4.68	5.58	4.05	11.07	5.67	10.76	7.26	5.63
Eu_N/Eu^*	0.71	0.74	0.74	0.84	0.77	0.80	0.81	0.87	0.77	0.73	0.78	0.64	0.82	1.04	0.64	0.80	0.90	0.72	0.86	0.85

Note: La to Lu are ICP-MS analyzed; N represents chondrite normalized values that are taken after Taylor and McLennan (1985); Eu*=(Sm).(Gd); REE=Sum of total REE.

TABLE 5.5
Major Element (wt. %), CIPW Norm (wt. %) and Chemical Index of Alteration (CIA) of Weathered Crust Profiles from Berinag Region

	1	2	3
Sample	**B2**	**B4**	**B5**
SiO_2	35.22	38.20	42.53
TiO_2	1.96	1.28	2.08
Al_2O_3	12.79	18.89	14.00
Fe_2O3^t	34.65	14.91	21.06
MnO	0.11	0.07	0.13
MgO	7.77	15.68	10.01
CaO	0.55	0.48	1.74
Na_2O	0.01	0.01	1.40
K_2O	1.80	4.23	3.04
P_2O_5	0.21	0.14	0.23
LOI	6.34	7.42	4.37
Total	101.41	101.31	100.59
CIPW Norms			
qtz	-	-	-
or	11.55	27.00	19.02
ab	0.09	0.11	12.52
an	1.46	1.55	7.60
crn	11.23	14.86	6.12
di	-	-	-
hy	51.77	22.38	27.36
ol	12.82	27.57	17.71
mt	6.49	3.56	4.93
Il	4.05	2.62	4.19
ap	0.53	0.35	0.55
Mg#	33.52	71.08	52.63
FeO^t/MgO	4.01	0.86	1.89
Salic	13.11	28.66	39.14
Femic	75.13	56.13	54.19
CI	27.43	45.96	26.10
DI	11.65	27.11	31.54
SI	18.87	46.71	29.68
AR	1.31	1.56	1.79
CIA	81	78	62

Note: Mg# = Mg number; t = total; CIPW normative minerals were calculated on anhydrous basis and normalizing the Fe2O3/FeO = 0.2 as recommended for basalt by Middlemost (1989) using the computer program SINCLAS (Verma et al., 2002). The standard mineral symbols are used after Kretz (1983) and are abbreviated in small letters as recommended for CIPW wt. % norm (McBirney, 1984). BA-basaltic andesite; B-basalt; subal-subalkaline; FOI-foidite; TEP-tephrite; LOI-loss on ignition; CI-crystallization index; SI-solidification index; AR-alkalinity ratio; CIA-chemical index of alteration.

TABLE 5.6
Trace eElements (ppm) and Important Ratios of Selected Samples of Weathered Crust Profiles from Berinag Region

	1	2	3
	B2	B4	B5
Cr	51.48	293.01	41.63
Ni	64.13	73.75	40.69
Co	92.19	81.55	55.45
Sc	35.58	42.79	36.20
V	328.32	294.45	325.54
Cu	180.44	50.36	165.48
Pb	37.69	4.37	6.42
Zn	106.42	40.54	97.48
Rb	44.50	116.60	43.60
Cs	0.56	1.59	1.74
Ba	139	335	168
Sr	5	8	16
Ga	17.91	26.00	20.96
Ta	0.647	0.343	2.391
Nb	10.1	5.6	11.1
Hf	0.08	0.28	0.13
Zr	163	123	191
Y	43.8	54.5	52.6
Th	2.22	1.74	3.66
U	0.17	5.61	0.23

Note: The elements Rb, Ba, Sr, Nb, Zr and Y are analyzed by both ICP-MS and XRF techniques, and values obtained by XRF are shown. All other elements are analyzed by ICP-MS only.

TABLE 5.7
Rare Earth Elements (ppm) and Important Ratios for Selected Samples of Weathered Crust Profiles of Berinag Region

	1	2	3
Sample No.	B3	B4	B5
La	19.011	2.564	35.226
Ce	48.639	6.978	88.725
Pr	5.159	0.758	9.336
Nd	27.292	4.101	49.134
Sm	5.536	1.221	8.598
Eu	1.622	0.399	2.071
Gd	6.868	1.146	8.605
Tb	1.033	0.146	0.822
Dy	5.353	0.608	2.612
Ho	1.054	0.098	0.326
Er	3.301	0.323	0.965

TABLE 5.7 (Continued)
Rare Earth Elements (ppm) and Important Ratios for Selected Samples of Weathered Crust Profiles of Berinag Region

	1	2	3
Sample No.	B3	B4	B5
Tm	0.495	0.054	0.093
Yb	2.573	0.319	0.52
Lu	0.362	0.052	0.07
∑REE	128.298	18.767	207.103
La_N	51.80	6.99	95.98
Ce_N	50.82	7.29	92.71
PrN	37.66	5.53	68.15
Nd_N	38.39	5.77	69.11
Sm_N	23.97	5.29	37.22
Eu_N	18.64	4.59	23.80
Gd_N	22.44	3.75	28.12
Tb_N	17.81	2.52	14.17
Dy_N	14.05	1.60	6.86
Ho_N	12.39	1.15	3.83
Er_N	13.26	1.30	3.88
Tm_N	13.90	1.52	2.61
Yb_N	10.38	1.29	2.10
Lu_N	9.50	1.36	1.84
Eu*	23.19	4.45	32.35
$(La/Lu)_N$	5.45	5.12	52.24
$(Ce/Yb)_N$	4.90	5.67	44.22
Eu_N/Eu*	0.80	1.03	0.74

Note: La to Lu are ICP-MS analyzed; N represents chondrite-normalized values where chondritic values are taken after Taylor and McLennan (1985); Eu*=√(Sm).(Gd); ∑REE=Sum of total REE.

$Fe_2O_3^t$ as total iron and CIPW normative minerals were calculated on an anhydrous basis and normalizing $Fe_2O_3/FeO = 0.2$ as recommended for basalt by Middlemost (1989) using the computer program SINCLAS (Verma *et al.*, 2002). The standard mineral symbols are used after Kretz (1983) and are abbreviated in small letters as recommended for CIPW wt.% norm (McBirney, 1984).

5.4.3.1 Major Element Geochemistry

Based on the CIA derived from major element of weathered crust profiles (WCP) (Nesbitt and Young, 1982, 1984, 1989), the WCP (N = 20, Bhowali-Bhimtal; N = 2, Berinag) have yielded elevated CIA values in the range of 48–80, which distinctly bear the normative corundum due to excess alumina enrichment and high K_2O content, ultimately shifting the altered BV compositions towards Al_2O_3 apex (Figure 5.7a). On a molar Al_2O_3-($CaO+Na_2O+K_2O$)-($FeO+MgO$) (A-CNK-FM) triangular plot of WCP and unweathered parent rock (Figure 5.7b), the WCP samples (N = 20) tend to be enriched into chlorite smectite composition. The removal of Na and Ca and enrichment of Al_2O_3 and K_2O in the solution during the breakdown of plagioclase and ferromagnesian silicates marks the chemical weathering trend of WCP from the study area.

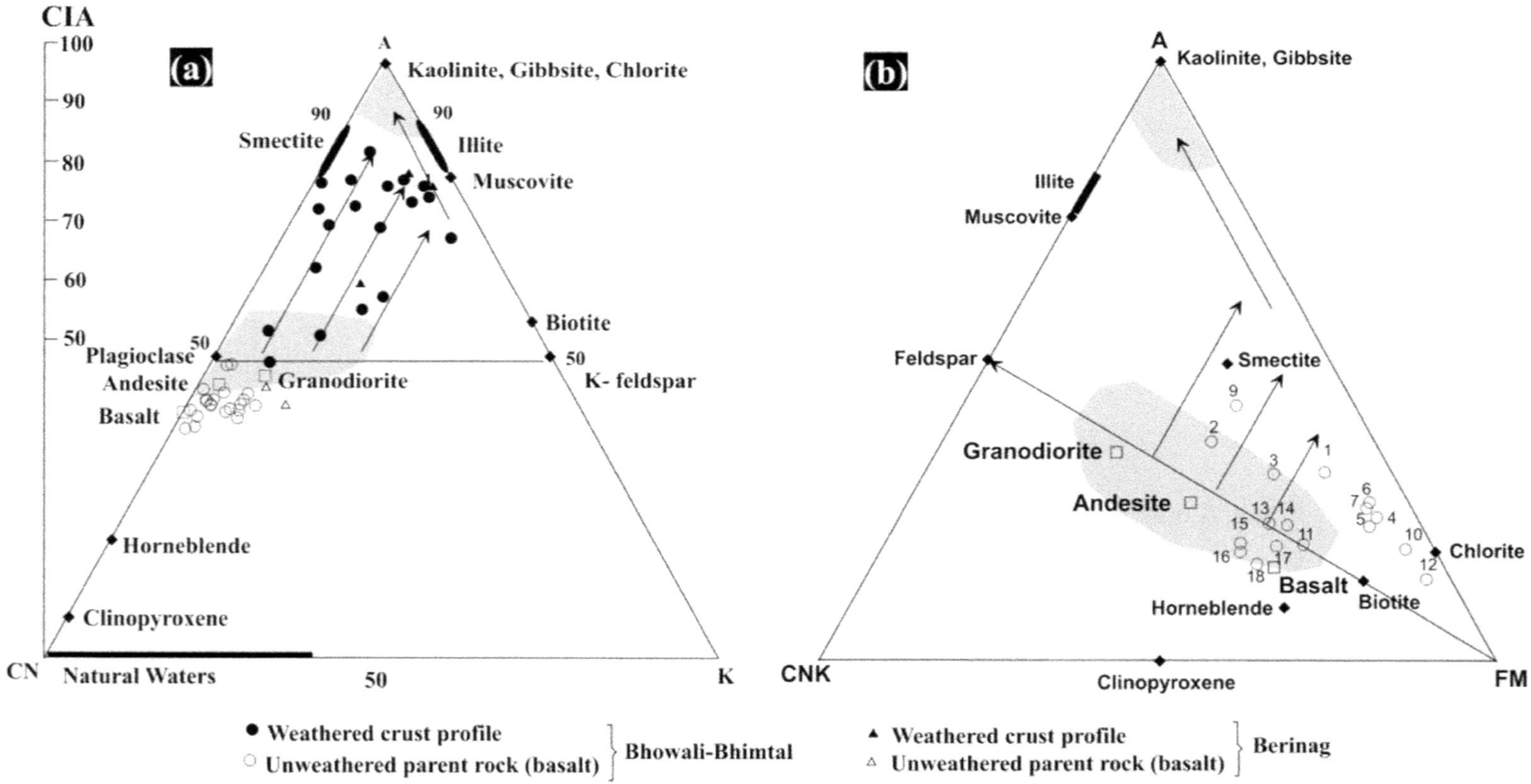

FIGURE 5.7 (a) A-CN-K and (b) A-CNK-FM diagram (Nesbitt and Young, 1984) showing the weathering trend of weathered crust profile in relation to its unweathered parent rocks from Bhowali-Bhimtal and Berinag locality. Compositions are plotted as molar proportions, and the average compositions of basalt, andesite and granodiorite along with the minerals (clinopyroxene, chlorite etc.) are shown for comparison. The arrow marks the trend of advancement of chemical weathering. The highest chemical index of alteration (CIA = 100) represents completely weathered product (e.g. clay minerals such as kaolinite, gibbsite, etc.).

5.4.3.2 Trace Element Geochemistry

Primitive mantle-normalized (Sun and McDonough, 1989) trace element spidergrams of weathered samples have been plotted and compared with those of unweathered samples of the Bhowali-Bhimtal area (Figure 5.8a). The LILE (Rb, K, La, Ce, Sr) and U have been affected to varying degrees during secondary processes, whereas HFSE (Ta, Nb, P, Zr, Ti, Y) and Ba, and to some extent Nd, remain intact, i.e. immobile, during prevailing chemical alteration environments. Similar features are also shown by the weathered crust of the Berinag region (Figure 5.8b), except for one sample showing U-anomaly, which could be related to an alteration mechanism.

5.4.3.3 Rare Earth Element Geochemistry

Chondrite normalized (after Taylor and McLennan, 1985) REE patterns of weathered samples are compared with unweathered samples of the Bhowali-Bhimtal region (Figure 5.9a) and the Berinag region (Figure 5.9b). The light rare earth element (LREE) patterns of weathered crust are inclined, restricted in a relatively narrower range compared with that of unweathered samples, whereas heavy rare earth element (HREE) patterns are comparatively more inclined and show a wide range of variations with the same degrees of Eu-anomalies as compared with that of unweathered samples. Two of the altered samples of Bhowali-Bhimtal have indeed shown much depletion and inclined

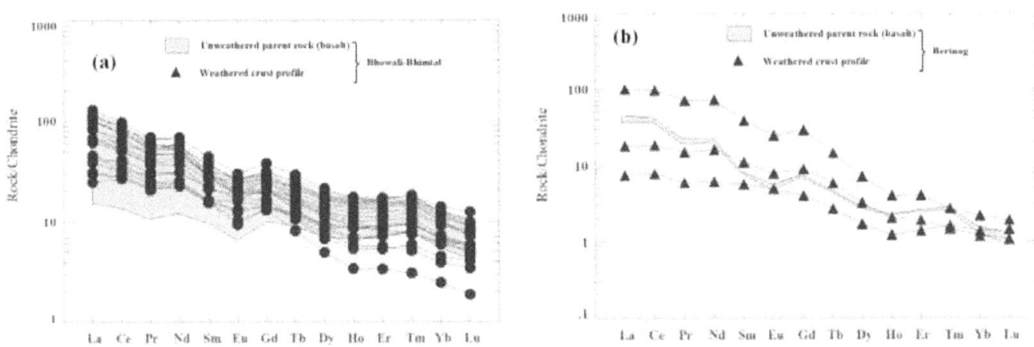

FIGURE 5.8 Chondrite normalized (after Taylor and McLennan, 1985) rare earth element patterns of weathered crust profiles of (a) Bhowali-Bhimtal area and (b) Berinag area. The shaded region marks the field occupied by unweathered parent rocks.

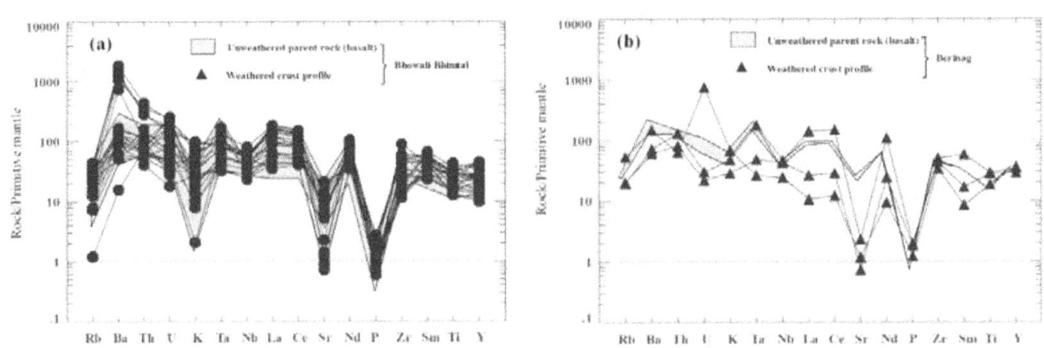

FIGURE 5.9 Primitive mantle (after Sun and McDonough, 1989) normalized trace elements spidergram of weathered crust profile from (a) Bhowali-Bhimtal area and (b) Berinag area. The stippled field marks the trend of unweathered parent rock (bedrock or protolith).

HREE patterns as compared with fresh samples. Although LREE are considered to be somewhat mobile during alteration, especially under seafloor conditions (Ludden and Thompson, 1979), the overall observed regular patterns of all weathered samples suggest that the REE were not significantly modified (altered or mobile) and retained their primary characteristics.

5.4.3.4 Implications for Paleoclimatic Conditions

Climate during the Precambrian can be deduced by understanding weathering, erosion and sedimentation cycles, for which lithological, lithochemical and geochemical data can be used as an indicative factor of paleoclimate (Timofeev, 1992, 2000; Maslov *et al.*, 2003). Lithological parameters act as direct evidence for paleoclimate in ultimate discharge basins, whereas lithogeochemical and other geochemical indicators usually provide insights on paleoclimate of provenance (Maslov *et al.*, 2003). The sedimentary records of Precambrian and Phanerozoic sedimentary rocks allow a more or less reliable reconstruction of climatic parameters (Negrusta, 1985; Negrusta and Negrusta, 2000a,b).

The field verification of the studied region in Bhowali-Bhimtal and Berinag suggests a penecontemporaneous nature of volcanism, i.e. concurrent episodes of volcanism along with sedimentation (Valdiya, 1980a; 1995), generating alternating volcano-sedimentary sequences. Mostly, such episodes suggest a steadily humid climate with enhanced greenhouse effect (Negrusta and Negrusta, 2000a). However, more studies are needed to confirm this.

The ternary A-CN-K diagrams (Nesbitt and Young, 1982) can be efficiently used for the estimation of weathering rate as well as paleoclimatic environments of sedimentation (Maslov *et al.*, 2002). On the A-CN-K diagram (Figure 5.7a), data points of Al-rich weathering products (kaolinite, gibbsite, chlorite and others) lie approximately parallel to the A-CN side. The shift toward A apex is proportional to the degree of protolith transformation. The shift toward K apex is commonly controlled by the post-sedimentary potassic metasomatism (Fedo *et al.*, 1995; Bhat and Ghosh, 2001), which is not seen in the present case. The observations are most likely indicating a humid climate; however, more studies would be needed to establish this.

5.5 CONCLUSIONS

- In the Bhowali-Bhimtal region of KLH, a total of seven spots of soil and weathered crust developed over the metabasalt (bedrock) were recognized. The weathered profiles are both oxidized ($c > 3 \times 10^{-3}$ SI unit) and reduced ($c < 3 \times 10^{-3}$ SI unit) in nature. The water seepage and degree of deformation are major factors in determining the intensity of chemical weathering that is commonly induced during alternately dry and humid climates.
- The ubiquitous occurrence of albitic plagioclase, chlorite and amphiboles in the unweathered (massive) parent rocks indicates different degrees of alteration of minerals without substantial changes in the bulk chemical compositions.
- Major elements in terms of molar A-CN-K and CIA suggest that the Ghorakhal region has suffered relatively intense chemical weathering as compared with the Jungaliagaon and Soangaon regions, and they have acquired aluminous composition. These chemical changes might be because of topographical differences of localities.
- The A-CNK-FM parameters indicate that parental basaltic rock tends more towards chloritization during the course of metamorphism and is subsequently chemically weathered to varying degrees.
- Sr, Rb, Ba and LREE appear slightly to moderately mobile, whereas the transition metals Co, Ni, V, Cr, Cu and HREE remain almost intact during the soil and weathered crust formation.

- During the weathering, the degree of elemental mobility can be recognized as follows: K, Rb and Ba are highly mobile; Pb, Th, Sr, Cu and HREE are moderately mobile; U, Ni, Zr, Ti, Y, Zn Ni, Cr and LREE are least mobile or immobile.
- Based on lithological and lithochemical (A-CN-K diagram) data of the study area, the most likely paleoclimate prevailing during the Precambrian appears to be humid with enhanced greenhouse effect.

ACKNOWLEDGEMENTS

We extend our sincere thanks to Padamshree Professor R. C. Sobti, FNA, for his invitation to contribute this chapter. The work is supported under a University Grants Commission (UGC)-funded major research project (No.F.5-7/2002/SR-1) awarded to SK. N. K Saini, Wadia Institute of Himalayan Geology (WIHG), Dehradun is thanked for WD-XRF analysis. D. R. Rao helped us with electron probe micro-analysis (EPMA) at WIHG, Dehradun. V. Balaram, Council of Scientific and Industrial Research-National Geophysical Research Institute (CSIR-NGRI), Hyderabad is thanked for extending ICP-MS facilities. The infrastructures developed under the Department of Science & Technology Fund for Improvement of S&T Infrastructure (DST-FIST) and UGC Centre of Advanced Study (CAS) programmes to Kumaun University were utilized in the present study.

REFERENCES

Auden, J.B. (1934) Geology of the Krol Belt. *Records of the Geological Survey of India,* 67, 141–256.

Babechuk, M.G., Widdowson, M., and Kamber, B.S. (2014). Quantifying chemical weathering intensity and trace element release from two contrasting basalt profiles, Deccan Traps, India. *Chemical Geology,* 363, 56–75.

Bartarya, S.K. (1988) Geohydrological and geomorphological studies of the Gaula river basin, district Nainital with special reference to the problem of erosion. Unpublished PhD thesis, Dept. of Geology, Kumaun University, Nainital, p. 266.

Bhat, M.I., and Ghosh, S.K. (2001) Geochemistry of the 2.51 Ga old Rampur group pelites, western Himalayas: implications for their provenance and weathering, *Precambrian Research,* 108, 1–16.

Chesworth, W., Dejou, J., and Larroque, P. (1981) The weathering of basalt and relative mobilities of the major elements at Belbex, France. *Geochimica et Cosmochimica Acta,* 45, 1235–1243.

Colman, S.M. (1982) Chemical Weathering of Basalts and Andesite: Evidence from Weathering Rinds. U.S. Geological Survey Professional Paper 1246, 43.

Condie, K.C., Dengate, J., and Cullers, R.L. (1995) Behavior of rare earth elements in a paleoweathering profile on granodiorite in the Front Range, Colorado, USA. *Geochimica et Cosmochimica Acta,* 59(2), 279–294.

Cramer, J.J., and Nesbitt, H.W. (1983) Mass-balance relations and trace element mobility during continental weathering of various igneous rocks. *Science Géologique Mémoire,* 73, 63–76.

Dearman, W.R. (1995) Description and classification of weathered rocks for engineering purposes: the background to the BS 5930: 1981 proposals. *Quarterly Journal of Engineering Geology,* 28(3), 267–276.

Duzgoren-Aydin, N.S., Aydin, A., and Malpas, J. (2002) Re-assessment of chemical weathering indices: case study on pyroclastic rocks of Hong Kong. *Engineering Geology,* 63, 99–119.

Eggleton, R.A., Foundoulis, C., and Varkevisser, D. (1987) Weathering of basalts: changes in rock chemistry and mineralogy. *Clays and Clay Minerals,* 35(3), 161–169.

Fedo C.M., Nesbitt H.W., and Young G.M. (1995) Unravelling the effects of potassium metasomatism in sedimentary rocks and palaeosoils, with implications for palaeoweathering conditions and provenance. *Geology,* 23, 921–924.

Fedo, C.M., Young, G.M., and Nessbit, H.W. (1997) Paleoclimatic control on the composition of the Paleoproterozoic Serpent Formation, Huronian Supergroup, Canada: a greenhouse to icehouse transition. *Precambrian Research,* 86, 201–223.

Fritz, S.J., and Mohr, D.W. (1984) Chemical alteration in the micro weathering environment within a spheroidally weathered anorthosite boulder. *Geochimica et Cosmochimica Acta,* 42, 417–424.

Fritz, S.J., and Ragland, P.C. (1980) Weathering rinds developed on plutonic igneous rocks in the North Carolina Piedmont. *American Journal of Science,* 280, 549–559.

Gavshin, V.M., Shcherbov, B.L., Bobrov, V.A., Solotchina, E.P., Sukhorukov, F.V., and Mel'gunov, M.S. (1997) Behavior of trace elements in the process of formation of weathering profile on granites. *Russian Geology and Geophysics,* 38(7), 1264–1276.

Goldich, S.S. (1938) A study of rock weathering. *Journal of Geology,* 46(1), 17–58.

Gouveia, M.A., Prudencio, M.I., Figueiredo, M.O., Pereira, L.C.J., Waerenborgh, J. C., Morjado, I., Pena, T., and Lopes, A. (1993) Behaviour of REE and other trace and major elements during weathering of granitic rocks Evora, Portugal. *Chemical Geology,* 107, 293–296.

Harris, R.C., and Adams, J.A.S. (1966) Geochemical and mineralogical studies on the weathering of granitic rocks. *American Journal of Science*, 264, 146–173.

Heim, A.A., and Gansser, A. (1939) Central Himalayas-geological observation of the Swiss expedition. *Mem. Soc. Helv. Sci. Nat.*, 73(1), 245.

Hencher, S.R., and McNicoll, D.P. (1995) Engineering in weathered rocks. *Quarterly Journal of Engineering Geology,* 28(3), 253–266.

Hill, I.G., Worden, R.H., and Meighan, I.G. (2000) Yttrium: the immobility mobility transition during basaltic weathering. *Geology,* 28(10), 923–926.

Kretz, R. (1983) Symbols of rock-forming minerals. *American Mineralogist,* 68(1–2), 277–279.

Kumar, S., Singh, B., Joshi, C.C., and Pandey, A. (2006) Magnetic susceptibility and biotite composition of granitoids of Amritpur region, Kumaun Lesser Himalaya: implications on granite series evaluation and nature of felsic magma. *Journal of Geological Society of India,* 68, 666–674.

Lucas-Tooth, H., and Pyne, C. (1964). *Advances in X-ray Analysis*, New York: Plenum Press, 7, 523.

Ludden, J. N., and Thompson, G. (1979) An evaluation of the behavior of the rare earth elements during the weathering of sea-floor basalt. *Earth and Planetary Science Letters*, 43(1), 85-92.

Maslov, A.V., Krupenin, M.T., and Gareev, E.Z. (2003) Lithological, lithochemical, and geochemical indicators of paleoclimate: evidence from Riphean of the Southern Urals. *Lithology and Mineral Resources*, 38(5), 427–446.

McBirney, A.R. (1984) *Igneous Petrology*, Oxford, UK: Oxford University Press, 509.A

Middlemost, E.A.K. (1989) Iron oxidation ratios, norms and the classification of volcanic rocks. *Chemical Geology,* 77, 19–26.

Minarik, L., Zigova, A., Skrian, P., and Stastny, M. (1998) Behaviour of rare-earth elements and Y during the rock weathering and soil formation in the Ricany granite massif, Central Bohemia. *Science of the Total Environment*, 215, 101–111.

Moon, V., and Jayavardane, J. (2004) Geomechanical and geochemical changes during early stages of weathering of Karamu Basalt, New Zealand. *Engineering Geology*, 74, 57–72.

Negrutsa, T.F. (1985) *Possibilities of the Reconstruction of Precambrian Paleoclimates, Sovremennye problemy paleoklimatologii i litologii (Modern Problems of Paleoclimatology and Lithology),* Leningrad: Leningrad. Gos. University, pp. 69–85.

Negrutsa, T.F., and Negrutsa, V.Z. (2000a) Lithogenetic formations and their implication for the reconstruction of Precambrian paleoclimates, Problemy litologii, geokhimii i rudogenezaosadochnogo protsessa (problems of lithology, geochemistry, and ore genesis in sedimentary process). *Moscow: GEOS,* 2, 66–70.

Negrutsa, V.Z., and Negrutsa, T.F. (2000b) Lithology of the Precambrian: achievements and perspectives, Problemy litologii, geokhimii i rudogeneza osadochnogo protsessa (problems of lithology, geochemistry, and ore genesis in sedimentary process), *Moscow: GEOS,* 2, 70–75.

Nesbitt, H.W. (1979) Mobility and fractionation of rare earth elements during weathering of a granodiorite. *Nature,* 279, 206–210.

Nesbitt, H.W. and Young, G.M. (1982) Early Proterozoic climates and plate motions inferred from major element chemistry of lutites. *Nature,* 299, 715–717.

Nesbitt, H.W., and Wilson, R.E. (1992) Recent chemical weathering of basalts. *American Journal of Science,* 292, 740–777.

Nesbitt, H.W., and Young, G.M. (1984) Prediction of some weathering trend of plutonic and volcanic rocks based on thermodynamic and kinetic consideration. *Geochimica Cosmochimica Acta,* 48, 1523–1534.

Nesbitt, H.W., Markovics, G., and Price, R.C. (1980) Chemical processes affecting alkali and alkaline earth during continental weathering. *Geochimica et Cosmochimica Acta,* 44, 1659–1666.

Norman, M.D., Leeman, W.P., Blanchard, D.P., Fitton, J.K., and James, D. (1989) Comparison of major and trace element analysis by ICP, XRF, INAA and ID methods. *Geostandards Newsletter,* 13, 283–290.

Ohlender, B., Land, M., Ingri, I., and Widerlund, A. (1996) Mobility of rare earth elements during weathering of till in northern Sweden. *Applied Geochemistry,* 11, 93–99.

Pandey, A., and Kumar, S. (2006) Geochemistry of metavolcanics of Berinag region, Kumaun Lesser Himalaya: a clue to Precambrian mafic magmatism in northwestern Himalaya. *Indian Journal of Geochemistry,* 21, 185–198.

Pandey, A., and Kumar, S. (2007) Clinopyroxene compositions from Bhowali-Bhimtal volcanics of Kumaun Lesser Himalaya, India: implication on mineral evolution and host magma type. In Saklani, P.S. (Ed.) *Himalaya (Geological Aspect),* vol. 5, Delhi: Satish Serial Publishing House, pp. 1–19.

Pant, C.C. (1985) Stromatolites from Middle-Riphean Gangolihat dolomites (Deoban), Berinag-Gangolihat area, Pithoragarh district, Kumaun Lesser Himalaya. *Journal of the Geological Society of India,* 26(1), 1–9.

Patino, L.C., Velbel, M.A., Price, J.R., and Wade, J.A. (2003) Trace element mobility during spheroidal weathering of basalts and andesites in Hawaii and Guatemala. *Chemical Geology,* 202, 343–364.

Price, D.G. (1995) Weathering and weathering processes. *Quarterly Journal of Engineering Geology,* 28(3), 243–252.

Rajamani, V., Tripathi, J.K., and Malviya, V.P. (2009) Weathering of lower crustal rocks in the Kaveri river catchment, southern India: implications to sediment geochemistry. *Chemical Geology,* 265, 410–419.

Ruxton, B.P. (1968) Measures of the degree of chemical weathering of rocks. *Journal of Geology,* 76, 518–527.

Saini, N.K., Mukherjee, P.K., Rathi, M.S., Khanna, P.P., and Purohit, K.K. (1998) A new geochemical reference sample of granite (DG-H) from Dalhousie, Himachal Himalaya. *Journal of Geological Society of India,* 52, 603–606.

Shankar, R., Kumar, G., and Saxena, S.P. (1989) Stratigraphy and sedimentation in Himalaya: a reappraisal. *Geological Survey of India Special Publication,* 26, 1–60.

Shankar, R., Nag, S., and Shome, S. (2002) Incidence of gold in subgreenschist to greenschist facies metabasalt of Bhowali, Nainital District, Uttranchal. *Journal of the Geological Society of India,* 59, 379–383.

Sharma, A., and Rajamani, V. (2000) Weathering of gneissic rocks in the upper reaches of Cauvery river, South India: implications to neotectonics of the region. *Chemical Geology,* 166, 203–223.

Sun, S.S., and McDonough, W.F. (1989) Chemical and isotopic systematics of oceanic basalts: implications for mantle composition and processes. In Saunders, A.D., and Norry, M.S. (Eds.) *Magmatism in Ocean Basins.* Geological Society of London Special Publication 42, pp. 313–345.

Tapia, J., Tosiani, T., and Loubet, M. (1998) Rare earth elements (REE) behaviour during gabbro weathering in tropical rain forests of south Venezuela. *Mineralogical Magazine,* 62A, 1493–1494.

Taylor, S.R., and McLennan, S.M. (1985) *The Continental Crust: Its Composition and Evolution.* Oxford: Blackwell, p. 295.

Timofeev, P.P. (1992) *Lithology – Fundamental Section of the Geological Science, Geonauki v SSSR (Geological Sciences in the USSR),* Moscow: Nedra, pp. 125–135.

Timofeev, P.P. (2000) *Genetic Science of Sedimentary Geological Formations (Theory of Formational Analysis), Sedimentary Basins of the Urals and Adjacent Regions: Regularities of Structure and Minerageny,* Yekaterinburg: Inst. Geol. Ural. Otd. Ross. Akad. Nauk, pp. 15–27.

Tripathi, J.K., and Rajamani, V. (1999) Geochemistry of the loessic sediments on Delhi ridge, eastern Thar Desert, Rajasthan: implications for exogenic processes. *Chemical Geology,* 155, 265–278.

Valdiya, K.S. (1980a) *Geology of Kumaun Lesser Himalaya.* Dehradun: Wadia Institute of Himalayan Geology, p. 291.

Valdiya, K.S. (1980b) The two intracrustal boundary thrusts of the Himalaya. *Tectonophysics,* 66, 323–348.

Valdiya, K.S. (1988) *Geology and Natural Environments of Nainital Hills, Kumaun Himalaya.* Nainital: Gyanodaya Prakashan, pp. 67–87.

Valdiya, K.S. (1995) Proterozoic sedimentation and Pan-African geodynamic development in the Himalaya. *Precambrian Geology,* 74, 35–55.

Verma, S.P., Torres-Alvarado, I.S., and Sotelo-Rodríguez, Z.T. (2002) SINCLAS: standard igneous norm and volcanic rock classification system. *Computers & Geosciences,* 28(5), 711–715.

Vyshnavi, S., Islam, R., and Sundriyal, Y.P. (2014) Role of physical and chemical weathering in development of soil profile in the Garhwal Lesser Himalaya. *Himalayan Geology*, 36(2), 111–117.

White, A.F., and Brantley, S.L. (1995) Chemical weathering rates of silicate minerals: an overview. *Reviews in Mineralogy,* 31, 1–21.

6 Ecological Restoration

Ecosystem Services and Conservation of Natural Heritage

Akshat Uniyal and N.S. Bisht
Department of Botany and Microbiology, HNB Garhwal (Central) University Srinagar (G), Uttarakhand

Isha Sharma
Department of Botany, School of Life Science, Cluster University Jammu, Jammu

Priyanka Uniyal
School of Pharmaceutical Sciences and Technology, Sardar Bhagwan Singh University, Uttarakhand

CONTENTS

DOI: 10.1201/9781003220824-7

Human use, population, and technology have reached that certain stage
Where mother Earth no longer accepts our presence with silence.

—The Dalai Lama

6.1 INTRODUCTION

6.1.1 Origin and Meaning of Ecological Restoration

Since our evolution, *Homo sapiens* have been exploiting Mother Nature for our own benefit, and now in the 21st century the ecosystem has reached an alarming condition of deterioration because of continuous exploitation. As aptly said by M. K. Gandhi, "There's enough on this planet for everyone's needs but not for everyone's greed". *Ecosystems* are dynamic communities where biotic components (plants, animals and microorganisms) interact (positively or negatively) with their abiotic environment as a functional unit to attain sustainability and can be *damaged*, *degraded* or *destroyed* by anthropogenic activity. Most of our ecosystems (whether terrestrial or aquatic) are witnessing degradation due to varying degrees of developmental pressure, population explosion, overexploitation, introduction of invasive alien species, climate change and yet more threats. Mother Nature, or we can say the ecosystem, continuously nurtures us in one way or another. In other words, we can say that human existence depends on various ecosystem benefits, and these benefits derived by us directly or indirectly from the ecosystem are referred as to "*ecosystem services*". But now, in the present era, these ecosystem services are reducing due to environment and ecosystem degradation by one way or another, and it is the need of the hour to restore the ecosystem. Ecological restoration is referred to as the process of recovering a 3D ecosystem, i.e., a damaged, degraded and destroyed ecosystem, and restoration ecology is the scientific study supporting the art and science of ecological restoration. The dictionary meaning of "restoration" is the act of bringing something back to its original or unscathed condition. Thus, the goal of ecological restoration is to bring back degraded ecosystems to imitate more closely (not necessarily to duplicate) conditions that prevailed before degradation and disruption, i.e., environmental conditions that had influenced native communities during evolutionary time. By the 1980s, the alliance between ecological restoration (the practice) and restoration ecology (the biological science) became so evident that it led to the recognition of restoration ecology keeping a focus on developing and testing ecological theory (Jordan et al., 1987). According to Nilsen (1991), the restoration process can be defined as a remedy for the ecosystem, where researchers are assisting nature to heal or retain its native form. Keeping this view in mind, the Ecological Restoration Society was founded in 1987, and its first annual congress was held in 1988, although its first scientific journal, entitled *Restoration Ecology*, was not initiated until 1993. A proper establishment of restoration is based on fundamental concepts of biodiversity conservation, which involve management strategies and actions designed to promote the recovery and rehabilitation of the ecosystem to its native form. The recommendations of the Society for Ecological Restoration (1993) provide the motive for ecosystem restoration (to restore the natural environmental conditions at par with evolutionary changes through time), which keeps a check on their further degradation along with conservation of native flora and fauna.

6.1.2 Why Restore the Ecosystem?

There are many reasons to restore ecosystems. A few are listed here:

- To restore natural capital like drinking water or wildlife populations
- To mitigate the effects of climate change (e.g., through carbon sequestration by organic farming or biochar application) on human population and ecosystems

- To conserve threatened and endangered species and prevent them from becoming extinct
- Aesthetic reasons
- Anthropogenic activities (causing habitat fragmentation and destruction)
- Regulated use/harvest, particularly for subsistence
- Cultural relevance of indigenous ecosystems to native people
- To improve environmental conditions

Christensen et al. (1996), in the report of the ESA (Ecological Society of America) entitled *The Scientific Basis for Ecosystem Management*, elaborate the role and importance of ecological restoration management. To support the maintenance and sustainability of the biosphere, Lubchenco et al. (1991), in the same report, focused on reclamation of ecosystem services out of 12 displayed topics for priority on the research agenda. The restoration of the ecosystem occupies a central position with various groups worldwide, like the Southwest Forest Alliance (1996) and the SAF (Society of American Foresters, 1993). In fact, restoring the ecosystem is an international concern, and it was considered as a theme on World Environment Day (5 June) 2021. It was also admitted as a centre of concern by the UN (1992) in the Rio Declaration (RDED). Principle 7 emphasized the cooperation of states globally to protect, conserve and restore our Earth's ecosystem (health and integrity). But, the question arises: What does ecosystem restoration mean exactly? The Society for Ecological Restoration (SER) defined ecological restoration as "the process of assisting the rehabilitation of an ecosystem that has been disturbed, damaged or degraded". This definition can simply be expressed as the process of recovering characteristics of ecosystems that have been degraded. However, ecosystem restoration can be applied to various restoration perspectives (from species to landscapes) and also the employment of diverse restoration approaches and actions to achieve the set goals. That is why another acceptable definition has been proposed by Martin (2017), which states: "Ecological restoration is the phenomenon to assist, facilitate and expedite the recovery of a degraded, damaged or destroyed ecosystem to reflect values regarded as inherent in the ecosystem".

An ecosystem is said to be degraded when there occurs disturbance in the structure, composition and function of ecosystem caused by anthropogenic activities or disturbances that are too severe to allow natural regeneration or recovery and lead to the loss of biodiversity. Generally, the factors driving degradation are interlinked directly or indirectly (whether anthropogenically induced exploitation, i.e., unsustainable use and management of land, water and other resources, or climate perturbations and extreme events like drought, fire and cyclones/hurricanes), which adversely affects the character and flow of ecosystem services as well as disruption and disturbance in our ecosystem. Therefore, restoration strategies for disturbed ecology become more necessary when these activities render the affected ecosystem incapable of self-repair. According to the Millennium Ecosystem Assessment (2005), ecosystem services are the benefits that humans derive naturally from the environment and ecosystems. They flow from natural capital stocks and are transformed by other forms of capital (e.g., human, financial, etc.) for the benefit of people.

6.1.3 Categories of Ecosystem Services

Ecosystem services are grouped into four broad categories:

(1) Provisioning (e.g., food and water production);
(2) Regulating (e.g., climate and pollutants control);
(3) Supporting (e.g., pedogenesis and primary production); and
(4) Cultural (e.g., recreational and spiritual benefits).

Following Rees (1995), the Millennium Ecosystem Assessment (2005) identified four partially overlapping types of natural capital:

(1) Renewable, i.e., well-functioning ecosystems and their living part, i.e., native biodiversity;
(2) Cultivated, defined as traditional crop varieties and livestock races;
(3) Replenishable (clean air, potable water and fertile soils); and
(4) Non-renewable (petroleum, copper, coal and diamonds). Ecological restoration primarily concerns itself with the first two types.

Regarding the typology, the Millennium Ecosystem Assessment proposed about 22 distinct services (Millennium Ecosystem Assessment, 2005; de Groot et al., 2010) in four main categories:

(1) Provisioning, e.g., food, fodder and firewood;
(2) Regulating, e.g., water regulation, waste recycling and flood prevention;
(3) Cultural, e.g., knowledge, recreation, aesthetics; and
(4) Supporting, e.g., soil formation, nutrient and water cycling and food chain dynamics.

According to the report of the Millennium Ecosystem Assessment (2005), ecosystem services are referred to as benefits which humans procure from ecosystems. The report distinguishes four classes of ecosystem services. However, the subsidiary services are considered on the basis of the other three services. Various ecosystem services derived directly or indirectly are listed in Figure 6.1.

The concept of ecosystem services has increasingly gained attention in decision-making processes, international studies, treaties and conventions over the last few decades (Alexander et al.,

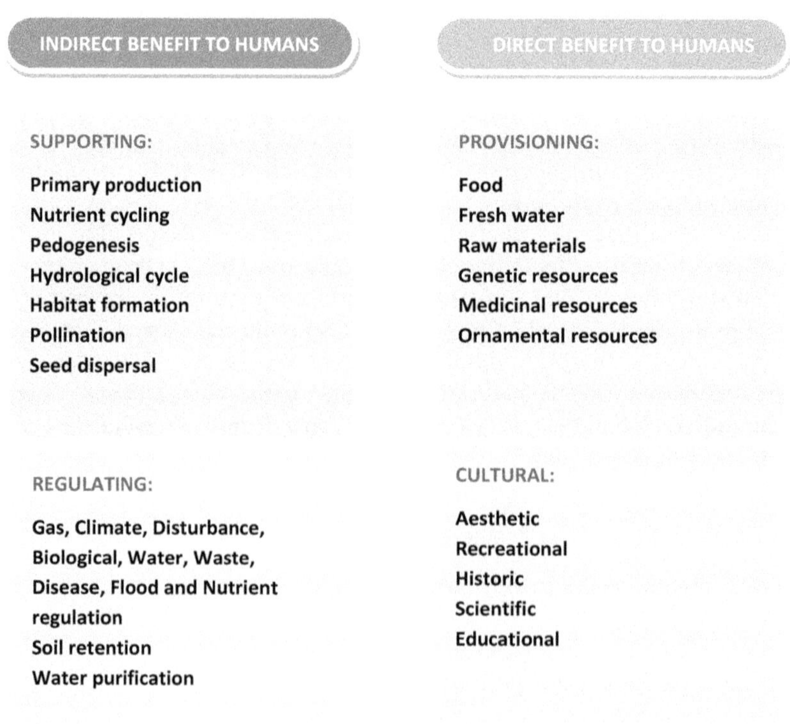

FIGURE 6.1 Direct and indirect benefits (ecosystem services) derived from ecosystem.

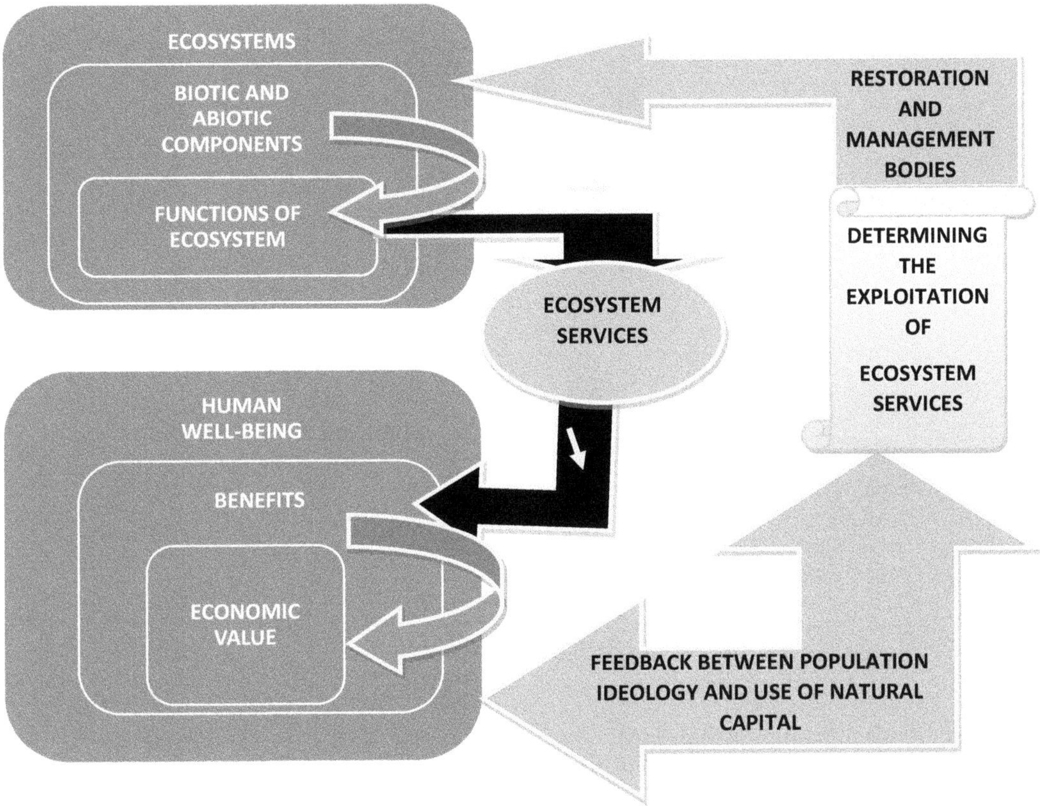

FIGURE 6.2 The ecosystem services flow illustrating the interconnections between ecosystem processes, functions, services, benefits and value.

2011; Diaz et al., 2015). The concept already had a profound impact on thinking, particularly when linked to the natural capital (Costanza and Daly, 1992; Millennium Ecosystem Assessment, 2005; de Groot et al., 2010; Guerry et al., 2015). The concept of restoration of natural capital (Aronson et al., 2007) is already attracting attention. Natural capital can be considered as an economic metaphor used for the fixed repositories of biological as well as physical elements found in the lithosphere, some of which are used directly by society and some are not. In a nutshell, this involves the maintenance of renewable and cultivated natural heritage (resources) to ensure the uniform and sustainable flow of ecosystem services to people. The ecosystem service concept and the efforts to value ecosystem services are attempts to create explicit and binding values related to monetary and non-monetary benefits derived from those services. Keeping this concern in mind, it is quite difficult to justify the possible roles, limitations and opportunities contributed by the research and implementation of both ecological restoration and ecosystem services for conservation of biodiversity and its proper functioning; well-integrated ecosystems in a limited and bounded globe (Convention on Biological Diversity (CBD), 2012; Aronson and Alexander, 2013).

The ecosystem services flow illustrating the interconnections between functions, processes, services, benefits and value is shown in Figure 6.2. Over the past 25–35 years, our capability to repair destroyed and damaged ecosystems along with managing ecosystem sustainability has vastly increased (Lamb, 2011; Holl et al., 2003; Suding, 2011). However, to achieve the set goals, i.e., re-plantation, rehabilitation of the aquatic ecosystem and reclamation of polluted terrestrial and marine ecosystems, and to accomplish restoration, much more multidisciplinary research and its

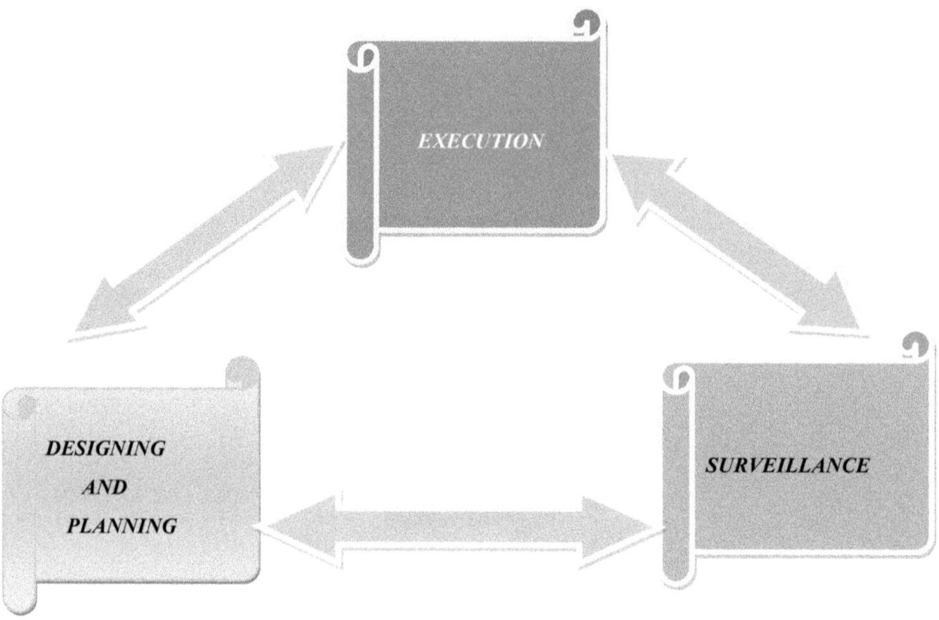

FIGURE 6.3 Conceptual diagram showing the three major restoration phases.

implementation is urgently required (SER (Society for Ecological Restoration) Science and Policy Working Group, 2004; Falk et al., 2006; Clewell and Aronson, 2013). So, policy reforms, extended research and development-related financing, and far-reaching ecological awareness are the foremost requirements to improve the efficiency of restoration processes (Aronson et al., 2007). Beyond conservation, our ecosystems need restoration (regeneration, rejuvenation and revamping). Ecological restoration is not a substitute for conservation. The basic principles, i.e., the fundamental ecological conservation upon which the restoration of a particular ecosystem is based, involve management practices-cum-actions required to facilitate the recovery of the indigenous ecosystem. The processes of restoration can be expressed as a combination of three phases: designing and planning, execution, and surveillance (Hobbs and Norton, 1996; Tischew et al., 2010), given in Figure 6.3. Alexander et al. (2016) studied four continents to demonstrate the utilization of the concept of ecological reconstruction/restoration and its services in land use planning and managing ecosystems. The restoration ecology comprises the reclamation of soil, water, forests and other natural capital, including the healing of damage and destruction due to climate change. According to Peter Barnes, "The problem of climate change is too large to be solved by voluntary individual responses and it requires an economy-wide solution, i.e. one that limits the total carbon intake of the economy" (Barnes, 2008).

6.1.4 Forest Ecosystem: Threats and Restoration Strategies

Plants in the form of forests cover nearly 31% of the land area worldwide, contribute over 80% of terrestrial biology and store 1 trillion tons of carbon. Moreover, the livelihood of around 1.6 billion people is totally dependent on forests' resources (United Nations International Year of Forests, 2011). A forest can be considered as a functional unit that encompasses soil, plants, insects, animals, birds and humans as its interacting units. Species composition and stratification are two main structural features of the forest ecosystem, including its components: productivity (gross primary productivity and net primary productivity), decomposition (fragmentation, humification and mineralization, leaching, catabolism, etc.), energy flow through different trophic levels, and most importantly, nutrient cycling.

Nevertheless, while forest resource practitioners are consistently focused on sustainable forest management and ecosystem conservation, forest habitat, particularly in the hot and humid regions, continues to become less comprehensive, mainly due to forest land conversion for agricultural purposes and in relation to urbanization (De Fries et al., 2010). Forest clearing for this purpose is one of the major causes reported earlier (Gibbs et al., 2010). In the last two decades of the 19th century, more than half of the new agricultural land across the tropics was constructed by clearing intact forests. Continuous anthropogenic deforestation also contributes to a total of 12–20% of the global greenhouse gas emissions that eventually lead to global warming (UN International Year of Forests, 2011). The boundaries and the attributes of the habitat continue to decline, and the associated loss of biodiversity threatens forest ecosystem functioning (processes, activities and properties like organic matter decomposition, nutrient cycling, water retention, etc.) as well as its ability to provide ecosystem services. However, it is much more stable and resistant to these deleterious changes as compared with other small ecosystems such as grasslands and wetlands. Restoration practices therefore might play a prominent role in enhancing or retaining the levels of biodiversity in anthropogenically disturbed ecosystems and also mitigate the changing climatic impact.

Several researchers associated with organizations at global, regional and national level set targets during the last two to three decades for large-scale forest landscape restoration strategies. The Sahara's great green wall and the African Sahel initiative aim to cover the region with vegetation. In addition, the Chinese government set a target to achieve the plantation of 6.7 million hectares of forest per annum (Cernansky, 2018). The Bonn Challenge, started in 2011, launched by the International Union for Conservation of Nature and Germany, aimed to restore 150 Mha of disturbed forest land by 2020 and 350 Mha by 2030.

Merlin et al. (2019), while presenting the theme of ecological basics that assist the forest restoration process, reported the experimental findings of physico-chemical properties of reconstructed soil of boreal forest in northern Alberta, Canada, following the mining of oil sand. Experimental data suggests that the limiting factors may differ by type of vegetation and shift through time. Soil, the sole limiting factor, plays a crucial role in early-stage restoration of vegetation. Research findings suggest that the enrichment and distribution of soil amendments directly affect the vegetation growth, succession, reproduction and net productivity (Alday et al., 2012). In particular, soil nutrient dynamics and water retention properties are the key regulating factors in vegetation development, confirmed by some experimental findings (Chang and Turner, 2019). In contrast, overgrowth of vegetation can govern the changes in maintenance along with micro-changes in soil (Huang et al., 2018) in the form of mass accumulation of organic content and nutrients, especially in the latter stages of the restoration process (Gu et al., 2019).

Furthermore, nutrient-rich soil follows a balancing pattern among the leading ecological processes, i.e., above-ground nutrient storage, decomposition and cycling; these cumulative effects may cause the involution of the interaction among abiotic (soil) and biotic (vegetation) factors (Huang et al., 2018). In conclusion, the knowledge of interactions between biotic and abiotic communities is important in restoring forest or any other ecosystem for predicting better ecological restoration and development practices.

6.1.5 Pedological Aspects

Soil-sequestered carbon works as a critical regulating component for water, nutrient cycling, temperature, waste material decomposition and filtration, eventually creating a suitable environment for biological interactions and existence (Dominati et al., 2010). We are totally dependent on these ecosystem services stored by soils, which are the product of interactions between abiotic and biotic factors and the backbone of self-maintenance in a well-established ecosystem.

Soil can be considered as a foundation to support all primary ecosystem functions, i.e., generation of biomass, regulation of the water cycle, a well-established nutrient pool and a better substrate for growth (Barrios, 2007: Bunemann et al., 2018). Restoration researchers often seek to restore a disturbed ecosystem along with its functions that are supported by soil and subject to a routine restoration process. However, scientists remain focused on manipulating the structure of an ecosystem, often focusing on reconstructing a specific suite of plants that is based on historical, reference or desirable conditions. In many situations, the emphasis has been only on the structural manipulation of flora and fauna, and this may limit a successful restoration, because the ecosystem may be faced with an unexpected shift due to sudden disturbance along with other adverse challenges (e.g., climatic factors), and in the later stages of the process, it is almost impossible to manage the native vegetation diversity (Hobbs et al., 2009).These vernal ecosystems can result from a shift in soil biotic factors and processes, most likely creating a mismatch between the soil ecology at the restoration site and the reference vegetation community that ecologists are attempting to restore. As natural resources are limited, it is very crucial to maintain and restore above-ground as well as below-ground processes.

Sustainable soil management is one of the prominent aspects, including the concepts of utilization, improvement and restoration of the native productive capacity and processes of the soil ecosystem (Lal and Stewart, 1992). We can use ecological restoration methods, which are linked with soil management, to improve degraded and disturbed ecology, reverse the trends of soil destruction, and enhance the soil's physico-chemical properties to regain ecosystem health. Ecological practitioners try to restore ecosystem health by means of applying various disciplines, from adaptation to indigenous strategies, from agronomical approaches to wildlife management, in terms of modern research-based implementations.

Here, we attempt to introduce a few important studies relevant to disturbed ecology and its restoration processes devoted to forest, forest soil, agricultural soil and aquatic ecosystems by means of various indigenous as well as recent approaches (e.g., bioremediation) to promote interdisciplinary collaboration and theory-informed restoration practices.

6.1.6 Forest Soil Management

The history of forests does not include a record of soil impacts, but we can infer some aspects of this from land use pattern and population factors. The management of forest soil has been characterized by avoidance of risk. Inattentiveness prevailed throughout the previous centuries in the absence of rules and regulations and conservation practices (MacCleery, 1993); this period of rapid land acquisition was also a time of high demand for food and fodder. Land clearing and overexploitation of forest resources resulted in an alarming 75% drawdown of saw timber stocks between 1800 and 1920 (Birdsey et al., 2006), leading to the implementation of the first forest conservation policy in the 1930s (MacCleery, 1993). Wood was considered as the sole source of energy and fuel for household or industrial uses. Moreover, until the early 19th century, it was the sole material utilized for railing purposes (Foster and Aber, 2004).

During the mid-20th century, the soil characteristics were recognized as a major component determining the forest quality; thus, during the last few decades, efforts have been aimed at identical forest species to a particular area. The adoption of a site index through time for measuring the production efficiency and yield led to the following facts: Soils may vary in their fertility parameters or their ability to produce and nourish plants; skilful management of forest along with other resources (i.e., wildlife) could make it more generative by understanding the soil.

6.1.7 Management and Restoration

Whenever we attempt to expose a soil ecosystem to multifarious management practices, it may lose or retain its capacity for sustaining biotic (plants, animals, microbes) as well as abiotic components (water retention, carbon sequestration, nutrient dynamics) or other functions (Heneghan et al., 2008).

6.1.8 Ecological Assessment of Interaction among Populations

In the past two and half decades, theories about biotic interaction (i.e., plant–plant) have been developed, with a strong potential for low-cost restoration process (Löf, 2017). In detail, it has been found that negative and positive interactions co-exist to a larger extent than previously believed (Bruno et al., 2003). Earlier, emphasis had been given to negative interactions, and to date, this is one of the most common management techniques of vegetation regeneration, which definitely reduces competition from undesired vegetation to a certain extent. We can explore this by taking the examples of forest fire, weed-killers and mechanical preparations. Conceptual application of the stress gradient hypothesis (He et al., 2013) and nurse plants (Gomez, 2009) has strong potential for restoring and improving the survival and growth of seedlings, keeping restoration costs as low as possible.

The concept of interaction between two contrasting biotic components facilitates the competition with the intensity of stress among biotic and abiotic, while negative interaction directs the growth and nourishment of one plant species by another species. In addition, management focused on positive or negative interactions for restoration purposes will be particularly compatible in scenarios of facing unfavourable challenges, such as drought due to adverse climatic conditions, with potential for an enhanced role of reinforcement to secure reclamation. However, the aforementioned restoration approaches are more suitable for a dry land ecosystem dominated by perennial vegetation (Gomez et al., 2004). Researchers are still trying to explore the ecosystems where floral diversity is more complex, i.e., tropical and temperate regions. To overcome this, not only is it necessity to select an appropriate combination of nurse and target vegetation species with the supporting species, but emphasis should be given to the community composition and densities, as covered canopy and competition increase through time. Biotic interactions between different communities may also play a vital role in low-cost forest reconstruction. A high population of herbivores, for example, one of the extreme challenges to forest reconstruction or reclamation, affects species dynamics and forest composition along with its functions (Cote et al., 2004).

A regeneration effort by means of physical protection by fences, chemical protection by repellents or shelters for individual trees seems more likely to be effective, but the costs of implementation and maintenance of these measures are quite high. In terms of low-cost strategies; such as providing physical protection to the target vegetation with the help of dense natural reclamation processor seeding browse tolerant plant species along with shrubs. Additionally, increased carrying capacity can be achieved by means of employing approaches like reducing the population of grazers or providing more trees/shrubs. Establishment of grasses, shrubs or lumber species may release the pressure of abundant herbivores from more suitable species on restoration sites, although supplementary feeding would be a good option if it improves the nourishment of the larger mammals (Felton et al., 2016).

6.1.9 Advancements in Restoration Techniques

Reclamation, reconstruction and regeneration practices continue to adapt to the current needs of society, disturbances (anthropogenic or natural), implementation of policies, climate change and recent technologies (Wagner et al., 2018). In an earlier point of view, the emphasis was given to reforestation followed by exploitation of forest resources for industrial or household purposes, but at the current time, a strong drive is required for the restoration of harsh, degraded environments by implementing new techniques of planting and seedling production (Oliet and Jacobs, 2012). In this regard, nursery systems are one of the major implementations; these aim to produce good-quality, stress-tolerant vegetation that can adapt in a harsh changing environment, followed by field emplacement (Haase and Davis, 2017). The nutrient content of plant biomass corresponds positively to seedling size, and stresses are mitigated by mobilization of these stored reserves on the seedling site (Uscola et al., 2015).

Technological advancements in programmed site construction have been observed with improved efficiency to control the site limiting factors (i.e., compaction, mounding and target vegetation), environmental impact, costs and safety parameters (Löf, 2017). Continuous application of soil fertilizers, organic input and tree residues (Oliet et al., 2019) will certainly help to improve the regeneration process. While the practical use of fencing is limited by its high installation cost, it is generally considered effective against ungulate damage. Thus, for a cost-effective restoration, it is necessary to evaluate alternatives such as nurse vegetation, tree shelters and afforestation schemes (Maltoni et al., 2019; Burney and Jacobs, 2018).

6.1.10 Biotic and Abiotic Disturbance

Disturbances are a considerable part of forest ecology. Disturbances by either natural or anthropogenic means lead to structural involution (Löf, 2017), as can wind throw and pest or disease outbreaks. Moreover, while forest resource economics play a central role in driving the management of commercial forests, there are still a lot of opportunities to use alternative methods of forestry, i.e., retaining the indigenous vegetation and prescribed burning, exactly mimicking the natural disturbances to promote diversity (Löf, 2017). Destruction of floral and faunal resources due to anthropogenic causes is a major concern in recent times; millions of hectares of degraded forests require to be restored globally (Lindenmayer, 2019).

In Asian regions, for example India, forest fire is a significant and prominent factor that contributes to the forest degradation process, although the extent of total damage needs to be verified.

According to a study during late 1900, during the sixth Five Year Plan (1980–85), 17,852 fire incidents were reported affecting an area of 5.7 million hectares (Saigal, 1989). However, the data provided by the forest survey of India suggest that the area of forest that is affected by consistent incidents of fires may be as high as 37 million hectares (Ministry of Environment, Forest and Climate Change, 1987). In this context, Rodgers (1986) concludes that the non-burning of *Sal* forest in the Dudhwa region of Uttar Pradesh since national park status was granted in 1974 has allowed a dense undergrowth of unpalatable shrubs to develop.

Many fact-finding studies have been done worldwide in the last few decades to explore the effects of fires on vegetation and loss of soil and biomass resources (Verma and Jayakumar, 2012), most of them are conducted in temperate regions. In the Indian context, studies reveal that 90% of forest fire cases are anthropogenic, and an estimated area of approximately 3.73 million hectares of vegetation is affected by forest fires (Srivastava and Garg, 2013).Tropical regions densely packed with dry deciduous forests are observed to be more vulnerable to fire than any other forest across the world (Janzen, 1988). The tropical dry deciduous forest of India is highly affected by fire, followed by moist and semi-evergreen forest.

In 2016, forest fire incidence in the Himalayan region of India, especially in Himachal and Uttarakhand, led to 3500 hectares of land losing its covering fauna and flora (Kavita, 2016). The main fact behind the forest fires in the Indian Himalayan region is the domination of forests by *Pinus* (locally called *Chir*) conifer, containing lignified leaves. The leaves deposited at ground level easily catch fire, though the bark of the tree is resistant to fire. Indian states experience the peak fire incidence during summer from April to June, though the extent and type vary from state to state and with type of forest as well as climatic conditions like a prolonged spell of dry conditions or delay in arrival of the monsoon, etc.(Chandran et al, 2011).

For reconstruction and regeneration, there are plenty of remedial measures and preventative programs for forest fires; these must be implemented by the government. In this regard, a joint forest management programme has also been introduced to counter forest fires using modern methods (Kumar, 2002). The conventional model of making fire lines criss-crossing the whole forest and clearing the deposited lignified biomass before the summer should be practised properly by forest authorities. People's understanding and awareness in this context is crucial. Assessing the impact of community efforts on fire prevention, a survey based on socio-economic and cultural activities demonstrated that people do not actually realize the economic as well as ecological disaster due to fires. Therefore, awareness programmes for motivation and educational strategy will be helpful for prevention. Proper and strategic management is essential for fire prevention. This is based on three common principles: engineering, enforcement and education (Nair, 1992).

Active community participation is important, and it needs to be further augmented with convenient fire prevention and planning techniques like implementing fire plans, generating fire maps, training-based capacity-building programmes followed by implementing mechanisms for fire suppression, and post-fire rehabilitation and management practices. In a soil-based management practice, a model for assessment and identification of plant species that are tolerant and resilient against fire threats along with invasive species should be developed.

According to Silva et al. (2019), new generation plantations have great potential to conserve native forests and their biodiversity; also providing food and fodder, assisting in the mitigation of climate change. A study of a *Eucalyptus regnans* plantation in southeastern Australia demonstrated a classical example of strategic conservation of biodiversity and establishment of effective restoration. The general principles of these studies focused on crucial vegetation and habitats, attributes of their structure, heterogeneity of landscape and ecological interactions (Lindenmayer, 2019). To restrict predation issues, repellents can be successfully implemented. This concept is supported by Villalobos et al. (2019), who revealed reduced predation of acorns and nuts by rodents. Meta-analyses give an exact idea and play a role in the identification of recent approaches across multidisciplinary research and development worldwide.

6.1.11 Soil Restoration in Agro-ecological System

As well as the ecological disturbance to forest soil, agro-ecology is also a major concern in recent times due to the overexploitation of soil for high productivity to feed the constantly growing population. Thus, it is mandatory for agro-ecological practitioners to attempt to implement techniques for sustainable soil health for better quality as well as quantity of crop yield. Soil destruction implies a gradual reduction of soil ecology in terms of its fertility and biology for proper functioning and services of the ecosystem.

Scientifically, we can explore the degradation (Figure 6.4) under the following four categories:

(1) Physical parameters
(2) Chemical composition of soil
(3) Soil biological activities
(4) Ecological factors

The physical degradation of soil results in loss of structural integrity, including pore size, which makes it susceptible to compaction, reduction in water infiltration rate, enhanced runoff, crusting, erosion and constant soil temperature fluctuations. Soil chemical degradation is defined by acidification, increased electrical conductivity (salinization), loss of nutrient dynamics, reduction in cation-exchange capacity value, accumulation of toxic chemicals as well as leachate loss of nitrate (NO_3^-), and Ca or Mg deficiencies. Soil biological components play a key role in nutrient bio-transformation mechanisms (i.e., conversion of unavailable nutrients to plant-available form or dissolved form). Disturbances in the biology of soil cause reduction in soil total organic carbon and soil carbon sink capacity, loss in soil microbial biodiversity, and emission of greenhouse gases into the atmosphere. Greenhouse gas emission (i.e., CO_2 and CH_4) from soil due to disturbed biology is one of the most severe consequences of soil degradation.

In addition, ecological degradation shows a combination of the three aforementioned processes, which leads to disorganization in ecosystem functions such as nutrient cycling, water infiltration and

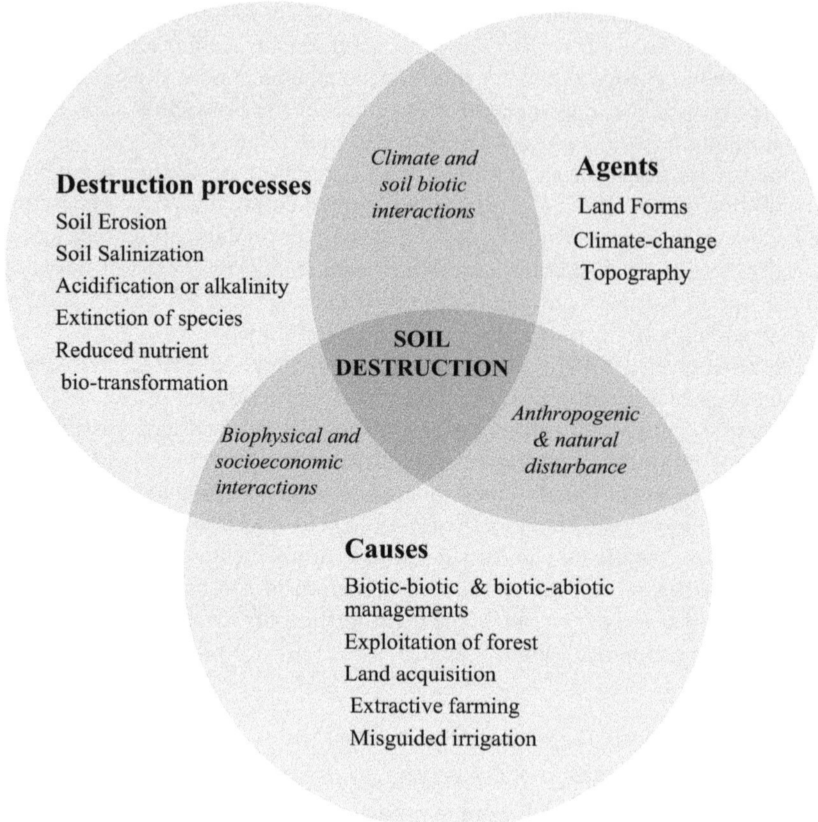

FIGURE 6.4 The processes, factors and causes as drivers of soil degradation.

purification capacity, disruption of the water cycle, and a decline in net biomass productivity. The eventual decline in soil health, either by means of anthropogenic disturbance or by natural causes, has a significant feedback effect, resulting in declining ecosystem services over time. The self-reinforcing soil destruction process is strongly intensified by the interaction among different factors, processes and causes of soil destruction (Figure 6.4). It includes the study of networks (types) of soil degradation. Factors include agents of degradation related to anthropogenic or natural drivers such as climatic conditions, geography and socio-economic parameters.

6.1.12 Soil Quality Index

The soil organic carbon (SOC) pool is considered as a sole indicator of soil characteristics and a prominent driver of sustainable agriculture. Additionally, other parameters, i.e., its distribution along with soil profile, its quality (biological, physical and chemical parameters) and the MRT (mean residence time) in the form of turnover rate are also crucial for determination.

Soil physical quality can be assessed by means of the stability and amount of the aggregate; crusting and compaction susceptibility; pore size and continuity; movement of water through different profiles (infiltration rate and amount); capability to hold water for plant availability; gaseous exchangeability; efficient depth of rooting; heat capacity and the temperature regime. In the same way, soil chemistry can be explored using appropriate indicators, including hydrogen ion concentration, cation exchange capacity, electric conductance, concentration of dissolved or available nutrients, and favourable elemental balance or elemental deficiency. Furthermore, soil biological parameters can be measured by microbial activity or microbial biomass (C, N); assessing the total microbial biomass carbon of soil gives an idea of soil biology, as microbes play a key role in nutrient transformation. Diversity of soil fauna and flora and absence of pathogens and pests indicate its disease-tolerant attributes.

Interestingly, quality indicators of soil may vary with soil types, climatic conditions and land uses. Soil type and climate are both important for favouring the growth of a certain type of organism; bacteria and fungi in particular have specific requirements or niche-micro differentiation.

6.1.13 Conservation Agriculture and Soil Quality

According to Lal (2015), the following are four basic principles of agro-ecosystem conservation:

(1) Application of integrated nutrient management, i.e., use of chemical and bio fertilizer amendments;
(2) Retention of crop waste;
(3) Application of crop rotation and crop cover; and
(4) Removal of soil mechanical disturbances.

In addition, the increased total organic content of soil under conservation agriculture prevents erosion and works as a good management strategy. Soil organic amendments such as farmyard manure and biochar implementation not only fulfil the nutrient requirement and enhance SOC for crop plants but also help in maintaining soil microbiota and other physical properties.

Site-specific management strategies can greatly enhance the SOC along with crop yield. Kimetu (2008), in his study in western Kenya, demonstrated an enhancement of SOC from 20 to 40 Mg/ha in 0.1 m depth, with a grain yield of 3.5 to 4.2 Mg/ha. Tillage systems efficiently restore soil health, reduce soil erosion, ameliorate the retention efficiency of fertilizers along with the rain water, and are more desirable (So et al., 2001), preventing the loss of SOC and nutrients. A curved arrow diagram is represented in Figure 6.5, demonstrating the impacts of conservation agriculture on soil health. The role of soil macro-organisms (e.g., bacteria, fungi, earthworms, termites) acting as a pioneer community to improve soil health has been widely acknowledged for centuries (Darwin, 1881). Conversion of plough tillage to conservation agriculture with crop residue and cover cropping (Figure 6.5) certainly enhances earthworm activity and helps to repair soil properties, with certain impacts on transportation of pollutants into the water reservoirs (Edwards et al., 1988).

6.2 AQUATIC RESTORATION

Due to very high anthropogenic disturbance to cater for the needs of the growing population, the destruction of freshwater as well as marine ecosystems by a variety of stressors has increased gradually over time. As a consequence, the aquatic systems are in urgent need of some strengthening measures for reclamation. This is an integrated process that could not be accomplished by manipulating a single or individual element.

The goal is to achieve a self-regulating and natural system that is associated ecologically with its native physiographical occurrence. However, more precisely, the restoration process must follow one of the following methods:

a) Reconstruction of native physical, chemical conditions;
b) Adjustment between the soil and water relationship; and
c) Manipulations to introduce native flora and fauna (Zedler, 1996).

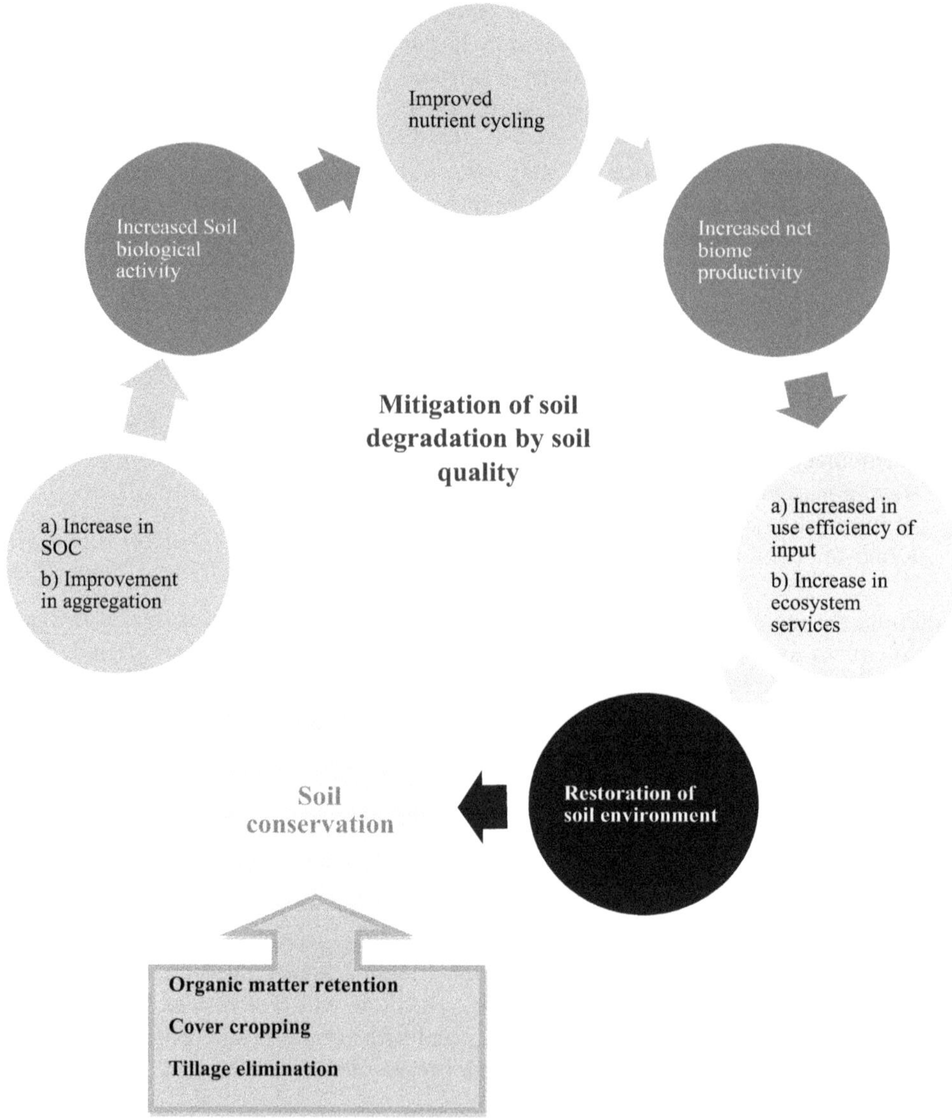

FIGURE 6.5 Sustainable soil adaptability and reduction in soil degradation by conservation agriculture.

Landscape water is considered to have poor self-purification characteristics, poor fluidity and semi-enclosed water, which easily precipitate and accumulate pollutants. According to relevant recent studies, about 93% of the landscape water in the globe experiences varying degrees of pollution and eutrophication (Ding et al., 2005).

Harmful algal blooms are one of the most notorious consequences of eutrophication of fresh water due to heavy contamination of nutrients by the leaching process. They are accompanied by serious issues related to human health, ecological balance, economic stability and landscape aesthetics (Conley, 2009).The ultimate source of landscape water comes from rivers, groundwater and tap water (Li, 2020). Pollutants of rivers and lakes enter the ground through osmosis, thereby polluting groundwater. However, much landscape water has a direct connectivity to the groundwater. In recent years, practitioners have stepped up efforts to control rivers and lakes, but there are still some areas, on which focus is still needed, that have not reached the standards for landscape water, which has caused landscape water to be polluted from the source.

In recent times, eutrophication of lakes, ponds and water reservoirs has caused most damage to freshwater. In addition to the commercial loading of pollutants (i.e., extensive use of fertilizers) from anthropogenic uses, the internal loading of pollutants from sediments is expected to further increase the occurrence of algal blooms and will consistently create pressure on river and lake ecosystems over the coming decades (Zhang et al., 2018).

Because of the lightweight spores of algae, they can be spread by air currents, and when sufficient nutrients are in available form or dissolved content, they grow and reproduce extremely fast. For their growth and reproduction, algae consume a lot of dissolved oxygen from the water body. This ultimately creates an anoxic environment with a black, smelly and muddy appearance, which directly affects the growth of other aquatic life. When the water body is covered by algae, it is quite impossible for aquatic vegetation to perform photosynthesis due to lack or absence of sunlight, and aquatic organisms suffocate due to lack of oxygen.

Algae, in addition to all these impacts, also produce various kinds of toxins, which accumulate in the water body and have harmful effects on human health, especially in young children.

6.2.1 Restoration Strategies

6.2.1.1 Control in External Loading

Exploring the exact point or non-point source of nutrients entering into the lakes is important for eutrophication control. Cao et al. (2018) studied the potency of contaminants in terms of how the quantity of contaminating nutrients affected the quality of water in different types of riparian buffers in a heavily eutrophicated water reservoir. Overall, the herb/shrub and grass riparian wetland construction is strongly recommended for effective removal of dissolved phosphorus and nitrogen.

6.2.1.2 Pre-capture from Water

The removal of dissolved salts, i.e., available phosphorus, from water bodies is a crucial step for the controlling eutrophication. For this purpose, adsorption is one of the most prominent and effective treatments. Lanthanum–chitosan-based magnetic spheres were synthesized by Cheng et al. (2018), who demonstrated their capability under various conditions. Due to their unique porous structures, high adsorption efficiency and low cost, magnetic spheres (lanthanum–chitosan based) are potentially applicable for eutrophicated water bodies.

Another lanthanum/aluminium hydroxide composite adsorbent was introduced and demonstrated by Pan et al. (2020) in a simulated pilot-scale river–lake system. This strategy to adsorb P from contaminated water was demonstrated to maintain its concentration below 10 μg/L.

6.2.1.3 Bioremediation Methods

The remediation process using microorganisms is based on the application of naturally occurring or easily cultivated microorganisms. Microbial-mediated transformation or degradation of pollutants within a controlled range helps to remove toxic substances from water or convert them into harmless substances, thereby repairing polluted water bodies (Li et al., 2009). This process involves careful screening and in-vitro culturing of the microbes in optimum growing conditions, followed by trials. These methods can be principally categorized into the following: preparing biofilms, inoculation, protozoan-based method, and immobilized microorganisms.

In practical models, the quality and quantity of inoculants is relatively standard and chiefly focused on additional functional inoculants that assist in complete degradation of pollutants in the water body (i.e., nitrifying bacteria, denitrifying bacteria, and lactic acid bacteria). Some researchers demonstrated that manipulating nitrification and denitrification processes with the help of microbes (i.e., *Bacillus subtilis*) eventually released the relevant pollutants from water bodies (Miuet al., 2013). Inoculating the microbes in water with low-carbon, nitrogen than landscape waters has observed with faster and better removal efficiency.

Gan et al. (2019), came up with amazing findings demonstrating the removal of nitrate from polluted water in about 5 hours. They used polyurethane foam as a carrier to immobilize microorganisms and demonstrated its denitrification capability in polluted water bodies.

Recently, the methods of native microorganism activation and immobilization have also been developed and implemented successfully. The activation of indigenous microbes will help the proliferation of native and beneficial microflora by adding some specific nutrients, biological enzymes, electrons or other suitable substrates to the water body, and provide optimum growing conditions for proper proliferation of microorganisms to degrade pollutants so as to achieve water body restoration (Li et al., 2009). The immobilized microbes method is based on biological materials, immobilizing bacterial groups on certain carriers to increase the density of the bacteria and enhance the protection of the functional bacteria.

Wang et al. (2018) isolated, screened and immobilized *p*-nitrophenol-degrading bacteria from activated sludge and used them to improve the degradation efficiency of *p*-nitrophenol in wastewater. The findings of this study suggest that at a controlled temperature (35°C), pH (9–12) and concentration of *p*-nitrophenol (£150 mg/L), the degradation rate of mixed contaminants is as high as about 80%. Polyurethane can also be used as a carrier for immobilization; in this regard, Anselmo and Novais (1992) used it to immobilize *Fusarium lanuginose* and carried out continuous in-vitro phenol degradation experiments.

6.3 CLIMATE CHANGE AND ECOLOGICAL RECONSTRUCTION APPROACHES

Profound ecological change is also caused by changes in the ocean and the atmosphere. The thin living film of life on the globe is intimately associated with the hydrosphere and atmosphere, which produce the sustaining fabric within which human society survives. Therefore, degradation and restoration of characteristics of the ecology are likely to have geographical or worldwide consequences. The resilience and viability of native ecosystems are increasingly threatened by anthropogenic greenhouse gas release, resulting in both ocean acidification and climate change, as are the human societies that depend upon them. The results of these threats can be extreme and in the last few years, have been observed to be gradually increasing. So far, Earth is committed to a significantly warmer climate, with an assumption of additional warming in forthcoming years, without a decisive change in carbon emissions trajectories(Hoegh-Guldberg, 2018). Scientific research continues to concentrate on understanding the climate system of Earth and its interrelationship with ecology. Impacts related to climate are being witnessed earlier than expected and appear to be increasing in frequency and severity. A number of potential climate tipping points are beginning to intimate the initial signs of activation (Lenton et al., 2019). According to the International Panel on Climate Change (IPCC) in the 2018 Special Report on 1.5°C (www.ipcc.ch/sr15/), allowing the Earth to warm beyond 1.5°C will result in climate change impacts, together with floods, heat waves, drought and increased sea-level, that are harmful for humankind and for the ecosystem. While the earliest internationally accepted target was 2°C, this difference of 0.5°C could decrease the threat of substantial degradation of coral reefs and Arctic ecosystems. A 1.5°C extreme warming target implies that the planet has approximately 12 years to reduce Earth's net carbon release by half to escape the most important impacts, but even if this purpose is accomplished, potential warming impacts may possibly extend for decades or even centuries (Nicholls et al., 2018), according to discussion at the Royal Society National Academy of Sciences Forum held in 2018 on Climate Change and Ecosystems. The Forum was mutually organized by the two societies, and the objectives were to

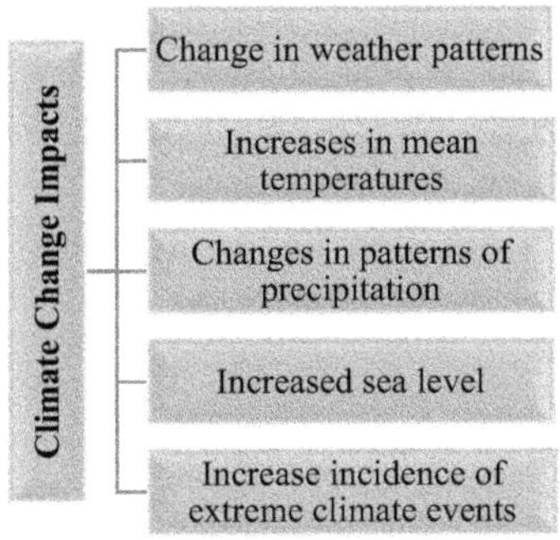

FIGURE 6.6 Major impacts of climate change.

create new opportunities for international collaboration, underline the current research findings on the main topics, analyse research gaps and future research preferences, and consider how research in this discipline may illuminate international policy (National Academy of Sciences, 2019). The Forum explores the current science on how a changing climate can influence aquatic, marine and terrestrial ecosystems, frequently in relation to other parameters. The Forum examined opportunities to serve and administer ecosystems to maximize both their resilience and societal resilience in a changing climate by researching a number of policy dimensions and sciences. This comprises how ecosystems can best be administered to enhance their resilience to a changing climate, their potential for modification under a changing climate, and how management of ecosystems can be an approach for more general adaptation to change. Therefore, a principal focus was to examine how to manage the ecosystem, and restoration has the ability to provide "nature-based solutions" (NBS) to tackle both the consequences and the causes of changing climate. However, the advantages, scalability and amplitude of various nature-based strategies need to be examined and better understood and evaluated (Bennett et al., 2016).

These changes will probably be sudden and uncertain as to timing and intensity (Figure 6.6). However, it is logical that even if instant, collaborative and decisive action is taken, substantial and remarkable changes are inevitable in the next two to three decades (Hulme, 2005; King, 2005).

6.3.1 Threats of Climate Change: A Challenge to Ecosystem Management

Changes in ecosystems threaten global biodiversity and have implications for worldwide food production. This section covers the advances in our understanding of the effects of climate change on ecosystem properties (trophic webs or energy flux, nutrient cycling or material flux, biological diversity) in various ecological communities (terrestrial soil microbes, terrestrial plants, invertebrates in marine sediments).

The constantly changing climate is responsible for the variation and instability of a particular ecosystem up to a certain level. However, it is quite difficult for practitioners to assess definite abrupt changes in ecological systems (ACES), because excessive interactions are, by their nature, probabilistic and seldom predictable. The current findings are directing researchers to explore and anticipate ACES with respect to climate change as an extreme priority (Turner et al., 2020). These analyses, associated with the main generalities that lead to examinations and hypotheses for upcoming research, are: Some aspects of ecological systems are more susceptible to abrupt change than others; climate change may be more likely than the desired trends to cause abrupt change (e.g., coral bleaching is instigated through maximum heat waves instead of moderate ocean warming); numerous drivers frequently combine to produce ACES (e.g., climate change-driven drought and intense fire can cause abrupt disturbance to communities of sensitive species); ancient abrupt events (ecological heritage, sequence and frequency of disturbance, conceptual context) are principal

ACES drivers due to ecosystem memory; and powerful productive feedbacks in an ecosystem can frequently cause constant state changes at critical transitions (tipping points).

Researchers also examined extreme climates and ancient contingencies, incorporating a modern understanding of aspects of subterrestrial ecological populations that make them resistant, vulnerable or flexible to climate extremes (Bardgett and Caruso, 2020). Soil microbiota are considered as a pioneer in moderating biogeochemical cycling. Central intrinsic features of these communities that possess resilience involve life-history strategy (regulation of resource use, growth rate) and diversity of microbial food web (slow and rapid forms of energy channels in fungal versus bacterial food webs). Rapid energy channels (e.g., bacteria in a soil context) allow fast nutrient recycling and rapid recovery from disturbances, thus allowing resilience to change, while slow energy channels (e.g., fungi) in a slow nutrient cycle lower the feedback to perturbations and thus confer resistance to change. The harmonizing functions of these two energy channels can simplify rapid yet steady restoration from perturbations, and contrarily, changes of the respective effects of these channels can cause impairment in the ecosystem. Extrinsic attributes involve changeability in an environment and contributions that the plant community makes to soil moisture, carbon and nutrients. Whereas the response of subterrestrial communities under sustained pressure is adequately understood, the authors determined the responses to extreme climates, and the potential for abrupt ecological change, as evaluative gaps in knowledge that should be addressed by experimentation.

Eventually, a changing climate drives loss of terrestrial biodiversity and influences the carbon storage in ecosystems either directly or indirectly through land use change, i.e., climate-driven changes in cropland development. The hypothesis is supported by Molotoks et al. (2020), who applied a modelling approach to survey variability in estimations of loss of carbon sinks and biodiversity. The findings suggest that large variability related to land use projections and future cropland development will probably have adverse impacts on biodiversity and carbon storage in various biodiversity hotspots, including Amazonia, the Congo Basin and Mexico. This study underlines the significance of involving indirect results through changes in land use when evaluating the total carbon and biodiversity impacts of climate change.

A thought-provoking paper was published by Harrison (2020), who anticipated that diversity in terrestrial plant community will be destroyed more than it is increased by a warmer climate, and carried out the experimental work to analyse this anticipation. This research alerts from the ongoing evidence, which indicates that a warm climate might generally increase diversity in temperate latitudes but this may hardly universalize because an emphasis of studies has arisen in the specific and uncommon context of north-temperate alpine ecosystems. She predicted that a net dropping of diversity will prevail in water-limited ecosystems, and that losses will also occur in temperature-limited systems without direct topographical gradients where potential replacement species of pools are not found nearby.

6.3.2 Opportunities to Improve Ecological Restoration

The key consideration is scientific observation of the opportunities to encourage and regulate ecosystems in order to increase societal and/or ecological resilience to ocean acidification and climate change, together with novel approaches to ecological restoration and conservation.

Experimental data from Thomas (2020) suggests a novel view of conservation approaches to biodiversity in a planet where the ecology is fundamentally transformed by human activity. Profound biological processes, unaffected by human actions, form a design for understanding the ecosystem responses to world change when human activity is quickly eliminated, included and shifted around populations, species and genes. These developing and ecological processes resume in an altered human world where novel ecological communities comprise populations, species and genes that are completely equivalent to the human-altered environment. The findings excitedly insist that facilitating instead of recapturing the entry of new species and genes that contribute advantages is an effectual conservation strategy in the Anthropocene. This research conclusively recommends greater concentration on connectivity or "trans-situ" conservation, facilitating the transport of species and genes to locations where they might be able to develop instead of in the rapidly changing world.

Much of the literature on environment-based approaches to the mitigation and adaptation of climate tends to focus on entirely terrestrial ecosystems (e.g., peatlands and forests) or terrestrial-coastal systems (e.g., salt marshes and mangroves). By contrast, few current studies are based on climate adaptation potential and mitigation strategies of soft-sediment benthic marine ecosystems (Solan et al., 2020). These are a huge habitat on the planet, they host most of the biodiversity, and the flora and fauna in the benthic zone can play a crucial role in modulating marine chemistry, climate-active gases and biogeochemical cycling, and subsequently remove the carbon from the marine-atmospheric system. Contrarily, vital conservation of the main areas in a network can extensively increase seascape-scale ecological conservation resilience and ecosystem services. Network integration of benthic covered zones is a principal aspect when considering extensive restoration to climate change, but questions remain regarding how to reach the measures scalability of benthic-based mitigation.

Likewise, Roberts et al. (2020) reported the collaborative potential between mitigation and conservation of marine biodiversity, including adaptability to climate change. Conservation frequently strengthens the capability to maintain the carbon level of ecosystems and also to sequester extra carbon, along with ecological restoration approaches to climate change. Current work has emphasized the role of marine megafauna in providing enhancement to the vertical transport of nutrients (physical mixing, cetacean deep-feeding and surface defecation) and thus, improvement in the fertility of the ocean and wide-ranging carbon sequestration. The authors call for an enlargement of marine covered zones from the recent 10% of sea zones in the Aichi biodiversity targets to 30% of sea zones to assist this phenomenon.

A case study by Lawler et al. (2020) also considered an ideally protected zone connectivity in status of climate change, evaluating the composition charges of a terrestrial conservation ecosystem network for the adjacent United States that appraise both projected and modern distributions of biodiversity on the basis of climate change scenarios. These findings suggest that the composition of the covered zone network changes significantly in the light of climate change, and also that the extra costs of designing for climate change may be trivial compared with the costs of developing the inventory network without considering climate change. Appropriately, conservation of some kinds of climate-change refugia may be a reasonable conservation strategy. It can be concluded that the maximum altitude influence of covered zones, which has been observed as uncertain for conservation, may contribute benefits with regard to climate change by conserving climate-change refugia (Lawler et al., 2020).

6.3.3 Microclimate: A Recent Adaptation Strategy

Microclimate is the local set of climatic conditions regulated in limited areas adjacent to the Earth's surface. These variables of the environment comprise wind speed, light, temperature, humidity and moisture content, which supply substantial indicators for selection of habitat and other ecological actions (Mislan and Helmuth, 2008). Wanga et al. (2020), using a satellite dataset, established a system to observe the changes in microclimate and vegetation in recovered landfills. Estimation of restoration process, its potency along with ecological indicators was innovated with endemic species distance and restoration. Collectively, with effective restoration, evapo-transpiration and vegetation were enhanced and surface temperature reduced.

The soil biotic components of the forest ecosystem experience variation in climatic exposure, which deviates considerably from the climatic conditions outside the forest (De Frenne, 2019). To a certain extent, changes in environmental conditions (i.e., under dense forest canopies, direct sunlight and wind speed are reduced) lead to a dampening of temperature and humidity variations. In a particular ecosystem, microclimate is of supreme importance and plays a key role in the regulation and functioning of the ecosystem. Microbial activity, i.e., rates of organic matter decomposition, carbon stock and nutrient dynamics, tends to be greater in forests than in neighbouring open habitats. Microclimate shows direct impacts on an organism's physiology, metabolism, behaviour and fitness. However, the impact may vary within the organism's community and depends on its mobility, lifespan, size and life cycle processes.

Recently, practitioners have paid much attention to the role of microclimate in the sustainability of local biodiversity and processes towards "microclimate forest restoration", or forest ecosystems restoration, with the obvious purpose of increasing their capacity to protect the local microclimates from climatic change. In this context, with gradually increasing sites of microclimate monitoring, novel microclimate modelling approaches have been introduced (Maclean, 2020). These vital recent advancements are intended to encourage the use of microclimate assessment studies instead of data-based long-term conditions. Once the observations of global microclimate variation are documented and analysed, more effort should be devoted to rolling out this information into further proper assessment of ecosystem functioning.

6.4 CONCLUSION

The comprehension of restoration processes is somewhat dependent on the recognition of services provided by an ecosystem, and accordingly, legal laws and reforms could be planned, implemented and executed to attain pre-determined goals. To achieve this, more advanced technology and practices should be applied to the restoration of soil, water, forest and other natural capital. At present, every form of natural capital is experiencing destruction and degradation, be it soil, forest, water, etc. Moreover, climate change, one of the major causes of this loss, in turn results from various anthropogenic activities. The changing climatic conditions in the past few decades have resulted in great devastation to natural capital and human life as well. This chapter elaborates a hypothetical model for ecosystem services and reconstruction of disturbed ecology. It clearly depicts the destruction caused to each and every form of natural capital and provides possible remedial measures to restore (rejuvenating, revegetating and revamping) our natural heritage. Large-scale and long-term restoration helps to improve the sustainability of ecological functioning, biodiversity and human well-being. Ecosystem services certainly provide a framework to access the types of restorative interventions required to target different forms and degrees of degradation and to achieve goals related to ecosystem management and its delivery to people. Moreover, we should adapt the concept of 3R, i.e., reduce, reuse and recycle, to avoid 3D, i.e., damage, destruction and degradation. To sum up, we can recall the quote of a great scholar and profound thinker:

> The natural world is the larger sacred community to which we belong. To be alienated from this community is to become destitute in all that makes us human. To damage this community is to diminish our own existence.
>
> Thomas Berry

ACKNOWLEDGEMENTS

We thank Dr. N. S. Bisht (Professor and Former Dean, school of life sciences), HNB Garhwal University, for their valuable constructive guidance and Dr. S. K. Malhotra (Professor and Head, Department of Zoology, Allahabad University) for sharing pearls of wisdom during the course of this study. We are immensely grateful to the professors, researchers and scientists working in the field of restoration ecology, whose splendid work is cited in this study.

REFERENCES

Alday JG, Marrs RH, and Martinez CR (2012). Soil and vegetation development during early succession on restored coal wastes: a six-year permanent plot study. *Plant Soil* 353:305–320. https://link.springer.com/article/10.1007/s11104-011-1033-2

Alexander S et al. (2011). Opportunities and challenges for ecological restoration within REDD. *Restor Ecol* 19:683–689. http://dx.doi.org/10.1111/j.1526-100x.2011.00822.x

Alexander SJ, Aronson, Whaley O, and Lamb D (2016). The relationship between ecological restoration and the ecosystem services concept. *Ecol Soc* 21:34. http://dx.doi.org/10.5751/ES-08288-210134

Anselmo AM and Novais JM (1992). Degradation of phenol by immobilized mycelium of *Fusarium flocciferum* in continuous culture. *Water Sci & Technol* 25:161–168. https://doi.org/10.2166/wst.1992.0024

Aronson J and Alexander S (2013). Ecosystem restoration is now a global priority: time to roll up our sleeves. *Restor Ecol* 21:293–296. http://dx.doi.org/10.1111/rec.12011

Aronson JS, Milton J, and Blignaut JN (2007). *Restoring Natural Capital: Definitions and Rationale*. Island Press, Washington, DC. http://hdl.handle.net/10019.1/41029

Bardgett RD and Caruso T (2020). Soil microbial community responses to climate extremes: resistance, resilience and transitions to alternative states. *Philos Trans R Soc Lond B Biol Sci* 375:1794. https://doi.org/10.1098/rstb.2019.0112

Barnes P (2008). *Climate Solutions: A Citizen's Guide*. Chelsea Green Publishers, White River Junction, VT. https://peter-barnes.org/book/climate-solutions/

Barrios E (2007). Soil biota, ecosystem services and land productivity. *Ecol Econ* 64:269–285. https://doi.org/10.1016/j.ecolecon.2007.03.004

Bennett EM et al. (2016). Bright spots: seeds of a good Anthropocene. *Front Ecol & Environ* 14(8):441–448. https://doi.org/10.1002/fee.1309

Birdsey RA, Pregitzer K, and Lucier A (2006). Forest carbon management in the United States. *J Environ Qual* 35(4):1461–1469. https://doi.org/10.2134/jeq2005.0162

Bruno JF, Stachowicz JJ, and Bertness MD (2003). Inclusion of facilitation into ecological theory. *Trends Ecol Evol* 18(3):119–125. https://doi.org/10.1016/S0169-5347(02)00045-9

Bunemann EK, Bongiorno G, Bai G, Creamer RE, Deyn GD, Goede RD, and Brussaard L (2018). Soil quality-a critical review. *Soil Biol & Biochem* 120:105–125. https://doi.org/10.1016/j.soilbio.2018.01.030

Burney OT and Jacobs DF (2018). Species selection—a fundamental silvicultural tool to promote forest regeneration under high animal browsing pressure. *Forest Ecol Manag* 408:67–74. www.forestfireindia.org/urls/fires_india.html

Cao X, Song C, Xiao J, and Zhou Y (2018). The optimal width and mechanism of riparian buffers for storm water nutrient removal in the Chinese eutrophic Lake Chaohu watershed. *Water* 10(10):1489. https://doi.org/10.3390/w10101489

Cernansky R (2018). How to rebuild a forest. *Nature* 560:542–544. https://doi.org/10.1038/d41586-018-06031-x

Chandran M, Sinha AR, and Rawat RBS (2011). Replacing controlled burning practice by alternate methods of reducing fuel load in the Himalayan long leaf pine (Pinus *r*oxburghii Sarg.) forests. 5th International Wildland Fire Conference, Sun City, South Africa, 9–13.

Chang CC and Turner BL (2019). Ecological succession in a changing world. *J Ecol* 107(2):503–509. https://link.springer.com/article/10.1007/s11629-017-4541-6

Cheng R, Shen LJ, Zhang YY, Dai DY, Zheng X, Liao LW, Wang L, and Shi L (2018). Enhanced phosphate removal from water by honeycomb-like microporous lanthanum-chitosan magnetic spheres. *Water* 10(11):1659. https://doi.org/10.3390/w10111659

Christensen NL et al. (1996). The report of the Ecological Society of America. Committee on the scientific basis for ecosystem management. *Ecol App* 6(3):665–691. https://doi.org/10.2307/2269460

Clewell AF and Aronson J (2013). *Ecological Restoration: Principles, Values, and Structure of an Emerging Profession.* Second edition. Island Press, Washington, DC. http://dx.doi. org/10.5822/978-1-59726-323-8

Conley DJ (2009). Controlling eutrophication: nitrogen and phosphorus. *Science* 323(5917):1014–1015. http://dx.doi.org/10.1126/science.1167755

Convention on Biological Diversity (CBD) (2012). UNEP/CBD/COP Decision XI/16. Ecosystem Restoration. Convention on Biological Diversity, Hyderabad, India. www.cbd.int/doc/decisions/cop-11/cop-11-dec-16-en.pdf

Costanza R and Daly HE (1992). Natural capital and sustainable development. *Conserv Biol* 6(1):37–46. http://dx.doi.org/10.1046/j.1523-1739.1992.610037.

Cote SD, Rooney TP, Tremblay JP, Dussault C, and Waller DM (2004). Ecological impacts of deer overabundance. *Annu Rev Eco Evol & Syst* 35:113–147. http://dx.doi.org/10.1146/annurev.ecolsys.35.021103.1005725

Darwin CR (1881). *The Formation of Vegetable Mould, through the Action of Worms, with Observations on their Habitats*. John Murray, London, UK.

De Frenne P (2019). Global buffering of temperatures under forest canopies. *Nat Ecol Evol* 3:744–749. www.nature.com/articles/s41559-019-0842-1

De Fries RS, Rudel T, Uriarteb M, and Hansen M (2010). Deforestation driven by urban population growth and agricultural trade in the twenty-first century. *Nature Geosci* 3(3):178–181. www.nature.com/articles/ngeo756

de Groot RB et al. (2010). Integrating the ecological and economic dimensions in biodiversity and ecosystem service valuation. In *The Economics of Ecosystems and Biodiversity: Ecological and Economic Foundations,* Ed. Pushpam Kumar, 2–37. Earthscan, London, UK.

Diaz S, Demissew S, Joly C, Lonsdale WM, and Larigauderie A (2015). A Rosetta stone for nature's benefits to people. *PLoS Bio* 13(1):e1002040. http://dx.doi.org/10.1371/ journal.pbio.1002040

Ding L, Shen YL, and Huang Y (2005). Rehabilitation technology and development status of water bodies in the park. *Journal of Suzhou University of Science and Technology (Engineering Technology Edition)*, 18(2):48–52.

Dominati E, Patterson M, and Mackay A (2010). A framework for classifying and quantifying the natural capital and ecosystem services of soils. *Ecol Econ* 69(9):1858–1868. https://doi.org/10.1016/j.ecolecon.2010.05.002

Edwards WM, Shipitalo MJ, and Norton LD (1988). Contribution of macroporosity to infiltration into a continuous corn no-tilled watershed: implications for contaminant movement. *J Contam Hydrol* 3(2–4):193–205. https://doi.org/10.1016/0169-7722(88)90031-9

Falk DA, Palmer M, and Zedler J (2006). *Foundations of Restoration Ecology*. Island Press, Washington, DC. https://islandpress.org/books/foundations-restoration-ecology

Felton AM, Felton A, Raubenheimer D, Simpson SJ, Krizsan SJ, Hedwall PO, and Stolter C (2016). The nutritional balancing act of a large herbivore: an experiment with captive moose (*Alces alces* L). *PLoS ONE* 11(3):e0150870. https://doi.org/10.1371/journal.pone.0150870

Foster DR and Aber JD (2004). *Forests in Time: The Environmental Consequences of 1,000 Years of Change in New England.* Yale University Press, New Haven, USA. www.cabdirect.org/cabdirect/abstract/20043159486

Gan Y, Zhao Q, and Ye Z (2019). Denitrification performance and microbial diversity of immobilized bacterial consortium treating nitrate micro-polluted water. *Biores Technol* 281:351–358. https://doi.org/10.1016/j.biortech.2019.02.111

Gibbs HK et al. (2010). Tropical forests were the primary sources of new agricultural land in the 1980s and 1990s. *Proc Natl Acad Sci USA* 107(38):16732–16737. https://doi.org/10.1073/pnas.0910275107

Gomez L (2009). The role of facilitative interactions in the restoration of degraded ecosystems: a meta-analysis across life-forms and ecosystems. *J Ecol* 97(6):1202–1214. https://doi.org/10.1111/j.1365-2745.2009.01573.x

Gomez L, Zamora R, Gomez JM, Hodar JA, Castro J, and Baraza E (2004). Applying plant facilitation to forest restoration: a meta-analysis of the use of shrubs as nurse plants. *Ecol Appl* 14(4):1128–1138. https://doi.org/10.1890/03-5084

Gu X et al. (2019). Vegetation restoration stimulates soil carbon sequestration and stabilization in a subtropical area of Southern China. *Catena* 181:104098. https://doi.org/10.1016/j.catena.2019.104098

Guerry AD et al. (2015). Natural capital and ecosystem services informing decisions: from promise to practice. *Proc Natl Acad Sci USA* 112(24):7348–7355. http://dx.doi.org/10.1073/pnas.1503751112

Haase DL and Davis AS (2017). Developing and supporting quality nursery facilities and staff are necessary to meet global forest and landscape restoration needs. *Reforesta* 4:69–93. https://doi.org/10.21750/REFOR.4.06.45

Haines YR and Potschin M (2009). *The Links between Biodiversity, Ecosystem Services and Human Well-being in Ecosystem Ecology: A New Synthesis,* 110–139. BES Ecological Reviews Series, Ed. Devid G Raffaeli, and Christopher LJ Frid, Cambridge University Press, Cambridge. https://doi.org/10.1017/CBO9780511750458.007

Harrison S (2020). Plant community diversity will decline more than increase under climatic warming. *Philos Trans R Soc Lond B Biol Sci* 375(1794):20190106. https://doi.org/10.1098/rstb.2019.0106

He Q, Bertness MD, and Altieri AW (2013). Global shifts towards positive species interactions with increasing environmental stress. *Ecol Lett* 16(5):695–706. https://doi.org/10.1111/ele.12080

Heneghan L, Miller SP, and Baer S (2008). Integrating soil ecological knowledge into restoration management. *Restor Ecol* 16(4):608–617. https://doi.org/10.1111/j.1526-100X.2008.00477.x

Hobbs RJ and Norton DA (1996). Towards a conceptual framework for restoration ecology. *Restor Ecol* 4(2):93–110. http://dx.doi.org/10.1111/j.1526-100X.1996.tb00112.x

Hobbs RJ, Higgs E, and Harris JA (2009). Novel ecosystems: implications for conservation and restoration. *Trends Ecol Evol* 24(11):599–605. https://doi.org/10.1016/j.tree.2009.05.012

Hoegh-Guldberg O (2018). *Global Warming of 1.5°C. An IPCC Special Report on the Impacts of Global warming of 1.5°C above Pre-industrial Levels and Related Global Greenhouse Gas Emission Pathways, in the Context of Strengthening the Global Response to the Threat of Climate Change, Sustainable Development, and Efforts to Eradicate Poverty*. IPCC, Geneva, Switzerland.

Holl KD, Crone EE, and Schultz CB (2003). Landscape restoration: moving from generalities to methodologies. *BioScience* 53(5):491–502. http://dx.doi.org/10.1641/0006-3568 (2003)053[0491:lrmfgt]2.0.co;2

Huang FF, Zhang WQ, Gan XH, Huang YH, Guo YD, and Wen XY (2018). Changes in vegetation and soil properties during recovery of a subtropical forest in South China. *J Mount Sci* 15:46–58. https://doi.org/10.1111/1365-2745.13132

Hulme PE (2005). Adapting to climate change: is there scope for ecological management in the face of a global threat? *J Appl Ecol* 42(5):784–794. https://doi.org/10.1111/j.1365-2664.2005.01082.x

Janzen DH (1988). Tropical dry forests: the most endangered major tropical ecosystem. In *Biodiversity*, Ed. EO Wilson, 130–137. National Academic Press, Washington.

Jordan WR, Gilpin ME, and Aber JD (1987). *Restoration Ecology: A Synthetic Approach to Ecological Restoration*. Cambridge University Press, New York.

Kavita U (2016). Uttarakhand battles fire crisis. The Hindu. Dehradun. 30 April 2016.

Kimetu J et al. (2008). Reversibility of soil productivity decline with organic matter of differing quality along a degradation gradient. *Ecosystems* 11:726–739. https://link.springer.com/article/10.1007/s10021-008-9154-z

King D (2005). Climate change: the science and the policy. *J Appl Ecol* 42(5):779–783. https://doi.org/10.1111/j.1365-2664.2005.01089.x

Kumar C (2002). Community involvement in forest fire prevention and control: lessons from Joint Forest Management (JFM). *Int Forest Fire News* 26:28–31. https://gfmc.online/iffn/country/in/in_6.html

Lal R (2015). Sequestering carbon and increasing productivity by conservation agriculture. *J Soil Water Conserv* 70(3):55A–62A. https://doi.org/10.2489/jswc.70.3.55A

Lal R and Stewart BA (1992). Need for land restoration. In *Soil Restoration*, Ed. R Lal and BA Stewart, 1–11. Springer, New York. https://link.springer.com/chapter/10.1007/978-1-4612-2820-2_1

Lamb D (2011). *Regreening the Bare Hills: Tropical Forest Restoration in the Asia–Pacific Region*. Springer, Dordrecht, Netherlands. http://dx.doi.org/10.1007/978-90-481-9870-2.

Lawler JJ et al. (2020). Planning for climate change through additions to a national protected area network: implications for cost and configuration. *Philos Trans R Soc Lond B Biol Sci* 375(1794):20190117. https://doi.org/10.1098/rstb.2019.0117

Lenton TM et al. (2019). Climate tipping points-too risky to bet against. *Nature* 575:592–595. www.nature.com/articles/d41586-019-03595-0

Li PP (2020). Analysis of water quality control strategy and application of landscape water body. *Housing & Real Est* 24:60–61.

Lindenmayer DB (2019). Integrating forest biodiversity conservation and restoration ecology principles to recover natural forest ecosystems. *New Forests* 50:169–181. https://doi.org/10.1007/s11056-018-9633-9

Löf M (2017). *Program and Book of Abstracts of the IUFRO 3rd Restoring Forest: Regeneration and Ecosystem Function for the Future*. Lund, Sweden, 12–14 September 2017. Report 51. Swedish University of Agricultural Sciences, Southern Swedish Forest Research Centre, Alnarp, 109.

Lubchenco J et al. (1991). The sustainable biosphere initiative: an ecological research agenda: a report from the Ecological Society of America. *Ecology* 72(2):371–412.

MacCleery DW (1993). *American Forests: A History of Resiliency and Recovery, FS-540*. U.S. Department of Agriculture, Forest Service; Forest History Society, Durham.

Maclean IMD (2020). Predicting future climate at high spatial and temporal resolution. *Glob Change Biol* 26(2):1003–1011. https://onlinelibrary.wiley.com/doi/abs/10.1111/gcb.14876

Maltoni A, Mariotti B, Tani A, Martini S, Jacobs DF, and Tognetti R (2019). Natural regeneration of *Pinus pinaster* facilitates *Quercus ilex* survival and growth under severe deer browsing pressure. *For Ecol Manag* 432:356–364. https://doi.org/10.1016/j.foreco.2018.09.045

Martin DM (2017). Ecological restoration should be redefined for the twenty-first century. *Restor Ecol* 25(5):668–673. https://doi.org/10.1111/rec.12554

Merlin M, Leishman F, Errington RC, Pinno BD, and Landhausser SM (2019). Exploring drivers and dynamics of early boreal forest recovery of heavily disturbed mine sites: a case study from a reconstructed landscape. *New Forests* 50(2):217–239. https://doi.org/10.1007/s1105 6-018-9649-1

Millennium Ecosystem Assessment (2005). *Ecosystems and Human Well-being*. Island Press, Washington, DC. www.bioquest.org/wp-content/blogs.dir/files/2009/06/ecosystems-and-health.pdf

Ministry of Environment, Forest and Climate Change (1987). Annual report 187-1988, Government of India. https://moef.gov.in/wp-content/uploads/2017/09/AR-87-88.pdf

Mislan KAS and Helmuth B (2008). Microclimate. In Jorgensen SE and Fath B (Eds.), *Encyclopedia of Ecology* 3:2389.

Molotoks A et al. (2020). Comparing the impact of future cropland expansion on global biodiversity and carbon storage across models and scenarios. *Philos Trans R Soc Lond B Biol Sci* 375(1794):20190189. https://doi.org/10.1098/rstb.2019.0189

Nair KK (1992). *Prevention of Forest Fires. National Seminar on Forest Fires Report.*

Nicholls RJ et al. (2018). Stabilization of global temperature at 1.5 C and 2.0 C: implications for coastal areas. *Philos Trans R Soc A* 376(2119):20160448. https://doi.org/10.1098/rsta.2016.0448

Nilsen R (1991). *Helping Nature Heal: An Introduction to Environmental Restoration/Whole Earth Catalog*. Ten Speed Press, Berkeley, CA. https://doi.org/10.1007/s1105 6-018-9649-1

Oliet JA and Jacobs DF (2012). Restoring forests: advances in techniques and theory. *New Forests* 43(5):535–541. https://link.springer.com/article/10.1007/s11056-012-9354-4

Oliet JA, Blasco R, Valenzuela P, Melero MB, and Puértolas J (2019). Should we use meshes or solid tube shelters when planting in Mediterranean semiarid environments? *New Forests* 50(2):267–282. https://doi.org/10.1007/ s1105 6-018-9659-z

Pan M et al. (2020). Synergistic recapturing of external and internal phosphorus for in situ eutrophication mitigation. *Water* 12(1):2. https://doi.org/10.3390/w12010002

Rees WE (1995). Cumulative environmental assessment and global change. *Environ Impact Assess Rev* 15(4):295–309. http://dx.doi.org/10.1016/0195-9255(95)00029-E.

Roberts CM, Leary BC, and Hawkins JP (2020). Climate change mitigation and nature conservation both require higher protected area targets. *Philos Trans R Soc Lond B Biol Sci* 375(1794):20190121. https://doi.org/10.1098/rstb.2019.0121

Rodgers WA (1986). The role of fire in the management of wildlife habitat – a review. *Indian Forester* 112(10):145–157.

Saigal PM (1989). A Suggested Classification of Forest Fires in India by Types and Causes. National Seminar on Forest Fire Fighting, Kulamaru (Kerala), 2–3 November 1989.

SER (Society for Ecological Restoration) Science and Policy Working Group (2004). *SER International Primer on Ecological Restoration*. Society for Ecological Restoration, Washington, DC.

Silva LN, Freer SP, and Madsen P (2019). Production, restoration, mitigation: a new generation of plantations. *New Forests* 50:153–168. https://doi.org/10.1007/s1105 6-018-9644-6

So H, Kirchhof G, Bakker R, and Smith G (2001). Low input tillage/cropping systems for limited resource areas. *Soil Tillage Res* 61(1–2):109–123. https://doi.org/10.1016/S0167-1987(01)00182-9

Society for Ecological Restoration (1993). Ecological restoration, conservation and environmentalism: mission statement and environmental policies of the Society for Ecological Restoration. *Restor Ecol* 1:206–207.

Society of American Foresters (1993). *Sustaining Long-Term Forest Health and Productivity*. Society of American Foresters, Bethesda, MD.

Solan M, Bennett EM, Mumby PJ, Leyland J, and Godbold JA (2020). Benthic-based contributions to climate change mitigation and adaptation. *Philos Trans R Soc Lond B Biol Sci* 375(1794):20190107. https://doi.org/10.1098/rstb.2019.0107

Southwest Forest Alliance (1996). *Forests Forever: A Plan to Restore Ecological and Economic Integrity to the Southwest National Forests and Forest Dependent Communities.* South West Forest Alliance, Flagstaff, AZ.

Srivastava P and Garg A (2013). Emissions from forest fires in India-as assessment based on MODIS fire and global land cover products. *Climate Change and Environmental Sustainability* 1(2):138–144. https://doi.org/10.5958/j.2320-642X. 1.2.013

Suding KN (2011). Toward an era of restoration in ecology: successes, failures, and opportunities ahead. *Annu Rev Ecol Evol Syst* 42:465–487. www.annualreviews.org/doi/pdf/10.1146/annurev-ecolsys-102710-145115

Thomas CD (2020). The development of Anthropocene biotas. *Philos Trans R Soc Lond B Biol Sci* 375(1794):20190113. https://doi.org/10.1098/rstb.2019.0113

Tischew S, Baasch A, Conrad MK, and Kirmer A (2010). Evaluating restoration success of frequently implemented compensation measures: results and demands for control procedures. *Restor Ecol* 18(4):467–480. http://dx.doi.org/10.1111/j.1526-100X.2008.00462.x

Turner MG et al. (2020). Climate change, ecosystems and abrupt change: science priorities. *Philos Trans R Soc Lond B Biol Sci* 375(1794):20190105. https://doi.org/10.1098/rstb.2019.0105

United Nations International Year of Forests (2011). www.un.org/en/events/iyof2011/

Uscola M, Salifu KF, Oliet JA, and Jacobs DF (2015). An exponential fertilization dose response model to promote restoration of the Mediterranean oak *Quercus ilex*. *New Forests* 46(5):795–812. https://link.springer.com/article/10.1007/s11056-015-9493-5

Verma S and Jayakumar S (2012). Impact of forest fire on physical, chemical and biological properties of soil: a review. *Proc Int Acad Ecol Environ Sci* 2(3):168–176. https://agris.fao.org/agris-search/search.do?recordID=CN2012200064

Villalobos A, Olsson G, Birkedal M, and Lof M (2019). The effects of four repellents on bank vole consumption and germination of beech nuts and acorns. *New Forests* 50(2):241–254. https://doi.org/10.1007/s1105 6-018-9660-6

Wagner RG, Gonzalez BCA, Nelson AS, and Jacobs DF (2018). Forest regeneration in changing environments. *New Forests* 49:699–703. https://link.springer.com/article/10.1007/s11056-018-9687-8

Wang QS, Han X, and Yu CL (2018). Study on the degradation of p-nitrophenol wastewater by CMC-bentonite immobilized microorganism. *Technology of Water Treatment* 318(7):67–87.

Wanga Y, Zhouyuan L, and Xiaodi Z (2020). The microclimatic effects of ecological restoration in brownfield based on remote sensing monitoring: the case studies of landfills in China. *Ecol Eng* 157:1–2. https://doi.org/10.1016/j.ecoleng.2020.105997

Zedler J (1996). Ecological issues in wetland mitigation: an introduction to the forum. *Ecol Appl* 6(1):33–37. https://doi.org/10.2307/2269550

Zhang H, Lyu T, Bi L, Tempero G, Hamilton D, and Pan G (2018). Combating hypoxia/anoxia at sediment-water interfaces: a preliminary study of oxygen nanobubble modified clay materials. *Sci Total Environ* 637:55. https://doi.org/10.1016/j.scitotenv.2018.04.284

7 Utilization of Sustainable Resources to Combat Challenges in Environment and Climate Change

Jyoti Rawat, Pooja Pant, and Veena Pande
Department of Biotechnology, Kumaun University, Sir J.C. Bose Technical Campus, Bhimtal (Nainital), Uttarakhand

CONTENTS

7.1 INTRODUCTION

Many environmental problems affect society, the economy, and the environment, such as unsustainable consumption and development, climate change, environmental displacement, loss of biodiversity, pollution, natural disasters such as fires and floods, etc. These are not being taken into account. Although some are persistent, others are emerging. Climate change, along with global population, poverty reduction, environmental degradation, and global stability, is one of the century's defining threats. The phrases "global warming" and "climate change" tend to be synonymous at times. However, global warming is one consequence of climate change. Climate change is a more precise definition of the issue facing the planet, since global warming can be followed by local or

DOI: 10.1201/9781003220824-8

temporary drops in average temperature. Human combustion of fossil fuels for oil, transportation, and industry, as well as forest clearing and livestock, are now generally acknowledged as major contributors to global average temperatures. Since the 19th century, human interaction with the climate system (primarily by greenhouse gas (GHG) emissions and land-use changes) has raised the global annual mean air temperature at the earth's surface by around 0.8°C (IPCC, 2013).Despite significant improvements in the production of systematic information on changing environments and their impacts on natural resources, policymakers continue to be hampered by a lack of confidence in most data and many of the assumptions on which they are based (Leyshon, 2014). The 2030 Agenda for Sustainable Development and its 17 Sustainable Development Goals (SDGs), which were adopted by world leaders in 2015, provide a roadmap for all nations' future development paths, with a focus on eradication of poverty, environmental sustainability, peace, and prosperity. The year 2014 was proclaimed to be the hottest on record, and 2015 appears to be on track to break that record as well (Feulner, 2017). This warming trend will continue: by 2100, the planet will have warmed by another 4°C or so if emissions are not significantly reduced in the coming decades (Stocker et al., 2014). It is generally recognized that a warming of this magnitude will have major implications for both the atmosphere and human communities (Field et al., 2014) and that mitigating climate change through the transition to decarbonized economies and societies should be a priority to mitigate these consequences (Edenhofer et al., 2015). Following a proposal from the International Civil Aviation Organization (ICAO) and the Parties to the Montreal Protocol on Substances that Deplete the Ozone Layer in 1999, the Intergovernmental Panel on Climate Change (IPCC) published a report on various aspects of the problem and its significance. The earth is increasingly warming, and climate change poses a significant challenge to our economies as well as the natural world on which we depend for food and other important resources. The accumulation of "greenhouse gases" emitted by human activities, especially the burning of fossil fuels (coal, oil, and gas), deforestation, and certain forms of agriculture, is primarily to blame. The global average surface air temperature rose by about 0.6°C during the 20th century, almost 1°C in Europe, and no less than 5°C in the Arctic. The effects of this artificial warming are already being felt all over the world. Changes in climatic conditions have affected energy use patterns as a result of climate change . Human habitats are threatened by the rate and severity of long-term global climate change. Society is becoming increasingly reliant on energy consumption, both at work and at home, as well as for mobility. Excessive usage depletes non-renewable resources because they are scarce and do not regenerate. The warming of the climate system is undeniable, as shown by the increasing global average ocean and air temperatures, widespread melting of snow and ice, and rising sea levels (global average sea level). The existence of the issue is undeniable, and it is growing in disturbing proportions with each passing day. As a result, immediate and vigorous action is needed to calibrate an appropriate response to the evolving climate change challenges. The current chapter discusses the environmental issues that exist for several causes, as well as how to resolve these challenges using renewable resources to address global concerns.

7.2 FACTORS RESPONSIBLE FOR CLIMATE CHANGE AND ENVIRONMENTAL POLLUTION

7.2.1 Depletion of Fossil Fuels

Fossil fuels, also known as mineral fuels, are hydrocarbons, mainly coal, fuel oil, or natural gas, formed from the remains of plants and animals. These are the driving force behind the industrialized world and its economic growth. At present, around 80% of all primary energy in the world is derived from fossil fuels, of which oils account for 32.8%, coal contributes 27.2%, and natural gas 20.9% (IEA, 2011). After fossil fuels, combustible biomass and waste, nuclear power, and hydroelectric dams are other contributors (around 18.3%), while only 0.8% of the world's primary energy is derived from geothermal, solar, and wind (Edenhofer et al., 2011). The world will remain reliant on fossil fuels for a long period due to their dominant contribution. It is also true that this reliance is causing an associated problem, that is, the emissions. Around 70% of all the anthropogenic GHG emissions are released by the energy sector, among which CO_2 makes the largest (70–75%) contribution from fossil fuel combustion (Halvorsen et al., 1989). About 30 billion tons of CO_2 emission was observed in 2008 from fossil fuel combustion, which was double the amount in 1970 (IEA, 2010). Special Report on Emissions Scenarios (SRES) has reported on future CO_2 emissions in 2000 from 1990 to 2100. The fossil fuel deposits are non-renewable and their use is limited. This is because the deposits take millions of years to accumulate but can be extracted rapidly. Despite awareness of fossil fuel depletion, discussions regarding this very crucial subject of oil, gas, and coal supply have long been overlooked. It is a great challenge now to use fossil fuel as a primary energy source for the next decades with their many associated problems. Thus, renewable resources need to be used more for combating future challenges (Kharecha and Hansen, 2008).

7.2.2 Greenhouse Gas Emission and Global Warming

A greenhouse gas (GHG) is a gas that absorbs and emits radiant energy, causing a greenhouse effect (IPCC, 2008). These are the gases responsible for making the earth warm; without these gases, the earth's average temperature would be −18°C rather than the present average of 15°C (Le Treut et al., 2007). The major GHGs are water vapour (H_2O), carbon dioxide (CO_2), methane (CH_4), nitrous oxide (N_2O), and ozone (O_3). Human activities since the beginning of the Industrial Revolution in 1750 have caused a 50% increase in the atmospheric concentration of CO_2, from 280 ppm in 1750 to 419 ppm in 2021. According to scientists, the last time the atmospheric concentration of CO_2 was this high was over 3 million years ago (Calma, 2021). The majority of anthropogenic (human-induced) CO_2 emission is caused by fossil fuel combustion and deforestation (Change, 2018). Methane gas emission is mainly caused by agriculture, gas venting, and fugitive emissions. Livestock primarily and traditional rice cultivation secondly are the major sources of agricultural methane (Reed, 2020). With the current emission rates, the temperature could be increased by 2°C, which the United Nations' IPCC has stated as the upper limit to avoid dangerous levels, by 2036. The increase in carbon stock in the atmosphere results largely from human activities such as use of fossil fuels, deforestation and forest degradation in tropical and boreal regions, and use of chemical fertilizers in agriculture. Earlier, CO_2 concentration in the atmosphere was maintained by natural sinks, photosynthesis of carbon compounds by plants and marine plankton, etc., which has been greatly disturbed after the start of the industrial era.

The global warming phenomenon explains the increase of the earth's temperature. This is a long-term process, but for the past hundred years it has become more rapid due to the burning of fossil fuels, increase in GHGs, more deforestation, and many other human activities. The increase in GHGs in the atmosphere causes the accumulation of ultraviolet radiation, which results in the earth's temperature rising above normal. Global warming has presented big climate change issues like sea-level rise, melting of glaciers, floods, droughts, etc. GHGs, ozone depletion, and global warming are different concepts but are in some way connected; the solution to one of the problems will automatically solve the others. To address this problem, the IPCC was formed in 1988 to report updates on global warming issues. It produced a report in 2018 stating that human activities have caused the temperature increase from 0.8 to 1.2°C since preindustrial times and also predicted that it would increase between 3 and 4°C by 2100 at this rate. The burning of fossil fuels, demolition of carbon sinks, forest burning, and other human activities have caused an accumulation of more CO_2 in the earth's atmosphere. This accumulation traps the heat in the lower atmosphere, causing global warming. As a result of this, we are facing a serious situation: the highest level of CO_2 in human history. The storms, droughts, floods, forest fires, and heat waves are anthropogenic calamities, occurring with increasingly high frequency due to climate change. If this overexploitation of natural resources occurs for a few more decades, life on earth will become miserable.

7.2.3 Ozone Layer Destruction

The ozone (O_3) layer or shield is a region of the earth's stratosphere that absorbs the harmful ultraviolet radiation coming from the sun. The maximum concentration of ozone is found in the stratosphere (mainly in the lower portion) as compared with other parts of the atmosphere (McElroy and Fogal, 2008). It absorbs approximately 97% to 99% of the sun's total UV radiation, which can cause huge damage to the living beings on earth. According to atmospheric research done in 1976, it was revealed that ozone layer depletion was occurring due to the chemicals released by industry, mainly chlorofluorocarbons (CFCs). The increase in UV radiation reaching the earth threatened life forms, causing skin cancer in humans and ecological problems, as a result of which many chemicals were banned. Man-made organohalogen compounds, especially CFCs and bromofluorocarbons (BFCs), are mainly responsible for depleting the ozone layer. CFCs and BFCs are highly stable compounds that survive in the stratosphere, where the chlorine and bromine radicals are liberated by the action of ultraviolet light. Each radical is then capable of initiating and catalysing a chain reaction that breaks down over 100,000 ozone molecules. Nitrous oxide was the ozone-depleting substance released in the highest concentration through human activities. The ozone level worldwide has dropped by about 4% since the 1970s. "Ozone holes" have occurred over approximately 5% of the earth's surface near the south and north poles, including the Arctic and Antarctic regions, which is a very serious issue that needs to be solved (Tabin, 2008).

7.2.4 Environmental Pollution

Any contaminant that causes adverse effects in the environment is termed a pollutant, and the phenomenon is known as pollution. Greater dependency of humans on technologies, natural resources, industrial products, etc. is causing enormous harm to the environment by introducing new pollutants. Pollution can be of any form – heat, air, water, thermal, radioactive, plastic, noise, etc. – and is divided into point source and nonpoint source pollution. According to a report, 9 million people worldwide were killed in 2015 due to pollution (Beil, 2017). The Industrial Revolution has not only caused global warming but also introduced different types of pollution into the world. The burning of coal and wood was the primary source of pollution. Harmful gases like carbon monoxide, CFCs, sulphur dioxide, and nitrogen oxide are released by industry and motor vehicles, which have contaminated fresh air for breathing, causing various lung diseases. Chemicals released from industries, hospital waste, etc. have contaminated fresh drinking water, which is already present in a very small amount on earth. Every year, 400 million tons of hazardous wastes is generated, and of this, 250 million metric tons comes from the United States alone. The plastics that we use on a daily basis are the worst contaminants; they are non-degradable and cause the death of many marine animals. The pesticides we use in agriculture contaminate the soil. Products like dichlorodiphenyltrichloroethane (DDT) and dieldrin are extremely hazardous for living organisms and human health; in many countries they are banned. We humans are developing our environment and ourselves, frequently in search of a better world, risking the life and health of all the life forms living on earth.

7.2.5 Human Interference

Environmental and climate change is a natural phenomenon that has occurred many times in earth's history and will occur more in the future, but human activities have increased its speed. Thus, it would not be incorrect to say that the present climate change is induced by humans and could be referred to as anthropogenic climate change. The earth's temperature has risen due to the increase of GHGs; the glaciers are melting, causing floods; the forest is being burned worldwide deliberately by humans for profit; the fertile lands are being used for constructing buildings and industries; the natural resources are being abused by humans for their benefit without maintaining sustainable development; drinking water has been polluted; no fresh air is to be found in metropolitan cities; these all are the results of human interference. The threat to the ecology and biodiversity of the world in the coming decades is disruptive climate change due to the build-up of human-generated GHGs in the atmosphere. All the nations in the world are trying to reduce their carbon footprints through lower consumption and better technology. But this will not be an easy step with a growing human population greatly dependent on technologies.

7.3 CHALLENGES DUE TO CLIMATE CHANGE AND ENVIRONMENTAL EFFECTS THAT THE WORLD MUST NEGOTIATE

7.3.1 Impact on Human Health

The climate change scenario is not only harming the environment and the other living creatures on earth but is also seriously affecting human health. The current situation has introduced new pests and pathogens into new regions. The most vulnerable people, like children, the poor, and the elderly, are suffering from serious health problems. The polluted air, the increased concentration of ground-level ozone, the fine particles and dust cause chest pains, throat irritation, congestion, asthma, and many more respiratory allergies. The increased CO_2 concentration and the temperature rise affect the aeroallergen distribution and amplify the allergenicity of pollen and mould spores. Depletion of ozone in the stratosphere is leading to an increase in UV exposure and causing the risk of skin cancer and cataracts. Cardiovascular disease is the third leading disease causing death. Heart attack, hypertension, stroke, etc. are triggered by the temperature extremes of cold and heat. Prolonged exposure to heat causes heat cramps, heatstroke, and death, which has become a common reason for deaths in poor and developing countries. Virulent pathogenic strains are becoming immune to the increased temperature; for example, the rise in sea surface temperature leads to the increase of *Vibrio* bacterial species, which cause seafood-borne disease and cholera. Heavy metals like lead cause a reduction in IQ level of children, birth defects, foetal loss, and sometimes long-term disability in those who are born without birth defects. Vector-borne and zoonotic diseases (VBZD) such as malaria, yellow fever, dengue, etc., which are spread by vectors such as insects, ticks, and mites, are mainly found in warmer climates. Most of them are climate sensitive, and their distribution and incidence are affected by the change in the climate. Extreme weather events such as floods, hurricanes, and drought cause a large number of deaths per year. Flooding leads to exposure to toxic chemicals in runoff, waterborne diseases, and ecosystem changes like the loss of wetlands.

7.3.2 Increased Risk of Tick-Borne Diseases

The extension of the range of many tick species into higher latitudes in North America has been helped by climate change and other environmental changes. Tick-borne diseases are likely to become more prevalent in Canada as temperatures rise and the climate becomes more desirable for ticks. As the season suitable for tick activity lengthens, tick-borne diseases may become more common. In addition to Lyme disease, four other tick-borne diseases (TBDs) have emerged and are expected to become more common: *Borrelia miyamotoi* disease, Anaplasmosis, Babesiosis, and Powassan virus. Increased temperature extends tick survival and activity periods, expands the range of tick reservoirs and tick hosts (e.g., mice and deer), and extends the season during which people may be exposed to ticks. As the climate changes, other ticks and TBDs may spread into Canada. Surveillance for existing and emerging TBDs, as well as public health actions to avoid infections by changing environmental and social-behavioural risk factors by increased public awareness, are among the public health measures to reduce the effects of all TBDs. Patient education, early detection, lab research, and treatment are all examples of clinical care methods.

7.3.3 Food and Water Scarcity

Water and food security are the most pressing concerns as a result of climate change, as both are extremely vulnerable to evolving weather patterns. According to studies, by the end of the 21st century, the average global temperature could rise by 1.4–5.8°C, resulting in a significant reduction in freshwater supplies and agricultural yield. Approximately 75% of Himalayan glaciers are retreating and will be gone by 2035. Furthermore, by 2050, rainfall in Africa (Sub-Saharan Africa) could drop by 10%, resulting in a 17% reduction in drainage. With an increase in population and food demand, the majority of freshwater supplies have already been exhausted, and agricultural production has decreased globally. Growing deserts, as well as a rise in the severity of floods and droughts, are some of the most visible climate change effects. Crop yields in arid and semi-arid areas around the world have plummeted, resulting in food shortages and a massive rise in food inflation. Natural resources and climate-dependent sectors such as forestry, agriculture, water, and fisheries have strong economic relations in Africa, the Middle East, the Arab countries, and Asia. To ensure water and food security under climate change and the formation of effective adaptation and mitigation policies, and to minimize the impact of climate change on water resources and irrigation, easy and economically feasible options are needed.

7.3.4 Loss of Biodiversity

The viewpoints on emerging issues include fishing, food security, climate change, and biodiversity. Fish production must increase by 50% from current levels to fulfil expected food requirements. The key pressures on marine biodiversity that are likely to occur as a result of the impacts of climate change on marine habitats could be resolved by implementing new conservation strategies and policy initiatives to counter those pressures. It should be emphasized that the majority of proposed actions to address pressures on marine biodiversity are incompatible with actions considered appropriate to meet future food security needs, especially in developing countries. To allow cohesive action on these critical and challenging issues, the two groups, experts and policymakers, must cooperate to find a single, compatible set of policies and management steps.

Throughout the evolutionary history of amphibians, global ecosystems have changed, often quickly and dramatically. As a result, modern amphibian populations are descended from those that have survived significant climatic shifts. Even though recent global climate change has caused warming in many areas, temperatures in some areas have remained stable or even decreased. Some amphibian population declines have been linked to climate events, but proof of clear causal relationships needs further study. There is evidence that climate change has an indirect impact on the initiation of breeding practices in certain amphibians, which happens earlier than in previous springs, although the costs and benefits of these changes are still being researched. Climate change can also play an indirect role in the spread of infectious disease epidemics. Whatever role climate change may have played in past and current amphibian declines, future climate changes, if they prove to be as drastic as expected, would undoubtedly pose challenges for surviving amphibian populations and active recovery attempts for species that have declined.

7.4 SUSTAINABLE RESOURCE MANAGEMENT TO TACKLE CHALLENGES ON THE PATH

Sustainable development is fuelled by green technology, which entails identifying eco-friendly sources of growth, developing new environmentally friendly industries, and creating jobs and technologies (Ghisetti and Quatraro, 2017). To achieve green growth, more investments and developments that serve as a foundation for long-term development must be made, as well as new economic opportunities (Przychodzen et al., 2020). Thus, the promotion of a green economy requires in-depth research on the conditions of its formation and its impact on national sustainable development.

Businesses (focusing on economic benefits), governments (setting environmental sustainability goals), and the general public (representing the interests of a social community) are all stakeholders in green economic development (Ramdhani et al., 2017). Green knowledge management processes play an important role in sustainable development, especially in the production, acquisition, sharing, and use of knowledge, as well as its effect on green technologies, eco-innovations, and the socio-economic dimension. (Abbas and Sağsan, 2019; Aldieri and Vinci, 2018). Innovation is required to achieve Sustainable Development Goals. Sustainable green technologies aim to create high-quality innovative goods that have a low environmental impact. The value of sustainable green technologies will be generally known as long as environmental problems exist (Hyung and Baral, 2019). The structural model's best indicator is compliance with environmental norms, which represents the effect of different environmental factors, such as developments, on the environmental situation. Environmental output is influenced by market orientation and innovations, with the latter having a positive effect.

7.4.1 Green Technology

Green technology is a rapidly growing field that focuses on new scientific and technical methods that benefit the earth. Environmentally friendly technology utilizes many methods for reducing the impact that various activities have upon the earth. A product or activity should be environmentally sustainable if it is long-lasting, creates as little waste and emissions as possible, and recycles and reuses materials wherever possible. Recycled, recyclable, and biodegradable content, plant-based products, reduction in pollution, GHG emissions reduction, renewable energy, and energy efficiency are all examples of eco-friendly technology. The eco-innovation principles empower businesses to face market challenges posed by competitors (Fernando and Wah, 2017). Green innovations that are both environmentally friendly and economically beneficial contribute significantly to a healthy society. Special attention should be paid to both the determinants of sustainable green technology invention as well as the variations in their growth goals as a result of this (Fujii and Managi, 2019).

India is a huge developing country with nearly 700 million rural residents whose livelihoods are directly dependent on climate-sensitive sectors (agriculture, forestry, and fisheries) and natural resources (such as water, biodiversity, mangroves and other coastal areas, grasslands). Furthermore, dry land farmers, forest dwellers, fishermen, and nomadic herders have limited adaptive potential (Ravindranath and Sathaye, 2002). According to India's national communications report to the United Nations Framework Convention on Climate Change, climate change is likely to affect all natural habitats as well as socioeconomic structures. Sustainable development is described as "development that meets current needs without jeopardizing future generations' ability to meet their own."

Once non-renewable resources are used, they cannot be used again for that purpose. Coal and oil, for example, are non-renewable resources. If the coal has been burned, the resource is no longer available. Coal and oil supplies are small, but new supplies can be discovered. The other major category of non-renewable resources is fossil fuels. They are primarily burned to generate energy for transportation and heating, converting their organic compounds to carbon dioxide and water, which are then released into the environment. Some of the CO_2 and H_2O can be absorbed by plants and other photosynthetic organisms and transformed back into organic materials, a process that could be interpreted as a net increase in CO_2 being a kind of recycling.

Semi-renewable resources are those that have the potential to be renewable if used properly. Farmland, for example, should be considered a sustainable resource. Crop rotation guarantees that the soil's fertility does not deteriorate. However, if farmland is intensively farmed with artificial fertilizers and pesticides, the soil's quality can begin to deteriorate.

Renewable resources are those that are not depleted as a result of their use. A wind generator, for example, does not decrease the amount of wind; it does not affect potential wind energy supplies.

Solar power does not deplete the sun's resources. Algal biofuels are gaining popularity as a domestic source of renewable energy. Scaling up the development of algal biofuels to meet even 5% of U.S. transportation fuel needs with current technologies, on the other hand, might generate unsustainable demands for electricity, water, and nutrient capital. Continued research and development may yield solutions to these problems, but evaluating whether algal biofuel is a viable fuel alternative will require a comparison of the environmental, economic, and social impacts of algal biofuel production and use with those of petroleum-based fuels and other fuel sources.

7.4.2 Biofuel Production

There is a need for new energy production technologies that not only minimize GHG emissions but also have fewer overall environmental impacts. In contrast to developing a set of criteria for guiding the identification of sustainable biofuel production alternatives as well as strategies for lowering the economic barriers that prevent the implementation of more sustainable biofuel, alternative approaches to biofuel production (i.e., first-, second-, and third-generation biofuels) with a focus on biodiversity and ecosystem services were developed. Recently, India's first biofuel-powered flight from Delhi to Dehradun was operated with a blend of 25% jatropha oil and 75% aviation turbine fuel. It is expected to reduce carbon emission and enhance fuel efficiency. Microalgae have recently sparked a lot of interest around the world because of their broad range of applications in the renewable energy, biopharmaceutical, and nutraceuticals industries. Microalgae are a source of biofuels, bioactive medicinal products, and food ingredients that are organic, nutritious, and cost-effective. Algae-based microalgae fuels are environmentally friendly, non-toxic, and have high potential for reducing global CO_2. The potential of microalgae as a source of renewable energy has attracted interest, but further optimization of mass culture conditions is required if microalgal biofuel production is to be economically viable and sustainable. The selection of biomass for biofuel production is also directly related to GHG emissions and environmental and economic sustainability (Cerri et al., 2017). There's also the possibility of integrating algae-based wastewater treatment, such as nutrient removal, with biofuel production. According to Gendy and El-Temtamy (2013), 1 kg of algal biomass could repair 1.83 kg of CO_2, as CO_2 makes up half of the dry weight of algal biomass. Several government mandates and incentives, as well as the creation of financial and market-based frameworks and applied research collaborations, will help to speed up the adoption of more sustainable biofuel production options.

7.4.3 Ecosystem Restoration

India's main ecological challenges are 96 million hectares of degraded land, unprecedented biodiversity loss, and extreme climatic events. There is a framework of different policies and strategies to overcome ecological and socioeconomic challenges. However, they are loosely linked, with separate aims and the participation of various government agencies and organizations. Multiple policies aimed at ecosystem restoration and biodiversity conservation will provide India with a cost-effective, productive, and long-term solution to its ecological and socioeconomic problems. Reduced GHG sources and increased GHG sinks through improved agriculture and terrestrial ecosystem stewardship are commonly cited as having the potential to provide approximately 30% of the CO_2 mitigation required to keep warming below 2°C by 2030 (Griscom et al., 2017; Erb et al., 2018; Le Quéré et al., 2018).

7.5 CURRENT POLICIES ON ENVIRONMENT AND CLIMATE PROTECTION

Climate change and global warming are two of the most serious threats society has ever faced, with consequences for human rights as well. Projections of future climate change are essential for planning measures to reduce impacts on agriculture (Michalak et al., 2022. In India, there is great uncertainty about implementing strict environmental laws and keeping them accountable for economic losses. Strong environmental regulations, both internationally and in India, are largely successful in minimizing long-term environmental degradation and promoting economic development. Biodiversity and public administration, in addition to the legal system, have had positive effects on environmental and climate goals. The Ministry of the Environment, Forests, and Climate Change ("MoEF&CC") in collaboration with the Central Pollution Control Board ("CPCB") and the National Pollution Control Boards ("SPCB") of each of the 28 states and nine Union Territories ("UT") in India administer and enforce environmental laws.

There are separate regulatory bodies for various environmental laws, such as the National Environmental Impact Assessment Authority, which oversees applications for environmental authorization and environmental assessment reports; the ozone cell, which oversees compliance with regulations on ozone-depleting substances; the forestry advisory committee ("FAC") for forest diversions; and Coastal Zone Management Authorities at the state level, overseeing the Coastal Regulatory Zone notification ("CRZ"), etc. Some of the more recent environmental laws, such as the E-waste Management rules, 2016 ("E-Waste Rules"), have incorporated a self-declaration mechanism, e.g., relating to the Reduction in the use of Hazardous Substances ("RoHS") requirements. Some states have also adopted an "auto-renewal" of consent orders (i.e., environmental permits) based on self- certification if certain criteria are met, such as when there is no increase in the overall production capacity and pollution load, or if there is only a marginal increase (up to a maximum of 10%) in the capital investment, etc. The most common consent orders or environmental permits to be obtained from the SPCBs/Union Territory Pollution Control Committee (UTPCC) (by manufacturing companies) are the consent to establish ("CTE") under the Water (Prevention and Control of Pollution) Act, 1974 ("Water Act") and the Air (Prevention and Control of Pollution) Act, 1981 ("Air Act"), in which a company submits its initial plans and shares its manufacturing capacity, pollution load, etc. for initial construction approval. In August 2018, a new online environmental portal was launched by the MoEF&CC, named "PARIVESH," which stands for "Pro-Active and Responsive facilitation by Interactive, Virtuous and Environmental Single-window Hub," to facilitate online submission and tracking of various environmental clearance applications.

In 2008, India adopted the National Climate Change Plan, which includes the National Mission to Improve Energy Efficiency. The latter contains a market-based energy efficiency improvement programme called the Perform, Achieve, and Trade (PAT) programme, under which industries must meet energy conservation targets by implementing energy efficiency measures or offsetting their excess consumption through the purchase of energy-saving certificates. India has implemented a hydrochlorofluorocarbon (HCFC) Phase-Out Management Plan, currently in stage 2, which aims to reduce the production and consumption of HCFCs. According to the Indian government, this plan will lead to cuts faster than required by the Kigali Amendment (Jaiswal and Deol, 2017).

7.6 CONCLUDING REMARKS

The current situation, in which climate models are primarily focused on pollution scenarios that are irrelevant to supply realities and issues, is troublesome. It is critical to recognize that the problems of fossil fuels and anthropogenic climate change are inextricably linked and that they should be viewed as two interconnected issues in need of a comprehensive solution. Eliminating carbon dioxide from the environment, protecting natural habitats, and sustainably maintaining and restoring forests are effective ways to minimize GHGs and delay short-term temperature rise. Simultaneously, we must drastically reduce global GHG emissions from fossil fuels such as coal, oil, and gas, since if we just do the first and neglect the second, we risk converting our carbon sinks into carbon sources as climate change continues.

REFERENCES

Abbas, J., & Sağsan, M. (2019). Impact of knowledge management practices on green innovation and corporate sustainable development: a structural analysis. *Journal of Cleaner Production, 229*, 611–620.

Aldieri, L., & Vinci, C. P. (2018). Green economy and sustainable development: the economic impact of innovation on employment. *Sustainability, 10*(10), 3541.

Beil, L. (2017). Pollution killed 9 million people in 2015. Retrieved from Science news.org.

Calma, J. (2021). The climate controversy swirling around NFTs. The Verge, 15. www.theverge.com/2021/3/15/22328203/nft-cryptoart-ethereum-blockchain-climate-change

Cerri, C. E. P., You, X., Cherubin, M. R., Moreira, C. S., Raucci, G. S., Castigioni, B. D. A., ... & Cerri, C. C. (2017). Assessing the greenhouse gas emissions of Brazilian soybean biodiesel production. *PloS One, 12*(5), e0176948.

Change, C. (2018). Atmospheric carbon dioxide. NOAA Climate.gov. Available online: www.climate.gov (accessed on 16 November 2020).

Edenhofer, O., Pichs-Madruga, R., Sokona, Y., Seyboth, K., Kadner, S., Zwickel, T., ... & Matschoss, P. (Eds.) (2011). *Renewable Energy Sources and Climate Change Mitigation: Special Report of the Intergovernmental Panel on Climate Change*. Cambridge University Press, Cambridge, UK.

Edenhofer, O., Pichs-Madruga, R., Sokona, Y., Farahani, E., Kadner, S., Seyboth, K., Adler, A., Baum, I., Brunner, S., Eickemeier, P., Kriemann, B., Savolainen, J., Schlömer, S., von Stechow, C., Zwickel, T., & Minx, J.C. (Eds.) (2015). *Climate Change 2014: Mitigation of Climate Change* (Vol. 3). Cambridge University Press, Cambridge, UK and New York.

Erb, K. H., Kastner, T., Plutzar, C., Bais, A. L. S., Carvalhais, N., Fetzel, T., ... & Luyssaert, S. (2018). Unexpectedly large impact of forest management and grazing on global vegetation biomass. *Nature, 553*(7686), 73–76.

Fernando, Y., & Wah, W. X. (2017). The impact of eco-innovation drivers on environmental performance: empirical results from the green technology sector in Malaysia. *Sustainable Production and Consumption, 12*, 27–43.

Feulner, G. (2017). Global challenges: climate change. *Global Challenges, 1*(1), 5.

Field, C. B., Barros, V. R., Dokken, D. J., Mach, K. J., Mastrandrea, M. D., Bilir, T. E., Chatterjee, M., Ebi, K. L., Estrada, Y. O., Genova, R. C., Girma, B., Kissel, E. S., Levy, A. N., MacCracken, S., Mastrandrea, P. R., & White, L. L. (2014). *Impacts, Adaptation, and Vulnerability. Part A: Global and Sectoral Aspects. Contribution of Working Group II to the Fifth Assessment Report of the Intergovernmental Panel on Climate Change*. Cambridge University Press, Cambridge, UK and New York, 1132 pp.

Fujii, H., & Managi, S. (2019). Decomposition analysis of sustainable green technology inventions in China. *Technological Forecasting and Social Change, 139*, 10–16.

Gendy, T. S., & El-Temtamy, S. A. (2013). Commercialization potential aspects of microalgae for biofuel production: an overview. *Egyptian Journal of Petroleum, 22*(1), 43–51.

Ghisetti, C., & Quatraro, F. (2017). Green technologies and environmental productivity: a cross-sectoral analysis of direct and indirect effects in Italian regions. *Ecological Economics, 132*, 1–13.

Griscom, B. W., Adams, J., Ellis, P. W., Houghton, R. A., Lomax, G., Miteva, D. A., ... & Fargione, J. (2017). Natural climate solutions. *Proceedings of the National Academy of Sciences, 114*(44), 11645–11650.

Halvorsen, B., Kverndokk, S., & Torvanger, A. (1989). Global, regional and national carbon dioxide emissions 1949-86-Documentation of a LOTUS Database. Working paper 59/89, Centre for Applied Research, Oslo.

Hyung, K., & Baral, P. (2019). Use of innovative public policy instruments to establish and enhance the linkage between green technology and finance. In *Handbook of Green Finance: Energy Security and Sustainable Development*, Sachs, J. D., Woo, W. T., Yoshino, N., & Taghizadeh-Hesary, F. (Eds.) Springer, Singapore; Asian Development Bank Institute, Tokyo (pp. 1–24).

IEA (2010). *Emissions from Fuel Combustion (2011 Edition)*. International Energy Agency (IEA), Paris.

IEA (2011). *Electricity Annual 2011*. Technical Report. International Energy Agency, Paris.

Jafari, M. (2013). Energy consumption impacted by climate change application: case study Astara, in power, control and optimization. In *Lecture Note in Electrical Engineering 239*, I. Zelinka, P. Vasant, and N. Barsoum (Eds.) Springer, Singapore; Asian Development Bank Institute, Tokyo (pp. 29–50). ISBN 978-3-319-00205-7 (Print) 978-3-319-00205-4.

Jaiswal, A., & Deol, B. (2017). *India Takes Major Step Towards Climate-Friendly Refrigerants*. www.nrdc.org/experts/anjali-jaiswal/india-takes-major-step-towards-climate-friendly-refrigerants

Kharecha, P. A., & Hansen, J. E. (2008). Implications of "peak oil" for atmospheric CO_2 and climate. *Global Biogeochemical Cycles*, *22*(3), 1–10.
Le Quéré, C., Andrew, R. M., Friedlingstein, P., Sitch, S., Pongratz, J., Manning, A. C., ... & Zhu, D. (2018). Global carbon budget 2017. *Earth System Science Data*, 10(1), 405–448.
Le Treut, H. (2007). Historical overview of climate change. *Climate Change 2007: The Physical Science Basis. Contribution of Working Group I to the Fourth Assessment Report of the Intergovernmental Panel on Climate Change*.
Leyshon, C. (2014). Critical issues in social science climate change research. *Contemporary Social Science*, *9*(4), 359–373.
McElroy, C. T., & Fogal, P. F. (2008). Ozone: From discovery to protection. *Atmosphere-ocean*, *46*(1), 1–13.
Michalak, J. L., Lawler, J. J., Gross, J. E., Agne, M. C., Emmet, R. L., Hsu, H. W., & Griffey, V. (2022). Climate-change vulnerability assessments of natural resources in US National Parks. *Conservation Science and Practice*, e12703.
Przychodzen, W., Leyva-de la Hiz, D. I., & Przychodzen, J. (2020). First-mover advantages in green innovation—Opportunities and threats for financial performance: a longitudinal analysis. *Corporate Social Responsibility and Environmental Management*, *27*(1), 339–357.
Quéré, C. L., Andrew, R. M., Friedlingstein, P., Sitch, S., Pongratz, J., Manning, A. C., ... & Zhu, D. (2018). Global carbon budget 2017. *Earth System Science Data*, *10*(1), 405–448.
Ramdhani, M. A., Aulawi, H., Ikhwana, A., & Mauluddin, Y. (2017). Model of green technology adaptation in small and medium-sized tannery industry. *Journal of Engineering and Applied Sciences*, *12*(4), 954–962.
Ravindranath, N. H., & Sathaye, J. A. (2002). Climate change: vulnerability, impacts and adaptation. In Ravindranath, N. H., & Sathaye, J. A. (Eds.) *Climate Change and Developing Countries* (pp. 63–95). Springer, Dordrecht.
Reed, J. (2020). Thai rice farmers step up to tackle carbon footprint. *Financial Times*. www.ft.com/content/8ff2b454-9390-11ea-899a-f62a20d54625. Accessed 28 July 2022.
Stocker, T., Qin, D., Plattner, G.K., Tignor, M., Allen, S.K., Boschung, J., Nauels, A., Xia, Y., Bex, V., Midgley, P.M. (Eds.) (2014). *Climate Change 2013: The Physical Science Basis: Working Group I Contribution to the Fifth Assessment Report of the Intergovernmental Panel on Climate Change*. Cambridge University Press, Cambridge, UK and New York, 1535 pp.
Tabin, S. (2008). *Global Warming: The Effect of Ozone Depletion*. APH Publishing, Delhi.

Part II

Palaeoclimate Studies and Computational Analysis of Climate Change

8 Mammalian Dental Enamel

An Archive for Palaeoecology and Palaeoclimate Studies

Rajeev Patnaik
Department of Geology, Panjab University, Chandigarh

CONTENTS

8.1 INTRODUCTION

Mammalian dental enamel, with high crystal density, low organic content and large-sized crystals, is the hardest biological tissue. In fossil mammals this tissue often remains intact for millions of years, preserving the original internal and external features, and in most cases, even the original composition. These properties of dental enamel make it very useful in reconstructing the past diet, habitat and environment of mammals. There are two widely used techniques to understand the diet and environment of ancient mammals. These are stable carbon and oxygen isotope composition and dental microwear pattern analyses. Stable carbon and oxygen isotopes are an integral part of the structure and composition of the enamel bioapatite ($Ca_{10}[PO_4, CO_3]_6[OH]_2$). Carbon isotopes of dental enamel depend upon the type of dominant vegetation the particular mammal consumes, which in turn helps in the reconstruction of the palaeohabitat (MacFadden et al., 2004; Hoppe et al., 1999; Cerling et al., 1997; Clementz et al., 2003). The dominant vegetation types are warm and dry season grasses (C4 plants) and trees, shrubs and cool season grasses (C3 plants). The dental enamel oxygen isotopes, on the other hand, depend upon the body water, which comes from drinking water, plant water and metabolically derived water (Kohn, 1996, Kohn et al., 1996, Sponheimer and Lee-Thorp, 1999).

The microwear pattern on the dental enamel surface forms due to the abrasion of the teeth by food items such as grasses, fruits, leaves, nuts, barks, etc. These patterns reflect the diet consumed by the individual during the last days prior to its death (Solounias et al., 1988; Teaford and Oyen, 1989; Fortelius and Solounias, 2000). Like the stable isotopes, these microwear features, when properly quantified, have the potential to reconstruct past habitats. Since most herbivorous mammals are terrestrial, they are either grass eaters (grazers) or leaf eaters (browsers), or in some cases mixed feeders (Hoffman, 1989). The browsers normally eat leaves and fruits from various dicotyledonous

DOI: 10.1201/9781003220824-10

species. On the other hand, the grazers feed mainly on grasses, sedges and rushes. Therefore, overall microwear pattern varies according to the properties of food and related items. On the whole, grasses and related plants make numerous scratches on the tooth surface because of the high concentration of silica phytoliths in their cell walls (MacNaughton et al., 1985; Robert and Roland, 1998). Leaf eaters usually have a very low percentage of scratches on their teeth, whereas fruit eaters show a high percentage of pits. Mixed feeders show microwear patterns similar to those of either grazers or browsers (Solounias and Moelleken, 1992, 1994; Solounias and Hayek, 1993).

Both these methods are independent of tooth morphology, which can sometimes mislead in the interpretation of palaeodiet. For example, in Africa, elephant ancestors with low-crowned teeth that are adapted to browsing were found to be grazing based on the carbon isotope composition (Lister, 2013). Similarly, Asian elephants are predominantly browsers in spite of having high-crowned teeth that are specialized to consume grasses (Patnaik et al., 2019).

8.2 STABLE ISOTOPE FRACTIONATION

8.2.1 Carbon Isotopes

The atmosphere consists of ~98.9% $^{12}CO_2$ (lighter carbon) and ~1.1% $^{13}CO_2$ (heavier carbon). The atmospheric CO_2 isotopic composition is at equilibrium with mainly the dissolved inorganic carbon (HCO_3) in sea water. Photosynthesis by plants results in biosynthetic chemical reactions leading to the conversion of the atmospheric carbon dioxide into carbohydrates, proteins and lipids. In photosynthesis, the compounds having the lighter isotopes react faster, using less energy and therefore enriching the ^{12}C values of the plant tissue relative to the atmospheric CO_2. The difference in $^{13}C/^{12}C$ ratio of different samples is very small; therefore, this ratio is expressed as $\delta^{13}C$ in ‰ or per mil. This ratio is usually negative, as it is always lower than the ratio found in the standard PDB (Cretaceous Belemnite *Belemnitella americana* from Peedee Formation of South Carolina). Further, photosynthetic pathways such as C3, C4 and CAM (crassulacean acid metabolism) produce different carbon isotope ratios. C3 plants (shrubs, trees and temperate grasses) depend upon the Calvin cycle and ribulose biphosphate carboxylase to fix CO_2 and produce depleted $\delta^{13}C$ values ranging between −35‰ and −25‰, averaging −27‰ (Marshall et al., 2007). On the other hand, C4 plants (warm season grasses) use the Hatch–Slack cycle and phosphoenol pyruvate carboxylase (PEP) to fix CO_2, enriching $\delta^{13}C$ values that have a range between −15‰ and −11‰, averaging −14‰ (Marshall et al., 2007).

Metabolism by herbivorous animals that consume these plants reverses the process of fractionation, leading to the enrichment of the heavier carbon isotope (^{13}C) by a factor of 14.1‰ ± 0.5‰ (Cerling and Harris, 1999). A recent study (Patnaik et al., 2014) on the Indian Siwalik mammals older than 8 million years show that most of them had a C3 diet. Most mammals younger than 3 million years were found to be grazers, but a few turned out to be browsers and mixed feeders as well (Figure 8.1).

8.2.2 Oxygen Isotopes

Palaeoecological and palaeoclimate interpretation based on $\delta^{18}O$ of dental enamel is rather complex. However, a careful analysis of the $\delta^{18}O$ values of enamel helps in reconstructing not only the diet but also seasonality conditions (Fricke and O'Neil, 1996; Longinelli, 1984). The $\delta^{18}O$ of tooth enamel of an animal depends upon its body water, which in turn, is controlled by the $\delta^{18}O$ values of ingested water. This $\delta^{18}O$ value of ingested water is further a result of the amount of precipitation of the region, latitude, altitude, aridity, evaporative processes, and behavioural water conservation factors as well as metabolic processes of the mammals (Kohn et al., 1996; Kohn, 1996; Luz and Kolodony, 1985; Luz et al., 1984). The $\delta^{18}O$ of large land mammals that frequently drink water depends on the $\delta^{18}O$ of rainfall; in general, drought-tolerant animals usually have relatively higher

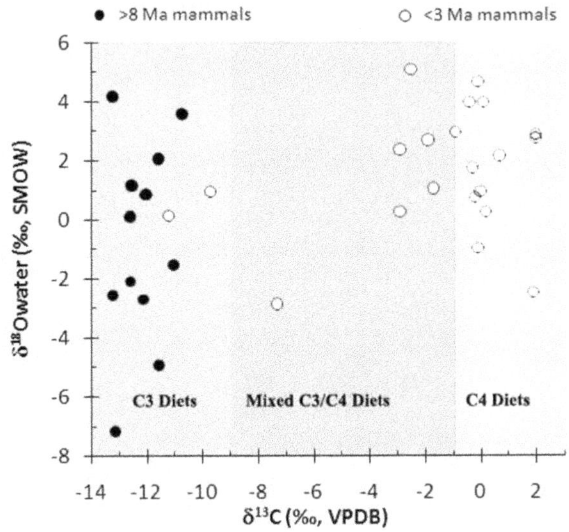

FIGURE 8.1 Stable carbon ($\delta^{13}C$) (VPDB) and oxygen ($\delta^{18}O$) (SMOW) values of some Siwalik mammalian dental enamel.

(Modified after Patnaik et al., 2014.)

$\delta^{18}O$ values because of $\delta^{18}O$ enrichment of environmental water due to evaporation (Ayliffe and Chivas, 1990; Levin et al., 2006). Bryant et al. (1996) observed that variation in $\delta^{18}O$ of water has largely been attributed to environmental temperature changes, which may lead to enrichment of $\delta^{18}O$ in warmer conditions and relative depletion in cooler conditions. Therefore, in temperate regions, summer would show a higher $\delta^{18}O$ in water, whereas in winter a lower $\delta^{18}O$ would be observed. In the tropical regions, where the environmental temperatures remain above 20°C, the "amount effect" dominates, whereby lower $\delta^{18}O$ values of water usually indicate periods of increased rainfall compared with higher values during drier conditions (Feranec and MacFadden, 2000; Dansgaard, 1964; Higgins and MacFadden, 2004). However, in a monsoonal system, this effect of temperature is dominated by the effect of precipitation amount. Due to kinetic fractionation, the more it rains, the less $\delta^{18}O$ is present in the atmospheric water, which in turn becomes more depleted with further precipitation (Dansgaard, 1964). Monsoon experiences both monthly variation in temperature as well as rainfall. The rainfall is typically $\delta^{18}O$ depleted during monsoonal months in spite of high summer temperatures (Rozanski et al., 1993; Araguas-Araguas and Froehlich, 1998). Dental enamel of amphibious mammals such as hippopotamus will generally have lower $\delta^{18}O$ values compared with their terrestrial counterparts (Bocherens et al., 1996; Cerling et al., 2003; Clementz et al., 2003). Warm season C4 grasses growing in open areas usually have higher $\delta^{18}O$ values due to loss of $\delta^{16}O$ by transpiration compared with C3 plants growing in wetter and closed areas, and therefore, grazers adapted to feeding on water-stressed grasses that grow in drier conditions may generally yield higher $\delta^{18}O$ values as compared with forest-dwelling browsers (Bocherens et al., 1996; Sternberg et al., 1989).

8.3 SAMPLING FOR STABLE ISOTOPE ANALYSES

8.3.1 Micro-drilling

Micro-drilling using diamond-studded drill bits yields powdered enamel samples. This is the most widely used method, as it generates enough enamel material for further analysis. However, a sample

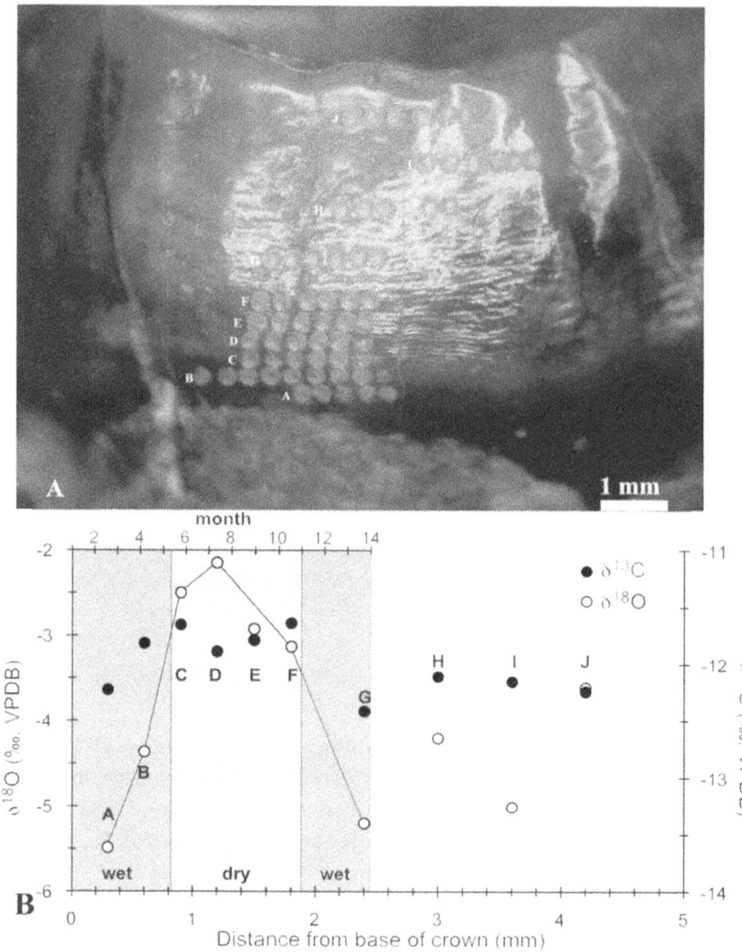

FIGURE 8.2 (A) Laser ablation pits on the enamel surface of a Miocene ape from the Siwaliks. (B) Stable carbon and oxygen isotope values giving the data on the diet and water intake during ~4 months of tooth growth. **(Modified from Patnaik et al., 2014.)**

micro-drilled from the enamel surface of a fossil mammal will yield an average value of $\delta^{13}C$ and $\delta^{18}O$, which would give us an idea of whether the herbivore was a grazer, browser or a mixed feeder and whether it lived in a dry or wet climate. Approximately 10 mg of enamel powder is removed from the front tooth perpendicular to the growth axis. The enamel powder is then pre-treated with 2.5% sodium hypochlorite (NaOCl) for 12 h, followed by 1 M acetic acid (pH 3.8) for 6 h to remove organics and secondary carbonates. Samples are then centrifuged at high speed and rinsed in distilled water to neutral pH before proceeding with the next solution. Samples are then freeze-dried for a maximum of 48 h. The pre-treated samples are then analysed using isotope ratio mass spectrometry (see Lee-Thorp et al., 1989).

8.3.2 Laser Ablation

Another method of sampling is the use of laser ablation, which allows the recovery of small amounts of sample from multiple levels within a single tooth (Cerling and Sharp, 1996; Passey and Cerling, 2006; Sponheimer et al., 2006). Basically, the enamel surface of the sample is laser beamed and is

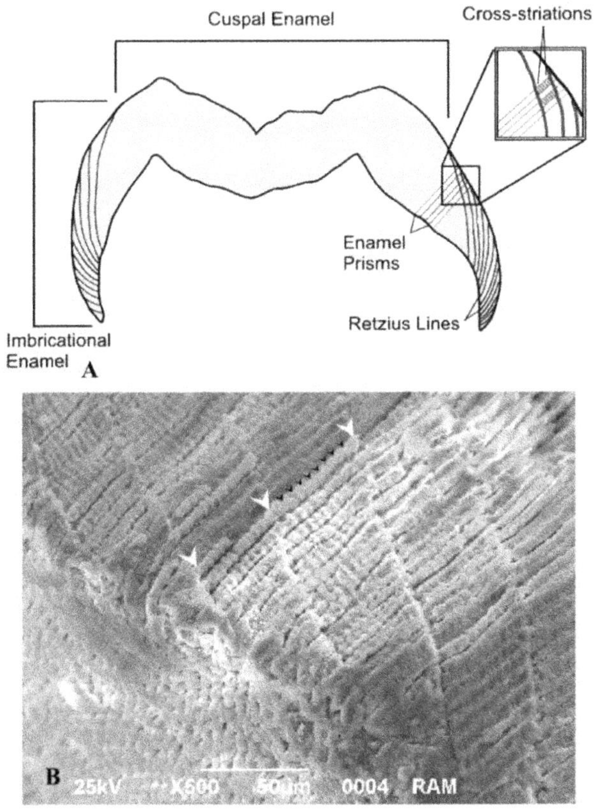

FIGURE 8.3 (A) Line drawing showing cuspal and imbricational enamel, orientation of prisms, Rhetzius lines and cross-striations in a hominid tooth. (B) Scanning electron microscopic image of internal structure of a Miocene ape enamel, showing Rhetzius lines (white arrows) and cross-striations (black arrows).

heated rapidly, producing CO_2 by thermal breakdown of the carbonate and phosphate components (Cerling and Sharp, 1996). Such a method is very useful in dealing with rare and very small specimens. For instance, specimens of fossil hominoids have yielded vital palaeodietary and palaeoecological information. Such sampling, if combined with information about enamel microstructure depicting the time required to grow each layer, provides robust dietary and seasonality data at the annual resolution (Figure 8.2) (Patnaik et al., 2014).

Dental enamel grows periodically, having rhythmic growth increments that are secreted on a daily and weekly basis. The daily growth increments are termed "cross-striations", and sub-millimetre growth lines that develop over a week or more are termed "Retzius lines". These Retzius lines run through the entire tooth and crop out on the tooth enamel surface as "perikymata", preserving a continuous record of temporal isotopic changes across the growth axis of the tooth (Figure 8.3).

Small mammals, particularly rodents, are difficult to use for stable isotope analyses using the conventional micro-drill method because of reduced sample size. The laser ablation technique helps in obtaining multiple small samples from a single tooth (Figure 8.4) (e.g. Kimura et al., 2013).

Teeth of high-crowned herbivorous mammals, particularly those of horses, bison, rhinocerotids and elephants, offer an excellent opportunity to study dietary and environmental signals at even inter- and intra-annual resolution, since they may preserve the isotopic record of 2 to 5 years, the time usually required by the tooth to mineralize (Feranec and MacFadden, 2000; Kohn et al., 1998;

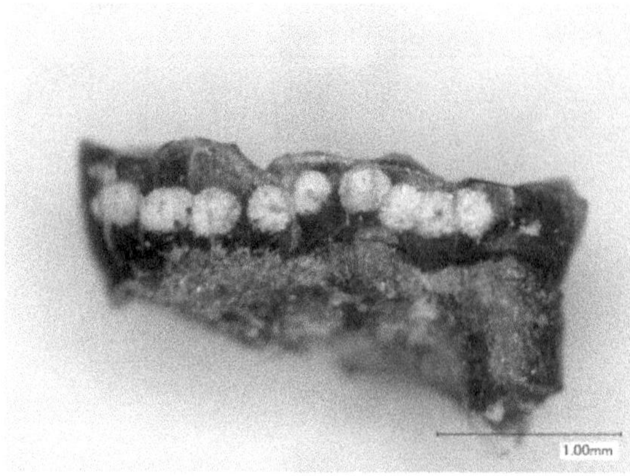

FIGURE 8.4 Laser ablation pits on a fossilised rodent tooth of Pliocene age (~3 million years old).

Sharp and Cerling, 1998). Therefore, serial $\delta^{18}O$ values extracted from powdered enamel along the growth axis of unworn teeth of a mammal may reflect the seasonal variation during the development of the tooth (Green et al., 2018). Such a serial sampling technique has been widely applied to extinct mammoths and mastodons to understand climate change, seasonal dietary shifts and tooth enamel growth rates (Koch et al., 1998; Feranec and MacFadden, 2000; Metcalfe and Longstaffe, 2012, 2014; Ma et al., 2019). Patnaik et al. (2019) applied a serial sampling method to extract $\delta^{18}O$ and $\delta^{13}C$ data from fossil and extant proboscideans and found a clear shift in diet during the Late Neogene (Figure 8.5).

8.4 DENTAL MICROWEAR ANALYSES

Herbivores' feeding habits usually reflect the surrounding vegetation. Grazers are known to eat mostly monocotyledons in an open landscape, whereas browsers prefer mostly dicotyledons, indicating the presence of primarily closed environments. Mixed feeders, on the other hand, may go for both browsing and grazing depending on the availability of the vegetation type. Dental microwear patterns result from the abrasion of the dental enamel by the last food items consumed (Hayek et al., 1992; Mainland, 2003; Merceron et al., 2004, 2005; Merceron and Ungar, 2005; Solounias et al., 1988; Walker et al., 1978). On an average, grazing taxa differ significantly from the browsing species, showing more scratches compared with the number of pits on the dental enamel surface. Within browsers, those that eat only leaves have a lower abrasion rate, exhibiting fewer pits compared with fruit and seed eaters (Figure 8.6). Mixed feeding results in a microwear pattern similar to that of either a grazer or a browser (Merceron et al., 2004, 2005; Solounias and Semprebon, 2002).

A lot of work has been done on the microwear patterns of extinct and extant primates (see El-Zaatari et al., 2005 for a review). First, Phase II or grinding wear facets (f9) are identified on the upper and lower molars of primates (Figure 8.7A). This is followed by a thorough cleaning of the enamel surface using acetone. A mould is prepared using a silicon rubber solution followed by the preparation of a cast using a polymer such as ARALDITE. This cast is mounted on the stub and viewed under a scanning electron microscope. At least two images at the magnification of ×500 are taken. These are then loaded on to software called MICROWARE 4 by Ungar (2002), which can be used to digitally count the number of pits and scratches.

A good example of a frugivore diet is shown by the microwear of *Indopithecus* (Figure 8.7B) and a more varied diet by the Late Miocene small primate *Sivaladapis* (Figure 8.8)

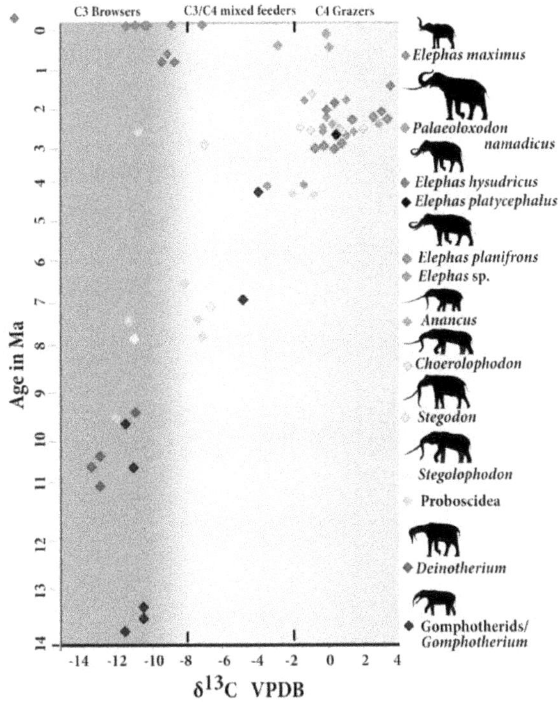

FIGURE 8.5 $\delta^{13}C$ values-based scatter plot of the proboscideans representing the last 14 Ma of the Indian subcontinent.

(Modified from Patnaik et al., 2019.)

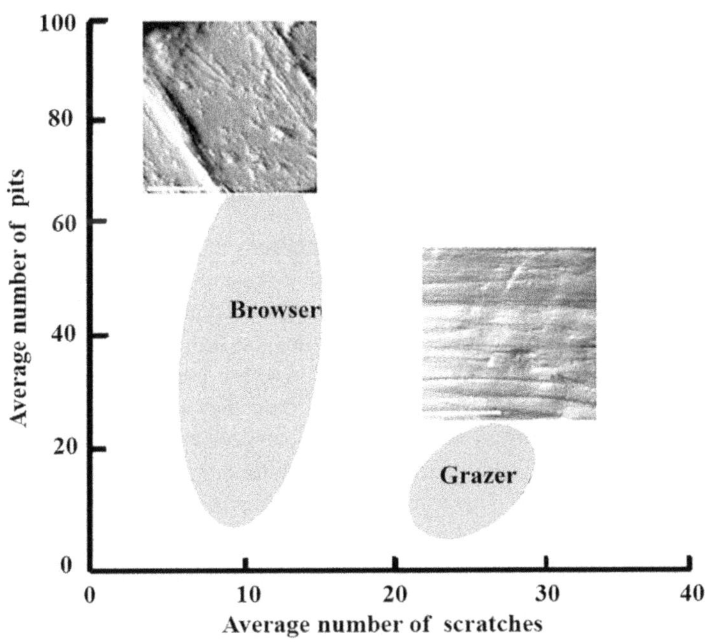

FIGURE 8.6 Graph showing the average number of pits and scratches found in grazer and browser mammals, with scanning electron microscopy of the tooth surface.

(Modified from Merceron et al., 2004.)

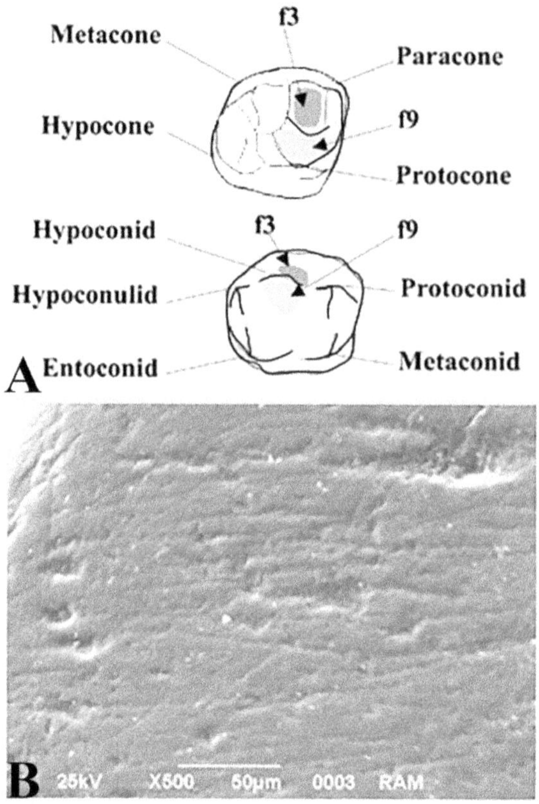

FIGURE 8.7 (A) Wear facets on upper and lower primate molars and (B) dental microwear pattern as seen on the enamel surface of the Miocene ape *Indopithecus*.

(Panel A modified from Merceron et al., 2005.)

8.5 CONCLUSION

There are several techniques available to decipher the diet and environment of extinct mammals. Most of these are based on the uniformitarianism principle, i.e., the present is the key to the past. The kind of diet and the type of habitat of present-day mammals can be used to reconstruct palaeodiet and palaeohabitat. Out of these, the two techniques that are independent of the ecological adaptation of species are the stable isotope studies and dental microwear analyses. Although a lot of work on these lines has been carried out internationally, from the Indian perspective, the data on fossil mammals is woefully lacking. At the national level, there are very few studies on stable isotopes of fossil terrestrial mammalian samples to understand their diet and habitat. However, new results on mammalian dental enamel isotopes and microwear are now being generated on a more regular basis to reconstruct past ecology and climate. These include new data on niche-partitioning among Neogene Siwalik rats and mice based on stable carbon isotopes that has come up very recently, and intra-tooth stable isotope profiles of Siwalik primates to infer intra-annual dietary variability and monsoonal conditions present some 9 Ma ago. Beside primates, both bulk and serial samples of other mammals such as proboscideans from the Middle Miocene to Pleistocene sediments have been carried out to evaluate their diet and habitat. Therefore, there is tremendous scope in this branch of palaeontology and geochemistry to infer the dietary evolution and palaeoclimate of fossil mammals of India.

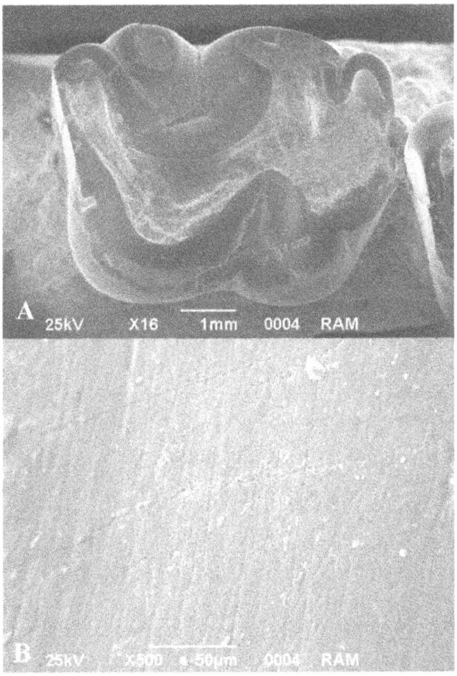

FIGURE 8.8 (A) *Sivaladapis* molar from Late Miocene Haritalyangar locality. (B) Dental microwear showing high number of scratches and low number of pits.

ACKNOWLEDGEMENTS

I would like to extend my sincere thanks to Prof. R.C. Sobti for inviting me to contribute this article. Funding by Science and Engineering Research Board (SERB) project HRR/2018/000063 is thankfully acknowledged.

REFERENCES

Araguas-Araguas L, and Froehlich K (1998). Stable isotope composition of precipitation over southeast Asia. *J. Geophys. Res.* 103(D22), 28721–28742.

Ayliffe L, and Chivas A (1990). Oxygen isotope composition of the bone phosphate of Australian kangaroos: potential as a paleoenvironmental recorder. *Geochim. Cosmochim. Acta* 54, 2603–2609.

Bocherens H, Koch P, Mariotti A, Geraads D, and Jaeger J-J (1996). Isotopic biogeochemistry (13C, 18O) of mammalian enamel from African Pleistocene hominoid sites. *Palaios* 11, 306–308.

Bryant JD, Froelich PN, Showers WJ, and Genna BJ (1996). A tale of two quarries: biologic and taphonomic signatures in the oxygen isotope composition of tooth enamel phosphates from modern and Miocene equids. *Palaios* 11, 397–408.

Cerling TE, and Harris JM (1999). Carbon isotope fractionation between diet and bioapatite in ungulate mammals and implications for ecological and paleoecological studies. *Oecologia* 120, 347–363.

Cerling TE, and Sharp Z (1996). Stable carbon and oxygen isotope analysis of fossil horse tooth enamel using laser ablation. *Palaeogeography, Palaeoclimatology, Palaeoecology* 126, 173–186.

Cerling TE, Harris JM, MacFadden MG, Leakey MG, and Quade J (1997). Global vegetation change through the Miocene/Pliocene boundary. *Nature* 389 , 153–158.

Cerling TE, Harris JM, Leakey MG, and Mudida N (2003). Stable isotope ecology of northern Kenya, with emphasis on the Turkana Basin. In: Leakey MG, Harris JM (Eds.), *Lothagam: The Dawn of Humanity in Eastern Africa*. Columbia University Press, New York, pp. 583–604.

Clementz MT, Hoppe KA, and Koch PL (2003). A paleoecological paradox: the habitat and dietary preferences of the extinct tethythere Desmostylus, inferred from stable isotope analysis. *Paleobiology* 29, 506–519.
Dansgaard W (1964). Stable isotopes in precipitation. *Tellus* 16, 436–468.
El-Zaatari S, Grine FE, Teaford MF, and Smith HF (2005). Molar microwear and dietary reconstructions of fossil Cercopithecoidea from the Plio-Pleistocene deposits of South Africa. *J. Hum. Evol.* 49, 180–205.
Feranec SR, and MacFadden BJ (2000). Evolution of the grazing niche in Pleistocene mammals from Florida: evidence from stable isotopes. *Palaeogeogr. Palaeoclimatol. Palaeoecol.* 162, 155–169.
Fortelius M, and Solounias N (2000). Functional characterization of ungulate molars using the abrasion – attrition wear gradient: a new method for reconstructing paleodiets. *Am. Mus. Novit.* 3301, 1–36.
Fricke HC, and O'Neil JR (1996). Inter- and intra-tooth variation in the oxygen isotope composition of mammalian tooth enamel phosphate: implications for palaeoclimatological and palaeobiological research. *Palaeogeogr. Palaeoclimatol. Palaeoecol.* 126, 91–99.
Green DR, Smith TM, Green GM, Bidlack FB, Tafforeau P, and Colman AS (2018). Quantitative reconstruction of seasonality from stable isotopes in teeth. *Geochim. Cosmochim. Acta* 235, 482–504.
Hayek L-AC, Bernor RL, Solounias N, and Steigerwald P (1992). Preliminary studies of hipparionine horse diet as measured by tooth microwear. *Ann. Zool. Fenn.* 28, 187–200.
Higgins P, and MacFadden BJ (2004). "Amount Effect" recorded in oxygen isotopes of late Glacial horse (*Equus*) and bison (*Bison*) teeth from the Sonoran and Chihuahuan deserts, southwestern United States. *Palaeogeogr. Palaeoclimatol. Palaeoecol.* 206, 337–353.
Hoffman RR (1989). Evolutionary steps of ecophysiological adaptation and diversification view of their digestive system. *Oecologia* 78, 443–457.
Hoppe KA, Koch PL, Carlson RW, and Webb SD (1999). Tracking mammoths and mastodons: reconstruction of migratory behavior using strontium isotope ratios. *Geology* 27, 439–442.
Kimura Y, Jacobs L, Cerling T, Uno K, Ferguson K, Flynn LJ, and Patnaik R (2013). Fossil mice and rats show isotopic evidence of niche partitioning and change in dental ecomorphology related to dietary shift in Late Miocene of Pakistan. *PloS One* 8(8), e69308.
Koch PL, Hoppe KA, and Webb SD (1998). The isotopic ecology of late Pleistocene mammals in North America. *Chem. Geol. Florida* 152, 119–138. doi: 10.1016/S0009-2541(98)00101-6.
Kohn M (1996). Predicting animal ^{18}O: accounting for diet and physiological adaptation. *Geochim. Cosmochim. Acta* 60, 4811–4829.
Kohn MJ, Schoeninger MJ, and Valley JW (1996). Herbivore tooth oxygen isotope compositions: effect of diet and physiology. *Geochim. Cosmochim. Acta* 60, 3889–3896.
Kohn MJ, Schoeninger MJ, and Valley JW (1998). Variability in oxygen isotope composition of herbivore teeth: reflections of seasonality or developmental physiology? *Chem. Geol.* 152, 97–112.
Lee-Thorp JA, Sealy JC, and van der Merwe NJ (1989). Stable carbon isotope ratio differences between bone collagen and bone apatite and their relationship to diet. *J. Archeol. Sci.* 16, 585e599.
Levin NE, Cerling TE, Passey BH, Harris JM, and Ehleringer JR (2006). A stable isotope aridity index for terrestrial environments. *Proc. Natl. Acad. Sci.* 103, 11201–11205.
Lister A (2013). The role of behaviour in adaptive morphological evolution of African proboscideans. *Nature* 500, 331–335.
Longinelli A (1984). Oxygen isotopes in mammal bone phosphate: a new tool for paleohydrological and paleoclimatological research. *Geochim. Cosmochim. Acta* 48, 385–390.
Luz B, and Kolodony Y (1985). Oxygen isotope variations in phosphate of biogenic apatites, IV: mammal teeth and bones. *Earth and Planet. Sci.* 75, 29–36.
Luz, B, Kolodony, Y, and Horowitz, M (1984). Fractionation of oxygen isotopes between mammalian bone-phosphate and environmental drinking water. *Geochim. Cosmochim. Acta* 48, 1689–1693.
Ma J, Wang Y, Jin CZ, Zhang HW, and Hu YW. (2019). A preliminary study of serial stable isotope analysis tracks foraging ecology of fossil Asian elephants in South China. *Vertebrata PalAsiatica* 57(3), 225–240. doi: 10.19615/j.cnki.1000-3118.190327.
MacFadden BJ, Higgins P, Clementz MT, and Jones DS (2004). Diets, habitat preferences, and niche differentiation of Cenozoic sirenians from Florida: evidence from stable isotopes. *Paleobiology* 30, 297–324.
MacNaughton SJ, Tarrants JL, Mac Naughton MM, and Davis RH (1985). Silica as a defence against herbivory and a growth promotor in African grasses. *Ecology* 66(2), 528–535.
Mainland, IL (2003). Dental microwear in grazing and browsing Gotland sheep (*Ovis aries*) and its implications for dietary reconstruction. *J. Archaeol. Sci.* 30, 1513–1527.

Marshall JD, Brooks JR, and Lajtha K (2007). Sources of variation in the stable isotopic composition of plants. *In:* Michener R, Lajtha K (Eds.), *Stable Isotopes in Ecology and Environmental Science*. 2nd ed. Blackwell Publishers, Boston, Massachusetts, pp. 22–60.
Merceron G, and Ungar P (2005). Dental microwear and palaeoecology of bovids from the Early Pliocene of Langebaanweg, Western Cape Province, South Africa. *S. Afr. J. Sci.* 101, 365–370. doi: 10.1007/s10914-007-9050-x.
Merceron G, Blondel C, Brunet M, Sen S, Solounias N, Viriot L, and Heintz E (2004). The Late Miocene paleoenvironment of Afghanistan as inferred from dental microwear in artiodactyls. *Palaeogeogr. Palaeoclimatol. Palaeoecol.* 207, 143–163.
Merceron G, Blondel C, de Bonis L, Koufos GD, and Viriot L (2005). A new method of dental microwear analysis: application to extant primates and *Ouranopithecus macedoniensis* (Late Miocene of Greece). *Palaios* 20, 551–561.
Metcalfe JZ, and Longstaffe FJ (2012). Mammoth tooth enamel growth rates inferred from stable isotope analysis and histology. *Quat. Res.* 77, 424–432.
Metcalfe JZ, and Longstaffe FJ (2014). Environmental change and seasonal behavior of mastodons in the Great Lakes region inferred from stable isotope analysis. *Quat. Res.* 82, 366–377.
Passey BH, and Cerling TE (2006). In situ stable isotope analysis (13C,18O) of very small teeth using laser ablation GC/IRMS. *Chem. Geol.* 235, 238–249.
Patnaik R, Cerling TE, Uno KT, and Fleagle JG (2014). Diet and habitat of Siwalik primates, *Indopithecus, Sivaladapis* and *Theropithecus*. *Ann. Zool. Fenn.* 51(1–2), 123–142.
Patnaik R, Singh NP, Paul D, and Sukumar R (2019). Dietary and habitat shifts in relation to climate of Neogene-Quaternary proboscideans and associated mammals of the Indian subcontinent. *Quat. Sci. Rev.* 224, 105968.
Robert D, and Roland JC (1998). *Organisation Cellulaire. Biologie Végétale, 1*. Douin, Vélizy-Villacoublay, p. 368.
Rozanski K, Araguas-Araguas L, and Gonfifiantini R (1993). Isotopic patterns in modern global precipitation, in climate change in continental isotopic records. *Geophys. Monogr.* 78, 1–36.
Sharp Z, and Cerling T (1998). Fossil isotope records of seasonal climate and ecology: straight from the horse's mouth. *Geology* 60, 219–222.
Solounias N, and Hayek CL-A (1993). New methods of tooth microwear analysis and application to dietary determination of two extinct antelopes. *J. Zool.* 229, 421–445.
Solounias N, and Moelleken SMC (1992). Dietary adaptation of two goat ancestors and evolutionary considerations. *Geobios* 25, 797–809.
Solounias N, and Moelleken SMC (1994). Differences in diet between two archaic ruminant species from Sansan, France. *Geobios* 7, 203–220.
Solounias N, and Semprebon G (2002). Advances in the reconstruction of ungulates ecomorphology with application to early fossil equids. *Am. Mus. Novit.* 3366, 1– 9.
Solounias N, Teaford MF, and Walker A (1988). Interpreting the diet of extinct ruminants: the case of a non-browsing giraffid. *Paleobiology* 14, 287–300.
Sponheimer M, and Lee-Thorp JA (1999). Oxygen isotopes in enamel carbonate and their ecological significance. *J. Archaeol. Sci.* 26, 723–728.
Sponheimer M, Passey BH, De Ruiter DJ, Guatellisteinberg D, Cerling TE, and Lee-Thorp JA (2006). Isotopic evidence for dietary variability in the early hominin *Paranthropus robustus*. *Science* 314, 980–982.
Sternberg LSL, Mulkey SS, and Wright SJ (1989). Oxygen isotope ratio stratification in a tropical moist forest. *Oecologia* 81, 51–56.
Teaford MF, and Oyen D (1989). *In vivo* and *in vitro* turnover in dental microwear. *Am. J. Phys. Anthropol.* 80, 447–460.
Ungar PS (2002). Microware software, version 4.0.2. A semi-automated image analysis system for the quantification of dental microwear. Fayetteville, Arkansas, USA.
Walker A, Hoeck HN, and Perez L (1978). Microwear of mammalian teeth as an indicator of diet. *Science* 201, 908–910.

9 Statistical Modeling for Climate Data

Minakshi Mishra and Pushpraj
Department of Statistics, Babasaheb Bhimrao Ambedkar University
Lucknow, Uttar Pradesh, India

CONTENTS

9.1 INTRODUCTION

Statistical modeling is the main perspective of any natural phenomenon due to the presence of random error (Rencher, 2000). Mathematical modeling does not incorporate the random error component, which makes mathematical modeling inferior to statistical modeling. Statistical modeling has its applications in weather forecasting, medical, agricultural, ecology, astronomy, and biological sciences (Agresti, 2015; Montgomery *et al.*, 2012).

Data and Software Used: The data used in the chapter is of Rainfall and Temperature in India from 1901 to 2020, which can be downloaded from the Climate Knowledge Portal of the World Bank. The URL is available in the Appendix. Various statistical software packages are available for modeling and analysis, like R, SPSS, SAS, Stata, Python, etc. In this chapter, R – Software is used to fit models, and codes are available in the Appendix. The reader should be familiar with R or an introductory course of R, which can be found in "An Introduction to R" by W. N. Venables and D. M. Smith, "Hands on Programming with R" by Hadley Wickham, "R in Action", etc. (Grolemund, 2014; R Core Team, 2021; Teetor, 2011; Wickham & Grolemund, 2016).

9.2 HISTORICAL DEVELOPMENTS

The term "regression" was coined by Sir Francis Galton during the late 19th century. For solving a regression equation, Gauss and Legendre developed the method of least squares, but they did not

DOI: 10.1201/9781003220824-11

coin the term regression; they solved it considering an equation of unknown constants, which is given by:

$$y = \alpha + \beta x + \varepsilon \ldots \quad (9.1)$$

Here, y is the dependent variable, x is the independent variable, α and β are constants that are to be estimated based on data, and ε is the random error. "Error" does not mean "mistake" but is a statistical term representing random fluctuations, measurement errors, or the effect of factors outside our control. For estimating α and β, certain assumptions are to be fulfilled. Then, various methods for solving Equation (9.1), e.g. method of maximum likelihood, method of generalized least squares, etc., were developed (Agresti, 2015; Draper & Smith, 2014; Montgomery *et al.*, 2012; Rencher, 2000).

9.3 LINEAR REGRESSION

An approach to modeling the relationship between a response variable (scalar) and one or more explanatory variables (also known as dependent and independent variables). The case of one explanatory variable is called simple linear regression; for more than one, the process is called multiple linear regression (Montgomery *et al.*, 2012).

9.3.1 Simple Linear Regression

In simple linear regression, an attempt is made to model the linear relationship between two variables, for example, rainfall and temperature, or the dose of a drug and response (Rencher, 2000). For a linear relationship, a model of the form as defined in Equation (9.1) is used. Certain assumptions are made about the distribution of the error terms, independence of the observed values of y, and so on. Using observed values of x and y, estimates of α and β are obtained, and inferences are made, such as confidence intervals and tests of hypotheses for α and β. We may also use the estimated model to forecast or predict the value of y for a particular value of x, in which case a measure of predictive accuracy may also be of interest. Considering the data described earlier, the simple linear regression model can be fitted for predicting average rainfall on the basis of average temperature in a year.

The regression equation is given by:

Average Rainfall = 164.933 – 3.147*Average Temperature.

Various estimates like standard error of α and β and their significance are shown in Table 9.1. The adjusted R-squared value is 0.01455, which implies that the model is poorly fitted.

Figure 9.1 shows the scatterplot and fitted line of Average Temperature versus Average Rainfall in a year. This plot was prepared using the ggplot2 package (Wickham, 2016).

9.3.2 Multiple Linear Regression

The response y is often influenced by more than one predictor variable. For example, the yield of a crop may depend on the amount of nitrogen, potash, and phosphate fertilizers used. These variables

TABLE 9.1
Regression Estimates and Their Coefficients

Coefficients	Estimates	Std. Error	t value	Pr(>|t|) [p – value]
Intercept (α)	164.933	45.971	3.588	0.000487
Slope (β)	−3.147	1.895	−1.660	0.0995

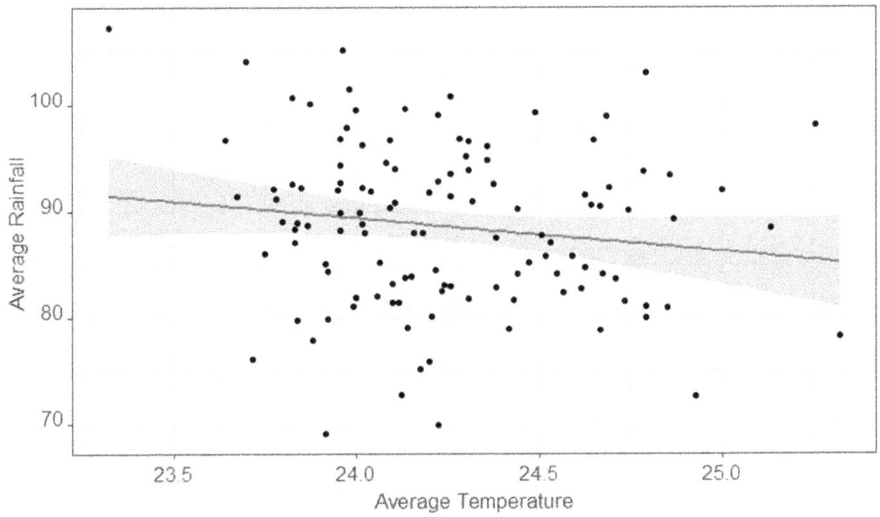

FIGURE 9.1 Scatter plot and fitted line of average temperature versus average rainfall in a year.

TABLE 9.2
Estimates and Their Coefficients

Coefficients	Estimates	Std. Error	t value	Pr(>|t|) [p – value]
Intercept (α)	133.061	36.956	3.601	0.000467
Slope (β_1)	2.769	1.259	2.199	0.029822
Slope (β_2)	–4.559	1.218	–3.742	0.000284

are controlled by the experimenter, but the yield may also depend on uncontrollable variables such as those associated with weather.

$$y = \alpha + \beta_1 x_1 + \beta_2 x_2 + \dots \beta_k x_k + \varepsilon \dots \tag{9.2}$$

Using observed values of x_1, x_2, ..., x_k, and y, estimates of α, β_1, β_2,..., β_k are obtained, and inferences are made such as confidence intervals and tests of hypotheses for α, β_1, β_2,..., β_k.

The multiple linear regression model may be fitted for the median rainfall as a dependent variable and minimum & maximum temperature as independent variables. The regression equation is given by:

Median Rainfall = 133.061 + 2.769*Minimum Temperature – 4.559*Maximum Temperature.

Various estimates like standard error of α, β_1, and β_2 and their significance are shown in Table 9.2. The adjusted R-squared value is 0.1013, which implies that the model is better fitted than the simple linear regression model.

Figure 9.2 shows the scatterplot and fitted line of minimum temperature versus median rainfall in a year. It can be seen that the fitted line shows an upward trend, i.e., slope is positive for the simple linear regression model, and it also has a positive slope with the multiple linear regression model, as shown in Table 9.2.

Figure 9.3 shows the scatterplot and fitted line of maximum temperature versus median rainfall in a year. It can be seen that the fitted line shows a downward trend, i.e., slope is negative for the simple linear regression model, and it also has a negative slope with the multiple linear regression model, as shown in Table 9.2.

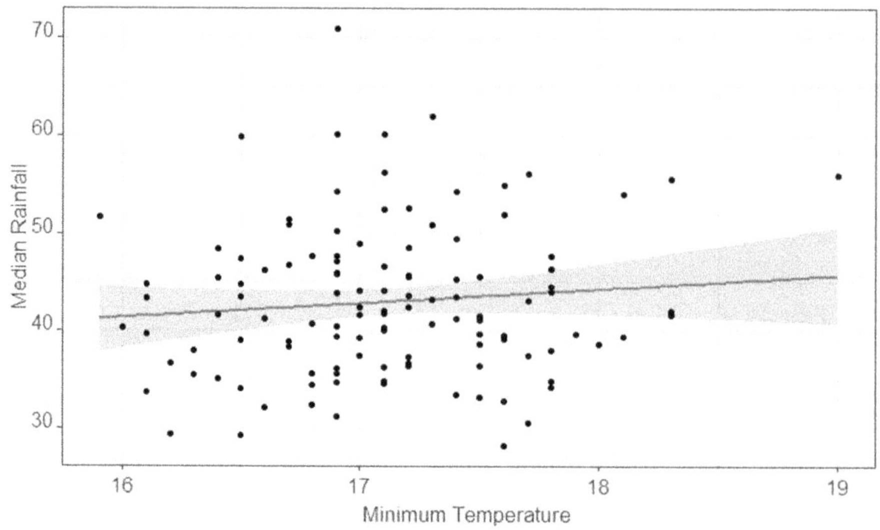

FIGURE 9.2 Scatter plot and fitted line of minimum temperature versus median rainfall in a year.

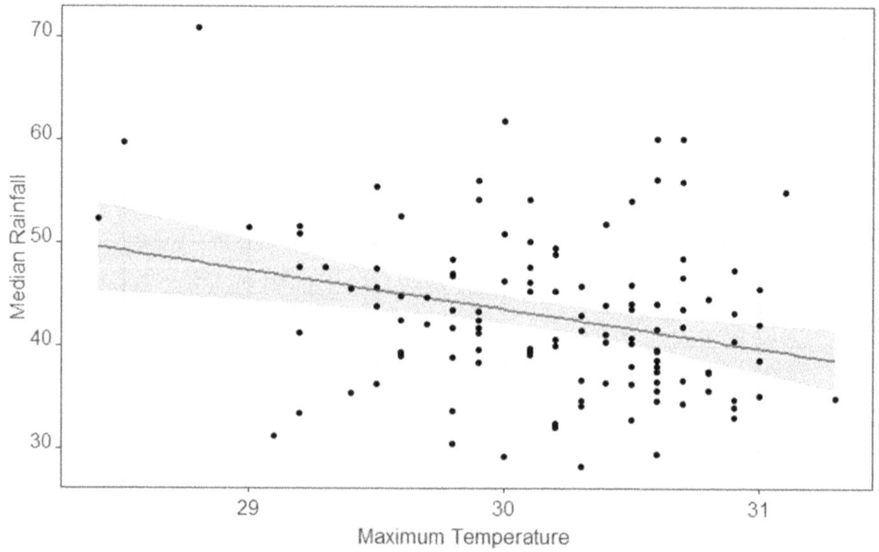

FIGURE 9.3 Scatter plot and fitted line of maximum temperature versus median rainfall in a year.

9.4 ARTIFICIAL NEURAL NETWORK

In the human body, the brain is the central processing unit for performing all the functions. It has around 86 billion neurons. A neuron may be defined as a cell transmitting nerve impulses or electrochemical signals. The brain is a complex network of neurons that process information through a system of several interconnected neurons (Liu et al., 2010; Salami & Ehteshami, 2016). An artificial neural network (ANN) uses statistical modeling techniques for fitting a model, but these techniques do not provide any parametric model, due to which these methods are called Black Box methods, and they can be used only in computer systems. With the help of an ANN, a model is fitted, and the value of a dependent variable is predicted by providing values of independent variables. ANNs

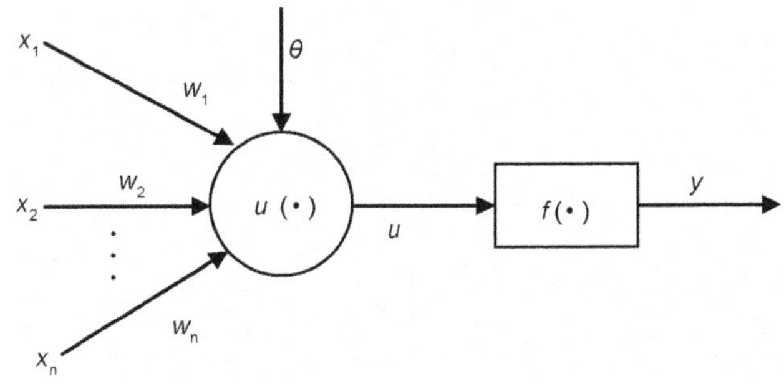

FIGURE 9.4 General neuron model.

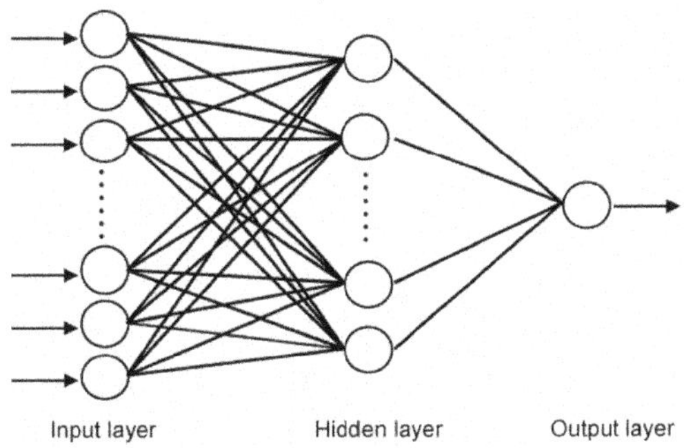

FIGURE 9.5 Neural network topology.

are mathematical algorithmic models that imitate the neural network behavior characteristics of animals and carry out parallel distributed information processing. These networks rely on the complexity of a system by adjusting large numbers of interconnected relationships between nodes for processing information. The general form of a neural network is shown in Figure 9.4. It is often used to model high-dimensional, nonlinear data. In the forecasting environment, neural networks are sometimes used to solve prediction problems instead of using a formal model building approach or the development of the underlying knowledge of the system that would be required to develop an analytical forecasting procedure. Multilayer feed forward ANNs are multivariate statistical models used to relate n predictor variables $X_1, X_2 \ldots X_n$ to one or more output or response variables y, as shown in Figure 9.4, where θ is the bias, $W_1, W_2, \ldots W_n$ are weights for $X_1, X_2 \ldots X_n$, respectively, and u(.) shows some linear or nonlinear combination of $X_1, X_2 \ldots X_n$. In a forecasting application, the inputs could be explanatory variables such as would be used in a regression model, and they could be previous values of the outcome or response variable. The model has several layers, each consisting of either the original or some constructed variables. The most common structure involves three layers: the inputs, which are the original predictors; the hidden layer, comprised of a set of constructed variables; and the output layer, made up of the responses, as shown in Figure 9.5. Each variable in a layer is called a node. The transformation functions are usually either sigmoidal (S-shaped) or linear and are usually called activation functions or transfer functions, shown in Figure 9.4 as f(.) (Liu *et al.*, 2010; Montgomery *et al.*, 2015). For determining the accuracy of a neural network model, the data is divided into two parts, namely, the training data set and testing. The difference between predicted and observed values of the testing data set will be considered for determining the accuracy of the model. A neural network model was fitted for the climate data as described earlier using a neural net package (Fritsch *et al.*, 2019), and the codes for the same are available in the Appendix. Figure 9.6 shows a neural network model fitted for median rainfall as a

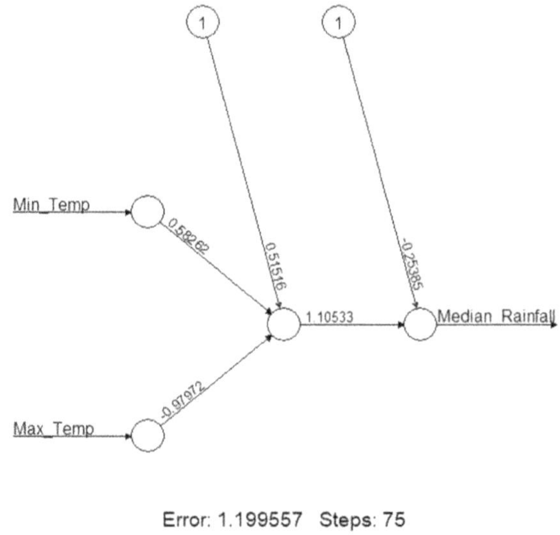

FIGURE 9.6 Neural network model fitted for median rainfall, minimum, and maximum temperature.

dependent variable and minimum and maximum temperature as independent variables. The results of the neural network model can be found in result.matrix of the print function and commands available in the Appendix.

9.5 TIME SERIES MODELING

Data that is observed at different time points leads to new and unique problems in statistical modeling and inference. Obviously, the correlation introduced by adjacent time points can severely restrict the applicability of the many conventional statistical methods traditionally dependent on the assumption that these adjacent observations are independent and identically distributed. The systematic approach of answering the mathematical and statistical questions posed by the time correlations is commonly referred to as time series analysis (Shumway & Stoffer, 2000). For detailed information, the reader should go through (Montgomery *et al.*, 2015; Shumway & Stoffer, 2000). The time series model in the terms of regression is represented by:

$$y_t = \beta_0 + \beta_1 x_{t1} + \beta_2 x_{t2} + \cdots + \beta_k x_{tk} + \varepsilon_t,\ t = 1, 2, \ldots, T \ldots \tag{9.3}$$

where T is the most recent time/period.

The trend of average temperature over the years for the climate data is represented by:

Average Temperature = 9.51 + 0.0075183 * Year.

The time series plot of average temperature with a linear fitted line is shown in Figure 9.7.

9.5.1 Autoregressive Models

The idea behind the autoregressive models is that the current value of the series, X_t, can be explained as a function of p past values, $X_{t-1}, X_{t-2}, \ldots, X_{t-p}$, where p is the number of steps required into the past to forecast the current value. It is represented by AR(p). The model takes the form:

$$X_t = \varphi_1 X_{t-1} + \varphi_2 X_{t-2} + \cdots + \varphi_p X_{t-p} + w_t, \ldots \tag{9.4}$$

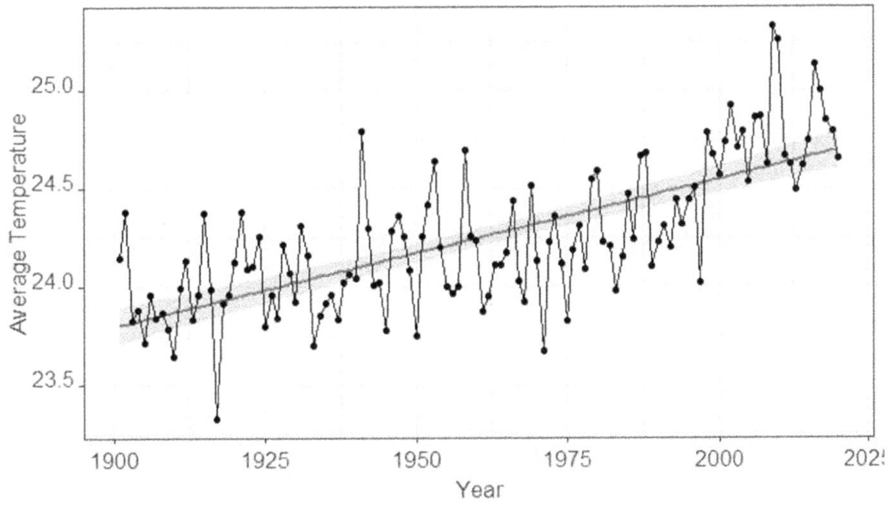

FIGURE 9.7 Time series plot of average temperature.

where X_t is stationary, $w_t \sim wn(0, \sigma^2_w)$, wn is white noise, and $\varphi_1, \varphi_2, \ldots, \varphi_p$ are constants ($\varphi_p \neq 0$). The mean of X_t in Equation (9.4) is zero. If the mean, μ, of X_t is not zero, replace X_t by $X_t - \mu$. From Equation (9.4),

$$(X_t - \mu) = \varphi_1(X_{t-1} - \mu) + \varphi_2(X_{t-2} - \mu) + \cdots + \varphi_p(X_{t-p} - \mu) + w_t$$

The autoregressive model of order 1, i.e., AR(1) for the average temperature in climate data, is given by:

$$(X_t - 24.2545) = 0.6587*(X_{t-1} - 24.2545) + w_t$$

9.5.2 Moving Average Models

The moving average model of order q is represented by MA(q). In this model, the X_t is assumed to be combined linearly with white noise w_t.

The model takes the form:

$$X_t = \mu + w_t - \theta_1 w_{t-1} - \theta_2 w_{t-2} - \cdots - \theta_q w_{t-q}, \ldots \quad (9.5)$$

where $w_t \sim wn(0, \sigma^2_w)$, wn is white noise, and $\theta_1, \theta_2, \ldots, \theta_q$ ($\theta_q \neq 0$) are parameters.

The moving average model of order 1, i.e., MA(1), for the average temperature in the climate data is given by:

$$X_t = 24.2493 + w_t - 0.6285 * w_{t-1}$$

9.5.3 Autoregressive Moving Average Models

A time series $\{X_t; t = 0, \pm1, \pm2, \ldots\}$ is termed ARMA(p, q) if it is stationary and takes the form:

$$X_t = \delta + \varphi_1 X_{t-1} + \varphi_2 X_{t-2} + \cdots + \varphi_p X_{t-p} + w_t - \theta_1 w_{t-1} - \theta_2 w_{t-2} - \cdots - \theta_q w_{t-q}, \ldots \quad (9.6)$$

with ($\varphi_p \neq 0$), ($\theta_q \neq 0$), and $\sigma^2_w > 0$. The parameters p and q are called the autoregressive and the moving average orders, respectively.

When q = 0, the model takes the form of an autoregressive model of order p, AR(p), and when p = 0, the model takes the form of a moving average model of order q, MA(q).

The autoregressive moving average model of order (1,1), i.e., ARMA(1,1), for the average temperature in the climate data is given by:

$$X_t = 24.3387 + 0.9898\, X_{t-1} + w_t - 0.8061\, w_{t-1}$$

9.6 CONCLUSION

This chapter is devoted to the modeling of climate data using R software. The methods discussed here are also applicable to a wide variety of problems occurring in engineering, management, biomedical, earth sciences, demography, etc. Step-by-step analyses along with R codes are suggested for data handling, plotting, modeling, and data fitting. Summing up, it can be concluded that this chapter would serve as a guide to scientists, engineers, researchers, and other practitioners to analyze real-life data sets using the important statistical software R.

9.7 APPENDIX

The data can be downloaded from the website https://climateknowledgeportal.worldbank.org/download-data. The data downloaded is of India for temperature (Celsius) and rainfall (mm) from 1901 to 2020 as csv files. Monthly average is available in the csv files as column name Statistics. Install R from the website www.r-project.org/ and R Studio from the URL www.rstudio.com/products/rstudio/download/ and run the R Studio so that the tasks become easier. Now, load csv files in R Studio by the following commands. Then, the analysis and graphs can be generated. Some packages are required, so first install the following packages and their dependencies:

1. Tidyverse
2. Rfit
3. grid
4. gridExtra
5. neuralnet

R – COMMANDS

```
## Represent Comments
rm(list=ls(all=TRUE)) ## Removing all objects
library(tidyverse)
library(grid)
library(gridExtra)
## Loading Data from Working Directory by Default(Documents)
data_tempreture <- read.csv(file = "tas_1901_2020_IND.csv")
data_rainfall <- read.csv(file = "pr_1901_2020_IND.csv")
## Merging datasets into one data frame and select only useful variables
data <- data.frame(Year = data_rainfall$Year, Rainfall = data_rainfall$Rainfall....MM.,
Temperature = data_tempreture$Temperature....Celsius.,
Statistics = data_tempreture$Statistics)
## Filtering data for a particular month
data1 = data[which(data$Statistics== " Jan Average"),]
```

```
## Summarizing some statistics like Average, Minimum, Maximum Rainfall and Temperature
# in a year
t <- data %>% group_by(Year) %>% summarise(Min_Rainfall = min(Rainfall),
   Max_Rainfall = max(Rainfall),
   Avg_Rainfall = mean(Rainfall),
   Median_Rainfall = median(Rainfall),
   Min_Temp = min(Temperature),
   Max_Temp = max(Temperature),
   Avg_Temp = mean(Temperature),
   Median_Temp = median(Temperature))
## Scatter Plot and Smooth Line Plot of Average Temperature v/s Average Rainfall in a Year
p=ggplot(t, aes(x=Avg_Temp,y= Avg_Rainfall)) + geom_smooth(method=lm,se = T) + theme_
   bw() + ylab("Average Rainfall") + xlab("Average Temperature")
p= p+ geom_point()
p=p+ theme(axis.text.x = element_text(color = "grey20", size = 12, angle = 0, hjust = .5, vjust = .5,
   face = "plain"),
   axis.text.y = element_text(color = "grey20", size = 12, angle = 0, hjust = 1, vjust = 0,
      face = "plain"),
axis.title.x = element_text(color = "grey20", size = 13, angle = 0, hjust = .5, vjust = 0, face = "plain"),
axis.title.y = element_text(color = "grey20", size = 13, angle = 90, hjust = .5, vjust = .5,
   face = "plain"))
grid.arrange(p, bottom = textGrob("Figure 9.1 Scatter Plot and Fitted line of Average Temperature
   v/s Average Rainfall in a Year", gp=gpar(fontsize=13)))
## Fitting a linear model i.e. Simple linear Regression model
# for Average Rainfall and Average Temperature with plots and summary
y=lm(Avg_Rainfall ~ Avg_Temp, data = t)
summary(y)
plot(y)
## Fitting a Multiple linear Regression model for Median Rainfall as dependent variable
# and Minimum and Maximum Temperature as independent variables with plots and summary
y=lm(Median_Rainfall ~ Min_Temp + Max_Temp, data = t)
summary(y)
plot(y)
## Scatter Plot and Fitted Line of Minimum Temperature v/s Median Rainfall in a Year
p=ggplot(t, aes(x=Min_Temp,y= Median_Rainfall)) + geom_smooth(method=lm,se = T) +
   theme_bw() + ylab("Median Rainfall") + xlab("Minimum Temperature")
p= p+ geom_point()
p=p+ theme(axis.text.x = element_text(color = "grey20", size = 12, angle = 0, hjust = .5, vjust = .5,
   face = "plain"),
   axis.text.y = element_text(color = "grey20", size = 12, angle = 0, hjust = 1, vjust = 0,
      face = "plain"),
axis.title.x = element_text(color = "grey20", size = 13, angle = 0, hjust = .5, vjust = 0, face = "plain"),
axis.title.y = element_text(color = "grey20", size = 13, angle = 90, hjust = .5, vjust = .5, face = "plain"))
grid.arrange(p, bottom = textGrob("Figure 9.2 Scatter Plot and Fitted Line of Minimum
   Temperature v/s Median Rainfall in a Year", gp=gpar(fontsize=13)))
## Scatter Plot and Fitted Line of Maximum Temperature v/s Median Rainfall in a Year
p=ggplot(t, aes(x=Max_Temp,y= Median_Rainfall)) + geom_smooth(method=lm,se = T) +
   theme_bw() + ylab("Median Rainfall") + xlab("Maximum Temperature")
p= p+ geom_point()
```

```
p=p+ theme(axis.text.x = element_text(color = "grey20", size = 12, angle = 0, hjust = .5, vjust = .5,
  face = "plain"),
  axis.text.y = element_text(color = "grey20", size = 12, angle = 0, hjust = 1, vjust = 0,
    face = "plain"),
axis.title.x = element_text(color = "grey20", size = 13, angle = 0, hjust = .5, vjust = 0, face = "plain"),
axis.title.y = element_text(color = "grey20", size = 13, angle = 90, hjust = .5, vjust = .5,
  face = "plain"))
grid.arrange(p, bottom = textGrob("Figure 9.3 Scatter Plot and Fitted Line of Maximum
  Temperature v/s Median Rainfall in a Year", gp=gpar(fontsize=13)))
## Non – Parametric Regression
library(Rfit)
y=rfit(Avg_Rainfall ~ Avg_Temp, data = t)
summary(y)
plot(y$fitted.values, y$residuals)
## Artificial Neural Network Modeling
library(neuralnet)
max_data <- apply(t, 2, max)
min_data <- apply(t, 2, min)
t_scaled <- scale(t,center = min_data, scale = max_data – min_data)
x <- nrow(t) ## Number of observations in a column of summarized dataset
y <- sample(1:x, 0.7*x, replace = F) ## Taking a sample of 70% observations
t1 <- t_scaled[y,] ## Training Dataset
t2 <- t_scaled[-y,] ## Testing Dataset
nn <- neuralnet(Median_Rainfall ~ Min_Temp + Max_Temp, data = t1, linear.output = T)
print(nn) ##Printing Neural Net Result
plot(nn)
predict(nn, t2, all.units = T)
prediction(nn)
## Time Series Modeling
##Linear Model
y=lm(Avg_Temp ~ Year, data = t)
summary(y)
## Time Series Plot
p=ggplot(t, aes(x=Year,y= Avg_Temp)) + geom_smooth(method=lm,se = T) +theme_bw() +
  ylab("Average Temperature") + xlab("Year")
p= p+ geom_point() + geom_line()
p=p+ theme(axis.text.x = element_text(color = "grey20", size = 12, angle = 0, hjust = .5, vjust = .5,
  face = "plain"),
axis.text.y = element_text(color = "grey20", size = 12, angle = 0, hjust = 1, vjust = 0,
  face = "plain"),
axis.title.x = element_text(color = "grey20", size = 13, angle = 0, hjust = .5, vjust = 0,
  face = "plain"),
axis.title.y = element_text(color = "grey20", size = 13, angle = 90, hjust = .5, vjust = .5,
  face = "plain"))
grid.arrange(p, bottom = textGrob("Figure 9.7 Time Series Plot of Average Temperature", gp=
  gpar(fontsize=13)))
## Auto REgressive model of order 1
arima(t$Avg_Temp, order = c(1,0,0))
## Moving Average Model of Order 1
```

```
arima(t$Avg_Temp, order = c(0,0,1))
## ARMA model of order(1,1)
arima(t$Avg_Temp, order = c(1,0,1))
```

REFERENCES

Agresti, A. (2015). *Foundations of Linear and Generalized Linear Models*. Wiley.

Draper, N. R., & Smith, H. (2014). *Applied Regression Analysis*. Wiley.

Fritsch, S., Guenther, F., & Wright, M. N. (2019). neuralnet: Training of Neural Networks. https://cran.r-project.org/package=neuralnet

Grolemund, G. (2014). *Hands-On Programming with R: Write Your Own Functions and Simulations*. O'Reilly Media.

Liu, Z. L., Peng, C. H., Xiang, W. H., Tian, D. L., Deng, X. W., & Zhao, M. F. (2010). Application of artificial neural networks in global climate change and ecological research: An overview. *Chinese Science Bulletin*, *55*(34), 3853–3863. https://doi.org/10.1007/s11434-010-4183-3

Montgomery, D. C., Jennings, C. L., & Kulahci, M. (2015). *Introduction to Time Series Analysis and Forecasting*. Wiley.

Montgomery, D. C., Peck, E. A., & Vining, G. G. (2012). *Introduction to Linear Regression Analysis*. Wiley.

R Core Team (2021). *R: A Language and Environment for Statistical Computing*. www.r-project.org/

Rencher, A. C. (2000). *Linear Models in Statistics*. Wiley.

Salami, E. S., & Ehteshami, M. (2016). Application of neural networks modeling to environmentally global climate change at San Joaquin Old River Station. *Modeling Earth Systems and Environment*, *2*(1), 1–9. https://doi.org/10.1007/s40808-016-0094-x

Shumway, R. H., & Stoffer, D. S. (2000). *Time Series Analysis and Its Applications*. Springer.

Teetor, P. (2011). *R Cookbook: Proven Recipes for Data Analysis, Statistics, and Graphics*. O'Reilly Media.

Wickham, H., & Grolemund, G. (2016). *R for Data Science: Import, Tidy, Transform, Visualize, and Model Data*. O'Reilly Media.

Wickham, H. (2016). *ggplot2: Elegant Graphics for Data Analysis*. Springer-Verlag. https://ggplot2.tidyverse.org

10 Knowledge Discovery Paradigms for Climate Change

Comparative Evaluation

Anu Gupta and Indu Chhabra
Department of Computer Applications, Panjab University, Chandigarh

CONTENTS

10.1 INTRODUCTION

The advent of data analytics undoubtedly contributed to the shift of research orientation. Though it may seem that they are applicable only for deriving valuable postulates and guidelines, they are more crucial for predictive analysis. Data mining, knowledge software, is a pattern analysis of unseen, interesting, but useful and actionable data from heterogeneous sources. It identifies interesting patterns using intelligent methods. The statistical description and data visualization filter out the quality data of the information industry.

The selection of data mining knowledge paradigm, interestingness measures, and their threshold ranges can simplify the knowledge derivation process, and hence, reliability and efficiencies are improved for the targeted predictions. Climate science, as information-concentrated learning, has primarily been influenced by the period of enormous information and its viability for the issue of environmental change [1]. The accessibility and collection of vast amounts of data have validated the technological advancements for their practical implementation. Earth is a complex dynamic system [2], and the development of data analytics is really a great challenge regardless of available data resources. Climate change, being knowledge-driven, has been a research focus over the past several decades [3]. The prime importance of knowledgeable data analytics in climate change is management and utilization of these resources for human benefit. Besides building up a mathematical multi-dimensional climate monitoring system, these data-intensive analytics have introduced dynamism in prediction and classification such as weather forecasting, early warning of natural

DOI: 10.1201/9781003220824-12

disasters, and traffic forecasting. Further, the application of an appropriate data mining technique [4] helps in the knowledge extraction of the potential associations and causal inferences.

The implementation of intelligent analytics through cloud computing [5] and Internet of Things have added significance. Weather forecasting can prepare human beings to ready themselves for any undesirable climate. Denissen et al. [6] have studied the effect of weather parameters like temperature, humidity, and wind speed extracted from various sensors to show the importance of weather to the daily life of human beings.

To identify trends embedded in the data, data mining uses different scientific and mathematical models for digging [7]. Being data-search proficient, it uses various statistical algorithms to find meaning in data and hence, to discover patterns and correlations within the databases. It actually balances the other data analysis techniques of statistics, on-line analytical processing (OLAP), spreadsheets, and the other basic data access tools for knowledge derivation. Its classification and association discovery techniques are utilized to make predictions as well as to describe the behavior of the patterns. However, these well-established techniques have yielded performance improvements for generalized applications, but ad hoc circumstances and the complexity of the situation require a customized architecture.

10.2 LITERATURE SURVEY

A huge, multi-dimensional information framework has been set up that makes it possible to screen the changes rapidly for different earth and environmental boundaries [8]. Dagade et al. [9] dealt with building an information paradigm utilizing Hadoop to investigate the climate boundary for extraction and investigation. Riyaz and Surekha [10] investigated the temperature-based climate National Climatic Data Center (NCDC) data through data mining program execution. They confirmed that Hadoop is useful for climate information due to technical significance. Katal, Wazid, and Goudar [11] discuss the different issues and implementation through well-known practices for big data. The specialized research-oriented difficulties, like internal failure handling, adaptability, and tackling heterogeneous information, are referenced. Equal programming models like MapReduce, Spark, and Distributed File System are proposed as decent instruments for big data.

The mining optimization requires comprehensive data analytics skills to frame the optimal infrastructure to maximize the inference performance over different cases. Schnase et al. [12] have explored Modern-Era Retrospective Analysis for Research and Applications (MERRA) analytics for climate science to target a broad range of users. It provides the complete life cycle management of large-scale scientific repositories targeting research applications. The ClimateSpark knowledge framework platform proposed by Hu et al. [13] aims to provide climate data analytics for computational efficiency. For energy efficiency management, a systematic review of data analytics is presented by Eggiman et al. [14] for rainfall data management, flood risk management, and water productivity optimization. Association Rule Mining [15] has been aimed at investigating co-occurrences among data objects. It is also confirmed as a practice to catalog the more frequently simultaneously occurring objects against the average co-occurring instances [16].

The main data analytical practices are grouped into four categories: clustering, classification, regression, and association rule mining [17]. A few of the most influential data mining algorithms, incorporated in almost all commercial and open source platforms (k-mean, support vector machine, apriori, kNN, and classification and regression trees) will be reviewed. Classification, based on the fundamental concept of distance measure, is one of the most fundamental big data analysis techniques with a collection of well-established methods [18]. Zakovorotnyi and Seerig [19] presented a practical and semi-automated artificial neural network method to group the daily energy consumption profiles for energy improvement. A decision tree, one of the best- known methods, relates a series of crafted rules for attribute classification. Later, Kamilaris and Prenafeta-Boldu [20] presented a review of big data analytics practices in agriculture, through the deep learning implementation in agriculture to minimize the classification error. A DSS-based novel energy architecture was developed by Marinakis et al. [21] for climate data analytics. It provided advanced analytics and dashboard interface services for multiple-sourced data access control. As a case study, the efficiency assessment of China's industry sectors can be found in Liu et al. [22], where the authors have verified the multi-dimensional association rules algorithms to seek possible energy leakages and faults to guide energy conservation. The significance of predictive analytics is thoroughly investigated by Poornima [23] for knowledge extraction and its usage in forecasting.

10.3 PARAMETRIC FRAMEWORK

Climatology, a science of weather prediction and climate change, is investigated through the established techniques of data mining to obtain its latest version of knowledge models as depicted in Figure 10.1. An adapted framework is modelled for different climate measurements through the case study of Practice-Scheduler.

Model Inputs are from major sources of abundant data related to
Business – Web, e-commerce, transactions, stocks
Science – remote sensing, bio informatics, scientific simulation
Society and everyone – news, digital cameras, YouTube

10.4 STRATEGIC ROADMAP

Through association and classification knowledge paradigms corresponding to the dedicated model, a pattern analyzer algorithm is developed that can learn how to classify the requested data into related clusters. In the present study, given the various records of weather parameters indicating the climate conditions for the players to practice, the system will predict the near-future weather conditions for the players to practice, or not, for various tournaments.

Case Study: Formulation and implementation of weather data analytics through the knowledge paradigms of market basket analysis and decision tree induction

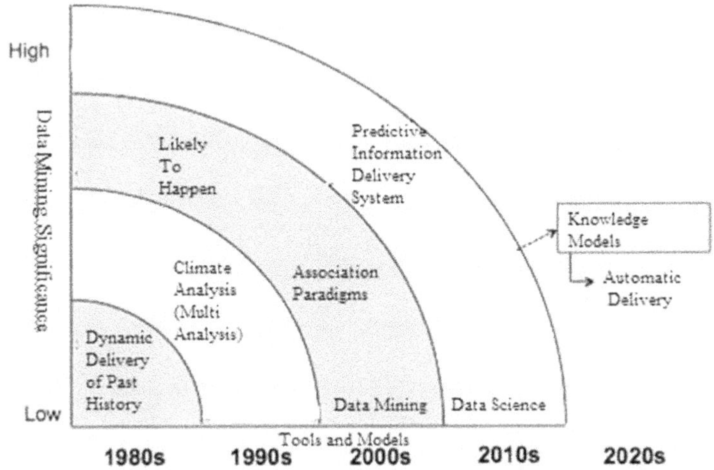

FIGURE 10.1 Parametric climate analysis.

Step 1: Learning through classifier model includes

- Developing classifier model through its algorithm
- Framing the classification rules
- Deciding the learning and threshold parameters according to the given database training tuples and associated class labels "play: Yes" and "play: no"

Step 2: Classifying and optimizing: In this step, the classifier is used for

- Estimating the accuracy through classification rules
- Tuning the training parameters for the test data
- Applying the software for the new data tuples if the accuracy is considered acceptable

10.5 IMPLEMENTATION

A data mining mathematical and logical paradigm, a representation of reality in the form of a model, is proposed. Figure 10.2 is utilized to navigate through a real-time weather database. To build a real-time solution, the associated patterns and relationships present in the data are converted into mathematical postulated logic of market basket analysis (MBA).This abstracted association, when combined with decision tree induction (DTI), allows us to partition feed-forward computations into meaningful classes.

The proposed Practice-Scheduler automatic system is implemented to decide on the appropriate days for practicing for national and international sports based on the analysis of historical weather data. Organizations that host other events such as concerts, festive shows, and bazaars may suffer a lot from frequent weather changes. They need to plan and choose the dates for their events months in advance. They choose the data to be analyzed based on many factors, and the most important factor is the weather, but which dates to choose? This chapter tries to answer that question. All collected data are stored as part of a data warehouse and are then processed and analyzed using MBA and classification DTI data mining techniques. As a result, useful information can be discovered about practice session planning and whether or not the players will be able to practice in such climate conditions, hence not compromising national pride.

The collected data are for three weather dimensions: windy, climate, and moisture. The initial design of the system is shown in Figure 10.2. The data are passed as records to the MBA and

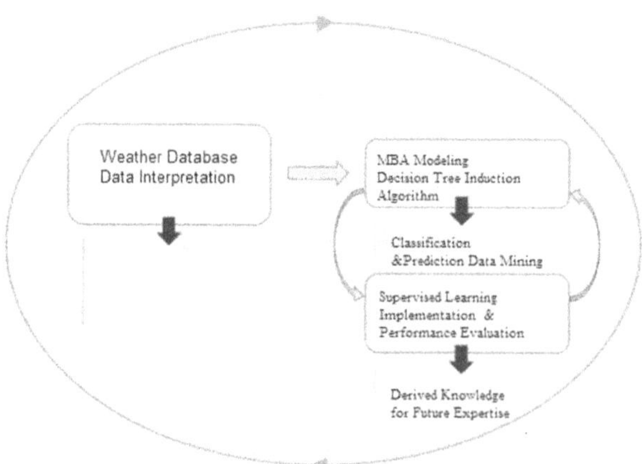

FIGURE 10.2 Implementation – climate prediction through association rule mining.

DTI algorithm. The algorithm applies the interestingness analytics to partition the data into various segments. To check whether or not a day is a good day for practice, the MBA algorithm makes its decision based on four factors. The four factors are temperature, humidity, overcast, and wind speed. The DTI algorithm has a number of cases that define whether or not a day is suitable, based on many conditions. The system investigates the various combinations of climate parameters through the predictive analytics.

10.5.1 Association Rule Framing-Go-Together Parameter Sets

$$\text{If } ((A_{i1} = 1) \wedge ((A_{i2} = 1) \ldots . \wedge ((A_{ik} = 1)) \Rightarrow A_{ik+1} = 1$$

With $1 \leq i_j \leq \rho$ for all j: On day i, the status of j weather parameters
ρ:threshold for the days to be analyzed

10.5.2 Frequency (Support) and Accuracy (Confidence) of Rules

Let Θ = Cloud affecting parameter-combinations
$f_\gamma(\Theta)$ = No. of cases which satisfy Θ
$f_\gamma(\Theta \wedge \Psi)$ = **Support** for $(\Theta \rightarrow \Psi)$
= No. of cases of Ψ in database that satisfy Θ also
Confidence = accuracy = $c\,(\Theta \Rightarrow \Psi)$
= No. of rows that satisfy Ψ among the no. of rows that satisfy Θ

Given a database $[A_{ij}]$: Indicator Matrix where rows are climate parameters, the "i-combinations" to be analyzed together and the columns are the parameter values themselves, (j) showing their varying range.

Postulates are:

1. Moisture <70: Low, Moisture 70–82: Normal, Moisture >82: High
2. Windy – True or False
3. Climate – Rainy, Sunny, Overcast

In the 12th case, though the climate is sunny, moisture is normal, and the player can play, sometimes, the classifier yields that the player cannot practice. The reason may be that the classifier is under-trained, the database is not sufficient, or this is the boundary case. So, exceptions should be removed by improving association mining through Hashing or advanced techniques of apriori or partition algorithms.

10.6 RESULTS AND OBSERVATIONS

10.6.1 MBA Analysis

Θ: climate, Ø: Windy, Ψ: Moisture

Confidence (Accuracy) = $f_\gamma(\Theta \wedge \Psi) / f_\gamma(\Theta)$

Rule 1: Climate → Moisture yields Practice "No"

If Θ = Sunny and Ψ = High

Then **Confidence** $c\,(\Theta \Rightarrow \Psi) = f_\gamma(\text{Sunny and High}) / f_\gamma(\text{Sunny}) = 5/8 = 0.63$

Results: 63% confidence after training that there is no chance of doing practice due to high humidity (as from Training database |D| in Table 10.1). Similarly for

TABLE 10.1
Training Dataset

MBA	Day	Windy	Climate	Moisture	Practice
	1	True	Rainy	70 (Normal)	No
A-1	2	False	Sunny	87 (High)	No
A-2	3	True	Sunny	96 (Very High)	No
	4	True	Overcast	87 (High)	Yes
	5	False	Rainy	70 (Normal)	Yes
A-3	6	False	Sunny	83 (High)	No
	7	False	Overcast	84 (High)	Yes
	8	True	Sunny	69 (Low)	Yes
	9	True	Rainy	84 (High)	No
	10	False	Rainy	82 (Normal)	Yes
	11	True	Sunny	74 (Normal)	Yes
A-4	12	False	Overcast	82 (Normal)	No
	13	False	Rainy	89 (High)	Yes
	14	True	Rainy	90 (High)	No
A-5	15	False	Sunny	90 (High)	No

Rule 2: Climate → Windy yielding Practice "Yes"
If Θ = Rainy and Ø = Windy = False (5th, 10th, and 13th Tuples)
Then **Confidence** c (Θ⇒Ø) = f_γ(Rainy and False) / f_γ(Rainy)= 3/84 = 0.75
Observation: 75% accuracy over trained database |D| that player can practice for nationals.

10.6.2 Decision Tree Induction Implementation

Climate = Sunny

- Moisture = High: No
- Moisture = Normal: Yes

Climate = Overcast : Yes
Climate = Rainy

- Windy = True: No
- Windy = False: Yes

Results: Here, **knowledge derivation** is that there are more "Yes" (three) chances of playing than "No" (two) chances.

10.7 COMPARATIVE ANALYSIS – PERFORMANCE EVALUATION OF MBA AND DTI

1. As compared to "MBA" yielding results based on interestingness measures of "Support" and "Confidence", only the DTI data mining results in the overall scenario, hence **reducing the generalization error**, which is desired from any customized application for reliable prediction forecasting.

2. Exception handling:
   Query is: Windy = False, Climate = Overcast having high humidity: a player can practice for national finals or not. Decision Tree is

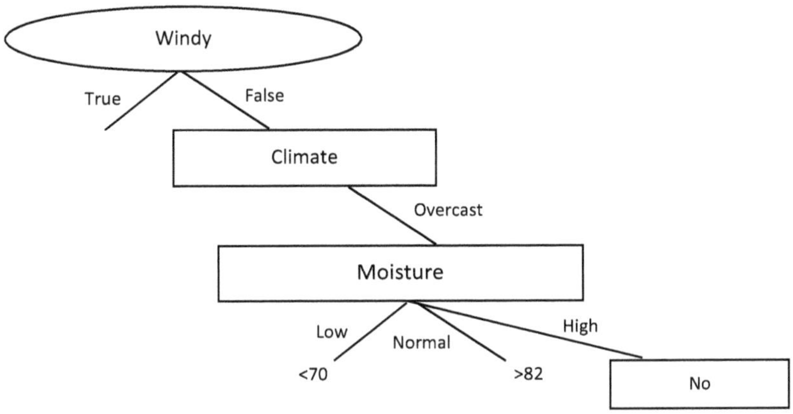

Similarly, though Windy = True, due to sunny climate and normal humidity, the player can practice, as from the 11th database record. So, the ID3 decision algorithm produces.
Windy = True

- Climate = Sunny
- Humidity = Normal: Yes

10.8 Discussion

It is observed through the literature survey that the temperature, pressure, humidity, and wind speed are some of the most important parameters for weather data analysis. Due to the tremendous speed at which weather data is being generated, this is a huge challenge in terms of storage and processing, and the feasible solution is cloud computing and storage.

The area of focus is to optimize the rules generated by association mining using intelligent algorithms. The core motivation is the discovery of globally searched prediction rules through the better understanding of attribute behavior. The optimization is demonstrated through the analysis of cloud parameter values, which definitely will help in framing the rule-based system for delivering better business forecasting for user satisfaction. Though MBA and DTI give good results for a limited-size database, voluminous data may result in latency of data access, causing slower performance. To avoid this latency, the more advanced apriori and partition algorithms may perform better for big data stored on disks, hence providing in-memory computing. Data mining technologies like Hadoop and Spark can be used efficiently to handle the unpredictable weather data generated daily. Also, a technology benchmarking comparison for Hadoop and Spark is important to study their suitability for weather data analysis.

10.9 Conclusion and Recommendations for Data Analysis

Data mining represents a link from the past history of data preserved for diverse situations. In order to unlock potential information, it performs knowledge analysis that would be too complicated and time-consuming for statisticians to derive previously unknown but useful facts. Weather analytics, being generated from various sensors across many locations simultaneously at great speed, is producing a challenge for storage and processing. Different technologies vary in terms of effectiveness and ease of use. The present research shows the execution and evaluation of association and classification data mining predictive analytics for weather datasets to forecast the climate conditions. The climate change mathematical and logical model is designed and implemented through the knowledge paradigms of MBA and DTI. For prediction and optimization, the interestingness parameters of support and confidence and decision tree branches are inspected for future prediction. The algorithms are implemented for a case study of predicting whether a player can practice for national and international tournaments depending on various weather conditions and hence, to plan the business shows to demonstrate the skills for national pride. The chapter concludes with the issues of these algorithms and futuristic approaches for knowledge discovery paradigms.

REFERENCES

[1] Hassani, H., Huang, X., and Silva, E., Big Data and Climate Change, *Big Data and Cognitive Computing*. 2019, 3(1), 12. https://doi.org/10.3390/bdcc3010012.

[2] Götz, M., Richerzhagen, M., Bodenstein, C., Cavallaro, G., Glock, P., Riedel, M., and Benediktsson, J.A., On Scalable Data Mining Techniques for Earth Science, *Procedia Computer Science*. 2015, 51, 2188–2197, ISSN 1877-0509, https://doi.org/10.1016/j.procs.2015.05.494.

[3] Almgren, K., Alshahrani, S., and Lee, J., *Weather Data Analysis Using Hadoop to Mitigate Event Planning Disasters*. UB Scholar Works, University of Bridgeport, Connecticut, USA, 2019.

[4] Witten, I.H., Frank, E., Hall, M.A., and Pal, C.J., *Data Mining: Practical Machine Learning Tools and Techniques*. Morgan Kaufmann: Burlington, MA, 2016.

[5] Khan, S., Shakil, K.A., and Alam, M., Cloud-Based Big Data Analytics—A Survey of Current Research and Future Directions. In Aggarwal, V.B., Bhatnagar, V., and Mishra, K.D. (eds) *Big Data Analytics*. Springer: Singapore, 2018, pp. 595–604.

[6] Denissen, J.J.A., et al., The Effects of Weather on Daily Mood: A Multilevel Approach, *Emotion*. 2008, 8(5), 662.

[7] Rotondo, A., and Quilligan, F., *Evolution Paths for Knowledge Discovery and Data Mining Process Models*, Springer Nature Singapore Pte Ltd, Published online: March 30, 2020.

[8] Faghmous, J.H., and Kumar, V., A Big Data Guide to Understanding Climate Change: The Case for Theory-hGuided Data Science, *Big Data*. 2014, 2, 155–163.

[9] Dagade, V., Lagali, M., Avadhani, S., and Kalekar, P., Big Data Weather Analytics Using Hadoop, *International Journal of Emerging Technology in Computer Science and Electronics*. April 2015, 14(02), 847–851. ISSN: 0976-1353.

[10] Riyaz, P.A., and Surekha, M.V., Leveraging Map Reduce with Hadoop for Weather Data Analytics, *IOSR Journal of Computer Engineering*. May–June 2015, 17(03), 6–12.

[11] Katal, A., Wazid, M., and Goudar, R.H., Big Data: Issues, Challenges, Tools and Good Practices, *Sixth International Conference on Contemporary Computing (IC3)*, IEEE, 2013, 404–409.

[12] Schnase, J.L., Duffy, D.Q., Tamkin, G.S., Nadeau, D., Thompson, J.H., Grieg, C.M., McInerney, M.A., and Webster, W.P., MERRA Analytic Services: Meeting the Big Data Challenges of Climate Science Through Cloud-Enabled Climate Analytics-As-A-Service, *Computer Environment Urban System*. 2017, 61, 198–211.

[13] Hu, F., Yang, C.P., Duffy, D., Schnase, J.L., and Li, Z., ClimateSpark: An In-memory Distributed Computing Framework for Big Climate Data Analytics, *Computer Geoscience*. 2018, 115, 154–166.

[14] Eggimann, S., Mutzner, L., Wani, O., Schneider, M.Y., Spuhler, D., Moy de Vitry, M., Beutler, P., and Maurer, M., The Potential of Knowing More: A Review of Data-Driven Urban Water Management, *Environment Science Technology*. 2017, 51, 2538–2553.

[15] Agrawal, R., Imielinski, T., and Swami, A., Mining Association Rules Between Sets of Items in Large Databases, *International Conference on Management of Data*, ACM, Washington, DC, 26–28 May 1993, 207–216.

[16] Yun, H., Ha, D., Hwang, B., and Ryu, K.H. Mining Association Rules on Significant Rare Data Using Relative Support, *Journal of System Software*. 2003, 67, 181–191.

[17] Chen, C.P., and Zhang, C.Y., Data-Intensive Applications, Challenges, Techniques and Technologies: A Survey on Big Data, *Information Science*. 2014, 275, 314–347.

[18] Kotsiantis, S.B., Zaharakis, I., and Pintelas, P., Supervised Machine Learning: A Review of Classification Techniques, *Emerging Artificial Intelligence Applications in Computer Engineering*. 2017, 160, 3–24.
[19] Zakovorotnyi, A., and Seerig, A., Building Energy Data Analysis by Clustering Measured Daily Profiles, *Energy Procedia*. 2017, 122, 583–588.
[20] Kamilaris, A., and Prenafeta-Boldu, F.X. Deep Learning in Agriculture: A Survey, *Computer Electronics Agriculture*. 2018, 147, 70–90.
[21] Marinakis, V., Doukas, H., Tsapelas, J., Mouzakitis, S., Sicilia, A., Madrazo, L., and Sgouridis, S., From Big Data to Smart Energy Services: An Application for Intelligent Energy Management. In Zissis et. al (eds) *Future Generations Computer Systems*, 2020, 110, 572–586.
[22] Liu, G., Yang, J., Hao, Y., and Zhang, Y., Big Data-Informed Energy Efficiency Assessment of China Industry Sectors Based on K-Means Clustering, *Journal of Cleaner Production*. 2018, 183, 304–314.
[23] Poornima, S., A Survey of Predictive Analytics Using Big Data with Data Mining, *International Journal of Bioinformatics Research and Applications*, 2018, 14(3), 269–282. Print ISSN: 1744-5485, Online ISSN: 1744-5493.

11 Optimization of Process Variables Using Central Composite Design for Heterotrophic Biological Denitrification

Archna Narula and Sravanthi Veluturla
M S Ramaiah Institute of Technology, Bangalore

R.C. Sobti
Department of Biotechnology, Panjab University, Chandigarh

CONTENTS

11.1 INTRODUCTION

The contamination of groundwater by agricultural practices and meteorological conditions has been rapidly increasing nitrate levels in the water. Nitrate pollution has been universally recognized to have a plethora of negative effects on human health. The possible health consequences of nitrate contamination include methemoglobinemia, blue baby syndrome in infants (Knobeloch et al. 2005),

DOI: 10.1201/9781003220824-13

and the possible formation of n-nitroso compounds in the gastric system (Sandor et al. 2001). The World Health Organization (WHO) has set a safe limit of 45 mg/l NO_3^- to regulate nitrate concentration in drinking waters. Several treatment methods with different performance levels are available for drinking water treatment to reach WHO standards (Choi and Batchelor 2008; Schoeman and Steyn 2001; Park et al. 2008; Robinson-Lora and Brennan 2009), but the most economical process has been biological denitrification.

The denitrifying bacteria utilize different energy sources, namely, organic carbon compounds (organotrophs), inorganic compounds (lithotrophs), and light (phototrophs). The dominant population of denitrifying bacteria appear to be organotrophs, mainly Pseudomonas and Alcaligenes. Bacterial denitrification, as a heterotrophic process, requires an external source of organic carbon to develop the metabolism of bacterial species (organotrophs). Various studies have been conducted to evaluate the potential use of solid natural carbon sources in the biological denitrification process (Soares and Abeliovich 1998). It has been found that the process has significant dependence on temperature, inlet nitrate concentration, flow rate through the reactor, and the microbial strain used (Ovez et al. 2006; Ovez 2006; Archna et al. 2011a; Archna et al. 2011b). Denitrification from nitrate to nitrogen requires four electron transport pathways: to nitrate reductase, to nitrite reductase, to nitric oxide reductase, and to nitrous oxide reductase. Most denitrifiers utilize aerobic respiration to obtain energy, but the processes of aerobic respiration and denitrification are normally distinct because of temperature and metabolic regulation; in the presence of oxygen, the genes encoding denitrifying enzymes are repressed (Rajta et al. 2019). In addition, nitrous oxide reductase is sensitive to the presence of molecular oxygen available during the process.

In view of the substantial dependence of the HBD process on temperature and flow rate, the optimization of these parameters at different levels for the strain of Pseudomonas requires many experiments for conventional techniques. Experimental optimization of multivariable systems follows the one-factor-at-a-time technique and requires more data to determine optimum levels, with prolonged time and low reliability. The primary purpose of the experimental design technique is to understand the interactions among the parameters, which could help in the optimization of experimental parameters and provide statistical models (Alam et al. 2007). An experimental statistical design technique was adopted to formulate the most appropriate temperature, flow rate, and nitrate loading for HBD.

11.2 HISTORICAL DEVELOPMENTS

Nitrate, being a stable and highly soluble ion with low potential for co-precipitation or adsorption, has posed challenges to conventional treatment technologies. The various techniques used for denitrification, in terms of their ease of operation, effectiveness, and cost, explored by various researchers are (Archna et al. 2012):

11.2.1 Chemical Denitrification

Nitrate reduction can be induced under basic pH according to the following reaction:

$$3NO_3^- + 8Fe\,(OH)_2 + 6H_2O \rightarrow NH_3 + 8Fe(OH)_3 + OH^-$$

Experimental data showed that a $Fe{:}NO_3^-$ ratio of about 15: 1 was required in the presence of copper catalyst for the reaction to proceed. This process generated a large quantity of iron sludge and formed ammonia, which required removal by air stripping. The process was associated with high costs.

In chemical denitrification using powdered aluminum and ammonia at a pH of 10.25, denitrification was explained on the basis of the following reactions:

$$3NO_3^- + 2Al + 3H_2O \rightarrow 3NO_2^- + 2Al(OH)_3$$

$$NO_2^- + 2Al + 5H_2O \rightarrow NH_3 + 2Al(OH)_3 + OH^-$$

$$2NO_2^- + 2Al + 4H_2O \rightarrow N_2 + 2Al(OH)_3 + 2OH^-$$

It was shown that 1.16 g of aluminum was required for the reduction of 1 g of nitrate.

Another removal technique used catalytic reduction of nitrate with Pd and Cu catalysts. It was found that Pd-Cu combined catalysts at a ratio of 4 could maximize the nitrate reduction into nitrogen with 80% nitrate removal efficiency, while 84% denitrification efficiency was achieved at ambient temperature and pressure using zero-valent magnesium (Mg(0)) for Mg(0):NO_3^--N molar ratio of 5.8 and pH of 2.

11.2.2 Reverse Osmosis for Denitrification

Reverse osmosis (RO) units can remove nitrates under pressures ranging from 300 to 1,500 psi to reverse the normal osmotic flow of water. Membranes used to date are made of cellulose acetate, polyamides, and composite materials. RO membranes present fouling, compaction, and deterioration with time as associated problems. These problems result from deposition of soluble materials, organic matter, suspended and colloidal particles, pH variations, and chlorine exposure. A pretreatment process is mandatory before RO.

A 15-gpm spiral wound cellulose acetate RO system was tested for 1,000 h, and 65% nitrate removal was observed. RO using both polyamide and cellulose triacetate membranes was tested, and sulfuric acid and sodium hexa-metaphosphate were added to feed water to prevent scaling. Polyamide membranes were found to be more effective than cellulose triacetate membranes.

11.2.3 Electrodialysis for Denitrification

Electrodialysis (ED) removed nitrate by selective removal of undesirable ions through a semi-permeable membrane. In ED, ions moved through membranes from a less concentrated to a more concentrated solution by application of direct electric current. An ED system required a supply of pressurized water (50–75 psi (345–578 kPa)) with pretreatment.

In the electrodialysis reversal (EDR) process, the polarity of the electrodes was reversed two to four times an hour to alter the direction of ion movement. The EDR process reduced scaling and chemical usage compared with conventional ED and was used for the production of drinking water from nitrate-rich water. The nitrate removal efficiency of ED and RO processes was almost the same.

NitRem, an ED-based process, was effective in reducing nitrate concentrations from 50 mg/l or more to less than 25 mg/l. The process had the attractive feature that it included the removal of nitrate without the addition of any chemicals.

11.2.4 Catalytic Denitrification

A catalytic process was developed for the removal of nitrite and nitrate from water. Palladium–alumina catalysts were effective in reducing nitrite to nitrogen (98%) and ammonia in the presence of hydrogen. The lead (5%), copper (1.25%), Al_2O_3 catalyst was found to completely remove nitrate from water having an initial nitrate concentration of 100 mg/l. The reaction was completed in 50 min. The process operated effectively at a temperature of 10°C and pH 6–8.

11.2.5 Electro Catalytic Reduction for Denitrification

NO_3^- from groundwater was removed using an electro catalytic reduction process. A commercially available carbon cloth, surface coated with 30% Rh (rhodium) (1 μg/cm), was used at an applied potential of –1.5 V versus standard calomel electrode (SCE) with a Pt auxiliary electrode. Electro catalytic reduction resulted in lowering of NO_3^- concentrations in groundwater from 73 to 39 mg/l on a timescale of 40–60 min.

11.2.6 Ion Exchange Process

The ion exchange process involved passage of nitrate water through a resin bed containing strong base anion (SBA) exchange resins, on which nitrate ions were exchanged for chloride or bicarbonate ions until the resin was exhausted. The exhausted resin was regenerated using a concentrated solution of sodium chloride or sodium bicarbonate.

Fifteen ion exchange plants used in the United States reduced nitrate from 18 to 6.8 mg/l. Addition of bentonite clay to the backwash water helped restore the resin, which was regenerated with 1 N NaOH and HCI.

11.2.7 Denitrification Using a Membrane Bioreactor

High-quality product water was produced using an immersed heterotrophic membrane bioreactor (MBR). NO_3^--contaminated water was made to flow through the lumen of tubular microporous membranes, where NO_3^- ions diffused through the membrane pores. Denitrification took place on the shell side of the membranes. The MBR achieved over 99% NO_3^-removal at an influent concentration of 200 mg/l NO_3^-.

11.2.7.1 Denitrification Using Nanofiltration

Nanofiltration (NF) made a breakthrough in drinking water production for the removal of nitrate in the last decade. NF70, NF45, UTC-20, and UTC-60 membranes have been experimentally tested for the removal of nitrate. The experimental results were not promising for most membranes, except for NF70, where a 76% nitrate removal was obtained.

11.2.7.2 Biological Denitrification

Anaerobic bacteria belonging to different genera grow by reducing ionic nitrogenous oxides to gaseous products. Nitrates or nitrites serve as the terminal electron acceptors instead of oxygen and result in generation of ATP. The transfer of electrons from the donor to the acceptor results in energy gain for the organism, which is applied for the synthesis of new cell mass and the maintenance of existing cell mass. The enzymes associated with denitrification are synthesized under anaerobic or partially aerobic conditions. Nitrate reduction to nitrogen occurs as:

$$NO_3^- \rightarrow NO_2^- \rightarrow NO \rightarrow N_2O \rightarrow N_2$$

Each step in this reaction is catalyzed by an enzyme system. Dissimilatory reduction of nitrate to nitrite is important for most bacteria, as the process involves energy conservation by increased substrate-level phosphorylation reaction.

Since biological denitrification can be heterotrophic, autotrophic, or a combination of both, it was observed that the process provided energy to the organism performing denitrification and resulted in enhanced cell mass. Anaerobic ciliates used denitrifying endosymbionts to gain energy, similarly to organisms using mitochondria in oxygen respiration.

The methods discussed have limited potential for full-scale application. The literature indicates wider application of HBD. The present study using a heterotrophic biological denitrification reactor (HBDR) was carried out to optimize nitrate removal using response surface methodology (RSM)–CCD with cotton as the carbon source and the strain of *Pseudomonas stutzeri* (NCIM 5136) at variable temperatures and flow rates. The experimental data was subjected to 2^3 full factorial CCD, and a second-order polynomial model was obtained. The R^2 value is 0.933, which showed that the values predicted by the statistical design model are in good agreement with experimental values. The model also predicted the maximum denitrification of 598.462 mg/day at a flow rate of 5 ml/min, and 40°C was the temperature for the strain *Pseudomonas stutzeri* (NCIM 5136).

11.3 MATERIALS AND METHODS

11.3.1 Organism

The lyophilized culture of *Pseudomonas stutzeri* (NCIM 5136) was procured from the National Chemical Laboratory, Pune, India. The organism was maintained on nutrient agar slants and sub-cultured every 2–3 weeks. The inoculum was prepared from the microbial strain.

11.3.2 Denitrification Process

The heterotrophic denitrification reactor consisted of a cylindrical PVC column 50 cm high and 10 cm in diameter (Soares 2000). This was followed by a trickling sand filter, which was packed with washed sand of 4 mm average diameter. The HDR column was filled with 180 g of cellulosic substrate, unprocessed short fiber cotton (*Gossypium hirsutum*), which was inoculated with the microbial strain. The configuration was selected based on factors such as bulkiness of the substrates, efforts to elude or at least diminish the entrapment of gases, and the point that to break down the substrate, bacteria need to attach themselves to the cellulose fibers. A thin layer of PVC fine net was placed at each end of the packing (Della Rocca et al. 2005). Water supplemented with 200 mg/l of nitrate (as KNO_3) and 3 mg/l of phosphate (as K_2HPO_4) was fed into the column.

The heterotrophic denitrification reactor was operated for 90 days continuously with a startup phase of 7 days (Figure 11.1). The column was replenished with inoculum every day to maintain an average dry cell mass (DCM) of 1.1 g/l. The flow to the HDR was varied from 2.5 to 12.5 ml/min using the peristaltic pump. The temperature of the nitrate-rich feed water was regulated between 20

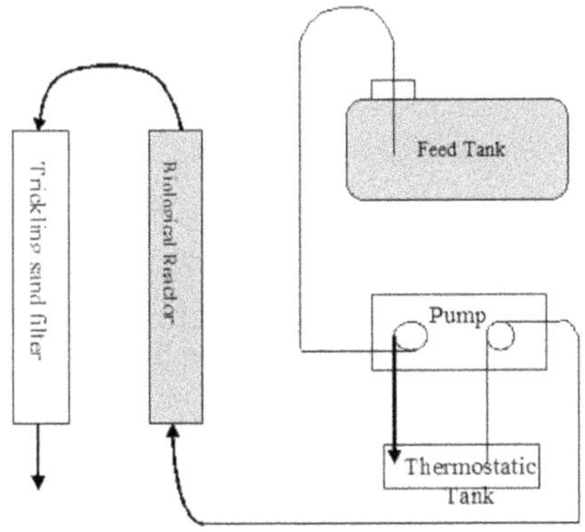

FIGURE 11.1 Heterotrophic denitrification reactor used for experimental study.

and 40°C for *Pseudomonas stutzeri* (NCIM 5136) using a thermostat. For the entire operation, the C:N ratio was maintained at 3.0. The water samples collected from the HDR were analyzed using a digital ion meter having a nitrate probe (ORION Benchtop ION Meters 720A, Thermo Fisher Scientific, Inc., USA).

11.3.3 Optimization Using CCD

To optimize the HBD process using cotton as the carbon source and the strain of *Pseudomonas stutzeri* (NCIM 5136), a 2^3-factorial central composite statistical design with 6 star-(α)-points ($\alpha = 1.682$), with 6 replicates at the center point and 8 cube points, all in duplicate, leading to 10 sets of experiments (Table 11.1), was applied.

The coding of the variables was in accordance with the following equation:

$$x_i = \frac{X_i - \overline{X_i}}{\Delta X_i} \tag{11.1}$$

The levels of independent variables, i.e., temperature and flow rate, selected for the study are presented in Table 11.2.

X_i (i = 1–2): real value of components of the independent variable

TABLE 11.1
Design of Experiments (Coded) for Optimization of Nitrate Removal for the Strain of *Pseudomonas stutzeri* (NCIM 5136)

Design No.	Variable 1	Variable 2
1	−1	−1
2	−1	+1
3	+1	−1
4	+1	+1
5	−α	0
6	+α	0
7	0	−α
8	0	+α
9	0	0
10	0	0

TABLE 11.2
Variables Studied in the Optimization Design for the Strains of *Pseudomonas stutzeri* (NCIM 5136)

			Levels of Variable				
Variable	Microbial Strain	Component	−α	−1	0	+1	+α
X_1	*Pseudomonas stutzeri*	Temperature (°C)	30	35	40	45	50
X_2	(NCIM 5136)	Flowrate (ml/min)	0	2.5	5	7.5	10

+α: highest values in the range studied for each variable
0: central value in the range studied for each variable
+1, −1: intermediate values of each variable
−α: lowest values in the range studied for each variable

11.4 RESULTS AND DISCUSSION

11.4.1 Optimization Study

CCD was used to determine the optimized reaction parameters for getting maximum nitrate removal using *Pseudomonas stutzeri* (NCIM 5136). The second-order polynomial of the same order was found for the HDR with the microbial strain *Pseudomonas stutzeri* (NCIM 5136).

$$Y = -9648.38 + 483.22X_1 + 138.789X_2 - 5.953X_1^2 - 13.733X_2^2 - 0.46X_1X_2$$

Where Y is the predicted response (nitrate removal), −9648.38 is the offset term, 483.22, 138.789 are linear effect terms, −5.95, −13.73 are squared effect terms, and −0.46 is the interaction effect. The +ve sign of coefficients indicates a synergistic effect, while the −ve sign reveals an antagonistic effect (Tekindal et al. 2012). From this equation, it was observed that the nitrate removal had a linear effect and the two process variables had quadratic effects. Temperature (X1) has a significant effect on the nitrate removal (response), since the coefficient of X1 (483.22) is the highest, followed by flowrate (X2). Significantly weaker interaction effects (X1X2) were observed. The multiple coefficient of determination, R^2, provides the proportion of variance for the polynomial models. Statistica software (version 6.0) was deployed for the present study.

The model was found to have a coefficient of multiple determination of 0.933 (Table 11.3 and Figure 11.2) for the strain *Pseudomonas stutzeri* (NCIM 5136), indicating that 93.3% of the variability in the response was supported by the model, i.e.. the results show 6.7% deviation between the experimental and the predicted values. The ideal value of the regression coefficient is 1, which indicates complete agreement between the experimental and predicted values. Since the R^2 value is near to 1, it can be concluded that the experimental value of nitrate removal is close to the predicted values for nitrate removal.

From the analysis of variance (ANOVA) for the strain *Pseudomonas stutzeri* (NCIM 5136) (Table 11.4), it can be observed that all the P values are <0.05, which signifies that the

TABLE 11.3
Experimental and Predicted Response of the Process of Denitrification for the Strain *Pseudomonas stutzeri* (NCIM 5136)

Experiment No.	Temperature (°C)	Flowrate (ml/min)	Experimental Nitrate Removed (mg/day)	Predicted Nitrate Removed (mg/day)
1	40.00000	1.464466	342	355.823
2	45.00000	2.50000	340	354.118
3	40.00000	5.00000	597	598.462
4	47.07107	5.00000	408	365.434
5	35.00000	7.50000	345	360.82
6	45.00000	7.50000	410	464.83
7	40.00000	8.53553	540	596.312
8	35.00000	2.50000	298	273.103
9	32.92893	5.00000	222	234.617
10	40.00000	5.00000	599.04	598.462

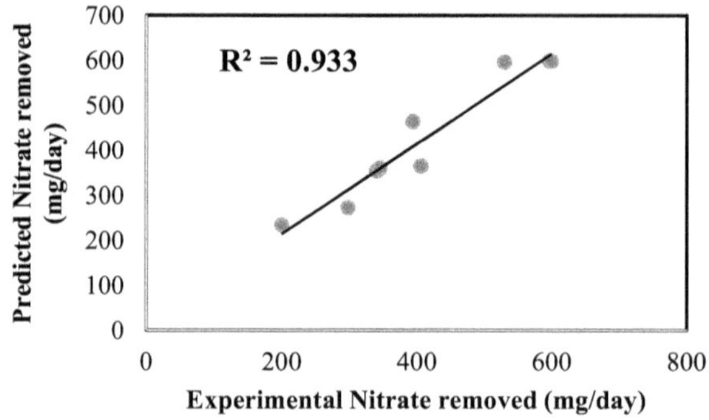

FIGURE 11.2 Predicted versus experimental values for the removal of nitrate using *Pseudomonas stutzeri* (NCIM 5136).

TABLE 11.4
ANOVA Table for the Strain *Pseudomonas stutzeri* (NCIM 5136)

	Sum of Squares (SS)	Degrees of Freedom (df)	Mean Squares (MS)	F Ratio	P Value
Temperature (X_1)	17116.5	1	17116.5	8.41354	0.044091
Temperature (X_1^2)	101418.8	1	101418.8	49.85182	0.002122
Flow Rate (X_2)	19702.5	1	19702.5	9.68467	0.035802
Flow Rate (X_2^2)	33769.0	1	33769.0	16.59897	0.015171
X_1 by X_2	132.3	1	132.3	0.06501	0.811323
Error	8137.6	4	2034.4		
Total SS	149246.0	9			

flowrate and the temperature play a significant role in the denitrification process. Figure 11.3 shows the surface plots representing the relative effect of temperature and flowrate on the amount of nitrate removal obtained per day for the microbial strain *Pseudomonas stutzeri* (NCIM 5136). The highest contour level was 598.462 mg/day when the flowrate was 5.0 ml/min at a temperature of 40°C for the strain *Pseudomonas stutzeri* (NCIM 5136). The comparison of F values of the model equation provided the significance of the regression analysis. These values were expressed as probabilities, given in the ANOVA (Table 11.4) for the microbial strain *Pseudomonas stutzeri* (NCIM 5136).

The advantage of biological denitrification of water is that the process is less cost intensive, as it uses cotton as the carbon source and the soil microbe *Pseudomonas stutzeri* (NCIM 5136). The optimum conditions for the process include variables like temperature, flow rate, and initial nitrate concentration. In the initial step of optimization, all these parameters have been studied separately. CCD was used as a statistical tool to define the optimum values of temperature and flow rate. The experiments showed that for *Pseudomonas stutzeri* (NCIM 5136), the amount of nitrate removed was 599.04 mg/day at a flowrate of 5.0 ml/min at 40°C. The experiments were conducted in an HDR for water having an inlet nitrate concentration of 100 mg/l at a pH of 7 ± 0.1.

11.5 COMPARISON WITH LITERATURE

The rates of denitrification obtained were comparable to those obtained in earlier studies. Studies were conducted for denitrification by Soares and Abeliovich (1998) using wheat straw. The highest

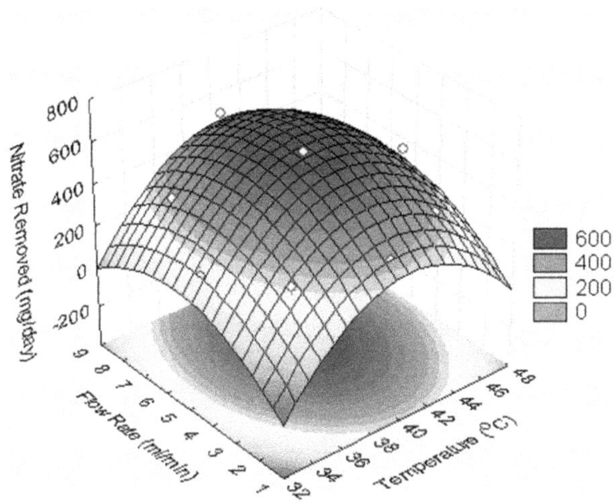

FIGURE 11.3 Response surface plot for continuous monitoring of process parameters in nitrate removal by the strain *Pseudomonas stutzeri* (NCIM 5136).

rate of denitrification of 0.053 g N removed per liter per day was observed with fresh reactors. The water velocity affected the rate of denitrification significantly. A few other works have been reported with a low-cost substrate for denitrification: Volokita et al. 1996a used unprocessed cotton, Volokita et al. 1996b employed shredded newspapers, and Soares et al. 2000 used wheat straw. Cotton has so far been the most efficient substrate (Volokita et al. 1996a) and was tested in a 9 m^3 packed reactor using 1200 kg of cotton (Soares et al. 2000). The system was unable to operate at full capacity due to compression of the bed by high water pressure. The system exhibited stability at feed rates of 0.8 and 1.5 m^3/h. The maximum rate of nitrate removal (denitrification) reported was 0.36 kg N/m^3 d, at a feed rate of 6 m^3/h, which lasted only for short periods.

11.6 CONCLUSION

The removal of nitrate was investigated using *Pseudomonas stutzeri* (NCIM 5136). The results obtained in the present work indicate that the maximum nitrogen removal using Pseudomonas species was 598.462 mg/day with a flowrate of 5.0 ml/min, at a temperature of 40°C, and using cotton as an electron donor. The optimum values for maximum nitrate removal were obtained using CCD. The polynomial model developed by the software was proved to be statistically valid by giving good correlation between the predicted and experimental values. The findings of this work throw light on the aspects of knowing how chitinous materials and various other alternate electron donor sources can sustain bioremediation processes.

REFERENCES

Alam M Z, Muyibi S A, and Toramae J (2007). Statistical optimization of adsorption processes for removal of 2,4-dichlorophenol by activated carbon derived from oil palm empty fruit bunches. *Journal of Environmental Science* 19:674–677.

Archna, Sobti R C, and Sharma S K (2011a). Microbial denitrification of ground water-batch study. *Asian Journal of Water, Environment and Pollution* 3:89–94.

Archna, Sobti R C, and Sharma S K (2011b). Biological denitrification of ground water using various carbon sources by *Pseudomonas fluorescens* and *Pseudomonas stutzeri* in a heterotrophic denitrification reactor. *Asian Journal of Water, Environment and Pollution* 4:89–94.

Archna, Sobti R C, and Sharma S K (2012). Nitrate removal from ground water: a review. *E-Journal of Chemistry* 9(4):1667–1675.

Choi J, and Batchelor B (2008). Nitrate reduction by fluoride green rust modified with copper. *Chemosphere* 70:1108–1116.

Della Rocca C, Belgirono V., and Meric S (2005). Cotton supported heterotrophic denitrification of nitrate-rich drinking water with a sand filtration post-treatment. *Water SA* 31(2):229–236.

Knobeloch L, Salna B, Hogan A, Postle J, and Anderson H. (2000). Blue babies and nitrate-contaminated well water. *Environmental Health Perspectives* 108(7):675–678.

Ovez B (2006). Batch biological denitrification using *Arundo donax*, *Glycyrrhiza glabra*, and *Gracilaria verrucosa* as carbon source. *Process Biochemistry* 41(6):1289–1295.

Ovez B, Ozgen S, and Yuksel M (2006). Biological denitrification in drinking water using *Glycyrrhiza glabra* and *Arunda donax* as the carbon source. *Process Biochemistry* 41:1539–1544.

Park J, Byun H, Choi W, and Kang W (2008). Cement paste column for simultaneous removal of fluoride, phosphate and nitrate in acidic wastewater. *Chemosphere* 70:1429–1437.

Rajta A, Bhatia R, Setia H, and Pathania P (2019). Role of heterotrophic aerobic denitrifying bacteria in nitrate removal from wastewater. *Journal of Applied Microbiology* 128(5):1261–1278.

Robinson-Lora M A and Brennan R A (2009). The use of crab-shell chitin for biological denitrification: batch and column tests. *Biosource Technology* 100(2):534–541.

Sandor J, Kiss I, Farkas O, and Ember I (2001). Association between gastric cancer mortality and nitrate content of drinking water: ecological study on small area inequalities. *European Journal of Epidemiology* 17:443–447.

Schoeman J J and Steyn A (2001). Nitrate removal with reverse osmosis in rural area in South Africa. *Desalination* 155:15–26.

Soares M I M (2000). Biological denitrification of ground water. *Water, Air, & Soil Pollution* 123:183–193.

Soares M I M and Abeliovich A (1998). Wheat straw as substrate for water denitrification. *Water* 32(12):3790–3794.

Soares M I M, Brenner A, Yevzori A, Messalem R, Leroux Y, and Abeliovich A (2000). Denitrification of groundwater: pilot-plant testing of cotton-packed bioreactor and post-microfiltration. *Water Science and Technology* 42(1–2):353–359.

Tekindal M A, Bayrak H, and Ozkaya Y (2012). Box-Behnken experimental design in factorial experiments: the importance of bread for nutrition and health. *Turkish Journal of Field Crops* 17:115–123.

Volokita M, Belkin S, Abeliovich A, and Soares M I M (1996a). Biological denitrification of drinking water using newspaper. *Water Research* 30(4):965–971.

Volokita M, Abeliovich A, and Soares M I M (1996b). Denitrification of groundwater using cotton as energy source. *Water Science and Technology* 34(1–2):379–385.

12 Effectiveness of Major Plant Components, Cellulose and Lignin, for Removal of Heavy Metal Ions from Industrial Wastewater

A Comparative Analysis

Manpreet Kaur
Energy Research Centre, Panjab University, Chandigarh

Gaurav Rattan and Anupama Kaushik
Dr. SSB University Institutes of Chemical Engineering and Technology, Panjab University, Chandigarh

Sonal Singhal
Department of Chemistry, Panjab University, Chandigarh

CONTENTS

DOI: 10.1201/9781003220824-14

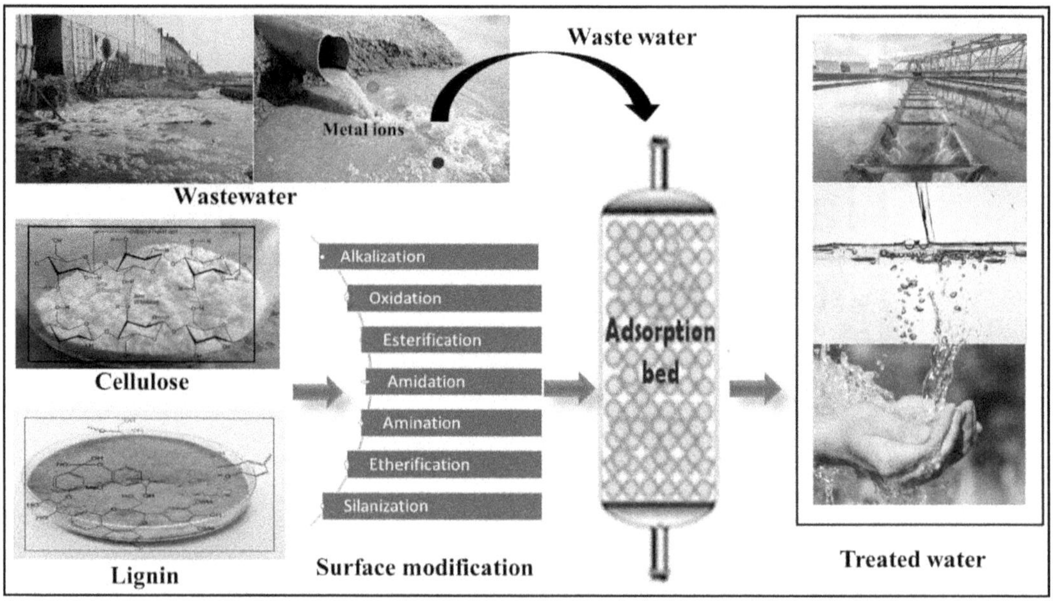

12.1 INTRODUCTION

Water is one of the basic necessities for the existence of life on earth, but sadly, fresh water sources are adversely affected by growing population, modernization and industrialization (Jamshaid et al., 2017). With the increase in human population worldwide, demand for products has also increased, which results in expansion of the industrial sector. As a consequence of this, water contamination through waste discharge from industries has become a serious environmental issue, not only to human beings but also to the aquatic system (Lofrano et al., 2016). According to National Institution for Transforming India (NITI), the rivers, streams, reservoirs, lakes as well as ground water of India are highly polluted as compared with other countries. About 70% of water (surface as well as ground water) in India is polluted, and the major water pollutants are fertilizers, pesticides, surfactants, textile dyes, plastics and heavy metal ions, which are directly discharged into water bodies every year without any treatment(Vyas & Nath, 2021). However, in this chapter, special emphasis has been put on heavy metal water pollution. As heavy metals are transported by runoff from various industries, such as steel industries, metal plating, textiles and mining operation, it has led to one of the biggest concerns in environmental contamination. Many heavy metals, such as iron (Fe), zinc (Zn) and copper (Cu), are micronutrients when present in low concentrations, but on overexposure to these, they become toxic. On the other hand, cadmium (Cd), lead (Pb), nickel (Ni), arsenic (As) and mercury (Hg) are heavy metals that are toxic even in low concentration (Wu et al., 2017). Basically, heavy metals are non-biodegradable, having a stable oxidation state, and accumulate for a long time in water as well as on the surface of soil (Rawat et al., 2003; Sharma et al., 2007).

Many large-scale industries have industrial effluent treatment plants, but small-scale industries do not have sufficient money for effluent treatment plants. Large numbers of small-scale industries are present in Delhi. Rawat and his co-workers reported the following fluxes of heavy metals ($*10^4$ tons per year): Mn 0.37, Pb 0.0146, Zn 0.04, Fe 2.23, Cr 0.45, Cu 0.20, Cd 0.003, Ni 0.10 and Co 0.01. Water contaminated with these metals is directly discharged into drains, which reach the Yamuna river, which is the major potable water source in Delhi (Rawat et al., 2003). Heavy metals were found in Brassicaceae (cabbage, turnip and radish) that were irrigated through the sewage disposal treatment plant of "Diesel Locomotive Works (DLW)" in Lohta village, Varanasi, India

(Rai & Tripathi, 2008). The quality of discharge water from the bleaching and dyeing industry of Tirupur, Tamilnadu, India was checked by a scientist, Azeez. The concentrations of heavy metals in water after washing dyed cloth were found to be Cu 4.80, Zn 4.20, Cr 1.68 and Cd 0.08 (μg/l) in these industries, which is a matter of concern (Azeez, 1999). The persistence and solubility of heavy metals in water lead to their transport into water bodies, where heavy metal ions interact with living and non-living bodies.

These metals then act as the sources for various disorders and diseases by gathering in the human body through the food chain (Dax et al., 2014). In the Taipei Province of China, people suffered from a disease of blood vessels named gangrene, also called black foot disease, when exposed to arsenic in drinking water (Gautam et al., 2014). In Japan, a disease named "Itai-Itai" was caused by excessive intake of cadmium through eating fish. This disease prevailed because of the discharging of cadmium waste by mining companies. People died due to renal dysfunction caused by cadmium (Nishijo et al., 2017). Various health effects of contaminated water and guidelines for drinking water standards given by the World Health Organization (WHO) are shown in Figure 12.1.

It is very important to treat industrial wastewater before discharge into water bodies. Various conventional methods are available for removing heavy metals from wastewater, including ion exchange, chemical precipitation, membrane process, nano-filtration, reverse osmosis, chemical reduction or oxidation, adsorption, electrochemical methods and chemical coagulation (Ahluwalia & Goyal, 2007). Adsorption is a widely used process for removing heavy metals from industrial wastewater. Adsorption has advantages over other conventional methods due to its low cost, easy availability, renewability, greater selectivity and high efficiency for removing heavy metals over a wide range of pH even very low concentration of metals (Chauhan, 2016).

Among various adsorbents, lignocellulose has much wider applications due to its abundance and easy availability from agricultural waste, plant waste and industrial by-products. Lignocellulose (from the Latin word *lignum* – wood) consists of three components, cellulose, hemicellulose and lignin, in which cellulose and hemicellulose form a frame in which lignin is joined as a connector,

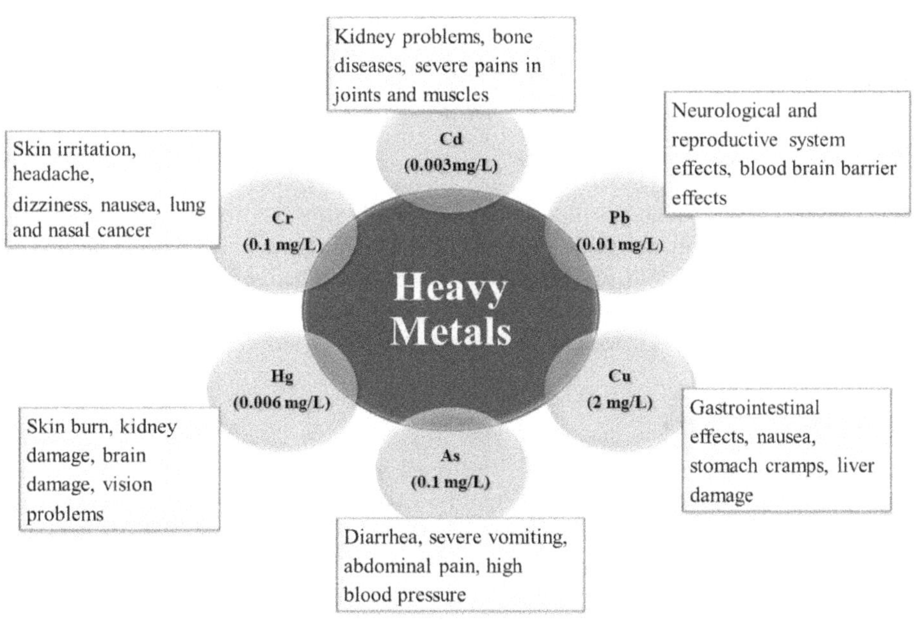

FIGURE 12.1 Effect of contaminated water and drinking water standards.

thus strengthening the cell wall and providing stability and shape to plants (Irmer, 2017). The lignocellulose materials used for removal of heavy metals from industrial wastewater are wheat straw, rice straw, sugarcane bagasse, hemp fibers, corn cob, rice husk, banana peel, corn stalks and Bel fruit shells (Malik et al., 2017). Dupont and co-workers studied the NICA–Donnan model for finding the interaction of heavy metals such as Pb (II), Cu (II) and Cd (II) with lignocellulose for removing metals from industrial effluents (Dupont et al., 2003).

This chapter discusses industrial wastewater discharge problems and their effect on human beings; the role of conventional methods in treating polluted water, with a special emphasis on adsorption processes; and the importance of agriculture waste as an adsorbent in wastewater treatment along with the content of cellulose, hemicellulose and lignin found in different agriculture wastes. Further, reagents used for modification of cellulose or lignin, such as organic acids, bases, minerals, oxidizing agents, organic compounds and binding functional groups are also mentioned in this chapter. This study is a review of current research literature and a comparison of the use of cellulose or lignin in its original form and in its modified form as an adsorbent for heavy metal removal from industrial wastewater.

12.2 CONVENTIONAL METHODS

The tremendous increase of noxious metal ions in the food chain as well in water bodies has put immense pressure on their separation techniques. Several conventional methods, such as chemical precipitation, ion exchange, electrochemical methods, ion exchange, membrane process, chemical coagulation or flocculation, and adsorption can be used for the treatment of industrial wastewater. These conventional methods are summarized as follows.

12.2.1 Chemical Precipitation

Chemical precipitation involves neutralization of ions, basically metallic cations, by combination with counter-ions. Precipitation is done by the addition of peroxide, lime, hydroxide, caustic soda, sodium sulfide and carbonate. In this method, a pH from 4.0 to 6.5 is required for removing heavy metals, which is done by using alkaline material. The drawback of this method is that it requires large amounts of alkaline solution in order to obtain the required pH range (Matlock et al., 2002).

12.2.2 Ion Exchange

This method involves the exchange of cations with metals in wastewater. Removal of all ions or selective removal of ions from solutions depends on size, charge, chemical structure and chemical composition (Dąbrowski et al., 2004). Different types of ion exchange material are used, either synthetic (resins, synthetic zeolites) or natural minerals (aluminosilicates, carbons, silicates, alumina). Zeolite is most effective in ion exchange (Fernandez et al., 2005).

12.2.3 Membrane Process

A membrane does the same work as a cell wall does in our body. It acts as a barrier that stops unwanted impurities passing through it. Different types of membrane processes used in wastewater treatment are nano-filtration, ultra-filtration, micro-filtration, reverse osmosis and forward osmosis. They differ only in the pore size of the membrane. Polymer material membranes are widely used, but some polymers are hydrophobic, due to which membrane pores are easily blocked, causing fouling and decreasing membrane efficiency (Kiran et al., 2019). Liquid membranes are also used for removing heavy metals from wastewater (Kitagawa et al., 1977).

12.2.4 Electrochemical Methods

These work on the principle of precipitation of metals in the elemental state on one electrode when electricity or potential is applied to another electrode. Metal reduction takes place to zero oxidation states from different oxidation states at the cathode through the flowing of electrons from the anode. There are three types of electrochemical methods: electro flotation, electro coagulation and electro deposition. In the electro flotation method, the cathode generates hydrogen gas, which floats the suspended heavy metal ions to the surface, where they are separated. Electro coagulation involves the release of coagulating ions on the electrode, which coagulates the metal, later separated by using centrifugation. Electro deposition involves recovery and removal at the same time by applying current (Koelmel et al., 2016).

12.2.5 Chemical Coagulation or Flocculation

Coagulation and flocculation methods are used for removing suspended material from wastewater. Oppositely charged coagulants are added to wastewater to neutralize the metal ions (Bratby, 2016). Large-size particles formed after neutralization are microflocs, which are not visible to the naked eye. In flocculation, small microflocs collide with each other to form large-size pinflocs, which are easily visible to the naked eye and settle in the water (Stechemesser, 2005). The merits and demerits of conventional methods are shown in Figure 12.2.

12.2.6 Adsorption

Adsorption is a surface phenomenon, which arises on the surface of the adsorbent due to a difference in residual surface energy because of unbalanced forces of atoms/ions/molecules present on the surface of a solution as compared with the energy of atoms/ions/molecules present in the bulk of the solution. In this process, adsorbate adsorbs on the surface of the adsorbent. In reverse, the removal of adsorbate from the surface of the adsorbent is called desorption, as shown in Figure 12.3. Adsorption processes are mainly of two types: physical adsorption corresponds to Van der Waals

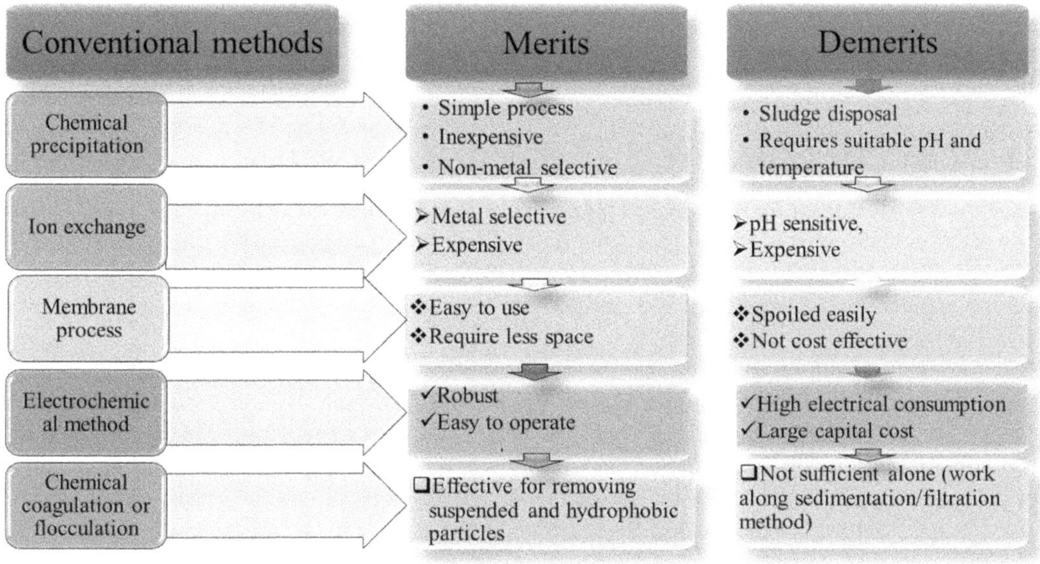

FIGURE 12.2 Merits and demerits of conventional methods.

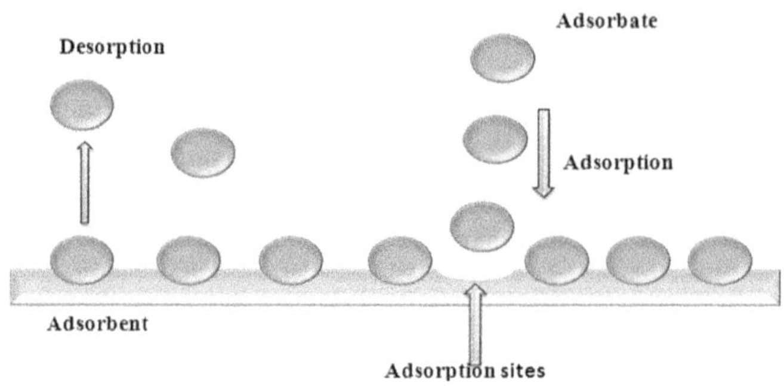

FIGURE 12.3 Adsorption process.

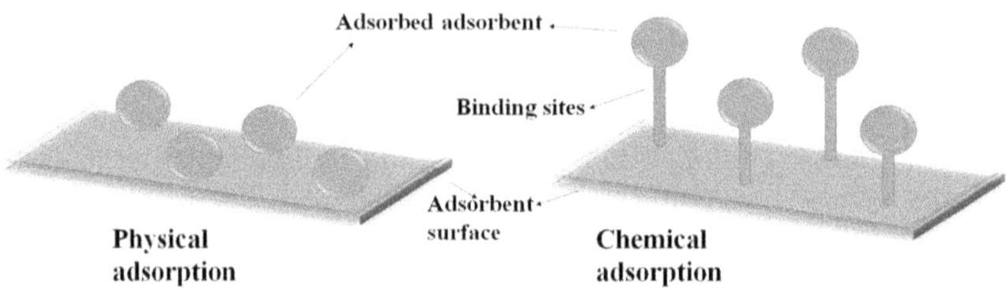

FIGURE 12.4 Physical and chemical adsorption.

force, and chemical adsorption corresponds to chemical bonds, as shown in Figure 12.4 (Wang & Guo, 2020).

Physical adsorption is fast and is carried out at low temperature. It is non-selective and has low adsorption energy because of weak forces of interaction. On the other hand, chemical adsorption requires large activation energy because it involves the formation and destruction of chemical bonds between the adsorbate and the adsorbent. Physical and chemical adsorption work together and cannot be isolated (Cossu & Stegmann, 2018; Hu & Xu, 2020). The adsorbent has a large surface area and minimum volume. Different types of materials used as adsorbents are activated carbon, clay, biochar, graphene, resins, zeolites, lignite coke, agricultural waste, activated alumina, silica, scavenges, etc. (Tareq et al., 2019).

12.3 AGRICULTURAL WASTE AS AN EFFICIENT ADSORBENT

Agricultural waste is a common term used for organic substances thrown away by human beings after the production of crops. It is a low-cost biodegradable material, which is found in abundant amounts in most of the northern parts of India (Wei et al., 2020). India, the second-highest agriculture-based economy, generates approximately 500 million tons per year of agriculture waste, including crop residues (Bhuvaneshwari et al., 2019).

The various agricultural material adsorbents are: wheat straw, rice straw, hemp stalk, rice bran, wheat husk, tree bark, groundnut shells, black gram husk, coconut shells, hazelnut shells, grape stalks, cotton, cotton stalks, sawdust of plants, walnut shells, cotton seed hulls, *Cassia fistula*, waste tea leaves, maize corncob, sugar cane bagasse, orange, apple, banana peels, soybean hulls, sugar beet pulp, arjun nuts, water hyacinth, coffee beans, etc. (Sud et al., 2008). Ninety-seven percent of Cr (VI) was removed at pH 3 using silver-impregnated ground husk carbon within 5 hours (Dai et al., 2018; Dubey & Gopal, 2007). Agricultural waste mainly consists of cellulose, hemicellulose, lignin (Figure 12.5) and extractives, many other small compounds like lipids, simple proteins, ash

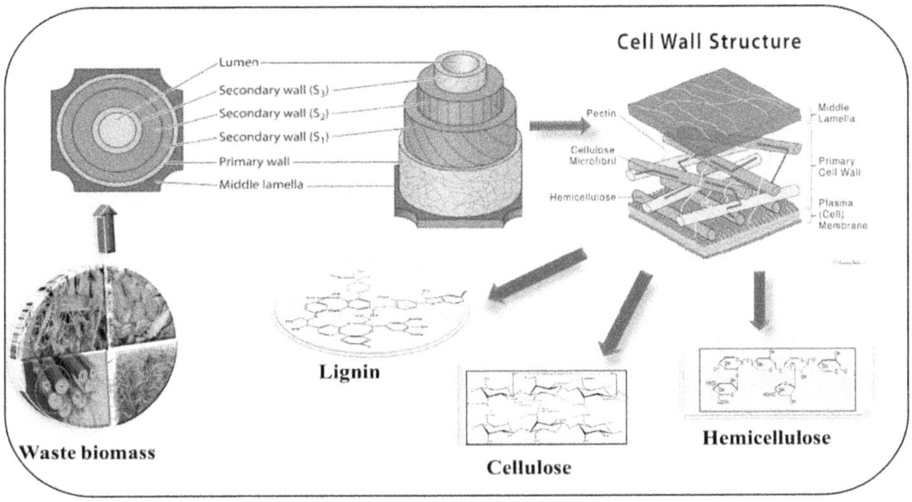

FIGURE 12.5 Distribution of cellulose, hemicellulose and lignin in plant cell wall.

and hydrocarbons. The cellulose, hemicellulose and lignin contents present in different agricultural wastes are given in Table 12.1.

12.3.1 Isolation of Cellulose, Hemicellulose and Lignin from Agricultural Waste

While dealing with agriculture waste, it is mandatory to break down the lignin barrier to separate cellulose, hemicellulose and lignin, which are further subjected to various modifications for removing heavy metals from industrial wastewater. Isolation of cellulose, hemicellulose and lignin from agricultural waste has been done using physical, chemical and combined processes as given in Figure 12.6.

12.3.1.1 Physical Pretreatment

Size reduction of different agricultural wastes such as rice straw, wheat straw, etc. is done before further treatment. This can be done by chipping, milling, grinding, shredding and high-pressure homogenization. These mechanical methods improve the digestibility of lignocellulose biomass and increase the porosity and availability of specific surface area. Physical treatment further decreases the cellulose crystallinity and degree of polymerization. Cellulose fibers are more effectively broken down with a vibratory ball mill instead of an ordinary mill. After milling, the size of cellulose fibers was reduced to 0.2–2 mm, and cellulose fibers 10–30 mm in size were found after chipping (Palmowski & Müller, 2000; Sun & Cheng, 2002). Milled biomass is either subjected to low-pressure steam or soaked in dilute acid for removing extractives.

12.3.1.2 Chemical Pretreatments

Chemical treatment, such as acidic and alkali hydrolysis, is done to simplify the disintegration of the fibers. These pretreatments helps in weakening hydrogen bonds and adding repulsive charge so that fibers easily separate and reduce the amorphous region.

12.3.1.2.1 Alkali Hydrolysis

Alkali hydrolysis is important for delignification of lignocellulosic material. Bases like sodium hydroxide, potassium hydroxide, ammonium hydroxide and calcium hydroxide are used in alkali hydrolysis treatment. During alkaline hydrolysis, structural changes occur inside lignocellulose,

TABLE 12.1
Lignocellulosic Contents in Plant Materials

Plant Materials	Cellulose (%)	Hemicellulose (%)	Lignin (%)	References
Aerial roots banana tree	67.32	13.46	15.62	(Ganapathy et al., 2019)
Amazonian litter leaves	9.1	18.3	21.2	(Bufacchi et al., 2020)
Apple wood	41.42	44.21	21.24	(Zhang et al., 2019)
Arundo donax	11	15	13	(Barana et al., 2016)
Bagasse	96	3		(Abou-Zeid et al., 2018)
Coccinia grandis	62.35	13.42	15.61	(Senthamaraikannan & Kathiresan, 2018)
Coffee parchment	22	18	53	(Reis et al., 2020)
Corn stalk	53	32.58	38	(Shui et al., 2016)
Date seeds	62.60			(Abu-Thabit et al., 2020)
Eggplant	22.7	24.8	11.7	(Aguilera-Sáez et al., 2019)
***Ficus religiosa* tree**	55.58	13.86	10.13	(Moshi et al., 2020)
Furcraea foetida	68.35	11.46	12.32	(Manimaran et al., 2019)
Garlic skin	25.03	30.94	19.53	(Ji et al., 2020)
Green bean	18.3	18	8.1	(Aguilera-Sáez et al., 2019)
Kegalia africana	61.5	12.42	20.94	(Siva et al., 2020)
Melon	18.6	15.3	6.6	(Aguilera-Sáez et al., 2019)
Mentha arvensis	38	20	8	(Prakash et al., 2018)
Miscanthus sacchariflorus	31.7	22.3	18.1	(Syaftika & Matsumura, 2018)
Miscanthus sinensis	30.5	29.9	14.7	(Syaftika & Matsumura, 2018)
Nerium oleander	45	15	21	(Sebeia et al., 2019)
Nerium oleander	45	–	21	(Jabli et al., 2018)
Oil palm frond	44.25	33.94	33.12	(Mohtar et al., 2015)
Oil palm empty fruit bunch	28.3	36.6	35.1	(Palamae et al., 2017)
Oil palm empty fruit bunch	32	24.4	34.9	(Karunakaran et al., 2020)
Oil palm trunk	22.4	41.3	36.1	(Mohtar et al., 2015)
Oil palm frond	45.2	17.5	22.2	(Hussin et al., 2018)
Pepper	21.8	20	10.9	(Aguilera-Sáez et al., 2019)
Pistachio shells	50			(Marett et al., 2017)
Poplar	55.78	14.94	20.20	(Gökkaya et al., 2020)
Poplar wood	41.1	21.9	24.4	(Wu et al., 2020)
Populus tremula	60	15	12	(Sebeia et al., 2019)
Rice husk	12		10	(Barana et al., 2016)
Rice straw	35	18	20.89	(Krishania et al., 2018)
	47.68	36.07	16.25	(Bisla et al., 2020)
Rose stem	58.3	39.3	2.51	(Ventura-Cruz et al., 2020)
Saharan aloe vera cactus leaves	60.2	14.2	13.7	(Balaji & Nagarajan, 2017)
Sander dust	28	10	29.1	(Vaidya et al., 2016)
Sugarcane bagasse	57.6	4.5	31.9	(Rocha et al., 2020)
Switch grass	35.8	21.5	21.1	(Syaftika & Matsumura, 2018)
***Tamarindus indica* L.**	72.84	11	15.38	(Binoj et al., 2018)
Tomato	23.3	16.8	6.6	(Aguilera-Sáez et al., 2019)
Triticale	75.13	13.93	8.42	(Tarrés et al., 2017)
Vine shoots	56.96		21.84	(Dávila et al., 2019)
Waste jute bags	53	58		(Ahuja et al., 2018)
Watermelon	14.5	18.6	9.8	(Aguilera-Sáez et al., 2019)
Wheat straw	40.5	20.5	1	(Li et al., 2019)

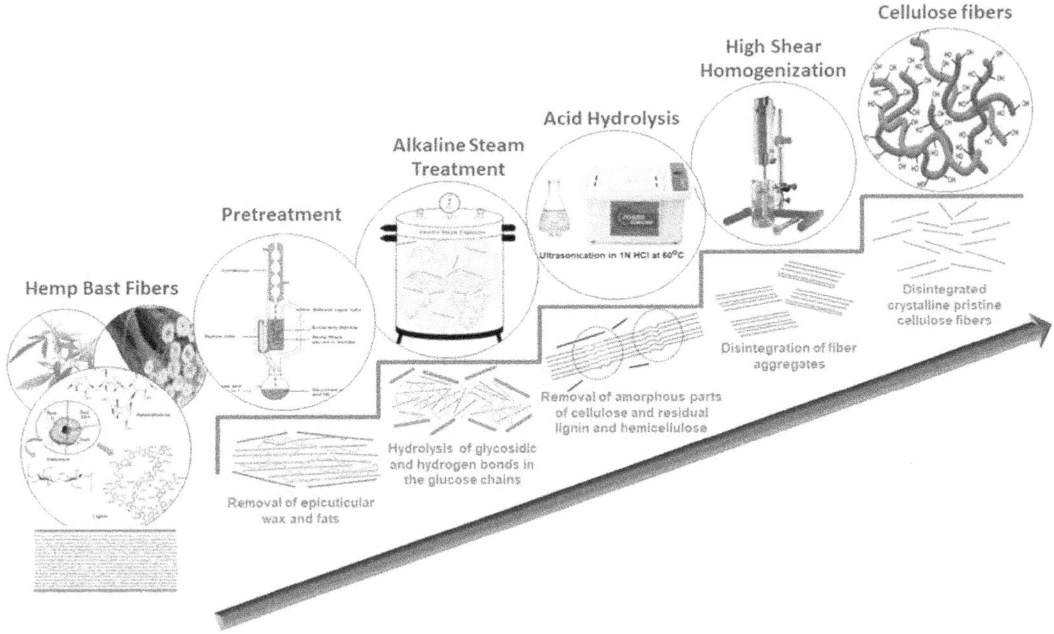

FIGURE 12.6 Extraction process of cellulose.

(Reprinted with permission from (Kaur et al., 2021). Copyright © Elsevier B.V.)

such as cellulose swelling, decrystallization, deacetylation of hemicelluloses, and lignin depolymerization. As a result, most of the lignin and hemicellulose is dissolved in alkaline solution. A study reported soda cooking to be an efficient treatment for separating cellulose and lignin, being economic, energy-efficient and environmentally friendly (Ahuja et al., 2018). Soda cooking includes saturation of lignocellulosic material with steam at elevated temperature and pressure followed by the instant release of pressure, causing sudden evaporation of water and resulting in rupture of lignocellulose material due to thermomechanical force (Kaushik & Singh, 2011). As a result of alkali hydrolysis, homolytic cleavage of β-aryl ether bonds and glycosidic bonds takes place, increasing the solubility of hemicellulose and lignin in alkaline medium and leaving behind cellulose pulp as residue. After treatment, the mother liquor is subjected to acid precipitation for extracting lignin. Swollen cellulose pulp is subjected to bleaching solution, such as hydrogen peroxide or sodium chlorite, for depolymerizing cellulose fibers and reducing their viscosity. Then, the pH of cellulose fibers is neutralized through washing, and finally, cellulose fibers are ready after drying in an oven.

The alkali treatment exhibits many advantages, including simplicity, low energy consumption, avoiding high pressure/temperature and lower capital investment, which leads to economical and efficient production of cellulose fibers by using easily available and less toxic chemicals (Garba et al., 2020).

12.3.1.2.2 Acid Hydrolysis

All agricultural waste requires some chemical pretreatment in order to maximize the yield of cellulose and minimize the degradation. Agriculture waste contains 40–60% cellulose and 15–30% lignin by weight . The linkages between anhydroglucopyranose units and H-bonding in cellulose make it highly rigid, and these are difficult to break. In microcrystalline cellulose production, acid hydrolysis of the amorphous region while preserving the crystalline part is still a scientific challenge. Acid concentration, hydrolysis time and hydrolysis temperature are major factors that affect the crystallinity of cellulose. The β-D-glycosidic bond is responsible for joining β-D-anhydroglucopyranose units in the long cellulose chain. In acidic solution, H_3O^+ moves towards the β-D-glycosidic bond, and at the same time the Cl^- ion weakens the glycosidic bond, which makes hydrolysis easier. Breakage of the β-D-glycosidic bond facilitates the opening of the H-bonded structure of cellulose (Garba et al., 2020; Li & Zhao, 2007).

Hydrochloric acid and sulfuric acid are the most popular reagents for hydrolysis of various cellulosic materials. El-Sakhawy and Hassan extracted microcrystalline cellulose from bagasse, cotton stalk and rice straw and studied the effect of hydrochloric acid (HCl) and sulfuric acid (H_2SO_4) on major cellulose properties like degree of polymerization, crystallinity index, particle size, bulk density and thermal stability. The level-off degree of polymerization was found higher for bagasse than for cotton stalk and rice straw using HCl as compared with H_2SO_4. The thermal degradation rate for the first stage was found to be higher for bagasse cellulose after hydrolysis using H_2SO_4 as compared with HCl (El-Sakhawy & Hassan, 2007). The higher degradation rate using H_2SO_4 makes it unsuitable for preparing composite materials. HCl is the most commonly used acid for hydrolysis of cellulose as compared with H_2SO_4, because H_2SO_4 introduces a sulfate ester group through an esterification reaction with cellulose. This sulfate group present on the surface of cellulose causes repulsion between cellulose chains, resulting in a non-flocculating suspension (Kaushik & Singh, 2011).

12.4 CELLULOSE

Cellulose is the most abundant organic compound that exists on earth. Annually, 10^{10}–10^{11} tons of cellulose is produced by nature. It is the chief structural material of cell walls of all plants. Bundles of cellulose molecules bundles are aggregated in the macro-fibril form, in which the amorphous part (less ordered) exists with a crystalline (highly ordered) part. The amounts of amorphous and crystalline portions depend upon the type of material and the measurement method. Hydrogen bonding is present in planes and Van der Waals interactions are present between planes in a crystalline section of cellulose. Cellulose has four different allomorphs: cellulose I(I_α and I_β), cellulose II, cellulose III (III_I and III_{II}), and cellulose IV (IV_I and IV_{II}) (Eyley et al., 2018). These allomorphs are easily converted into one another by chemical treatments. Cellulose I is the allomorph that is mostly found in nature, and cellulose III is thermodynamically the most stable form. I_α is the chief form in most organisms, while I_β is the chief form in plants. I_α can be easily converted into I_β by water treatment at 260–280°C for about 15–30 minutes (Collinson & Thielemans, 2010).

12.4.1 Structure and Adsorption Characteristics of Cellulose

The molecular formula of cellulose is $(C_6H_{10}O_5)_n$. Cellulose is a linear homopolysaccharide polymer having β-D-glucopyranose repeating units that are covalently bonded through an acetal group between the –OH group of C_1 and C_4 carbon atoms (β-1,4-glucosidic linkage). Because of its linear structure, it can be easily converted into fibers (Chauhan, 2016).

X-ray analysis has shown that cellulose is made up of 12000 or more glucose units, in which each glucose unit is at right angles to the previous one. Preferred bond angles between every anhydroglucone unit are accommodated through rotation of 180° in plane. These chains lie side by side in bundles held together by H-bonding between C_6-CH_2OH of one chain and C_2-OH of other chains. These bundles are twisted to form rope-like structures. These ropes are themselves grouped to form fibers or may be embedded in lignin, as in the case of wood. The structure of cellulose and the hydroxyl groups present on it are responsible for properties like chirality, hydrophilicity and degradability. Figure 12.7 shows intramolecular and intermolecular H-bonding between glucose units of cellulose. Due to H-bonding, cellulose is insoluble in most solvents, and the rigidity of cellulose is also increased.

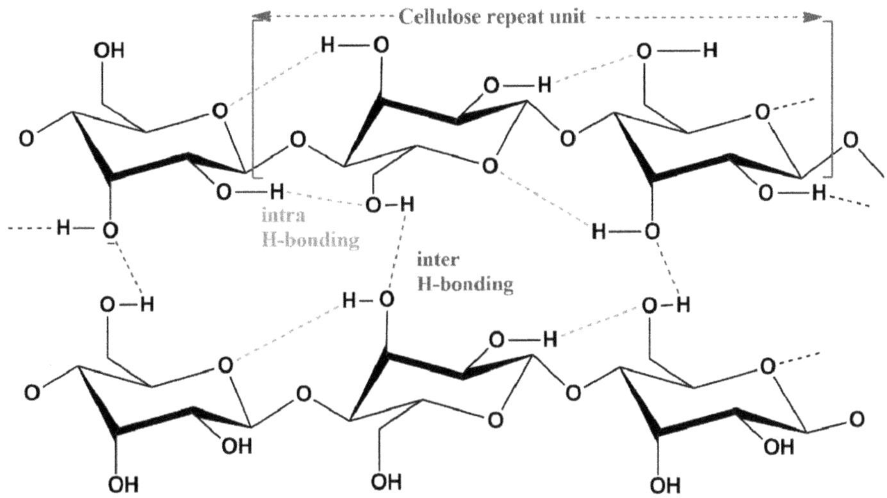

FIGURE 12.7 Cellulose structure.

The adsorption characteristics of cellulose are low due to the deficiency of binding sites for specific molecules (Tang et al., 2019). That is why modification of cellulose is done before using it as an adsorbent for removal of heavy metals from industrial wastewater.

12.4.2 Cellulose Surface Modification

The three hydroxyl groups present on each glucon unit of cellulose make it highly reactive for various modifications. Graft polymerization is a widely used method for the modification of cellulose. To increase the adsorption capacity and make cellulose fibers more efficient adsorbents, several researchers reported different modifying methods, such as animation, carboxylation, amidoximation, esterification, etherification and sulfonation, as shown in Figure 12.8.

Garcia-Reyes and Rangel-Mendez studied the contribution of lignocelluloses material for removal of Cr (III) from water. Sorghum straw, oats straw and agave straw removed 11.60, 6.26 and 12.08 mg/g of Cr (III), respectively (Garcia-Reyes & Rangel-Mendez, 2009).

12.5 Lignin

Lignin is the second most abundant naturally occurring biomass present on earth after cellulose. It accounts for about 15–30% by weight of lignocellulosic material. Lignin is a heterogeneous, long-chain polymer consisting of phenylpropanoid monomers connected through ether bonds. Lignin is found as a cell wall component in the woody stems of hardwoods and softwoods as well as in all vascular plants. Lignin works as a glue that fills the gap between cellulose and hemicellulose polymers. It increases the rigidity, strengthens the cell wall and increases the hydrophobic property of plant material. It also permits flow of water and fluids through the plant's vascular system, as well as providing a physical barrier to phytopathogens and other environmental exposures (Figueiredo et al., 2018). Annually 70 million tons of lignin is produced. Sulfite and kraft pulping processes in the paper and pulp industries are used to produce lignin, called black liquor (Lu et al., 2017).

12.5.1 Structure of Lignin

Lignin is an asymmetrical, three-dimensional, amorphous and highly cross-linked phenolic polymer. After extraction from different types of waste biomass, typically, we get a yield of molecular masses

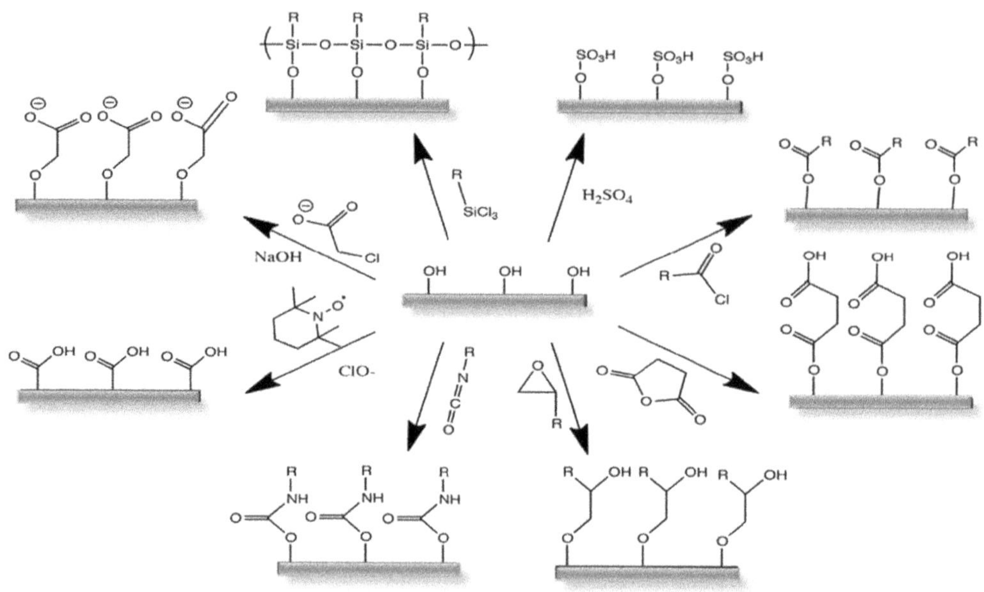

FIGURE 12.8 Some surface modifications of cellulose.

(From Patel et al., 2019.)

varying between 1000 and 2000 g/mol. As a result, the physical and chemical characteristics of lignin will differ depending on the initial source and separation technique performed (Upton & Kasko, 2016). The chemical structure of lignin is complex, formed through a biosynthesis process from three phenylpropanoid monomers, guaiacyl (G), p-hydroxyphenyl (H) and syringyl (S) units as shown in Figure 12.9, derived from coniferyl alcohol, p-coumaryl alcohol and sinapyl alcohol, respectively (Ge & Li, 2018).

β-O-4′ ether bonds are present between most of the monomers of lignin, which account for more than half of the linkage structures in lignin and are a key target for most degradation processes β-5 phenylcoumaran, 5-5′ biphenyl, 4-O-5′ diphenyl ether, β-β′ resinol, β-10 diphenyl methane and α-O-4′ ether are the other bonds present in smaller amounts in lignin (Upton & Kasko, 2016; Zakzeski et al., 2010). Different linkages present in lignin structure are shown in Figure 12.10.

12.5.2 Adsorption Capacity of Lignin

Lignin can also be directly used for heavy metal removal from wastewater without any modification. Hydroxyl, phenolic and methoxyl groups on lignin make it more favorable for heavy metal adsorption. Guo in 2008 studied the adsorption of heavy metals using lignin extracted from black liquor of the paper industry. The affinity of lignin for various metals was found to be in the following order: Pb(II)>Cu(II)>Cd(II)>Zn(II)>Ni(II). It was concluded that phenolic and carboxylic acid sites present on lignin were involved in bonding with heavy metals (Ge et al., 2015). Similarly, Šćiban and co-workers reported kraft lignin from paper industries as an efficient and abundant waste material for the adsorption of heavy metal ions in the order of Cr (VI)>Cd(II)>Cu(II)>Zn(II). They also studied the effect of interference of another metal ion with the adsorption of Cu(II), which was found in decreasing order as Ni(II)>Cd(II)>Pb(II) (Šćiban et al., 2011). Albadarin et al. used alkaline lignin with low sulfonate content for removal of toxic Cr(VI) and checked the effects on adsorption by varying pH of the solution, ionic strength, initial concentration of metal solutions and

TABLE 12.2
Metal Adsorption Capacity of Various Modified Celluloses

Cellulose Starting Material	Modifying Agents	Type: Functional Group	Metal Impurity	Adsorption Capacity (mg/g)	Reference
Cellulose	Tris(hydroxymethyl)aminomethane, polyethyleneimine, dopamine hydrochloride	Aerogel: hydroxyl	Cu(II)	103.5	(Tang et al., 2019)
Cellulose	N-methylolacrylamide, carbonyl diimidazole, triethylenetetramine	Porous: carboxyl, amide, carbonyl sulfide, secondary amino group	Cd(II)	401.1	(Chen et al., 2019)
Cellulose membrane	Alpha-zirconium phosphate, graphene oxide	Ion exchange membrane	Cu(II)	97	(Ibrahim et al., 2019)
			Zn(II)	98	
			Ni(II)	99	
Cotton linters	TEMPO, branched-polyethyleneimine	Sponge: amide	Cd(II), Cr(III)	84, 94	(Fiorati et al., 2020)
			Cu(II), Zn(II)	194, 125	
Cellulose	Acrylic acid, potassium fulvate, N,N-methylenebisacrylamide	Graft copolymer: carboxyl, hydroxyl	Pb(II)	232.6	(Essawy et al., 2017)
Bagasse	Sodium metaperiodate, TEMPO, sodium hypochlorite	Oxidized fiber: carboxyl	Cu(II), Ca(II), Pb(II)	92.23, 97.34, 82.19	(Abou-Zeid et al., 2018)
Cellulose filament fibers	Bisacrylamide	Aerogel: amide	Cu(II)	51.3	(J. Liu et al., 2020)
Dialdehyde cellulose	Graphene oxide	Composite heterogeneous (GTD-1), homogeneous (GTD-2): carboxyl	Cu(II)	65.1	(Wang et al., 2020)
			Pb(II)	80.9	
			Cu(II)	74.2	
			Pb(II)	91.7	
Cellulose acetate	3,3-Dithiodipropionic acid, ammonium thioglycolate,1,1-carbonyldiimidazole	Membrane: thiol	Cu(II)	49	(Choi et al., 2020)
			Cd(II)	45.9	
			Pb(II)	22	
Carboxymethyl cellulose	Graphene oxide	Composite fibers: carboxyl	Pb(II)	101.4	(Yu et al., 2020)
Carboxymethyl cellulose	Acrylamide	Composite hydrogel: amide, hydroxyl, carboxyl	Cu(II)	227.3	(Godiya et al., 2019)
			Pb(II)	312.5	
			Cd(II)	256.4	
Bamboo	Glutamic acid, thiourea, epichlorohydrin, sodium hydroxide	Porous cellulose beads: hydroxyl, ethoxy (-O-)	Cu(II)	98.86	(Qiao et al., 2020)
			Co(II)	102.29	

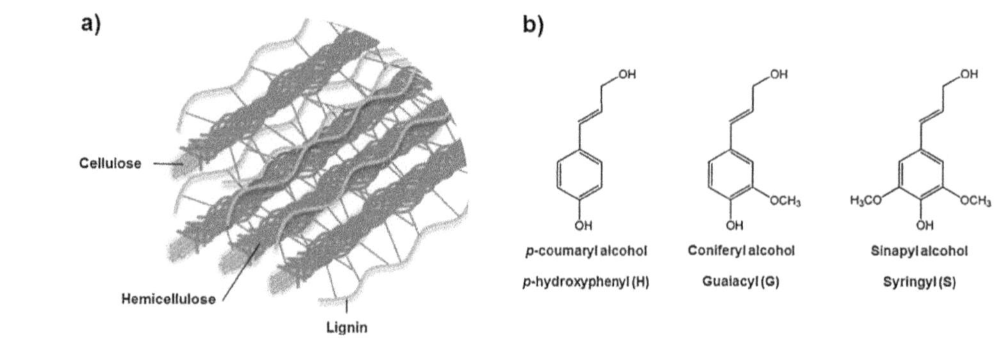

FIGURE 12.9 (a) Lignin present in lignocellulose, (b) structure of monomer units present in lignin.

(Reprinted with permission from (Figueiredo et al., 2018). Copyright © Elsevier Ltd.)

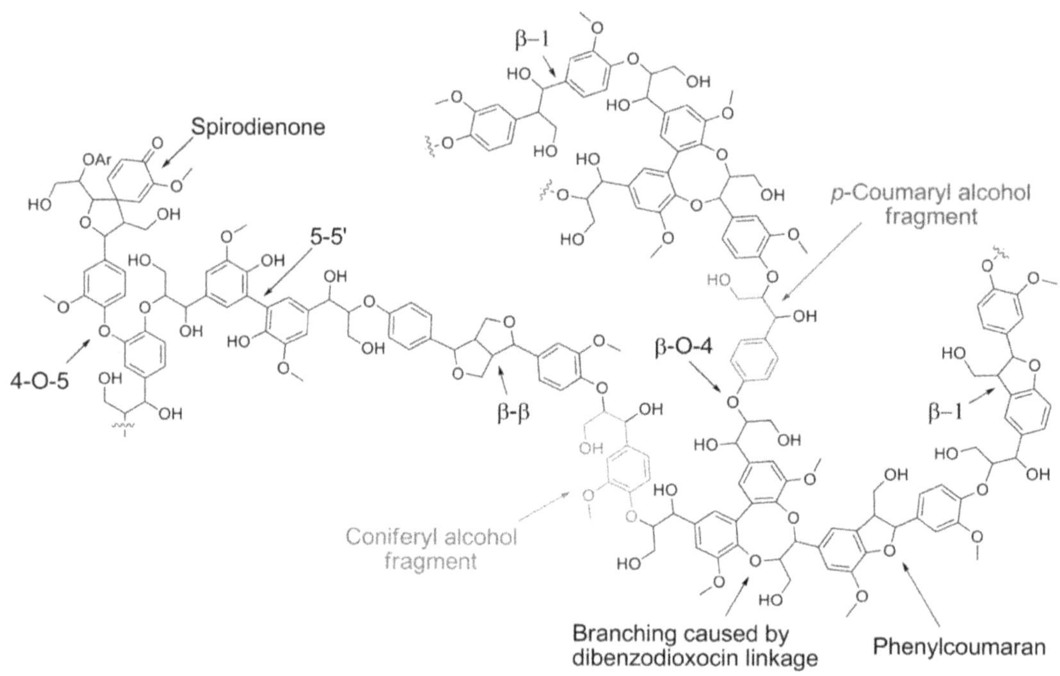

FIGURE 12.10 Different linkages present in lignin structure.

(Reprinted with permission from (Zakzeski et al., 2010). Copyright © 2010, American Chemical Society.)

adsorbent dose (Albadarin et al., 2011). Maximum adsorption of Cr(VI) (60 mg/g) was found at pH 2 with a metal solution concentration of 100 mg/l.

12.5.3 Surface Modification of Lignin

Unmodified lignin has a non-specific nature and low adsorption capacity for heavy metal ions. The hydroxyl, methoxyl, carboxyl and phenolic groups present on lignin can be easily modified for the production of new adsorbents. These surface modifications include enhancing the reactivity of hydroxyl groups or modifying the nature of reactive chemical sites to create new adsorbents that are

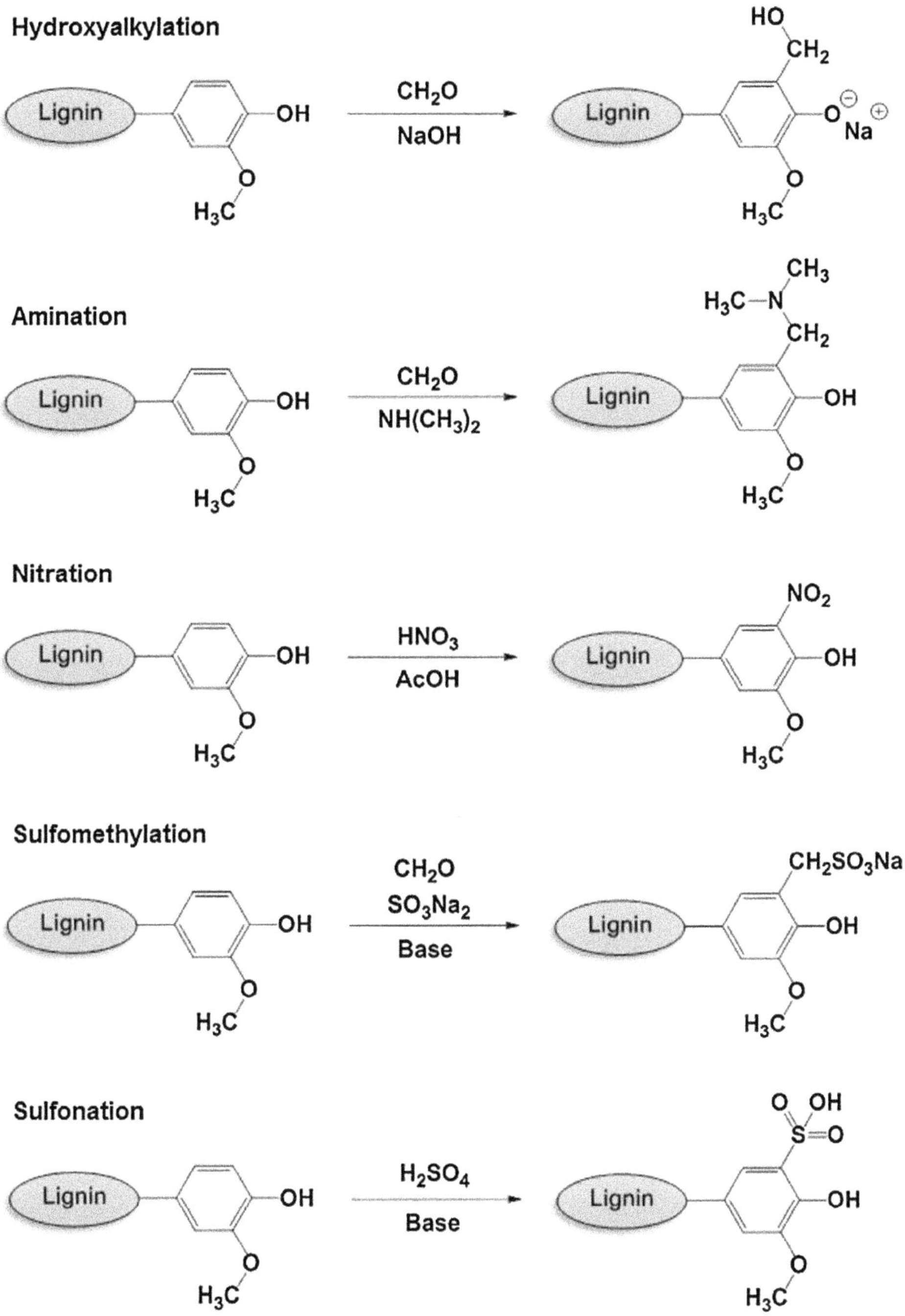

FIGURE 12.11 Modifications of lignin.

(Reprinted with permission from (Figueiredo et al., 2018). Copyright © Elsevier Ltd.)

even more efficient and reactive. Various modifications are possible on the surface of lignin, such as amination, sulfomethylation, hydroxyalkylation, nitration, etc., as shown in Figure 12.11 and Figure 12.12, respectively, and the metal adsorption capacities of various modified lignins are given in Table 12.3.

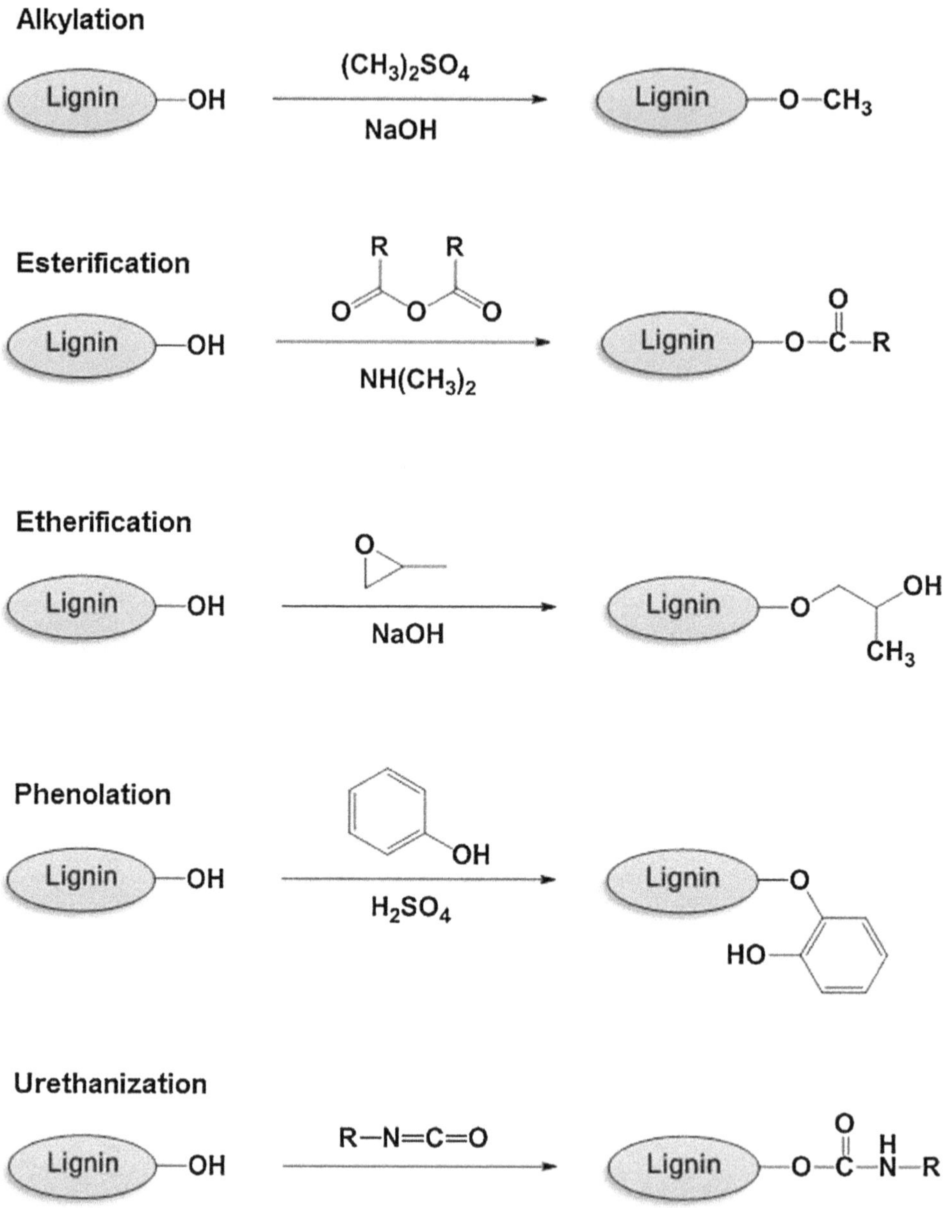

FIGURE 12.12 Modifications of hydroxyl groups on lignin.

(Reprinted with permission from (Figueiredo et al., 2018). Copyright © Elsevier Ltd.)

TABLE 12.3
Metal Adsorption Capacities of Various Modified Lignins

Material	Modifying/Grafting Agent	Type: Binding Functional Group	Metal Impurity	Adsorption Capacity (mg/g)	References
Kraft lignin	Poly(ethylene imine), sodium alginate	Porous Microspheres: amine	Cd(II)	74.84	(Popovic et al., 2020)
			Cr(VI)	54.20	
			As(V)	53.12	
			Ni(II)	49.42	
Rice straw	Sulfur trioxide	Mesoporous: sulfoxide	Pb(II)	952	(Xu et al., 2017)
Alkaline lignin	Carbon nanotubes	Nanotubes: polyhydric phenol	Pb(II)	235	(Li et al., 2017)
Lignin	Diethylenetriamine, sodium dodecanesulfonate, $FeCl_3 \cdot 6H_2O$	Magnetic composite	Cr(VI)	123	(Song et al., 2015)
Wheat straw		Carboxylic, aromatic-OH, hydroxyl	Cu(II)	26	(Todorciuc et al., 2015)
Lignosulfonate	Sodium alginate	Porous sphere: sulfonic, hydroxyl, carboxyl	Pb(II)	27.1	(Li, Ge, et al., 2015)
Alkaline lignin	Methylamine	Amine	Pb(II)	60.5	(Ge et al., 2015)
Lignin	Polyethyleneimine	Microsphere: amine	Pb(II)	33.9	(Ge et al., 2016)
Alkaline lignin	Xanthate, carbon disulfide	Porous resin: xanthate	Pb(II)	64.9	(Li, Xiao, et al., 2015)
Alkaline lignin	Polyethyleneimine, formaldehyde, carbon disulfide	Porous: dithiocarbamate	Pb(II)	188	(Li, Xiao, et al., 2015)
Bamboo	Propargyl bromide, 1,2,4-triazole-3-thiol	Azomethine, imino, thioether	Cd(II)	72.4	(Jin et al., 2017)
Lignin	Poly(ethylene imine)	Composite: dithiocarbamate	Cu(II)	98.0	(Qin et al., 2017)
			Zn(II)	78.0	
			Ni(II)	67.0	
Lignin	$MgO\text{-}SiO_2$	Hybrid sorbent	Cu(II)	83.9	(Ciesielczyk et al., 2017)
Bamboo lignin	1-Aza-18-crown-6	Crown ether units	Pb(II)	91.4	(Jin et al., 2020)
3Kraft lignin	Polyethyleneimine	Spherical particles: amine	Cr(VI)	657.9	(Kwak et al., 2020)
Alkali lignin	Polyacrylic acid	Lignin composite hydrogel	Pb(II)	1.076	(M. Liu et al., 2020)
			Cu(II)	0.323	
			Cd(II)	0.059	
Kraft lignin	Poly(ethylene imine)	Microsphere: amine	Cd(II)	74.84	(Popovic et al., 2020)
			Cr(VI)	54.20	
			As(V)	53.12	
			Ni(II)	49.42	
Wheat straw lignin	Alkali		Cu(II)	26.0	(Todorciuc et al., 2015)

12.6 CONCLUSION

Heavy metal contamination of water is a worldwide problem. The influence of these toxic metal ions in water is minimized by various methods such as chemical precipitation, ion exchange, membrane process, chemical coagulation/flocculation and adsorption. Among of these methods, the most effective method was found to be adsorption due to its economic feasibility, eco-friendly nature, reusability and low energy consumption properties. Agricultural waste lignocellulosic materials are widely used adsorbents due to their easy availability, cheapness and lower economic value. Straw, stalks, roots, shells, bagasse, husks, leaves and seeds are the major sources of cellulose, hemicellulose and lignin. Of cellulose and lignin, cellulose is the better and more efficient adsorbent. Natural and modified cellulose materials are used to detoxify various metals in wastewater. Lignin has limited use because of its complex molecular structure. The repeating units of the lignin polymer chain are lacking in stereo-regularity and heterogeneously interpenetrated by the cellulose polymer chain. Hence, very strong reagents are required to isolate hydrophobic lignin from the hydrophilic polymer chain of cellulose, giving the complex structure of lignin by self-condensation.

REFERENCES

Abou-Zeid, R. E., Dacrory, S., Ali, K. A., & Kamel, S. (2018). Novel method of preparation of tricarboxylic cellulose nanofiber for efficient removal of heavy metal ions from aqueous solution. *International Journal of Biological Macromolecules, 119*, 207–214.

Abu-Thabit, N. Y., Judeh, A. A., Hakeem, A. S., Ul-Hamid, A., Umar, Y., & Ahmad, A. (2020). Isolation and characterization of microcrystalline cellulose from date seeds (*Phoenix dactylifera* L.). *International Journal of Biological Macromolecules, 155*, 730–739.

Aguilera-Sáez, L. M., Arrabal-Campos, F. M., Callejón-Ferre, Á. J., Medina, M. D. S., & Fernández, I. (2019). Use of multivariate NMR analysis in the content prediction of hemicellulose, cellulose and lignin in greenhouse crop residues. *Phytochemistry, 158*, 110–119.

Ahluwalia, S. S., & Goyal, D. (2007). Microbial and plant derived biomass for removal of heavy metals from wastewater. *Bioresource Technology*, *98*(12), 2243–2257.

Ahuja, D., Kaushik, A., & Singh, M. (2018). Simultaneous extraction of lignin and cellulose nanofibrils from waste jute bags using one pot pre-treatment. *International Journal of Biological Macromolecules, 107*, 1294–1301.

Albadarin, A. B., Ala'a, H., Al-Laqtah, N. A., Walker, G. M., Allen, S. J., & Ahmad, M. N. (2011). Biosorption of toxic chromium from aqueous phase by lignin: mechanism, effect of other metal ions and salts. *Chemical Engineering Journal*, *169*(1–3), 20–30.

Azeez, P. (1999). Influence of dyeing and bleaching industries on ground water of Tirupur, Tamilnadu, India. *Bulletin of Environmental Contamination and Toxicology, 62*, 330–335.

Balaji, A., & Nagarajan, K. (2017). Characterization of alkali treated and untreated new cellulosic fiber from Saharan aloe vera cactus leaves. *Carbohydrate Polymers, 174*, 200–208.

Barana, D., Salanti, A., Orlandi, M., Ali, D. S., & Zoia, L. (2016). Biorefinery process for the simultaneous recovery of lignin, hemicelluloses, cellulose nanocrystals and silica from rice husk and *Arundo donax*. *Industrial Crops and Products, 86*, 31–39.

Bhuvaneshwari, S., Hettiarachchi, H., & Meegoda, J. N. (2019). Crop residue burning in India: policy challenges and potential solutions. *International Journal of Environmental Research and Public Health, 16*(5), 832.

Binoj, J., Raj, R. E., & Indran, S. (2018). Characterization of industrial discarded fruit wastes (*Tamarindus indica* L.) as potential alternate for man-made vitreous fiber in polymer composites. *Process Safety and Environmental Protection, 116*, 527–534.

Bisla, V., Rattan, G., Singhal, S., & Kaushik, A. (2020). Green and novel adsorbent from rice straw extracted cellulose for efficient adsorption of Hg (II) ions in an aqueous medium. *International Journal of Biological Macromolecules, 161*, 194–203.

Bratby, J. (2016). *Coagulation and Flocculation in Water and Wastewater Treatment*. IWA Publishing.

Bufacchi, P., Bizzo, W. A., Buckeridge, M. S., Franco-Jacome, D. L., Grandis, A., Cambler, A. B., & Krieger Filho, G. C. (2020). Thermal degradation of leaves from the Amazon rainforest litter considering non-structural, structural carbohydrates and lignin composition. *Bioresource Technology Reports, 11*, 100490.

Chauhan, S. (2016). Use of cellulose and its derivatives for metal ion sorption. *Journal of Chemical and Pharmaceutical Research*, *8*(4), 416–420.

Chen, Q., Zheng, J., Wen, L., Yang, C., & Zhang, L. (2019). A multi-functional-group modified cellulose for enhanced heavy metal cadmium adsorption: performance and quantum chemical mechanism. *Chemosphere, 224*, 509–518.

Choi, H. Y., Bae, J. H., Hasegawa, Y., An, S., Kim, I. S., Lee, H., & Kim, M. (2020). Thiol-functionalized cellulose nanofiber membranes for the effective adsorption of heavy metal ions in water. *Carbohydrate Polymers, 234*, 115881.

Ciesielczyk, F., Bartczak, P., Klapiszewski, Ł., & Jesionowski, T. (2017) Treatment of model and galvanic waste solutions of copper (II) ions using a lignin/inorganic oxide hybrid as an effective sorbent. *Journal of Hazardous Materials*, *328*, 150–159.

Collinson, S., & Thielemans, W. (2010). The catalytic oxidation of biomass to new materials focusing on starch, cellulose and lignin. *Coordination Chemistry Reviews*, *254*(15–16), 1854–1870.

Cossu, R., & Stegmann, R. (2018). *Solid Waste Landfilling: Concepts, Processes, Technology*. Elsevier.

Dąbrowski, A., Hubicki, Z., Podkościelny, P., & Robens, E. (2004). Selective removal of the heavy metal ions from waters and industrial wastewaters by ion-exchange method. *Chemosphere*, *56*(2), 91–106.

Dai, Y., Sun, Q., Wang, W., Lu, L., Liu, M., Li, J., Yang, S., Sun, Y., Zhang, K., & Xu, J. (2018). Utilizations of agricultural waste as adsorbent for the removal of contaminants: a review. *Chemosphere, 211*, 235–253.

Dávila, I., Remón, J., Gullón, P., Labidi, J., & Budarin, V. (2019). Production and characterization of lignin and cellulose fractions obtained from pretreated vine shoots by microwave assisted alkali treatment. *Bioresource Technology, 289*, 121726.

Dax, D., Chávez, M. S., Xu, C., Willför, S., Mendonça, R. T., & Sánchez, J. (2014). Cationic hemicellulose-based hydrogels for arsenic and chromium removal from aqueous solutions. *Carbohydrate Polymers, 111*, 797–805.

Dubey, S. P., & Gopal, K. (2007). Adsorption of chromium (VI) on low cost adsorbents derived from agricultural waste material: a comparative study. *Journal of Hazardous Materials*, *145*(3), 465–470.

Dupont, L., Bouanda, J., Dumonceau, J., & Aplincourt, M. (2003). Metal ions binding onto a lignocellulosic substrate extracted from wheat bran: a NICA–Donnan approach. *Journal of Colloid and Interface Science*, *263*(1), 35–41.

El-Sakhawy, M., & Hassan, M. L. (2007). Physical and mechanical properties of microcrystalline cellulose prepared from agricultural residues. *Carbohydrate Polymers*, *67*(1), 1–10.

Essawy, H. A., Mohamed, M. F., Ammar, N. S., & Ibrahim, H. S. (2017). The promise of a specially-designed graft copolymer of acrylic acid onto cellulose as selective sorbent for heavy metal ions. *International Journal of Biological Macromolecules, 103*, 261–267.

Eyley, S., Schütz, C., & Thielemans, W. (2018). Surface chemistry and characterization of cellulose nanocrystals. In Rosenau, T., Potthast, A., & Hell, J. (eds) *Cellulose Science and Technology* (pp. 223–252). John Wiley & Sons.

Fan, H., Ma, X., Zhou, S., Huang, J., Liu, Y., & Liu, Y. (2019). Highly efficient removal of heavy metal ions by carboxymethyl cellulose-immobilized Fe_3O_4 nanoparticles prepared via high-gravity technology. *Carbohydrate Polymers, 213*, 39–49.

Fernandez, Y., Maranon, E., Castrillón, L., & Vázquez, I. (2005). Removal of Cd and Zn from inorganic industrial waste leachate by ion exchange. *Journal of Hazardous Materials*, *126*(1–3), 169–175.

Figueiredo, P., Lintinen, K., Hirvonen, J. T., Kostiainen, M. A., & Santos, H. A. (2018). Properties and chemical modifications of lignin: towards lignin-based nanomaterials for biomedical applications. *Progress in Materials Science, 93*, 233–269.

Fiorati, A., Grassi, G., Graziano, A., Liberatori, G., Pastori, N., Melone, L., Bonciani, L., Pontorno, L., Punta, C., & Corsi, I. (2020). Eco-design of nanostructured cellulose sponges for sea-water decontamination from heavy metal ions. *Journal of Cleaner Production, 246*, 119009.

Ganapathy, T., Sathiskumar, R., Senthamaraikannan, P., Saravanakumar, S., & Khan, A. (2019). Characterization of raw and alkali treated new natural cellulosic fibres extracted from the aerial roots of banyan tree. *International Journal of Biological Macromolecules, 138*, 573–581.

Garba, Z. N., Lawan, I., Zhou, W., Zhang, M., Wang, L., & Yuan, Z. (2020). Microcrystalline cellulose (MCC) based materials as emerging adsorbents for the removal of dyes and heavy metals–a review. *Science of the Total Environment, 717*, 135070.

Garcia-Reyes, R. B., & Rangel-Mendez, J. R. (2009). Contribution of agro-waste material main components (hemicelluloses, cellulose, and lignin) to the removal of chromium (III) from aqueous solution. *Journal of Chemical Technology & Biotechnology*, *84*(10), 1533–1538.

Gautam, R. K., Sharma, S. K., Mahiya, S., & Chattopadhyaya, M. C. (2014). *Heavy Metals in Water: Presence, Removal and Safety*. The Royal Society of Chemistry.

Ge, Y., & Li, Z. (2018). Application of lignin and its derivatives in adsorption of heavy metal ions in water: a review. *ACS Sustainable Chemistry & Engineering*, *6*(5), 7181–7192.

Ge, Y., Song, Q., & Li, Z. (2015). A Mannich base biosorbent derived from alkaline lignin for lead removal from aqueous solution. *Journal of Industrial and Engineering Chemistry, 23*, 228–234.

Ge, Y., Qin, L., & Li, Z. (2016). Lignin microspheres: an effective and recyclable natural polymer-based adsorbent for lead ion removal. *Materials & Design, 95*, 141–147.

Godiya, C. B., Cheng, X., Li, D., Chen, Z., & Lu, X. (2019). Carboxymethyl cellulose/polyacrylamide composite hydrogel for cascaded treatment/reuse of heavy metal ions in wastewater. *Journal of Hazardous Materials, 364*, 28–38.

Gökkaya, D. S., Sert, M., Sağlam, M., Yüksel, M., & Ballice, L. (2020). Hydrothermal gasification of the isolated hemicellulose and sawdust of the white poplar (*Populus alba* L.). *The Journal of Supercritical Fluids, 162*, 104846.

Hamad, A. A., Hassouna, M. S., Shalaby, T. I., Elkady, M. F., Abd Elkawi, M. A., & Hamad, H. A. (2020). Electrospun cellulose acetate nanofiber incorporated with hydroxyapatite for removal of heavy metals. *International Journal of Biological Macromolecules*, 151, 1299–1313.

Hu, H., & Xu, K. (2020). Physicochemical technologies for HRPs and risk control. In Ren, H. (ed.) *High-Risk Pollutants in Wastewater* (pp. 169–207). Elsevier.

Hussin, M. H., Samad, N. A., Latif, N. H. A., Rozuli, N. A., Yusoff, S. B., Gambier, F., & Brosse, N. (2018). Production of oil palm (*Elaeis guineensis*) fronds lignin-derived non-toxic aldehyde for eco-friendly wood adhesive. *International Journal of Biological Macromolecules, 113*, 1266–1272.

Ibrahim, Y., Abdulkarem, E., Naddeo, V., Banat, F., & Hasan, S. W. (2019). Synthesis of super hydrophilic cellulose-alpha zirconium phosphate ion exchange membrane via surface coating for the removal of heavy metals from wastewater. *Science of the Total Environment, 690*, 167–180.

Irmer, J. (2017). Lignin – a natural resource with huge potential. www.biooekonomie-bw.de/en/articles/dossiers/lignin-a-natural-resource-with-huge-potential

Jabli, M., Tka, N., Ramzi, K., & Saleh, T. A. (2018). Physicochemical characteristics and dyeing properties of lignin-cellulosic fibers derived from *Nerium oleander*. *Journal of Molecular Liquids, 249*, 1138–1144.

Jamshaid, A., Hamid, A., Muhammad, N., Naseer, A., Ghauri, M., Iqbal, J., Rafiq, S., & Shah, N. S. (2017). Cellulose-based materials for the removal of heavy metals from wastewater–an overview. *ChemBioEng Reviews*, *4*(4), 240–256.

Ji, Q., Yu, X., Yagoub, A. E.-G. A., Chen, L., & Zhou, C. (2020). Efficient removal of lignin from vegetable wastes by ultrasonic and microwave-assisted treatment with ternary deep eutectic solvent. *Industrial Crops and Products, 149*, 112357.Jin, C., Zhang, X., Xin, J., Liu, G., Wu, G., Kong, Z., & Zhang, J. (2017). Clickable synthesis of 1, 2, 4-triazole modified lignin-based adsorbent for the selective removal of Cd (II). *ACS Sustainable Chemistry & Engineering*, *5*(5), 4086–4093.

Jin, C., Liu, G., Wu, G., Huo, S., Liu, Z., & Kong, Z. (2020). Facile fabrication of crown ether functionalized lignin-based biosorbent for the selective removal of Pb (II). *Industrial Crops and Products, 155*, 112829.

Karunakaran, V., Abd-Talib, N., & Yong, T.-L. K. (2020). Lignin from oil palm empty fruit bunches (EFB) under subcritical phenol conditions as a precursor for carbon fiber production. *Materials Today: Proceedings, 31*, 100–105.

Kaur, M., Tewatia, P., Rattan, G., Singhal, S., & Kaushik, A. (2021). Diamidoximated cellulosic bioadsorbents from hemp stalks for elimination of uranium (VI) and textile waste in aqueous systems. *Journal of Hazardous Materials, 417*, 126060.

Kaushik, A., & Singh, M. (2011). Isolation and characterization of cellulose nanofibrils from wheat straw using steam explosion coupled with high shear homogenization. *Carbohydrate Research, 346*(1), 76–85.

Kiran, S., Nosheen, S., Abrar, S., Anjum, F., Gulzar, T., & Naz, S. (2019). Advanced approaches for remediation of textile wastewater: a comparative study. In ul-Islam, S., & Butola, B.S. (eds) *Advanced Functional Textiles and Polymers: Fabrication, Processing and Applications* (pp. 201–264). Scrivener Publishing LLC.

Kitagawa, T., Nishikawa, Y., Frankenfeld, J., & Li, N. (1977). Wastewater treatment by liquid membrane process. *Environmental Science & Technology, 11*(6), 602–605.

Koelmel, J., Prasad, M., Velvizhi, G., Butti, S., & Mohan, S. V. (2016). Metalliferous waste in India and knowledge explosion in metal recovery techniques and processes for the prevention of pollution. In Prasad, M.N.V., & Shih, K. *Environmental Materials and Waste* (pp. 339–390). Elsevier.

Kong, Q., Preis, S., Li, L., Luo, P., Hu, Y., & Wei, C. (2020). Graphene oxide-terminated hyperbranched amino polymer-carboxymethyl cellulose ternary nanocomposite for efficient removal of heavy metals from aqueous solutions. *International Journal of Biological Macromolecules, 149*, 581–592.

Krishania, M., Kumar, V., & Sangwan, R. S. (2018). Integrated approach for extraction of xylose, cellulose, lignin and silica from rice straw. *Bioresource Technology Reports, 1*, 89–93.

Kwak, H. W., Lee, H., & Lee, K. H. (2020). Surface-modified spherical lignin particles with superior Cr (VI) removal efficiency. *Chemosphere, 239*, 124733.

Li, C., & Zhao, Z. K. (2007). Efficient acid-catalyzed hydrolysis of cellulose in ionic liquid. *Advanced Synthesis & Catalysis, 349*(11–12), 1847–1850.Li, Z., Ge, Y., & Wan, L. (2015). Fabrication of a green porous lignin-based sphere for the removal of lead ions from aqueous media. *Journal of Hazardous Materials, 285*, 77–83.

Li, Z., Xiao, D., Ge, Y., & Koehler, S. (2015). Surface-functionalized porous lignin for fast and efficient lead removal from aqueous solution. *ACS Applied Materials & Interfaces, 7*(27), 15000–15009.

Li, Z., Chen, J., & Ge, Y. (2017). Removal of lead ion and oil droplet from aqueous solution by lignin-grafted carbon nanotubes. *Chemical Engineering Journal, 308*, 809–817.

Li, J., Feng, P., Xiu, H., Li, J., Yang, X., Ma, F., Li, X., Zhang, X., Kozliak, E., & Ji, Y. (2019). Morphological changes of lignin during separation of wheat straw components by the hydrothermal-ethanol method. *Bioresource Technology, 294*, 122157.

Liu, J., Chen, T.-W., Yang, Y.-L., Bai, Z.-C., Xia, L.-R., Wang, M., Lv, X.-L., & Li, L. (2020). Removal of heavy metal ions and anionic dyes from aqueous solutions using amide-functionalized cellulose-based adsorbents. *Carbohydrate Polymers, 230*, 115619.

Liu, M., Liu, Y., Shen, J., Zhang, S., Liu, X., Chen, X., Ma, Y., Ren, S., Fang, G., & Li, S. (2020). Simultaneous removal of Pb^{2+}, Cu^{2+} and Cd^{2+} ions from wastewater using hierarchical porous polyacrylic acid grafted with lignin. *Journal of Hazardous Materials, 392*, 122208.

Lofrano, G., Carotenuto, M., Libralato, G., Domingos, R. F., Markus, A., Dini, L., Gautam, R. K., Baldantoni, D., Rossi, M., & Sharma, S. K. (2016). Polymer functionalized nanocomposites for metals removal from water and wastewater: an overview. *Water Research, 92*, 22–37.

Lu, Y., Lu, Y.-C., Hu, H.-Q., Xie, F.-J., Wei, X.-Y., & Fan, X. (2017). Structural characterization of lignin and its degradation products with spectroscopic methods. *Journal of Spectroscopy, 2017*, 1–15.

Malik, D., Jain, C., & Yadav, A. K. (2017). Removal of heavy metals from emerging cellulosic low-cost adsorbents: a review. *Applied Water Science, 7*(5), 2113–2136.

Manimaran, P., Saravanan, S., Sanjay, M., Siengchin, S., Jawaid, M., & Khan, A. (2019). Characterization of new cellulosic fiber: *Dracaena reflexa* as a reinforcement for polymer composite structures. *Journal of Materials Research and Technology, 8*(2), 1952–1963.

Marett, J., Aning, A., & Foster, E. J. (2017). The isolation of cellulose nanocrystals from pistachio shells via acid hydrolysis. *Industrial Crops and Products, 109*, 869–874.

Matlock, M. M., Howerton, B. S., & Atwood, D. A. (2002). Chemical precipitation of heavy metals from acid mine drainage. *Water Research, 36*(19), 4757–4764.

Mohtar, S., Busu, T. T. M., Noor, A. M., Shaari, N., Yusoff, N., Bustam, M., Mutalib, M. A., & Mat, H. (2015). Extraction and characterization of lignin from oil palm biomass via ionic liquid dissolution and non-toxic aluminium potassium sulfate dodecahydrate precipitation processes. *Bioresource Technology, 192*, 212–218.

Moshi, A. A. M., Ravindran, D., Bharathi, S. S., Indran, S., Saravanakumar, S., & Liu, Y. (2020). Characterization of a new cellulosic natural fiber extracted from the root of *Ficus religiosa* tree. *International Journal of Biological Macromolecules, 142*, 212–221.

Nishijo, M., Nakagawa, H., Suwazono, Y., Nogawa, K., & Kido, T. (2017). Causes of death in patients with Itai-itai disease suffering from severe chronic cadmium poisoning: a nested case–control analysis of a follow-up study in Japan. *BMJ Open*, *7*(7), e015694.

Palamae, S., Dechatiwongse, P., Choorit, W., Chisti, Y., & Prasertsan, P. (2017). Cellulose and hemicellulose recovery from oil palm empty fruit bunch (EFB) fibers and production of sugars from the fibers. *Carbohydrate Polymers, 155*, 491–497.

Palmowski, L., & Müller, J. (2000). Influence of the size reduction of organic waste on their anaerobic digestion. *Water Science and Technology*, *41*(3), 155–162.

Patel, D. K., Dutta, S. D., & Lim, K.-T. (2019). Nanocellulose-based polymer hybrids and their emerging applications in biomedical engineering and water purification. *RSC Advances*, *9*(33), 19143–19162.

Popovic, A. L., Rusmirovic, J. D., Velickovic, Z., Radovanovic, Z., Ristic, M., Pavlovic, V. P., & Marinkovic, A. D. (2020). Novel amino-functionalized lignin microspheres: high performance biosorbent with enhanced capacity for heavy metal ion removal. *International Journal of Biological Macromolecules, 156*, 1160–1173.

Prakash, O., Naik, M., Katiyar, R., Naik, S., Kumar, D., Maji, D., Shukla, A., Nannaware, A. D., Kalra, A., & Rout, P. K. (2018). Novel process for isolation of major bio-polymers from *Mentha arvensis* distilled biomass and saccharification of the isolated cellulose to glucose. *Industrial Crops and Products, 119*, 1–8.

Qiao, L., Li, S., Li, Y., Liu, Y., & Du, K. (2020). Fabrication of superporous cellulose beads via enhanced inner cross-linked linkages for high efficient adsorption of heavy metal ions. *Journal of Cleaner Production, 253*, 120017.

Qin, L., Ge, Y., Deng, B., & Li, Z. (2017). Poly (ethylene imine) anchored lignin composite for heavy metals capturing in water. *Journal of the Taiwan Institute of Chemical Engineers, 71*, 84–90.

Rai, P. K., & Tripathi, B. (2008). Heavy metals in industrial wastewater, soil and vegetables in Lohta village, India. *Toxicological and Environ Chemistry*, *90*(2), 247–257.

Rani, K., Gomathi, T., Vijayalakshmi, K., Saranya, M., & Sudha, P. (2019). Banana fiber Cellulose Nano Crystals grafted with butyl acrylate for heavy metal lead (II) removal. *International Journal of Biological Macromolecules, 131*, 461–472.

Rawat, M., Moturi, M. C. Z., & Subramanian, V. (2003). Inventory compilation and distribution of heavy metals in wastewater from small-scale industrial areas of Delhi, India. *Journal of Environmental Monitoring*, *5*(6), 906–912.

Reis, R. S., Tienne, L. G., de HS Souza, D., Maria de Fátima, V. M., & Monteiro, S. N. (2020). Characterization of coffee parchment and innovative steam explosion treatment to obtain microfibrillated cellulose as potential composite reinforcement. *Journal of Materials Research and Technology, 9*(4), 9412–9421.

Rocha, G., Andrade, L., Martin, C., Araujo, G., Mouchrek Filho, V., & Curvelo, A. (2020). Simultaneous obtaining of oxidized lignin and cellulosic pulp from steam-exploded sugarcane bagasse. *Industrial Crops and Products, 147*, 112227.

Šćiban, M. B., Klašnja, M. T., & Antov, M. G. (2011). Study of the biosorption of different heavy metal ions onto Kraft lignin. *Ecological Engineering*, *37*(12), 2092–2095.

Sebeia, N., Jabli, M., Ghith, A., El Ghoul, Y., & Alminderej, F. M. (2019). *Populus tremula*, *Nerium oleander* and *Pergularia tomentosa* seed fibers as sources of cellulose and lignin for the bio-sorption of methylene blue. *International Journal of Biological Macromolecules, 121*, 655–665.

Senthamaraikannan, P., & Kathiresan, M. (2018). Characterization of raw and alkali treated new natural cellulosic fiber from *Coccinia grandis* L. *Carbohydrate Polymers, 186*, 332–343.

Sharma, R. K., Agrawal, M., & Marshall, F. (2007). Heavy metal contamination of soil and vegetables in suburban areas of Varanasi, India. *Ecotoxicology and Environmental Safety*, *66*(2), 258–266.

Shui, T., Feng, S., Yuan, Z., Kuboki, T., & Xu, C. C. (2016). Highly efficient organosolv fractionation of cornstalk into cellulose and lignin in organic acids. *Bioresource Technology, 218*, 953–961.

Siva, R., Valarmathi, T., Palanikumar, K., & Samrot, A. V. (2020). Study on a novel natural cellulosic fiber from *Kigelia africana* fruit: characterization and analysis. *Carbohydrate Polymers, 244*, 116494.

Song, Z., Li, W., Liu, W., Yang, Y., Wang, N., Wang, H., & Gao, H. (2015). Novel magnetic lignin composite sorbent for chromium (VI) adsorption. *RSC Advances*, *5*(17), 13028–13035.

Stechemesser, H. (2005). *Coagulation and Flocculation*. CRC Press.
Sud, D., Mahajan, G., & Kaur, M. (2008). Agricultural waste material as potential adsorbent for sequestering heavy metal ions from aqueous solutions–a review. *Bioresource Technology*, *99*(14), 6017–6027.
Sun, Y., & Cheng, J. (2002). Hydrolysis of lignocellulosic materials for ethanol production: a review. *Bioresource Technology*, *83*(1), 1–11.
Syaftika, N., & Matsumura, Y. (2018). Comparative study of hydrothermal pretreatment for rice straw and its corresponding mixture of cellulose, xylan, and lignin. *Bioresource Technology, 255*, 1–6.
Tang, J., Song, Y., Zhao, F., Spinney, S., da Silva Bernardes, J., & Tam, K. C. (2019). Compressible cellulose nanofibril (CNF) based aerogels produced via a bio-inspired strategy for heavy metal ion and dye removal. *Carbohydrate Polymers, 208*, 404–412.
Tareq, R., Akter, N., & Azam, M. S. (2019). Biochars and biochar composites: low-cost adsorbents for environmental remediation. In *Biochar from Biomass and Waste* (pp. 169–209). Elsevier.
Tarrés, Q., Ehman, N. V., Vallejos, M. E., Area, M. C., Delgado-Aguilar, M., & Mutjé, P. (2017). Lignocellulosic nanofibers from triticale straw: the influence of hemicelluloses and lignin in their production and properties. *Carbohydrate Polymers, 163*, 20–27.
Todorciuc, T., Bulgariu, L., & Popa, V. I. (2015). Adsorption of Cu (II) from aqueous solution on wheat straw lignin: equilibrium and kinetic studies. *Cellulose Chemistry and Technology, 49*, 439–447.
Upton, B. M., & Kasko, A. M. (2016). Strategies for the conversion of lignin to high-value polymeric materials: review and perspective. *Chemical Reviews*, *116*(4), 2275–2306.
Vaidya, A. A., Gaugler, M., & Smith, D. A. (2016). Green route to modification of wood waste, cellulose and hemicellulose using reactive extrusion. *Carbohydrate Polymers, 136*, 1238–1250.
Ventura-Cruz, S., Flores-Alamo, N., & Tecante, A. (2020). Preparation of microcrystalline cellulose from residual Rose stems (Rosa spp.) by successive delignification with alkaline hydrogen peroxide. *International Journal of Biological Macromolecules, 155*, 324–329.
Vyas, J. N., & Nath, S. (2021). The role of Government and the public in water resource management in India. In Singh, P., Singh, R., Singh, V.P., & Bhadouria, R. (eds) *Pollutants and Water Management: Resources, Strategies and Scarcity* (pp. 399–415). John Wiley & Sons Ltd.
Wang, J., & Guo, X. (2020). Adsorption isotherm models: classification, physical meaning, application and solving method. *Chemosphere, 258*, 127279.
Wang, J., Liu, M., Duan, C., Sun, J., & Xu, Y. (2019). Preparation and characterization of cellulose-based adsorbent and its application in heavy metal ions removal. *Carbohydrate Polymers, 206*, 837–843.
Wang, Z., Yao, M., Wang, X., Li, S., Liu, Y., & Yang, G. (2020). Influence of reaction media on synthesis of dialdehyde cellulose/GO composites and their adsorption performances on heavy metals. *Carbohydrate Polymers, 232*, 115781.
Wei, J., Liang, G., Alex, J., Zhang, T., & Ma, C. (2020). Research progress of energy utilization of agricultural waste in China: Bibliometric analysis by citespace. *Sustainability*, *12*(3), 812.
Wu, S.-P., Dai, X.-Z., Kan, J.-R., Shilong, F.-D., & Zhu, M.-Y. (2017). Fabrication of carboxymethyl chitosan–hemicellulose resin for adsorptive removal of heavy metals from wastewater. *Chinese Chemical Letters*, *28*(3), 625–632.
Wu, X., Zhang, T., Liu, N., Zhao, Y., Tian, G., & Wang, Z. (2020). Sequential extraction of hemicelluloses and lignin for wood fractionation using acid hydrotrope at mild conditions. *Industrial Crops and Products, 145*, 112086.
Xu, F., Zhu, T.-T., Rao, Q.-Q., Shui, S.-W., Li, W.-W., He, H.-B., & Yao, R.-S. (2017). Fabrication of mesoporous lignin-based biosorbent from rice straw and its application for heavy-metal-ion removal. *Journal of Environmental Sciences, 53*, 132–140.
Yao, M., Wang, Z., Liu, Y., Yang, G., & Chen, J. (2019). Preparation of dialdehyde cellulose grafted graphene oxide composite and its adsorption behavior for heavy metals from aqueous solution. *Carbohydrate Polymers, 212*, 345–351.
Yu, H., Hong, H.-J., Kim, S. M., Ko, H. C., & Jeong, H. S. (2020). Mechanically enhanced graphene oxide/carboxymethyl cellulose nanofibril composite fiber as a scalable adsorbent for heavy metal removal. *Carbohydrate Polymers, 240*, 116348.
Zakzeski, J., Bruijnincx, P. C., Jongerius, A. L., & Weckhuysen, B. M. (2010). The catalytic valorization of lignin for the production of renewable chemicals. *Chemical Reviews*, *110*(6), 3552–3599.

Zhang, J., Liu, J., Kou, L., Zhang, X., & Tan, T. (2019). Bioethanol production from cellulose obtained from the catalytic hydro-deoxygenation (lignin-first refined to aviation fuel) of apple wood. *Fuel, 250*, 245–253.

Zhao, B., Jiang, H., Lin, Z., Xu, S., Xie, J., & Zhang, A. (2019). Preparation of acrylamide/acrylic acid cellulose hydrogels for the adsorption of heavy metal ions. *Carbohydrate Polymers, 224*, 115022.

Part III

Climate Change: Challenges and Management Strategies

13 Emerging Issues of Climate Change

Global Perspective, Ecosystem, and Health

U.R. Zargar
Department of Zoology, Government Degree College, Anantnag

M.I. Rather
Department of Environmental Sciences, University of Kashmir, Kashmir

CONTENTS

13.1 INTRODUCTION

Climate change is a global phenomenon that is currently felt by different countries across the globe at different levels. In a real sense, climate alteration is now observed as an actual phenomenon, and there are ample chances that both people and the ecosystem will bear the brunt of the changing behavior of the earth's climate. Alternatively, governments and the general public need to initiate a universal procedure to combat warming and climate change. There is substantial evidence from scientific research that highlights the potential negative impact of altered climate on the human population and different ecosystems. Challenging impacts of climate change are being observed, and their severity and frequency have increased manifold. Recently, several tipping points related to climate change have been identified, and they are showing early signs of activation (Lenton et al., 2019). Pertinently, the idea of tipping points was initially set in motion by the IPCC (Intergovernmental Panel on Climate Change) at the end of the 20th century. Massive discontinuities in various weather

DOI: 10.1201/9781003220824-16

and climate systems were previously predicted to occur if there was a rise of global warming above 5°C (above the pre-industrial level). The current data, however, does not conform very well with early expectations, and it has been revealed that tipping points can be increased even with 1–2°C of global warming (IPCC, 2018; 2019). Furthermore, the IPCC (2018) believes that global warming exceeding 1.5°C will have great implications for the earth and will lead to various catastrophes in the form of flash floods, drought, heatwaves, etc., which could in turn impact biodiversity and human life (www.ipcc.ch/sr15/).

. There is another aspect to climate change that is very much linked to the economy of the nations. The current estimate suggests that the average global economy will shrink by 23% towards the end of the current century. Consequently, the world will face more conflicts, poverty, and other socio-economic issues. The earth is committed to facing more intense climate warming, which is forecast to continue in the next decades unless there is a sharp decline in carbon emissions (Hoegh-Guldberg et al., 2018).

The current knowledge and evidence reveal that the climate of the planet is showing a rapid spike as compared with the past, which is primarily due to excess dependence on fossil fuels as a prime source of energy. The world is currently 80% dependent on fossil fuel for the continuation of industries. Although there are huge benefits of using fossil fuels as far as the economy is concerned, at the same time, there are severe concerns over accumulating greenhouse gases in the atmosphere, which has the potential to alter global climate.

The temperature at the global level showed an increase of from 0.3–0.6°C during the previous century, and this trend is expected to show further acceleration (Bhutiyani et al., 2007; 2009). There is a profound impact on the snow cover in the mountainous regions. It has been reported that about 10% of snow cover has been lost in the last 50 years. As per estimates, the average temperature for the Indian subcontinent is predicted to increase by between 3.5 and 5.5°C by the end of the 21st century, and a greater temperature rise is expected in the Tibetan region of the Himalaya (IPCC, 2007). The last century witnessed warm winters and a rise of temperature up to 1.6°C in the north-western Himalayan region (Bhutiyani et al., 2007; 2009).

It is pertinent to mention that if we are considering the human contribution to climate change, then other predictors that include various climatic variabilities known to us and external natural factors should be exempted. To understand the human pressures on climate change, determination of all potential factors or fingerprints and then, comparison with observed patterns is needed. There are various physical drivers of climate change, which include greenhouse gases, land surface changes, clouds and aerosols, and solar and volcanic activity (fifth IPCC assessment report, 2011).

In the current scenario, the major issue to be perceived is how sustainable development can be achieved under such circumstances (Yang et al., 2013) Climate change has a drastic impact on the environment and other human development indices, as is reflected by research authentic data.

In general, simulation and forecast data suggest that there will be a rise of approximately 2.5–10 degrees Fahrenheit on a global scale in the next century, which may have an immense socio-economic impact on developing and low-income countries. For instance, climate change and global warming will have a significant effect on high-impact areas, especially coastal and island countries, where there will be a large rise in sea level. It has been estimated that the global sea level rose by 8 inches from 1880 onwards, and a rapid rise of 1–8 feet is expected after the end of the current century. This projected change in sea level is attributed to seawater expansion and melting glaciers and ice caps.

Climate change can pose different challenges to the environment, ecosystems, health, and people. Climate change was thought to have a meager impact on biodiversity and the environment when compared with the direct effects of more intense anthropogenic activities such as habitat fragmentation and other human activities. However, this relative perception is now declining, and the detrimental environmental effects of altered climatic effects are more evident and are expected to increase in the future (Hoffmann and Beierkuhnlein, 2019; Ohashi et al., 2019). Similarly, climate change, together with anthropogenic health stressors, influences the diseases of humans and other animals in innumerable ways. There are ample chances that some current health issues will escalate and will lead to emerging health threats. This chapter aims to highlight the emerging issues of climate change and its effects on the environment/ecosystem and health.

13.2 PARIS AGREEMENT AND CLIMATE CHANGE

The Paris Climate Agreement (COP21) is fundamentally a legal procedure for the implementation of certain rules on climate change. This agreement was a major agreement in history adopted by 196 parties in Paris (on December 12, 2015) and came into existence from November 2016 onwards. The main outcome of the Paris climate change conference was that the world needs to cut carbon emissions to reduce global warming below 2 degrees Centigrade by the year 2021 and to mitigate the adverse impacts of temperature rise on ecologically fragile regions of the earth.

The general expectation of climate scientists in the last decade was that a reduction in greenhouse gases (by up to 40–70%) could reduce the challenges caused by climate alterations. The crux of the Paris Agreement was that the temperature milestone was mutually agreed upon by member countries after hectic and complex political deliberations (Hallegatte et al., 2016).

The climate agreement signed in Paris will benefit climate-susceptible and island countries if proposed temperature targets are kept below 2°C. Two areas remain unaddressed: one, the issues of developed and underdeveloped countries related to climate change are still ignored, and two, the framework for the effective implementation of the climate agreement is still lacking on the ground.

Three major objectives of the Paris Climate Agreement were:

- To limit the global temperature below 2°C (or limit it to 1.5°C)
- To facilitate and intensify recovery efforts against growing challenges of climate change
- To facilitate financial assistance between member countries for the implementation of various objectives

The climate agreement has been generally appreciated across the globe, with some reservations. The general opinion is that this agreement will enhance significant change in global warming, which will ease the pressure from growing temperature rise. The fact is that some objectives are overambitious and are barely achievable for some countries. However, a common expectation is that such a global effort will enhances cooperation in taking prompt action against the climate threat.

The climate agreement of 2015 has significant relevance for the Indian sub-continent, which is currently facing a climate crisis in different forms. Clear signs of climate alterations have been observed across different areas of the Himalayan region, with Himalayan glaciers showing huge retreats. The climate impacts are also evident in the north-western Himalayan region, the Hindu Kash, and the Tibetan plateau, which have resulted in unusual climate changes including flash floods, landslides, and other natural disasters, affecting both human and wildlife populations. The most vulnerable parts of the sub-continent, where climatic impacts are more intense and drastic, are coastal and island areas. Some of the countries in the sub-continent, including Bangladesh and Pakistan, have been recognized as the most vulnerable to climate change. Therefore, any reduction in warming will have a significant impact on ecologically fragile regions of the sub-continent.

The climate conference did not sit very well with developing countries, and some countries show reluctance to accept some of the objectives/approaches proposed by the climate meeting. Most developing countries argue that highly developed countries should adopt a holistic approach for carrying forward common goals against climate change and global warming. Furthermore, they advocate that rich nations should show flexibility as far as their contribution towards technology and money is concerned to counter rising temperatures.

Econometric analyses suggest deleterious effects at higher temperatures due to climate change; nevertheless, very little data is available to show the relative economic outcome of securing these climate milestones. Glanemannn et al. (2020) incorporated a damage–cost curve extracting the observed relation between economic prosperity and rising temperatures into the integrated and holistic Dynamic Integrated Climate-Economy model (DICE). The results reveal that the Paris Climate Agreement, which has political significance, also constitutes an economically beneficial roadway if realized in a judicious manner. In an attempt to ascertain the direct impact of rising temperatures on gross domestic product (GDP), Ueckerdt et al. (2019) observed that any limit to global temperature rise between 1.9 and 2°C is a convenient temperature regime for economic activities and will have no major impact on overall economic costs.

13.3 CLIMATE CHANGE AND HIMALAYAN REGION – RECENT TRENDS

The Himalayan region of India is located in a strategic position, giving shelter to 51 million people, who generally practice agriculture in eco-fragile zones. The Himalayan region is a perennial reservoir of freshwater for millions of people in the sub-continent and has great potential for hydroelectricity (Government of India, 2010; Department of Science and Technology, 2012). The importance of the Himalayas lies in the fact that it has been divided into well-defined 34 'biological hot spots' (Gautam et al., 2013) for delineation of prominent ecological regions.

A recent update on climatic alterations in the Himalayan region shows that the eco-fragile regions of the Himalayan Hotspot are prone to natural disasters, and there is great apprehension about significant climate change impacts, which may lead to major natural disasters (Shugar et al., 2021), loss of precious biological diversity, and great risk of food shortages. India initiated robust planning in 2008, known as the National Action Plan on Climate Change, to counter impacts of climate change in the country, and one of the objectives is to develop the sustainable Himalayan ecosystem and increase strategic know-how about climate change.

Climatic alteration has been observed in different regions of the Himalayas. For instance, the climate of the Hindu Kash Himalaya (HKH) has shown remarkable alterations in temperature in the 20th century; it is estimated that the warming trend has changed since the 20th century, and it is expected that the rate of global warming of the earth will be doubled to 0.32°C per decade at the beginning of the 21st century (Yan and Liu 2014). The estimated warming rate is significantly more severe in the winter season across different areas of HKH (Bhutiyani et al., 2007; Shrestha and Aryal, 2010). There are clear signs that the Himalayan mountains are showing a significantly more rapid increase than Indian landmass, which is also confirmed by Pepin et al. (2015), showing that temperatures at high altitude are increasing at a faster pace.

The main highlights of climatic impact on HKH are as follows (Sabin et al., 2020; Figure 13.1):

- One of the key observations with relation to climate alteration in HKH is the declining pattern of snowfall or retreating glaciers in the past couple of decades. However, high-altitude regions of the Karakoram range of the Himalayas have shown an increased level of precipitation, partly due to western disturbances.
- The predicted model reveals that the range of temperature increase in the HKH will be 2.6–4.6°C by the end of the 21st century. These predictive models further emphasize that there will be a substantial decrease in snowfall, while high-elevation areas with a height greater than 4000 m will experience an overall increase in annual precipitation during the 21st century.

13.3.1 Climate Change and North-Western Himalaya

Many studies on the north-western Himalaya confirm that climatic and warming effects were more prominent from the last quarter of the 20th century (i.e., 1975 onwards) (Dimri and Dash, 2012; Negi et al., 2018). The rapid growth of tree ring formation over the years authenticates the signs of climate change (Borgaonkar et al., 2009). Tewari et al. (2017), while reviewing the effects of climate change on the ecosystem of the north-western Himalayan region, highlighted that there is significant

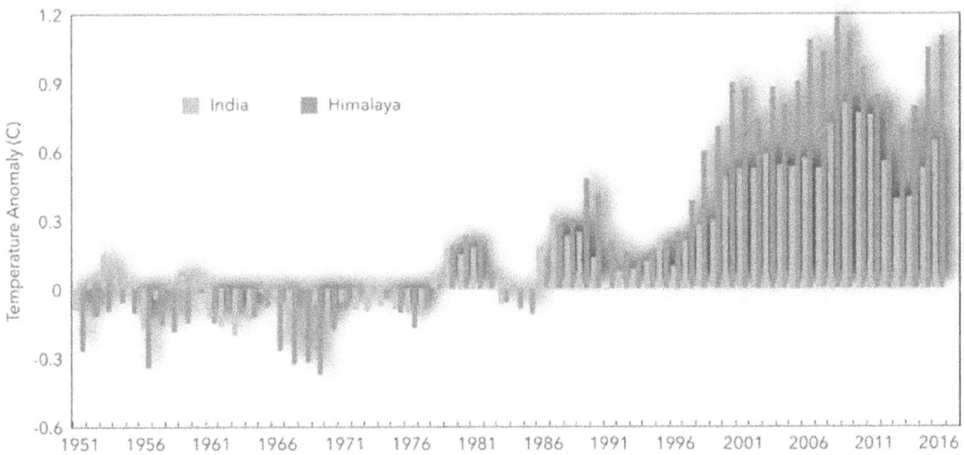

FIGURE 13.1 Comparison of mean annual average temperature between Indian landmass and HKH from 1951–2018.

(Adapted from Sabin et al., 2020.)

evidence of the global warming phenomenon, variability in precipitation, and extreme climate. Massive human-induced changes have already altered the physicochemical features of water in river systems of the north-west Himalayan region, including the river Jhelum (Rather et al., 2016).

Climate change in the north-western Himalaya, including Jammu and Kashmir (Union Territory), causes a plethora of complicated problems, encompassing biodiversity, endangered species, agriculture, livelihood, food security, floods, droughts, landslides, and human health. As per the United Nations Environmental Programme (UNEP) report, some areas of the north-western Himalaya are vulnerable (mild–high). According to Indian Network for Climate Change Assessment (INCCA) predictions, minimum temperatures are expected to rise by 1°C to 4.5°C, and maximum temperatures by 0.5°C to 2.5°C. It is projected that rainy days in the Himalayan region in the 2030s may surge by 5–10 days on an average, with an increase of more than 15 days in the eastern part of the Jammu and Kashmir region, and the intensity is expected to increase by 1–2 mm/day. Glacier retreat in the Himalayan region is attributed to three factors: rising global temperature, winter precipitation, and human-induced influence. Changes in temperature, precipitation, and cold waves will have a major effect on the agriculture sector.

The steady rise in the minimum and maximum winter temperatures is another concern in the north-western Himalaya. This temperature is not conducive to snow formation; therefore, the north-western Himalaya experiences precipitation mainly in the form of rain in peak winters. In Gulmarg, the average maximum winter temperature has risen from 1.92°C in 2001 to 2.17°C in 2010. It has been estimated that on average, the temperature in Kashmir Himalaya increased by 1.2°C from 1980 to 2010 (Romshoo and Rashid, 2014).

The Himalayan region is very sensitive to climate change. Climate change will cause a significant rise in temperature across all the three climatic zones, i.e., subtropical (Jammu), temperate (Kashmir), and cold desert (Ladakh), by the end of the 21st century, with the subtropical belt extending into the temperate climatic zone and the cold desert being invaded by the temperate zone. This massive change will result in substantial changes in environmental and socio-economic factors, such as changes in glacier dynamics, frequency of climate extremes, changes in streamflows, changes in biodiversity, increasing incidence of plant invasions, changes in land use pattern, specifically agriculture and horticulture, and above all, huge loss of hydroelectricity and other valuable ecosystem services (Romshoo et al., 2020).

The problem of food security is growing in Jammu and Kashmir. Rain-fed agriculture, which is dependent on natural precipitation, will suffer the most. Due to the decline in snowfall in recent years, there is a drastic reduction in horticultural crops like apples. Due to growing urbanization, there will be a massive reduction in forest cover in the future, which will significantly impact the ecological condition of the state. Climate change will increase the incidence of vectors and water-borne diseases (cholera, malaria, etc.) and impact the water, air, and soil quality. Abrupt flood due to sudden breakdown of glacial lakes (GLOF) has the potential to increase the possibility of massive landslides in higher reaches, which can indirectly affect food security in these areas.

13.3.1.1 Climate Change and Forests

The climate of a given region determines the vegetation patterns (distribution, structure, and ecology of forests). It is projected that there will be a large-scale change in forest distribution due to climate change (Romshoo and Rashid, 2010). Climate change will also have a substantial impact on the distribution of vegetation across the Kashmir Himalayan region. Applying various modeling techniques, it has been predicted that by the end of the century, deciduous forests and grasslands will be eliminated from the region, and temperate evergreen broadleaf forest, boreal evergreen forest, mixed forest types, and savannah will emerge. There will also be huge repercussions on livelihood and food security as a result of less streamflow and agricultural production due to the substantial decrease of ice cover by the end of the century (Rashid and Romshoo, 2020).

Due to global warming, there will be both poleward and upward movement of plants (Lenihan et al., 2003). With this change in the temperature and vegetation pattern, there are more chances of extinction of species of ecological significance. It is also projected that the yield of non-timber forest products (NTFP) will be significantly decreased due to the recurrence of droughts and forest fires. The services provided by the forests will drastically decrease. Due to climate change in Jammu and Kashmir, there will be a reduction of alpine pastures, reduction in important tree species like fir, spruce, and deodar, a massive spread of alien species like *Parthenium*, *Lantana*, *Xanthium*, etc., increased incidence of forest fire, and increased forest cover reduction.

13.3.1.2 Climate Change and Wetlands

Wetlands are considered one of the important ecosystems of the world, providing water for domestic and irrigation purposes, fisheries, buffering capacity, refuge for biodiversity, and climate stability. Wetlands will be drastically impacted due to rapid climate change (MEA, 2005). A decrease in the snow cover and consequent reduced discharge will seriously impact the aquatic biodiversity and hydrology of the wetlands (Johnson et al., 2005). Due to enhanced global warming, wetlands can become potential sources of carbon dioxide and methane, which are both potent greenhouse gases (Clement and Aidoud, 2007). In the Himalayan mountainous region of Kashmir, there are 3,813 wetlands and water bodies (Romshoo and Rashid, 2010). Many important species of Kashmir wetlands are reported to have been lost in the last few decades, primarily due to climate change (Kaul and Zutshi, 1967; Khan, 2000; Pandit and Kumar, 2006).

As well as climate change, wetlands are subjected to human-induced changes, which alter the ecosystem services provided by these important ecosystems (Dar et al., 2020). Some of the prominent threats to this ecosystem are land use/land cover (LULC) changes, pollution, changes in hydrological patterns, and encroachments, mainly through urbanization (Rather et al., 2016; Rashid and Aneaus, 2020). Wetlands need to be protected from these threats to secure the ecological functions and livelihood of the community.

13.3.1.3 Climate Change and Agriculture

Climate change, besides having impacts on the various aspects of the environment, has broad impacts on crop productivity and other associated sectors on both regional and global scales. As a consequence, there have been changes in cropping patterns, especially in mountainous regions, particularly the Himalayan region. One such study was done by Batool et al., 2019, in which they analyzed the impact of rainfall variability on crop productivity and changes in cropping patterns in the north-western Himalayas. There has been a reduction in rainfall patterns in recent decades, which has led to a reduction in crop yield/production and massive changes in the cropping pattern in the region, and people have shifted towards the horticulture sector.

Changes in climatic and weather patterns will drastically alter the agriculture sector of the Jammu and Kashmir state (Batool et al., 2019). There will be a massive reduction in crop production and hence, a big jolt to food distribution. Due to erratic rainfall patterns and less soil moisture, regular drought-like conditions will lead to crop failure. Crops like rice, wheat, and mustard, which are heavily dependent on irrigation water, will be severely reduced. There will be a reduction in oil-seed and vegetable production due to climate change, posing a threat to food security. Increased incidence of disease-causing pests and appearance of new alien weeds will be another big issue regarding climate change. Due to climate change, paddy fields are being converted into orchards and drylands in various districts of Kashmir Himalaya. There has been a reduction in crop yields, a change in cropping patterns, and a loss of agrobiodiversity.

13.3.1.4 Climate Change and Water Resources

The valley of Kashmir is bestowed with abundant water resources and glaciers, and the majority of the people are dependent on it for sustenance. But unfortunately, this vital natural resource is facing diminution due to climate change (Dar et al., 2014). There has been a large-scale change in the hydrological pattern of water bodies, alterations in water supply systems, and incremental pressure on water resources (Shafiq et al., 2019). Furthermore, there have been sudden changes in precipitation patterns, enhanced rates of evapotranspiration, the recession of glaciers, and lower availability of freshwaters. According to the State Action Plan on Climate Change (SAPCC), the quantity of glaciers is decreasing day by day due to climate change, and melting will continue even without further temperature increase. As a result of glacier mass loss, there will be a reduction in streamflows, which will impact not only irrigation but power generation facilities as well (Romshoo et al., 2020). It is not only the surface water that is facing shortage; the groundwater level also is receding very fast in the valley. It has also been observed that due to climate change, water yield in springs has declined or has gone dry, mainly due to the erratic rainfall in recent decades (Bhat et al., 2020).

13.3.1.5 Climate Change and Tourism

The feasibility of tourism at a particular location is very much dependent on the climatic/seasonal variations. Thus, alteration in the duration and quality of climate-dependent tourism seasons could severely impact the success of the tourism enterprise. The decline in snowfall has a big impact on the tourism sector, one of the biggest sectors of Kashmir. Snow in Kashmir is the biggest attraction for tourists around the globe. Without it, winter tourism is bound to suffer.

The tourism industry, which contributes much to the GDP of the state, will face drastic changes due to climate change. It will have implications for snow-associated sports activities and winter tourism, like snowboarding and skiing, as winters are becoming prolonged with little snowfall on account of climate change. This problem, compounded by the excessive melting of glaciers and ice covers due to increased temperature in the winter months, poses a grave threat to the future and sustainability of winter tourism in the Kashmir region (Dar et al., 2014).

Climate change may cause a significant impact on snow-linked winter tourism (Bourdeau, 2009; Surugiu et al., 2010) in the valley of Kashmir. The impact will be felt in ski resorts like Gulmarg, which will obviously lose their characteristics and features due to the unavailability of snow, which is an attraction for tourists. Climate change will worsen the already damaged tourism industry as well as having large-scale consequences for agriculture, horticulture, and other important sectors of the economy (Muslim, 2012). There has been an unprecedented reduction rate of Himalayan glaciers in the recent past (Dobhal et al., 2004; Murtaza et al., 2021; Rashid et al., 2021). With this speed, there are bound to be some complications, like change in the river water flow and impacts on the discharges, which will eventually have consequences for the distribution of drinking water supplies. Typically, the spring months in Kashmir Himalaya are characterized by heavy precipitation in the form of rainfall (Bhutiyani et al., 2009), but this pattern is also changing due to climate change (Panwar, 2020; Silwal et al., 2020). Due to climate change, snow is melting much earlier, leaving the rest of the summer (June, July, and August) with low or no snow. The problem is further compounded in the summer months by the temperature increase and massive reduction in water discharge. Furthermore, there is an increase in the minimum temperature in the winter months (Hunt et al., 2020). Climate change has impacted river/spring ecosystems as well as other ecosystems, as their source is snow and glaciers, which either have dried up or are very likely to be dry in the near future, and this will have serious consequences for water availability (Jeelani et al., 2012). All these changing scenarios in the climatic setup in the Himalayas have decreased the very essence of the various unique climatic features, which has recently impacted the tourism industry, and climate change will continue to be a challenge for policymakers and tourism administrators.

13.3.1.6 Adaptive Measures

The various adaptive measures that need to be undertaken in the Northwest Himalayan region in response to climate change (JK SAPCC, 2013) are as follows:

- Massive afforestation programs in river catchments
- Rehabilitation of degraded forests on a war footing
- Ecotourism promotion and generation of sustainable tourism
- Use of innovative methods for water conservation, including sprinkle irrigation
- Promotion of organic farming and focus on traditional methods of farming
- Use of pest- and drought-resistant crops
- Introduction of advanced water harvesting methods, including rainfall-runoff harvesting and rooftop rainwater harvesting
- Public awareness about the possible impacts of climate change
- Establishment of proper weather forecasts and early warning systems

13.4 CLIMATE CHANGE AND DISEASE

Climate change is considered one of the major threats to plants and animals, as it will act as a major predictor for increasing disease burden and is forecast to play an important role in the resistance of species against new and emerging diseases. There are convincing empirical data that show the long-term impact of climate change on diseases. Pathogens and diseases show sensitivity towards climatic changes, and most of these organisms perform optimally under specific temperature regimes. Climate change may induce both direct or indirect impacts on the distribution and epidemiological parameters of parasites. Anthropogenic climate disruptions are a major concern for ecologists, as they pose a significant influence on various organisms, including pathogens. Climatic disturbances like drought, flood, heatwave, and drought present a direct and significant risk of infectious diseases in humans (Watts et al., 2015). Climate alterations can modulate changes in disease patterns in the marine ecosystem and could lead to changes in the diversity pattern of parasites in the marine environment (Byers, 2020).

Climate change can have a detrimental impact on animal health, including the health of human beings (Epstein, 1999; Costello et al., 2009; Willox et al., 2015), particularly when infectious diseases are the main concern (Altizer et al., 2013). It is now established that three components are necessary for the spread of infectious diseases: a pathogen, a host, and the environment (Epstein, 2001). The interaction between different epidemiological, ecological, environmental, and climatic alterations is well documented (Teklehaimanot et al., 2004; Greer et al., 2008). Basic information on these links is essential for understanding the disease ecology of various infections, the transmission of disease agents, vectors (intermediate hosts), reservoir hosts, and definitive hosts (Medone et al., 2015). Numerous pathogens need an intermediate host for the completion of their life cycle, and perfect climatic and weather conditions are pivotal for the distribution and transmission of pathogens, vectors, and hosts. The changes in climate or weather are directly or indirectly related to the transmission of infection between different hosts, and the role of the environment is very significant (Epstein, 2001, Wu et al., 2014).

Extended climate change will lead to the shift of many deadly diseases (Ostfeld and Brunner, 2015, Rodó et al., 2013), and these altered climatic patterns may enhance the chance to create new clustered infections at non-specific places (Epstein, 2000). Climate change has the potential to force the distributional pattern, inducing seasonal and geographical patterns, of different infections, and weather has the main influence on intensity and timing of disease occurrence (Kuhn et al., 2005, Wu et al., 2014). The research data reveal that diseases transmitted by insects are highly sensitive to climate change (Kuhn et al., 2005, Tian et al., 2015a). Vector-borne diseases like hantavirus, malaria, dengue, and cholera have already shown a response to climate change (Tian et al., 2015b; Watson et al., 1998; Yu et al., 2015). Furthermore, other diseases such as giardiasis, salmonellosis, and cholera are expected to show a surge in outbreaks in response to an increase in temperature and flooding (Chretien et al., 2014). Moreover, new emerging climate-generated diseases may have a long-lasting impact on society.

Climate change can modulate pathogens, hosts/vectors, and the environment in many ways. For instance, climate variables alter infection patterns both temporally and spatially, and such change has a drastic effect on various developmental stages of pathogens, hosts, and the way they interact with the human host. Recent thematic models confirm the impact of severe weather and meteorological hazards on many disease systems (Wu et al., 2016; Liang and Gong, 2017).

It is clear from the research that climate change has a profound impact on the virulence, distribution, transmission, and abundance of different parasites in wild as well as in domestic animals (Moller et al., 2013). Lindgren (1998), for instance, found a relationship between alterations in climate and the spread of tick-borne diseases in the human host.

There is still scanty empirical analysis of the effects of the impact of climate change on various host–parasitic interactions, and the results of many research studies are yet to come. The main reasons for the lack of large-scale studies on the effect of climate change on many parasitic diseases (both micro- and macroparasites) are the uneven distribution and over disperse nature of parasites. However, many studies confirm the impact of climate change on host–parasite interaction. For example, Møller (2010), while evaluating the prevalence and intensity of many parasites of the barn swallow, *Hirundo rustica*, observed that the number of parasites that do not complete their life cycle on hosts was significantly affected by climate change. Ectoparasites like mites may have reduced their infection rate in response to climate change. Alterations in the prevalence of brood parasitic cuckoo have been linked to climate change (Saino et al., 2009; Møller et al., 2011). A meta-analysis by Garamszegi (2011) confirmed that the incidence of blood parasites and avian malaria has increased during recent decades, which may be linked to climate change. Bedsides, significant changes in the phenology of both parasites and hosts have been observed. The parasite-mediated mortality and changes in host resistance have been related to changing patterns of climate change (Garamszegi, 2011; Møller et al., 2011; Møller and Erritzøe, 2003). Although there is convincing evidence that confirms the impacts of climate change on host–parasite interaction, nevertheless, there are some unresolved research questions, including unexplained heterogeneity among parasite taxa.

The most obvious impact of altered climate will be the spread of pathogens and diseases to a diverse geographical area. It has been predicted that many global diseases will be shifted to other regions that previously had no occurrence of these diseases. The spread and distribution of malaria towards northern and southern regions of the world is an example of the poleward shift of diseases

(MEA, 2005). A similar change is expected with tick-borne diseases if there is no let-up in climate change in the future. It has been reported that erratic climate warming may affect the vertical distribution of tick-borne encephalitis in Central Europe. Climate change can trigger cascading effects on the various host–parasite systems that could have an unusual effect on the spread of diseases.

The rising temperatures may affect pathogens and parasites in many ways, including growth and development, prolonged transmission, and unusual expansion to new areas. For example, it has been observed that species of parasitic worms are using the rising temperatures to increase their chances of survival at the cost of the fish populations they use as hosts. In addition, there are chances of cascading effects of increased temperature on the host–parasite system due to climate change. Furthermore, disease response has increased over the last 20 years (Liang and Gong, 2017).

The shifting nature of diseases and parasites may show a ripple effect on humans. The infection load in humans will show an incremental increase across the globe in the coming decades due to climate change. The infection burden will increase significantly in low-economic countries located in Asia and Africa and will have an additional burden on their economies, as more funds will be needed for the management of these diseases.

There are many examples at the global level that suggest climate change is inducing biological invasions in areas where the climatic effect was probably not predicted. The climatic shifts have posed concern over lemur parasites in Madagascar, which are showing a massive change in distribution across this area (Barrett et al., 2013). The massive distribution of lemur parasites across Madagascar is one of the examples of how parasites are showing invasion across remote areas. They have spread to about 12–26% of surface land area and 40–86% of forest area. This expansion is expected to increase, and lemur hosts will more vulnerable to new, emerging parasites. Predicting changing patterns of parasites in remote areas will help assess the overall health of the ecosystem, risk factors for wild animals and humans, and the overall impact assessment.

13.4.1 Climate Change and Emerging Diseases

The major concern for epidemiologists in recent years has been the rise of new emerging diseases, and there is evidence showing that climate change plays a pivotal role in the shaping and emergence of these diseases. Currently, the world is facing an unprecedented deadly emerging disease in the form of Coronavirus Disease 2019 (COVID-19), which was first reported from Wuhan, China at the end of 2019 (Huang et al., 2021) More such episodes of emerging diseases with greater intensity of virulence are expected in the future.

There is a possibility that climate change may have acted as a trigger for the emergence of Covid-19; nevertheless, research is still in its infancy to ascertain the fundamental causes for the origin and transmission of these diseases. Climate change has been considered as an important predictor for the transmission of zoonotic and newly emerging diseases (Naicker, 2011). The alterations in climate and weather adversely impact on the various epidemiological features of many diseases. The substantial increase in the temperature has a possible role in the transmission of many mosquito-borne diseases, such as dengue transmission in Dhaka (Banu et al., 2014). Morin et al. (2019) predicted that warming at higher latitudes is expected to alter the ecology of the avian influenza virus (AIV) and help to sustain the virus in the atmosphere for a long period.

13.4.2 Biodiversity–Disease Relationship in Relation to Climate Change

It is now established that climate change has a differential influence on various diseases, with some being more sensitive while others are less responsive (Zargar, 2011). Climate change may expand the range of vector-borne diseases from the tropical zone, where the species diversity of hosts is comparatively high, in contrast to the temperate climatic zone, where species diversity is very low (Dobson and Carper, 1992; Harvell et al., 2002). It is, however, too early to predict the impact of biodiversity and global warming on the propagation of vector-borne disease, as the vector behavior and transmission mechanisms of the host differ (Miller and Huppert, 2013). However, it is pertinent to elaborate that the changing pattern of biodiversity may stimulate changes in the host population, which may indirectly shape the disease pattern in the wild as well as in the human population.

Current knowledge on disease ecology predicts that climate change modulates disease patterns through biodiversity alteration, although the mechanism behind these changes is still unknown (Zargar et al., 2015). Any negative effects of climate change on biodiversity may affect disease transmission, because biodiversity may act as a buffer against disease propagation. The fundamental relationship between biodiversity and diseases has been explained by two theories, i.e., amplification and dilution effect theories. The dilution effect theory predicts an inverse relationship between biodiversity and disease, while the amplification theory predicts a direct relationship between disease and biodiversity (Zargar et al., 2015).

The paucity of information on this aspect suggests that the biodiversity–disease relationship needs to be verified regarding climatic alteration. The response to climate change at small spatial scales as well as at the global scale is still unknown. It is interesting to see how climatic alteration affects biodiversity and then, how a change in biodiversity affects the disease outcome. It has been argued that ecologists should analyze the infection pattern in the context of biodiversity and ecological interaction (Lafferty, 2009).

13.4.3 Eradication of Disease Linked with Global Warming and Climate Change

Health scientists point out that global warming may affect patterns of disease, and this could have a dramatic impact on public health strategies for disease prevention and control. The variable effects of climate change on different diseases will also complicate eradication programs. Many microbial and vector-borne diseases will be sensitive to climatic changes, but some disease-causing organisms or vectors will be less responsive (Slenning, 2010). For example, initiating a campaign to eradicate a disease in a tropical region without knowing that it has already shifted to other areas will render control efforts ineffective.

General reservation is shown by parasite ecologists, who suggest that disease eradication is not a good idea, as nature has a well-organized mechanism to maintain ecological balance and the co-existence of living organisms. This suggests that eradicating one organism may change that balance, with negative impacts on the whole ecosystem.

The general conclusions from recent understanding of climate change and disease are as follows:

- There are convincing pieces of evidence which suggest that there is a link between outbreaks of different diseases and climate change.
- Shifting of disease towards the poleward direction could have a great impact on the whole ecosystem.
- The effects of climate change will be superimposed on a multitude of other anthropogenic environmental changes.
- The threat of emerging diseases and their possible link with climate change is badly affecting global health.
- The entire ecosystem will be badly affected if climatic effects are felt by parasites and diseases of key species.

In the past, climate change and human health have been treated separately by governments and other funding agencies. However, the two are intimately related to each other and need greater understanding by all stakeholders in the future. There has been considerable progress in the development of various models, leading us to the conclusion that climatic change modulates the transmission of diseases.

Future suggestions for improving our understanding of the disease–climate change relationship are highlighted here (Rodó et al., 2013):

- A long-term and holistic approach where key indices of human health and disease transmission are regularly monitored and other factors like parasite epidemiology, disease–parasite relationship, vector populations, and other demographic characterizations of the local human population are taken into consideration
- Undertaking the comparative analysis of different climate-driven disease models to understand various aspects of disease patterns and future predictions
- Spatial and temporal investigations on different disease models in relation to climate change, which could increase our level of understanding about the impact of climate variability on various epidemiological parameters
- Exhaustive studies on various diseases to generate seasonal and decadal predictions
- Investigation of strategies to generate skillful seasonal-to-decadal disease predictions
- Development of baseline predictive models for the future of various infectious diseases that are driven by climate change, also including different levels of mitigation and adaptation strategies

13.5 CONCLUSION

Climate change is a global phenomenon that has potential to alter the environment, ecosystems, and people across the globe. Climate change has been predicted to have a significant effect on the stability of the environment/ecosystems and biodiversity in the future, especially those ecosystems that are highly fragile and sensitive. The global impact of climate change on the health of animals is now a fact, and if climate-related effects are not mitigated quickly, there is a chance that humans may have to face more pandemics in the form of new, emerging diseases.

Climate change has a major impact on socio-economic development in developed, developing, and low-income countries. Global efforts are in full swing to reverse climatic alterations. From this perspective, the Paris Climate Agreement has great significance and can act as a cornerstone for mitigating the global impacts of climate change. The Paris climatic pact is significant for India and other sub-continental countries that are facing either direct or indirect impacts of climate change.

Climatic alterations in the Himalayan region show that the eco-fragile regions of the Himalayan Hotspot are prone to natural disasters, and therefore, there is an urgent need to understand the various impacts of global warming and climate change. In recent years, more focus has been given to the Northwest Himalayan regions to understand various facets of climate change, which will also help to predict future changes. More holistic and collaborative action is needed to understand key factors related to climate change in the Northwest Himalayan region and the possible mitigation plan for the future.

The impact of climate change on disease is growing, and health scientists point out that global warming may affect patterns of disease, which could have a dramatic impact on public health strategies for disease prevention and control. So, we need a long-term and holistic approach where key indices of human health and disease transmission are regularly monitored and other factors like parasite epidemiology, disease–parasite relationship, vector populations, and other demographic characterizations of the local human population are taken into consideration.

REFERENCES

Altizer S, et al. (2013). Climate change and infectious diseases: from evidence to a predictive framework. *Science,* 341: 514–519.

Banu S, et al. (2014). Dynamic spatiotemporal trends of dengue transmission in the Asia-Pacific region, 1955–2004. *PLoS ONE,* 9(2): e89440. DOI: 10.1371/journal.pone.0089440

Barrett M, et al. (2013) Climate change, predictive modeling and lemur health: Assessing impacts of changing climate on health and conservation in Madagascar. *Biological Conservation,* 157: 409–422.

Batool N, et al. (2019). Rainfall variability and dynamics of cropping pattern in Kashmir Himalayas: a case study of climate change and agriculture. *SN Applied Sciences*, 1(6): 1–9.

Bhat SU, et al. (2020). Water quality scenario of Kashmir Himalayan Springs—a case study of Baramulla District, Kashmir Valley. *Water, Air, & Soil Pollution*, 231(9): 1–17.

Bhutiyani MR, et al. (2007). Long-term trends in maximum, minimum, and mean annual air temperatures across the northwestern Himalaya during the 20th century. *Climatic Change,* 85: 159–177.

Bhutiyani MR, et al. (2009). Climate change and the precipitation variations in the northwestern Himalaya: 1866–2006. *International Journal of Climatology,* 30(4): 535–548.

Borgaonkar HP, et al. (2009). Assessment of tree-ring analysis of high-elevation *Cedrus deodara* D. Don from western Himalaya (India) in relation to climate and glacier fluctuations. *Dendrochronologia,* 27(1): 59–69.

Byers JE (2020). Effects of climate change on parasites and disease in estuarine and nearshore environments. *PLoS Biol,* 18(11): e3000743.

Chretien JP, et al. (2014). Global climate anomalies and potential infectious disease risks. *PLoS Currents*, 7: 2014–2015.

Clement B and Aidoud, A (2007). The hypothesis of changes in palustrian plant communities under climate change. *Ecohydrology,* 2: 117.

Costello A, et al. (2009). Managing the health effects of climate change. *Lancet,* 373: 1773–1964.

Dar RA, et al. (2014). Sustainability of winter tourism in a changing climate over Kashmir Himalaya. *Environmental Monitoring and Assessment,* 186(4): 2549–2562.

Dar SA, et al. (2020). Current status of wetlands in Srinagar City: threats, management strategies, and future perspectives. *Frontiers in Environmental Science,* 7: 199.

Department of Science and Technology (2012). *Dynamics of Glaciers in the Indian Himalaya: Science Plan. Himalayan Glaciology Technical Report No. 2.* Science and Engineering Research Board (SERB), Department of Science and Technology, New Delhi, p. 125.

Dimri A and Dash SK (2012) Wintertime climatic trends in the Western Himalayas. *Climatic Change,* 111: 775–800.

Dobhal DP, et al. (2004). Recession and morphogeometrical changes of Dokriani glacier (1962–1995) Garhwal Himalaya, India. *Current Science*, 86(5): 692–696.

Dobson A and Carper R (1992). *Global Warming and Potential Changes in Host–Parasite and Disease–Vector Relationship.* Yale University Press, Connecticut, pp. 201–217.

Epstein PR (1999). Climate and health. *Science*, 285: 347–348.

Epstein PR (2000). Is global warming harmful to health? *Scientific American*, 283: 50–57.

Epstein PR (2001). Climate change and emerging infectious diseases. *Microbes and Infection,* 3: 747–754.

Garamszegi LZ (2011). Climate change increases the risk of malaria in birds. *Global Change Biology,* 17: 1751–1759.

Gautam MR, et al. (2013). Climate change in the Himalayas current state of knowledge. The World Bank Development Research Group Environment and Energy Team, Policy research working paper WP-6516, p. 47.

Government of India (2010). National Mission for sustaining the Himalayan eco-system under National Action Plan on climate change. Draft Report. Department of Science and Technology, Government of India, p. 51.

Greer A, et al. (2008). Climate change and infectious diseases in North America: the road ahead. *Canadian Medical Association Journal*, 178: 715–722.

Hallegatte S, et al. (2016). *Shock Waves: Managing the Impacts of Climate Change on Poverty*. Climate Change and Development Series. World Bank, Washington, DC.

Harvell CD, et al. (2002). Ecology – climate warming and disease risks for terrestrial and marine biota. *Science*, 296: 2158–2162

Hoegh-Guldberg O, et al. (2018). *Global warming of 1.5°C. An IPCC Special Report on the impacts of global warming of 1.5°C above pre-industrial levels and related global greenhouse gas emission pathways, in the context of strengthening the global response to the threat of climate change, sustainable development, and efforts to eradicate poverty*. IPCC, Geneva, Switzerland.

Hoffmann S, et al. (2019). Predicted climate shifts within terrestrial protected areas worldwide. *Nature Communications,* 10, 4787.

Huang D, et al. (2021). Efficacy and safety of umifenovir for coronavirus disease 2019 (COVID-19): A systematic review and meta-analysis. *Journal of Medical Virology*, 93(1): 481–490.

Hunt KM, et al. (2020). The impacts of climate change on the winter water cycle of the western Himalaya. *Climate Dynamics,* 55(7): 2287–2307.

IPCC (2007) *Climate Change 2007 – The Physical Science Basis Contribution of Working Group I to the Fourth Assessment Report of the IPCC.* (ISBN 978 0521 88009-1 Hardback; 978 0521 70596-7 Paperback)

IPCC (2018) *Global Warming of 1.5°C.An IPCC Special Report on the Impacts of Global Warming of 1.5°C above Pre-industrial Levels and Related Global Greenhouse Gas Emission Pathways, in the Context of Strengthening the Global Response to the Threat of Climate Change, Sustainable Development, and Efforts to Eradicate Poverty.* Masson-Delmotte, V., et al. (eds.). In Press.

Jeelani G, Feddema JJ, van der Veen CJ, and Stearns L. (2012). Role of snow and glacier melt in controlling river hydrology in Liddar watershed (western Himalaya) under current and future climate. *Water Resource Research*, 48: W12508.

JK SAPCC (2013). www.jkenvis.org/pdf/jkenvis_sapcc.pdf

Johnson WC, et al. (2005). Vulnerability of northern prairie wetlands to climate change. *Bioscience,* 55: 863–872.

Kaul V and Zutshi DP (1967). A study of aquatic and marshland vegetation of Srinagar. *Proceedings of the Natural Institute of Science, India, B,* 33(3–4): 111–127.

Khan MA (2000). Wetland biodiversity in the Kashmir Himalaya: assessment and conservation strategies. In Khan MA (ed.) *Environmental Biodiversity and Conservation.* APH Publishing, New Delhi, pp. 69–93.

Kuhn K, et al. (2005). *Using Climate to Predict Infectious Disease Epidemics.* World Health Organization, Geneva, Switzerland.

Lenihan JM, et al. (2003). Climate change effects on vegetation distribution, carbon, and fire in California. *Ecological Applications,* 13(6): 1667–1681.

Lenton TM, et al. (2019). Climate tipping points—too risky to bet against. *Nature,* 575: 592–595.

Liang L and Gong P (2017). Climate change and human infectious diseases: A synthesis of research findings from global and spatio-temporal perspectives. *Environment International,* 103*:* 99–108.

Lindgren E (1998). Climate and tick-borne encephalitis. *Conservation Ecology* [serial online], 2(1): 5.

Medone P, et al. (2015). The impact of climate change on the geographical distribution of two vectors of Chagas disease: implications for the force of infection. *Philosophical Transactions of the Royal Society B: Biological Sciences*, 370(1665): 20130560.

Miller E and Huppert A. (2013) The effects of host diversity on vector-borne disease: the conditions under which diversity will amplify or dilute the disease risk. *PLoS ONE*, 8: e80279.

Møller AP (2010). Host-parasite interactions and vectors in the barn swallow in relation to climate change. *Global Change Biology,* 16: 1158–1170.

Møller AP and Erritzøe J (2003). Climate, body condition and spleen size in birds. *Oecologia,* 137: 621–626.Møller AP, et al. (2011) Rapid change in host use of the common cuckoo *Cuculus canorus* linked to climate change. *Proceedings of the Royal Society B: Biological Sciences,* 278: 733–738.

Morin JG, et al. (2019) Population-based analysis of the relationship between substance use and adolescent cognitive development. *American Journal of Psychiatry,* 176(2): 98–106.

Murtaza KO, et al. (2021). Glacial geomorphology and recent glacial recession of the Harmukh Range, NW Himalaya. *Quaternary International,* 575: 236–248.

Negi HS, et al. (2018). Recent wintertime climatic variability over the North West Himalayan cryosphere. *Current Science,* 114(4): 25.

Ohashi H, et al. (2019). Biodiversity can benefit from climate stabilization despite adverse side effects of land-based mitigation. *Nature Communications,* 10: 5240.

Ostfeld RS and Brunner JL (2015). Climate change and Ixodes tick-borne diseases of humans. *Philosophical Transactions of the Royal Society B: Biological Sciences*, 370(1665): 20140051.

Pandit AK and Kumar R (2006). Comparative studies on ecology of Hokersar wetland, Kashmir: present and past. *Journal of Himalayan Ecology and Sustainable Development*, 1: 73–81.

Pepin N, et al. (2015) Elevation-dependent warming in mountain regions of the world. *Nature Climate Change,* 5: 424–430.

Rashid I and Aneaus S (2020). Landscape transformation of an urban wetland in Kashmir Himalaya, India using high-resolution remote sensing data, geospatial modeling, and ground observations over the last 5 decades (1965–2018). *Environmental Monitoring and Assessment,* 19210: 1–14.
Rashid I and Romshoo SA (2020) Impact of climate change on vegetation distribution in the Kashmir Himalaya. In Dar GH and Khuroo AA (eds) *Biodiversity of the Himalaya: Jammu and Kashmir State*. Topics in Biodiversity and Conservation. Springer, Singapore, pp. 1029–1047. (ISBN 978-981-329-174-4)
Rashid I, et al. (2021). Retreat of Machoi Glacier, Kashmir Himalaya between 1972 and 2019 using remote sensing methods and field observations. *Science of the Total Environment,* 785: 147376.
Rather MI, et al. (2016). Massive land system changes impact water quality of the Jhelum River in Kashmir Himalaya. *Environmental Monitoring and Assessment,* 188: 185.
Rodó X, et al. (2013). Climate change and infectious diseases: can we meet the needs for better prediction? *Climatic Change,* 118: 625–640.Romshoo SA and Rashid I (2010). Potential and constraints of geospatial data for precise assessment of the impacts of climate change at landscape level. *International Journal of Geomatics and Geosciences,* 1(3): 386–405.
Romshoo SA and Rashid I (2014) Assessing the impacts of changing land cover and climate on Hokersar wetland in Indian Himalayas. *Arabian Journal of Geosciences,* 7(1): 143–160.
Romshoo SA, et al. (2020). Twenty-first century-end climate scenario of Jammu and Kashmir Himalaya, India, using ensemble climate models. *Climatic Change,* 162(3): 1473–1491.
Sabin TP, et al. (2020). Climate change over the Himalayas. In: Krishnan R, et al. (eds) *Assessment of Climate Change over the Indian Region*. Springer, Singapore.
Saino N, et al. (2009). Climate change effects on migration phenology may mismatch brood parasitic cuckoos and their hosts. *Biology Letters,* 5: 539–541.
Shafiq M, et al. (2019). Assessment of present and future climate change over Kashmir Himalayas, India. *Theoretical and Applied Climatology,* 137(3): 3183–3195.
Shrestha AB and Aryal R (2010). Climate change in Nepal and its impact on Himalayan glaciers. *Regional Environmental Change,* 11: 65–77.
Shugar, et al. (2021). A massive rock and ice avalanche caused the 2021 disaster at Chamoli, Indian Himalaya. *Science*, 376: 300–306.
Silwal CB, et al. (2020). Climate change and its possible impact in groundwater resource of the Kankai River Basin, East Nepal Himalaya. *Climate*, 8(11): 137.
Surugiu C, et al. (2010). Tourism destinations vulnerable to climate changes: an econometric approach on predeal resort. *Buletinul Universităţii Petrol – Gaze din Ploieşti, Seria Ştiinţe Economice,* 62(1): 111–120.
Teklehaimanot HD, et al. (2004). Alert threshold algorithms and malaria epidemic detection. *Emerging Infectious Diseases,* 10: 1220–1226.
Tewari VP, et al. (2017). Climate change effects in the Western Himalayan ecosystems of India: evidence and strategies. *Forest Ecosystems, 4:* 13.
Tian HY, et al. (2015a). Avian influenza H5N1 viral and bird migration networks in Asia. *Proceedings of the National Academy of Sciences of the United States of America,* 112: 172–177.
Tian HY, et al. (2015b) How environmental conditions impact mosquito ecology and Japanese encephalitis: an eco-epidemiological approach. *Environment International,* 79: 17–24.
Ueckerdt F, et al. (2019). The economically optimal warming limit of the planet. *Earth System Dynamics,* 10: 741–763.
Watson RT, et al. (1998). *The Regional Impacts of Climate Change: an Assessment of Vulnerability*. Cambridge University Press, Cambridge, UK.
Watts A, et al. (2015). Health and climate change: policy responses to protect public health. *Lancet,* 386: 1861–1914.
Willox AC, et al. (2015). Examining relationships between climate change and mental health in the Circumpolar North. *Regional Environmental Change,* 15: 169–182.
Wu XX, et al. (2014). Impact of global change on transmission of human infectious diseases. *Science China Earth Sciences,* 57: 189–203.
Wu X, et al. (2016). Impact of climate change on human infectious diseases: empirical evidence and human adaptation. *Environment International,* 86: 14–23.
Yale Sustainability (2021). https://macmillan.yale.edu/news/refresher-paris-climate-agreement. Accessed January 22, 2021.

Yan LB and Liu XD (2014). Has climatic warming over the Tibetan Plateau paused or continued in recent years? *J. Earth Ocean Atmos. Sci.*, 1(2014): 13–28.

Yang P, et al. (2013). Spatial and diurnal characteristics of summer rainfall over Beijing Municipality based on a high-density AWS dataset. *International Journal of Climatology,* 33: 2769–2780.

Yu P, et al. (2015). Hantavirus infection in rodents and hemorrhagic fever with renal syndrome in Shaanxi Province, China, 1984–2012. *Epidemiology & Infection,* 143: 405–411.

Zargar UR (2011). Proceed with caution on disease eradication. Science and Development Network. www.academia.edu/11134720/Proceed_with_caution_on_disease_eradicationletters/proceed-with-caution-on-disease eradication-1.html. Accessed August 18, 2022.

Zargar UR, et al. (2015). Does alteration in biodiversity really affect disease outcome? – A debate is brewing. *Saudi Journal of Biological Sciences,* 22: 14–18.

14 Climate Change and Agriculture

Shaista Mehraj, M.R. Mir Mehreen Manzoor, R.K. Sharma, Aadil Hussain Mir, Irfan Latief Khan, and Sameena Maqbool
College of Temperate Sericulture, Sher-e-Kashmir University of Agricultural Sciences and Technology, Jammu and Kashmir

CONTENTS

DOI: 10.1201/9781003220824-17

14.1 INTRODUCTION

Climate change refers to changes beyond the average atmospheric conditions that are caused both by natural factors, such as the orbit of earth's revolution, volcanic activities and crustal movements, and by artificial factors, such as the increase in the concentration of greenhouse gases (GHGs) and aerosols. Climate change due to global warming, which refers to the average increase in global temperature, has become a megatrend that will lead to significant global changes in the future. Concerning its impacts, the UN Intergovernmental Panel on Climate Change (IPCC) presented considerable scientific evidence in its sixth report on climate change (2019), and the impacts have become clearly recognized worldwide. In addition, people have become more aware of the fact that global warming cannot be avoided due to the continued increase in GHG emissions and the changes in the climate system.

According to the sixth report of UN IPCC (2019) on climate change, it is indisputable that global warming has serious impacts on the earth, and it is very likely that the increase in GHG emissions caused by anthropogenic activities has caused global warming since the mid-20th century. Especially, this report warns us that if mankind continues its present level of consumption of fossil fuels (e.g., oil and coal), the average temperature of the earth will rise by up to 6.4°C by the end of the 21st century (2001–2100) and the sea level will rise by 59 cm. In fact, the average temperature of the earth has risen 0.74°C over the past 100 years (1906–2005) (Korea Meteorological Agency, 2008). Global warming not only causes a change in average temperature and precipitation but also increases the frequency of floods, droughts, and heat waves, and the intensity of typhoons and hurricanes following the change in temperature and precipitation patterns. The impacts of climate change are also shown in various other forms throughout the world, including the rise of sea level, decrease in glaciers, northward movement of plant habitats, changes in animal habitats, rise of ocean temperature, shortened winter and early arrival of spring.

14.2 GLOBAL GREENHOUSE GAS EMISSIONS

Global GHG emissions continued to grow for the third consecutive year in 2019, reaching a record high of 52.4 Gt CO_2 without land use change emissions and 59.1 Gt CO_2 when including land use change . Fossil carbon dioxide emissions dominate total GHG emissions, including land use change (65%), and consequently, the growth in GHG emissions. Although 2020 emissions will be lower than in 2019 due to the COVID-19 crisis and associated responses, GHG concentrations in the atmosphere continue to rise, with the immediate reduction in emissions expected to have a negligible long-term impact on climate change. However, the unprecedented scale of COVID-19 economic recovery measures presents the opening for a low-carbon transition that creates the structural changes required for sustained emissions reductions. Seizing this opening will be critical to bridging the emissions gap.

The United Nations Secretary-General is calling on governments to use COVID-19 recovery as an opportunity to create more sustainable, resilient and inclusive societies. Aligned with this, the United Nations Framework Convention on Climate Change (UNFCCC) has stressed that governments could integrate and specify some of their post-COVID-19 recovery plans and policies in their new or updated Nationally Determined Contributions and long-term mitigation strategies, both of which countries are requested to submit in 2020. The most significant and encouraging development in terms of climate policy in 2020 is the growing number of countries that have committed to achieving net zero emissions goals by around mid-century. These commitments are broadly consistent with the Paris Agreement temperature goal provided they are achieved globally. The litmus test of these announcements will be the extent to which they are reflected in near-term policy action and in significantly more ambitious NDCs for the period to 2030.

14.3 CLIMATE CHANGE AND AGRICULTURE

Based on some of the past experiences indicated earlier, the impact of climate change on agriculture will be one of the major deciding factors influencing the future food security of mankind on the earth. Agriculture is not only sensitive to climate change but also one of the major drivers for climate change. Understanding the weather changes over a period of time and adjusting the management practices towards achieving a better harvest are challenges to the growth of the agricultural sector as a whole. The climate sensitivity of agriculture is uncertain, as there is regional variation in rainfall, temperature, crops and cropping systems, soils and management practices. The inter-annual variations in temperature and precipitation were much higher than the predicted changes in temperature and precipitation. Crop losses may increase if the predicted climate change increases the climate variability. Different crops will respond differently, as global warming will have a complex impact. The tropics are more dependent on agriculture, as 75% of the world population lives in the tropics, and for two-thirds of these people, agriculture is the main occupation. With low levels of technology, a wide range of pests, diseases and weeds, land degradation, unequal land distribution and rapid population growth, any impact on tropical agriculture will affect their livelihood. Rice, wheat, maize, sorghum, soybean and barley are the six major crops in the world, grown in 40% of cropped area, and contribute 55% of non-meat calories and over 70% of animal feed (FAO, 2006). Consequently, any effect on these crops would adversely affect food security.

14.4 LIKELY EFFECTS OF CLIMATE CHANGE ON KEY SECTORS AT GLOBAL LEVEL

The IPCC Fourth Assessment Report of the Working Group II: Impacts, Adaptation and Vulnerability described the likely effects of climate change, including from increases in extreme events. The effects on key sectors, in the absence of countermeasures, are summarized as follows.

14.4.1 Water

Drought-affected areas are likely to be more widely distributed. Heavier precipitation events are very likely to increase in frequency, leading to higher flood risks. By mid-century, water availability is likely to decrease in mid-latitudes, in the dry tropics and in other regions supplied by melted water from mountain ranges. More than one-sixth of the world's population is currently dependent on melt water from mountain ranges.

14.4.2 Food

While some mid-latitude and high-latitude areas will initially benefit from higher agricultural production, for many others at lower latitudes, especially in seasonally dry and tropical regions, the increase in temperature and the frequency of droughts and floods are likely to affect crop production negatively, which could increase the number of people at risk from hunger and increased levels of displacement and migration.

14.4.3 Industry, Settlement and Society

The most vulnerable industries, settlements and societies are generally those located in coastal areas and river flood plains, and those whose economies are closely linked with climate-sensitive resources. This applies particularly to locations already prone to extreme weather events and especially to areas undergoing rapid urbanization. Where extreme weather events become more intense or more frequent, the economic and social costs of those events will increase.

14.4.4 Health

The projected changes in climate are likely to alter the health status of millions of people, including increased deaths, disease and injury due to heat waves, floods, storms, fires and droughts. Increased malnutrition, diarrhoea and malaria in some areas will increase vulnerability to extreme public health events, and development goals will be threatened by long-term damage to health systems from disasters.

14.5 OBSERVED CHANGES IN CLIMATE AND WEATHER EVENTS IN INDIA

14.5.1 Surface Temperature

At the national level, an increase of 0.4°C has been observed in surface air temperatures over the past century. A warming trend has been observed along the west coast and in central India, the interior peninsula and north-eastern India. However, cooling trends have been observed in northwest India and parts of south India.

14.5.2 Rainfall

While the observed monsoon rainfall at the all-India level does not show any significant trend, regional monsoon variations have been recorded. A trend of increasing monsoon seasonal rainfall has been found along the west coast, northern Andhra Pradesh and north-western India (+10% to + 12% of the normal over the last 100 years), while a trend of decreasing monsoon seasonal rainfall has been observed over eastern Madhya Pradesh, north-eastern India, and some parts of Gujarat and Kerala (−6% to −8% of the normal over the last 100 years).

14.5.3 Extreme Weather Events

Trends of extreme weather events have been observed in multi-decadal periods of more frequent droughts followed by less severe droughts. There has been an overall increasing trend in severe storm incidence along the coast at the rate of 0.011 events per year. While the states of West Bengal and Gujarat have reported increasing trends, a decline has been observed in Orissa. Scientists analysing a daily rainfall data set have shown

(i) a rising trend in the frequency of heavy rain events
(ii) a significant decrease in the frequency of moderate events over central India from 1951 to 2000

14.5.4 Rise in Sea Level

Using the records of coastal tide gauges in the north Indian Ocean for more than 40 years, scientists have estimated that sea level rise (SLR) was between 1.06 and 1.75 mm per year. These rates are consistent with the 1–2 mm per year global SLR estimates of the IPCC.

14.5.5 Impacts on Himalayan Glaciers

The Himalayas possess one of the largest resources of snow and ice, and their glaciers form a source of water for perennial rivers such as the Indus, the Ganga and the Brahmaputra. Glacial melt may impact their long-term lean-season flows, with adverse impacts on the economy in terms of water availability and hydropower generation. The available monitoring data on Himalayan glaciers indicates that while recession of some glaciers has occurred in some Himalayan regions in recent years, the trend is not consistent across the entire mountain chain.

14.6 SOME PROJECTIONS OF CLIMATE CHANGE OVER INDIA FOR THE 21ST CENTURY

Some modelling and other studies have projected the following changes due to increase in atmospheric GHG concentrations arising from increased global anthropogenic emissions.

14.6.1 Annual Mean Surface Temperature

Simulation studies by Indian Institute of Tropical Meteorology (IITM), Pune, estimated that the annual mean surface temperature rise by the end of century is expected to be in the range from 3 to 5°C, with warming more pronounced in the northern parts of India.

14.6.2 Impacts on Water Resources

Changes in key climate variables, namely, temperature, precipitation and humidity, may have significant long-term implications for the quality and quantity of water. The river systems of the Brahmaputra, the Ganga and the Indus, which benefit from melting snow in the lean season, are likely to be particularly affected by the decrease in snow cover. Due to SLR, fresh water sources near the coastal regions will suffer salt intrusion.

14.6.3 Impacts on Agriculture and Food Production

Food production in India is sensitive to climate change, such as variability in monsoon rainfall and temperature changes within a season. Studies by Indian Agricultural Research Institute (IARI) and others indicate greater expected loss in the rabi crop. Every 1°C rise in temperature reduces wheat production by 4–5 million tonnes. Small change in temperature and rainfall have significant effects on the quality of fruits, vegetables, tea, coffee, aromatic and medicinal plants, and basmati rice. Pathogens and insect populations are strongly dependent upon temperature and humidity, and changes in these parameters may change their population dynamics. Other impacts on agricultural and related sectors include lower yields from dairy cattle and decline in fish breeding, migration and harvests. Global reports indicate a loss of 10–40% in crop production by 2100.

14.6.4 Impacts on Forests

Climate projections indicate that the country is likely to experience a shift in forest types with consequent changes in forest produce and, in turn, livelihoods based on those products. Correspondingly, the associated biodiversity is likely to be adversely impacted.

14.6.5 Vulnerability to Extreme Events

Heavily populated regions such as coastal areas are exposed to climatic events such as cyclones, floods and drought, and large declines in sown areas in arid and semiarid zones occur during climate extremes. Large areas in Rajasthan, Andhra Pradesh, Gujarat and Maharashtra and comparatively small areas in Karnataka, Orissa, Madhya Pradesh, Tamil Nadu, Bihar, West Bengal and Uttar Pradesh are frequented by drought. About 40 million hectares of land is flood prone, including most of the river basins in the north and the north-eastern belt, affecting about 30 million people on average each year. Such vulnerable regions may be particularly impacted by climate change.

14.6.6 Impacts on Coastal Areas

A mean SLR of 15–38 cm is projected along India's coast by the mid-21st century and a rise of 46–59 cm by 2100. In addition, a projected increase in the intensity of tropical cyclones poses a threat to the heavily populated coastal zones in the country (NATCOM, 2004).

14.6.7 Impacts on Biodiversity

The IPCC has projected that global average temperature increase during the 21st century will range from 1.4 to 4°C. Research by the Consultative Group on International Agricultural Research based on distribution models of wild relatives of three staple crops of the poor, i.e., peanuts, cowpea and potato, suggests that 16–22% per cent of wild species will be threatened with extinction by 2055. Loss of genetic diversity can have serious long-term consequences globally.

14.6.8 Impact on Pests

Some of the most dramatic effects of climate change on pests and diseases are likely to be seen among arthropod insects like mosquitoes, midges, ticks, fleas and sand flies, and the viruses they carry. With changes in temperature and humidity levels, the populations of these insects may expand their geographic range and expose animals and humans to diseases to which they have no natural immunity. Plant pests, which include insects, pathogens and weeds, continue to be one of the biggest constraints to food and agricultural production. Fruit flies, for instance, cause extensive damage to fruit and vegetable production. Controlling such pests often requires the use of pesticides, which can have serious side effects for human health and the environment. Climate change may also play a role in food safety. A growing number of pests and diseases could lead to higher and even unsafe levels of pesticide residues and veterinary drugs in local food supplies. And changes in rainfall, temperature and relative humidity can readily contaminate foods like groundnuts, wheat, maize, rice and coffee with fungi that produce potentially fatal mycotoxins.

14.6.9 Effect on Insecticide Use Efficiency

A warmer temperature requires a greater number of insecticide applications (i.e., three, more than normal) for controlling corn pests. Entomologists predict more generations of insects in a warm climate, which will necessitate a greater number of insecticide applications. It will increase the cost of protection and environmental pollution. Synthetic pyrethroids and Naturalyte Spinosad will be less effective in higher temperatures. Therefore, it is advisable for farmers not to use insecticides with a similar mode of action frequently to avoid development of resistance in the case of a greater number of applications. Cultural management practices, e.g., early planting, may not be helpful because of early emergence of pests due to warmth.

14.6.10 Effect on Natural Pest Control

Global warming is expected to make regional climates more varied and unpredictable, which could affect the relationship between insects and their natural enemies. In years of most variable rainfall, caterpillars have a significantly lower number of parasitoids. This could be because the parasitoids use cues, e.g., change in local climate, to determine the best time for laying eggs. Unpredictable rains might disrupt the parasitoids' ability to track their caterpillar hosts. The wasps use the start of the rain as a cue to hatch out of their cocoons and look for a caterpillar to lay their eggs. If the rains are late, they emerge late and may not find the larval stage of the host, resulting in reduced natural pest control.

14.6.11 Impact of Climate Change on Disease

Any direct yield gains caused by increased CO_2 or climate change could be offset partly or entirely by losses caused by phytophagous insects, plant pathogens and weeds. It is, therefore, important to consider these biotic constraints on crop yields under climate change. Climate change has the potential to modify host physiology and resistance and to alter stages and rates of development of the pathogen. The most likely impacts would be shifts in the geographical distribution of host and pathogen, changes in the physiology of host–pathogen interactions and changes in crop loss. Another important impact may be through changes in the efficacy of control strategies.

14.7 RESEARCH FINDINGS OF ICAR (INDIAN COUNCIL OF AGRICULTURAL RESEARCH) ON CLIMATE CHANGE

To meet the challenges posed by climate change to the agricultural system, ICAR has accorded high priority to understanding the impacts of climate change and developing adaptation and mitigation strategies through its network research programmes.

Some of the research findings are as follows:

- Significant negative rainfall trends were observed in the eastern parts of Madhya Pradesh, Chhattisgarh and parts of Bihar, Uttar Pradesh, parts of north-west and north-east India, and also a small pocket in Tamil Nadu.
- A significant increase in rainfall has also been noticed in Jammu and Kashmir and in some parts of the southern peninsula.
- The maximum and minimum temperature (1960–2003) analysis for the north-west region of India showed that the minimum temperature is increasing at annual, kharif and rabi season time scales. The rate of increase of minimum temperature during rabi is much higher than during kharif. The maximum temperature showed an increasing trend in annual, kharif and rabi time scales, but a very sharp rise was observed from the year 2000 onwards.
- It was observed from experiments on the impact of high temperature on pollen sterility and germination in rice that maximum temperature above 35°C and minimum temperature 23°C at flowering stage increased the pollen sterility in two normal and three basmati varieties of rice, and the effect was more profound in basmati cultivars.
- Biological yields were reduced drastically with elevated ambient temperature in tunnel experiments. The degrees of reduction in grain yield were enhanced with a rise in ambient temperature of 1, 2 and 3°C. The grain yield was reduced by 60, 64 and 70% in Pusa Sugandh-2 and 45, 52 and 54% in Pusa 44 variety, which was mainly attributed to maximum reduction in number of panicles/m^2 followed by the number of panicles/m^2 and 1000 grain weight.
- High thermal stress during post-flowering duration manifested 18, 60 and 12% reduction in economic yield of wheat, mustard and potato, respectively.
- Coconut yields were not affected with the increase of maximum temperature up to 44°C, but above that, the yield was reduced.

14.8 SOME CURRENT ACTIONS FOR ADAPTATION AND MITIGATION IN INDIA

Adaptation, in the context of climate change, comprises the measures taken to minimize the adverse impacts of climate change, e.g., relocating communities living close to the sea shore, for instance, to cope with the rising sea level or switching to crops that can withstand higher temperatures. Mitigation comprises measures to reduce the emissions of GHGs that cause climate change in the first place, e.g., by switching to renewable sources of energy, such as solar energy or wind energy or nuclear energy, instead of burning fossil fuel in thermal power stations. Current government expenditure in India on adaptation to climate variability exceeds 2.6% of the gross domestic product (GDP), with agriculture, water resources, health and sanitation, forests, coastal-zone infrastructure and extreme weather events being specific areas of concern.

14.8.1 Crop Improvement Programmes

Programmes address measures such as development of arid land crops and pest management as well as capacity building of extension workers and non-governmental organizations (NGOs) to support better vulnerability-reducing practices. Regarding drought proofing, the current programmes seek to minimize the adverse effects of drought on production of crops and livestock and on productivity of land, water and human resources, so as to ultimately lead to drought proofing of the affected areas. They also aim to promote overall economic development and improve the socio-economic conditions of the resource poor and disadvantaged.

14.8.2 Forestry

India has a strong and rapidly growing afforestation programme. The afforestation process was accelerated by the enactment of the Forest Conservation Act of 1980, which aimed at stopping the clearing and degradation of forests through a strict, centralized control of the rights to use forestland and mandatory requirements of compensatory afforestation in case of any diversion of forestland for any non-forestry purpose. In addition, an aggressive afforestation and sustainable forest management programme resulted in annual reforestation of 1.78 Mha during 1985–1997, and the figure is currently 1.1 Mha annually. Due to this, the carbon stocks in Indian forests have increased over the last 20 years to 9–10 gigatons of carbon (Gt C) during 1986 to 2005.

14.8.3 Water

The National Water Policy (2002) stresses that non-conventional methods for utilization of water, including inter-basin transfers, artificial recharge of groundwater and desalination of brackish or sea water, as well as traditional water conservation practices like rainwater harvesting, including rooftop rainwater harvesting, should be practised to increase the utilizable water resources. Many states now have mandatory water harvesting programmes in several cities.

14.9 INDIA'S POLICY STRUCTURE RELEVANT TO GHG MITIGATION

India has in place a detailed policy, regulatory and legislative structure that relates strongly to GHG mitigation. The Integrated Energy Policy was adopted in 2006.

Some of its key provisions are:

- Promotion of energy efficiency in all sectors
- Emphasis on mass transport
- Emphasis on renewables, including biofuel plantations
- Accelerated development of nuclear and hydropower for clean energy
- Focused R&D on several clean energy-related technologies.

14.10 COVID-19 AND CLIMATE CHANGE

CO_2 emissions could decrease by about 7% in 2020 (range: 2–12%) compared with 2019 emission levels due to COVID-19, with a smaller drop expected in GHG emissions as non-CO_2 emissions are likely to be less affected. However, atmospheric concentrations of GHGs continue to rise. The reduction in GHG emissions in 2020 due to COVID-19 is likely to be significantly larger than the 1.2% reduction during the global financial crises in the late 2000s. Studies indicate that the biggest changes have occurred in transport, as COVID-19 restrictions were targeted to limit mobility, though reductions have also occurred in other sectors.

14.11 CONCLUSION

From the preceding, it is clear that the occurrence of floods and droughts, heat and cold waves is common across the world due to climate change. Their adverse impact on the livelihood of farmers is tremendous, all the more so in India, as our economy is more dependent on agriculture. Interestingly, weather extremes of opposite nature, such as cold and heat waves, floods, and droughts, are noticed within the same year over the same region or in different regions, and this is likely to increase in the coming decades. Human and crop losses are likely to be heavy. Climate change as a whole is associated with increasing GHGs and human-induced aerosols, and the imbalance between them may lead to uncertainty even in year-to-year monsoon behaviour over India. Therefore, there should be a determined effort from developed and developing countries to make industrialization environment friendly by reducing GHGs being pumped into the atmosphere. In the same fashion, awareness programmes on climate change and its effects on various sectors, i.e., agriculture, health, infrastructure, water, forestry, fisheries, land and ocean biodiversity, and sea level, and the role played by human interventions in climate change need to be taken up on a priority basis. In the process, the lifestyles of people should also be changed so as not to harm the earth–atmosphere continuum by pumping GHGs and chlorofluorocarbons (CFCs) into the atmosphere. From the agricultural point of view, the effects of extreme weather events on crops are to be documented on a regional scale, which will be handy to planners in such re-occurrence events for mitigating the ill effects. Also, there is a need to guide farmers on the projected impact of climate change and make them aware of probable mitigation and adaptation options to minimize the risk in the agricultural sector. COVID-19 has provided insight into how rapid lifestyle changes can be brought about by governments (who must create conditions that make lifestyle changes possible), civil society actors (who must encourage positive social norms and a sense of collective agency for lifestyle changes) and infrastructure (which must support behaviour changes). The lockdown period in many countries may be long enough to establish new, lasting routines if supported by longer-term measures. In planning the recovery from COVID-19, governments have an opportunity to catalyse low-carbon lifestyle changes by disrupting entrenched practices.

REFERENCES

Emissions Gap Report. 9 December 2020. UNEP, UNEP Copenhagen, Climate Centre (UNEP-CCC).

FAO. 3–12 October, 2006. Climate change adaptation and mitigation in agriculture. Course: Climate Change and Food Security. Rome, Italy.

IPCC (Intergovernmental Panel on Climatic Change). 2019. Climate Change: The Physical Science Basis. Extracts from the IV Assessment Report. Survey of the Environment 2019, *The Hindu,* pp. 147–155.

Korea Meteorological Agency. 2008. *Understanding of Climate Change and Scenario Application.*

NATCOM. 2004. Initial National Communication to the United Nations Framework Convention on Climate Change, Ministry of Environment & Forests, Government of India.

15 Climate Change
Mitigation and Adaptation

Seema Singh
Department of Zoology, Nari Shiksha Niketan (PG) College, Lucknow

Ruchi Saxena
Department of Chemistry, Nari Shiksha Niketan (PG) College, Lucknow

CONTENTS

DOI: 10.1201/9781003220824-18

15.1 INTRODUCTION

Climate change is a significant and lasting modification in the statistical distribution of weather patterns of the earth, over periods ranging from a decade to millions of years at local, regional or global scales as a consequence of global warming. It is characterized by an increased frequency of natural calamities such as cyclones, thunderstorms, landslides, tornadoes, precipitation, droughts, floods, heatwaves, wildfires and expansion of the deserts. Global warming is caused naturally by certain primordial factors like oceanic circulation, solar radiation, plate tectonics, volcanic eruptions, biotic processes and by human-induced emissions of greenhouse gases (GHGs). These GHGs emanate from burning fossil fuels (coal, oil and natural gas) for energy in homes, industries, agriculture and transportation. The vital records of the planet earth exhibit warming for the past 40 years with an average surface temperature rise of about 1.18°C from the late 19th century onwards. The years 2016 and 2020 have been observed to be the hottest years on record to date (NASA, NOAA Data, 2017). The mid-tropospheric CO_2 was found to have increased from the pre-industrial era of 280 ppm to an alarming level of 417 ppm in June, 2021 (National Oceanic and Atmospheric Administration, 2021). Glaciers have retreated in the mountain ranges of the Alps, the Himalayas, the Andes, the Rockies, Rwenzori, Kenya, Kilimanjaro and Alaska (Kaser et al., 2004). The Arctic ice too, has decreased by 278 billion tons/year and Antarctic ice by 149 billion tons/year (World Glacier Monitoring Service, 2021, Velicogna et al., 2020). As a consequence of melting glaciers, the global sea level has risen by about 20 cm in the last century and is still increasing at the rate of 3.4 mm/year (Nerem et al., 2018). Since the oceans absorb 90% of the extra heat of the earth, their heat has also increased by 326 zettajoules since 1955, with a net 0.33°C rise in temperature (National Centres for Environmental Information NOAA, 2021). Further, the increase in ocean temperature has resulted in the absorption of about 7.2 to 10.8 billion metric tons/year (Sabine et al., 2004) of $CO_{2,}$ which is approximately 20%–30% of the total CO_2 emitted globally.

15.1.1 Chronology of Climate Change

Climate change is a dynamic process that has been experienced by the earth since its origin. According to the nebular theory, the earth was first formed approximately 4.5 billion years ago by the gradual condensation of interstellar dust, when its temperature was estimated to be 5000–6000°C and the atmosphere was composed of NH_3, CH_4, H_2O (steam), CO_2, nascent H and cyanides. Thereafter, for about half a billion years, the earth cooled and solidified, and its steam condensed due to volcanic outgassing, which probably created the primordial atmosphere consisting of minerals, salts and gases: H_2S, CH_4, NH_3, nascent H, excessive CO_2 and no O_2. The earth cooled further to 1000°C with interactions between the solids, liquids, gases and high-energy solar radiations creating a "hot dilute soup" containing hydrocarbons and complex organic compounds. The earth continued to cool; its temperature decreased to 7°C, acquiring extreme greenhouse conditions with excessive CO_2, and initiating the evolution of anaerobic organisms, which was approximately 4 billion years ago in the Precambrian eon and also in the Phanerozoic eon. A major diversification of life began in the Palaeozoic era of the Phanerozoic eon, when the Great Oxidation Event (GOE) occurred approximately 2.7–2.8 billion years ago. Then, 2.45 billion years ago, geological, isotopic and chemical evidence suggests that the biologically produced molecular dioxygen (O_2) changed the existing weak reducing atmosphere with practically no oxygen into an oxidizing atmosphere with abundant oxygen. This enabled the subsequent development of multicellular life forms. The earth experienced a massive extinction event in the Tertiary period of the Cenozoic era, around 55 million years ago, when it became very hot and humid with a temperature of 28°C and the concentration of CO_2 at 1400–4000 ppm, causing the ice on the polar caps to melt completely. During this period, five ice ages known as the *Snowball earth* also occurred, with intermittent warm periods called *the Greenhouse earth*. The five ice ages were the Huronian Ice Age (200 million years period), the Cryogenian Ice Age (215 million years period), the Andean-Saharan Ice Age (30 million years period), the late Palaeozoic Ice Age (100 million years period) and the present Quaternary Ice Age, which started 2.58 million years ago (Earle, 2019). In the interglacial quaternary period, complex life forms evolved throughout the Mesozoic and the Cenozoic eras. This period also saw the origin of man, evolving through the Pliocene, Pleistocene and the current Holocene epoch. Man started living in a stable climate from the past 11,700 years, and with civilization, agriculture developed 11,000 years ago, with the emission of GHGs like CO_2 and CH_4 in low concentrations, followed by the domestication of animals 8000 years ago, which further increased the concentration of CH_4. Both CO_2 and CH_4 showed an exponential rise over the last 5000 years, going through the industrial revolution in 1750, and are still on the increase due to the expansion of agriculture by deforestation, burning of fossil fuels and heavy industrialization. Between 1850 and 1890, the mean global temperature was 13.6°C and the CO_2 in ice was 290 ppm, but by the middle of the 20th century the temperature rose to 13.9°C, CO_2 level rose to 315 ppm and chlorofluorocarbons (CFCs) were also detected for the first time in the atmosphere, causing the *Greenhouse Effect.* Hence, by the year 1950, large-scale degradation of the environment by human interventions was observed, which marked the beginning of the Anthropocene epoch (Farquhar et al., 2000).

15.1.2 Science of Climate Change

It is hypothesized that the following physical-geographical factors of the earth are responsible for climatic change:

a) The length of the elliptical orbit of the earth changes every 100,000 years.
b) The tilt of the earth's axis changes during rotation every 40,000 years.
c) The axis of the earth wobbles slightly every 26,000 years.
d) Eleven-year sunspot cycles and 22-year solar magnetic cycles change the energy output of the earth.
e) Shifting tectonic plates cause volcanic eruptions and continental drift.
f) Strikes by large meteors.
g) Anthropogenic interference.

The three major determinants of the earth's climate are the sun, which is the source of energy for the survival of the planet, the oceans, where there is a decrease in pH by carbon sinks affecting the water cycle, and the greenhouse effect, which is a dynamic process in the lower and mid-troposphere increasing the temperature.

15.1.3 Vulnerability

Both the terrestrial and marine ecosystems are highly vulnerable to the adverse effects of climate change. From the climate perspective, the word "vulnerability" is the degree to which a system is susceptible to, or unable to cope with, the adverse effects of climate change, variability and extremes. There are five aspects to vulnerability: physical, economic, social, ecological and geographical.

Excessive exposure or sensitivity and low adaptive capacity cause high vulnerability, with the relocation or extinction of many species expected to affect the tundra, mangroves, coral reefs and also the cave ecosystem in the near future. Consequences of global warming are frequent storms, droughts, wildfires, landslides, bleaching of coral reefs, and rise in sea levels leading to scarcity of food and water, ultimately causing distress migration.

In Mozambique, 68.5 million people in 2017 were forced to leave their homes due to drought, and approximately 22.5 to 24 million were forced to move by *sudden-onset* weather events throughout the world (Podesta, 2019). Half of the world's children are vulnerable to such climate distress, with little access to nutrition, education and safety. The United Nations High Commission for Refugees (UNHCR) finds that 80% of displaced people worldwide live in areas with acute food insecurity fuelling social conflicts. The World Bank estimates that by 2050, 1.2 billion people will become climate refugees, with 143 million emerging from sub-Saharan Africa, 17 million from Latin America and 40 million from South Asia (*The Times of India*, 2021). Impacts of climate change are assessed by seven vulnerability criteria: magnitude, timing, persistence and reversibility, likelihood and frequencies, potential for adaptation, geographical distribution, and importance of the system at risk.

15.2 GLOBAL CLIMATE – A PUBLIC GOOD (BENCHMARK INTERNATIONAL CONFERENCES)

Climate degradation affects everyone, as it has no boundaries; therefore, the sum of all the GHGs (CH_4, N_2O, SF_6, hydrofluorocarbon (HFC), perfluorocarbon (PFC) and CO_2) emitted should be tackled by international cooperation to strategize mitigation and adaptation.

15.2.1 The First World Climate Conference (FWCC or WCC-1) 1979, Geneva; Participating Countries: 53

A world conference of climate experts was convened by the World Meteorological Organization (WMO), the United Nations Educational, Scientific and Cultural Organization (UNESCO), the Food and Agriculture Organization (FAO), the World Health Organization (WHO), United Nations Environment Programme (UNEP) and the International Council of Scientific Unions (ICSU) to foresee and prevent potential man-made changes in climate that might be adverse to the well-being of humanity and endorsed plans to establish a World Climate Programme (WCP) (WCC-1, E-Library, 2009).

15.2.2 The Intergovernmental Panel on Climate Change (IPCC) 1988, Geneva; Participating Countries: 195

To provide regular scientific assessments to the policymakers on the current state of climate change and prepare a comprehensive review with recommendations on mitigation and adaptation, IPCC was established in 1988 and endorsed by the UN General Assembly. The given recommendations were to be used by the parties to reduce risks due to climate change. Since 1988, IPCC has done five assessment cycles with reports.

First IPCC Assessment Report (FAR, 1990)
Climate change is a challenge with serious consequences; therefore, international cooperation is needed, and the United Nations Framework Convention on Climate Change (UNFCCC) was established in 1992. It was enforced in 1994 and made an international treaty to reduce global warming to cope with the consequences of climate change.

The Second Assessment Report (SAR, 1995)
Anthropogenic activities were declared as the critical cause of global warming in the coming century.

The Third Assessment Report (TAR, 2001)
States that global warming is unprecedented and very likely, with highly damaging future effects, which need urgent adaptation.

The Fourth Assessment Report (AR4, 2007)
Laid the groundwork for a post-Kyoto agreement, focusing on limiting global warming to 2°C.

The Fifth Assessment Report (AR5, 2013 and 2014)
Provided scientific input in the Paris Agreement.

The Sixth Assessment Report (AR6, 2021)
Climate change 2021: The Physical Science Basis
The current state of the climate system is based on physical science, according to the AR6, which claims that the world has already warmed by about 1.07°C since pre-industrial times. In 2019, the annual average concentration of CO_2 was 410 ppm, CH_4 was 1866 ppb and N_2O was 332 ppb, and these are on the increase, mainly due to anthropogenic activities. Global warming is predominantly driven by CO_2 and relatively less by the *short-lived climate forces*: CH_4 and N_2O. CO_2 is mostly generated from fossil fuel burning and industrial processes emanating from the industries of developed countries and China. The emissions of the other GHGs in the atmosphere do not make a very significant contribution to global warming. IPCC has made predictions on climate change for shared socioeconomic pathways in the near and long-term future for five different emission and global temperature scenarios, given in the following Table 15.1 (AR 6 Climate change 2021).

The predictions according to the Shared Socioeconomic Pathways in the IPCC-AR6, 2021 for the near future, i.e. 2021–2040, are: increase in global temperature of about 1.5–1.6°C, 4.1 times more heatwaves, 1.5 times more heavy precipitation events and 2.0 times more drought than normal (once in 10 years). Similarly, a rise of 2°C in temperature is predicted in the mid-term (2041–2060), with 5.6 times more heatwaves, 1.7 times more heavy precipitation events and 2.4 times more drought than the expected normal (AR6 Climate change 2021, the physical basis – IPCC). Therefore, policies should be strictly followed for mitigations and adaptations so that the world can restrict the emission of CO_2 below 500 GT and the temperature below 1.5°C.

TABLE 15.1
Shared Socioeconomic Pathways in the IPCC in SAR, 2021

		Estimated Warming (°C)	
SSP	Scenario	Near Future 2021–2040	Mid Term 2041–2060
SSP1-1.9	Very low GHG emissions: CO_2 emissions cut down to net zero around 2050	1.5	1.6
SSP1-2.6	Low GHG emissions: CO_2 emissions cut down to net zero around 2075	1.5	1.7
SSP 2-4.5	Intermediate GHG emissions: CO_2 emissions around current levels until 2050, then falling but not reaching net zero by 2100	1.5	2.0
SSP 3-7.0	High GHG emissions: CO_2 emissions double by 2100	1.5	2.1
SSP 5-8.5	Very high GHG emissions: CO_2 emissions triple by 2075	1.6	2.4

15.2.3 The Second World Climate Conference (SWCC or WCC-2) WMO, UNESCO, UNEP, FAO and ICSU, 1990, Geneva; Participating Countries: 116

The SWCC supported the principles of the Climate Change Convention – *common concern of humankind* and *common but differentiated responsibilitie*s with equity among nations. An integrated approach was given based on scientific capabilities, social and economic factors, natural variability of climate and environmental challenges for application through a global framework with a declaration for implementation. It did not specify any international targets for reducing emissions.

15.2.4 RIO Earth Summit 1992, Rio (Environment and Development), Participating Countries: 195

Only 154 countries signed to adopt the UNFCCC for the stabilization of GHG emissions within a timeframe. Food production was to be ensured with sustainable economic development. Guidelines were given on forest development and biological diversity, but nothing was legally binding. The cardinal body: Conference of Parties (COP) was created for drafting the climate policy, legislation and framing guidelines necessary to promote its effective implementation. The presidency of COP rotates among five recognized UN regions: i) Africa, ii) Asia, iii) Latin America and the Caribbean, iv) Central and Eastern Europe, and v) Western Europe. It meets annually in its secretariat in Bonn to review the national communications and emission inventories submitted by party nations and assesses the progress made in achieving the ultimate objective of the Convention (United Nations Conference on Environment and Development (UNCED), Earth Summit, 1992).

COP-1 Berlin, Germany (1995) was the first international joint venture against climate change. The Berlin Mandate allowed parties to make specific commitments for the activities to be implemented jointly. Following COP-2 Geneva, Switzerland (1996), the Geneva Ministerial Declaration called upon party nations to accelerate negotiations on a legally binding protocol.

15.2.5 Kyoto Protocol 1997 (COP-3) (Commitments for Common but Differentiated Responsibilities), Kyoto; Signatory Countries: 192

In the first commitment, guidelines were given to reduce GHG emissions by 5.2% between 2008 and 2012 from the baseline year of 1990 for the gases CO_2, CH_4 and N_2O and from the baseline year of 1995 for HFC, PFC and SF_6. This target was legally binding for the 39 most developed nations of the world, known as Annex I Parties under the UNFCCC. A series of four workshops known as *the Convention Dialogue* were held between 2005 and 2009 for long-term negotiations for the target by a market-based mechanism called Emission Trading:

1. *Joint implementation is where one nation gets credit for implementing a project in another nation to reduce carbon emissions.*
2. *The Clean Development Mechanism is undertaking of joint projects in developing nations by developed nations through a special payment called the administrative fee.*
3. *Emission trading between nations is done to fulfil their commitments with the flexibility of substituting one gas with another* (UNFCCC, 2021a).

COP-4 Buenos Aires, Argentina (1998): In the Buenos Aires Action Plan, a time period of 2-years was given for the complete implementation of the Kyoto Protocol by the year 2000 and it was decided to review the financial mechanism after every 4 years. COP-5 Bonn, Germany (1999): negotiations were done on the adoption of the guidelines on climate change. COP-6 part I; Hague, Netherlands (2000); all negotiations failed. COP-6 part II Bonn, Germany (2000): A consensus was reached on the Bonn Agreements. COP-7 Marrakesh, Morocco (2001): the Marrakesh Accord endorsed the guidelines for implementation of the Kyoto Protocol. The Special Climate Change Fund (SCCF) and the Least Developed Countries Fund (LDCF) were established. COP-8 Delhi, India (2002): the Delhi Ministerial Declaration called for the developed countries to transfer technology to developing countries. COP-9 Milan, Italy (2003): IPCC emission reporting guidelines were adopted. The Special Climate Change Fund SCCF and the Least Developed Countries Fund (LDCF) were increased. COP-10 Buenos Aires, Argentina (2004): future strategies were developed on the post-Kyoto protocol, which are sustained development pathways, transfer of technologies, land-use change and reforestation. The allocation of emission reduction obligations was to be done by 2012, at the end of the first commitment period. It also adopted UNFCCC's financial mechanism for developed countries and education, training and public awareness for developing countries. COP-11/CMP-1 (Conference of Parties serving as meeting of parties) Montreal, Canada (2005): The Montreal Action Plan was an agreement to extend the life of the Kyoto Protocol beyond 2012 and negotiate deeper cuts in greenhouse-gas emissions with a map for the future (CBC News, 2005; UNFCCC, 2006). Capacity-building, development and transfer of technologies, adverse effects of climate change on developing and least developed countries were discussed. Finance and budget guidelines were given for the Global Environment Facility (GEF). The Kyoto Protocol was implemented with amendments on carbon trading, Certified Emission Reductions (CERs), carbon sinks, industrial tree plantation, and clean energy projects like solar, wind, biomass, and hydropower generation, which were ratified by 141 countries. In the second commitment, an international legal system was established in which developed countries were to commit to greenhouse gas cuts with accountability and models on international cooperation for future efforts.

COP-12/CMP-2 Nairobi, Kenya (2006): Five-year plan for adaptation on a clean development mechanism (CDM) by developing countries with the procedures and modalities of the Special Climate Change Fund (SCCF).

15.2.6 The Bali Road Map 2007 (COP-13/CMP-3) (Shared Vision for Long-Term Cooperative Action) Bali; Participating Countries: 187

The Bali Road Map established the Ad Hoc Working Group on Long-term Cooperative Action with a mandate on mitigation for Reducing Emissions from Deforestation and forest Degradation (REDD). Finance for long-term cooperative action between developed and developing nations was also established for the climate-resilient development globally. Nationally Appropriate Mitigation Actions (NAMAs) and National Adaptation Programme of Action (NAPA) were established for finance mobilization of public/private capital (UNFCCC, 2021b). This would help to implement National Mitigation and Adaptation. It was mandatory for all developed countries to have deep cuts in global emissions from 10% to 40% by 2020 and 100% by 2050.
(COP-14/CMP-4, Poznan, Poland (2008) Development of Global environment facility; Participating countries: 180. Assessment and transfer of appropriate technology to the developing countries from the global environment technology facility.

15.2.7 Copenhagen Accord 2009 (COP-15) Copenhagen; Participating Countries: 183

The Copenhagen Accord was to limit the global temperature rise below 2°C. Mid-term commitments for economy-wide emission reduction targets for developed countries and quantified mitigation by major developing countries were established with no legal binding. Voluntary GHG reduction targets were given by Australia, Brazil, United States, China and India till 2020. The Copenhagen Green Climate Fund was established with $100 billion as an operating entity to support projects in developing countries. Adaptation in forestry, technology and capacity-building activities should have a *Prompt Start*. The deadline for assessment for implementation of the accord was 6 years (i.e., 2015).

COP-16 Cancún, Mexico (2010) adopted major tenets of the Copenhagen Accord, especially limiting global warming to 2°C, protecting vulnerable forests, and establishing a framework for a Green Climate Fund for developing countries for mitigation and adaptation; generating US$100 billion per annum as Green Climate Fund with fast-start finance of $30 billion for 2010–2012 for adaptation and mitigation in developing countries, with priority to the least developed country; and transferring clean-green technologies at the right place and time for the maximum benefit of vulnerable people, educating the masses in developing countries, to meet the challenges of natural disasters. A networked *Climate Technology Centre* was also set up.

COP-17 Durban, South Africa, 2011: the Durban Platform established a new international legally binding emissions reduction protocol and enhanced action on the second commitment period of the Kyoto Protocol with deadlines for post-2020 mitigation, beginning in 2013.

15.2.8 Doha Climate Gateway 2012 (COP-18/CMP-8)

The Doha Amendment to the Kyoto Protocol was adopted and, together with the Bali Action Plan, transferred to the SBI (Subsidiary Body for Implementation) and SBSTA (Subsidiary Body for Scientific and Technological Advice). The concept of *loss and damage* was introduced as developed countries pledged to help developing countries and small island nations (UNFCCC, 2021c).

15.2.9 Warsaw Climate Change Conference 2013 (COP-19/CMP-9)

At this COP, the Warsaw REDD+ framework of finance, institutional arrangements and methodological issues of loss and damage came into existence, including a Nationally Determined Contribution (NDC) for transfer of technology with finance for implementation by civil societies in cities, ensuring full implementation of the Bali Action Plan. All nations in unison to build adaptation capacity for climate change (UNFCCC, 2021d).

COP-20 Lima, Peru (2014): Directions for creating a climate for health, nutrition, gender equity and inter-generational climate justice for achieving food security and agricultural sustainability. Preparedness for El Nino and its impact. A non-binding agreement for REDD+ was reached among countries to set up a system tackling loss and damage due to climate change.

15.2.10 Paris Agreement 2015 (COP-21), Paris, Participating Countries: 196

The Paris Agreement is a legally binding international treaty on climate change, which sets long-term goals to guide all countries for NDCs to reduce global greenhouse gas emissions and limit the global temperature to 2°C for a climate-neutral world by 2050. Nations have to pledge their own targets by 2020 on long-term low GHG emission development strategies (LT-LEDS) and also the finance (UNFCCC, 2021e).

COP-22, 2016, Marrakech: Emphasis on water scarcity, water cleanliness, and water-related sustainability, a major problem in the developing world. Paris Agreement enforced.

15.2.11 Bonn Climate Change 2017 (COP-23) Bonn

Reviewed the Paris Agreement and aimed for *Further, Faster Ambition Together* implementation by establishing the Koronivia Joint Work on Agriculture for mitigation and adaptation. Phasing out coal from power generation by 2030 by Past Coal Alliance of 30 countries. A Gender Action Plan was designed for the participation of local communities, Indigenous Peoples Platform and traditionally marginalized groups (UNFCCC, 2021f).

COP-24, 2018 Katowice: Paris Agreement was ratified again. COP-25 Chile 2019 Madrid; Spain.

Even after the NDCs and existing climate pledges are met, the target of limiting temperature rise to 1.5°C cannot be achieved. Therefore, there should be a 38% rise in the NDC to meet the target by 2030. Emission from fossil fuels are expected to rise continuously till 2020.

COP-26, 2021 Glasgow: Due to the COVID-19 pandemic, COP-26 is re-scheduled to November 2021.

To preserve the ecosystems, a two-pronged approach has been proposed by assessing their vulnerability (IPCC, 2014):

1. Mitigation: reducing and stabilizing the emissions of heat-trapping greenhouse gases in the atmosphere
2. Adaptation: adjusting to life in a changing climate for the future.

15.3 MITIGATION

Mitigation is the moderation of the GHG emissions from the sources and enhancing the carbon sinks.

Power, industry, transportation and buildings are the four main sectors for GHG emissions; therefore, mitigation becomes mandatory for all nations and can be approached in the following two ways:

1. Conventional mitigation technologies
2. Nonconventional mitigation technologies

15.3.1 Conventional Mitigation Technologies

Decarbonization by conventional mitigation technologies is realized through the following: (i) renewable energy, (ii) nuclear power, (iii) carbon trapping and storage, and (iv) shifting to low-carbon emission fuels and deploying energy-efficient processes and technologies to reduce energy consumption.

15.3.1.1 Renewable Energy

Decarbonization can be achieved in the energy production sector by the use of solar power, e.g., photovoltaic devices, thermal devices for heating and cooling applications, wind power on the shore and deserts, hydropower, marine power, geothermal power, biomass power, biofuels and natural gas. Additionally, the use of electric vehicles and power generated by the electrolysis of hydrogen are potential sources for renewable energy. One of the main associated technological challenges is the diverse power production and intermittent supply, which is overcome by integrating renewable and grid technologies. The total share of renewable energy in global power was approximately 33% in 2018, with global electricity production as follows: hydropower: 26.2%, wind power: 15.8%, photovoltaic solar power: 5.5%, biopower: 2.4%, geothermal: 2.2%, concentrated solar power and marine power: 0.46% (Fawzy et al., 2020). Technology-push, demand-pull and systemic policy are the instruments for the successful deployment of renewable energy solutions.

15.3.1.2 Atomic Energy

Generating power by use of atomic energy has played a significant role in global energy production by preventing 1.2–2.4 Gt CO_2 emissions annually. Globally, 450 nuclear energy plants are operational, producing 396.4 GW power, which is estimated to reach 748 GW by 2050. Power production should be increased to 930 GW by 2050 to meet the target of limiting the increase in temperature below 2°C as per the Paris Agreement. Therefore, enhanced fusion-based nuclear technology with zero-emission technology can contribute to mitigation.

15.3.1.3 Energy in Industrial Sector

The three carbon-capturing technologies are pre-combustion, post-combustion and oxyfuel combustion. Decarbonization of power as well as of the industrial sectors is done by separating and capturing CO_2 gas, transporting it and storing it in geological reservoirs for very long periods. Two carbon capture and storage projects were in operation in 2018, with an annual capture capacity of 2.4 Mt CO_2. According to the International Energy Agency (IEA), the targeted capacity is 1488 Mt C for 2040. Presently, nine carbon capture projects are under development, with a projected increase in capacity to 11 Mt CO_2 by 2025.

15.3.1.4 Energy Efficiency in Industrial Sector

Efficiency in the energy or power sector can be brought about through streamlining thermal power plants by improving fuel combustion, increasing the efficiency of turbine generators and harnessing waste heat for extra electricity production. The use of backpressure steam turbines where steam pressure reduction is required can enhance the efficiency many times. Similarly, turbo-expanders can be deployed for gas pressure. Waste heat can be recovered from steel and cement on construction sites by using micro and small gas turbines. Combining thermal and solar power units increases efficiency and reduces losses with the use of standard technologies. Energy-efficient buildings with improved insulation can be constructed by using glazed tiles painted white to reflect sunlight. Increasing energy efficiency can be achieved in the home by using induction lighting and cooking, LED lighting, improved water mills and metal cooking stoves. Transport can be made efficient by using hybrid, electric and hydrogen (H_2) vehicles with updated thermal engines, ropeways, cable cars and electric railways. Managing the distance and frequency of travel, sharing vehicles and switching to rail additionally increase efficiency, as the railways are currently the most energy-efficient mode of transport.

15.3.2 Nonconventional (Negative Emissions) Technologies

The IPCC has adopted bioenergy carbon capture and storage, biochar, enhanced weathering, direct air carbon capture and storage, ocean fertilization, ocean alkalinity enhancement, soil carbon sequestration, afforestation and reforestation, wetland construction and restoration for the removal of CO_2.

15.3.2.1 Carbon Capture and Storage – Biochar

Biochar is a negative emission technology in which the carbon captured by green plants during photosynthesis is processed into a char and mixed in soils for long periods. Biochar is produced from dedicated crops and agricultural and forest residues through pyrolysis, gasification or hydrothermal carbonization. It is estimated that by 2050, a potential global carbon removal in the range of 0.3–2 $GtCO_2$/year can be achieved through biochar, though biomass availability is one of the limiting factors for successful large-scale deployment of biochar projects.

15.3.2.2 Weathering

Enhanced weathering is the sequestration of atmospheric carbon in both inorganic and organic forms by accelerating the CO_2 uptake by rocks on a shorter timescale. Inorganic carbon is sequestered by the production of carbonates and organic carbon by enhanced biomass production through photosynthesis and nutrients from the rocks. Biomass production by enhanced weathering has a favourable impact on soil hydrology too, as it increases the pH as well as nutrients in the soil, decreasing the dependence on fertilizers and pesticides. Enhanced weathering removes 135 kg of CO_2/ton of rocks from the atmosphere.

15.3.2.3 Direct Air Carbon Capture and Storage (DACCS)

In DACSS, atmospheric CO_2 is directly removed from the air by synthesizing chemical sorbents (LiOH, NaOH and KOH), which then regenerate the CO_2 for storage in geological reservoirs. The global potential for carbon dioxide removal has been estimated to be in the range of 0.5 to 5 $GtCO_2$/year by 2050, and this may increase to 40 $GtCO_2$/year by the end of the century.

15.3.2.4 Ocean Alkalinity Enhancement and Fertilization

The concept of ocean fertilization is to create a carbon-negative atmosphere by adding extra nutrients in the ocean to increase photosynthetic plants for the uptake of CO_2. Phytoplankton present on the surface of the oceans sequester CO_2 from the air to the ocean floor as a biological pump. It is estimated that ocean fertilization can potentially sequester up to 3.7 $GtCO_2$/year by 2100. The ocean has stored approximately 140,000 Gt of CO_2 in the form of carbonic acid till date.

15.3.2.5 Agriculture, Forestry and Land-Use (AFOLU) Sector

15.3.2.5.1 Soil Carbon Sequestration

Carbon sequestration in the soil can be increased by changing land management practices, like rotation of crops, zero-tillage, conservation tillage, nutrients and composting, to trap CO_2 in the soil. By 2050, the global carbon dioxide removal by soil sequestration is estimated to be between 2.3 and 5.3 $GtCO_2$/year.

15.3.2.5.2 Afforestation and Reforestation

Afforestation is the conversion of non-forest areas into forests, and reforestation is replanting trees in former forest areas that have undergone deforestation. Remission through the development of forests will preserve biodiversity, control floods, and improve the quality of soil gradually by the accumulation of organic CO_2 to delay climate change. The projected land use for afforestation is 500 Mha for the removal of 0.5 to 3.6 Gt CO_2/year globally by 2050.

15.3.2.5.3 Wetland Restoration and Construction

Wetland ecosystems have an enormous capacity to sequester atmospheric CO_2 through photosynthesis from biomass and store it both above and below the ground in the form of soil organic matter. These wetland habitats also emit non-CO_2 GHGs like CH_4 and N_2O. The accumulation of organic carbon in the topsoil and the sequestration rate were found to be 1060 tCO_2/ha/year and 6.3 ± 4.8 tCO_2/ha/year in mangroves, 917 tCO_2/ha/year and 8.0 ± 8.5 tCO_2/ha/year in salt marshes, and 500 tCO_2/ha/year and 4.4 ± 0.95 tCO_2/ha/year in seagrass meadows, respectively,. The peatlands and coastal wetlands also store CO_2, approximately between 44% and 71% of the total terrestrial biological carbon.

Negative emission technologies should be considered as complementary to conventional decarbonization methods, not as a substitute.

15.3.2.5.4 Alternative Negative Emissions Utilization and Storage Techniques

(i) Mineral carbonation is a method in which carbonates are formed by the combination of CO_2 with minerals and stored or used for natural weathering. The two main methods for mineral carbonation are (i) an ex-situ process above ground by grinding and pre-treatment of minerals before the reaction, (ii) an in-situ process of directly injecting the CO_2 in silicate rocks below ground. Silicate rocks composed of Ca, Mg and Fe react with CO_2 to form stable carbonates. The mineral carbonates so formed can replace the conventional aggregates in concrete for construction of buildings. The potential CO_2 removal is estimated at approximately 0.5–1 $GtCO_2$/year on replacing conventional construction materials such as steel and cement.

(ii) Radiative forcing geoengineering technologies reduce global temperatures by increasing the earth's reflectivity through increasing shortwave solar radiation, which is reflected back into space, or by terrestrial radiation, i.e., by enhancing longwave radiation, which is emitted by the earth's surface into space.

(iii) Stratospheric aerosol injection is a solar radiation management technique that mimics the cooling effect of a volcanic eruption. The temperature of the stratosphere is reduced temporarily by artificially injecting aerosols, which reflect the solar radiation; e.g., in 1991, large amounts (between 15 and 30 million tons) of sulphur dioxide gas (SO_2) were ejected during the volcanic eruption in the Philippines (Mount Pinatubo) and formed aerosols, which reflected the sunlight. The stratospheric aerosol injection reduced global temperatures by 0.4–0.5°C. This technique has certain drawbacks, as it alters the hydrological cycle and depletes ozone from the stratosphere.

(iv) Absorbing CO_2 with homemade charcoal. Charcoal made from vegetation is able to lock in CO_2, effectively storing this greenhouse gas. But, more space is needed for cultivation of plants.

(v) Particles in the stratosphere. Climate control by chemicals is done by launching a balloon into the air, which releases chemicals into the stratosphere to turn down the heat. These particles create a global dimming effect and gradually reduce the amount of direct irradiation of the earth's surface.

(vi) Marine sky brightening. To reduce or limit temperature increase by global warming, solar management can be done by marine sky brightening or the cloud albedo enhancement technique. The solar rays are reflected back into the atmosphere through cloud seeding with seawater particles or chemicals (Fawzy et al., 2020).

In spite of making arduous mitigations, climate change is inevitable. Thus, adaptation is necessary to overcome the vulnerabilities.

15.4 ADAPTATION

Adaptation is the adjustment of species in response to actual or expected climatic variations and their effects, which threaten their survival. It is a process that includes learning about risk, evaluating responses, fostering options, and creating the conditions that enable mobilizing of resources, implementation and revising choices with the new learning. Thus, based on the vulnerability due to the effects of climate change, three types of adaptation can takes place:

1. *Proactive adaptation* takes place before impacts of climate change are observed.
2. *Spontaneous adaptation* does not constitute a conscious response to climatic stimuli but is triggered by ecological changes in natural systems and by welfare schemes for society.
3. *Planned adaptation* is a planned policy decision based on assessment reports.

Planned adaptation can be approached in two ways.

15.4.1 Bottom-Up Approach (Community-Based Adaptation (CbA))

The bottom-up approach relies on the past knowledge of the local stakeholders to cope with or respond to impacts of climate change. CbA can be defined as a community-led process, based on priorities, needs, knowledge and capacities of the people who plan and cope with the impacts of climate change. The approach is vital to the threat climate change poses to the poor. CbA is a human-centred approach, which aims to support the most vulnerable people living in high-risk environments. CbA was implemented by the United Nations Development Programme (UNDP) in

Bangladesh, Bolivia, Guatemala, Jamaica, Kazakhstan, Morocco, Namibia, Niger, Samoa and Viet Nam during 2008–2013.

15.4.2 Top-Down Approach (Ecosystem-Based Adaptation (EbA))

The top-down approach is governed by scientific research and climate model projections to assess the risks of future climate change for action under the plans made by the UN. EbA is a human-centred approach to adaptation for reducing vulnerabilities by planned ecological development. Ecosystems should be regulated to help societies to adapt to the present natural calamities caused by climate change and future climate change by equipping the ecosystem with various solutions, e.g., developing mangrove forests to protect coastal areas against storms and waves, forest products as food by local communities when agriculture is affected by climate, and forests regulating water quality and river flows.

On the basis of their response to climate change, species can be classified into the following functional groups:

1. *Hardy Species* that have the ability to cope with the new climate in their current location
2. *Dependent Species* that will have to relocate with the change in climate
3. *Transformer Species* that can adapt quickly to a new climatic habitat
4. *Tender Species* that cannot adapt to any of the changed climatic situations and will become prematurely extinct.

The National Adaptation Plan Action (NAPA) was established for universal implementation together with a specifically devised need-based adaptation plan for least developed countries (LDCs). The Support Adaptation Committee (AC) was also formed to promote speedy implementation with a review on adaptation. The AC should provide technical support by sharing knowledge, experience, and good practices to increase the ability to adapt against destructive impacts of climate change and the capability to adjust to the changing climate. The technical examination process on adaptation (TEP-A) has been laid down to identify opportunities for strengthening resilience, reducing vulnerabilities and increasing the understanding and implementation of adaptation actions. The Green Climate Fund will help financially to expedite the formulation for implementation of national adaptation plans.

15.4.3 The Adaptation Cycle under the UN

15.4.3.1 Assess Impacts, Vulnerability and Risks

An initial assessment of vulnerabilities and current risk of those who are affected or will be affected by natural calamities due to climate change has to be made for the adaptation plans.

15.4.3.2 Plan for Adaptation

The Warsaw International Mechanism for Loss and Damage enumerates the workplan on events, non-economic losses, risk management, displacement and transformational approaches related to the adverse impacts of climate change:

- *A new transparency framework* should be recorded in a public registry to provide information on climate change impacts and adaptation.
- *Establishing a Clearing House for Risk Transfer* with the help of the insurance industry. A repository should be created for information on risk transfer to develop and implement comprehensive risk management strategies.

- *A Task Force on Displacement* was created to recognize the needs of countries regarding displacement and to develop integrated approaches by improving natural resources with the diffusion of technologies.
- *Identification of adaptation* activities and their appraisal should be taken by assessing costs and benefits between the available options.
- *Comprehensive planning* is required to avoid duplication and maladaptation of activities.
- *Integrated long-term national and local development* with sustainable strategies for adaptation for poverty alleviation.
- *Management of climate risk* with rapid scaling up of adaptation through local governance, civil society organizations and the private sector.
- Development of robust *resource mobilization mechanisms* to ensure a continuous flow of funds by diversifying economic activities towards less climate-sensitive and less hazardous areas.

15.4.3.3 Implementation of Adaptation Measures

Implementation should be at national, regional and local levels by multiple projects, programmes, policies or strategies. It could be a stand-alone process that should be fully integrated into the mainstream with sustainable development plans. Preference should be given to communities with high vulnerabilities. Leverage in disaster prevention and response ranges from improved early warning systems, through contingency planning and integrated response, to adaptation for risk reduction.

15.4.3.4 Monitor and Evaluate Adaptation

The adaptation process should be monitored and evaluated by keeping a record of the progress made in implementation to determine the effectiveness of adaptation for future ventures like protection of coastlines together with sea-level encroachments, managing land and forests, planning for reduced water availability, and development of resilient crop varieties to conserve energy and public infrastructure.

15.5 CONCLUSION

Society should be made more climate-resilient through proactive adaptations to extreme climate variations in which sustained development prospects are less threatened by climate hazards. Adaptations are not possible as a stand-alone function without the collaboration of agencies responsible for both economic and technological resources. It is mandatory for the world community to invest in expanding knowledge that is relevant in reducing climatic risks by strategically conserving the biodiversity by translocation of species into new habitats, ex-situ conservation by gene banking, cryopreservation, and nature reserves and zoological gardens. There is an urgent need to adapt by narrowing the existing adaptation deficit to preserve the planet earth.

REFERENCES

AR 6 Climate change (2021). *The Physical Science Basis-IPCC.* www.ipcc.ch/report/ar6/wg1/

CBC News (2005). The climate-change conference ends with key deals. 10 December.

Earle S (2019). *Glaciation: Glacial Periods in Earth's History, Physical Geology, 2nd Edition.* BC campus OpenEd CA. https://opentextbc.ca/geology/chapter/16-1-glacial-periods-in-earths-history/

Farquhar J, Bao H and Thiemens M (2000). Atmospheric influence of Earth's earliest sulfur cycle. *Science* 289(5480): 756–758.

Fawzy S, Osman AI, Doran J and Rooney DW (2020). Strategies for mitigation of climate change: a review. *Environ Chem Lett* 18: 2069–2094. https://link.springer.com/article/10.1007/s10311-020-01059-w

IPCC (2014). *Climate Change 2014: Synthesis Report. Contribution of Working Groups I, II and III to the Fifth Assessment Report of the Intergovernmental Panel on Climate Change* (Core Writing Team, RK Pachauri and LA Meyer (eds), IPCC, Geneva, Switzerland, pp. 151.

Kaser G, Hardy DR, Molg T and Bradley RS (2004). *International Journal of Climatology* 24(3): 329–339.

Katowice Climate Change Conference (2018). https://unfccc.int/process-and-meetings/conferences/katowice-climate-change-conference-december-2018/katowice-climate-change-conference-december-2018

NASA, NOAA Data (2017). www.giss.nasa.gov/research/news/20170118/

National Centres for Environmental Information NOAA (2021). Global Ocean Heat and Salt Content. www.nodc.noaa.gov/OC5/3M_HEAT_CONTENT/index3.html

National Oceanic and Atmospheric Administration (2021). Summer 2021 Neck and Neck with Dust Bowl Summer. www.noaa.gov

Nerem RS, Beckley BD, Fasullo JT, Hamlington BD, Masters D and Mitchum GT (2018). Climate-change–driven accelerated sea-level rise detected in the altimeter era. *PNAS* 115(9) pp. 2022–2025.

Podesta J (2019) The climate crisis, migration and refugees. Brookings. www.brookings.edu/research/the climate-crisis-migration-and-refugees/

Sabine CL, Feely RA, Gruber N, Key RM, Lee K, Bullister JL, Wanninkhof R, Wong CS, Wallace DWR, Tilbrook B, Millero FJ, Peng T-H, Kozyr A, Ono T and Rios AF (2004). The oceanic sink for anthropogenic CO_2. *Science* 305: 367–371.

UNFCCC (2006). Decision 1: Consideration of commitments for subsequent periods for Parties included in Annex I to the Convention under Article 3, paragraph 9, of the Kyoto Protocol (PDF). Retrieved 22 February 2013. https://unfccc.int/kyoto_protocol

UNFCCC (2021a). https://unfccc.int/kyoto_protocol

UNFCCC (2021b). https://unfccc.int/process/conferences/the-big-picture/milestones/bali-road-map

UNFCCC (2021c). Doha Climate Gateway. https://unfccc.int/process/conferences/the-big-picture/milestones/the-doha-climate-gateway

UNFCCC (2021d). Warsaw Framework for REDD-Plus (2021). https://unfccc.int/topics/land-use/resources/warsaw-framework-for-redd-plus

UNFCCC (2021e). The Paris Agreement. https://unfccc.int/process-and-meetings/the-paris-agreement/the-paris-agreement

UNFCCC (2021f). Bonn Climate Change Conference. https://unfccc.int/process-and-meetings/conferences/bonn-climate-change-conference-june-2019/bonn-climate-change-conference-june-2019

United Nations Conference on Environment and Development (UNCED), Earth Summit (1992). Sustainable Development Knowledge Platform. www.un.org/en/conferences/environment/rio1992

Velicogna I, Mohajerani Y, Geruo A, Landerer F, Mouginot J, Noel B, Rignot E, Sutterly T, Broeke MVD, Wessem MV and Wiese D (2020). Continuity of ice sheet mass loss in Greenland and Antarctica from the GRACE and GRACE Follow-On missions. *Geophysical Research Letters,* 47(8) pp. 1–8.

WCC-1, E-Library (2009). https://library.wmo.int/index.php?lvl=author_see&id=5288#.YT-niYbhUW4

World Glacier Monitoring Service (2021). http://wgms.ch/

16 Climate Change and Its Impact on Poultry Production

Tufail Banday, Shaista S. Nissa, Syed Shanaz, and Shahana Riyaz Tramboo
Sher-e-Kashmir University of Agricultural Science and Technology, Srinagar, Jammu and Kashmir

CONTENTS

16.1 INTRODUCTION

The climate crisis is the general process of climate variation, especially caused by human activities, which is characterized by changes in the normal climate of the planet (wind, precipitation, and temperature). The earth's climate and humanity's future are in jeopardy due to the unbalanced sustainability of the planet's biological systems. Global warming, on the other hand, may be a steady increase in the average temperature of the earth's climate produced by increased radiation impacting the earth from the sun being trapped inside the environment rather than being released. The total amount of warm inner nursery soil is so large that the temperature of the soil is rising at a considerably faster pace than in the past.

Global warming leads to climate change, since all frameworks within the worldwide climate framework are connected, and consequently, an increase of heat energy naturally increases the temperature of soil over the globe. A hotter climate causes ice sheets and mountain snow to soften, resulting in a rise in ocean levels. Global change is primarily driven by the spread of nursery gases or greenhouse gases (GHGs) across the planet, which results in climate warming (IPCC, 2013).

DOI: 10.1201/9781003220824-19

16.2 CAUSES OF CLIMATE CHANGE

The major causes of climate change are classified into two groups: naturally and human caused.

Naturally caused: The planet has gone through warm and cool periods in the past, which is a natural process, but warming has been quicker than ever before since the previous century, which cannot be explained only by natural reasons. The effect of these natural factors that are still in play today is too minor or they occur too slowly to explain the fast warming witnessed in recent decades, according to NASA.

Human caused: According to the 2010 World Development Report, global warming and climate change are caused by human-induced activities that result in the creation of GHGs and their buildup in the earth's atmosphere, causing an increase in atmospheric temperature. Human-caused GHG emissions are the primary cause of the earth's quickly changing climate. They perform a vital role in keeping the planet warm, but their levels have risen dramatically in recent decades. The burning of fossil fuels such as gas, oil, and coal is the primary source of human-generated emissions. Deforestation also contributes to climate change by releasing trapped carbon into the atmosphere.

16.3 GREENHOUSE GASES (GHGS)

Nursery gases or greenhouse gases are characterized as the gases that trap warmth within the environment, which results in change within the climatic radiation balance of the earth. GHGs prevent warmth from emanating or reflecting away from the soil, and in this way, may cause warming of the climate. Different evidence has appeared that GHG levels have grown since the Industrial Revolution. GHGs of specific concern include carbon dioxide (CO_2), nitrous oxide, methane, hydrofluorocarbons, and sulfur hexafluoride. The seas retain a significant quantity of carbon in the form of CO_2, which is later transmitted into the air. CO_2 concentrations have increased by about 36% across the planet since the Industrial Revolution, with the majority of this increase being attributable to the burning of fossil fuels. The following are the most often studied GHGs:

Carbon dioxide: This is the most important heat-trapping gas, accounting for 55% of the total amount of long-term GHGs. It is released in a variety of ways, including through breathing, volcanic eruptions, and human activities such as deforestation, land use, and fossil fuel burning. The United States Natural Security Organization uses it as a marker because of its inclusivity.

Methane: This is the primary component of common gas produced by anaerobic workouts like arrival use, mechanical exercises, fossil fuel combustion, and powerful wastes. Methane can be a precursor to ozone, which is a critical GHG.

Nitrous oxide: This may be a byproduct of mechanical processes, fertilizer generation, and the combustion of different materials. Nitrous oxide remains in the environment for a very long time.

Fluorinated gases: These are strong GHGs that are produced in smaller amounts but have been demonstrated to be both long-lasting and amazingly warming GHGs. They are completely human-caused and have no natural sources.

Sulfur hexafluoride: This is utilized basically in dielectric materials, particularly dielectric fluids, and as insulation in high-voltage applications such as transformers and grid switching gear.

16.4 LIVESTOCK AND CLIMATE CRISIS

16.4.1 Effect of Livestock on Climate Crisis

Animal products provide 17% of worldwide kilocalorie utilization and 33% of worldwide protein utilization and are therefore vital agrarian products for worldwide food security (Rosegrant *et al.*,

2008). In response to global population increase and due to expanding requirements, animal production is changing drastically. Animal farming contributes to climate change. Of total anthropogenic GHG emissions, 14.5% come from the animal supply chain, which adds up to 7.1 gigatons of carbon dioxide equivalents per year. Of this, 3.2 gigatons is contributed by food production and preparation, 2.8 gigatons by enteric aging, and 0.71 gigatons by excrement, and the remaining 0.42 gigatons of CO_2 emission equivalents is due to the preparation and transportation of animal products (Gerber *et al.*, 2013). Methane and nitrous oxide are the most important GHGs from animal farming. Methane, which is primarily delivered by enteric aging and excrement, has a 28 times higher impact on worldwide warming than CO_2. Nitrous oxide, which is produced by the use of organic/inorganic fertilizers and feces, has a 265-fold greater impact on global warming than CO_2 (IPCC, 2013).

Along with the associated soil carbon dioxide and nitrous oxide, the animal sector contributes to the nursery gas spread increasing from feed era. Soil carbon dioxide levels rise when plant structures deteriorate, normal soil matter mineralizes, and land use changes, among other factors. Aside from the development of custom fertilizers and pesticides, the use of fossil fuels in on-farm agricultural activities also contributes to the rise in GHG levels in the air (Goglio *et al.*, 2018). Organic and inorganic fertilizer application also results in nitrous oxide emissions.

16.4.2 Effect of the Climate Crisis on Livestock

Climate change has possible consequences for livestock both directly and indirectly. These include changes in the creation, characteristics of feed and forages, milk production and animal growth, water availability, diseases, and reproduction. These impacts are primarily due to changes in the normal temperature of the earth and high CO_2 concentration, precipitation variability, and a combination of these components (Nardone *et al.*, 2010). Most of the imperative variables for animal production, such as forage amount and quality, water accessibility, creature generation, propagation, and wellbeing, are influenced by increased temperature and high CO_2 concentrations.

16.5 POULTRY AND GREENHOUSE GASES

Poultry, which includes guinea fowl, fowl, ostrich, duck, turkey, goose, etc., makes an astounding contribution to individual nourishment as a main provider of eggs, meat, crude substances to businesses, source of salary, and business to individuals other than giving financial administrations (Demeke, 2004). The poultry division is an important section of the agricultural industry. It plays a vital part in the economy, sustenance, and employment of destitute country families in numerous developing nations. The poultry sector has grown over the last decade due to the expanding human population and thus, the requirement for animal protein.

Worldwide poultry supply chains contribute GHG emissions of 0.6 Gg CO_2-eq/year, representing 8% of emissions from the animal sector (Gerber *et al.*, 2013). The two essential sources of GHGs from poultry cultivation are:

Fossil fuels: These sources of energy are especially utilized to warm the poultry and control ventilation and lighting. Much of the CO_2 created by the poultry industry comes from the misuse of fossil fuels from different sources, including power, propane utilized in stationary combustion units (such as heaters or incinerators), and diesel utilized in portable combustion units such as trucks, tractors, and generators that are used for cultivation. Propane burning which needs heat produces 68% of emanations from broiler and pullet ranches and 20% from electricity respectively.

Manure and its treatment: The amount of GHGs transmitted from excrement (litter) depends on how it is treated and disposed of. During handling and storing, various GHGs like nitrous oxide and methane are emitted.

Poultry flocks are especially vulnerable to thermal conditions, which ends up affecting the behavioral and physiological activities of birds (Ayo-Enwerem, 2017). Birds can only tolerate narrow temperature ranges due to their thermoregulatory mechanism. Modern poultry genotypes are thought to create higher body heat as a result of their increased metabolic activity (Settar et al., 1999; Deeb and Cahaner, 2002).

16.6 THERMOREGULATORY MECHANISM OF POULTRY

Birds (such as chickens) have a higher internal body temperature than mammalian animals, ranging from 106 to 108 degrees Fahrenheit. The temperature at which enzymatic proteins are destroyed and the living cells and tissues progressively fail to operate starts to occur in the region of 46 degrees Celsius, and thus, poultry have considerably less scope when suffering from heat stress compared with other animals and quickly succumb to a higher temperature. Due to lack of sweat glands, domestic poultry are less tolerant to heat than to cold, due to which they die from heat stress.

16.7 HEAT LOSS IN POULTRY

To keep their body temperature at 105 degrees Fahrenheit, fowl use a variety of methods to disperse heat. Radiation, conduction, convection, and evaporation all help to dissipate body heat to the surrounding environment (Mustaf *et al.*, 2009). When the ambient temperature is below or within the bird's thermoneutral zone (55 to 75 degrees Fahrenheit), the first three techniques, known as sensible heat loss, are effective. The amount of perceptible heat loss is proportional to the temperature differential between the bird and its surroundings (Mack *et al.*, 2013). When the ambient temperature drops below 77 degrees Fahrenheit, the heat loss mode switches from sensible to evaporative (Fedde, 1998).

16.8 EFFECT OF THE CLIMATE CRISIS ON POULTRY

The climate crisis is one of the main big environmental stressors affecting animal production, as it negatively affects their health and performance, particularly for chickens. Global temperatures are continuously rising, challenging various agricultural patterns, including poultry (Moorhead, 2009). Infections, pests, and parasites, as well as exercise and nourishment, all influence the welfare of poultry, and the climate crisis might have a critical effect on poultry efficiency and profitability (Lay *et al.*, 2011). Elevated temperature and water scarcity affect chicken production by placing stress on the birds' homeostasis as a result of climate change leading to extreme events such as drought or flood (Tiruneh and Tegene, 2018). Increased global temperature is one of them. Bird performance and output suffer as a result of behavioral, physiological, and immunological responses to high ambient temperatures. The effects of a hot climate on poultry performance can be significant (Bhadauria *et al.*, 2014). Heat stress can result in significant financial losses for poultry producers and have an impact on every element of their performance.

16.8.1 Growth and Production

Heat exhaustion causes a drop in voluntary feed intake in birds and hence, depresses growth rate and production (Sahin *et al.*, 2001). Vital development promoters like plasma triiodothyronine and thyroxine are seriously affected in heat-stressed broiler chickens. Heat stress results in expanded water utilization and diminished booster utilization. In order to preserve the balance between heat generation and heat stress, fowls reduce their booster utilization to decrease warmth from the digestive system. Research has illustrated that for each degree rise in temperature between 32 and 38°C, booster utilization is decreased by 5% (Balogun *et al.*, 2013). Heat stress results in diminished food intake, body weight, and food effectiveness. It also causes delayed development and influences the digestibility of supplements.

16.8.2 Egg Quality and Reproductive Performance

Hyperthermia reduces the efficiency of laying hens, diminishing egg number and quality. In order to maintain a steady body temperature, birds use their food for metabolic endurance, which results in lower egg generation and egg quality. At high temperatures, the respiratory rate of birds increases, resulting in hyperventilation. This hyperventilation causes a reduction in blood CO_2 levels, which is afterward compensated by kidney take-up of bicarbonate, subsequently influencing eggshell quality. Investigation has moreover found that warm stretch altogether diminishes bolster admissions, egg generation, cruel egg weight, and live weight, conjointly impact a few egg quality characteristics, such as eggshell thickness and egg specific gravity. Warm stretch diminishes the quality of eggs and the execution of laying eggs consequently diminishing their regenerative execution. Increased temperature diminishes eggshell thickness and increases egg breakage (Lin *et al.*, 2004). Heat stress has appeared to cause a critical diminution of eggshell weight, egg weight, eggshell thickness, and eggshell percentage (Ebeid *et al.*, 2012). In breeder cocks, high temperature influences semen generation and semen characteristics such as consistency, spermatozoa concentration, and seminal volume. High temperature diminishes the seminiferous epithelial cell separation, which is shown in diminished semen quality and amount with time (Obidi *et al.*, 2008).

16.8.3 Immunity

When the ambient temperature of birds increases, they get uncomfortable, which becomes quite evident from their behavior. They are under constant stress, and any kind of stress reduces the bird's immune response and its competence to mount a response to infection. Likewise, heat stress has been found to have an immunosuppressing impact on broilers and layers. It decreases the relative weights of resistant organs and liver in broilers and laying hens (Felver-Gant *et al.*, 2012).

16.8.4 Meat Quality

Warm push can influence the organ and muscle digestion system which can hold on indeed after the butcher. It has been detailed that long-term warm conditions adversely influence fat deposition and meat quality in broilers, in a breed-dependent way (Lu *et al.*, 2007). Later research illustrated that heat stress is related to deterioration of meat quality and its chemical composition in broilers (Dai *et al.*, 2012). Constant heat stress has been found to diminish the extent of breast muscle while expanding the extent of thigh muscle in broilers, with moo protein substance and higher fat deposition (Zhang *et al.*, 2012). Discoloration, dry muscle, increase in blood thickness, and unpleasantness of the skin are shown by heat-stressed birds (Soliman and Safwat, 2020).

16.8.5 Disease Incidence

The global spread and re-emergence of vector-borne diseases has demonstrated the link between the climate crisis and animal–human interactions (Bett *et al.*, 2017). In terms of parasite growth and dispersion, generation, destructiveness, and transmission, the climatic crisis can have an impact on irresistible diseases and their vectors. Moreover, it could potentially condition the reasons such as diseases, pests, and parasites which can end up affecting the wealth of birds and also severely affecting poultry production. The prevalence of zoonotic diseases is also increased at high temperatures. The biodiversity, distribution, and migratory pattern of birds are also affected by climate change, which can lead to various disease outbreaks. High temperature increases the incidence of respiratory infections in poultry. Infectious illnesses such as infectious bronchitis, Newcastle disease, infectious bursal disease, and avian influenza are becoming more common as a result of changes in precipitation patterns and rising temperatures (Mitchell, 2017).

16.9 MODERATING THE IMPACTS OF CLIMATE CHANGE

Breeding: This is predicated on the disclosure and choice of qualities included in heat resistance and disease resistance in hot climatic zones around the globe, and it incorporates unmistakable evidence and determination of these qualities (Ranjan *et al.*, 2019).

Nutrition: The harmful effects of increasing temperature have been mitigated by dietary modification. Feed additives (electrolytes, phytobiotics, fat, protein, natural antioxidants, and minerals), feed limitation, drinking cool water, and feed shape, among others, have all been shown to decrease the negative impacts of heat stress.

Housing: Buildings should be designed in such a way that they can better withstand new climatic and weather extremes. Approaches to deal with current climatic extremes include using bio-fuel geyser or fermentative assimilation of chicken debris, adding renewable energy such as solar or wind energy to control poultry house shelters. The housing system should be built in such a way that the pen has enough ventilation. Proper ventilation makes it easier to regulate heat stress in the chicken house and eliminates moisture-laden air (Ranjan *et al.*, 2019). The use of coolers such as fans, cooling pads, and static pressure controllers in the pen house helps mitigate the effect of the increase in temperature on the birds (Sinha *et al.*, 2018).

16.10 CONCLUSION

Climate change affects poultry in both direct and indirect ways. It has a detrimental impact on all areas of chicken production, lowering productivity and profitability. The detrimental effects of climate change on poultry can be moderated via adequate nutrition and environmental adjustment, but the most significant strategy for reducing the long-term effects of climate crisis on poultry birds is the genetic selection of heat-tolerant birds.

REFERENCES

Ayo-Enwerem, M. C., Ahaotu, E. O., Nwogu, C. M. and Opara, J. 2017. Growth performance of starter broilers fed diets containing red sandalwood (*Pterocarpus santolinoides)* leaf meal. *Direct Research Journal of Veterinary Medicine and Animal Science* 2(4): 106–109.

Balogun, A. A. B., Akinseye, F. M. and Agbede, J. O. 2013. Water and feed consumption in broiler birds during a typical hot weather condition in Akure, Ondo State, Nigeria. *International Journal of Biological and Chemical Sciences* 7(3): 1119–1125.

Bhadauria, P., Kataria, J. M., Majumdar, S., Bhanja, S. K., Divya and Kolluri, G. 2014. Impact of hot climate on poultry production system – A review. *Journal of Poultry Science and Technology* 2(4): 56–63.

Bett, B., Kiunga, P., Gachohi, J., Sindato, C., Mbotha, D., Robinson, T., Lindahl, J. and Grace, D. 2017. Effects of climate change on the occurrence and distribution of livestock diseases. *Veterinary Medicine* 137: 119–129.

Dai, S. F., Gao, F., Xu, X. L., Zhang, W. H., Song, S. X. and Zhou, G. H. 2012. Effects of dietary glutamine and gammaaminobutyric acid on meat colour, pH, composition and water-holding characteristic in broilers under cyclic heat stress. *British Poultry Science* 53(4): 471–481.

Deeb, N. and Cahaner, A. 2002. Genotype-by-environment interaction with broiler genotypes differing in growth rate. 3. Growth rate and water consumption of broiler progeny from weight-selected versus non selected parents under normal and high ambient temperatures. *Poultry Science* 81: 293–301.

Demeke, S. 2004. Egg production and performance of local white leghorn hens under intensive and rural household conditions in Ethiopia. *Livestock Research for Rural Development* 16(2).

Ebeid, T. A., Suzuki, T. and Sugiyama, T. 2012. High temperature influences eggshell quality and calbindin-D28k localization of eggshell gland and all intestinal segments of laying hens. *Poultry Science* 91: 2282–2287.

Fedde, M. R. 1998. Relationship of structure and function of the avian respiratory system to disease susceptibility. *Poultry Science* 77: 1130–1138.

Felver-Gant, J. N., Mack, L. A., Dennis, R. L., Eicher, S. D., and Cheng, H. W. 2012. Genetic variations alter physiological responses following heat stress in 2 strains of laying hens. *Poultry Science* 91: 1542–1551.

Gerber, P. J., Steinfeld, H., Henderson, B., Mottet, A., Opio, C., Dijkman, J., Falcucci, A. and Tempio, G. 2013. *Tackling Climate Change through Livestock: A Global Assessment of Emissions and Mitigation Opportunities*. Food and Agriculture Organization of the United Nations (FAO), Rome, Italy.

Goglio, P., Smith, N., Grant, B. B., Desjardins, R. L., Gao, X., Hanis, K., Tenuta, M., Campbell, C. A., McConkey, B. G., Nemecek, T., Burgess, P. J. and Williams, A. G. 2018. A comparison of methods to quantify greenhouse gas emissions of cropping systems in LCA. *Journal of Cleaner Production* 172: 4010–4017.

IPCC (Intergovernmental Panel on Climate Change). 2013. *Climate Change 2013: The Physical Science Basis. Contribution of Working Group I to the Fifth Assessment Report of the Intergovernmental Panel on Climate Change*. Stocker, T. F., Qin, D., Plattner, G.-K. Tignor, M., Allen, S. K., Boschung, J., Nauels, A., Xia, Y., Bex, V. and Midgley, P. M. (Eds.). Cambridge University Press, Cambridge, United Kingdom and New York, p. 1535.

Lay, D. C., Fulton, R. M., Hester, P. Y., Karcher, D. M., Kjaer, J. B., Mench, J. A., Mullens, B. A., Newberry, R. C., Nicol, C. J., O'sullivan, N. P. and Porter, R. E. 2011. Hen welfare in different housing systems. *Poultry Science* 90(1): 278–294.

Lin, H., Mertens, K., Kemps, B., Govaerts, T., De Ketelaere, B., De Baerdemaeker, J., Decuypere, E. and Buyse, J. 2004. New approach of testing the effect of heat stress on eggshell quality: mechanical and material properties of eggshell and membrane. *British Poultry Science* 45: 476–482.

Lu, Q., Wen, J. and Zhang, H. 2007. Effect of chronic heat exposure on fat deposition and meat quality in two genetic types of chicken. *Poultry Science* 86: 1059–1064.

Mack, L. A., Felver-Grant, J. N., Dennis, R. L. and Cheng, H. W. 2013. Genetic variations alter production and behavioral responses following heat stress in 2 strains of laying hens. *Poultry Science* 92: 285–294.

Mitchell, A. 2017. How Will Climate Change Affect Poultry Disease Problems? The Poultry Site. www.thepoultrysite.com/articles/how-will-climate-change-affect-poultry-disease-problems

Moorhead, A., 2009. *Climate, Agriculture and Food Security: A Strategy for Change*. Alliance of the CGIAR Centers.

Mustaf, S., Kahraman, N. S. and Firat, M. Z. 2009. Intermittent partial surface wetting and its effect on body-surface temperatures and egg production of white brown domestic laying hens in Antalya (Turkey). *British Poultry Science* 50: 33–38.

Nardone, A., Ronchi, B., Lacetera, N., Ranieri, M. S. and Bernabucci, U. 2010. Effects of climate change on animal production and sustainability of livestock systems. *Livestock Science* 130: 57–69.

Obidi, J. A., Onyeanusi, B. I., Ayo, J. O., Rekwot, P. I. and Abdullahi, S. J. 2008. Effect of timing of artificial insemination on fertility and hatchability of Shika brown breeder hens. *International Journal of Poultry Science* 7(12): 1224–1226.

Ranjan, A., Ranjana, S., Indu, D., Abdul, R. and Shiwani, T. 2019. Effect of heat stress on poultry production and their managemental approaches. *International Journal of Current Microbiology and Applied Sciences* 8(2): 1548–1555.

Rosegrant, M, W., Ewing, M., Yohe, G., Burton, I., Saleemul, H. and Valmonte-Santos, R. 2008. Climate change and agriculture: threats and opportunities. *Asian Journal of Agriculture and Development* 7(1): 41–82.

Sahin, N., Sahin, K. and Kucuk, O. 2001. Effects of vitamin E and vitamin A supplementation on performance, thyroid status and serum concentrations of some metabolites and minerals in broilers reared under heat stress (32°C). *Veterinary Medicine* 46(11–12): 286–292.

Settar, P., Yalcin, S., Turkmut, L., Ozkan, S. and Cahanar, A. 1999. Season by genotype interaction related to broiler growth rate and heat tolerance. *Poultry Science* 78: 1353–1358.

Sinha, R., Kamboj, M. L., Lathwal, S. S. and Ranjan, A. 2018. Effect of housing management on production performance of crossbred cows during hot-humid season. *Indian Journal of Animal Research* 52(7): 1091–1094.

Soliman, A. and Safwat, A. M. 2020. Climate change impact on immune status and productivity of poultry as well as the quality of meat and egg products. In Omran, E.-S.E. and Negm, A. M. (eds) *Climate Change Impacts on Agriculture and Food Security in Egypt*, Springer Water 481–498.

Tiruneh, S. and Tegene, F. 2018. Impacts of climate change on livestock production and productivity and different adaptation strategies in Ethiopia. *Journal of Applied and Advanced Research* 3(3): 52–58.

Zhang, Z. Y., Jia, G. Q., Zuo, J. J., Zhang, Y., Lei, J., Ren, L. and Feng, D.Y. 2012. Effects of constant and cyclic heat stress on muscle metabolism and meat quality of broiler breast fillet and thigh meat. *Poultry Science* 91: 2931–2937.

17 Climate Change and Renewable Energy

Improvements and Interpretations for Sustainable Development

Richa Kothari, Gagandeep Kour, Shubham Raina, Anita Singh, S. Dhar, and D. Pathania
Department of Environmental Sciences, Central University of Jammu, Samba

Har Mohan Singh and V.V. Tyagi
School of Energy Management, Shri Mata Vaishno Devi University, Jammu

Arjun Tyagi
Department of Electrical Engineering, Netaji Subhas University of Technology, New Delhi

CONTENTS

DOI: 10.1201/9781003220824-20

17.1 INTRODUCTION

Energy demand is rapidly increasing at the present time due to economic activities. Fossil fuels are the main source of energy, and transport and power production mainly depend on coal-based thermal power plants and liquid petroleum products. The combustion of solid, liquid, and gas fossil fuels produces a huge amount of energy and gaseous emissions. Oxides of carbon (COx), sulfur (SOx), and nitrogen (NOx) comprise a large portion of gaseous emissions, and suspended particles (PM), ash, and metal compounds are also integrated parts of the burning of fossil fuels (Singh et al., 2019). These waste materials create challenges in the form of deterioration of the environment and climate changes. Depletion of the ozone layer, melting of glaciers, rising of sea level, global warming, and extinction of flora and fauna are repercussions of climate change. The USA, China, and India are the largest primary energy consumers in the world, and India produced 809.2 Mtoe in 2018. India is the third largest carbon dioxide (CO_2) emitter, with a 40% share of power generation in 2018 (Jain et al., 2020). The International Energy Agency (IES) estimated that the energy demand will increase by 30% in the years 2016–2040 (Kafka and Miller, 2019). To fulfill the demand for energy, more fossil fuel is required, which may have a catastrophic impact on the earth. Therefore, the need for renewable energy sources is increasing in the contemporary world for clean energy production.

Solar, wind, biomass, small hydro, and geothermal energy sources are major renewable energy sources that have low emissions. Although renewable energy sources showed high cost in their initial developing years, this is continuously reducing, and renewable energy technologies have reached a cost-competitive era in which the levelized cost of electricity from wind power has reached 2.4 INR/kWh, while coal power has 3.7 INR/kWh (Das et al., 2020). The solar photovoltaic (PV) cost was 6.17 INR/Unit in 2014, which was reduced to 2.44 INR/Unit in 2018 (MNRE, 2019). This shows that the cost of electricity production by renewable technologies is lower than with conventional technologies like coal-based thermal power plants.

Renewable energy plays a vital role in boosting power production to fill the gap between demand and supply and mitigate the impact of climate change on the earth. In this direction, the United Nations (UN) and various other international organizations are working to ensure an affordable power supply sustainably. Each year, the United Nations organizes a Conference of Parties (COP) to ensure emission reduction (Das et al., 2020). Similarly, the International Renewable Energy Agency (IRENA), Renewable Energy and Energy Efficiency Partnership (REEEP), International Solar Alliance (ISA), and the IEA are working for sustainable clean energy development. According to the IEA, 70% of world energy investment is driven by governments. This shows that world energy policy is dependent on government policies and directions (IEA, 2018). The development of renewable energy depends on climate change policies that mention the need for clean and green energy for sustainable development, and the consequences of climate change policies catalyze the making of policies favorable to renewable energy throughout the world.

17.2 ENERGY AND CLIMATE CHANGE

Changes in global climatic patterns are regarded as climate change. Climate change is real, and one piece of evidence of climate change is the rising global mean temperature, called global warming. This is evident from a rise of 0.85°C in average surface air temperature from 1880 to 2012 (Mei et al., 2020). Climate change is mainly attributed to greenhouse gases (GHGs) emitted through anthropogenic activities. The GHGs mainly responsible for global warming and the resulting climate change are carbon dioxide (CO_2), methane (CH_4), and nitrous oxide (N_2O) with CO_2 as the main contributing gas.

Energy plays a crucial role in boosting the economic growth of a country, as economic growth instigates energy demand due to increased consumption. Fossil fuels are the backbone of the energy sector, as they contribute more than 80% of primary energy demand. Energy and climate are bilaterally related. Most of the GHGs emissions are contributed by the energy sector. These GHGs are released during the burning of fossil fuels. According to the IEA, fossil fuels will remain in a dominant position in the energy supply market because of rapid industrialization, urbanization, and lack of development of renewable energy technologies (Jin et al., 2017; Ucal and Xydis, 2020). Fossil fuels are responsible for more than 90% of the CO_2 emissions in the energy production sector. This CO_2 emission is expected to further increase in the future, thereby accelerating the pace of climate change. N_2O and CH_4 are the other two GHGs released from energy production facilities and contribute to 10% of emissions during oil and gas extraction, processing, and distribution (Ucal and Xydis, 2020).

Climate change will affect energy production as well as energy consumption. For forecasting peak electricity demands for Los Angeles while considering rising temperature, a bottom-up building energy model was adopted, and it predicted that peak demand would increase by a value of 25.5% by 2060 (Burillo et al., 2019). To estimate the impact of temperature, sunshine duration, and rainfall on the power demand of China, a fixed-effect regression feedback model was used, which showed that electricity demand would be significantly affected by temperature and rainfall (Fan et al., 2019).

Econometric analysis of the climate change scenario depicts that future energy demand will rise. This will further increase GHGs in the atmosphere, thereby adding to climate change. To achieve the low-carbon future while at the same time meeting the growing energy demand requires innovation in energy systems, and this has already attracted researchers' interest in opting for alternative options to mitigate climate change and transform the pathways of energy systems. To combat climate change and meet the growing energy demand, dependence on renewable energy sources has attracted much attention from researchers and policymakers. Renewable energy sources are regarded as cleaner fuels, as they cause minimal damage to the environment during processing and extraction, and these are among suitable remedies to deal with the problems of climate change.

International organizations have come together in a joint effort to combat climate change and reduce GHGs emissions and have opted for sustainable development. The IEA has projected lowering the various emissions related to energy sector using a bridge scenario consisting of five strategies (Kang et al., 2020). These strategies are as follows:

Strategy 1: Increasing energy efficiency in all types of development activities, such as industry, construction, and transportation

Strategy 2: Minimal use/lessening the use of coal-burning power plants with lowest efficiency and outlawing further construction

Strategy 3: Increasing investment in different renewable energy technologies

Strategy 4: A hierarchical termination of fossil fuel subsidies to the end-users

Strategy 5: Cutting down methane emissions from oil and gas manufacturing facilities

Existing studies also give considerable hope for modern energy technologies like carbon capture and storage (CCS), electric vehicles, smart grid systems, etc. Several types of research have explored the integration of new ideas and technologies with the traditional ones to improve energy efficiency and reduce GHG emissions. Renewable resources like solar, wind, hydropower, and thermal energy have gained attention. In recent times, biohydrogen production and other biofuels; waste-to-energy approach have attracted the interest of many researchers. The cost–benefit and the mitigation potential of these renewable energy resources are most important to consider.

17.3 RENEWABLE ENERGY

Renewable energy is often referred to as the source of clean energy that comes from various natural resources or processes and is naturally replenished easily on earth. Sunlight and wind are the best examples of renewable sources, which keep shining and blowing on the earth. Renewable energy is harnessed from natural sources and utilized in various lighting, heating, and transport applications. The ultimate source of energy on the earth is the sun, which creates the various direct and indirect forms of energy (Figure 17.1). Table 17.1 describe renewable energy production and its advantages and disadvantages. Life cycle assessment of renewable energy technologies for GHGs emissions is lower than for fossil fuel technologies. It is estimated that renewable energy technologies have a range of 4 to 46 g CO_2eq/kWh, whereas 469 to 1001 g CO_2eq/kWh emission is estimated from fossil fuel-based technologies (IPCC, 2011).

17.3.1 Solar Power Generation

Solar energy technologies are a wide range of technologies utilizing various methods such as heating and electricity production. Solar PV is the direct conversion of solar radiation into electricity. Concentrator solar power (CSP) is the conversion and storage of solar radiation thermal energy into high-temperature fluids/materials to run an electrical generator for power production. Solar photolysis is a chemical process in which chemical bonds are broken by solar radiation to produce energy; the best example is solar water photolysis for hydrogen production. Solar PV-assisted

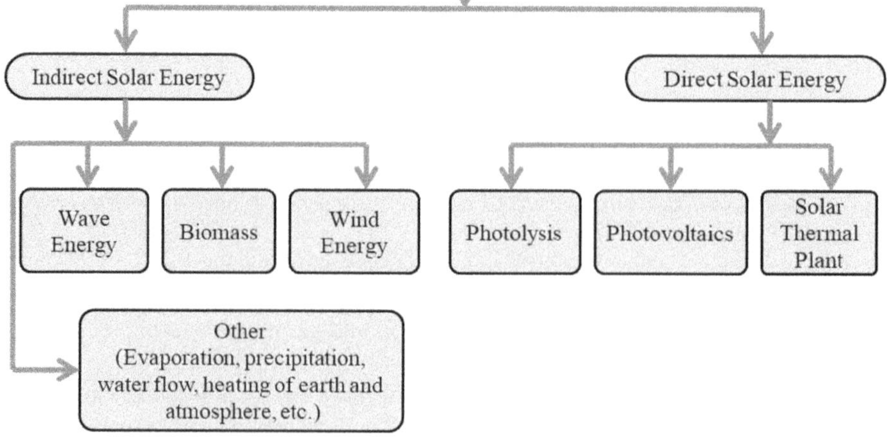

FIGURE 17.1 Renewable energy categorization.

TABLE 17.1
Renewable Energy: At a Glance

Renewable Technologies	Sources	Principle of Generation	Advantages	Limitations
Solar thermal	Sunlight	Absorption of radiated heat and light of sun	Available in abundance and infinite source of energy	Requires large installation space
Photovoltaic	Sunlight	Conversion of sunlight into electricity	Abundant natural source; can make homes self reliant	Requires large area and discarded panels are hazardous; high investment cost
Wind energy	Wind	Use of wind turbines for the generation of electrical energy	Low maintenance cost	High investment cost; high wind flux required; site specific
Geothermal energy	Magmatic heat	Trapping of heat energy of thermal sites	Infinite; negligible operational emissions	Limited area; high installation cost; requires electricity/fuel wood for running heat pump
Bioenergy	Biomass	Conversion of biomass into desired fuels/concentrated form	Abundantly available resource; carbon neutral	Constant supply of biomass is an issue; hydrolysis of complex biomass is difficult
Hydroelectricity	Potential energy of water	Water is allowed to fall on turbines to generate electricity	No emissions; long-term benefits	High construction cost; building of dams disturbs local eco-system; potential hazard to low-lying areas during floods and accidental damage

electrolysis for hydrogen production is also a form of solar energy for solar hydrogen/fuel production (Pathak et al., 2019). Solar energy is a source of radiation from the sun, and an average of 1367 W/m^2 solar energy is received on the earth, which varies from place to place. Global solar radiation is measured on the horizontal surface, while solar energy technologies such as PV and CSP are fixed at specific inclined surfaces so that these technologies trap maximum solar radiation (Mousavi Maleki et al., 2017).

17.3.1.1 Solar Photovoltaic

Solar cells are PV cells that convert sunlight (photons) directly into electricity (voltage); the conversion of photons into electricity is called the photovoltaic effect. Solar PV cells produce direct current (DC) that can be transformed to alternative current (AC) and stored in batteries for future use. The quantity of electricity depends on the intensity of sunlight striking the PV cell surface. A large number of PV cells make a PV module (panel), and a large number of PV panels make a PV array. PV panels can easily be installed on the roofs of buildings for power production. PV cells are made of semiconductor materials, mainly silicons. Single crystalline, multi-crystalline, and amorphous silicon are the main types of solar cell that are utilized in commercial applications. Nanocrystal-based solar cells and dye-sensitized solar cells (perovskite) are emerging technologies for solar cells that are in the developmental stage (Rathore et al. 2021). Silicon PV cells were first introduced by Bell Laboratories in 1955 with 2% efficiency, which is continuously improving, and

15–22.5% operational efficiency is achieved by commercial solar plants, whereas 46% efficiency has been achieved by research laboratories. As the demand for clean energy increases, governments provide subsidies and various lucrative incentives associated with clean energy that help to trim down the cost of solar PV power (Kafka and Miller, 2019).

17.3.1.2 Solar Energy Collectors

Solar thermal energy conversion technologies are the primary applications of solar energy utilization in which solar collectors are utilized to trap solar radiation. Generally, flat-plate collectors and evacuated tube collectors are used currently; flat-plate collectors are utilized for low-thermal applications that range from 60 to 100°C, while evacuated tube collectors are utilized for high-temperature applications. Flat-plate collectors are glazed box-like structures, and evacuated tube collectors are similar to flat-plate collectors but with a vacuum space between the heat pipe and glass cover, which makes them suitable for high-heat applications. Concentrator solar collectors are designed for higher-temperature applications than the flat-plate and evacuated tube collectors. Concentrator solar collectors redirect solar radiation passing through an aperture into an absorber, by which the solar energy is highly concentrated before being transformed into heat. The integration of sun-tracking systems makes these collectors more efficient. Parabolic dish, parabolic trough, and linear Fresnel reflector are well-known solar concentrators that are utilized at the commercial level (Olczak et al., 2020; Pathak et al., 2019).

Wiser et al. (2016) analyzed various environmental and public health benefits of solar energy in the USA. They suggested that solar power reduced GHGs and air contaminants up to 10% from 2015 to 2050, and solar power has also decreased water consumption and withdrawals by 9% and 4%, respectively, in various drought-prone states of the USA. The life cycle assessment of GHGs emissions of solar PV and solar concentrator for electricity production ranges from 30 to 80 g of CO_2eq/kWh and from 14 to 32 g of CO_2eq/kWh, respectively, which has lower intensity than the natural gas-fired power plants (IPCC, 2011). This shows that solar power is a lower-emission renewable technology than conventional energy technologies.

17.3.2 Hydropower

Hydropower is the leading renewable energy source, which contributes two-thirds of global electricity production from all renewable sources combined. Many countries, such as Nepal, Paraguay, Norway, and Ethiopia, have produced more than 70% of their energy from hydropower, and China is one of the largest hydropower producers, increasing its production more than four times from 2000 to 2017. It is expected to achieve 45 to 70% energy production from hydropower by 2040 (IEA, 2018). Africa, Brazil, India, China, and southeastern Asia are the hot spots for hydropower growth. Run-of-river plants, plants with significant storage, and pumped storage are three main types of hydropower systems, which can be classified on the basis of whether the flow of water at a given site is controlled or modified. Small run-of-river hydropower plants are considered most significant for renewable power because they can be easily developed at low cost, and construction and maintenance are also easy for lower-income countries (Kuriqi et al., 2021).

The USA and some developed countries of Europe attempted to study the impact of hydropower development on the ecological environment for the first time in the 1980s. Small hydropower plants are considered as clean and green power because of their low impact on the environment. Large hydropower plants need a large land area for the water reservoir and other requirements, seriously disturbing the ecological balance by submerging vegetation, excavation of canals, and various restoration and resettlement activities (Shiji et al., 2021). The water reservoirs of hydropower are emission sources of CO_2 because a large vegetative flora is found in the water reservoir that needs to perform photosynthesis and produce organic biomass. The degradation of the organic biomass creates other GHGs such as nitrous oxide and methane. Commonly, hydropower is considered a low-carbon energy option but a hydropower plant is a vast civil and mechanical structure with many parts, and 4 to 14 g of CO_2eq/kWh emission is estimated during construction, operation, and dismantling, but sometimes GHG emissions may be higher (IPCC, 2011).

17.3.3 Bioenergy

It is the first kind of traditional fuel used for heating and lighting applications. Bioenergy is considered a carbon-neutral energy option because it is produced from biological materials that have almost equal carbon emission and carbon sequestration. Green plants take CO_2, oxygen (O_2), and sunlight from the atmosphere and convert them into biomass materials that are utilized for energy production (Sinha et al., 2019). Bauen et al. (2009) reported that 1500 EJ/year biomass potential was estimated on the basis of the diverse range of feedstock by 2050. Agriculture, forestry, and municipal solid waste are the main feedstocks utilized for biopower and heat production. Sugar cane waste (bagasse), agriculture residues, grains, and vegetable oils are utilized for biofuel production. Marginal and waste lands of roadsides, railway tracks, and parks can be utilized for energy crops such as maize and sweet sorghum and herbaceous plants such as miscanthus, canary grass, kenaf, and eucalyptus for lignocellulosic biomass (Lamichhane et al., 2021). Surplus and waste food materials, unused agriculture grains, and residues are also notable feedstocks for bioenergy production (Kothari et al., 2020). Aquatic plant biomass is also an important feedstock, which has high biomass productivity compared with terrestrial plants. Due to their high photosynthesis rate, the high biomass productivity of algae is considered as a potential source of bioenergy production, which consumes 1.8 kg of CO_2 from the atmosphere for the production of 1 kg of biomass. It shows that algae can significantly help to reduce atmospheric CO_2. Algal biomass can be utilized for biodiesel, biohydrogen, and bioethanol production, that are renewable options for biofuels (Singh et al., 2019).

Combustion, gasification, anaerobic digestion, and fermentation are the main thermochemical and biochemical routes of bioenergy production. Biomass cookstoves and gasifiers are utilized for thermal and power production, whereas biogas digesters and fermenters are utilized for biogas and bioethanol production. The commercial applications of bioenergy options, such as anaerobic digestion for biomass to power and direct combustion of biomass, are considered efficient heating technologies. Biopower mainly deals with biomass gasification and bagasse-based cogeneration at the commercial level, whereas biodiesel blending for transport vehicles has an ambitious mandate to reduce the consumption of petroleum fuels. IPCC (2011) suggested 16 to 74 g of CO_2/kWh emission from most of the commercial bioenergy technologies.

17.3.4 Ocean Energy

Ocean energy refers to consolidated forms of renewable energy obtained from the sea. Wave, tidal, and ocean temperature are the main types of ocean technologies that are utilized for electricity production at the current time. Tidal energy conversion is somewhat similar to traditional hydroelectric power plants in which the difference between the low and high tides is utilized for electricity production. Tidal energy production needs a dam or barrage in the shallow area of the seashore (bay, estuary) where the low and high tide difference is at least 5 m. Ocean thermal energy is produced by converting the temperature difference between the upper layer of the sea and the deep layer of the sea, because the temperature difference (20°C) generates a potential that can be utilized to run a generator for electricity production. The ocean thermal energy conversion plant may be land based or floating on the sea (Polis et al., 2017; Melikoglu, 2018). The 254 MW tidal power station in Sihwa Lake of South Korea, the La Rance tidal power plant in France (240 MW), Annapolis Royal in Nova Scotia, Canada (20 MW), and the Jiangxia tidal power plant in China (4.1 MW) are the main tidal power plants in various parts of the globe (Li and Pan, 2017). The theoretical annual capacity of the ocean or marine energy is 4 to 18 million tonnes of oil equivalent per year, and the global exploitation potential of ocean power is 337 GW (Derakhshan et al., 2017; Melikoglu, 2018). The cost of ocean energy conversion systems is high compared with other renewable energy technologies, which is hindering the commercialization of ocean energy technologies (Contestabile et al., 2017). Direct emission of GHGs is not seen with ocean energy, just as with other renewable energy technologies (solar and wind), but life cycle assessment studies have suggested that the raw material extraction, component manufacturing, construction, maintenance, and dismantling activities have recorded GHGs emissions of less than 23 g CO_2eq/kWh with wave and tidal energy systems (IPCC, 2011).

17.3.5 Geothermal Energy

Geothermal energy is thermal energy that is present in the geothermal fluids in the form of heat (Dincer and Ozcan, 2018). Geothermal energy can be directly utilized or can be further processed for the generation of electricity. With every 100 m descent into the earth's crust, there is an average temperature rise of 2–3°C. A geothermal system comprises three main elements: the first is the heat source, the second is the reservoir, and the third is the heat fluid or geothermal liquid. The heat source is present at a 5–10 km depth having a temperature greater than 600°C. Geothermal water contains a marginal fraction of dissolved minerals and ions such as calcium, iron, chlorides, sulfates, etc. The concentration of these minerals depends on the underlying rock type (Sanyal, 2005).

As a novel energy source, geothermal energy has various advantages and disadvantages. It is seen that geothermal systems emit few emissions during the extraction of energy and are constant and reliable sources of thermal energy for practical use. The pumps used for energy extraction require only 25–50% of the energy for heating and cooling in comparison to conventional energy systems, require less space for installation, and have a minimal maintenance cost. Various disadvantages include the release of GHGs and others such as CO_2, ammonia, methane, and hydrogen sulfide. Installation costs are high, and the technology is still in the developmental phase for tapping deep underground reservoirs of magmatic rock fluids. The Geysers Complex in the USA is the world's largest geothermal site, having 22 operational plants with a combined capacity of 1520 MW. Lardarello Complex, Italy has a combined capacity of 790 MW. This plant generates 10% of the world's total geothermal energy. In Mexico, Cerro Prieto Geothermal Complex is in third place, producing 720 MW of geothermal energy. This is the largest geothermal power plant in the world (Soltani et al., 2019).

17.3.6 Wind Energy

Wind energy is among the primitive forms of energy that can be harnessed by the utilization of technology. Sailing and flying a kite are conventional applications of wind energy, but now the wind energy application scenario is changed; it has various modern applications, such as wind energy farms for power production and water pumping with the help of electrical and mechanical tools. The utilization of wind energy for power production is the main application at the current time. Wind energy harnesses the kinetic energy of air, and the prime application of energy for mitigation of climate change is to produce electricity, because the GHG emission intensity of wind energy ranges from 8 to 20 g CO_2/kWh, which is very low, while payback times are 3.4 to 8.5 months (IPCC, 2011). Land (onshore) and sea (offshore) wind turbines are heavily deployed worldwide for energy production. Wind energy power generation has ambitious expansion plans, and the wind power production scenario will range from 251 to 392 GW by 2030 (EWEA, 2015). The top ten leading countries share 90% of total wind power installation in the world, as is mentioned in Figure 17.2, and the remaining 10% is shared by the rest of the countries in 2017. China has the largest (41%) wind power installation, and the USA, Germany, India, Spain, and the UK share 20%, 12%, 7%,

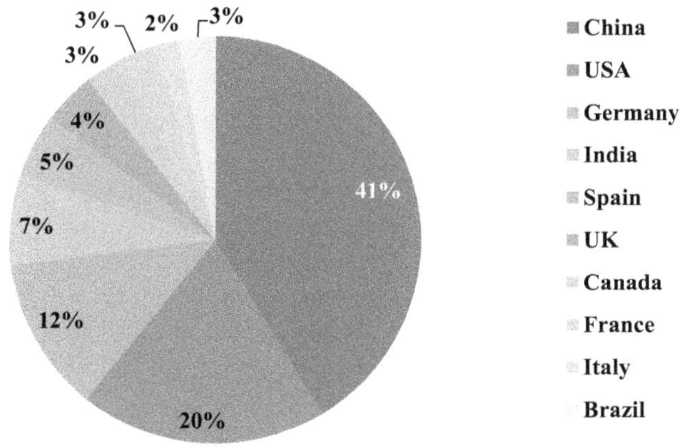

FIGURE 17.2 Wind power installation share of top 10 countries.

(From Chaurasiya et al., 2019.)

5%, and 4%, respectively, of wind energy installation in 2017 (Chaurasiya et al., 2019). Germany is the leading country in Europe with 6.5 GW of wind power. India is the leading country in Asia after China. Africa, Latin America, Brazil, and the Middle East are other important wind power producers.

Wind power production depends on the atmospheric conditions, and climate change can easily alter the large-scale flow of wind at the local level, such as stratification or surface roughness. For the planning of wind power production, it is important to know about the complete climate profile of wind resources of a particular place and the need for energy and storage. The climate profile has temperature, rain, solar radiation, and wind flow records that help to analyze the suitability of a particular place for wind power production.

Wind power production does not significantly impact the environment, because wind turbines do not release GHGs, which are the main cause of climate change. Wind turbines do not use water for cooling, so they have a very low water footprint. Regional noise pollution is created by wind turbines, which has a low impact. However, ecological concerns are also raised due to wind power plants. Bird collision with wind turbines is a major ecological challenge, and electrocution by transmission lines is another challenge seen in the various region of the world. Ecologically sensitive sites and hot spots of migratory birds must be avoided when installing wind turbines. Great Indian Bustard collisions with transmission lines are seen in the Indian states of Gujarat, Rajasthan, and Maharashtra (Wildlife Institute of India, 2018). Burning wind turbines that create fire in the fields and forest is a major emerging ecological concern.

17.4 IMPACT OF CLIMATE CHANGE ON RENEWABLE ENERGY GENERATION

In order to mitigate the impacts of climatic change, there is a need to lower the CO_2 concentration, and renewable energy resources would play a crucial role in the low-carbon future. To achieve a rise of below 2°C compared with pre-industrial levels, countries need to shift towards renewable sources, with the contribution of renewables to the energy sector to as high as 65% by 2050 (Solaun and Cerdá, 2019). But there are certain implications that climate change poses for renewables, including reliability and performance of the energy system (Cronin et al., 2018).

Climate change will affect demand and supply in the energy sector. Renewable resources can also be impacted by different aspects of regional climatic change such as rainfall, wind speed, and air temperature. Climate change is already impacting storm frequencies, drought frequencies, etc. Climate change will affect electricity generation, as increasing temperatures will affect the cooling and heating loads. Electricity generation will also be affected by increased temperature and a decrease in rainfall. Globally, the growing energy demands and their relationship to climate change have motivated many researchers to investigate the impacts of climate change on energy systems.

Among the renewables, the energy or electricity produced by PV or wind turbines is greatly affected by meteorological parameters such as the amount of available solar radiation, cloud cover, wind speed, and wind direction. Future climate change will affect these meteorological conditions. The amount of these meteorological factors will increase or decrease, and this will directly affect renewable energy, which will change in line with these changes (Oka et al., 2020). The impact of climate change and the subsequent changes in the meteorological conditions is discussed in detail in the following subsections.

17.4.1 Impact of Climate Change on Hydroelectric Power Generation

Globally, hydropower generation accounts for roughly 16% of the energy sector. Climate change is expected to impact hydropower generation, as it is very likely to affect the water availability for power generation. Moreover, at present, power generator companies are generally not accounting for the impact of climate change in their development plans. That means they could be overestimating their ability to meet future electricity needs, and that mismatching may create a crisis in the power sector.

The future climate change scenarios will have different effects on water availability, as they will greatly affect rainfall in different regions of the globe. One study has stated that regional climate change has resulted in the decline of hydroelectric power production in Brazil's Sao Francisco basin (de Jong et al., 2019). In regard to hydropower generation in future climatic scenarios in Brazil, a hydropower model was developed for 52 climate scenarios in terms of changes in precipitation and temperature patterns. The results showed that a decrease in rainfall, independently of the temperature, will compromise the hydroelectric plant operation (Hidalgo et al., 2020).Water availability affects electricity production in both hydroelectric and thermal power plants. The extent to which water availability altered by climate change will affect the power generation in the western USA was evaluated. The outcome of the study suggested that the inter-annual changes in net power generation are directly connected to local changes in water availability (Voisin et al., 2020). In order to assess the impact of climate change on hydropower generation in the USA, a monthly water systems model was employed, based on multiple climate models and the emission scenarios. The results showed an increase in future hydroelectric power production due to climate change (Boehlert et al., 2016). A study by Turner et al. (2017) also states that the hydropower potential will increase by 5–20% in most of the high-latitude countries like Canada, Russia, northern Europe, and northeast China, while a decrease of the same percentage is predicted for southern Europe, the southern USA, southeast China, and southern South America.

In India, all six units of the Parli thermal power plant in Beed district of Maharashtra have been shut down because of severe water shortage in the Marathwada region. Four hydroelectric power plants along with two mega hydroelectric plants of 1500 MW at NathpaJhakri and 412 MW at Rampur, besides 300 MW at Baspa and 120 MW at Bhabha, have been temporarily shut down due to a rise in silt levels in the Satluj river (CEA ANNUAL REPORT 2019–20).

India's hydropower plants have been badly stricken as well in past years due to climate change. Figures 17.3 and 17.4 show the yearly growth of installed capacity of hydroelectric power and yearly energy generated by them in India. From 2014 to 2021, even as 12.67% of new hydro capacity was installed, hydropower generation increased by around 10% only. Moreover, between 2014 and 2018, even as 10.84% of new hydro capacity was installed, the hydropower generation dropped by around 7% (CEA ANNUAL REPORT 2019–20).

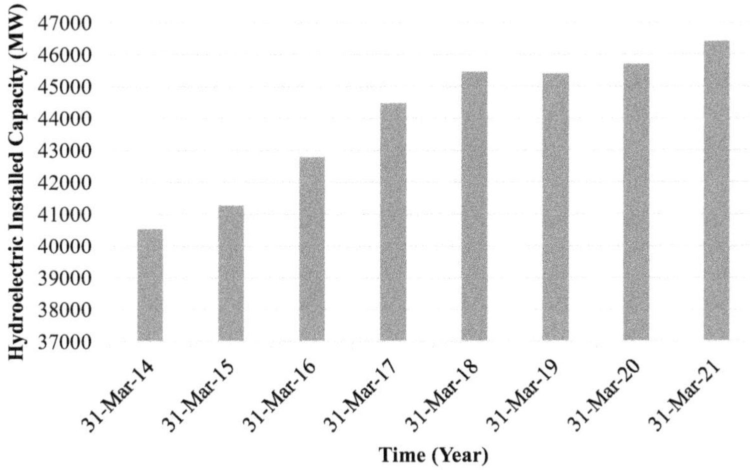

FIGURE 17.3 Yearly growth in hydroelectric installed capacity in India.
(From CEA ANNUAL REPORT 2019–20.)

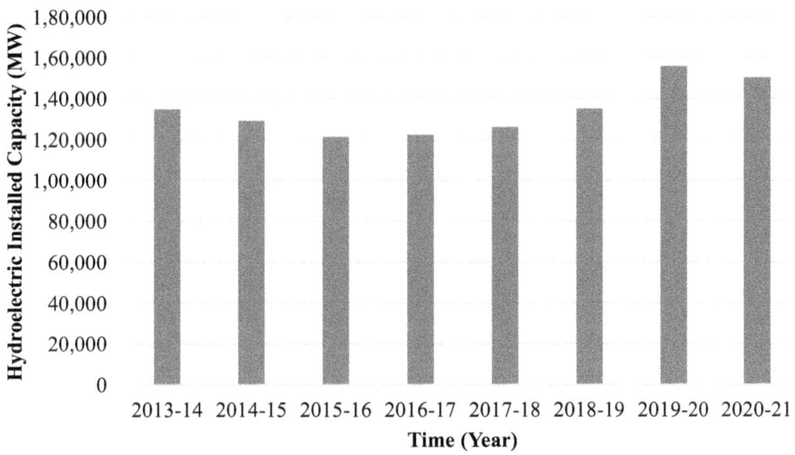

FIGURE 17.4 Yearwise hydroelectric power generation in India.
(From CEA ANNUAL REPORT 2019–20.)

17.4.2 Impact of Climate Change on Solar Energy Generation

Climate change may impart positive effects on renewable resources depending upon the geographic conditions. A South America-based regional climate model was used to estimate the impact of climate change on solar and wind energy in Brazil. The solar energy potential could also experience the same trend at some locations, particularly in the northeast and southeast regions (de Jong et al., 2019). As per the study carried out by Zhao et al. (2020), climate change is expected to badly affect PV energy potential in China, prompting the Chinese government to pay more attention towards the development of PV power generation through improvements in PV technologies and wise site selection for the PV-based power generation. To estimate the impact of climate change on the renewable energy generation sector in Japan, three Representative Concentration Pathways (RCP) and seven Global Climate Models (GCM) were analyzed. The results showed an average increase in PV energy generation by an estimated value of about 1.7% in 2030. This value showed a further increase in average PV energy generation by 3.9% in 2050 and 4.9% in 2070 as a result of the ongoing climate change scenario (Oka et al., 2020). As far as the European power system is concerned, the share of solar power in the cost-optimal power system would increase significantly by around 5% (Schlott et al., 2018).

17.4.3 Impact of Climate Change on Wind Energy Production

Expanding wind energy generation is a key factor to mitigate climate change and to make a low-carbon future. There is a growing interest in how the changing climate impedes wind power generation in the energy sector. The future climate change scenario is projected to affect wind energy resources both regionally and seasonally. Climate change is projected to cause positive as well as negative impacts on wind energy generation. The installed capacity of wind turbines grew at a rate of more than 20% between 2000 and 2019 and is further projected to rise by 50% by 2023. There is evidence of increased wind power generation in Europe and in the great plains of the USA (Pryor et al., 2020). A South American regional climate model used to assess the impact of climate change on wind energy in Brazil showed that wind power generation could increase by 40% at some wind farm locations (de Jong et al., 2019). To explore the effect of climate change on wind power generation across the Caspian Sea, climate modelling using an adaptive neuro-fuzzy inference system was employed, showing a slight decrease in the projected future (Nabipour et al., 2020). Temporal variability is also projected, with the wind power generation to decrease in the winter and summer, while for spring and autumn, these changes remain negligible. The projected annual wind power generation in Japan showed a decrease of ~5% in response to climate change (Ohba, 2019).

17.5 IMPROVEMENTS IN RENEWABLE ENERGY TECHNOLOGIES FOR SUSTAINABLE DEVELOPMENT

Environmental issues such as climate change and global warming have triggered the major focus on renewable energy. Renewable energy sources including biomass, wind, solar, tidal, hydro, etc. have shown technological advancements in recent years. Different strategies are being employed for improvement in the renewable energy sector. Strategies include energy saving, increasing efficiency, and replacing fossil fuels with renewable sources like biofuels (Vivek et al., 2021). There has been a sharp rise in the demand for crude oil to meet the ever-growing energy usage. The production of liquid biofuels such as bioethanol, biodiesel, and gaseous biofuels like biogas has emerged as a better, cleaner, and sustainable alternative that has the potential to eliminate maximum consumption of fossil-based fuels. Different technologies for the hydrolysis of lignocellulosic biomass are used that can efficiently digest cellulose and hemicellulose. Apart from acid or alkali hydrolysis, enzymatic hydrolysis is being encouraged for the dissolution of polysaccharides into their respective monosaccharides. Hydrolytic enzymes are also engineered to enhance hydrolysis. Agricultural waste such as bagasse, rice husk, corn straw, etc. is being worked upon at lab scale to produce bioethanol (Rastogi and Shrivastava, 2017). The efficiency of vehicular engines is being enhanced for running on biofuels. There is a strong race going on to develop more efficient electric vehicles: they are being equipped with modern batteries that last longer in a single charge; hybrid technologies such as PV systems are being used to aid battery charging; heat pump technology is being utilized for heating modern cabins. It is expected that the use of these hybrid systems can reduce vehicular emissions by 75% (Tomura and Nakagawa, 2019). A hybrid solar and wind power plant may be more reliable than a single power plant because the sun is absent during night time and wind flow also fluctuates. Thus, in the hybrid power plant, each may cover the absence of the other, reducing the energy fluctuations. However, hybrid wind and solar power plant development largely depend on government policies.

The wind–solar hybrid power production in Romania has achieved 35% to 65% renewable energy utilization. The addition of battery storage is an advanced way to store surplus energy (Das et al., 2020). The stored energy can be utilized during high load demand.

Hydrogen fuel is being produced using different wastewaters; it is also produced through water splitting via different routes including electrolysis, thermolysis, photolysis, photo-fermentation, dark-fermentation, thermochemical routes such as pyrolysis, gasification, and liquefaction. However, at present, electrolysis is the only well-established hydrogen-generating technology available for commercial use (Martinez-Burgos et al., 2020). In the solar sector, most of the PV panels used presently are silicon based and show an average efficiency of only 20%. In 2019, data reveals the generation of 720 TWh of global electricity through PV systems, and this is expected to reach above 1900 TWh by the end of the year 2025. Thin solar films are emerging as the most promising PV systems; they are lightweight, flexible, and about 350 times smaller in comparison to traditional silicon-based panels. Thermo-radiative PV devices or reverse solar panels are among the recent advancements in the solar sector, which have the potential of generating electricity at night. These thermo-radiative panels work through exploiting the radiated heat of solar panels. Transparent solar windows installed on green buildings can provide 10% conversion efficiency.

Furthermore, the key developments in the field of biofuels, including bioethanol, biodiesel, biohydrogen, biogas, wind, and solar energy, are also discussed in the upcoming section.

17.5.1 Advances in the Solar Energy Sector

Solar energy is utilized as photothermal energy for cooking, drying, and space/water heating, while in PV systems, the energy from the sun is directly converted into electricity. In PV, there are four mature technologies based on silicon-based semiconductors, thin-film cells, multijunction solar cells, and next-gen solar cells. Silicon-based solar panels are highly efficient and are used in monocrystalline as well as in multi-crystalline form. Film solar cells are a low-cost alternative to silicon cells. Multijunction solar cells can have a conversion efficiency of up to 40% but are costly in terms of manufacturing. Next-generation PV systems comprise kesterite cells, dry sensitized solar cells, perovskite, and quantum dot solar cells. Green buildings being equipped with PV systems on rooftops and window glasses is also a new trend (Rashmi et al., 2021).

17.5.2 Biofuels

Advancements in bioethanol, biohydrogen, biodiesel and biogas are discussed as:

17.5.2.1 Recent Advances in Bioethanol

Recent advancements in bioethanol production have primarily focused on technological development to achieve an increase in ethanol yield and to minimize the processing cost as well as the steps used in the process (Zabed et al., 2017). Many pre-treatment methods are applied for better hydrolysis to produce fermentable sugars, but these methods are very costly. Recently, an advancement in pre-treatment was carried out on the destruction of knotty links between cellulose, hemicellulose, and lignin by the liquid hot water treatment (LHW) method, which has a low cost and proved to be effective in hydrolysis yield (Jiang et al., 2016). Ammonia and steam pre-treatment are also effective in producing fermentable sugars. The chemical hydrolysis that leads to the formation of fermentable sugars is basically a toxic, corrosive, and dangerous method. Therefore, the advancement of hydrolysis is important for the bioethanol industry. Advancement in the fermentation of hexose and pentose sugars is quite challenging, since most of the fermenting microbes have a temperature range of 28 to 30°C, while cellulolytic enzyme activity is highest around 50°C. Therefore, genetically modified microbes growing at 50–55°C could enhance the process of simultaneous saccharification and fermentation.

17.5.2.2 Recent Advances in Biodiesel

Biodiesel is a promising fuel having the potential to replace fossil-based diesel. Various ionic liquids, being non-volatile and soluble in organic as well as inorganic materials and environmentally friendly, are the best alternatives to replace alkali catalysts. Therefore, the recent development in biodiesel production concentrates on using ionic liquids by deposition on nanoporous materials as catalysts for biodiesel production. Various hybrids of ionic liquids and nanoporous materials such as nanoporous silica, polymers, and carbonaceous materials have been used for catalyzing the esterification reactions. The results have been satisfactory. However, despite the high yield of biodiesel, their high cost in comparison to conventional catalyst remained a challenge.

17.5.2.3 Recent Advances in Biohydrogen

Recent advancement in biohydrogen production mainly concentrates on hydrogen production through the photo-biological route, which has various advantages as compared with gasification and thermochemical techniques. Photo-biological hydrogen production is a capable method for the generation of clean biohydrogen that utilizes different microbial strains (algae and bacteria) for converting solar energy to hydrogen (Sharma and Arya, 2019). It is being examined as a promising technique, but it still needs more advancements and scientific research.

17.5.2.4 Recent Advances in Biogas

The generation of biogas achieved via the anaerobic route provides sustainable bioenergy but contains different impurities in the form of nitrogen, hydrogen, H_2S, CO_2, and water vapor, which lead to a lower calorific value in comparison to natural gas. So, the recent advancement in biogas production focuses on the different cleaning as well as upgrading techniques to enhance the quality of biogas. The advanced biogas upgrading techniques used are hydrate and cryogenic separation, membrane enrichment, multistage high-pressurized anaerobic digestion, and the in-situ upgrading route.

17.5.3 Advances in the Wind Energy Sector

At present, power generation from wind is chiefly based on ground-gen and fly-gen concepts; in ground-gen, processed chemical energy is subjected to power generation on a ground-mounted dynamo, while in the case of fly-gen, the dynamo is placed just back of the rotor. These concepts are proven to be low cost and easy to maintain (Watson et al., 2019). Drone and multi-drone technologies have emerged recently that can generate power at higher elevations due to fast winds. Studies suggest sustained drone models can generate up to 1 MW of energy (Wijnja et al., 2018). The multirotor wind turbine system diffuses load over a single turbine by the placement of multiple rotors on a single turbine. Magnetic levitation models use permanent magnets to raise the blades of the rotor for reducing friction (Kozlov et al., 2016).

17.6 CHALLENGES AND PROSPECTS

Policies to mitigate climate change directly or indirectly emphasize the utilization of renewable energy along with phasing out the use of fossil fuels. The main targets of any climate change policy are phasing out fossil fuels, carbon pricing, emission trading, and reduction in GHG emission. Different countries have launched different policies regarding climate change and renewable energy utilization. Various countries have launched a program to enhance the use of renewable energy and to achieve net zero emissions, like Costa Rica, France, and the UK by 2050. Different countries have also announced the shutting down of coal-fired power plants to mitigate the climate change problem. Germany announced that it would shut down coal-fired power plants by 2038, Chile by 2040, and the US states of New Mexico and Washington by 2031 and 2025, respectively (Renewables 2020 Global Status Report). Worldwide, more focus is on the generation of renewable energies to achieve the

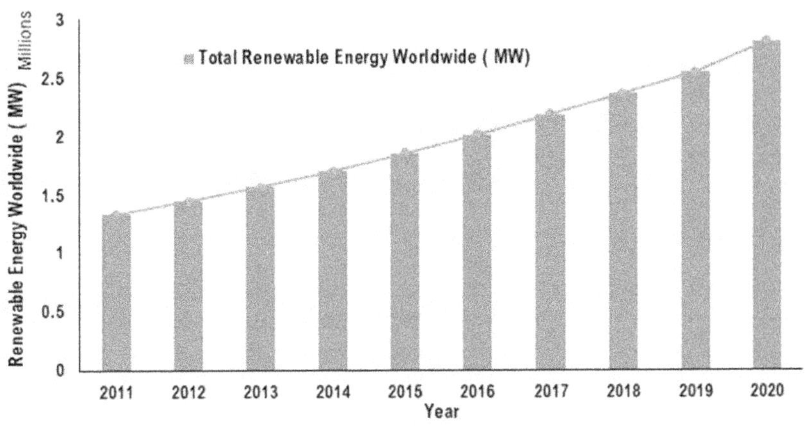

FIGURE 17.5 Total renewable energy production (MW) from 2011 to 2020.
(From IRENA, 2021.)

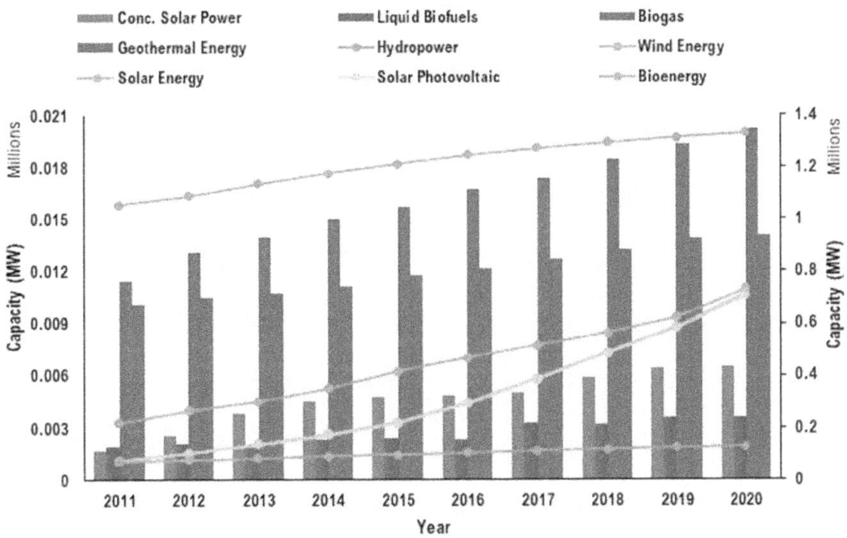

FIGURE 17.6 Global status of different forms of renewable energy production (MW) from 2011 to 2020.
(From IRENA, 2021.)

goals of climate change policies, as shown by Figure 17.5. Much focused research is going towards enhancement of the efficiency and efficacy of different types of renewable energies. Figure 17.6 shows the global status of solar energy, geothermal energy, hydropower, liquid biofuels, bioenergy, wind energy, biogas, etc.

There are various challenges associated with renewable energy development and its applications. The complete reliance on renewable energy is still a critical issue, because a consistent supply of solar and wind resources is not possible; this can be solved by the integration of renewable energy on one platform or an integrated renewable energy grid that can supply consistent power with high power quality. Due to lack of information, the benefits of renewable energy do not reach remote and rural areas, which requires the dissemination of proper information in society. Social challenges are observed in developing countries like India, where biogas from human waste is not accepted for domestic applications, especially cooking use. High cost is a major issue with renewable energy development that needs to be reduced by innovation and lucrative government policies. Renewable energy technologies have various advantages as well as various challenges that have hindered the development of renewable energy sources.

17.7 CONCLUSION

The utilization of fossil fuels creates various environmental problems on the earth and leads to climate change, which can be seen in the form of global warming, floods, and droughts. The importance of renewable energy sources cannot be ignored, because they have low GHGs emissions that can help to reduce the climate change impact in the current time when the demand for clean energy is high. Energy options like solar, hydro, wind, ocean, and biomass are developed at the commercial level to supply reliable power. Cost is a major challenge at the current time for the development of various renewable energy generation facilities, but solar and wind power prices are showing that cost will not be a big challenge in the future. The advancement in technologies and renewable energy governing policies may be the main solutions for the sustainable adoption of renewable energy.

ACKNOWLEDGEMENT

The authors are grateful to the Department of Science and Technology (Government of India), for proving financial support of project (DST/INT/AUS/P-76/2020) "Thermal Energy Storage for Food/ Grain Drying with CST/RE to Lower Pollution".

REFERENCES

Bauen, A., Berndes, G., Junginger, M., Londo, M., Vuille, F., Ball, R., Bole, T., Chudziak, C., Faaij, A. & Mozaffarian, H., 2009. Bioenergy: A Sustainable and Reliable Energy Source. A Review of Status and Prospects. www.ieabioenergy.com/wp-content/uploads/2013/10/MAIN-REPORT-Bioenergy-a-sustainable-and-reliable-energy-source-A-review-of-status-and-prospects.pdf

Boehlert, B., Strzepek, K. M., Gebretsadik, Y., Swanson, R., McCluskey, A., Neumann, J. E., McFarland, J., & Martinich, J. 2016. Climate change impacts and greenhouse gas mitigation effects on U.S. hydropower generation. *Applied Energy,* 183, 1511–1519. https://doi.org/10.1016/j.apenergy.2016.09.054.

Burillo, D., Chester, M. V., Pincetl, S., Fournier, E. D., & Reyna, J. 2019. Forecasting peak electricity demand for Los Angeles considering higher air temperatures due to climate change. *Applied Energy*, 236, 1–9. https://doi.org/10.1016/j.apenergy.2018.11.039

CEA ANNUAL REPORT. 2019–20. Central Electricity Authority. Government of India, Ministry of Power. https://cea.nic.in/annual-report/?lang=en

Chaurasiya, P. K., Warudkar, V., & Ahmed, S., 2019. Wind energy development and policy in India: a review. *Energy Strategy Reviews*, 24, 342–357.

Contestabile, P., Iuppa, C., Di Lauro, E., Cavallaro, L., Andersen, T. L., & Vicinanza, D. 2017. Wave loadings acting on innovative rubble mound breakwater for overtopping wave energy conversion. *Coastal Engineering,* 122, 60–74. https://doi.org/10.1016/j.coastaleng.2017.02.001.

Cronin, J., Anandarajah, G., & Dessens, O., 2018. Climate change impacts on the energy system: a review of trends and gaps. *Climatic Change*, 151, 79–93.

Das, A., Jani, H. K., Nagababu, G., & Kachhwaha, S. S. 2020. A comprehensive review of wind–solar hybrid energy policies in India: barriers and recommendations. *Renewable Energy Focus*, 35, 108–121.

de Jong, P., Barreto, T. B., Tanajura, C. A., Kouloukoui, D., Oliveira-Esquerre, K. P., Kiperstok, A., & Torres, E. A. 2019. Estimating the impact of climate change on wind and solar energy in Brazil using a South American regional climate model. *Renewable Energy*, 141, 390–401. https://doi.org/10.1016/j.renene.2019.03.086

Dincer, I., & Ozcan, H. 2018. 1.17 Geothermal Energy. *Comprehensive Energy Systems*, 1, 702–732. https://doi.org/10.1016/B978-0-12-809597-3.00119-X

EWEA. August 2015. *Wind Energy Scenarios for 2030*. European Wind Energy Association, Brussels.

Fan, J. L., Hu, J. W., & Zhang, X. 2019. Impacts of climate change on electricity demand in China: an empirical estimation based on panel data. *Energy*, 170, 880–888. https://doi.org/10.1016/j.energy.2018.12.044

Hidalgo, I. G., Paredes-Arquiola, J., Andreu, J., Lerma-Elvira, N., Lopes, J. E. G., & Cioffi, F. 2020. Hydropower generation in future climate scenarios. *Energy for Sustainable Development*, 59, 180–188. https://doi.org/10.1016/j.esd.2020.10.007

IEA. 2018. *World Energy Outlook 2018*. www.iea.org/reports/world-energy-outlook-2018

IPCC. 2011. Summary for Policymakers. In *IPCC Special Report on Renewable Energy Sources and Climate Change Mitigation* (O. Edenhofer, R. Pichs-Madruga, Y. Sokona, K. Seyboth, P. Matschoss, S. Kadner, T. Zwickel, P. Eickemeier, G. Hansen, S. Schlomer & C. von Stechow, eds), Cambridge University Press, Cambridge, UK and New York.

IRENA. 2021. *Renewable Capacity Statistics 2021*. International Renewable Energy Agency (IRENA), Abu Dhabi. www.irena.org/-/media/Files/IRENA/Agency/Publication/2021/Apr/IRENA_RE_Capacity_Statistics_2021.pdf

Jain, A., Das, P., Yamujala, S., Bhakar, R. & Mathur, J., 2020. Resource potential and variability assessment of solar and wind energy in India. *Energy*, 211, 118993.

Jiang, W., Chang, S., Qu, Y., Zhang, Z., & Xu, J. 2016. Changes on structural properties of biomass pretreated by combined deacetylation with liquid hot water and its effect on enzymatic hydrolysis. *Bioresource Technology*, 220, 448–456. https://doi.org/10.1016/j.biortech.2016.08.087

Jin, S. W., Li, Y. P., Nie, S., & Sun, J. 2017. The potential role of carbon capture and storage technology in sustainable electric-power systems under multiple uncertainties. *Renewable and Sustainable Energy Reviews*, 80, 467–480. https://doi.org/10.1016/j.rser.2017.05.230.

Kafka, J. L., & Miller, M. A. 2019. A climatology of solar irradiance and its controls across the United States: implications for solar panel orientation. *Renewable Energy*, 135, 897–907.

Kang, J. N., Wei, Y. M., Liu, L. C., Han, R., Yu, B. Y., & Wang, J. W. 2020. Energy systems for climate change mitigation: a systematic review. *Applied Energy*, 263, 114602.

Kothari, R., Vathistha, A., Singh, H. M., Pathak, V. V., Tyagi, V. V., Yadav, B. C., Ashokkumar, V., & Singh, D. P. 2020. Assessment of Indian bioenergy policy for sustainable environment and its impact for rural India: strategic implementation and challenges. *Environmental Technology & Innovation*, 20, 101078.

Kozlov, S. V., Sirotkin, E. A., & Solomin, E. V. May 2016. Wind turbine rotor magnetic levitation. In *2016 2nd International Conference on Industrial Engineering, Applications and Manufacturing (ICIEAM)* (pp. 1–4). IEEE Xplore, INSPEC Accession Number: 16838924, Conference Location: Chelyabinsk, Russia. https://doi.org/10.1109/ICIEAM.2016.7911477

Kuriqi, A., Pinheiro, A. N., Sordo-Ward, A., Bejarano, M. D., & Garrote, L. 2021. Ecological impacts of run-of-river hydropower plants—current status and future prospects on the brink of energy transition. *Renewable and Sustainable Energy Reviews*, 142, 110833.

Lamichhane, G., Acharya, A., Poudel, D. K., Aryal, B., Gyawali, N., Niraula, P., … & Parajuli, N. 2021. Recent advances in bioethanol production from lignocellulosic biomass. *International Journal of Green Energy*, 18(7), 731–744. https://doi.org/10.1080/15435075.2021.1880910

Li, Y., & Pan, D.-Z. 2017. The ebb and flow of tidal barrage development in Zhejiang Province, China. *Renewable and Sustainable Energy Reviews*, 80, 380–389. https://doi.org/10.1016/j.rser.2017.05.122.

Martinez-Burgos, W. J., Sydney, E. B., de Paula, D. R., Medeiros, A. B. P., de Carvalho, J. C., Soccol, V. T., de Souza Vandenberghe, L. P., Woiciechowski, A. L., & Soccol, C. R. 2020. Biohydrogen production in cassava processing wastewater using microbial consortia: process optimization and kinetic analysis of the microbial community. *Bioresource Technology*, 309, 123331. https://doi.org/10.1016/j.biortech.2020.123331

Mei, H., Li, Y. P., Suo, C., Ma, Y., & Lv, J. 2020. Analyzing the impact of climate change on energy-economy-carbon nexus system in China. *Applied Energy*, 262, 114568.

Melikoglu, M. 2018. Current status and future of ocean energy sources: a global review. *Ocean Engineering*, 148, 563–573.

MNRE. 2019. MNRE Annual Report 2018–2019. https://mnre.gov.in/img/documents/uploads/file_f-1608040317211.pdf, accessed on September 10, 2021.

Mousavi Maleki, S. A., Hizam, H., & Gomes, C. 2017. Estimation of hourly, daily and monthly global solar radiation on inclined surfaces: models re-visited. *Energies*, 10(1), 134.

Nabipour, N., Mosavi, A., Hajnal, E., Nadai, L., Shamshirband, S., & Chau, K. W. 2020. Modeling climate change impact on wind power resources using adaptive neuro-fuzzy inference system. *Engineering Applications of Computational Fluid Mechanics*, 14(1), 491–506. https://doi.org/10.1080/19942060.2020.1722241

Ohba, M. 2019. The impact of global warming on wind energy resources and ramp events in Japan. *Atmosphere*, 10(5), 265.

Oka, K., Mizutani, W., & Ashina, S. 2020. Climate change impacts on potential solar energy production: a study case in Fukushima, Japan. *Renewable Energy*, 153, 249–260. https://doi.org/10.1016/j.renene.2020.01.126

Olczak, P., Matuszewska, D., & Zabagło, J., 2020. The comparison of solar energy gaining effectiveness between flat plate collectors and evacuated tube collectors with heat pipe: case study. *Energies*, 13(7), 1829.

Pathak, A. K., Chopra, K., Singh, H. M., Tyagi, V. V., Kothari, R., Anand, S., & Pandey, A. K. 2019. Role of Solar Energy Applications for Environmental Sustainability. In *Environmental Biotechnology: For Sustainable Future* (pp. 341–374), R. C. Sobti, N. K. Arora & R. Kothari (eds), Springer, Singapore.

Polis, H. J., Dreyer, S. J., & Jenkins, L. D. 2017. Public willingness to pay and policy preferences for tidal energy research and development: a study of households in Washington State. *Ecological Economics,* 136, 213–225. https://doi.org/10.1016/j.ecolecon.2017.01.024.

Pryor, S. C., Barthelmie, R. J., Bukovsky, M. S., Leung, L.R., & Sakaguchi, K. 2020. Climate change impacts on wind power generation. *Nature Reviews Earth & Environment,* 1, 627–643. https://doi.org/10.1038/s43017-020-0101-7

Rashmi, W., Mahesh, V., Anirban, S., Sharnil, P., & Khalid, M. 2021. Hybrid solar PVT systems for thermal energy storage: role of nanomaterials, challenges, and opportunities. In: *Energy Systems and Nanotechnology. Advances in Sustainability Science and Technology*, D. Tripathi & R. K. Sharma (eds), Springer, Singapore. https://doi.org/10.1007/978-981-16-1256-5_9

Rastogi, M., & Shrivastava, S. 2017. Recent advances in second generation bioethanol production: an insight to pretreatment, saccharification and fermentation processes. *Renewable and Sustainable Energy Reviews*, 80, 330–340. https://doi.org/10.1016/j.rser.2017.05.225

Rathore, N., Panwar, N.L., Yettou, F., & Gama, A. 2021. A comprehensive review of different types of solar photovoltaic cells and their applications. *International Journal of Ambient Energy*, 42, 1200–1217.

Renewables. 2020. Global Status Report. www.ren21.net/wp-content/uploads/2019/05/gsr_2020_full_report_en.pdf, accessed on September 21, 2021.

Sanyal, S. K. January 2005. Classification of geothermal systems–a possible scheme. In *Proceedings of the 13th Workshop on Geothermal Reservoir Engineering, Stanford, California* (pp. 85–88).

Schlott, M., Kies, A., Brown, T., Schramm, S., & Greiner, M. 2018. The impact of climate change on a cost-optimal highly renewable European electricity network. *Applied Energy*, 230, 1645–1659. https://doi.org/10.1016/j.apenergy.2018.09.084

Sharma, A., & Arya, S. K. 2019. Photobiological production of biohydrogen: recent advances and strategy. In: Rastegari, A., Yadav, A., Gupta, A. (eds) *Prospects of Renewable Bioprocessing in Future Energy Systems*, Springer, Cham, 10, 89–116. https://doi.org/10.1007/978-3-030-14463-0_3

Shiji, C., Dhakal, S., & Ou, C. 2021. Greening small hydropower: a brief review. *Energy Strategy Reviews*, 36, 100676.

Singh, H. M., Kothari, R., Gupta, R., & Tyagi, V. V. 2019. Bio-fixation of flue gas from thermal power plants with algal biomass: overview and research perspectives. *Journal of Environmental Management*, 245, 519–539.

Sinha, S. K., Subramanian, K. A., Singh, H. M., Tyagi, V. V., & Mishra, A. 2019. Progressive trends in bio-fuel policies in India: targets and implementation strategy. *Biofuels*, 10(1), 155–166.

Solaun, K., & Cerdá, E. 2019. Climate change impacts on renewable energy generation. A review of quantitative projections. *Renewable and Sustainable Energy Reviews*, 116, 109415.

Soltani, M., MoradiKashkooli, F., Dehghani-Sanij, A. R., Nokhosteen, A., Ahmadi-Joughi, A., Gharali, K., … & Dusseault, M. B. 2019. A comprehensive review of geothermal energy evolution and development. *International Journal of Green Energy*, 16(13), 971–1009. https://doi.org/10.1080/15435075.2019.1650047

Tomura, Y., & Nakagawa, T. 2019. Effective utilization system of renewable energy through the use of vehicle. *Energy Procedia*, 158, 3000–3007. https://doi.org/10.1016/j.egypro.2019.01.972

Turner, S. W. D., Yi, J., & Galelli, S. 2017. Examining global electricity supply vulnerability to climate change using a high-fidelity hydropower dam model. *Science of The Total Environment,* 591, 663–675.

Ucal, M., & Xydis, G. 2020. Multidirectional relationship between energy resources, climate changes and sustainable development: technoeconomic analysis. *Sustainable Cities and Society*, 60, 102210. https://doi.org/10.1016/j.scs.2020.102210

Vivek, C. M., Ramkumar, P., Srividhya, P. K., & Sivasubramanian, M. 2021. Recent strategies and trends in implanting of renewable energy sources for sustainability–A review. *Materials Today: Proceedings*, 46, 8204–8208. https://doi.org/10.1016/j.matpr.2021.03.208

Voisin, N., Dyreson, A., Fu, T., O'Connell, M., Turner, S. W., Zhou, T., & Macknick, J. 2020. Impact of climate change on water availability and its propagation through the Western US power grid. *Applied Energy*, 276, 115467. https://doi.org/10.1016/j.apenergy.2020.115467

Watson, S., Moro, A., Reis, V., Baniotopoulos, C., Barth, S., Bartoli, G., … & Wiser, R. 2019. Future emerging technologies in the wind power sector: a European perspective. *Renewable and Sustainable Energy Reviews*, 113, 109270. https://doi.org/10.1016/j.rser.2019.109270

Wijnja, J., Schmehl, R., De Breuker, R., Jensen, K., & Vander Lind, D. 2018. Aeroelastic analysis of a large airborne wind turbine. *Journal of Guidance, Control, and Dynamics*, 41(11), 2374–2385. https://doi.org/10.2514/1.G001663

Wildlife Institute of India. 2018. Power-Line Mitigation Measures. Second edition (2020). https://wii.gov.in/images/images/documents/species_recovery_programme/gib_power_line_mitigation.pdf, accessed on September 9, 2021.

Wiser, R., Millstein, D., Mai, T., Macknick, J., Carpenter, A., Cohen, S., Cole, W., Frew, B., & Heath, G. 2016. The environmental and public health benefits of achieving high penetrations of solar energy in the United States. *Energy*, 113, 472–486.

Zabed, H., Sahu, J. N., Suely, A., Boyce, A. N., & Faruq, G. 2017. Bioethanol production from renewable sources: current perspectives and technological progress. *Renewable and Sustainable Energy Reviews*, 71, 475–501. https://doi.org/10.1016/j.rser.2016.12.076

Zhao, X., Huang, G., Lu, C., Zhou, X., & Li, Y. 2020. Impacts of climate change on photovoltaic energy potential: a case study of China. *Applied Energy*, 280, 115888.

18 Technology and Innovation to Tackle Climate Challenges

Reshu Chaudhary
Department of Biotechnology, KVSCOS, Swami Vivekanand Subharti University, Meerut

R.S. Sengar
Department of Agriculture Biotechnology, College of Agriculture, Sardar Vallabhbhai Patel University of Agriculture and Technology, Meerut

CONTENTS

18.1 INTRODUCTION

Global warming and climate have emerged as a major global concern, cutting across geographical and national boundaries. There are several physical and biological evidences of climate change and variability. The number and frequencies of extreme rainfall, floods, droughts and cyclone have increased substantially, especially during the last 50 years. The Intergovernmental Panel on

DOI: 10.1201/9781003220824-21

Climate Change (IPCC) in its fourth assessment report has clearly and confidently identified several anthropogenic activities as governing factors for the substantial increase in the concentrations of greenhouse gases (GHGs) in the atmosphere, which are responsible for the rising global atmospheric temperature. Even if all future emissions of GHGs are taken care of, still a further warming of about 0.1°C per decade would be expected through the 'enhanced greenhouse effect'. The last 50 years have been reported to be the warmest in the last 1300 years, and 11 of the last 12 years have been reported to be the warmest since 1980. The average temperature of the Earth is 15°C, which has increased by 0–5°C in the last 100 years. Accordingly, it should have increased by 0.005°C per year in this century, but in 1998 it saw an increase of 0.17°C. At a temperature of 32°C, life becomes difficult. According to a report by the United Nations Educational, Scientific and Cultural Organization (UNESCO)'s IPCC, by 2050 the Earth's temperature may increase by 1–3°C. This is a slight relief, but the problem should not be underestimated. On the one hand, man has created an infinite chain of industries and is doing this day by day. At the same time, he is rapidly cutting down the forests and filling the land, water and air with various types of pollution.

Global warming may cause melting of glaciers, which could cause floods followed by droughts in the near future. Water availability may become another severe social and agricultural problem due to greater loss of moisture through evapo-transpiration as well as more competition among the public, industrial and agriculture sectors. Coastal inundation due to sea level rise and water expansion may lead to greater loss of coastal biodiversity, especially mangrove plant/animal species responsible for coastal ecosystem protection.

Global warming may be useful for the polar region but harmful for temperate, subtropical and tropical regions. The increase in atmospheric CO_2 level has been reported to be beneficial for C_3 crops, but this does not show a positive effect of rising CO_2 level. Spread of endemic disease and insect pests, emergence of new biotypes, and disease and greater crop/weed competition may be other consequences of warming. All these agro-biological changes may result in a marked reduction in agricultural income due to low productivity and higher cost of production, which may lead to the migration of rural populations from vulnerable areas to other areas for their livelihood.

It is an established fact that various anthropogenic activities have given impetus to the rate of global warming, directly or indirectly. Thus, the option for reducing or minimizing the detrimental impact of global warming and climate change is adaptation, which refers to the ability of crops and cultivators to modify the existing biological and non-biological systems by, for example, shifting the optimum thermal range, inducing heat shock protein, development of heat-tolerant crop cultivars, rescheduling irrigation and fertilization, enhancing the support price of produce, crop insurance, susceptibility against climatic risk, etc., where the term 'susceptibility' refers to exposure, sensitivity and adaptive capacity (Metzger *et al.*, 2005). But all these modifications would lead to additional costs of adaptation for maintaining the ceiling of crop productivity and production.

Climate change today is given a high priority on the agenda of every government, non-governmental organization (NGO), corporate organization, scientist, civil society group, creative person, young farmer and ordinary person. This is reflected in the national and international media highlighting the various issues concerning the phenomenon of climate change and global warming. Considering the various apprehensions about the likely impacts of climate change on people, food security, ecosystem and livelihood, almost all nations are preparing to adapt themselves to this unprecedented global threat. The government of India has already formed the National Council for Climate Change to prepare a roadmap for energy efficiency and sustainable development and coordinate national action plans for impact assessment, adaptation and mitigation of climate change without compromising our food sovereignty.

18.2 GREENHOUSE EFFECT

Before the industrial revolution, climate change was a gradual process, but in the recent past, the rate of change has been triggered due to a tremendous increase in pollution, found to be a consequence of industrialization. Human-made GHGs (CH_4, O_3, CO_2, NO_x, H_2O and CFCs) have caused the largest change in climate forcings in recent centuries. These gases absorb the Earth's infra-red (heat) radiation. The temperature of the Earth is regulated by the greenhouse effect. Global warming is also known as the greenhouse effect. The greenhouse Earth is surrounded by a shield of atmospheric gases rather than a glass or plastic cover. The greenhouse effect is a fundamental property of the Earth's atmosphere. The atmospheric greenhouse effect warms the Earth's surface by reflecting back down to it a portion of the solar heat energy that the Earth's surface sends skyward. The atmosphere is transparent to the sun's visible light (short-wave), which is absorbed by the Earth's surface. The Earth then slowly expels this stored heat in the form of invisible infra-red radiation (long-wave radiation). But, the atmosphere acts somewhat like a one-way mirror, reflecting some but not all of this radiated heat energy. It comes back down to Earth to warm its surface further. Thus, the Earth gradually warms. Due to the large heat capacity of the oceans, the Earth requires about a century to return to equilibrium. The chief GHGs are water vapour, CO_2, CH_4, nitrous oxide and various complex human-made compounds. They are entirely responsible for reflecting downward heat that the Earth's surface would otherwise emit into space. The gases that make up the rest of the atmosphere, nitrogen (78%), oxygen (21%) and argon (0.9%), are transparent to heat as well as to light and contribute nothing to the greenhouse effect. Without this effect, the Earth's surface on an average would be 33°C colder, yielding an ice-covered planet inhospitable to life as we know it. Having some GHGs in the air thus makes life possible. But if their concentrations exceed critical levels, conditions could become less favourable for human beings and other species. Carbon dioxide (CO_2) gas generated by man's burning of fossil fuels and the forests is responsible for about half the GHG warming. Other gases (chlorofluorocarbons (CFCs), nitrous oxide, methane and ozone) are responsible for the rest. Increases in all these gases are due to mankind's explosive population growth over the last century and increased industrial expansion. Approximately 80% of the atmospheric CO_2 increase is due to man's use of fossil fuels: oil, coal and gas.

Aerosols (fine particles in the air) comprise the other main human-made climate forcing besides GHGs. Some aerosols, such as black carbon (soot), a product of incomplete combustion of fossil fuels, biofuels and outdoor biomass burning, absorb sunlight and thus heat the atmosphere. Exploitation of natural resources by the increasing population is no doubt one of the reasons for accelerated global warming. CFCs are used for refrigeration, as aerosol propellants and for insulation. They are responsible for 14% of the greenhouse effect. CFCs have 15,000 times greater potential for global warming than that of CO_2, while oxides of nitrogen (gaseous form) have 10 to 1000 (around 230) times greater effect on global warming than that of carbon dioxide. Ozone is responsible for 15% of global warming (Figure 18.1).

Although there are dozens of GHGs, four of them are most important and are closely linked to common human activities. These are as follows:

Greenhouse Gases	Greenhouse Effect
Carbon dioxide	55%
Methane	15%
CFCs	24%
Nitrous Oxide	6%

18.2.1 Carbon Dioxide

Carbon dioxide in our atmosphere acts like a light filter, allowing certain wavelengths (those of visible light) to pass through but absorbing others (especially infra-red light). An important balance exists between the atmospheric concentration of carbon dioxide and life. With less

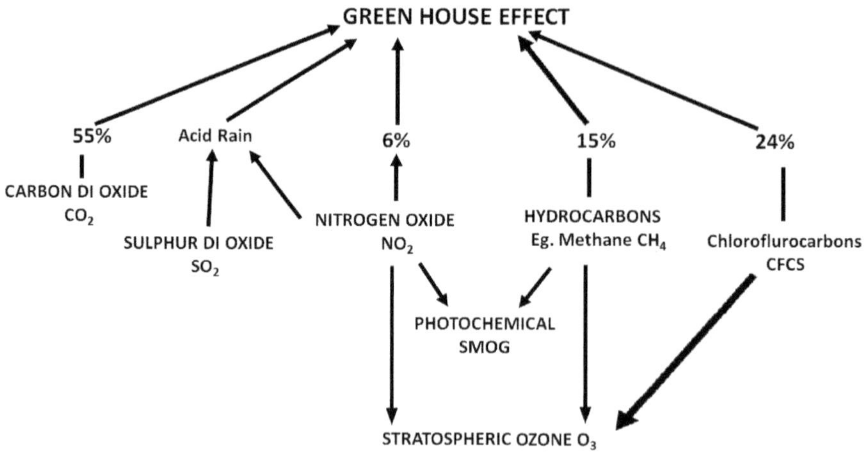

FIGURE 18.1 Greenhouse effect.

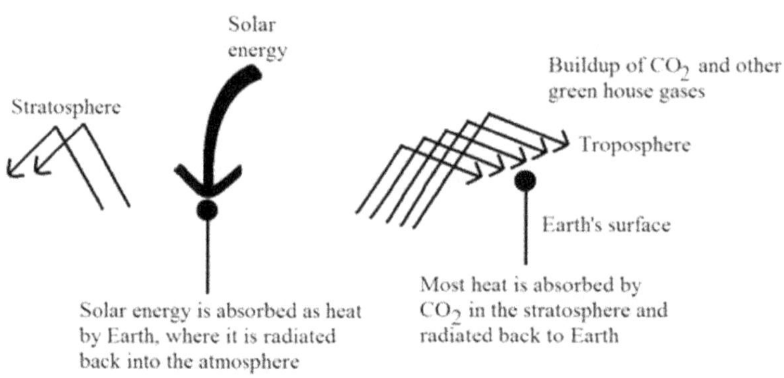

FIGURE 18.2 CO_2 production.

carbon dioxide, more heat would be lost, and Earth would be frozen, like Mars. With more, more heat would be trapped, and our world would be as hot as Venus, at 800°C, with lakes of molten lead.

Burning of petrol alone releases a huge quantity of carbon dioxide into the atmosphere; for every 1000 litres of petrol consumed, automobile exhaust releases nearly 320 kg of carbon dioxide and 2–8 kg of nitrogen oxide, besides various other air pollutants, into the atmosphere. CO_2 concentration has increased from 280 ppm (parts per million) at the dawn of the industrial revolution to around 355 ppm today. The destruction of forests and the degradation of soils add an estimated 5–9 billion tonnes of CO_2 to the atmosphere (Figure 18.2).

18.2.2 Nitrous Oxide

Nitrous oxide (N_2O) is produced by a variety of biological processes in soils and water. Its concentration has increased by 8% since pre-industrial times (from 288 to 310 ppb). Bringing new land into cultivation may be the largest source of the gas. Two recently identified major sources are the production of nylon and nitrogen fertilizers. It is estimated that a 70 to 80% reduction in human emission of N_2O would be needed in order to stabilize concentrations at their present high level.

18.2.3 Methane

Methane (CH_4) is released as a result of the combustion of carbon and microbial decay in the absence of oxygen. Methane is also emitted from coalmines and the production and distribution networks of natural gas. Methane is 50 times more powerful in trapping heat than CO_2. Human activity has more than doubled its concentration from 0.7 ppm in pre-industrial times to more than 1.7 ppm today. Because it is removed from the atmosphere relatively rapidly by chemical reactions in the air, emissions reductions of only about 15 to 20% would be required to stabilize methane as its current high concentration.

18.2.4 Chlorofluorocarbons

CFCs are powerful heat trappers. But the net impact of CFCs on climate change may be neutral, because they also contribute to the depletion of the upper atmospheric ozone layer.

18.3 PREDICTING CHANGES IN CLIMATE

We use climatic models to predict the change and variability of the climatic system. These models simulate the behaviour of the atmosphere, oceans and ice caps to examine their effects on climate. The interactions range from very simple to extremely complex. The most complex models are known as General Circulation Models (GCMs). To date, five GCMs have been used to predict climate change, and there are significant differences between them in the magnitude of changes in temperature that they predict. As temperature change also brings change in precipitation, which is the single most important factor for agriculture, these predictions are not in good agreement. There are a few areas where these agreements on possible impacts of climate change on agriculture hold true. These are: warming of higher latitudes, northern advances of monsoon rainfall, and reduced soil water availability.

18.3.1 Warming of Higher Latitudes

With the increase in temperature in the higher latitudes, agriculture can flourish, and the southern region of the boreal forest will retreat northward due to competition for land to cultivate.

18.3.2 Northern Advances of Monsoon Rainfall

As the Intertropical Convergence Zone (ITCZ) is likely to be advanced northward with global warming, India is expected to receive more rainfall, causing floods and erosion.

18.3.3 Reduced Soil Water Availability

GCMs predict a reduction in soil water in Africa, parts of South East Asia and the Arabian peninsula, eastern Australia and the southern half of North America (between December and February); in west Africa, western Europe, China and Soviet Central Asia, Mexico and Central America, Brazil, and north-eastern and western Australia (June to August).

18.4 SOURCES OF GHG EMISSIONS

Global GHG emissions can be roughly traced back to four broad categories: energy, agriculture, industry and waste. Overwhelmingly, almost three-quarters of GHG emissions come from our energy consumption.

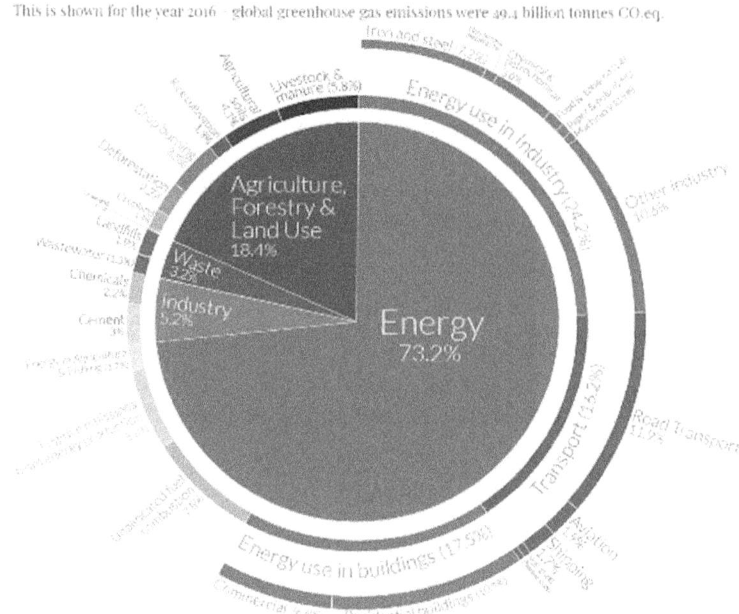

FIGURE 18.3 Global greenhouse gas emission by sector.

Sector	Global GHG Emissions Share
Energy Use	73.2%
Agriculture, Forestry and Land Use	18.4%
Industrial Processes	5.2%
Waste	3.2%

There are even more granular breakdowns to consider within each area. We'll focus on the top two, which together account for more than 91% of worldwide GHG emissions (Figure 18.3).

18.4.1 Sectors That Make Use of Energy

We can further split things down into sub-categories within this broad area, such as transportation, buildings and industry-related energy usage, to mention a few.

Sub-sector	Share of GHG Emissions	Further Breakdown
Transportation	16.2%	• Road 11.9% • Aviation 1.9% • Rail 0.4% • Pipeline 0.3% • Ship 1.7%
Infrastructures	17.5%	• Residential 10.9% • Commercial 6.6%
Industrial energy	24.2%	• Iron and steel 7.2% • Non-ferrous metals 0.7% • Machinery 0.5% • Food and tobacco 1.0% • Paper, pulp and printing 0.6% • Chemical and petrochemical (energy) 3.6% • Other industries 10.6%

Sub-sector	Share of GHG Emissions	Further Breakdown
Agriculture and fishing both need a lot of energy	1.7%	-
Combustion of unallocated fuel	7.8%	-
Emissions from energy generation that are fugitive	5.8%	• Coal 1.9% • Oil and natural gas 3.9%
Total	73.2%	

Hundreds of millions of people rely on gasoline and diesel vehicles to get around. As a result, they account for over 12% of world emissions.

18.4.2 But This Challenge Is Also an Opportunity

The consumer adoption of electric vehicles (EVs) could significantly help shift the world away from fossil fuel use, both for passenger travel and for freight – although there are still speedbumps to overcome. Meanwhile, buildings contribute 17.5% of energy-related emissions overall – which makes sense when you realize the stunning fact that cities use 60–80% of the world's annual energy needs.

18.4.3 Agriculture, Forestry and Land Use

The second biggest category of emissions is the sector that we rely on daily for the food we eat.

Perhaps unsurprisingly, methane from cows and other livestock contributes the most to emissions, at 5.8% total. These foods also have some of the highest carbon footprints from farm to table.

Sub-sector	GHG Emissions Share
Livestock and Manure	5.8%
Agricultural Soils	4.1%
Crop Burning	3.5%
Forest Land	2.2%
Cropland	1.4%
Rice Cultivation	1.3%
Grassland	0.1%
Total	**18.4%**

Another essential factor to consider is the amount of land required for our overall farming needs. There is a definite relation between human land usage and rising global emissions when large tracts of forest are destroyed for grazing and crops. The global energy mix is ripe for change, despite the fact that many of these energy systems are still in place. As the globe marches toward a green energy revolution, the potential areas of disruption have become increasingly evident, as shown by the data.

18.5 TECHNOLOGICAL ADVANCEMENTS THAT MIGHT POTENTIALLY SAVE US FROM THE CONSEQUENCES OF CLIMATE CHANGE

Any successful approach to significantly reduce GHG emissions will necessitate efforts that not only deploy low-emission technologies that are currently available, but also promote innovation on new technologies that are required. As a result, there has been a surge of interest in recent years over how to encourage such innovation, particularly the role that governments may and should play in the process. From a source perspective, we are still, despite all of our efforts, in an age of oil. To get to a net zero position by 2050, the economy needs to dramatically shift to an age of electricity. Any technology or invention that will facilitate or drive that transition will do well. More broadly, at least some adaptation to climate change will almost certainly be necessary, and such adaptations also will require some degree of technological change (NRC, 2010).

18.6 THE TRANSFORMATION WAS LED BY TECHNOLOGY

Technologies and ideas that increase battery efficiency and longevity for storage, which smooths out intermittent generation from renewable sources, are examples of this. Clean hydrogen – which is currently being used to power buses in Northern Ireland – energy storage, and carbon capture, storage and use are among the other priority topics. Digital innovations are transforming the world.

18.6.1 A Meal

The world's 7 billion people account for around a quarter of all global emissions, with meat consumption accounting for a portion of that. In a 2013 blog post, Bill Gates stated: 'There is no way to produce enough meat for 9 billion people' (Gates, 2013). That is why we require more meat-producing solutions that do not deplete our natural resources. Scholze *et al.* (2006), Sutherland (2006) and Schröter *et al.* (2005) have looked at the effects of global warming on key ecosystem processes (change in forest cover/carbon storage, wildfire frequency and freshwater availability) for the world ecosystems using a Dynamic Global Vegetation Model (DGVM). One option is to begin manufacturing lab-grown meat and meat substitutes that look, taste and feel like the real thing (Figure 18.4). It may appear to be science fiction, but businesses and investors alike are taking it seriously. The start-up Beyond Meat, which is already supported by Bill Gates, has developed the world's first plant-based meat burger. It's largely made of pea protein, which is a type of vegetable protein.

18.6.2 Transportation

Transportation (Figure 18.5) accounts for 23% of worldwide CO_2 emissions due to energy use. However, the demand for transportation is only going to grow. We've already discovered alternate ways to power vehicles, such as using electricity, but we'll need considerably more efficient batteries and battery-charging technology to do it on a large scale. Researchers at the University of

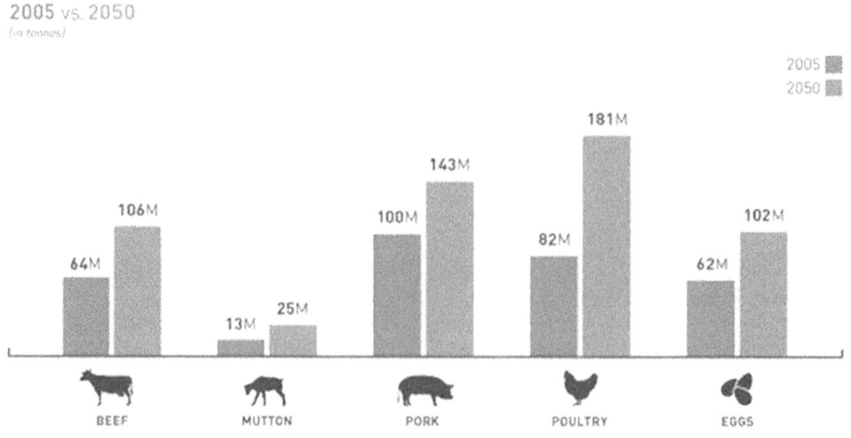

FIGURE 18.4 Global demand for meat.

FIGURE 18.5 Different modes of transport.

Surrey claim to have made a scientific breakthrough in this area. They claim to have discovered new materials that may be used as an alternative to batteries and have been shown to be 1000–10,000 times more potent than the current battery replacement, the supercapacitor.

18.6.3 Manufacturing Electric Power

We already know that nuclear power produces electricity without emitting carbon dioxide, but we have yet to harness it in a safe and cost-effective manner. General Fusion, a Canadian firm, is developing a commercially viable nuclear-fusion energy generating plant. Fusion produces no greenhouse gas emissions and only emits helium as a byproduct. Fusion also requires less land than other renewable technologies, according to the company. Fusion energy is intrinsically safe, with no risk of a meltdown and no long-term waste, and enough fusion fuel exists to power the globe for hundreds of millions of years (Figure 18.6).

Smart metering will usher in a slew of innovative energy-saving ideas. It will influence many decisions and allow businesses to customize products to meet specific needs. It's also about empowering individuals to be prosumers rather than consumers, because when combined with microgeneration systems like solar panels, it can help decarbonize the power grid. Smart meters can also assist in converting the electric vehicle parked in the driveway into a battery that can store and feed electricity back into the grid.

18.6.4 City Structures

Buildings produce a large amount of GHG emissions. Whether at home, at the office, at school or in a hospital, we require lights, power, heating and cooling. The emissions from these sources account for over 20% of total global emissions. Building smarter cities is one part of the solution. Sidewalk Labs (part of Alphabet Inc., the parent company of Google) is doing just that, using digital technology to solve today's critical urban problems. One of their current studies examines how traffic

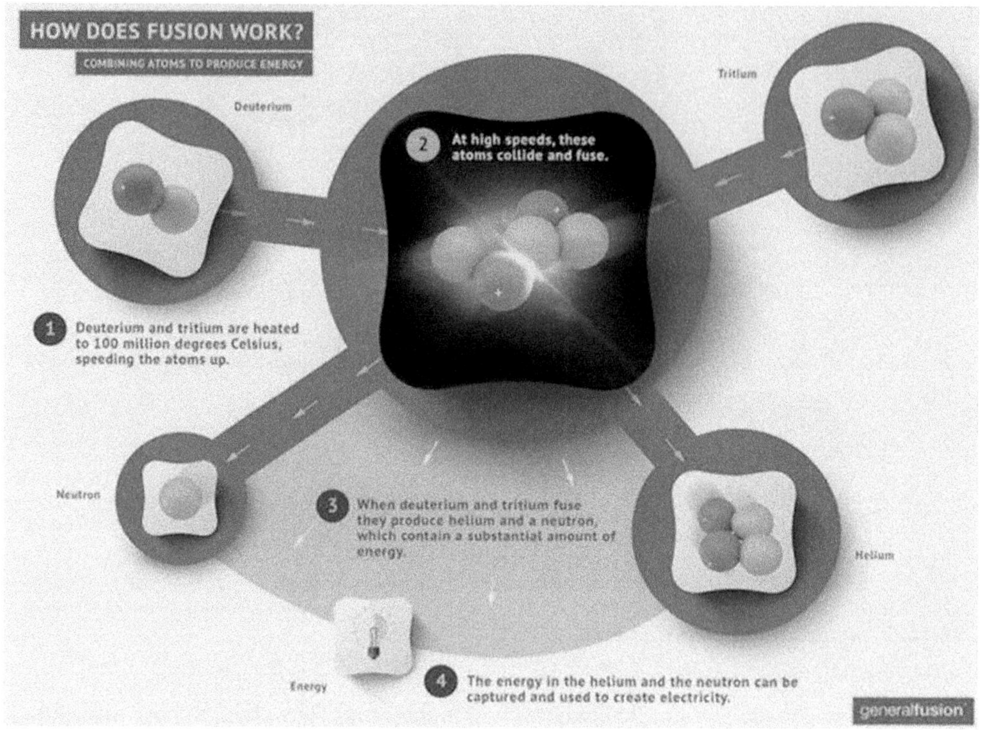

FIGURE 18.6 Nuclear fusion energy power plant (general fusion).

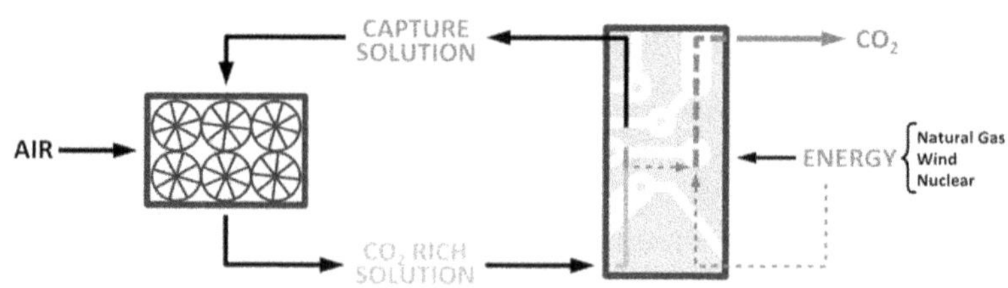

FIGURE 18.7 Carbon engineering.

flows through a city and how congestion hotspots might be addressed. This has the potential to significantly reduce air pollution in our cities.

Members of the Breakthrough Energy Coalition (BEC), which includes Amazon's Jeff Bezos, Ali Baba's Jack Ma and Richard Branson, have pledged to invest more than $1 billion in new technologies over the next 20 years.

18.6.5 Industrial Production

Making the goods we use every day puts a huge pressure on the environment – industry accounts for around 30% of all emissions. But what if we could eliminate CO_2 emissions from the atmosphere? Carbon Engineering (Figure 18.7), a Canadian start-up, is working on precisely that – extracting carbon dioxide from the atmosphere and using it to generate fuel. 'Direct air capture can remove significantly more CO_2 per acre of land footprint than trees and plants', the company claims.

In Squamish, British Columbia, the business is already operating a demonstration facility that removes 1 tonne of CO_2 from the atmosphere every day.

18.6.6 Incentives That Are Unique

We trained people on how to separate rubbish and rewarded them with free bins and recycling collections when recycling was initially implemented. Climate change requires a programme similar to this. Technology isn't the only solution. We don't want people to believe they can sit back and wait for technology to solve the problem. We must build the mechanisms necessary for people to believe that their individual activities have a purpose and an impact on climate change (www.irishtimes.com).

18.7 CONCLUSION

We should be more passionate about land conservation, forest conservation, pollution prevention and use of solar energy than we are in terms of industrialization and ruthless use of agricultural land. If we commit our forestry to the pledge of planting ten trees for cutting one tree; if we curb the industries emitting carbon dioxide, the things that produce carbon gas (including refrigerators, perfumes, air conditioners etc.) to pollute rivers and reservoirs; avoid these things or develop a method of purifying the water; avoid the use of nitrogen and artificial fertilizers in agriculture and use a carbon dioxide-reducing electricity system, then perhaps we will do good for ourselves and the country. Otherwise, this global warming will be accompanied by all the calamities that starvation, malnutrition, disease and social human disturbances. will give birth to. The changing mood of the weather has forced countries that are avoiding restraint under the guise of growth to accept the bondage. Whether it is the US alone emitting 25% of carbon dioxide in the atmosphere, or the fast-paced economies like India and China, which have tasted development, all are beginning to recognize that there is a need for joint efforts to tackle the challenge of climate change. But efforts are also under way to put the gamble of responsibility on each other's shoulders when it comes to shared funds. According to a cost estimate by the World Bank, if the damage done so far to the environment is to be reversed, then the world will have to achieve 5% of its economic growth using pollution-free technologies and conservation efforts. Developing countries are apprehensive about this; they do not want their economies to be prevented from taking off in the name of pollution-free technologies. However, under the guise of development, the third world countries, whose population will have to suffer the most, are turning their backs on the dreadful future. It is clear from the latest report of the IPCC, set up by the United Nations, that the biggest price for the changing mood of the environment will have to be paid by the populations of Asia and Africa, who are doing it earlier. Recently, eight of the world's richest countries gathered in Germany's Hailey Gendem, despite all the differences, have agreed that if the Earth is to be saved, the amount of carbon dioxide reaching the atmosphere every year will have to be halved by 2050. They also believed that due to the indiscriminate race for industrial development in the world, it is his first responsibility to carry its ill-effects, but not every proposed platform to save the environment should be like the Kyoto Protocol; it is also a great responsibility of India and China. This agreement is a good opportunity for developing countries, including India, which can not only help in saving the environment but can also save their development from being affected.

REFERENCES

Gates B (2013) Future of Food. Gate Notes. *The Bill Gates Blog*.www.gatesnotes.com/About-Bill-Gates/Future-of-Food

Metzger MJ, Leemans R and Schroter D (2005) A multidisciplinary multi-scale framework for assessing vulnerabilities to global change. *International Journal of Applied Earth Observation and Geoinformation* 7, 253–267.

NRC, National Research Council (2010), *Adapting to the Impacts of Climate Change*, Washington, DC: The National Academies Press.

Scholze M, Knorr W, Arnell NW and Prentice IC (2006) A climate-change risk analysis for world ecosystems. *Proceedings of the National Academy of Sciences of the United States of America* 103, 13116–13120.

Schröter D, Cramer W, Leemans R, Prentice IC, Araújo MB, Arnell NW, Bondeau A, Bugmann H, Carter TR, Gracia CA, de la Vega-Leinert AC, Erhard M, Ewert F, Glendining M, House JI, Kankaanpää S, Klein RJ, Lavorel S, Lindner M, Metzger MJ, Meyer J, Mitchell TD, Reginster I, Rounsevell M, Sabaté S, Sitch S, Smith B, Smith J, Smith P, Sykes MT, Thonicke K, Thuiller W, Tuck G, Zaehle S and Zierl B (2005) Ecosystem service supply and vulnerability to global change in Europe. *Science* 310(5752), 1333–1337. doi: 10.1126/science.1115233

Sutherland WJ (2006) Predicting the ecological consequences of environmental change: a review of the methods. *Journal of Applied Ecology* 43, 599–616.

www.irishtimes.com (2020) How technology and innovation can help to tackle climate change. *The Irish Times*, Friday 2 April 2020.

19 Role of Bioenergy in Climate Change, Food, Energy and Rural Development

Omkar Verma and M. Prashanth
Geology Discipline Group, School of Sciences, Indira Gandhi National Open University, New Delhi

Ashu Khosla
Department of Geology, Panjab University, Chandigarh

Nisha Changotra
GPS Nihalpur, Department of School Education, Government of Jammu and Kashmir, Kathua

CONTENTS

19.1 INTRODUCTION

Energy is the basic lifeline of global socio-economic development, as it has been fuelling our day-to-day lives since the beginning of human history. At present, non-renewable (fossil fuels and nuclear power) and renewable (biomass, hydropower, solar, wind and tidal) are the two main sources of energy. Non-renewable energy sources have been fulfilling nearly 82% of total global energy consumption, whilst the remaining 18% are coming from the renewable sources (Figure19.1; REN21, 2020). Among non-renewable energy sources, the fossil fuel petroleum is the largest source of commercial energy, followed by coal. The main concerns associated with utilization of fossil fuel include (i) depletion of their reserves, because they are non-renewable energy sources, which could lead to an energy crisis, (ii) their burning increases the greenhouse gas (GHG) concentration in the atmosphere, which is accelerating global warming and ultimately, contributing to climate change, and (iii) deteriorating air quality and posing serious threats in human health (Hoekman et al., 2018; Wu

DOI: 10.1201/9781003220824-22

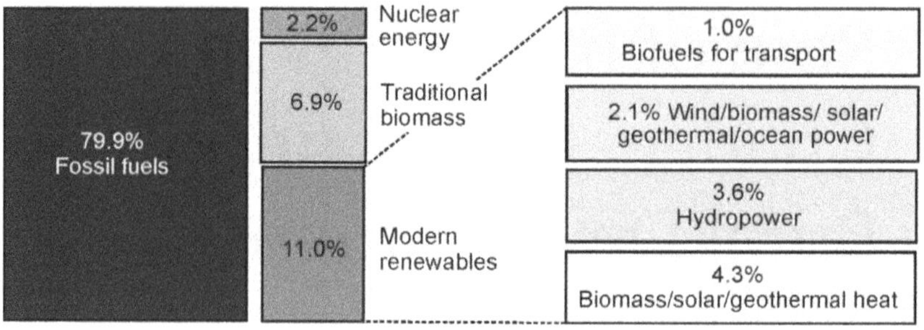

FIGURE 19.1 Total energy consumption in 2018.

(Redrawn after REN21, 2020.)

et al., 2018). It is noted that public reliance on fossil fuels is not sustainable, as the projected energy demand will grow more than 50% in the next two decades (Ozturk et al., 2017; Wu et al., 2018). Additionally, the rising cost of fossil fuels because of limited supply and their drastic ill effects on the environment and climate system, along with continuously growing concern over future energy security, are prompting many countries to harvest energy from alternate, economically efficient and environmentally friendly sources. In this context, bioenergy has emerged as one such option, as it provides an alternative to fossil fuels and is being globally promoted as a cleaner and greener source of energy (Sims, 2003; Miller, 2007; Reid et al., 2019).

Bioenergy (also known as biomass energy) is defined as energy produced from modern/living biomass sources. There are several types of biomass, such as ligno-cellulosic materials, agricultural crops and organic wastes (municipal, kitchen, agricultural and animal), that can be used for bioenergy production (Demirbas, 2009). The traditional uses of bioenergy include burning of unprocessed biomass to generate heat energy for cooking and heating purposes, primarily in rural areas of developing nations. The modern bioenergy uses comprise biofuels for transport and processed biomass for heat and electricity generation (Bringezu et al., 2009). The shares of traditional and modern bioenergy accounted for nearly 6.9% and 5%, respectively, of the total share (i.e., 18%) of renewable energy in 2018 (REN21, 2020). A considerable increase in the consumption of modern bioenergy is evident, from 1.8% to nearly 5% from 2007 to 2018 (REN21, 2008; REN21, 2020). The belief that bioenergy, especially biofuels, can reduce our dependency on fossil fuels, mitigate climate change and create millions of rural jobs has driven governments around the globe to promote the production and use of sustainable biofuels, particularly biodiesel and bioethanol, through various governmental policies. This chapter presents an account of bioenergy versus fossil fuels, types of biofuels, and threat and benefit perceptions of bioenergy with special reference to climate change.

19.2 BIOENERGY VERSUS FOSSIL FUEL

The history of bioenergy is as old as humankind, and its origin was noticed when the earliest humans discovered fire by burning wood. Apart from language, fire is considered as the most significant discovery, because it has several uses in modern society (Gowlett, 2016). Wood was the first form of bioenergy, and it still accounts for nearly 75% of the global energy supply (http://biofuel.org.uk/history-of-biofuels.html). The initial development in bioenergy, especially biofuels for transport, mainly happened in Europe (Patil, 2007; Singh & Walia, 2016). In addition, the earliest developments, consisting of the invention of charcoal, vegetable oil, whale oil, ethanol and biogas, took place prior to the twentieth century. It is to be noted that the twentieth century witnessed some instances of energy crises, particularly related to fossil fuel, for example the 1973 world-wide fuel crises, and the role of bioenergy in climate change mitigation and rural development alternatively gave a boost to the technological development of bioenergy (Pfeiffer & Thrän, 2018). As a consequence, considerable growth in bioenergy, including biofuels, can be seen from the twentieth to the early twenty-first century (REN21, 2020). Now, bioenergy includes highly commercialized liquid (bioethanol, biodiesel, bio-oil), solid (wood pellets, charcoal) and gaseous (biogas, syngas) biofuels, and bioelectricity, which have been used in several economic sectors (Singh & Walia, 2016; Xiao-Yang, 2016).

It is noted that bioenergy developments has faced many ups and downs during its long history, which are directly associated with the rate of supply of fossil fuel. In 1960, a group of five oil-producing countries comprising Iran, Iraq, Kuwait, Saudi Arabia and Venezuela formed a permanent intergovernmental organization known as the Organization of Petroleum Exporting Countries (OPEC) (www.opec.org/opec_web/en/about_us/24.htm). At present, OPEC is a group of 13 oil-producing countries; 8 more have been added to the existing original 5 members, namely, Algeria, Angola, Congo, Equatorial Guinea, Gabon, Libya, Nigeria and United Arab Emirates. The OPEC has 78% of the world's oil reserves and acts as a major controlling body over the supplies and prices of the world's oil (Miller, 2007). The main objectives of the OPEC are to "coordinate and unify petroleum policies among member countries to secure fair and stable prices for petroleum producers; an efficient, economic and regular supply of petroleum to consuming nations; and a fair return on capital to those investing in the industry" (www.opec.org/opec_web/en/about_us/24.htm). It is seen that OPEC often sets a high oil price and reduces oil production. As a consequence, oil-importing (i.e., non-OPEC) nations are forced to invest more in oil imports, making development more expensive and also putting them into energy insecurity (Demirbas, 2009). Indeed, the 2003 Iraq war was considered by some as the war for oil, because Iraq has huge oil deposits (Gray, 2004). Additionally, over the last few decades, fossil fuel reserves have been depleted at a very fast rate, and it has been predicted that global petroleum supply will fall rapidly by 2030, generating a huge energy gap that will perhaps be impossible to fill. The Intergovernmental Panel on Climate Change (IPCC) also has concluded that intensive burning of fossil fuels is mainly responsible for the current global warming, as this alone accounts for about 90% of GHG emissions into the atmosphere (IPCC, 2014). As a result, today's world is facing three major interrelated challenges: energy insecurity, climate change and environmental degradation (REN21, 2020). The main problems with fossil fuels are their uneven distribution (with major deposits are in politically unstable regions of the globe), non-renewable nature and excessive GHG emissions, causing the climate to change.

In fact, the non-OPEC nations first realized the importance of bioenergy (of which biofuels are a subset) during the 1973 energy crisis, when OPEC suddenly disrupted oil supply. Visualizing the significance of using bioenergy at a global level instead of succumbing to OPEC, the member countries of the Organization for Economic Co-operation and Development (OECD) in 1974 established the International Energy Agency (IEA). The main functions of the IEA are to improve co-operation and information exchange between countries that have national programs in bioenergy, to shape a secure and sustainable energy future for all, and to help to avoid future energy crises (www.iea.org/about). It has been observed that bioenergy, particularly biofuels, has been playing an important role in addressing several global challenges, such as reducing dependency on fossil fuels and controlling their prices, reducing GHG emissions, mitigating climate change, increasing energy security and creating rural jobs, which encourages countries across the globe to introduce and support policies related to bioenergy production and use (Rai & Ingle, 2019). As a consequence, many countries of the world, notably the US, Canada, Brazil, Argentina, the EU, China, Japan, India, South Africa and others, are actively promoting technology development for sustainable bioenergy use (Reid et al., 2019). Therefore, bioenergy is now becoming a popular and important renewable energy source.

19.3 TYPES OF BIOFUELS

Biofuels are just one form of bioenergy. Making biofuels from biomass is not a new process, as it was first done in Germany by Rudolph Diesel in 1900 by producing biodiesel from vegetable oils (Singh & Walia, 2016). At present, biofuels have three generations. First-generation biofuels (biodiesel, ethanol, vegetable oils) are mainly derived from food-grains, and conventional technology has been used for their production. The second-generation biofuels (biohydrogen, methanol, Fischer-Tropsch diesel) are derived from crops specifically grown for fuels and non-grain parts of plants through the process of fermentation and gasification or pyrolysis. But, more research and development work is required to make second-generation technologies economically viable. Algal fuel, derived from aquatic algae, is a third-generation biofuel; other biofuels of the generation are biopropanol and biobutanol. Unfortunately, the technology for production of these biofuels is yet not fully developed.

Biofuels may be in liquid, solid and gaseous forms. The liquid or transport biofuels such as biodiesel (methyl esters), biomethanol (methyl alcohol) and bioethanol (ethyl alcohol) produced from plants and animal wastes have the potential to replace conventional gasoline and diesel fuel. Biodiesel and bioethanol are dominant biofuels for use as road fuel (Bringezu et al., 2009). The main biofuel-producing nations are the US, Brazil and the EU. Brazil and US are the leading producers of ethanol. Brazil produces ethanol from bagasse, a residue of sugarcane, while corn is the feedstock (source) of choice in the US. About 40% of Brazilian motor vehicles are already running on cane-derived ethanol, and it is further planned that within a decade all Brazilian vehicles will run on this biofuel. Three nations in the EU, Germany, France and Italy, produce around 95% of the world's biodiesel. Among these, Germany is producing more than half of the world's biodiesel. Rapeseed and sunflower seeds are primary feedstocks in the EU (Miller, 2007). The global production of ethanol has increased threefold from 17 billion to over 52 billion litres and biodiesel production expanded around 11-fold from less than 1 billion to about 11 billion litres between 2000 and 2007. Both these biofuels contributed 1.8% to global road fuel (Bringezu et al., 2009).

19.4 BIOENERGY: THREAT AND BENEFIT PERCEPTIONS

The production and use of bioenergy present opportunities to tackle major issues such as reducing greenhouse gases emissions, mitigating climate change, providing energy security, saving foreign exchange and creating jobs, and also offer a unique solution to waste management by converting organic wastes into usable energy (Miller, 2007; Bringezu et al., 2009; Reid et al., 2019). However, at the same time, there are some serious risks to land use, food, environment and biodiversity associated with bioenergy (Berndes et al., 2011; Popp et al., 2014; Singh & Walia, 2016). Therefore, an appropriate understanding of the benefits and threats of bioenergy among stakeholders and the general public is very important for the future growth of the bioenergy industry.

19.4.1 Land Use Change

Land use change is defined as a process by which humans alter the natural landscape/land cover by various means such as clearing forests for agriculture activities, increasing areas for a particular crop, urbanization or changing management for existing land cover (Davis et al., 2019). It is noted that both direct and indirect land use changes are associated with bioenergy feedstock production and the bioenergy industry (Berndes et al., 2011). Direct land use change commonly occurs on the locations used for feedstock production by the introduction of new activity related to bioenergy, and this is an observable as well as a measurable change (Berndes et al., 2011; Ernst & Young, 2011). For example, Brazil has been converting grazing land into agricultural land for cultivating sugarcane to meet its growing demand. On the other hand, indirect land use change generally occurs elsewhere, not in situ, as a result of bioenergy production, and thus, it cannot be directly observable and measurable (Berndes et al., 2011; Ernst & Young, 2011). Take an example; the increased demand for sugarbeet for biofuel production in Europe caused Brazil to produce more sugar from sugarcane (Ernst & Young, 2011). The direct and indirect land use changes are interrelated and mainly involve conversion of forest, pasture and wetlands together with natural landscape into agricultural land for production of bioenergy feedstocks. It also involves changes in cropping patterns, for example shifting food crops to bioenergy crops. Land use change results in changes in landscape, ecosystems, biodiversity, soil, environment, waters, food, energy, life and climate (Popp et al., 2014).

Bioenergy biomass production depends upon the availability of arable land. Growing demand for bioenergy crops may only be met through a wide expansion of cropland. This may also lead to loss of biodiversity, deforestation and an increase in GHG emissions. Production–consumption chain analyses of bioenergy show a number of serious impacts on the environment. Currently, cropland has been expanded in the US, EU and Latin America for the production of energy crops. The US has been using genetically modified varieties and does not need rapid expansion of cropland. According to European bioenergy mandates, the EU demand for bioenergy crops is mostly met by imports from developing countries. In other words, the European threat to the expansion of cropland is being diverted to developing nations. Brazil is the leading producer of modern bioenergy (bioethanol and biodiesel) and plans to increase the production of bioenergy crops. As a result, Brazil has been reducing larger areas of the Amazon rainforest and pasture land (Welfle, 2017). The tropical nations, specifically Malaysia and Indonesia, are exporting palm oil to the growing bioenergy markets in Europe, the US and China. This results in destruction of rainforest in Southeast Asia (Khatiwada et al., 2021). Overall, expansion of palm oil plantations has imposed severe environmental damages and biodiversity losses by converting rainforests into arable land in tropical regions, particularly Malaysia and Indonesia (Hooijer et al., 2006). The biodiversity loss is costly and irreversible. While there is hope that climate change can be reversed, as has already happened many times in the geological past, there is no way to bring back an extinct species (Verma, 2021).

In the context of climate change, the impacts of these changes can have both positive and negative feedbacks on climate in term of GHG emissions. The expansion of vegetative cover for growing bioenergy crops, especially in the unused marginal, degraded and barren areas, will considerably reduce GHG emissions and help in mitigating climate change. Additionally, the cultivation of ligno-cellulosic bioenergy feedstocks comprising perennial grasses and rotational woody crops has great potential for mitigating climate change by reducing GHG emissions (Popp et al., 2014). The soil cover and plants store a significant amount of carbon. The clearing of trees for converting forest cover into agricultural or pasture lands, burning of biomass and ploughing of grasslands for bioenergy production can release more GHGs into the atmosphere and also make the soil carbon depleted. Therefore, a proper understanding of bioenergy-driven impacts on land use change is useful for developing strategies for promoting bioenergy as an alternative energy source to fossil fuel and reducing GHG emissions.

19.4.2 Climate Change Mitigation

Most climatologists are of the view that the earth's temperature is increasing at an unprecedented rate (Figure19.2). Even a small increase in the earth's average temperature could melt glaciers and polar ice caps, raise sea level, flood low-lying areas, increase frequency of extreme weather events, shift vegetation zones and collapse ecosystems, thereby threatening species habitat. Besides, the rapid spread of infectious diseases, increasing intensity of natural hazards, reduced crop yields, and change in population dynamics because of climate change will also cause misery to human beings. A primary cause of global warming is the build-up of GHGs, especially carbon dioxide, released by

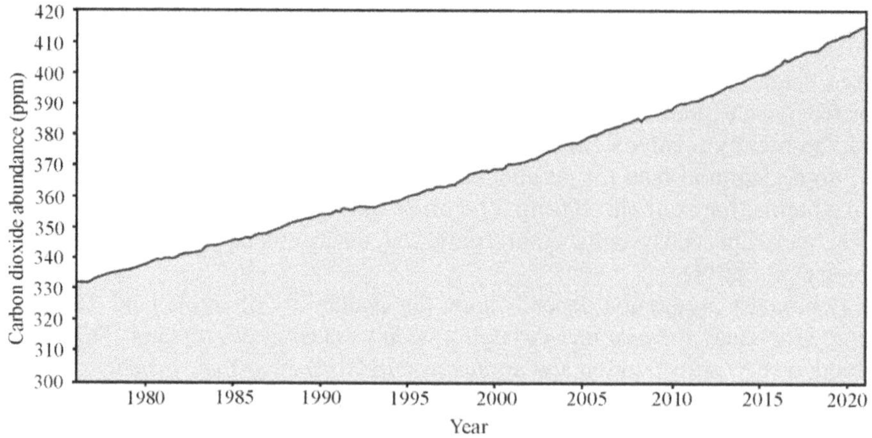

FIGURE 19.2 Increase in carbon dioxide from fossil fuel consumption.

(From www.climate.gov/news-features/understanding-climate/climate-change-atmospheric-carbon-dioxide.)

the burning of fossil fuels (oil, gas, diesel and coal), in the atmosphere. Little or no carbon dioxide is emitted during the production and use of bioenergy provided the biomass is sustainably generated. In fact, it is thought that bioenergy may play an important role in reducing temperature to well under 2°C and thus, helping to mitigate climate change (Rogelj et al., 2018; Reid et al., 2019). As a result, many governments across the world are looking to bioenergy, especially biofuels, as a way of reducing reliance on oil imports and slowing GHG emissions. Some signatories of the Kyoto Protocol (the EU and Japan) have taken biofuels as a key tool for meeting their GHG emissions reduction levels. Hence, bioenergy is often called "carbon neutral energy".

Initially, it was proposed that carbon sequestration occurs during the growth of bioenergy crops by absorbing carbon dioxide emitted during the burning of biofuels and may help to mitigate GHG emissions. But this belief proved wrong, because this study did not take any account of various inputs used from cultivation of bioenergy crops to consumption of biofuels. A complete life-cycle assessment of GHG emissions for bioenergy reported only modest carbon savings associated with the first generation of biofuels. While higher carbon savings may be achieved from a second generation of biofuels, the technologies for this generation are beyond our reach. A positive greenhouse balance can resulted when energy crops are grown on marginal and infertile land with minimum land use change and organic wastes are used for the production of bioenergy crops. Otherwise, intensive agricultural production and conversion of natural land to arable land may lead to negative feedback.

Fossil fuel deposits, ocean, atmosphere, biosphere and soil are main five global storehouses of carbon. Among them, the flow of carbon from fossil fuel into the atmosphere by its burning is a unidirectional response, and the remaining four have bi-directional interrelationships as far as the flow of carbon is concerned (Berndes et al., 2011). In other words, the carbon dioxide produced by the consumption of fossil fuel is adding more to what is already existing in the atmosphere and causing the temperature to rise. The other storehouses of carbon transfer carbon from one to another, but the net carbon content in the system remains the same. In this context, bioenergy has the advantage of reducing carbon dioxide from the atmosphere by capturing and storing carbon. Further, studies around the globe have shown that the net carbon emissions resulting from the use of bioenergy are much lower than the consumption of fossil fuel (e.g., Wu et al., 2018). In addition, bioenergy has the ability to minimize air pollution.

19.4.3 Energy and Food Security

It is believed that production of modern bioenergy, particularly biofuels, can improve the energy security of oil-importing nations by generating a substitute for fossil fuel. Biomass is used to produce bioenergy, and it is the world's most widely distributed and plentifully available resource on the surface of the earth; thus, no organization (e.g., OPEC) or no one nation (e.g., Saudi Arabia) will be able to dominate the global supply of modern bioenergy (Miller, 2007). In this way, economics of governments of bioenergy-producing nations will not be held hostage to the OPEC and whims of other oil-exporting nations. It is considered that the ability of biofuels to replace convectional oil has not been properly assessed. The world consumed 83.94 million barrels of oil per day in 2005, of which US consumption was 20.6 million barrels of oil per day. On the other hand, global total production of transport biofuels was around 325 million barrels in 2005. This production of biofuels would have met at least 3 weeks' requirement for transportation fuels in the US. If present biofuel production can only meet a few weeks' oil requirement for a nation, it would seem impossible that biofuels will meet the current rapidly rising demand for energy.

Food insecurity is another major concern associated with production and use of biofuels, because it involves the use of agricultural lands and products for non-food purposes (Popp et al., 2014). The increasing demand for energy feedstock could be generating competition between land use to grow food crops and bioenergy feedstocks. Since the price of biofuel is related to the price of conventional oil, as a result, the rising price of biofuels motivates farmers to produce bioenergy crops on arable land, because these crops become more profitable than food crops; this can be a threat to global food security. If the production of food declines, prices of food will go up, which may cause poor rural people to go hungry, starve and die. A study has shown that an increase of about 1%in food prices may put 16 million people into food insecurity. Furthermore, rising demand for biofuels is also likely to increase the cost of land, labour and agricultural inputs and also limit the ability of bioenergy to improve welfare for rural people (Souza et al., 2017).

19.4.4 Rural Development

About three-quarters of the poor population in developing nations live in rural areas (Sheingate, 2008). Bioenergy production also has a huge potential to reduce poverty in these areas by creating new employment opportunities and increasing economic growth along with meeting their energy demands (GNESD, 2011). The production of bioenergy and biofuels requires biomass feedstocks (energy and food crops, organic wastes and aquatic plants). This creates additional demand for cultivation of bioenergy crops. Growing feedstock crops for bioenergy could help increase farmers' income. Energy crops can be grown anywhere, but tropical regions characterized by high rainfall, longer growing seasons and higher productivity are more suitable for their production. There are certain bioenergy crops, such as miscanthus and jatropha, that can be grown on marginal and infertile land. Further, producing biofuels helps rural residents to save the money being spent on buying conventional fuel for agricultural and other activities. Bioethanol production has created around 1 million direct and 3.5 million indirect rural jobs in Brazil, while the biodiesel industry is expected to create 17 million rural jobs (de Castro, 2007).

Rural areas are hotspots for the production of biomass for bioenergy. Today, numerous bioconversion technological platforms, such as biological (e.g., anaerobic fermentation) and thermal (e.g., combustion, pyrolysis and gasification), are available, by which rural people are generating biogas, biofuels and electricity from agricultural products and wastes for various purposes (GNESD, 2011). At present, bioenergy is playing an important role in providing clean energy to rural people, creating suitable new jobs in both agricultural and bioenergy-related industry and finally, helping to alleviate poverty in rural areas.

19.5 CONCLUSIONS

Despite many potential benefits, one must not take it for granted that bioenergy will automatically contribute to climate change mitigation, environmental quality, energy security and socio-economic development of a region. Recently, a number of issues related to the production and use of bioenergy have been discussed in many scientific journals and forums. The main concerns of these issues are shifting from fossil fuels to bioenergy, the food versus fuel versus fibre debate, GHG emission levels, sustainable bioenergy production, deforestation, impact on water and soil resources, poverty reduction potential, bioenergy polices and prices, energy balance and efficiency, and centralized and decentralized production models. No doubt, bioenergy may provide an alternative to fossil fuel to some extent and mitigate climate change. But, this may only be achieved with (i) sustainable growth of biomass feedstock, (ii) implementation of bioenergy certification programmes, (iii) no/minimal natural land conversion to arable land, (iv) using more biowaste for bioenergy production and (v) technologies required for the production of second- and third-generation biofuels being made available in the market. Since bioenergy is emerging as a new field, the "Bioenergy Revolution", further research and development work is required to determine the full potential of global biomass for making bioenergy a reality.

ACKNOWLEDGEMENTS

One of us (Ashu Khosla) is thankful to Professor R.C. Sobti for inviting him to contribute the article for CRC Press.

REFERENCES

Berndes G, Bird N, & Cowie A (2011) *Bioenergy, Land Use Change and Climate Change Mitigation*. IEA Bioenergy, New Zealand. www.osti.gov/etdeweb/servlets/purl/22110326

Bringezu S, Schütz H, O'Brien M, Kauppi L, Howarth RW, & McNeely J (2009) Towards sustainable production and use of resources: Assessing biofuels. *United Nations Environment Programme*, pp. 1–118. https://wedocs.unep.org/handle/20.500.11822/8680

Davis KF, Dalin C, DeFries R, Galloway JN, Leach AM, & Mueller ND (2019) Sustainable pathways for meeting future food demand. *Encyclopedia of Food Security and Sustainability* 3: 14–20.

de Castro JFM (2007) *Biofuels – An Overview*, Final Report. DGIS/DMW/IB. pp. 1–42. https://np-net.pbworks.com/f/BZOS+(2007)+Biofuels+in+Africa_overview.pdf

Demirbas A (2009) *Biofuels – Securing the Planet's Future Energy Needs*. Springer-Verlag, London. Ernst & Young (2011) Biofuels and indirect land use change: The case for mitigation. www.iucn.org/sites/dev/files/content/documents/biofuels_and_indirect_land_use_change.pdf

GNESD (2011) Bioenergy: The potential for rural development and poverty alleviation. Global Network on Energy for Sustainable Development (GNESD). Summary for policy-makers. GNESD-SPM-BET-11/2011, pp. 1–29. https://backend.orbit.dtu.dk/ws/portalfiles/portal/7590946/Bioenergy.pdf

Gray M (2004) *Geodiversity – Valuing and Conserving Abiotic Nature*. John Wiley and Sons Ltd, West Sussex, UK.

Gowlett JAJ (2016) The discovery of fire by humans: Along and convoluted process. *Philosophical Transactions of the Royal Society B* 371: 20150164.http://dx.doi.org/10.1098/rstb.2015.0164

Hoekman SK, Broch A, & Liu X (2018) Environmental implications of higher ethanol production and use in the U.S.: A literature review. Part I – Impacts on water, soil, and air quality. *Renewable and Sustainable Energy Reviews* 81(2): 3140–3158.

Hooijer A, Silvius M, Wösten H, & Page S (2006) Peat-CO2. Assessment of CO2 Emissions from Drained Peatlands in SE Asia. Delft Hydraulics Report Q3943, Amsterdam. https://inis.iaea.org/search/search.aspx?orig_q=RN:42022627

http://biofuel.org.uk/history-of-biofuels.html (Retrieved on or before April 2021).

IPCC (2014) Climate Change 2014: Mitigation of Climate Change. In: *Contribution of Working Group III to the Fifth Assessment Report of the Intergovernmental Panel on Climate Change*, eds. Edenhofer O,

Pichs-Madruga R, Sokona Y, Farahani E, Kadner S, Seyboth K, Adler A, Baum I, Brunner S, Eickemeier P, Kriemann B, Savolainen J, Schlömer S, von Stechow C, Zwickel T,&Minx JC. Cambridge University Press, Cambridge. www.ipcc.ch/site/assets/uploads/2018/02/ipcc_wg3_ar5_full.pdf

Khatiwada D, Palmén C, & Silveira S (2021) Evaluating the palm oil demand in Indonesia: Production trends, yields, and emerging issues. *Biofuels* 12(2): 135–147.

Miller GT (2007) *People and Environment, Indian Edition.* Cengage Learning India Private Limited, New Delhi.

Ozturk M, Saba N, Altay V, Iqwal R, Hakeem KR, Jawaid M, & Ibrahim FH (2017) Biomass and bioenergy: An overview of the development potential in Turkey and Malaysia. *Renewable and Sustainable Energy Reviews* 79: 1285–1302.

Patil V (2007) The relevance of biofuels. *Current Science* 96(6): 707.

Pfeiffer D, & Thrän D (2018) One century of bioenergy in Germany: Wildcard and advanced technology. *Chemie IngenieurTechnik* 90(11): 167611698.

Popp J, Lakner Z, Harani-Rakos M, & Fart M (2014) The effect of bioenergy expansion: Food, energy, and environment. *Renewable and Sustainable Energy Reviews* 32: 559–578.

Rai M, & Ingle AP (2019) *Sustainable Bioenergy: Advances and Impacts.* Elsevier, Netherlands.

Reid WV, Ali MK, & Field CB (2019) The future of bioenergy. *Global Change Biology* 26: 274–286.

REN21 (2008) Renewables 2007 Global Status Report. Deutsche Gesellschaft für Technische Zusammenarbeit (GTZ) GmbH. http://rael.berkeley.edu/old_drupal/sites/default/files/Kammen%202008-%20REN%2021%20Renewables%202007.pdf

REN21 (2020) *Renewables 2020 Global Status Report.* UN Environment Programme, Paris, France. www.ren21.net/wp-content/uploads/2019/05/gsr_2020_full_report_en.pdf

Rogelj J, Shindell D, Jiang K, Fifita S, Forster P, Ginzburg V, Handa C, Kheshgi H, Kobayashi S, Kriegler E, Mundaca L, Séférian R, & Vilariño MV (2018) Mitigation pathways compatible with 1.5°C in the context of sustainable development. In*: Global warming of 1.5°C. An IPCC Special Report on the Impacts of Global Warming of 1.5°C above Pre-industrial Levels and Related Global Greenhouse Gas Emission Pathways, in the Context of Strengthening the Global Response to the Threat of Climate Change, Sustainable Development, and Efforts to Eradicate Poverty*, eds. Masson-Delmotte V, Zhai P, Pörtner H-O, Roberts D, Skea J, Shukla PR, Pirani A, Moufouma-Okia W, Péan C, Pidcock R, Connors S, Matthews JBR, Chen Y, Zhou X, Gomis MI, Lonnoy E, Maycock T, Tignor M, Waterfield T, Intergovernmental Panel on Climate Change, Geneva, Switzerland, pp. 313–443.

Sheingate A (2008) *Agrarian Social Pacts and Poverty Reduction.* United Nations Research Institute for Social Development, Switzerland, pp. 1–57. www.unrisd.org/80256B3C005BCCF9/%28httpAuxPages%29/13F125EBBAEC4FF9C1257AE5004E75B0/$file/Sheingate.pdf

Sims R (2003) Bioenergy to mitigate for climate change and meet the needs of society, the economy and the environment. *Mitigation and Adaptation Strategies for Global Change* 8: 349–370.

Singh RS, & Walia AK(2016) Biofuels: Historical perspectives and public opinions. In: *Biofuels: Production and Future Prospectus*, eds. Singh RS, Pandey A, & Gnansounou E, CRC Press, Boca Raton, Florida, pp. 3–23.

Souza GM, Ballester MVR, Cruz CHB, Chum H, Dale B, Dale VH, Fernandes ECM, Foust T, Karp A, Lynd L, Filho RM, Milanez A, Nigro F, Osseweijer P, Verdade LM, Victoria RL, & Wielen LW (2017) The role of bioenergy in a climate-changing world. *Environmental Development* 23: 57–64.

Verma O (2021) Climate change and its impacts with special reference to India. In *Water, Cryosphere, and Climate Change in the Himalayas*, eds. Talool AK, Kotlia BS, & Kumar K, Springer Nature, Switzerland, pp. 37–55.

Welfle A (2017) Balancing growing global bioenergy resource demands – Brazil's biomass potential and the availability of resource for trade. *Biomass Bioenergy* 105: 83–95.

Wu Y, Zhao F, Liu S, Wang L, Qiu L, Alexandrov G, & Jothiprakash V (2018) Bioenergy production and environmental impacts. *Geoscience Letters* 5: 14. https://geoscienceletters.springeropen.com/track/pdf/10.1186/s40562-018-0114-y.pdf

www.climate.gov/news-features/understanding-climate/climate-change-atmospheric-carbon-dioxide (Retrieved on or before April 2021).

www.opec.org/opec_web/en/about_us/24.htm (Retrieved on or before April 2021).

Xiao-Yang Z (2016) Developing bioenergy to tackle climate change: bioenergy path and practice of Tianguan group. *Advances in Climate Change Research* 7: 17–25.

20 A Review of Fluoride Contamination in Groundwater and Its Removal by Adsorption Process

Chandra Bhan and Jiwan Singh
Laboratory of Environmental Nanotechnology and Bioremediation, Department of Environmental Science, Babasaheb Bhimrao Ambedkar University, Lucknow

CONTENTS

20.1 INTRODUCTION

Water is a natural gift on the earth that makes life possible. About a third of the earth is covered with water, but only a little of this is freshwater, which is used for different purposes. Due to the rapidly growing population and industrialization, water has been contaminated by various foreign chemical compounds, and the load of these pollutants is increasing continuously nowadays. For the existence of life on earth, water is a vital resource. Our earth contains 1386 million km^3 of water in total. Of the 10.63 million km^3 that is freshwater, only 30.1% occurs as groundwater. Due to the growth of the population, the requirement of groundwater for drinking purposes has increased (Mukherjee et al., 2018). Groundwater is considered as 'a hidden sea'; this source of water has been extracted by 50% of the population of the world, and of this, 70% of water is used for agricultural practices (Jadhav et al., 2015). Natural and anthropogenic inputs have altered the quality of water. Fluoride is the most prevalent pollutant in terms of health effects and the number of populations affected worldwide (Sahu et al., 2018). Groundwater contamination with fluoride is a universal issue. Fluoride is an essential component for human teeth and bone development. However, high concentrations cause serious and long-lasting health problems. About 85 million tons of fluoride are present in the earth's crust, while 12 million tons are found in the geographical region of India. The rift system of Africa, including Jordan, Kenya, and Tanzania, the Middle East region (Iraq, Iran, Syria), and south Asia (India, Sri

DOI: 10.1201/9781003220824-23

Lanka, Pakistan) contain a high concentration of fluoride in groundwater (Yadav et al., 2019).The level of fluoride concentration in the Kenyan rift valley is 11 mg/L, while the Tanzanian rift valley contains 4.6 mg/L (Demelash et al., 2019). Fluorite, fluorspar, topaz, mica, cryolite, apatite, sellaite, and other minerals contain a large quantity of fluoride (Banerjee, 2015; Ali et al., 2019). Igneous and sedimentary rock sand mineralized veins are the main and abundant sources of these minerals (Ali et al., 2016). Different factors such as pH, ionic concentration, carbonate and bicarbonate ion-based mobilization, rock–water interactions, atmospheric deposition, and contact time are responsible for the dissolution of fluoride in groundwater (Jha and Tripathi, 2021). Calcium fluoride is the natural source of reactive fluorine. Granite has five times higher concentrations compared with basalt rocks. However, a more abundant concentration of fluoride occurs in shale compared with limestone and sandstone. About 1200 to 8500 mg/kg is available in alkaline rocks (Bibi et al., 2017). This book chapter covers the background on fluoride contamination in groundwater, main sources of fluoride, health effects, and mitigation strategies. The adsorption technique is the main focus of this chapter, including several adsorbents, preparation methods, and adsorption capacity.

20.1.1 Fluoride Contamination in Groundwater of Different Regions of the World

Weathering and leaching of fluoride-containing rocks are the main sources of fluoride in the groundwater of the Pakistan region (Rahman et al., 2018). Fluoride-bearing minerals, alkaline conditions, and dry climate are the controlling factors for fluoride in groundwater in the area of Prakasam district, Andhra Pradesh (Rao, 2017). Ion exchange and mineral dissolution related to water–mineral interactions are responsible for the increasing amount of fluoride in groundwater (Yuan et al., 2020).

Huaibei plain of Northern China is a fluoride-rich region due to the weakly alkaline conditions, where sodium and calcium ions are in excess (Hu et al., 2017). Fluoride concentration in the groundwater of Yuncheng Basin, China was found to be up to 14.1 mg/L (Li et al, 2018). Safe drinking water is not available for 680 million people in 13 countries, including Uganda, Sudan, Malawi, Kenya, Ethiopia, Tanzania, Eritrea, and Mozambique in Sub-Sahara Africa. Moreover, in the Rift Valley of east Africa, a high fluoride content has been reported in groundwater (Kut et al., 2016). Okala et al. (2016) have reported fluoride concentrations in the range of 0.22 to 74.98 mg/L. In muscovite, cordierite, and villiaumite minerals, the highest fluoride contents of 308 to 6366 mg/L were reported. Brindha and Elango (2011) reported a high level of fluoride in several regions of the world, such as Ghana (282. 29 mg/L), Sri Lanka (5 mg/L), Iran (0.2 to 9.2 mg/L), Pakistan (0.11 to 22.8 mg/L), Korea (40.8 mg/L), Ethiopia (0.01 to 13 mg/L), and Tanzania (15 to 63 mg/L). The plain area of Balod, Birbhum, and Brahmputra, India, reported fluoride concentration in the range from 10 to 20 mg/L (Yadav et al., 2016; Batabyal and Gupta, 2017). Jirsa et al. (2013) reported the maximum level of fluoride concentration in the Bogoria lakes, Africa to be in the range of 530 to 1310 mg/L. The thermal water of South Korea contained 40.8 mg/L of fluoride (Chae et al., 2007), and thermal springs in Ethiopia also have fluoride levels of 78 mg/L (Ayenew, 2008). Fluoride concentrations in different areas of the world are reported in Table 20.1.

20.2 FLUORIDE CONTAMINATION IN DIFFERENT AREAS OF INDIA

The groundwater of the semi-arid region of Nalgonda, Andhra Pradesh has been contaminated with a high concentration of fluoride, and 5.9 mg/L of fluoride was found in the Shaslar Vagu watershed (Kumar et al., 2017a). Fluoride in elevated levels has been reported in several states of India, including Rajasthan (0.2 to 69 mg/L), Gujarat (1.58 to 31 mg/L), Assam (0.2 to 23 mg/L), Haryana (0.17 to 48 mg/L), and Delhi (Kisku and Sahu, 2020). Keesari et al. (2021) have reported fluoride content ranging from 0.04 to 8.2 mg/L in the north-eastern area of Rajasthan. Raj and Shaji (2017)

TABLE 20.1
Fluoride Concentration in Groundwater across the World

Area/Country	Fluoride Concentration (mg/L)	Reference
Hail, Saudi Arabia	2.8	Al-Khateeb et al., 1991
Badghis, Afghanistan	0.02–63	Hayat and Baba, 2017
Durgai, Pakistan	0.7–4.9	Rashid et al., 2020
Sindh, Pakistan	0.23–6.8	Talpur et al., 2020
Punjab, Pakistan	0.3–2.5	Masood et al., 2021
Middle and Eastern region of Turkey	13.7	WHO, 2002
Savelugu-Nanton, Ghana	0.1–4.1	Tay, 2017
Texas, USA	0.3–4.3	Reardon and Wang, 2000
Munster, Germany	8.8	Queste et al., 2001
Laxemar, Sweden	0.3–74	Berger et al., 2016
Hordaland, Norway	0.02–9.48	BArdsen et al., 1999
Finland	More than 3	Azbar and Turkman, 2000
Banyuputih, Indonesia	0.1–4.2	Heikens et al., 2005
Indonesia	0.1–4.2	WHO, 2006
Estonia	0.01–7.2	Indermitte et al., 2009
Southeastern Tunisia	29	Melki et al., 2020
Kutain, China	21.5	Wang et al., 1997
Shaanxi, China	1.08	Gao et al., 2020
Sri Lanka	0.01–4.34	Young et al., 2011
Japan	1.4	Tsutsui et al., 2000
Quasim province	5.4	Yadav et al., 2019

investigated the presence of fluoride in groundwater of the highly populated town Alleppey of Kerala, where fluoride concentration was found in the range of 0.68 to 2.88 mg/L; fluorapatite minerals are a rich source of fluoride in this region. Das and Bhattacharjya (2020) found fluoride content from 0.04 to 4.7 mg/L in groundwater and 0.02 to 3.75 mg/L in the case of Bharalu river water. The surface water of the Dungarpur, Udaipur, and Banswara region of Rajasthan is contaminated with fluoride having a concentration up to 3.05 mg/L (Choubisa and Choubisa, 2015). Chouhan (2016) investigated the presence of fluoride in a range from 0.5 to 8.4 mg/L in 25 different villages of Sirohi district, Rajasthan.

Several regions of the Unnao and Kanpur districts have been investigated for high concentrations of fluoride in groundwater. It was found in the range of 0.8 to 13.9 mg/L in Unnao and 0.14 to 5.34 mg/L in Kanpur (Ali et al., 2019). Dutta et al., (2019) found a high level of fluoride in Chaohatta and Chitisapur villages of Fatehpur, Uttar Pradesh, which had 0.83 to 7.6 mg/L of fluoride concentration present in groundwater. About 60% of the population of Chaohatta suffer from dental and skeletal fluorosis, and many people are found physically handicapped. Fluoride concentrations in different regions of India are listed in Table 20.2.

20.3 SOURCES OF FLUORIDE IN GROUNDWATER

Fluorine-bearing rocks are the primary and geogenic sources of fluoride ions in the groundwater. Geogenic minerals such as fluorite, fluorspar, cyolite, mica, and topaz contain a large quantity of fluoride (Banerjee, 2015). Igneous rocks, sedimentary structures, and mineralized veins are rich sources of fluoride (Ali et al., 2016). Table 20.3 shows the amount of fluorine present in the minerals. Industries such as iron smelters, phosphate fertilizers, pesticides, ceramics, and coal-based thermal power stations are secondary and artificial sources of fluoride, which is released with

TABLE 20.2
Fluoride Concentration in Groundwater of Different Areas of India

Local Area/State	Fluoride Concentration (mg/L)	References
Chandrapur, Maharashtra	0.44 to 7.52	Dubey et al., 2018
Yavatmal, Maharashtra	0.3 to 11.9	Madhnure et al., 2007
Karbi Anglong, Assam	0.15 to 17.13	Hanse et al., 2019
Bhagalpur, Bihar	1.93 to 2.98	Gouri and Choudhary, 2017
Koshi, Bihar	0.6 to 1.6	Singh and Singha, 2009
Jamui district	3.6 to 5.8	Kumar et al., 2018
Gaya, Bihar	0.19 to 14.4	Yasmin et al., 2011
Balod, Chhattishgarh	1.5 to 14.0	Yadav et al., 2016
Chouhatta, Uttar Pradesh	0.83 to 7.6	Dutta et al., 2019
Varanasi, UP	0.28–2.01	Chaurasia et al., 2018
Sonbhdra, UP	0.08–6.7	Dey and Raju, 2016
Unnao, UP	0.06–1.83	Ansari and Umar, 2019
Banda, UP	0.32–3.5	Singh, 2016
Agra, UP	0.14–4.88	Ali et al., 2017
Pratapgarh, UP	0.41–3.99	Tiwari et al., 2017
Patan, Gujarat	0.4 to 4.8	Kumar et al., 2017b
Chirala and Ongole, Andhra Pradesh	0.6 to 1.8	Rao et al., 2017
Basara, Telangana	0.06 to 4.3	Narsimha and Sudarshan, 2017
Madurai, Tamil Nadu	1.8	Thivya et al, 2017
Mahendergarh, Haryana	0.3 to 16.0	Yadav et al., 2019
Bankura, West Bengal	11.51	Rudra and Khan, 2018
Guntur, Andhra Pradesh	1.3 to 12.9	Subba Rao et al., 2020
Vijayapura, Karnatka	0.3 to 4.8	Ugran et al., 2017

TABLE 20.3
Fluoride-Containing Minerals and Their Chemical Formula

Minerals	Chemical Formula	Fluorine %
Sellite	MgF_2	61
Fluorspar	CaF_2	49
Cryolite	Na_3AlF_6	45
Fluorapatite	$Ca_5(PO_4)_3F$	3–4
Svabite	$Ca_5(AsO_4)_3F$	2.99
Topaz	$Al_2SiO4(F,OH)_2$	11.47
Villianmite	NaF	55

effluents into the water body and subsequently enters the groundwater (Kumar, 2017; Preethi and Meenakshi, 2018).Groundwater contamination with fluoride and its effects on humans are presented in Figure 20.1.

20.3.1 Risk of Fluoride to Health

Accumulation of an excessive concentration of fluoride in bones causes skeletal fluorosis. Hardening of bones happens due to the fluoride presence, and the elasticity of bones is decreased. A higher

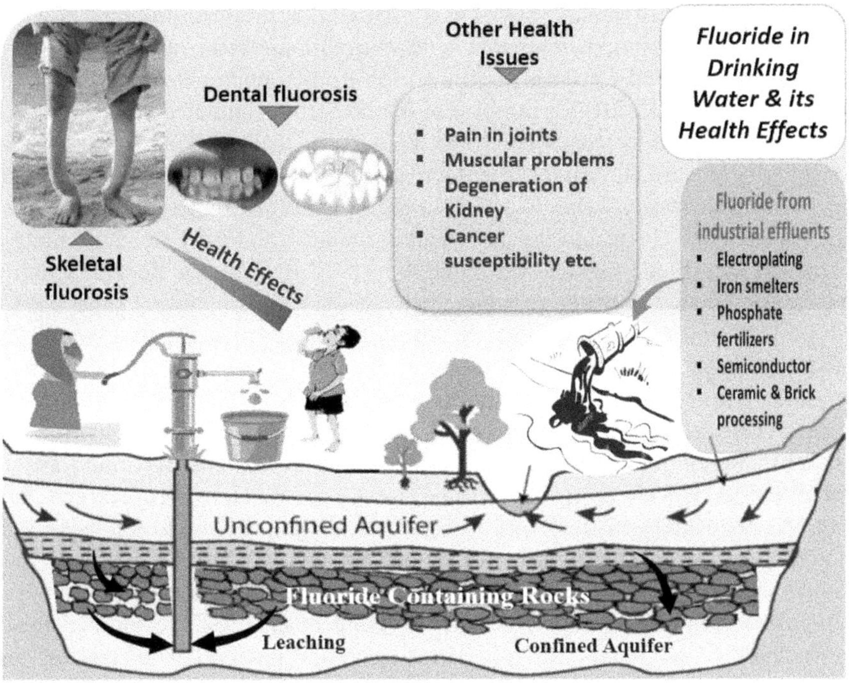

FIGURE 20.1 Fluoride contamination in water and its effects on human health.

TABLE 20.4
Health Effect of Fluoride

Concentration of Fluoride in Water (mg/L)	Health Effect
Less than 0.5	Beneficial for dental caries
0.5 to 1.5	Strengthens teeth and bones
1.5 to 4.0	Causes dental fluorosis
More than 4	Causes skeletal fluorosis
More than 10	Crippling fluorosis in bones and causes cancer

concentration of fluoride in bone tissues disrupts the mobility of joints. The function of ligaments and cartilage can be obstructed (Kalia et al., 2010). Excessive intake of fluoride causes dysfunction of the thyroid gland (Kheradpisheh et al., 2018). Naseem et al. (2016) reported fertility rate reduction in females. Moreover, it causes various health issues such as anorexia, dyspnea, restlessness, gastroenteritis, sweating, stiffness, tachycardia, and muscle weakness (Sahu et al., 2017). Pranesh et al. (2019) reported compressive cervical myeloradiculopathy and neurological complications in a 40-year-old man due to skeletal fluorosis. Different concentration levels in drinking water and their health effects are described in Table 20.4.

20.4 APPLICATION OF DIFFERENT TYPES OF ADSORBENTS

Elimination of excess fluoride from drinking water and wastewater is currently performed by methods such as membrane technique, coagulation and precipitation, electrochemical process, ion exchange, and their advances (Habuda-Stanić et al., 2014). But, the adsorption method is considered a cost-effective, simple operation compared with conventional removal methods. A number of adsorbents have been reported for fluoride elimination from contaminated water. Lai et al. (2018) applied zeolite modified by La(III) for removal of fluoride from simulated zinc sulfate solution and obtained the highest adsorption capacity of 20.83 mg/g. Biftu et al. (2020) carried out defluorination of water by applying adsorbent derived from activation of *Leucaena leucocephala* by nitric acid, and the adsorption capacity was found to be 1.16 mg/g. Chen et al. (2009) synthesized the adsorbent by using sand and polymer latex and the adsorbent modified with Fe-Al-Ce nano-material, and the adsorption capacity of coated granules was found to be 2.22 mg/g for the fluoride elimination.

Activated carbon having magnetic properties was developed using sweet lime waste and $FeCl_3$ at 250°C and 500°C and used for fluoride removal. The adsorption capacity was 7.75 mg/g for ACP-250 and 12.6 mg/g for ACP-500 (Ibrahim et al., 2019). Iron-infused carbon was prepared from *Pisum sativum* peel and applied for fluoride adsorption from aqueous solution by a fixed-bed column process (Sahu et al., 2020). Bhan et al. (2021) synthesized adsorbent from mentha ash with sodium and aluminum and used it for fluoride ion removal. The maximum adsorption capacity was observed to be 87% at neutral pH and normal temperature. Mullick and Neogi (2019) prepared Mg-Mn-Zr-loaded activated carbon by ultrasonic exposure and applied it for effective fluoride removal from water. The highest adsorption capacity was found to be 26.27 mg/g at equilibrium. Mullick and Neogi (2018) synthesized zirconium-activated adsorbent by an ultrasonication process and observed 94.4% removal of fluoride. Pang et al. (2020) examined the adsorption capacity of Zr-modified activated carbon by the drop-casting method. The removal capacity was 28.5 mg/g at normal temperature. Velazquez-Jimenez et al. (2014) also used a zirconium–carbon hybrid adsorbent for fluoride removal. Ghosh et al. (2014) reported various composite materials such as iron-activated attapulgite, La-loaded chitosan beads, Al(III)–Ce(IV) combined adsorbent, CaO nanoparticles, and $La3^{+}$, $Al3^{+}$ $ZrO2^{+}$ activated zeolite with good fluoride adsorption capacity. Carbon nanotubes and hydroxyapatite composite were prepared by co-precipitation method and applied for fluoride adsorption, and a good adsorption capacity of 11.05 mg/g was observed (Tang et al., 2018). Inaniyan and Raychoudhury (2019) used activated carbon impregnated with cerium nitrate ($CeNO_3$) to eliminate fluoride from groundwater. The highest removal capacity of adsorbent (4.6 mg/g) was evaluated by an adsorption isotherm study. Talat et al. (2018) synthesized activated carbon by using coconut husk at 700°C in the presence of N_2 gas. The prepared carbon was modified by KOH and dehydrated at 350°C for 3 hours in a nitrogen environment. The maximum removal capacity of activated carbon was reported to be 6.5 mg/g with adsorbent amount 1.4 g/L at pH 5. Activated carbon was derived from *Crocus sativus* leaves and modified by calcium chloride. The highest removal percentage of fluoride was found 85.43% at pH 4, adsorbent amount 15 g, and contact time 70 min. The Freundlich isotherm was best correlated for fluoride adsorption on adsorbent (Dehghani et al., 2018). Corn husk and rice husk were used to prepare activated carbon treated with hydrochloric acid and sodium hydroxide, respectively. Maximum fluoride uptake capacity was observed to be 5.8 mg/g and 7.9 mg/g for corn husk and rice husk, respectively (Gebrewold et al., 2018). Bibi et al. (2017) conducted a fluoride adsorption study using an adsorbent prepared from rice husk and potato peel. Raw material was converted into an activated carbon in a muffle furnace at 700°C for 120 min. Fluoride adsorption capacity was found to be 2.91 mg/g. Mohanta and Ahmaruzzaman (2018) reported a nanocomposite with carbon and SnO_2 for fluoride removal. Sawdust was converted into carbon by a chemical vapor deposition process, and it was impregnated with SnO_2 nanoparticles by sonication treatment. Sivarajasekar et al. (2017) reported spirogyra biomass activated carbon synthesized by a thermal treatment process and applied to remove fluoride. The highest percentage of fluoride removal was found to be 95% under optimum conditions.

Morinda tinctoria activated carbon modified by aluminum hydroxide was used to eliminate fluoride from contaminated water. The intraparticle diffusion model and Freundlich isotherm model best described the adsorption of fluoride on aluminum activated carbon, and the maximum removal

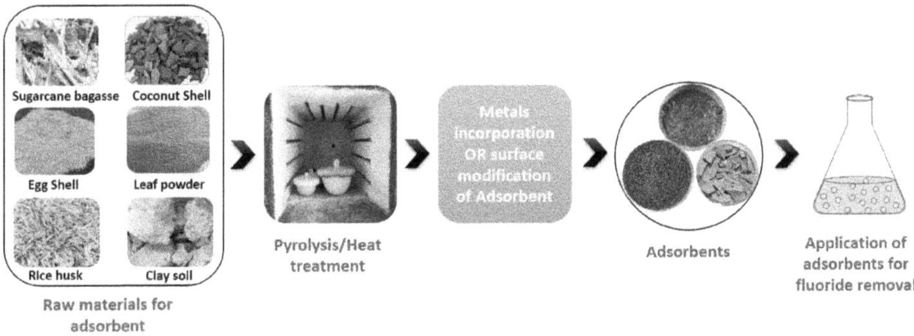

FIGURE 20.2 Common process of adsorbent preparation.

percentage achieved was 26.03 mg/g (Amalraj and Pius, 2017). Palm shell activated carbon was modified by chitosan from prawn shell and steam treatment. Maximum adsorption was 55% at neutral pH, and pseudo-second-order kinetics fitted well for this adsorption mechanism (Issabayeva et al., 2021). Mukherjee et al. (2021) applied the adsorbent for fluoride elimination from water. Activated carbon was prepared from paper sludge generated by industry. Dehydrated paper sludge was pyrolyzed at 950°C for 60 min. Pyrolyzed material was modified by orthophosphoric acid. The highest adsorption percentage observed was 99% at optimum conditions (0.3 g/L dose, pH 5, 25°C, and 12 mg/L fluoride concentration). Activated carbon from waste tires modified by bimetals (iron and cobalt) was applied for fluoride elimination from wastewater. Maximum fluoride adsorption was 85.34% at adsorbent dose 1.25 g/L, pH 2, and 30.9 min contact time (Bose et al., 2021). Sewage sludge was used to prepare activated carbon and modified by trivalent lanthanum ion. The prepared adsorbent was applied for fluoride removal. Lanthanum-modified activated carbon showed 90% removal of fluoride and after regeneration of adsorbent, fluoride removal was 73.8% (Zhang et al., 2021). Takmil et al. (2020) reported the synthesis of activated carbon/Fe_3O_4 composites from oak shells. The oak shell was converted into activated carbon at 800°C and subsequently, an AC/Fe_3O_4 composite was prepared. The synthesized composite showed 454.54 mg/g adsorption capacity during fluoride adsorption.

Nagaraj et al. (2020) synthesized adsorbent from bentonite modified with tri-metals (aluminum, lanthanum, and cerium) and also decorated it with chitosan. Fluoride elimination efficiency of 9.87 mg/g was achieved from groundwater. Fluoride concentration in groundwater decreased to the level of 0.63 mg/L from 2.08 mg/L after adsorption. Many more alumina and clay adsorbents have been used for defluoridation of water and are listed in Table 20.5. The common process of adsorbent synthesis for fluoride removal is shown in Figure 20.2.

20.5 CONCLUSION

The literature has reported that most areas of the world are contaminated with a high level of fluoride. Africa, Tanzania, Ghana, Pakistan, India, and other regions of the world are badly affected. About 530 mg/L to 1310 mg/L was reported in Bogoria lakes of Africa. It is revealed that different adsorbents with efficient adsorption capacity have been synthesized; however, there is a high-priority need for fluoride removal on a large scale. There is a need to enhance the economic feasibility of the adsorption process by regeneration and removal of adsorbents. Regeneration and reutilization of adsorbent can reduce the cost of adsorbent and its process time of development. Based on the literature review, various types of adsorbents have been developed from a large number of waste materials and biomass by surface modification. Therefore, it's an urgent need to implement the application of adsorbents for large-scale water treatment. The majority of researchers applied the

TABLE 20.5
Fluoride Removal with Different Types of Adsorbents

Adsorbent	Modification	Adsorption capacity/ Percentage	References
Red mud	Hydrochloric acid	3.95 mg/g	Guo et al., 2014
Natural diatomite and ignimbrite	Sulfuric acid	12.03 mg/g and 8.82 mg/g	Kır et al., 2016
Activated alumina	Aluminum tri-sec butoxide	90%	Rafique et al., 2013
Activated alumina	Lanthanum	6.7 mg/g	Cheng et al., 2014
Magnetite	Lanthanum oxide and aluminum oxide	96 and 90%	Garcia-Sanchez et al., 2016
Alumina	Manganese		You et al., 2021
Diatomite	Aluminum hydroxide	89.4%	Akafu et al., 2019
Hydroxyapatite–carbon nanotube	Carbon nanotubes	11.05 mg/g	Tang et al., 2018
Laterite soil		0.48 mg/g	Iriel et al., 2018
Crocus sativus leaves	Calcium chloride	2.01 mg/g	Dehghani et al., 2018
Activated carbon	Sulfuric acid	98.36%	Mihayo et al., 2021

prepared adsorbent for the removal of fluoride from aqueous solution. Adsorbent should be applied for groundwater treatment. A wide range of adsorbents have been synthesized by different processes and from various materials. Particularly, carbon materials such as activated carbon/nanocomposites and carbon nanotubes have shown efficient removal capacity, but these adsorbents were relatively high cost, not suitable for large-scale employment. Activated carbon, clay, and zeolite-based adsorbents were observed to be better than chemically synthesized composites. Waste biomass has the advantage of low cost and is easily and locally available in huge amounts. Modification of adsorbents by metals is attracting a lot of attention due to being non-toxic, environmentally friendly, and commercially cheap.

REFERENCES

Akafu, T., Chimdi, A., and Gomoro, K. 2019. Removal of fluoride from drinking water by sorption using diatomite modified with aluminum hydroxide. *J Anal Methods Chem* 2019:1–11.

Ali, S., Thakur, S. K., Sarkar, A., and Shekhar, S. 2016. Worldwide contamination of water by fluoride. *Environ Chem Lett* 14:291–315.

Ali, S., Kumari, M., Gupta, S. K., Sinha, A., and Mishra, B. K. 2017. Investigation and mapping of fluoride-endemic areas and associated health risk – a case study of Agra, Uttar Pradesh, India. *Hum Ecol Risk Assess* 2:590–604.

Ali, S., Fakhri, Y., Golbini, et al. 2019. Concentration of fluoride in groundwater of India: a systematic review, meta-analysis and risk assessment. *Groundw Sustain Dev* 9:100224.

Al-Khateeb, T. L., Al-Marsafi, A. I., and O'Mullane, D. M. 1991. Caries prevalence and treatment need amongst children in an Arabian community. *Commun Dent Oral Epidemiol* 19:277–280.

Amalraj, A., and Pius, A. 2017. Removal of fluoride from drinking water using aluminum hydroxide coated activated carbon prepared from bark of *Morinda tinctoria*. *Appl Water Sci* 7:2653–2665.

Ansari, J. A., and Umar, R. 2019. Evaluation of hydrogeochemical characteristics and groundwater quality in the quaternary aquifers of Unnao District, Uttar Pradesh, India. *Hydro Res* 1:36–47.

Ayenew, T. 2008. The distribution and hydrogeological controls of fluoride in the groundwater of central Ethiopian rift and adjacent highlands. *Environ Geol* 54:1313–1324.

Azbar, N. U. R. I., and Turkman, A. 2000. Defluoridation in drinking waters. *Water Sci Technol* 42:403–407.

Banerjee, A. 2015. Groundwater fluoride contamination: a reappraisal. *Geosci Front* 6:277–284.

Bardsen, A., Klock, K. S., and Bjorvatn, K. 1999. Dental fluorosis among persons exposed to high-and low-fluoride drinking water in western Norway. *Community Dent Oral Epidemiol* 27:259–267.

Batabyal, A. K., and Gupta, S. 2017. Fluoride-contaminated groundwater of Birbhum district, West Bengal, India: interpretation of drinking and irrigation suitability and major geochemical processes using principal component analysis. *Environ Monit Assess* 189:1–24.

Berger, T., Mathurin, F. A., Drake, H., and Astrom, M. E. 2016. Fluoride abundance and controls in fresh groundwater in Quaternary deposits and bedrock fractures in an area with fluorine-rich granitoid rocks. *Sci Total Environ* 569:948–960.

Bhan, C., Singh, J., and Sharma, Y. C. 2021. Development of adsorbent from Mentha plant ash and its application in fluoride adsorption from aqueous solution: a mechanism, isotherm, thermodynamic, and kinetics studies. *Int J Phyto Remediation* 23:1–11.

Bibi, S., Farooqi, A., Yasmin, A., Kamran, M. A., and Niazi, N. K. 2017. Arsenic and fluoride removal by potato peel and rice husk (PPRH) ash in aqueous environments. *Int J Phyto Remediation* 19:1029–1036.

Biftu, W. K., Ravulapalli, S., and Kunta, R. 2020. Effective de-fluoridation of water using *Leucaena leucocephala* active carbon as adsorbent. *Int J Environ Res* 14:415–426.

Bose, S., Mukherjee, T., and Rahaman, M. 2021. Simultaneous adsorption of manganese and fluoride from aqueous solution via bimetal impregnated activated carbon derived from waste tire: response surface method modeling approach. *Environ Prog Sustain Energy* 40:13600.

Brindha, K., and Elango, L. 2011. Fluoride in groundwater: causes, implications and mitigation measures. *Fluoride Proper Appl Environ Manage* 1:111–136.

Chae, G. T., Yun, S. T., Mayer, B., Kim, K. H., Kim, S. Y., Kwon, J. S., and Koh, Y. K. 2007. Fluorine geochemistry in bedrock groundwater of South Korea. *Sci Total Environ* 385:272–283.

Chaurasia, A. K., Pandey, H. K., Tiwari, S. K., Prakash, R., Pandey, P., and Ram, A. 2018. Groundwater quality assessment using water quality index (WQI) in parts of Varanasi district, Uttar Pradesh, India. *J Geol Soc India* 92:76–82.

Chen, L., Wu, H. X., Wang, T. J., Jin, Y., Zhang, Y., and Dou, X. M. 2009. Granulation of Fe–Al–Ce nano-adsorbent for fluoride removal from drinking water by spray coating on sand in a fluidized bed. *Powder Technol* 193:59–64.

Cheng, J., Meng, X., Jing, C., and Hao, J. 2014. $La3^{+}$-modified activated alumina for fluoride removal from water. *J Hazard Mater* 278:343–349.

Choubisa, S. L., and Choubisa, D. 2015. Neighbourhood fluorosis in people residing in the vicinity of superphosphate fertilizer plants near Udaipur city of Rajasthan (India). *Environ Monit Assess* 187:1–6.

Chouhan, H. S. 2016. Physico–chemical investigation of ground water of Sirohi District (Raj.). *J Environ Sci Comput Eng Technol* 5:64–74.

Das, M., and Bhattacharjya, R. K. 2020. A regression-based analysis to assess the impact of fluoride reach river water on the groundwater aquifer adjacent to the river: a case study in Bharalu River Basin of Guwahati, India. *Pollution* 6:637–650.

Dehghani, M. H., Farhang, M., Alimohammadi, M., Afsharnia, M., and Mckay, G. 2018. Adsorptive removal of fluoride from water by activated carbon derived from $CaCl_2$-modified *Crocus sativus* leaves: equilibrium adsorption isotherms, optimization, and influence of anions. *Chem Eng Commun* 205:955–965.

Demelash, H., Beyene, A., Abebe, Z., and Melese, A. 2019. Fluoride concentration in ground water and prevalence of dental fluorosis in Ethiopian Rift Valley: systematic review and meta-analysis. *BMC Public Health* 19:1–9.

Dey, S., and Raju, N. J. 2016. Hydrogeochemical assessment to explore the extent, nature and source of fluoride contamination within the groundwaters of the Panda River Basin, Sonbhadra District, Uttar Pradesh, India. In: Raju, N. (ed.) *Geostatistical and Geospatial Approaches for the Characterization of Natural Resources in the Environment* 221–229. Springer: Cham.

Dubey, M. S., Deshpande, S. M., Gaikwad, G. D., and Dongre, A. 2018. Fluoride contaminated groundwater from Warora area of Chandrapur District, Central India: groundwater geochemistry and comments on the source of fluoride. *J Appl Geochem* 20:123–132.

Dutta, V., Fatima, N., and Kumar, N. 2019. Excessive fluoride in groundwater of Central Ganga Alluvial Plain: a case study of Fatehpur, North India. *Int J Sci Technol* 16:7791–7798.

Gao, Y., Qian, H., Wang, H., Chen, J., Ren, W., and Yang, F. 2020. Assessment of background levels and pollution sources for arsenic and fluoride in the phreatic and confined groundwater of Xi'an city, Shaanxi, China. *Environ Sci Pollut Res* 27:34702–34714.

Garcia-Sanchez, J. J., Solache-Rios, M., Martinez-Gutierrez, J. M., Arteaga-Larios, N. V., Ojeda-Escamilla, M. C., and Rodriguez-Torres, I. 2016. Modified natural magnetite with Al and La ions for the adsorption of fluoride ions from aqueous solutions. *J Fluor Chem* 186:115–124.

Gebrewold, B. D., Kijjanapanich, P., Rene, E. R., Lens, P. N., and Annachhatre, A. P. 2018. Fluoride removal from groundwater using chemically modified rice husk and corn cob activated carbon. *Environ Technol* 40:2913–2927.

Ghosh, A., Chakrabarti, S., Biswas, K., and Ghosh, U. C. 2014. Agglomerated nanoparticles of hydrous Ce (IV)+ Zr (IV) mixed oxide: preparation, characterization and physicochemical aspects on fluoride adsorption. *Appl Surf Sci* 307:665–676.

Gouri, K., and Choudhary, S. K. 2017. Fluoride contamination in groundwater sources of Bhagalpur Municipal Corporation Area, Bhagalpur, Bihar. *IOSR J Environ Sci Toxicol Food Technol* 11:45–49.

Guo, H., Yang, L., and Zhou, X. 2014. Simultaneous removal of fluoride and arsenic from aqueous solution using activated red mud. *Sep Sci Technol* 49:2412–2425.

Habuda-Stanić, M., Ravancic, M. E., and Flanagan, A. 2014. A review on adsorption of fluoride from aqueous solution. *Materials* 7:6317–6366.

Hanse, A., Chabukdhara, M., Baruah, S. G., Boruah, H., and Gupta, S. K. 2019. Fluoride contamination in groundwater and associated health risks in Karbi Anglong District, Assam, Northeast India. *Environ Monit Assess* 191:1–17.

Hayat, E., and Baba, A. 2017. Quality of groundwater resources in Afghanistan. *Environ Monit Assess* 189:1–16.

Heikens, A., Sumarti, S., Van Bergen, M., et al. 2005. The impact of the hyperacid Ijen Crater Lake: risks of excess fluoride to human health. *Sci Total Environ* 346:56–69.

Hu, Y., Xia, C., Dong, Z., and Liu, G. 2017. Geochemical characterization of fluoride in the groundwater of the Huaibei Plain, China. *Anal Lett* 50:889–903.

Ibrahim, M., Siddique, A., Verma, L., Singh, J., and Koduru, J. R. 2019. Adsorptive removal of fluoride from aqueous solution by biogenic iron permeated activated carbon derived from sweet lime waste. *Acta Chim Slov* 66:123–136.

Inaniyan, M., and Raychoudhury, T. 2019. Application of activated carbon–metal composite for fluoride removal from contaminated groundwater in India. *Int J Environ Sci Technol* 16:7545–7554.

Indermitte, E., Saava, A., and Karro, E. 2009. Exposure to high fluoride drinking water and risk of dental fluorosis in Estonia. *Int J Environ Res Public Health* 6:710–721.

Iriel, A., Bruneel, S. P., Schenone, N., and Cirelli, A. F. 2018. The removal of fluoride from aqueous solution by a lateritic soil adsorption: kinetic and equilibrium studies. *Ecotoxicol Environ Saf* 149:166–172.

Issabayeva, G., Wong, S. H., Pang, C. Y., Wong, M. C., and Aroua, M. K. 2021. Fluoride removal by low-cost palm shell activated carbon modified with prawn shell chitosan adsorbents. *Int J Environ Sci Technol* 19:1–10.

Jadhav, S. V., Bringas, E., Yadav, G. D., Rathod, V. K., Ortiz, I., and Marathe, K. V. 2015. Arsenic and fluoride contaminated groundwaters: a review of current technologies for contaminants removal. *J Environ Manage* 162:306–325.

Jha, P. K., and Tripathi, P. 2021. Arsenic and fluoride contamination in groundwater: a review of global scenarios with special reference to India. *Groundw Sustain Dev* 13:1–27.

Jirsa, F., Gruber, M., Stojanovic, A., et al. 2013. Major and trace element geochemistry of Lake Bogoria and Lake Nakuru, Kenya, during extreme draught. *Geochemistry* 73:275–282.

Kalia, L. V., Lee, L., Kalia, S. K., et al. 2010. Thoracic myelopathy from coincident fluorosis and epidural lipomatosis. *Canad J Neurol Sci* 37:276–278.

Keesari, T., Pant, D., Roy, A., et al. 2021. Fluoride geochemistry and exposure risk through groundwater sources in northeastern parts of Rajasthan, India. *Arch Environ Contam Toxicol* 80:294–307.

Kheradpisheh, Z., Mirzaei, M., Mahvi, A. H., et al. 2018. Impact of drinking water fluoride on human thyroid hormones: a case-control study. *Sci Rep* 8:1–7.

Kır, E., Oruc, H., Kır, I., and Sardohan-Koseoglu, T. 2016. Removal of fluoride from aqueous solution by natural and acid-activated diatomite and ignimbrite materials. *Desalin Water Treat* 57:21944–21956.

Kisku, G. C., and Sahu, P. 2020. Fluoride contamination and health effects: an Indian scenario. In: Shukla, V., Kumar, N. (eds) *Environmental Concerns and Sustainable Development Vol. 1* 213–233. Springer: Singapore.

Kumar, M. S., Dhakate, R., Yadagiri, G., and Reddy, K. S. 2017a. Principal component and multivariate statistical approach for evaluation of hydrochemical characterization of fluoride-rich groundwater of Shaslar Vagu watershed, Nalgonda District, India. *Arabian J Geosci* 10: 1–17.

Kumar, P., Singh, C. K., Saraswat, C., Mishra, B., and Sharma, T. 2017b. Evaluation of aqueous geochemistry of fluoride enriched groundwater: a case study of the Patan district, Gujarat, Western India. *Water Sci* 31:215–229.

Kumar, P. S. 2017b. Fluoride in groundwater-sources, geochemical mobilization and treatment options. *Int J Environ Sci Nat Resour* 1:106–108.

Kumar, S., Venkatesh, A. S., Singh, R., Udayabhanu, G., and Saha, D. 2018. Geochemical signatures and isotopic systematics constraining dynamics of fluoride contamination in groundwater across Jamui district, Indo-Gangetic alluvial plains, India. *Chemosphere* 205:493–505.

Kut, K. M. K., Sarswat, A., Srivastava, A., Pittman Jr, C. U., and Mohan, D. 2016. A review of fluoride in African groundwater and local remediation methods. *Groundw Sustain Dev* 2:190–212.

Lai, Y. Q., Yang, K., Yang, C., Tian, Z. L., Guo, W. C., and Jie, L. I. 2018. Thermodynamics and kinetics of fluoride removal from simulated zinc sulfate solution by La (III)-modified zeolite. *Trans Nonferrous Met Soc China* 28:783–793.

Li, D., Gao, X., Wang, Y., and Luo, W. 2018. Diverse mechanisms drive fluoride enrichment in groundwater in two neighboring sites in northern China. *Environ Pollut* 237:430–441.

Madhnure, P., Sirsikar, D. Y., Tiwari, A. N., Ranjan, B., and Malpe, D. B. 2007. Occurrence of fluoride in the groundwaters of Pandharkawada area, Yavatmal district, Maharashtra, India. *Curr Sci* 92:675–679.

Masood, N., Hudson-Edwards, K. A., and Farooqi, A. 2021. Groundwater nitrate and fluoride profiles, sources and health risk assessment in the coal mining areas of Salt Range, Punjab Pakistan. *Environ Geochem Health* 44:1–14.

Melki, S., Mabrouk El Asmi, A., Ould Baba Sy, M., and Gueddari, M. 2020. Actual and predictive transport modeling of fluoride contamination of the Sfax-Agareb Coastal Aquifer in the Mediterranean Basin. *J Chem* 2020:1–14.

Mihayo, D., Vegi, M. R., and Vuai, S. A. H. 2021. Defluoridation of aqueous solution using thermally activated biosorbents prepared from *Adansonia digitata* fruit pericarp. *Adsorp Sci Technol* 2021:1–16.

Mohanta, D., and Ahmaruzzaman, M. 2018. Bio-inspired adsorption of arsenite and fluoride from aqueous solutions using activated carbon@ SnO_2 nanocomposites: isotherms, kinetics, thermodynamics, cost estimation and regeneration studies. *J Environ Chem Eng* 6:356–366.

Mukherjee, A., Bhanja, S. N., and Wada, Y. 2018. Groundwater depletion causing reduction of baseflow triggering Ganges river summer drying. *Sci Rep* 8:1–9.

Mukherjee, S., Kamila, B., Paul, S., Hazra, B., Chowdhury, S., and Halder, G. 2021. Optimizing fluoride uptake influencing parameters of paper industry waste derived activated carbon. *Microchem J* 160:105643.

Mullick, A., and Neogi, S. 2018. Acoustic cavitation induced synthesis of zirconium impregnated activated carbon for effective fluoride scavenging from water by adsorption. *Ultrason Sonochem* 45:65–77.

Mullick, A., and Neogi, S. 2019. Ultrasound assisted synthesis of Mg-Mn-Zr impregnated activated carbon for effective fluoride adsorption from water. *Ultrason Sonochem* 50:126–137.

Nagaraj, A., Pillay, K., Kumar, S. K., and Rajan, M. 2020. Dicarboxylic acid cross-linked metal ion decorated bentonite clay and chitosan for fluoride removal studies. *RSC Adv* 10:16791–16803.

Narsimha, A., and Sudarshan, V. 2017. Assessment of fluoride contamination in groundwater from Basara, Adilabad district, Telangana state, India. *Appl Water Sci* 7:2717–2725.

Naseem, M., Khurshid, Z., Khan, H. A., Niazi, F., Zohaib, S., and Zafar, M. S. 2016. Oral health challenges in pregnant women: recommendations for dental care professionals. *Saudi J Dent Res* 7:138–146.

Olaka, L. A., Wilke, F. D., Olago, D. O., Odada, E. O., Mulch, A., and Musolff, A. 2016. Groundwater fluoride enrichment in an active rift setting: Central Kenya Rift case study. *Sci Total Environ* 545:641–653.

Pang, T., Chan, T. S. A., Jande, Y. A. C., and Shen, J. 2020. Removal of fluoride from water using activated carbon fibres modified with zirconium by a drop-coating method. *Chemosphere* 255:126950.

Pranesh, M. B., Arjundas, G., Kalyanaraman, S., and Bharati, R. S. 2019. Autopsy study of a case of skeletal fluorosis (1977). *Neurol India* 67:643.

Preethi, J., and Meenakshi, S. 2018. Fabrication of La^{3+} impregnated chitosan/β-cyclodextrin biopolymeric materials for effective utilization of chromate and fluoride adsorption in single systems. *J Chem Eng Data* 63:723–731.

Queste, A., Lacombe, M., Hellmeier, W., et al. 2001. High concentrations of fluoride and boron in drinking water wells in the Muenster region-results of a preliminary investigation. *Int J Hyg Environ Health* 203:221–224.

Rafique, A., Awan, M. A., Wasti, A., Qazi, I. A., and Arshad, M. 2013. Removal of fluoride from drinking water using modified immobilized activated alumina. *J Chem* 2013:1–7.

Rahman, Z. U., Khan, B., Ahmada, I., et al. 2018. A review of groundwater fluoride contamination in Pakistan and an assessment of the risk of fluorosis. *Fluoride* 51:171–181.

Raj, D., and Shaji, E. 2017. Fluoride contamination in groundwater resources of Alleppey, southern India. *Geosci Front* 8:117–124.

Rao, N. S. 2017. Controlling factors of fluoride in groundwater in a part of South India. *Arabian J Geosci* 10:1–15.

Rao, N. S., Rao, P. S., Dinakar, A., Rao, P. N., and Marghade, D. 2017. Fluoride occurrence in the groundwater in a coastal region of Andhra Pradesh, India. *Appl Water Sci* 7(3):1467–1478.

Rashid, A., Farooqi, A., Gao, X., Zahir, S., Noor, S., and Khattak, J. A. 2020. Geochemical modeling, source apportionment, health risk exposure and control of higher fluoride in groundwater of sub-district Dargai, Pakistan. *Chemosphere* 243:125409.

Reardon, E. J., and Wang, Y. 2000. A limestone reactor for fluoride removal from wastewaters. *Environ Sci Technol* 34:3247–3253.

Rudra, S., and Khan, U. 2018. Fluoride contamination of ground water at Jamda Village of Bankura District, West Bengal: a geographical analysis. *Indian J Geo Env* 15–16:107–114.

Sahu, B. L., Banjare, G. R., Ramteke, S., Patel, K. S., and Matini, L. 2017. Fluoride contamination of groundwater and toxicities in Dongargaon block, Chhattisgarh, India. *Expos Health* 9:143–156.

Sahu, N., Bhan, C., and Singh, J. 2020. Removal of fluoride from an aqueous solution by batch and column process using activated carbon derived from iron infused *Pisum sativum* peel: characterization, isotherm, kinetics study. *Environ Eng Res* 26:1–11.

Sahu, P., Kisku, G. C., Singh, P. K., Kumar, V., Kumar, P., and Shukla, N. 2018. Multivariate statistical interpretation on seasonal variations of fluoride-contaminated groundwater quality of Lalganj Tehsil, Raebareli District (UP), India. *Environ Earth Sci* 77:1–11.

Singh, R. K., and Singha, M. M. 2009. Fluoride contamination in water of Koshi region (Bihar). *Int J Chem Sci* 7:919–922.

Singh, S. 2016. Fluoride contamination in groundwater in some villages of Banda District, Uttar Pradesh, India. *Int J Innov Res Sci Technol* 3:180–183.

Sivarajasekar, N., Paramasivan, T., Subashini, R., Kandasamy, S., and Prakash Maran, J. 2017. Central composite design optimization of fluoride removal by spirogyra biomass. *Asian J Microbiol Biotechnol Environ Sci* 19:130–137.

Subba Rao, N., Ravindra, B., and Wu, J. 2020. Geochemical and health risk evaluation of fluoride rich groundwater in Sattenapalle Region, Guntur district, Andhra Pradesh, India. *Hum Ecol Risk Assess* 26:2316–2348.

Takmil, F., Esmaeili, H., Mousavi, S. M., and Hashemi, S. A. 2020. Nano-magnetically modified activated carbon prepared by oak shell for treatment of wastewater containing fluoride ion. *Adv Powder Technol* 31:3236–3245.

Talat, M., Mohan, S., Dixit, V., Singh, D. K., Hasan, S. H., and Srivastava, O. N. 2018. Effective removal of fluoride from water by coconut husk activated carbon in fixed bed column: experimental and breakthrough curves analysis. *Groundw Sustain Dev* 7:48–55.

Talpur, S. A., Noonari, T. M., Rashid, A., et al. 2020. Hydrogeochemical signatures and suitability assessment of groundwater with elevated fluoride in unconfined aquifers Badin district, Sindh, Pakistan. *SN Appl Sci* 2:1–15.

Tang, Q., Duan, T., Li, P., Zhang, P., and Wu, D. 2018. Enhanced defluoridation capacity from aqueous media via hydroxyapatite decorated with carbon nanotube. *Front Chem* 6:1–9.

Tay, C. K. 2017. Hydrogeochemical framework and factor analysis of fluoride contamination in groundwater within the Savelugu-Nanton district, northern Ghana. *West Afr J Appl Ecol* 25:33–55.

Thivya, C., Chidambaram, S., Rao, M. S., Thilagavathi, R., Prasanna, M. V., and Manikandan, S. J. A. W. S. 2017. Assessment of fluoride contaminations in groundwater of hard rock aquifers in Madurai district, Tamil Nadu (India). *Appl Water Sci* 7:1011–1023.

Tiwari, A. K., Singh, A. K., and Mahato, M. K. 2017. GIS based evaluation of fluoride contamination and assessment of fluoride exposure dose in groundwater of a district in Uttar Pradesh, India. *Hum Ecol Risk Assess* 23:56–66.

Tsutsui, A., Yagi, M., and Horowitz, A. M. 2000. The prevalence of dental caries and fluorosis in Japanese communities with up to 1.4 ppm of naturally occurring fluoride. *J Public Health Dent* 60:147–153.

Ugran, V., Desai, N. N., Chakraborti, D., et al. 2017. Groundwater fluoride contamination and its possible health implications in Indi taluk of Vijayapura District(Karnataka State), India. *Environ Geochem Health* 39:1017–1029.

Velazquez-Jimenez, L. H., Hurt, R. H., Matos, J., and Rangel-Mendez, J. R. 2014. Zirconium–carbon hybrid sorbent for removal of fluoride from water: oxalic acid mediated Zr (IV) assembly and adsorption mechanism. *Environ Sci Technol* 48:1166–1174.

Wang, G. Q., Hang, Y. Z., Xiao, B. Y., et al. 1997. Toxicity from water containing arsenic and fluoride in Xinjiang. *Fluoride* 30:81–84.

WHO 2002. World Health Organization, Fluorides, Geneva, Environmental Health Criteria, 227.

WHO 2006. *Fluoride in Drinking Water*, Fawell, J., Bailey, K., Chilton, J., Dahi, E., Fewtrell, L., and Magara, Y. (eds), 41–75. IWA Publishing, London, UK.

Yadav, A., Sahu, Y. K., Rajhans, K. P., et al. 2016. Fluoride contamination of groundwater and skeleton fluorosis in Central India. *J Environ Prot* 7:784.

Yadav, S., Bansal, S. K., Yadav, S., and Kumar, S. 2019. Fluoride distribution in underground water of district Mahendergarh, Haryana, India. *Appl Water Sci* 9:1–11.

Yasmin, S., Monterio, S., Ligimol, P. A., and D'Souza, D. 2011. Fluoride contamination and fluorosis in Gaya Region of Bihar, India. *Curr Biot* 5:232–236.

You, K., Li, P., Fu et al. 2021. Fabrication of manganese-supported activated alumina adsorbent for defluoridation of water: a kinetics and thermodynamics study. *Water* 13:1219.

Young, S. M., Pitawala, A. H. M. T. G., and Ishiga, H. 2011. Factors controlling fluoride contents of groundwater in north-central and northwestern Sri Lanka. *Environ Earth Sci* 63:1333–1342.

Yuan, L., Fei, W., Jia et al. 2020. Health risk in children to fluoride exposure in a typical endemic fluorosis area on Loess Plateau, north China, in the last decade. *Chemosphere* 243:125451.

Zhang, Y., He, L., Wang, G., Zhang, X., Liu, Y., and Chen, Y. 2021. Defluorination and regeneration study of lanthanum-doped sewage sludge-based activated carbon. *J Environ Chem Eng* 9:1–10.

21 Himalaya in a Rapidly Changing World

Climate Change Impacts and Conservation Implications on the Montane Freshwater Fauna

Aashna Sharma, Vineet Kumar Dubey,
Jeyaraj Antony Johnson, and Kuppusamy Sivakumar
Wildlife Institute of India, Dehradun

Yogesh Kumar Rawal
Department of Zoology, Panjab University, Chandigarh

CONTENTS

21.1 INTRODUCTION

The montane ecosystems are conspicuous in their capacity to hold tremendous biodiversity despite human intrusions and expansions into wilderness. It does not thus come as a surprise that a large percentage of global montane regions are nature reserves, idiosyncratic to the endemism they hold. The increasing greenhouse gas emissions and pollution, nonetheless, are rising threats to the biodiversity of these landscapes. Of these, climate change raises most concern. Mainly attributed to anthropogenic climate change, the world has entered the sixth mass extinction, with a colossal part of montane biosphere already lost to the rapidly warming world.

DOI: 10.1201/9781003220824-24

Climate, which as per definition is the long-term average of the temperature, precipitation and wind velocity, is, in point of fact, mainly driven by temperature, which in turn governs the latter two factors. A global average of multiple measurements taken over the world's land and oceans indicates that human activities have caused an estimated 1°C of global warming above the pre-industrial levels (IPCC, 2018). Seemingly trivial, nonetheless, this small temperature rise has resulted in devastating ecological implications and biome shifts. The projected climatic change in the 21st century is equivalent to the largest global changes the Earth has witnessed in the past 65 million years (Diffenbaugh and Field, 2013; Kemp et al., 2015). Additionally, various global climate models (GCMs) project plausible average increases in temperature up to 4.8°C by 2100 (IPCC, 2013). Even supposing adherence by all nations to agreements on curbing greenhouse gas emissions, the climate would regardless continue to change at a similar pace for at least a few centuries (Pecl et al., 2017; IPCC, 2018). While the future climate trajectories will depend on global political decisions, the situation is already critical for the biodiversity, which would have to track phenology, disperse at unrealistic scales and alter their thermal niches to avert extinction (Settele et al., 2015; IPCC, 2018; Urban, 2019).

A global-scale biological impact is evident, with burgeoning research indicating that species across biomes are 'on the move' to track their niche (preferred environmental conditions) in a changing climate (Pacifici et al., 2015; Isaak et al., 2016; Cotto et al., 2017; Pecl et al., 2017; Morrison et al., 2018; Littlefield et al., 2019; Troia and Giam, 2019; Fandos et al., 2020; Herrera-R et al., 2020). Although the pace of range shifts and adaptive responses is unprecedented, so is the rate of global warming, portending massive shifts in ecosystem types altogether (Hoegh-Guldberg et al., 2018). This is critical, as the species' rate of responses often does not keep pace with the local climate velocities – the rate of shift in species distribution necessary to track their thermally 'suitable' niche (Isaak et al., 2016).

Trends of species range shift in the past decade based on multi-species metanalyses (Chen et al., 2011; Moritz and Agudo, 2013; Ockendon et al., 2014; Diamond, 2018; Freeman et al., 2018) indicate distinctive trends in species ranges shifting either poleward (latitudinally) or mountainward (altitudinally) for the terrestrial and freshwater taxa (Bobrowski et al., 2018; Lu et al., 2021; Mamantov et al., 2021), or downwards (towards greater depths) in the marine realms (Perry et al., 2005; Engelhard et al., 2014; Thatje, 2021). Nonetheless, there is no ubiquitous pattern recorded for species globally, as individual species are likely to respond idiosyncratically, based on their phenological attributes, migratory abilities and thus, resistance or resilience (Rapacciuolo et al., 2014; Rowe et al., 2015; Campos-Cerqueira et al., 2017; Chardon et al., 2020).

Climate change together with habitat alterations in an overriding Anthropocene forms a daunting concoction, limiting the species' inherent capacities to migrate and search for climate refugia (Yalcin and Leroux, 2018). Extinctions are, thus, the most dreaded aftermaths of climate change (Hannah, 2012), more so where a major chunk of biodiversity is constrained by geography, as in high-elevation biodiversity hotspots (Malcolm et al., 2006; Pimm et al., 2014; Rahbek et al., 2019; Trew and Maclean, 2021). This is predicted to cause a staggering instability in the ecosystem functioning of our planet, apart from the overhead release of more greenhouse gases (Hannah, 2012), making the climate change cycle uncontrollably vicious. With our planet already having lost 84% of the freshwater biodiversity since 1970, the faster pace of climate change raises serious issues for the remnant freshwater biota in high-elevation biodiversity hotspots (WWF, 2020).

Although climate change impacts have been documented for varied taxa globally, the research has been largely focussed on the terrestrial realm (Mantyka-Pringle et al., 2016), barricading the policymakers from designing climate-adaptation strategies for the freshwater taxa. However, some evidence exists on responses of freshwater taxa to environmental perturbations that occurred over the past centuries and decades (Isaak et al., 2016). Although not substantial, research on freshwater taxa has witnessed an upsurge in the past decade (Kovach et al., 2016; Ruiz-Navarro et al., 2016;

Radinger et al., 2017; Troia and Giam, 2019; Mota-Ferreira et al., 2021). The detected patterns of species responses, nevertheless, are deduced from a limited number of studies focussed on specific regions or taxa, thus hindering the understanding of the envelope of responses freshwater species can exhibit to climate change.

Freshwater ecosystems, and the fauna they harbour, are specifically sensitive to the changing climate (Sala et al., 2000; Markovic et al., 2017), particularly due to a strong association of their habitats with the climatic and hydrologic regimes (Ruiz-Navarro et al., 2016) and thus, the air temperature and precipitation patterns. This is more so for the lotic habitats, as rivers, being network-constrained environments, are predominantly isolated within the larger landscapes (Rubenson and Olden, 2017), such that they function as 'biogeographic islands' (Dudgeon, 2011; Olden et al., 2011). This results in species movements being delimited by the dendritic structuring of the river and thereby, directionally restricted (Connolly et al., 2008; Lois and Cowley, 2017). Tracking their niche under climate change is thus a daunting task for riverine taxa, more so for the montane riverscapes, where inaccessibility and rigorous terrains have kept the major part of the biodiversity still poorly inventoried.

Montane stream networks ubiquitously hold unique biodiversity due to long-term isolation of the species and their adaptive specialization for the topology of mountain drainages (Isaak et al., 2016). Strong evidence of climate-driven range contraction, declining numbers and local extinction of montane fish fauna is, however, considered a heightening risk amongst the majority of high-altitude environments worldwide (Comte and Grenouillet, 2013; Comte et al., 2013; Troia and Giam, 2019). With research evidencing faster temperature rise in higher elevations (Mountain Research Initiative EDW Working Group, 2015; Wang et al., 2016; Lamprecht et al., 2018; Hock et al., 2019), the future persistence of the native freshwater taxa is under jeopardy. Cold- and cool-water organisms are presumed to be at particular risk due to the highest climate sensitivity predictions for riverscapes fed by glaciers and snow (Null et al., 2013; Lee et al., 2020).

Limited exploration of the montane freshwater biodiversity has brought forth mixed opinions on the impacts of climate change therein. While the majority of research conducted portends adversities for the freshwater fish under climate change (Comte and Grenouillet, 2013; Filipe et al., 2013; Eby et al., 2014; Ruiz-Navarro et al., 2016; Lopes et al., 2017; Troia and Giam, 2019), there are a few who estimate climate velocities to be slower in mountains, thus enabling the species to cope (Isaak et al., 2016). Regardless, these doubts and mixed opinions need to be verified with in-depth field surveys and robust ecological niche models built for montane fishes to better inform management decisions.

21.2 HIMALAYA

Himalaya, the highest mountain chain on the globe, represents an extraordinary diversity and endemism (Myers et al., 2000; Mittermeier et al., 2004) and runs as a 2,400-km long arc from west-northwest to east-southeast (Wadia, 1931). It includes in it the Pamir, Hindu Kush and the Karakoram ranges apart from the Himalaya proper. The Indian Himalaya region, which falls in the political boundaries of India, stands tall in six Indian states, i.e., Himachal Pradesh, Uttarakhand, Sikkim, Arunachal Pradesh and hilly regions of West Bengal and Assam, as well as two union territories, Jammu and Kashmir, and Ladakh. It falls closely into the Himalaya proper and has a strict geographical extent, finely defined between 75° and 95° E and from 27° to 35° N. The westernmost extents are distinctly bounded by the river Indus and anchored by the Nanga Parbat. The easternmost limits, on the other hand, are demarcated by the Brahmaputra, anchored by Namcha Barwa. To the north and south are the lower limits of the Tibetan High Plateau and the upper limits of the Gangetic plains. The region is predicted to be warming at a rate much higher than the global average rate of 0.4°C, which is also evident from a six times higher pace of glacier retreat (Solomon et al., 2007; Dimri et al., 2020).

21.3 MAJOR RIVERS OF HIMALAYA

With the majority of freshwater stored as ice in the Himalayan glaciers, the Himalaya is aptly named the 'third pole' of the globe. This vast reserve of freshwater, nonetheless, is under severe threat as Himalaya continues to warm at an unprecedented pace. Three major rivers, the Indus, Ganga and Brahmaputra, drain the Himalayan region and flow into the Indian subcontinent. These rivers are significantly important to the livelihoods of many, and often bear the brunt of overexploitation through rampant damming and illegal fishing activities (Pandit and Grumbine, 2012).

The hydrological attributes of these basins, especially the lower-order headwaters and springs, is as yet undeciphered, with studies largely concentrated on the lower-elevation downstream stretches (Dimri et al., 2020). Additionally, most of the information on these river basins is inadequate or is not collated into published form, thus restricting the understanding we currently have on the hydrology in Himalaya. A detailed review by Qazi et al. (2020) highlights that although high flows attributable to climate change have been investigated with a smattering of intensive research, climate-induced low flows in the high elevation are still not understood. Additionally, the estimates of the ice mass of the Himalayan glaciers are still rudimentary. Nonetheless, a substantial ice mass feeding the major Himalayan rivers is predicted with continued loss and ice thinning, eventually impacting the biodiversity these freshwater rivers hold.

Very recently, many researchers have cautioned about the potential impact of climate change on terrestrial plants and animals in this landscape (Bhattacharyya et al., 2019; Hamid et al., 2019; Bagaria et al., 2020, 2021; Mainali et al., 2020; Singh et al., 2020); however, no study has been devoted to understanding the impact of climate change on its freshwater species to date (but see Sharma et al., 2021a). Identifying stream reaches, riverscapes and the species with a higher risk of extinction thus stands out as an urgent priority in view of ongoing climate modifications in the fragile landscape of Himalaya.

Enhancing the research base on fish diversity estimations in the Indian Himalaya (Menon, 1954; 1962; Sehgal, 1999), an escalation in ichthyological investigations has been recorded in the past two decades (Karmakar, 2000; Badola, 2009; Uniyal, 2010; Bhatt et al., 2012). To name a few, the Bhagirathi river basin in the western Himalayan region has been explored for its fish fauna by many researchers (Rawal, 2002; Johal and Rawal, 2005; Joshi and Tyagi, 2008; Agarwal and Singh, 2012; Rajvanshi et al., 2012; Sarkar et al., 2012; Singh and Agarwal, 2014; Agarwal et al., 2018; Johnson et al., 2020), and similarly, the Beas and Sutlej river basins in the north-western Himalayan region have also been explored for their fish and fisheries (Johal et al., 1999; 2001; Sivakumar, 2002; 2008; Mehta and Sharma, 2008; Negi and Banyal, 2017). This uniqueness in freshwater fishes is ascribable to different geomorphic conditions, changing thermal regimes and fast water currents of the Himalayan rivers.

21.4 A 'YET TO EMBARK' CLIMATE RESEARCH IN HIMALAYAN FRESHWATERS

An intensive exploration on the research conducted on the Indian Himalayan ichthyofauna indicates the climate change studies to be at a relatively naïve stage. As it stands, a meagre 1.02% (n=14) of the total studies published till the end of the year 2020 are focussed on the impacts of climate on the freshwater fishes. Nonetheless, what is interesting is that multifarious aspects other than climate change have been intensively researched. The ichthyological investigations in Himalaya span wide-ranging themes such as biodiversity and distribution, biology, ecology, evolution, taxonomy, disease, fish farming, aquatic conservation management, population biology, molecular taxonomy and population genetics, followed by climate change and altitudinal distribution patterns. Noticeably, an appreciable part of the studies conducted so far have focussed on taxonomy, thereby contributing to the fish species database of this biodiversity hotspot. This is followed by their diversity and distribution apart from ecology, as seen in Figure 21.1.

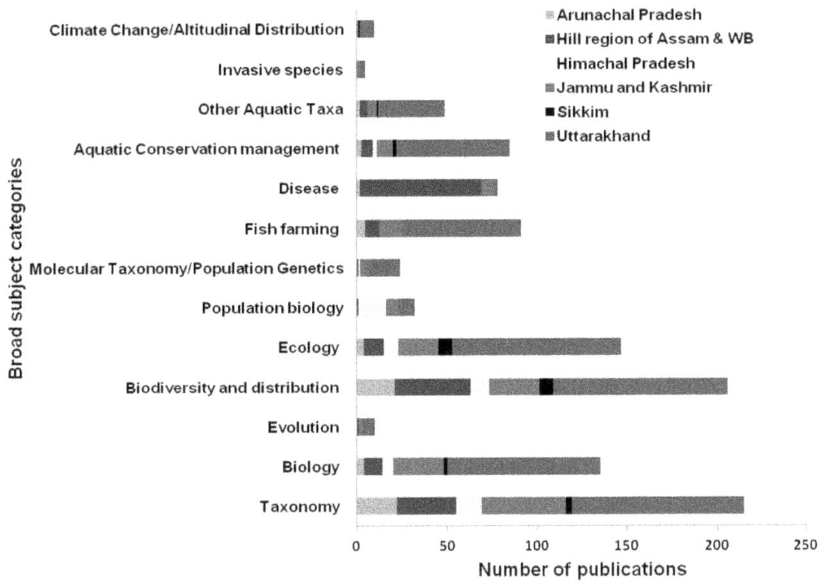

FIGURE 21.1 Trends in publications on the freshwater ichthyofauna of the Indian Himalayan region under different broad subject categories.

If teased out, so far, a total of 846 studies have been conducted on the freshwater fishes of Himalaya. These studies, dating from the first published research by Bloch (1785) to the year 2020, are available as a detailed account collated as a database (see Sharma et al., 2016). The review suggests that the number of studies on fishes of Indian Himalaya followed a slow growth trajectory after the 1930s and peaked during 1991 to 2000, with almost 21% of all studies published in this period. However, the number of studies has comparatively plummeted since the 1991–2000 peak.

Knowledge of the occurrence of fish in India dates back to three millennia BC, with documentations by pioneers like Hamilton (1822) and Hora (1930), which were appreciably continued by Menon (1954; 1962), Jayaram (1977), Johal and Tandon (1981) and Jhingran (1982). Since then, many documentations on the fish diversity and community patterns of Himalayan fishes have been conducted and have been carried forward to date (Sehgal, 1999; Raina and Petr, 1999; Nautiyal, 2006; Bhatt et al., 2012; Johnson et al., 2020).

Records till the year 2005 suggest that the Indian Himalaya is known to be home to around 266 fish species, which is about 27% of India's total freshwater fish diversity, including a vast variety of threatened, migratory and endemic species (Nautiyal, 2005). Nonetheless, isolated studies on the enumeration of species have been ongoing, and the current numbers can be assumed to be much higher. A pan-Himalayan taxonomic richness of the fish fauna has been reviewed by Nautiyal (2005), who found the central (Nepal) Himalaya to be richest in fish diversity (181 species), followed by the western (167 species) and the eastern Himalaya (159). Another exploration on the fishes of north-east Himalaya by Vishwanath et al. (2007) further surpassed this record (296).

Studies on the pan-Himalayan distribution of fishes are rare, with only a few species explored for their broader range, e.g., *Tor* and *Schizothorax* (Nautiyal, 2005; Pinder et al., 2019; Regmi, 2019). With climate change and biological invasions cumulatively acting as overarching deterrents to conserving fish diversity, what raises concern is that research on fish invasions in Himalaya is equivalently benign. Furthermore, the available studies on the impacts of climate change and invasions on the fish fauna of Indian Himalaya are anecdotal; they warrant ground-truthing and confirmation through empirical field design, survey and ecological data sets on local and broader spatial scales.

21.5 ONGOING CHALLENGES

With the inaccessibility and complex high-altitude terrains hindering survey efforts, a substantial part of Himalayan riverscapes is still uncharted, and the extant distribution of the majority of freshwater species still in its exploratory stage (Sharma et al., 2016). Maintaining the extent and ecosystem functionality of the aquatic ecosystems in Himalaya and preventing any further retrogression is, therefore, important as a strategy to address climate change.

With escalating concern to save the extraordinary diversity and endemism of Himalaya (Myers et al., 2000; Mittermeier et al., 2004), dozens of studies have been conducted in recent decades, predicting potential distribution or future range shifts as a consequence of climate change. These include burgeoning studies encompassing a range of taxa, including Himalayan flora (Yang et al., 2013; Bobrowski et al., 2018; Manish et al., 2016; Mungi et al., 2018; Hamid et al., 2019; Rana et al., 2021; Rathore et al., 2019), fungi (Shrestha and Bawa, 2014) amphibians (Subba et al., 2018), birds (Bagaria et al., 2020) and mammals (Forrest et al., 2012; Escobar et al., 2015; Su et al., 2018; Bhattacharyya et al., 2019; Bagaria et al., 2021).

Regardless, the freshwater realm remains entirely omitted from the Himalayan climate research. This mirrors the global pattern of disproportionately low understanding of freshwater organisms' responses to climate change. This is particularly grim when montane freshwater ecosystems have already been predicted to be much more sensitive to climate change than the other realms (Lee et al., 2020). Thus, while climate studies strongly prognosticate the altered distribution of other taxa in Himalaya, the impact on its cold-water fauna is unclear.

A major challenge to decrypting the impacts of climate change on Himalayan freshwater biota is the lack of an overarching research approach. Although the past decade has seen an upsurge of studies on climate change, hydrology and policy measures (Pratap et al., 2016; Maurer et al., 2019; Bobrowski et al., 2018; Su et al., 2018; Bhattacharyya et al., 2019; Bagaria et al., 2020b; Qazi et al., 2020), the lack of a multidisciplinary approach to substantiating inferences on climate biology seemingly is the major deterrent to freshwater biodiversity conservation. Lack of such collated research hinders a holistic understanding of the impacts of climate change on the hydrological regimes in the Himalayan riverscape and eventually, the plausible responses of species they harbour.

Additionally, climate science in Himalaya presents the greatest uncertainties on how the glacier meltdown will transform river flows and discharges in the years to come (Qazi et al., 2020). This comes as another major challenge, because ascertaining the species responses in dynamic systems like rivers becomes complex when the future trajectories of flow and discharge patterns are still uncertain. Regardless, the impacts of global warming on the Himalayan glaciers are evidently colossal, thus heightening the climate-induced future vulnerabilities of the Himalayan freshwater biodiversity.

21.6 CLIMATE VULNERABILITIES AND HIMALAYA

Climate change is causing the poles and mountains to warm faster than the rest of the globe (Overland et al., 2019). Himalaya, the highest mountain chain on Earth, is conspicuously sensitive to these ongoing climate alterations due to its sixfold quicker glacial mass retreat as compared with glaciers elsewhere (Singh et al., 2017). This definitely is a cause of concern when most of the Himalayan nations are still developing economically (UN DESA, 2020) and are thus still coming to terms with climate change research, which necessitates allocation of sufficient funds.

Geodetic studies suggest an overall vast thinning of the whole western, central and eastern Indian Himalayan glaciers in the last decade (Pratap et al., 2016; Maurer et al., 2019). This is evidenced by two massive flash flood events that occurred in 2013 and 2021 in the western Himalaya. Being geotectonically active, Himalayan river basins are much more prone to slope instabilities and landslides, which are further aggravated by the glacial lake outbursts (Dimri et al., 2020). While the 2013 Kedarnath disaster was attributable to the Chorabari glacier lake outburst (Bhambri et al., 2015), the Chamoli disaster in 2021 was another catastrophe, believed to be a glacial lake outburst of the Nanda Devi glacier (ISRO, 2021).

While aberrant flash floods are expected in some riverscapes of Himalaya, spring-fed streams are expected to show speeded-up drying due to the predicted intensified pre-monsoonal drought (Grover et al., 2015; Dimri et al., 2020), thus destroying important spawning grounds for the Himalayan fishes. Climate change would thus create an intricate and complex chain of events in the Himalayan riverscapes, which is expected to cause large-scale loss of freshwater biodiversity. Further adding to the threat is the presence of non-native exotics in the Himalayan riverscape, as invasions are forecast to increase substantially in a changing climate.

Species' vulnerability to changing climate is ascertained based on the habitat available to support their expansion as well as the species phenology, physiology and biotic interactions, all of which play a cumulative role (McKelvey and Buotte, 2018). Understanding the species responses to climate change thus necessitates knowledge of these covariates to support climate mitigation. Conservation planning in Himalayan freshwaters nonetheless remains adjourned, given the complete lack of such a knowledge base. In fact, this condition resonates with that of the other global montane riverscapes with noticeably sparse data on climate change–species relationships. This warrants mechanistic climate change predictions with a holistic inclusion of phenology and life-history traits in the species distribution models.

21.7 HIMALAYAN FRESHWATER FAUNA UNDER THREAT

Climate change is expected to change the distribution and govern the survival, extinction or spread of species across global montane systems. Although the impacts of climate will be species and taxa specific, there are distinct global parallelisms observed for various taxa (Guisan et al., 2019). Of all these, the freshwater fauna faces the greatest threat due to their ectothermic physiology and thus, a strong dependence on the thermal conditions of the environment for their metabolic functions (James et al., 2017). The riverine networks of Himalaya are in particular jeopardy due to the harsh climate and geomorphological constraints. Furthermore, the rampant damming and hydrological alteration in Himalayan river networks (Pandit and Grumbine, 2012) limits the dispersal abilities of species, specifically the ichthyofauna, thus elevating the cacophony of climate-induced local extirpations. This concurs with the escalating evidence on biodiversity degradation at genetic, species and habitat levels when climate change and land use co-occur (Mantyka-Pringle et al., 2014).

The natural flow regime and spatio-temporal connectivity in the downstream areas, particularly in the Ganges and Indus basin, have been predicted to be disturbed as a consequence of climate change (Immerzeel et al., 2010). Furthermore, these rivers are also subjected to extreme events, such as intense rainfall and flash floods, besides a multitude of anthropogenic stressors like pollution and channel modification. Therefore, given the consideration of various threats posed to the aquatic ecosystems of Himalaya, it is important to develop effective monitoring tools followed by rigorous sampling in high-altitude rivers and streams for planning relevant conservation and mitigation measures to save them from potential loss in future environments.

Recent decades have seen an upsurge of research describing the potential impact of climate change on freshwater organisms worldwide (Sharma and Jackson, 2008; Buisson and Grenouillet, 2009; Ruesch et al., 2012; Furniss et al., 2013; Isaak et al., 2015; 2016; Troia and Giam, 2019). Research further suggests that freshwater species with restricted range sizes are most vulnerable to changing climate (Mantyka-Pringle et al., 2014). While evidence of climate change impacts continues to be documented from montane riverscapes globally, its implications for the Indian freshwater taxa are still not explicit, more so for the Himalayan riverscape (Sarkar et al., 2017). Building an understanding with available studies from other high-altitude montane stream networks, this section provides a synopsis of possible climate change impacts on the Himalayan fishes, freshwater

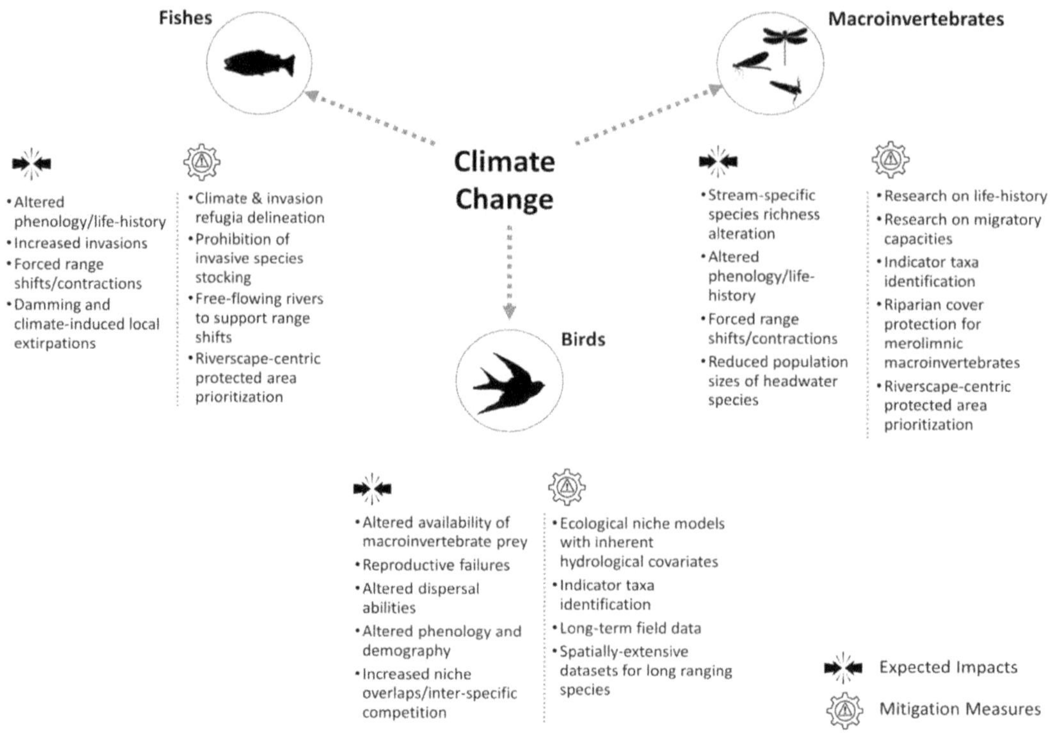

FIGURE 21.2 A representation of expected impacts and possible mitigation measures under future climate change for fishes, macroinvertebrates and birds across Himalayan stream networks.

macroinvertebrates and riverine birds, with probable mitigation measures and adaptation strategies representatively presented in Figure 21.2.

21.7.1 Fishes

Himalayan freshwater fish diversity is structured by diverse geomorphic conditions, thermal regimes and rapid water current. Due to their distinctive adaptation to cool- and cold-water conditions and restricted dispersal abilities to track their thermal environmental niche (Comte and Grenouillet, 2013; Isaak et al., 2016), it is important to assess the potential impact of climate change on these organisms to prevent them from future extinction.

Typical of the Himalayan cold-water river systems are two species, snow trout (*Schizothorax richardsonii*) and golden mahseer (*Tor putitora*). Most primitive amongst all the Schizothoracines, snow trout is specialized for torrential streams and has co-evolved with the changing patterns of Himalayan geomorphology (He and Chen, 2006). Idiosyncratically representing the Himalayan coldwaters (Regmi, 2019; Vishwanath, 2010), it is often regarded as a potential model organism for climate change studies (Kamalam et al., 2019). However, its potential as a climate change indicator taxon has been tapped with just one available study to date (see Sharma et al., 2021a). Both the golden mahseer and snow trout are typical of the Himalayan streams and migrate long distances for spawning and/or wintering (Sehgal, 1999; Pinder et al., 2019).

While species distribution models built for the snow trout indicate a high-altitude squeeze with evident range contraction in the lagging edges of its distributions (Sharma et al., 2021a), the golden mahseer is expected to face similar threats. Additionally, the rampant damming in the Himalayan watersheds is posing further threats to the existing range shift pathways as well as blocking the migratory routes necessary for their breeding (Bhatt et al., 2012). There are thus high threats of local extinction for most Himalayan cold-water fishes, further exacerbated by the continuous stocking of invasive species like the brown trout under the auspices of various angling associations (Sharma et al., 2021b).

Identification of climate and invasion refugia to support current and future sustenance of high-altitude Himalayan fishes is a sought-for mitigation measure, which can benefit from detailed investigations on life-history plasticity and adaptive capacities of the populations in the wild as well as their phenological dependence on the climate. This can help inform species-specific policy decisions. Additionally, an immediate halt to enhancing invasive species through increasing population size and frequency is highly recommended. This needs to be supported by the prioritization of 'riverscapes' rather than 'landscapes' to design climate mitigation measures for riverine fauna. Furthermore, as the cold-water Himalayan fishes tend to migrate to spring-fed tributaries for spawning, the dams need to be planned to ensure there is no existing mainstem–tributary disconnect, as well as ensuring that some rivers in Himalaya are maintained in their free-flowing state.

21.7.2 Freshwater Macroinvertebrates

While the Himalayan riverscapes remain understudied for the macroinvertebrate diversity they hold, the Nepal Himalaya has been extensively explored for this faunal group. Shah et al. (2015; 2020a,b) have investigated the impacts of run-off pollutants as well as using the macroinvertebrates as potential organisms to identify the 'climate sensitive zones' or the bands of ecotonal transition in the dynamic river systems. Scattered records from other parts of the Himalayan range present community assemblage structures of the benthic macroinvertebrates. Nonetheless, the phenology, life-history, temperature tolerances, distributional ranges and thus, the climate change impacts on the Himalayan macroinvertebrate fauna are still largely unascertained, which is definitely a matter of concern.

Climate change models for the macroinvertebrates of other montane stream networks predict that species currently preferring cool to warm waters would expand into higher elevations, while the cold-water specialist macroinvertebrates would undergo massive range contractions (Domisch et al., 2011). The increasing global temperatures are further predicted to reduce the dissolved oxygen available for aquatic insect respiration, translating into reduced thermal tolerance (Verberk and Bilton, 2013). Temperature, oxygen and salinity act in congruence under global warming to further exacerbate the stressors on freshwater macroinvertebrate bioenergetics, thus dictating their growth and ability to survive (Verberk et al., 2020).

Other studies report that cold-water specialists migrate further upstream and occupy novel areas created by the retreating glaciers, with concurrent range contractions at the lagging edges (Lencioni, 2018). Dispersal and climate-induced range shifts in macroinvertebrate taxa can occur through their terrestrial adults shifting overland or through the aquatic life stages drifting upstream (Brittain and Eikeland, 1988; Lencioni et al., 2015). This focuses attention on maintenance of free-flowing rivers as well as protection of riparian vegetation cover to mitigate the climate-forced migrations of the narrow-ranging benthic macroinvertebrates. Regardless, the climate change impacts for the Himalayan taxa still need to be understood with intensive long-term field observations and life-stage-specific growth and phenological responses.

21.7.3 Riverine Birds

The impacts of climate change on higher trophic levels such as riverine birds are still the least explored globally, with no study so far conducted on the climate-induced responses of Himalayan riverine birds. The majority of Himalayan river birds are insectivores, dependent on the freshwater macroinvertebrates. As climate change is expected to alter the macroinvertebrate abundance, the riparian bird densities would invariably change (Trevelline et al., 2018). Furthermore, climate-induced river acidification poses serious implications for the macroinvertebrates (Weiss et al., 2018) that form the major diet of most Himalayan riverine birds (Buckton and Ormerod, 2008). The altered flow and stream acidification in other montane stream networks have been documented to reduce the macroinvertebrate prey base, thus affecting provisioning during nesting and ultimately impacting the nestlings as well as the adults (Strasevicius et al., 2013; Trevelline et al., 2018).

Himalayan riverine birds like the brown dipper (*Cinclus pallasii*), little forktails (*Enicurus scouleri*), spotted forktails (*E. maculatus*) and plumbeous water redstarts (*Rhyacornis fuliginosus*) are mainly dependent on the stream flows for foraging. They segregate their niches based on the habitat complexity and flow variations but nonetheless, prefer similar diets. This enables their co-existence. However, changing climate and damming across Himalaya is expected to reduce the habitat complexity and alter flow regimes, thereby increasing the chances of higher niche overlaps amongst these species.

Furthermore, flow variability during the breeding season is expected to have serious implications on the bank-nesting species due to possible nest inundations (Strasevicius et al., 2013), as well as altering the dispersal decisions in others (Roche et al., 2012), thus altering both phenology and demography (Jenouvrier, 2013; Warren et al., 2013). Nonetheless, with no study available on their responses to climate change through plasticity either in foraging behaviour or in life-history attributes, these presumptions warrant validation through empirical testing. Long-term field data is thus warranted on their behaviour, diet and breeding biology to build mechanistic species distribution models in Himalaya.

21.8 POSSIBLE PHENOLOGICAL AND LIFE-HISTORY CHANGES

Climate change has been documented to elicit changes in breeding phenology and life-history attributes of freshwater species, more so in climate-resilient species (Crozier and Hutchings, 2014). Understandably, the variations in breeding phenology and life-history of freshwater fish can indicate their ability to respond to climate change, a useful management intervention (Hoxmeier and Dieterman, 2012; Al-Chokhachy and Sepulveda, 2019). This is specifically pertinent for riverscapes like Himalaya, where empirical research on climate impacts has just germinated, with no temporal accounts of the phenology and life-history of native fish species (but see Sharma et al., 2021a,b), and no records on river thermal profiles to compare with.

A species' life-history plasticity to climate change proffers a measure of its abilities to persist or expand and the possibilities of its local extinction (Shuter et al., 2012; Lefevre et al., 2021). Furthermore, the changes in phenology and life-history are climate-adaptive strategies, which might be similar if the species is faced with invasion threats. Competitive interactions with the exotic-invasives create an environmental stress, which triggers plasticity in life-history attributes if a species is resilient (Crozier and Hutchings, 2014). This plasticity then establishes as '*acclimation*', a phenomenon that happens when a history of continuous exposure to changed environmental conditions elicits the species' response to a stressor (Angilletta, 2009). As climate change entails a prolonged exposure to a changed environment, phenotypes and life-history responses will largely be influenced by acclimation, which depends on the species' inherent resilience to change (Hofmann and Todgham, 2010; Crozier and Hutchings, 2014). In fact, comparisons of the physiologies of different populations of a species have been documented to play a major role in the pursuit of understanding climate change impacts on wildlife (Pörtner and Farrell, 2008; Jarić et al., 2019).

The impacts of changing environments on freshwater fishes over a long temporal window of 150 years (Olden et al., 2006) proffer a better interpretability of results from the present study.

Olden et al. (2006) suggested that triggered by a changing environment, certain fish species exhibit an '*opportunistic strategy*' over time, where through traits of fast life-history, they are noticed to enhance their fecundity at the cost of smaller maturation sizes, which, however, if given the chance of a favourable environment, can transition back to 'normal' life-history or 'equilibrium strategy'.

Based on what the climate biologists have so far proposed, an understanding of the cumulative impacts of invasion and climate change on the native freshwater biota of Himalaya can be built. Climate change and invasions act synergistically to present similar stressors for the natives in Himalaya, while the plasticity in traits dictates their survival ability and vulnerability to these stressors. This finds common ground with research suggesting that freshwater fishes cope with climate change by advancing their spawning and migration timings. In fact, studies of durations as short as 35 years indicate the warming temperatures to be a proximate cue triggering plasticity in the timings of these life cycle events (Kennedy and Crozier, 2010; Russell et al., 2012).

As it stands, climate research on Himalayan freshwater biota lags far behind. Nonetheless, what cannot wait is the development of policy measures targeting their sustenance under increasingly rapid global warming. It thus seems most expedient to understand the native species' phenology and life-history plasticity to build on their climate-adaptive abilities. The inferences drawn from a series of research by Sharma et al. (2021b,c), are compelling to fit this hypothesis. They conducted an intensive study on the snow trout sympatric with the invasive brown trout (*Salmo trutta*) in the watersheds of western Himalaya to decrypt the plastic capacities of the former. The native snow trout was found to mature at a much earlier size in the watersheds invaded by the brown trout. This early maturity inflicted by the brown trout is a definitive indication of plasticity in breeding, which is also known to be genetically controlled in different populations if the plasticity is long displayed (Hutchings, 2002), but is prodigiously dictated by the common stressors of invasion and climate change (Crozier and Hutchings, 2014; Rogers and Dougherty, 2019). What attracts attention is the early maturity in sympatric snow trout with a higher number of lower-length individuals, which is a unanimous plastic response of freshwater fishes to globally warming temperatures (Daufresne et al., 2009; Gardner et al., 2011; Audzijonyte et al., 2020). Although this seems to be a definitive indication of the snow trout's inherent capacity for survival in warming temperatures in the future, the possibilities of their survival or vulnerability to extirpation will be chiefly dictated by the ability to keep pace with the velocity of stressors, which are often intricate (Comte and Grenouillet, 2013).

Climate change is further expected to change the phenological responses of freshwater species; however, the patterns of change are uncertain, given the dearth of research in Himalaya. The climate-induced glacier meltdown is predicted to alter the river flow with increased variability in discharge. These changes are predicted to be more aberrant in the pre-monsoons, with the possibility of rampant cloud bursts, flash floods and droughts (Qazi et al., 2020). With the seasonal changes in river flow acting as the major cue for species' reproductive and migratory cycles, the unpredictable flow patterns are expected to play havoc with the species' permanence and viability.

Furthermore, climate predictions on precipitation patterns indicate a longer window of monsoons in Himalaya with expected erratic rainfall (Qazi et al., 2020). Monsoons being the major spawning season for most Himalayan freshwater fishes, the predicted aberrant patterns will have far-reaching consequences for their breeding cycles. Himalayan fishes tend to undergo a spawning migration from larger mainstems into lower-order tributaries (Sehgal, 1999). This comes with their preference for low-flow spring-fed tributaries, which provide ambient temperatures and flow for their eggs to hatch. With invasive species having highly plastic temporal windows of breeding, they are facultatively dependent on the seasonal cue and are expected to proliferate under altered temperature and precipitation by utilizing empty niches (Loehman et al., 2018). Given the severity of climate change impacts expected on the Himalayan biota, studies are warranted on mapping the current and future suitable environmental niches for scientifically backed policy decisions.

21.9 *SCHIZOTHORAX RICHARDSONII* – A CASE STUDY

The distributional ranges of species under the current and future climatic conditions can be detected using various species distribution models (SDMs) under multiple Global Circulation Models (GCMs) and Representative Concentration Pathways (RCPs). The recent years have seen a noteworthy upsurge in biodiversity scenario building and SDMs (Elith et al., 2010; Merow et al., 2013; Thuiller et al., 2016; Guisan et al., 2017), yet advanced modelling techniques keep non-aficionados unclear as to which methodology can be the best (Guillera-Arroita et al., 2015; Thuiller et al., 2019). Inter-model comparison studies have revealed high degrees of variability based on model selection, so much so that the choice of SDMs, GCMs and RCPs has an overwhelming influence on the future biodiversity projections (Thuiller et al., 2019; Hao et al., 2020). A single model provides an incomplete description of the predicted future of a species and the uncertainties involved (Benestad et al., 2017). Currently, the ensemble projections based on multiple SDMs and GCMs prove to be the best option available to provide stringent and close-to-reality projections (Araújo and New, 2007; Araújo et al., 2019; Hao et al., 2019).

SDM studies in Himalaya have witnessed an impetus in recent decades, beginning with an initial set of studies broadly ignoring the single-model prediction caveats, i.e., those on Himalayan flora (Yang et al., 2013; Manish et al., 2016; Bobrowski et al., 2018; Mungi et al., 2018), fungi (Shrestha and Bawa, 2014), amphibians (Subba et al., 2018) and mammals (Escobar et al., 2015; Khadka and James, 2017; Su et al., 2018). Nevertheless, the most recent few years have seen studies addressing the errors and uncertainties in SDMs by incorporating multi-model ensembles for a range of taxa in Himalaya (Chhetri et al., 2018; Devkota et al., 2019; Srivastava et al., 2018; Ahmad et al., 2020; Bagaria et al., 2020b). Surprisingly, though, the freshwater fishes of Himalaya still await future research, with no models projected so far to assess their climate change responses.

To address the issue, a recent study by Sharma et al. (2021a) used snow trout as a model organism for predicting the current suitability and climate-driven potential range shift of Himalayan cold-water specialists. The authors used a multi-model ensemble with a comprehensive input data set for accurate predictions (Jarnevich et al., 2015; Ruaro et al., 2019).They collated occurrence data from long-term, extensive field-based sampling with additional records derived from an in-depth literature survey.

A comprehensive input data set including topographic, hydrogeomorphic and climatic variables was used to build correlative SDMs for the current and future time periods (2050 and 2070) under three carbon emission scenarios (RCPs) by integrating five GCMs. Overall, they built an ensemble modelling framework for the current and future distribution of snow trout across the core Himalaya or the Himalaya proper, evaluated the performance and response of topographic, hydrogeomorphic and climatic variables, and quantified the future range shifts by using a multi-projection ensemble (incorporating multiple GCMs, SDMs and RCPs). Furthermore, the range shifts were specifically investigated in the context of elevation as probable 'mountainward' movement of the suitable environmental niche for the snow trout in the future scenarios.

The study offers scientific backing on managing high-elevation riverscapes across the Himalayan nations. The SDM results invariably suggest a wide geographic distribution of snow trout in the mid-elevation catchments of the world's major river basins: Indus, Ganges and Brahmaputra. The predicted wide distribution of this species is corroborated by its long history of origin in the Himalaya, with its primitive ancestors diversifying and spreading out through vicariant-based speciation pertaining to a series of geological events dating back to the late Miocene (He and Chen, 2006; Li et al., 2009; Yang et al., 2012; Regmi, 2019).

Specialized with morphological traits like muscular and cylindrical body shape, minute scales, and a modified lower lip for adhesion, the species is well adapted to the torrential flow, cold waters and heterogeneous substrate conditions that are the major characteristics of the Himalayan rivers (Menon, 1962; Sehgal, 1999; Rajput et al., 2013).

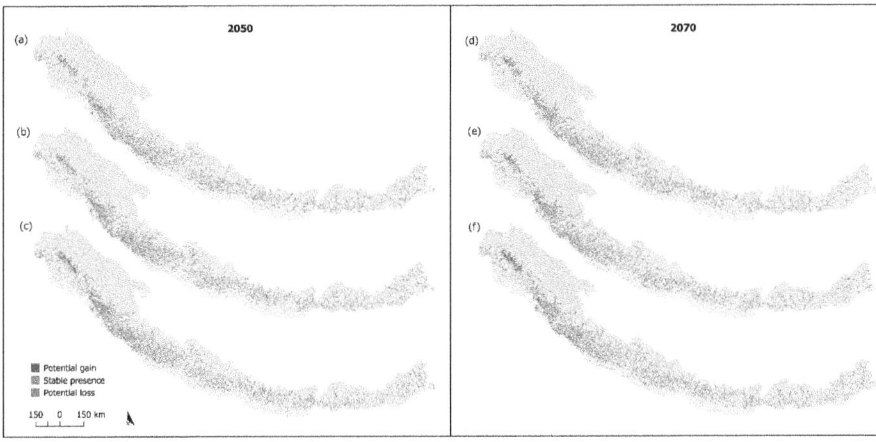

FIGURE 21.3 Predicted range shift of the snow trout under future climate change based on the ensemble of five general circulation models (GCMs), RCPs 2.6 (a) 4.5 (b) and 8.5 (c) for the year 2050; and RCPs 2.6 (d) 4.5 (e) and 8.5 (f) for the year 2070.

(From Sharma et al., 2021a.)

The higher probability of occurrence of snow trout in the main channel as well as in the tributaries as per the study predictions possibly hint at its potamodromous migratory behaviour. Nonetheless, their predictions for the future scenarios indicate that the snow trout will have to move towards higher elevations in search of future refugia in parallel with a predominant loss from most of its lagging edges in the future time periods as a consequence of climate change, and that a significant part of its current distributional range will be lost over time, as evident in Figure 21.3. The models further predict that snow trout will expand their range upwards into the high-altitude streams with a concurrent predominant range contraction in most of their lagging edges, ultimately creating a high-altitude squeeze. The net habitat loss under three RCP scenarios (RCPs 2.6, 4.5 and 8.5) was estimated to range from 7.41% to 16.29% for the year 2050, which would further increase in the year 2070, ranging from 9.46% to 26.56% (Figure 21.4).

The habitat suitability predictions that the authors present for some of the major river basins (for, e.g., tributaries of upper Ganges and Beas) showed higher probabilities of snow trout occurrence. With scarce abundance and population-level information available for these basins, these predictions can aid targeted species surveys and documentation in future studies. Their study provides a strong basis for climate-adaptive actions by identifying the high-altitude rivers and their tributaries that can act as climate refugia for the snow trout in future. The modelling framework they use also provides a foremost basis not only for conserving the snow trout but also for evaluating the climate impact on several other cold-water genera, which are equally vulnerable and ecologically important in the Himalaya.

Such results call for conservation efforts beyond political boundaries by combined decisions of the policymakers of Himalayan countries like India, Nepal and Bhutan to maintain the fragile riverine ecosystems of Himalaya. Reducing the unsustainable harnessing of rivers for hydropower development projects and energy efficiency by improving green energy potential could be the decisions needed.

21.10 DATA LIMITATIONS AND CHALLENGES

Although it can be ascertained that climate change will bring major perturbations in the seasonal flow regimes and basin hydrologies in Himalaya, what hinders the development of robust SDMs is

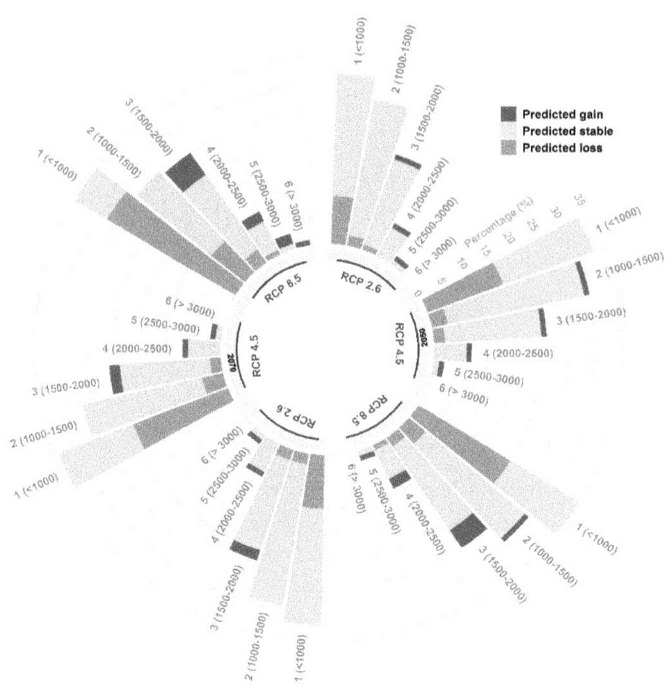

FIGURE 21.4 Circular stacked bar chart representing the potential elevational shift of the snow trout under RCPs 2.6, 4.5 and 8.5 for the years 2050 and 2070. The elevation ranges are represented in six categories as 1, <1000 m; 2, 1000–1500 m; 3, 1500–2000 m; 4, 2000–2500 m; 5, 2500–3000 m and 6, >3000 m. The percentage (%) corresponds to the predicted percentage gain, stable and loss of snow trout habitat at different elevational gradients. The dashed circular lines represent the percentage next to the bars.

(From Sharma et al., 2021a.)

the inherent inability of the present climate models and downscaling techniques to capture and predict the hydrological changes with precision (Hasson et al., 2019). Additionally, the current climate models built for Himalaya do not consider the aerosol-based reduced albedo, which speeds up the surface warming and thus, the glacier meltdown (Sabin et al., 2020).

The Himalayan orography poses major challenges to simulating the future hydroclimatic conditions (Dimri et al., 2020). The currently available spatial resolution of the climate data for the GCMs is coarse, which is unsuitable to model the distribution of most freshwater species in Himalaya which have home ranges smaller than the highest resolution of the available data. In fact, a study by Palazzi et al. (2015) highlighted the noticeably differing predictions for precipitation climatology in various sub-regions of the Hindu Kush Himalaya, indicative of the need to incorporate an ensemble of GCMs for climate models with reduced uncertainties.

Furthermore, the density of ground stations for climate data monitoring needs to be substantially enhanced to result in better inter-grid interpolations to build better climate layers in Himalaya (Dimri et al., 2020). Finer-scale climate models for Himalaya can aid in enhanced model parameterization, which can holistically account for local processes like convection, thereby improving the hydrological predictions (Rangwala et al., 2020). It thus warrants enhanced participation of multiple organizations and research institutes to come up with statistically downscaled climate data that can be utilized for building SDMs for Himalayan freshwater fauna.

Furthermore, the climate models that can be built using the fine-scale data would still hold major uncertainties and challenges for the SDMs of Himalayan freshwater fauna. These projections cannot overcome the uncertainties involved in a complex interplay of climatic conditions, species response, several anthropogenic drivers and time-related demographic conditions. For instance, the future distribution of the species resulting from the SDM correlative approach can differ based on the rate of climate warming and species' potential to adapt to new environmental conditions (Filipe et al., 2013; Urban, 2019).

Furthermore, the high-altitude Himalayan rivers are subjected to extreme events, such as intense rainfall and flash floods, besides a multitude of anthropogenic stressors like pollution and channel modification, which might further alter the future distribution of the species being modelled. Nevertheless, the currently available multi-model ensemble framework provides the foremost opportunity of transferability to develop more robust models in the future by incorporating several drivers and stressors that can influence the distribution freshwater species of Himalaya.

21.11 MITIGATION MEASURES

Policy measures designed to mitigate climate change impacts on the most wide-ranging and well-studied freshwater species in Himalaya can in turn act as '*umbrella*' conservation strategies for many other co-occurring freshwater fauna, which are yet to be explored and studied. Riverscape prioritization thus warrants a delineation of prime habitats for high-altitude freshwater specialists like the snow trout, where climate refugia are precisely delineated.

Protected area design and management across the Himalayan riverscapes is targeted based on the species' current range sizes. Climate change, however, threatens major perturbations to their existing geographical ranges, which are expected to shift towards climatically suitable habitats (Isaak et al., 2016; Hoegh-Guldberg et al., 2018). This collapses the effectiveness of existing protected areas, as considerable habitats falling within their boundaries might not stay optimal for species survival under continued global warming (Marquet et al., 2019).

Effective management of biodiversity under climate change thus warrants conjugation with habitat protection and restoration (Roberts et al., 2020). The globally protected land and inland water area coverage was 'targeted' at 17% by 2020 (Aichi target 11, Convention on Biological Diversity, 2010). To be reassessed by June 2021, the current protected area ratio is noticeably meagre, given the endemic biodiversity many landscapes and riverscapes abound in. What attracts interest is that while 10% of the marine area is also targeted for conservation prioritization, the freshwaters have yet not been considered as a separate ecosystem. Merging the management measures for the terrestrial and freshwater regimes is thus commonplace.

Although the issue is expected to be a future conservation concern across the montane habitats of the globe, the condition is Himalaya is even more momentous due to the existential conservation policies being largely 'landscape' centric. Currently, the policies implemented in the Indian part of Himalaya, i.e., the Indian Fisheries Act 1897, the Wildlife Protection Act 1972, the Environment Protection Act 1986 and the Wetlands (Conservation and Management) Rules 2017, although safeguard certain species, are indissolubly linked to 'landscapes' and thus self-defeating for 'riverscape' management.

Targeting climate mitigation for Himalayan freshwater fauna thus warrants an overarching approach of building future SDMs and riverscape-centric management policies based on flow-dictated species' phenologies and life-history traits. This can aid in identification of the species-specific current and future ecological niches for informed expansion of existing protected areas with concurrent designation of new protected watersheds.

REFERENCES

Agarwal, N.K. and Singh, G., 2012. Documentation of fishes and physico-chemical characters of a stream Indrawati – a spring fed tributary of river Bhagirathi at Uttarkashi (Central Himalaya, Garhwal). *Environment Conservation Journal*, 13(3), pp. 117–124.

Agarwal, N.K., Singh, G., Singh, H., Kumar, N. and Rawat, U.S., 2018. Ecological impacts of dams on the fish diversity of Bhagirathi river in central Himalaya (India). *Coldwater Fisheries Society of India*, 1(1), pp. 74–84.

Ahmad, S., Yang, L., Khan, T.U., Wanghe, K., Li, M. and Luan, X., 2020. Using an ensemble modelling approach to predict the potential distribution of Himalayan gray goral (*Naemorhedus goral bedfordi*) in Pakistan. *Global Ecology and Conservation,* 21, p. e00845.

Al-Chokhachy, R. and Sepulveda, A.J., 2019. Impacts of non-native Brown Trout on Yellowstone Cutthroat Trout in a tributary stream. *North American Journal of Fisheries Management*, 39(1), pp. 17–28.

Angilletta Jr, M.J., 2009. *Thermal Adaptation: A Theoretical and Empirical Synthesis*. Oxford University Press Inc., New York., p. 285.

Araújo, M.B. and New, M., 2007. Ensemble forecasting of species distributions. *Trends in Ecology & Evolution,* 22(1), pp. 42–47.

Araújo, M.B., Anderson, R.P., Barbosa, A.M., Beale, C.M., Dormann, C.F., Early, R. et al., 2019. Standards for distribution models in biodiversity assessments. *Science Advances,* 5(1), p. eaat4858.

Audzijonyte, A., Richards, S.A., Stuart-Smith, R.D., Pecl, G., Edgar, G.J., Barrett, N.S. et al., 2020. Fish body sizes change with temperature but not all species shrink with warming. *Nature Ecology and Evolution,* 4(6), pp. 809–814.

Badola, S.P., 2009. *Ichthyology of the Central Himalaya*. Transmedia Publication, Uttarakhand, p. 206.

Bagaria, P., Sharma, L.K., Joshi, B.D., Kumar, H., Mukherjee, T., Thakur, M. et al., 2020. West to east shift in range predicted for Himalayan Langur in climate change scenario. *Global Ecology and Conservation,* 22, p. e00926.

Bagaria, P., Thapa, A., Sharma, L.K., Joshi, B.D., Singh, H., Sharma, C.M. et al., 2021. Distribution modelling and climate change risk assessment strategy for rare Himalayan Galliformes species using archetypal data abundant cohorts for adaptation planning. *Climate Risk Management*, 31,p. 100264.

Benestad, R., Sillmann, J., Thorarinsdottir, T.L., Guttorp, P., Mesquita, M.D.S., Tye, M.R. et al., 2017. New vigour involving statisticians to overcome ensemble fatigue. *Nature Climate Change,* 7(10), pp. 697–703.

Bhambri, R., Mehta, M., Dobhal, D.P., and Gupta, A.K., 2015. *Glacier Lake Inventory of Uttarakhand*. Wadia Institute of Himalayan Geology, Dehradun, pp. 1–78.

Bhatt, J.P., Manish, K., and Pandit, M.K., 2012. Elevational gradients in fish diversity in the Himalaya: water discharge is the key driver of distribution patterns. *PloS ONE*, 7(9), p. e46237.

Bhattacharyya, S., Mungi, N.A., Kawamichi, T., Rawat, G.S., Adhikari, B.S. and Wilkening, J.L., 2019. Insights from present distribution of an alpine mammal Royle's pika (*Ochotona roylei*) to predict future climate change impacts in the Himalaya. *Regional Environmental Change*, 19(8), pp. 2423–2435.

Bloch, M.E., 1785. *Naturgeschichte der auslandischen Fische*, Berlin; 9 parts in 2, and atlas (3 vols.). Atlas.

Bobrowski, M., Bechtel, B., Böhner, J., Oldeland, J., Weidinger, J. and Schickhoff, U., 2018. Application of thermal and phenological land surface parameters for improving ecological niche models of *Betula utilis* in the Himalayan region. *Remote Sensing*, 10(6), p. 814.

Brittain, J.E. and Eikeland, T.J., 1988. Invertebrate drift—a review. *Hydrobiologia*, 166(1), pp. 77–93.

Buckton, S.T. and Ormerod, S.J., 2008. Niche segregation of Himalayan river birds. *Journal of Field Ornithology*, 79(2), pp. 176–185.

Buisson, L. and Grenouillet, G., 2009. Contrasted impacts of climate change on stream fish assemblages along an environmental gradient. *Diversity and Distributions,* 15, pp. 613–626.

Chardon, N.I., Pironon, S., Peterson, M.L. and Doak, D.F., 2020. Incorporating intraspecific variation into species distribution models improves distribution predictions, but cannot predict species traits for a wide-spread plant species. *Ecography*, 43(1), pp. 60–74.

Chen, I.C., Hill, J.K., Ohlemüller, R., Roy, D.B. and Thomas, C.D., 2011. Rapid range shifts of species associated with high levels of climate warming. *Science*, 333(6045), pp. 1024–1026.

Chhetri, B., Badola, H.K. and Barat, S., 2018. Predicting climate-driven habitat shifting of the near threatened Satyr Tragopan (*Tragopan satyra*; Galliformes) in the Himalayas. *Avian Biology Research*, 11(4), pp. 221–230.

Comte, L. and Grenouillet, G., 2013. Do stream fish track climate change? Assessing distribution shifts in recent decades. *Ecography*, 36(11), pp. 1236–1246.

Comte, L., Buisson, L., Daufresne, M. and Grenouillet, G., 2013. Climate-induced changes in the distribution of freshwater fish: observed and predicted trends. *Freshwater Biology*, 58(4), pp. 625–639.

Connolly, P.J., Jezorek, I.G., Martens, K.D. and Prentice, E.F., 2008. Measuring the performance of two stationary interrogation systems for detecting downstream and upstream movement of PIT-tagged salmonids. *North American Journal of Fisheries Management,* 28(2), pp.402–417.

Convention on Biological Diversity. 2010. *COP Decision. X/2. Strategic Plan for Biodiversity 2011–2020.* Secretariat of the Convention on Biological Diversity, Montreal, Canada. www.cbd.int/decision/cop/?id=12268

Cotto, O., Wessely, J., Georges, D., Klonner, G., Schmid, M., Dullinger, S. et al., 2017. A dynamic eco-evolutionary model predicts slow response of alpine plants to climate warming. *Nature Communications,* 8(1), pp.1–9.

Crozier, L.G. and Hutchings, J.A., 2014. Plastic and evolutionary responses to climate change in fish. *Evolutionary Applications*, 7(1), pp.68–87.

Daufresne, M., Lengfellner, K. and Sommer, U., 2009. Global warming benefits the small in aquatic ecosystems. *Proceedings of the National Academy of Sciences*, 106(31), pp. 12788–12793.

Devkota, S., Dymytrova, L., Chaudhary, R.P., Werth, S. and Scheidegger, C., 2019. Climate change-induced range shift of the endemic epiphytic lichen *Lobaria pindarensis* in the Hindu Kush Himalayan region. *The Lichenologist*, 51(2), pp. 157–173.

Diamond, S.E., 2018. Contemporary climate-driven range shifts: putting evolution back on the table. *Functional Ecology*, 32(7), pp. 1652–1665.

Diffenbaugh, N.S. and Field, C.B., 2013. Changes in ecologically critical terrestrial climate conditions. *Science,* 341, pp. 486–492.

Dimri, A.P., Bookhagen, B., Stoffel, M. and Yasunari, T., Eds., 2020. *Himalayan Weather and Climate and Their Impact on the Environment.* Springer Nature, Cham, Switzerland, p. 577.

Domisch, S., Jaehnig, S.C. and Haase, P., 2011. Climate-change winners and losers: stream macroinvertebrates of a submontane region in Central Europe. *Freshwater Biology*, 56(10), pp. 2009–2020.

Dudgeon, D., 2011. Asian river fishes in the Anthropocene: threats and conservation challenges in an era of rapid environmental change. *Journal of Fish Biology*, 79(6), pp. 1487–1524.

Eby, L.A., Helmy, O., Holsinger, L.M. and Young, M.K., 2014. Evidence of climate-induced range contractions in bull trout *Salvelinus confluentus* in a Rocky Mountain watershed, USA. *PLoS ONE*, 9(6), p. 98812.

Elith, J., Kearney, M. and Phillips, S., 2010. The art of modelling range-shifting species. *Methods in Ecology and Evolution*, 1(4), pp. 330–342.

Engelhard, G.H., Righton, D.A. and Pinnegar, J.K., 2014. Climate change and fishing: a century of shifting distribution in North Sea cod. *Global Change Biology*, 20(8), pp. 2473–2483.

Escobar, L.E., Awan, M.N. and Qiao, H., 2015. Anthropogenic disturbance and habitat loss for the red-listed Asiatic black bear (*Ursus thibetanus*): using ecological niche modelling and nighttime light satellite imagery. *Biological Conservation,* 191, pp. 400–407.

Fandos, G., Rotics, S., Sapir, N., Fiedler, W., Kaatz, M., Wikelski, M. et al., 2020. Seasonal niche tracking of climate emerges at the population level in a migratory bird. *Proceedings of the Royal Society B,* 287(1935), p. 20201799.

Filipe, A.F., Markovic, D., Pletterbauer, F., Tisseuil, C., De Wever, A., Schmutz, S. et al., 2013. Forecasting fish distribution along stream networks: brown trout (*Salmo trutta*) in Europe. *Diversity and Distributions,* 19(8), pp. 1059–1071.

Forrest, J.L., Wikramanayake, E., Shrestha, R., Areendran, G., Gyeltshen, K., Maheshwari, A. et al., 2012. Conservation and climate change: assessing the vulnerability of snow leopard habitat to treeline shift in the Himalaya. *Biological Conservation,* 150(1), pp. 129–135.

Freeman, B.G., Lee-Yaw, J.A., Sunday, J.M. and Hargreaves, A.L., 2018. Expanding, shifting and shrinking: the impact of global warming on species' elevational distributions. *Global Ecology and Biogeography,* 27(11), pp. 1268–1276.

Furniss, M. J., Roby, K. B., Cenderelli, D., Chatel, J., Clifton, C.F., Clingenpeel, A. et al. (2013). *Assessing the Vulnerability of Watersheds to Climate Change: Results of National Forest Watershed Vulnerability Pilot Assessments (General Technical Report PNW-GTR-884).* U.S. Forest Service, Pacific Northwest Research Station, Portland.

Gardner, J.L., Peters, A., Kearney, M.R., Joseph, L. and Heinsohn, R., 2011. Declining body size: a third universal response to warming? *Trends in Ecology & Evolution*, 26(6), pp. 285–291.

Grover, V., Borsdorf, A., Breuste, J.H., Tiwari, P.C. and Frangetto, F., 2015. *Impacts of Global Change on Mountains: Adaptation and Responses.* CRC Press, Taylor and Francis group, New York, p. 527.

Guillera-Arroita, G., Lahoz-Monfort, J.J., Elith, J., Gordon, A., Kujala, H., Lentini, P.E. et al., 2015. Is my species distribution model fit for purpose? Matching data and models to applications. *Global Ecology and Biogeography,* 24(3), pp. 276–292.

Guisan, A., Thuiller, W. and Zimmermann, N.E., 2017. *Habitat Suitability and Distribution Models: With Applications in R.* Cambridge University Press, UK, p. 462.

Guisan, A.A., Broennimann, O., Buri, A., Cianfrani, C., D'Amen, M., Di Cola, V. et al., 2019. Chapter seventeen: Climate change impacts on mountain biodiversity. In *Biodiversity and Climate Change* (Lovejoy, T.E. and Hannah, L., Eds.) Yale University Press, USA, pp. 229–241.

Hamid, M., Khuroo, A.A., Charles, B., Ahmad, R., Singh, C.P. and Aravind, N.A., 2019. Impact of climate change on the distribution range and niche dynamics of Himalayan birch, a typical treeline species in Himalayas. *Biodiversity and Conservation*, 28(8), pp. 2345–2370.

Hamilton, G., 1822. *Water Quality Criteria and Nomenclature for Freshwater Fish.* Food and Agriculture Organization of the United Nations, FAO, Rome, 7, pp. 6–9.

Hannah, L., 2012. Saving a million species. In *Saving a Million Species: : extinction risk from climate change* (Hannah, L., Ed.) Island Press, Center for Resource Economics, Washington, pp. 389–394.

Hao, T., Elith, J., Guillera-Arroita, G. and Lahoz-Monfort, J.J., 2019. A review of evidence about use and performance of species distribution modelling ensembles like BIOMOD. *Diversity and Distributions*, 25(5), pp. 839–852.

Hao, T., Elith, J., Lahoz-Monfort, J.J. and Guillera-Arroita, G., 2020. Testing whether ensemble modelling is advantageous for maximising predictive performance of species distribution models. *Ecography*, 43(4), pp. 549–558.

Hasson, S., Böhner, J. and Chishtie, F., 2019. Low fidelity of CORDEX and their driving experiments indicates future climatic uncertainty over Himalayan watersheds of Indus basin. *Climate Dynamics*, 52(1), pp. 777–798.

He, D. and Chen, Y., 2006. Biogeography and molecular phylogeny of the genus Schizothorax (Teleostei: Cyprinidae) in China inferred from cytochrome b sequences. *Journal of Biogeography*, 33(8), pp. 1448–1460.

Herrera-R, G.A., Oberdorff, T., Anderson, E.P., Brosse, S., Carvajal-Vallejos, F.M., Frederico, R.G., et al., 2020. The combined effects of climate change and river fragmentation on the distribution of Andean Amazon fishes. *Global Change Biology,* 26(10), pp. 5509–5523.

Hock, R., Rasul, G., Adler, C., Cáceres, B., Gruber, S., Hirabayashi, Y. et al., 2019. High mountain areas. In *IPCC Special Report on the Ocean and Cryosphere in a Changing Climate* (Pörtner, H.O., Roberts, D.C., Masson-Delmotte, V., Zhai, P., Tignor, M., Poloczanska, E., et al., Eds.). IPCC – Intergovernmental Panel on Climate Change, Cambridge University Press, Cambridge, UK and New York, pp. 131–202.

Hoegh-Guldberg, O., Jacob, D., Bindi, M., Brown, S., Camilloni, I., Diedhiou, A. et al., 2018. Impacts of 1.5 C global warming on natural and human systems. In *Global Warming of 1.5 C. An IPCC Special Report.* (Masson-Delmotte, V., Zhai, P., Pörtner, H.-O., Roberts, D., Skea, J., Shukla, P.R. et al., Eds). Cambridge University Press, Cambridge, UK and New York, pp. 175–312.

Hofmann, G.E. and Todgham, A.E., 2010. Living in the now: physiological mechanisms to tolerate a rapidly changing environment. *Annual Review of Physiology,* 72, pp. 127–145.

Hora, S.L., 1930. V. Ecology, bionomics and evolution of the torrential fauna, with special reference to the organs of attachment. *Philosophical Transactions of the Royal Society of London. Series B.*, 218(450–461), pp. 171–282.

Hoxmeier, R.J.H. and Dieterman, D.J., 2012. Demographic Responses of Brook Trout to Removal of Brown Trout from a Driftless Area Stream in Minnesota. Final Report, https://files.dnr.state.mn.us/areas/fisheries/lanesboro/research/677report.pdf. Accessed on 16 March 2020, p. 677.

Hutchings, J.A., 2002. Life histories of fish. In *Handbook of Fish Biology and Fisheries* (Hart, P.J.B. and Reynolds, J.D., Eds.). Blackwell Science, Fish Biology, Oxford, pp. 149–174.

Immerzeel, W.W., Van Beek, L.P. and Bierkens, M.F., 2010. Climate change will affect the Asian water towers. *Science*, 328(5984), pp. 1382–1385.

IPCC (Intergovernmental Panel for Climate Change), 2013. *The Physical Science Basis. Working Group I Contribution to the Fifth Assessment Report of the Intergovernmental Panel on Climate Change.* Cambridge University Press, Cambridge, UK.

IPCC (Intergovernmental Panel for Climate Change), 2018. Global warming of 1.5°C. *An IPCC Special Report on the Impacts of Global Warming of 1.5°C above Pre-industrial Levels and Related Global Greenhouse Gas Emission Pathways, in the Context of Strengthening the Global Response to the Threat of Climate Change, Sustainable Development, and Efforts to Eradicate Poverty* (Masson-Delmotte, V., Zhai, P., Pörtner, H.O., Roberts, D., Skea, J., Shukla, P.R. et al., Eds.) Cambridge University Press, Cambridge, UK and New York, 616 pp.

Isaak, D.J., Young, M.K., Nagel, D.E., Horan, D.L. and Groce, M.C., 2015. The cold-water climate shield: delineating refugia for preserving salmonid fishes through the 21st century. *Global Change Biology*, 21(7), pp. 2540–2553. https://doi.org/10.1111/gcb.12879

Isaak, D.J., Young, M.K., Luce, C.H., Hostetler, S.W., Wenger, S.J., Peterson, E.E. et al., 2016. Slow climate velocities of mountain streams portend their role as refugia for cold-water biodiversity. *Proceedings of the National Academy of Sciences,* 113(16), pp. 4374–4379.

ISRO, Indian Space Research Organisation, 2021. www.iirs.gov.in. Accessed on 15 April 2021.

James, C.S., Reside, A.E., VanDerWal, J., Pearson, R.G., Burrows, D., Capon, S.J. et al., 2017. Sink or swim? Potential for high faunal turnover in Australian rivers under climate change. *Journal of Biogeography,* 44(3), pp. 489–501.

Jarić, I., Lennox, R.J., Kalinkat, G., Cvijanović, G. and Radinger, J., 2019. Susceptibility of European freshwater fish to climate change: species profiling based on life-history and environmental characteristics. *Global Change Biology*, 25(2), pp. 448–458.

Jarnevich, C.S., Stohlgren, T.J., Kumar, S., Morisette, J.T. and Holcombe, T.R., 2015. Caveats for correlative species distribution modeling. *Ecological Informatics,* 29, pp. 6–15.

Jayaram, K.C., 1977. Zoogeography of Indian freshwater fishes. *Proceedings of the Indian Academy of Sciences-Section B*, 86, pp. 265–274.

Jenouvrier, S., 2013. Impacts of climate change on avian populations. *Global Change Biology*, 19(7), pp. 2036–2057.

Jhingran, V.G., 1982. *Fish and Fisheries of India.* Hindustan Publishing Corporation (India), Delhi. p. 666.

Johal, M.S. and Rawal, Y.K., 2005. General ecology of some hillstreams of Himachal Pradesh with special reference to fish communities. In *Biological Diversity in Freshwater Environments* (Nautiyal, P., Bhatt, P., Gussian, J.P., and Dobriyal, A.K., Eds.). Transmedia, Srinagar (Uttaranchal), India. pp. 36–56.

Johal, M.S. and Tandon, K.K., 1981. Age, growth and length–weight relationship of *Tor putitora* from Gobindsagar, Himachal Pradesh India. *Fishery Bulletin, Special Publication on Coldwater Fisheries Seminar Organized by CIFE, Chandigarh, India*, pp. 43–48.

Johal, M.S., Tandon K.K. and Sandhu G.S., 1999. Age and growth of an endangered cold water fish golden mahseer *Tor putitora* (Ham.) from Gobindsagar, Himachal Pradesh, India. In *Recent Research Advances in Ichthyology* (Saksena, D.N. Ed.). Oxford and IBH Publishing Co., New Delhi, pp. 59–73.

Johal, M.S., Tandon, K.K., Rawal, Y.K., Tyor, A.K., Banyal, H.S. and Rumana, H.S., 2001. Species richness of fish in relation to environmental factors. *Current Science*, 80(4), pp. 499–501.

Johnson, J.A., Sharma, A., Rajput, V., Dubey, V.K. and Sivakumar, K., 2020. Taxonomic and guild structure of fish assemblages in the streams of Western Himalaya, India. *Community Ecology*, 21(3), pp. 239–257.

Joshi, K.D. and Tyagi, B.C. 2008. Rearing of carp fry under polyhouse ponds at high altitudes of Uttaranchal Himalaya, during winters. *Journal of Inland Fisheries Society of India*, 40(1), pp. 78–81.

Kamalam, B.S., Mahija, J., Baral, P., Pandey, A., Akhtar, M.S., Ciji, A. et al., 2019. Temperature and oxygen related ecophysiological traits of snow trout (*Schizothorax richardsonii*) are sensitive to seasonal changes in a Himalayan stream environment. *Journal of Thermal Biology,* 83, pp.22–29.

Karmakar, K.A., 2000. Fish communities and their distribution in Himalayan drainage system. *Records of the Zoological Survey of India,* 98, pp. 25–37.

Kemp, D.B., Eichenseer, K. and Kiessling, W., 2015. Maximum rates of climate change are systematically underestimated in the geological record. *Nature Communications*, 6(1), pp.1–6.

Kennedy, R.J. and Crozier, W.W., 2010. Evidence of changing migratory patterns of wild Atlantic salmon *Salmo salar* smolts in the River Bush, Northern Ireland, and possible associations with climate change. *Journal of Fish Biology*, 76(7), pp. 1786–1805.

Khadka, K.K. and James, D.A., 2017. Modeling and mapping the current and future climatic-niche of endangered Himalayan musk deer. *Ecological Informatics,* 40, pp. 1–7.

Kovach, R.P., Muhlfeld, C.C., Al-Chokhachy, R., Dunham, J.B., Letcher, B.H. and Kershner, J.L., 2016. Impacts of climatic variation on trout: a global synthesis and path forward. *Reviews in Fish Biology and Fisheries,* 26(2), pp. 135–151.

Lamprecht, A., Semenchuk, P.R., Steinbauer, K., Winkler, M. and Pauli, H., 2018. Climate change leads to accelerated transformation of high-elevation vegetation in the central Alps. *New Phytologist*, 220(2), pp. 447–459.

Lee, S.Y., Fullerton, A.H., Sun, N. and Torgersen, C.E., 2020. Projecting spatiotemporally explicit effects of climate change on stream temperature: a model comparison and implications for coldwater fishes. *Journal of Hydrology,* 588, p. 125066.

Lefevre, S., Wang, T. and McKenzie, D.J., 2021. The role of mechanistic physiology in investigating impacts of global warming on fishes. *Journal of Experimental Biology,* 224, p. jeb238840.

Lencioni, V., 2018. Glacial influence and stream macroinvertebrate biodiversity under climate change: lessons from the Southern Alps. *Science of the Total Environment,* 622, pp. 563–575.

Lencioni, V., Jousson, O., Guella, G. and Bernabò, P., 2015. Cold adaptive potential of chironomids overwintering in a glacial stream. *Physiological Entomology*, 40(1), pp. 43–53.

Li, J., He, Q., Hua, X., Zhou, J., Xu, H., Chen, J. and Fu, C., 2009. Climate and history explain the species richness peak at mid-elevation for *Schizothorax* fishes (Cypriniformes: Cyprinidae) distributed in the Tibetan Plateau and its adjacent regions. *Global Ecology and Biogeography,* 18(2), pp. 264–272.

Littlefield, C.E., Krosby, M., Michalak, J.L. and Lawler, J.J., 2019. Connectivity for species on the move: supporting climate-driven range shifts. *Frontiers in Ecology and the Environment,* 17(5), pp. 270–278.

Loehman, R.A., Bentz, B.J., DeNitto, G.A., Keane, R.E., Manning, M.E., Duncan, J.P. et al., 2018. Effects of climate change on ecological disturbance in the Northern Rockies. In *Climate Change and Rocky Mountain Ecosystems* (Halofsky, J.E., Peterson, D.L., Dante-Wood, S.K., Hoang, L., Ho, J.J., Joyce, L.A., Eds.) Springer, Cham, pp. 115–141.

Lois, S. and Cowley, D.E., 2017. Conservation of interacting species in network-constrained environments. *Diversity and Distributions,* 23, pp. 1235–1245.

Lopes, T.M., Bailly, D., Almeida, B.A., Santos, N.C., Gimenez, B.C., Landgraf, G.O. et al., 2017. Two sides of a coin: Effects of climate change on the native and non-native distribution of *Colossoma macropomum* in South America. *PloS ONE,* 12(6), p. e0179684.

Lu, X., Liang, E., Wang, Y., Babst, F. and Camarero, J.J., 2021. Mountain treelines climb slowly despite rapid climate warming. *Global Ecology and Biogeography,* 30(1), pp. 305–315.

Mainali, K., Shrestha, B.B., Sharma, R.K., Adhikari, A., Gurarie, E., Singer, M. et al., 2020. Contrasting responses to climate change at Himalayan treelines revealed by population demographics of two dominant species. *Ecology and Evolution*, 10(3), pp. 1209–1222.

Malcolm, J.R., Liu, C., Neilson, R.P., Hansen, L. and Hannah, L.E.E., 2006. Global warming and extinctions of endemic species from biodiversity hotspots. *Conservation Biology*, 20(2), pp. 538–548.

Mamantov, M.A., Gibson-Reinemer, D.K., Linck, E.B. and Sheldon, K.S., 2021. Climate-driven range shifts of montane species vary with elevation. *Global Ecology and Biogeography,* 30(4), pp. 784–794.

Manish, K., Telwala, Y., Nautiyal, D.C. and Pandit, M.K., 2016. Modelling the impacts of future climate change on plant communities in the Himalaya: a case study from Eastern Himalaya, India. *Modeling Earth Systems and Environment*, 2(2), pp. 1–12.

Mantyka-Pringle, C.S., Martin, T.G., Moffatt, D.B., Linke, S. and Rhodes, J.R., 2014. Understanding and predicting the combined effects of climate change and land-use change on freshwater macroinvertebrates and fish. *Journal of Applied Ecology*, 51(3), pp. 572–581.

Mantyka-Pringle, C.S., Martin, T.G., Moffatt, D.B., Udy, J., Olley, J., Saxton, N. et al., 2016. Prioritizing management actions for the conservation of freshwater biodiversity under changing climate and land-cover. *Biological Conservation,* 197, pp. 80–89.

Markovic, D., Carrizo, S.F., Kärcher, O., Walz, A. and David, J.N., 2017. Vulnerability of European freshwater catchments to climate change. *Global Change Biology*, 23(9), pp. 3567–3580.

Marquet, P.A., Lessmann, J. and Shaw, M.R., 2019. Chapter twenty-two: Protected-area management and climate change. *In Biodiversity and Climate Change* (Lovejoy, T.E. and Hannah, L., Eds.) Yale University Press, USA, pp. 291–301.

Maurer, J.M., Schaefer, J.M., Rupper, S. and Corley, A., 2019. Acceleration of ice loss across the Himalayas over the past 40 years. *Science Advances*, 5(6), p. eaav7266.

McKelvey, K.S. and Buotte, P.C., 2018. Chapter nine: Climate change and wildlife in the Northern Rockies region. In *Climate Change Vulnerability and Adaptation in the Northern Rocky Mountains [Part 2]* (Halofsky, J.E.; Peterson, D.L.; Dante-Wood, S.K.; Hoang, L.; Ho, J.J.; Joyce, L.A., Eds.). Gen. Tech. Rep. RMRS-GTR-374. US Department of Agriculture, Forest Service, Rocky Mountain Research Station, Fort Collins, CO, pp. 353–397.

Mehta, H.S. and Sharma, I., 2008. Pisces. In *Fauna of Pin Valley National Park*. Zoological Survey of India, Kolkata (Director, ZSI, Kolkata, Ed.), 34, pp. 75–84.

Menon, A.G.K., 1954. Fish geography of the Himalayas. *Proceedings of National Academy of Sciences, India,* 20(4), pp. 467–493.

Menon, A.G.K., 1962. A distributional list of fishes of the Himalayas. *Journal of Zoological Society of India*, 14(1–2), pp. 23–32.

Merow, C., Smith, M.J., and Silander Jr, J.A., 2013. A practical guide to MaxEnt for modeling species' distributions: what it does, and why inputs and settings matter. *Ecography,* 36, pp. 1058–1069.

Mittermeier, R.A., Robles-Gil, P., Hoffmann, M., Pilgrim, J.D., Brooks, T., Mittermeier, C. et al., 2004. *Hotspots Revisited: Earth's Biologically Richest and Most Endangered Terrestrial Ecoregions*. Cemex, Mexico City, Mexico, p. 390.

Moritz, C. and Agudo, R., 2013. The future of species under climate change: resilience or decline? *Science*, 341(6145), pp. 504–508.

Morrison, L., Estrada, A. and Early, R., 2018. Species traits suggest European mammals facing the greatest climate change are also least able to colonize new locations. *Diversity and Distributions*, 24(9), pp. 1321–1332.

Mota-Ferreira, M., Filipe, A.F., Filomena Magalhães, M., Carona, S. and Beja, P., 2021. Spatial modelling of temporal dynamics in stream fish communities under anthropogenic change. *Diversity and Distributions,* 27(2), pp. 313–326.

Mountain Research Initiative EDW Working Group, 2015. Elevation-dependent warming in mountain regions of the world. *Nature Climate Change,* 5, pp. 424–430.

Mungi, N.A., Coops, N.C., Ramesh, K. and Rawat, G.S., 2018. How global climate change and regional disturbance can expand the invasion risk? Case study of *Lantana camara* invasion in the Himalaya. *Biological Invasions*, 20(7), pp. 1849–1863.

Myers, N., Mittermeier, R.A., Mittermeier, C.G., Da Fonseca, G.A. and Kent, J., 2000. Biodiversity hotspots for conservation priorities. *Nature*, 403(6772), p. 853.

Nautiyal, P., 2005. Taxonomic richness in the fish fauna of the Himalaya, Central Highlands and Western Ghats (Indian Subcontinent). *International Journal of Ecology and Environmental Sciences*, 31(2), pp.73–92.

Nautiyal, P., 2006. Assessment of ecological health of fish populations of the upper reaches of Ganga (Garhwal Himalaya). In *Advances in Fish Research* (Dutta Munshi, J.S., Singh, H.R. Eds.), vol 4. Narendra Publishing House, Delhi, pp. 211–238.

Negi, R.K. and Banyal, H.S., 2017. Ichthyofaunal study in Trans-Himalayan Rakchham-Chhitkul Wildlife Sanctuary in Baspa (Sangla) valley, district Kinnaur, Himachal Pradesh, *India. International Journal of Biology*, 9(1), pp. 36–40.

Null, S.E., Viers, J.H., Deas, M.L., Tanaka, S.K. and Mount, J.F., 2013. Stream temperature sensitivity to climate warming in California's Sierra Nevada: impacts to coldwater habitat. *Climatic Change*, 116(1), pp. 149–170.

Ockendon, N., Baker, D.J., Carr, J.A., White, E.C., Almond, R.E., Amano, T. et al., 2014. Mechanisms underpinning climatic impacts on natural populations: altered species interactions are more important than direct effects. *Global Change Biology,* 20(7), pp. 2221–2229.

Olden, J.D., Kennard, M.J., Lawler, J.J. and Poff, N.L., 2011. Challenges and opportunities in implementing managed relocation for conservation of freshwater species. *Conservation Biology*, 25(1), pp. 40–47.

Olden, J.D., Poff, N.L. and Bestgen, K.R., 2006. Life-history strategies predict fish invasions and extirpations in the Colorado River Basin. *Ecological Monographs*, 76(1), pp. 25–40.

Overland, J.E., Hanna, E., Hanssen-Bauer, I., Kim, S.J., Walsh, J.E., Wang, M. et al., 2019. Surface air temperature. Arctic Report Card, p. 5.

Pacifici, M., Foden, W.B., Visconti, P., Watson, J.E., Butchart, S.H., Kovacs, K.M. et al., 2015. Assessing species vulnerability to climate change. *Nature Climate Change,* 5(3), pp. 215–224.

Palazzi, E., von Hardenberg, J., Terzago, S. and Provenzale, A., 2015. Precipitation in the Karakoram-Himalaya: a CMIP5 view. *Climate Dynamics*, 45(1–2), pp. 21–45.

Pandit, M.K. and Grumbine, R.E., 2012. Potential effects of ongoing and proposed hydropower development on terrestrial biological diversity in the Indian Himalaya. *Conservation Biology*, 26(6), pp.1061–1071.

Pecl, G.T., Araújo, M.B., Bell, J.D., Blanchard, J., Bonebrake, T.C., Chen, I.C. et al., 2017. Biodiversity redistribution under climate change: impacts on ecosystems and human well-being. *Science,* 355 (6332), p. eaai9214.

Perry, A.L., Low, P.J., Ellis, J.R. and Reynolds, J.D., 2005. Climate change and distribution shifts in marine fishes. *Science*, 308(5730), pp. 1912–1915.

Pimm, S.L., Jenkins, C.N., Abell, R., Brooks, T.M., Gittleman, J.L., Joppa, L.N. et al., 2014. The biodiversity of species and their rates of extinction, distribution, and protection. *Science,* 344, p. 6187.

Pinder, A.C., Britton, J.R., Harrison, A.J., Nautiyal, P., Bower, S.D., Cooke, S.J. et al., 2019. Mahseer (Tor spp.) fishes of the world: status, challenges and opportunities for conservation. *Reviews in Fish Biology and Fisheries,* 29(2), pp. 417–452.

Pörtner, H.O. and Farrell, A.P., 2008. Physiology and climate change. *Science*, 322(5902), pp. 690–692.

Pratap, B., Dobhal, D.P., Bhambri, R., Mehta, M. and Tewari, V.C., 2016. Four decades of glacier mass balance observations in the Indian Himalaya. *Regional Environmental Change*, 16(3), pp. 643–658.

Qazi, N.Q., Jain, S.K., Thayyen, R.J., Patil, P.R. and Singh, M.K., 2020. Hydrology of the Himalayas. In *Himalayan Weather and Climate and their Impact on the Environment.* (Dimri, A., Bookhagen, B., Stoffel, M., Yasunari, T., Eds.), Springer Nature, Cham, Switzerland, pp. 419–450.

Radinger, J., Essl, F., Hölker, F., Horký, P., Slavík, O. and Wolter, C., 2017. The future distribution of river fish: the complex interplay of climate and land use changes, species dispersal and movement barriers. *Global Change Biology*, 23(11), pp. 4970–4986.

Rahbek, C., Borregaard, M.K., Colwell, R.K., Dalsgaard, B.O., Holt, B.G., Morueta-Holme, N. et al., 2019. Humboldt's enigma: what causes global patterns of mountain biodiversity? *Science*, 365(6458), pp. 1108–1113.

Raina, H.S. and Petr, T., 1999. Coldwater fish and fisheries in the Indian Himalayas: lakes and reservoirs. In Fish and fisheries at higher altitudes: Asia. FAO Fisheries Technical Paper, 385, pp. 64–88.

Rajput, V., Johnson, J.A. and Sivakumar, K., 2013. Environmental effects on the morphology of the Snow Trout *Schizothorax richardsonii* (Gray, 1832). *Taprobanica: The Journal of Asian Biodiversity*, 5(2), pp. 102–110.

Rajvanshi, A., Arora, R., Mathur, V.B., Sivakumar, K., Sathyakumar, S., Rawat, G.S. et al., 2012. *Assessment of Cumulative Impacts of Hydroelectric Projects on Aquatic and Terrestrial Biodiversity in Alaknanda and Bhagirathi Basins*. Technical Report, Wildlife Institute of India, Uttarakhand, p. 203.

Rana, S.K., Luo, D., Rana, H.K., O'Neill, A.R. and Sun, H., 2021. Geoclimatic factors influence the population genetic connectivity of *Incarvillea arguta* (Bignoniaceae) in the Himalaya–Hengduan Mountains biodiversity hotspot. *Journal of Systematics and Evolution,* 59(1), pp. 151–168.

Rangwala, I., Palazzi, E. and Miller, J.R., 2020. Projected climate change in the Himalayas during the twenty-first century. In *Himalayan Weather and Climate and Their Impact on the Environment.* (Dimri, A., Bookhagen, B., Stoffel, M., Yasunari, T., Eds.), Springer Nature, Cham, Switzerland, pp. 51–71.

Rapacciuolo, G., Maher, S.P., Schneider, A.C., Hammond, T.T., Jabis, M.D., Walsh, R.E. et al., 2014. Beyond a warming fingerprint: individualistic biogeographic responses to heterogeneous climate change in California. *Global Change Biology,* 20(9), pp. 2841–2855.

Rathore, P., Roy, A. and Karnatak, H., 2019. Assessing the vulnerability of Oak (*Quercus*) forest ecosystems under projected climate and land use land cover changes in Western Himalaya. *Biodiversity and Conservation*, 28(8), pp. 2275–2294.

Rawal, Y.K., 2002. Ecology and abundance of fish communities in hillstreams of Western Himalayas and analysis of factors responsible for their distribution. Doctoral dissertation, PhD Thesis, Panjab University, Chandigarh.

Regmi, B., 2019. Phylogenomics and geometric morphometrics define species flocks of snow trout (Teleostei: Schizothorax) in the Central Himalayas. Dissertation, University of Arkansas, Fayetteville.

Roberts, C.M., O'Leary, B.C. and Hawkins, J.P., 2020. Climate change mitigation and nature conservation both require higher protected area targets. *Philosophical Transactions of the Royal Society B,* 375(1794), p. 20190121.

Roche, E.A., Gratto-Trevor, C.L., Goossen, J.P. and White, C.L., 2012. Flooding affects dispersal decisions in piping plovers (*Charadrius melodus*) in prairie Canada. *The Auk*, 129(2), pp. 296–306.

Rogers, L.A. and Dougherty, A.B., 2019. Effects of climate and demography on reproductive phenology of a harvested marine fish population. *Global Change Biology*, 25(2), pp. 708–720.
Rowe, K.C., Rowe, K.M., Tingley, M.W., Koo, M.S., Patton, J.L., Conroy, C.J. et al., 2015. Spatially heterogeneous impact of climate change on small mammals of montane California. *Proceedings of the Royal Society B: Biological Sciences,* 282(1799), p. 20141857.
Ruaro, R., Conceição, E.O., Silva, J.C., Cafofo, E.G., Angulo-Valencia, M.A., Mantovano, T. et al., 2019. Climate change will decrease the range of a keystone fish species in La Plata River Basin, South America. *Hydrobiologia*, 836(1), pp. 1–19.
Rubenson, E.S. and Olden, J.D., 2017. Dynamism in the upstream invasion edge of a freshwater fish exposes range boundary constraints. *Oecologia*, 184(2), pp. 453–467.
Ruesch, A.S., Torgersen, C.E., Lawler, J.J., Olden, J.D., Peterson, E.E., Volk, C.J. et al., 2012. Projected climate-induced habitat loss for salmonids in the John Day River network, Oregon, USA. *Conservation Biology,* 26(5), pp. 873–882.
Ruiz-Navarro, A., Gillingham, P.K. and Britton, J.R., 2016. Predicting shifts in the climate space of freshwater fishes in Great Britain due to climate change. *Biological Conservation,* 203, pp. 33–42.
Russell, I.C., Aprahamian, M.W., Barry, J., Davidson, I.C., Fiske, P., Ibbotson, A.T. et al., 2012. The influence of the freshwater environment and the biological characteristics of Atlantic salmon smolts on their subsequent marine survival. *ICES Journal of Marine Science,* 69(9), pp. 1563–1573.
Sabin, T.P., Krishnan, R., Vellore, R., Priya, P., Borgaonkar, H.P., Singh, B.B. et al., 2020. Climate change over the Himalayas. In *Assessment of Climate Change over the Indian Region* (Krishnan, R., Sanjay, J., Gnanaseelan, C., Mujumdar, M., Kulkarni, A., Chakraborty, S., Eds.) Springer, Singapore, pp. 207–222.
Sala, O.E., Chapin, F.S., Armesto, J.J., Berlow, E., Bloomfield, J., Dirzo, R. et al., 2000. Global biodiversity scenarios for the year 2100. *Science*, 287(5459), pp. 1770–1774.
Sarkar, U.K., Pathak, A.K., Sinha, R.K., Sivakumar, K., Pandian, A.K., Pandey, A. et al., 2012. Freshwater fish biodiversity in the River Ganga (India): changing pattern, threats and conservation perspectives. *Reviews in Fish Biology and Fisheries,* 22(1), pp. 251–272.
Sehgal, K.L., 1999. Coldwater fish and fisheries in the Indian Himalayas: rivers and streams. Fish and fisheries at higher altitudes: Asia. Food and Agriculture Organization of the United Nations, Technical Paper, 385, pp. 41–63.
Settele, J., Scholes, R., Betts, R.A., Bunn, S., Leadley, P., Nepstad, D. et al., 2015. Terrestrial and inland water systems. In *Climate Change 2014 Impacts, Adaptation and Vulnerability: Part A: Global and Sectoral Aspects*. (Field, C.B., Barros, V.R., Dokken, D.J., Mach, K.J., Mastrandrea, M.D., Bilir, T.E. et al., Eds.) Cambridge University Press, Cambridge, UK and New York, pp. 271–360.
Shah, R.D.T., Sharma, S., Haase, P., Jähnig, S.C. and Pauls, S.U., 2015. The climate sensitive zone along an altitudinal gradient in central Himalayan rivers: a useful concept to monitor climate change impacts in mountain regions. *Climatic Change*, 132(2), pp. 265–278.
Shah, R.D.T., Sharma, S. and Bharati, L., 2020a. Water diversion induced changes in aquatic biodiversity in monsoon-dominated rivers of Western Himalayas in Nepal: implications for environmental flows. *Ecological Indicators,* 108, p. 105735.
Shah, R.D.T., Sharma, S., Narayan Shah, D. and Rijal, D., 2020b. Structure of benthic macroinvertebrate communities in the rivers of Western Himalaya, Nepal. *Geosciences,* 10(4), p. 150.
Sharma, S. and Jackson, D.A., 2008. Predicting smallmouth bass *Micropterus dolomieu* occurrence across North America under climatic change: a comparison of statistical approaches. *Canadian Journal of Fisheries and Aquatic Sciences,* 65, pp. 471–481.
Sharma, A., Dubey, V.K., Johnson, J.A. and Sivakumar, K., 2016. Fishes of the Indian Himalayan Region. In *Bibliography on the Fauna and Micro flora of the Indian Himalayan Region* (Sathyakumar, S., Ed.). ENVIS Bulletin: Wildlife and Protected Areas. Vol. 17. Wildlife Institute of India, Dehradun, India. pp. 155–207.
Sharma, A., Dubey, V.K., Johnson, J.A., Rawal, Y.K. and Sivakumar, K., 2021a. Is there always space at the top? Ensemble modeling reveals climate-driven high-altitude squeeze for the vulnerable snow trout *Schizothorax richardsonii* in Himalaya. *Ecological Indicators,* 120, p. 106900.
Sharma, A., Dubey, V.K., Johnson, J.A., Rawal, Y.K. and Sivakumar, K., 2021b. Introduced, invaded and forgotten: allopatric and sympatric native snow trout life-histories indicate brown trout invasion effects in the Himalayan hinterlands. *Biological Invasions*, 23(5), pp. 1497–1515.

Shrestha, U.B. and Bawa, K.S., 2014. Impact of climate change on potential distribution of Chinese caterpillar fungus (*Ophiocordyceps sinensis*) in Nepal Himalaya. *PLoS ONE*, 9(9), p. e106405.

Shuter, B.J., Finstad, A.G., Helland, I.P., Zweimüller, I. and Hölker, F., 2012. The role of winter phenology in shaping the ecology of freshwater fish and their sensitivities to climate change. *Aquatic Sciences*, 74(4), pp. 637–657.

Singh, G. and Agarwal, N.K., 2014. Fish assemblage structure and habitat use of the snow fed stream Assiganga-a major tributary of river Bhagirathi in Central Himalaya (India). *International Journal of Aquatic Biology*, 2(6), pp. 305–312.

Singh, P.B., Mainali, K., Jiang, Z., Thapa, A., Subedi, N., Awan, M.N. et al., 2020. Projected distribution and climate refugia of endangered Kashmir musk deer *Moschus cupreus* in greater Himalaya, South Asia. *Scientific Reports,* 10(1), pp. 1–13.

Singh, R.K., Mishra, N. and Sharma, A.K., 2017. Effect of climate change on Himalayan regions. *International Engineering and Technological Applied Research Journal*, 1(1), pp. 7–9.

Sivakumar, K., 2002. Fishes of trans-Himalayas. *Newsletter, WII,* 8(2), pp. 12–13.

Sivakumar, K., 2008. Species richness, distribution pattern and habitat use of fishes in the Trans-Himalayas, India. *Electronic Journal of Ichthyology,* 1, pp. 31–42.

Solomon, S., Manning, M., Marquis, M. and Qin, D., 2007. *Climate Change 2007 – The Physical Science Basis: Working Group I Contribution to the Fourth Assessment Report of the IPCC*, Vol. 4, Cambridge University Press, Cambridge, UK and New York.

Srivastava, V., Griess, V.C. and Padalia, H., 2018. Mapping invasion potential using ensemble modelling. A case study on *Yushania maling* in the Darjeeling Himalayas. *Ecological Modelling,* 385, pp. 35–44.

Strasevicius, D., Jonsson, M., Nyholm, N.E.I. and Malmqvist, B., 2013. Reduced breeding success of Pied Flycatchers *Ficedula hypoleuca* along regulated rivers. *Ibis*, 155(2), pp. 348–356.

Su, J., Aryal, A., Hegab, I.M., Shrestha, U.B., Coogan, S.C., Sathyakumar, S., Dalannast, M., Dou, Z., Suo, Y., Dabu, X. and Fu, H., 2018. Decreasing brown bear (*Ursus arctos*) habitat due to climate change in Central Asia and the Asian Highlands. *Ecology and Evolution*, 8(23), pp. 11887–11899.

Subba, B., Sen, S., Ravikanth, G. and Nobis, M.P., 2018. Direct modelling of limited migration improves projected distributions of Himalayan amphibians under climate change. *Biological Conservation,* 227, pp. 352–360.

Thatje, S., 2021. Climate warming affects the depth distribution of marine ectotherms. *Marine Ecology Progress Series,* 660, pp. 233–240.

Thuiller, W., Georges, D., Engler, R. and Breiner, F., 2016. Package "biomod2". Ensemble platform for species distribution modeling. Version 3.

Thuiller, W., Guéguen, M., Renaud, J., Karger, D.N. and Zimmermann, N.E., 2019. Uncertainty in ensembles of global biodiversity scenarios. *Nature Communications*, 10(1), pp. 1–9.

Trevelline, B.K., Nuttle, T., Porter, B.A., Brouwer, N.L., Hoenig, B.D., Steffensmeier, Z.D. et al., 2018. Stream acidification and reduced aquatic prey availability are associated with dietary shifts in an obligate riparian neotropical migratory songbird. *PeerJ,* 6, p. e5141.

Trew, B.T. and Maclean, I.M., 2021. Vulnerability of global biodiversity hotspots to climate change. *Global Ecology and Biogeography*, 30(4), pp. 768–783. https://doi.org/10.1111/geb.13272.

Troia, M.J. and Giam, X., 2019. Extreme heat events and the vulnerability of endemic montane fishes to climate change. *Ecography*, 42(11), pp. 1913–1925.

UN DESA (United Nations, Department of Economic and Social Affairs), 2020. World Economic Situation and Prospects: August 2020 briefing, No. 140. www.un.org/development/desa/dpad/publication/world-economic-situation-and-prospects-august-2020-briefing-no-140/.

Uniyal, D.P., 2010. Pisces. In *Fauna of Uttarakhand*, State Fauna Series (Director, Zoological Survey of India, Ed.), 18, pp. 533–621.

Urban, M.C., 2019. Projecting biological impacts from climate change like a climate scientist. *Wiley Interdisciplinary Reviews: Climate Change*, 10(4), p. e585.

Verberk, W.C. and Bilton, D.T., 2013. Respiratory control in aquatic insects dictates their vulnerability to global warming. *Biology Letters*, 9(5), p. 20130473.

Verberk, W.C., Buchwalter, D.B. and Kefford, B.J., 2020. Energetics as a lens to understanding aquatic insect's responses to changing temperature, dissolved oxygen and salinity regimes. *Current Opinion in Insect Science*, 41, pp. 46–53.

Vishwanath, W., 2010. *Schizothorax richardsonii* (errata version published in 2018). The IUCN Red List of Threatened Species 2010. https://dx.doi.org/10.2305/ IUCN.UK.2010-4.RLTS.T166525A6228314.en. Assessed 16 March 2020.

Vishwanath, W., Lakra, W.S. and Sarkar, U.K., 2007. *Fishes of North East India*. National Bureau of Fish Genetic Resources, Indian Council of Agricultural Research, Lucknow, UP, India, p. 264.

Wadia, D.N., 1931. The syntaxis of the northwest Himalaya: its rocks, tectonics and orogeny. *Records of the Geological Survey of India*, 65(2), pp. 189–220.

Wang, Q., Fan, X. and Wang, M., 2016. Evidence of high-elevation amplification versus Arctic amplification. *Scientific Reports*, 6(1), pp. 1–8.

Warren, R., VanDerWal, J., Price, J., Welbergen, J.A., Atkinson, I., Ramirez-Villegas, J. et al., 2013. Quantifying the benefit of early climate change mitigation in avoiding biodiversity loss. *Nature Climate Change*, 3(7), pp. 678–682.

Weiss, L.C., Pötter, L., Steiger, A., Kruppert, S., Frost, U. and Tollrian, R., 2018. Rising pCO_2 in freshwater ecosystems has the potential to negatively affect predator-induced defenses in Daphnia. *Current Biology*, 28(2), pp. 327–332.

WWF, 2020. *Living Planet Report 2020 – Bending the curve of biodiversity loss*. (Almond, R.E.A., Grooten, M., Petersen, T., Eds.). WWF, Gland, Switzerland, p. 159.

Yalcin, S. and Leroux, S.J., 2018. An empirical test of the relative and combined effects of land-cover and climate change on local colonization and extinction. *Global Change Biology*, 24(8), pp. 3849–3861.

Yang, J., Yang, J.X. and Chen, X.Y., 2012. A re-examination of the molecular phylogeny and biogeography of the genus *Schizothorax* (Teleostei: Cyprinidae) through enhanced sampling, with emphasis on the species in the Yunnan–Guizhou Plateau, China. *Journal of Zoological Systematics and Evolutionary Research*, 50(3), pp. 184–191.

Yang, X.Q., Kushwaha, S.P.S., Saran, S., Xu, J. and Roy, P.S., 2013. Maxent modeling for predicting the potential distribution of medicinal plant, *Justicia adhatoda* L. in Lesser Himalayan foothills. *Ecological Engineering*, 51, pp. 83–87.

22 A Solution to Healthy Living in Relation to Vector-Borne Encephalitis

Goutam Chandra and Sunanda Burman
Mosquito, Microbiology and Nanotechnology Research Units, Parasitology Laboratory, Department of Zoology, The University of Burdwan, Purba Bardhaman, India

CONTENTS

22.1 INTRODUCTION

Different forms of encephalitis are dreadful diseases, involving human beings and some animals of economic importance. Japanese encephalitis virus (JEV), an arthropod-borne virus, has a wide distribution including Asia, the western Pacific and northern Australia. The World Health Organization (WHO) has declared Japanese Encephalitis (JE) to be the most important cause of mosquito-borne paediatric viral encephalitis in endemic Asian countries, mostly affecting the paediatric age group of 0–14 years, which accounts for about 75% of JE cases. Globally, it accounts for about 68,000 clinical cases per year [1].

In India in the year 2005, an outbreak of JE resulted in a majority of paediatric deaths among 1700 total casualties [2, 3]. The virus belongs to the family *Flaviviridae*, which contain a positive-sense RNA genome within a host-derived membrane. They are distributed throughout the world,

DOI: 10.1201/9781003220824-25

crossing national boundaries, and therefore need global surveillance. JE is considered to be an emerging zoonotic disease as well as a major public health problem in India [4]. Symptomatic cases are rare, and only 1 in 250 subclinical cases were reported. In the case of symptomatic cases, about 30% fatality can be encountered in infected patients [5]. A spectrum of clinical features is found, including flu-like symptoms, stiffness in the neck, seizures, spastic paralysis, disorientation, coma and even death of the infected patients. JEV is considered to be one of the major public health problems due to its neuro-psychiatric effects, which lead to morbidity and mortality in a large number of infected patients, high epidemic potential, high fatality rate, and sequelae among the survivors, and this also imparts a socio-economic burden due to the increase in disability in affected patients [1, 6]. However, JE may affect adults, particularly people who come to JE endemic areas and lack childhood immunization against JEV. JE mainly occurs in rural areas, mainly in association with rice cultivation and irrigated rice fields and ditches associated with rice fields, where vector mosquitoes breed profusely in close proximity with domestic pigs, wading birds and ducks. Like most of the flaviviruses, its transmission occurs through mosquitoes, especially by the genus *Culex*, in a natural cycle between birds and some species of livestock [5]. Reservoir hosts include pigs, birds, rodents, squirrels and also bats, while humans and horses (equines) are the dead-end hosts. Ardeid birds play a considerable role in the long-distance propagation of JEV and also act as a reservoir of this viral disease [7]. Domestic pigs, having long-lasting viraemia, are the key amplifiers of the virus and are mainly responsible for the disease in humans [8]. Insufficient viraemia in horses and other vertebrates other than birds makes them incidental dead-end hosts [9].

Pigs are the biological amplifiers of JE and act as reservoirs of the virus [10]. Interestingly, in countries like Pakistan, where piggeries can rarely be found, JE outbreaks are reported only very rarely [11]. Disease manifestation in the livestock varies from asymptomatic infection in some species to acute neurological disorder in others. Pigs are among the animals that are mainly affected by this viral infection, which leads to abortion, stillbirth and birth defects, while the adult swine remains asymptomatic. Thus, this virus is responsible for reproductive problems in pigs, which in turn causes great economic loss to farmers. Infected piglets face neurological disease, ultimately leading to significant mortality in pigs as well [12–14].

Besides pigs, the domestic animals that can be infected by JEV are equines. Infected horses show non-specific signs, including pyrexia, but this can sporadically lead to major neurological disorders. Syndromic manifestations in horses include a hyper-excitable type syndrome, which results in high fevers (41°C or higher) along with profuse sweating and tremors in muscle. Behavioural changes include aimless wandering, aggression, vision problems, collapse, coma and finally death. Neurological sequelae may result in 5% mortality in horses, which may increase to 30% [15]. In extreme cases, where disease manifestation is severe, the death of infected horses may occur. Though veterinary vaccination is available for both pigs and horses, JEV disease in livestock exists as a great problem, resulting in great economic loss for caretakers, as detection and treatment of the disease in these domestic animals are not always available, or the costs of the treatment are not affordable by the farmers. Control and prevention of JEV in livestock, i.e., within the veterinary sector, is important not only to prevent economic loss to the farmers but also to prevent transmission of the viral infection from livestock to man. Such measures are essential in breaking the zoonotic transmission cycle, i.e., association of JEV with domestic animals, consequently entering the human population, where humans act as a dead-end host [16].

JEV outbreaks significantly affect agricultural economy and public health, particularly in areas of low vaccine coverage and where there is a lack of proper diagnostic facilities.

As viral outbreaks cannot completely be avoided, preventive strategies should be taken to control JE incidence. Management strategies for JE eradication do not work fully, as the pathogen cannot be eradicated from a population due to its zoonotic potential. Thus, re-evaluation of encephalitis infection control in man, pigs and horses is greatly needed. The present discussion focuses on topics including types of encephalitis, reservoirs, vector–host relationship, geographical distribution, zoophilic and zoo-anthropophagic cycles in relation to transmission of JEV, and most importantly, the solution to control vector-borne encephalitis.

22.2 DIFFERENT VECTOR-BORNE ENCEPHALITIDES

Symptomatic human arboviral diseases are divided into three different syndromes: i. febrile systemic illness, e.g., Dengue fever; ii. haemorrhagic fever, e.g., Dengue haemorrhagic fever and yellow fever; and iii. Encephalitis, which includes Japanese Encephalitis, Eastern Equine Encephalitis, Western Equine Encephalitis, Venezuelan Equine Encephalitis, and La Crosse Encephalitis [17], which are vector-borne.

22.2.1 Japanese Encephalitis (JE)

JE is endemic to the areas of South Asia, East Asia, Southeast Asia and the Pacific region. JE is caused by JEV, a zoonotic flavivirus. Infection of this virus occurs in frogs, snakes, egrets, bats and most domestic animals like cattle, poultry and ducks. Pigs and Ardeid birds are considered as the reservoir and amplifier hosts and are unaffected by the infection of this virus [18, 15, 19, 20]. Horses and donkeys also suffer from Japanese encephalitis as well as humans. JE does not spread from man to man because of the low and insufficient viraemia in man. Man is an accidental, dead-end host.

22.2.2 Eastern Equine Encephalitis (EEE)

Eastern equine encephalitis virus (EEEV) is an alphavirus of the family Togaviridae. In the United States, about six to eight affected persons are reported annually, occurring mainly in the season of May through October, mostly in the provinces of Florida, Georgia, Wisconsin, New Jersey and Maryland [21–23]. Though passerine birds are considered to be the principal reservoir hosts of EEEV in North America, rodents can also amplify the virus in a natural cycle. The primary reservoir hosts for Madariaga virus (South American EEEV) are still uncertain, but small mammals may play a prominent role. Some experiments have suggested that reptiles (especially snakes) may help maintain EEEV over the winter. EEEV mainly affects horses and other equid species. However, disease manifestations have been seen in sheep, cattle, dogs, llamas, alpacas, pigs, deer and nonhuman primates. As per reports, other susceptible animals include goats, moose, certain rodents, bats, reptiles and even amphibians.

22.2.3 Western Equine Encephalitis (WEE)

Western equine encephalitis virus also belongs to the genus *Alphavirus* of the family Togaviridae. It was first discovered from a horse in 1930, and it is mainly a disease of equines [24]. WEE is reported from the states in the Mississippi River of the United States and also in some western Canadian states [25]. The virus, like other encephalitis viruses, is maintained by a bird–mosquito–bird life cycle. Birds and humans are incidental hosts. Passerine birds are the main amplifying reservoir host of the virus, and often the natural viral cycle involves blacktail jackrabbit, *Lepus californicus* [26]. Infections in mules, horses and pheasants more often precede epidemics in humans. Infections lead to fever, chills and malaise, and high viral load often results in translocation of virus into the central nervous system across the blood brain barrier, causing cerebral as well as meningeal inflammation and necrosis [27]. Asymptomatic infections in cattle, squirrels, other rodents, snowshoe hare (*Lepus americanus*), opossums, snakes, tortoises and frogs may occur [28].

22.2.4 Tick-Borne Encephalitis (TBE)

Tick-borne encephalitis, caused by tick-borne encephalitis virus (TBEV), is also an arbovirus belonging to the family Flaviviridae. It mainly occurs in middle and eastern European countries and some parts of Asia and is considered to be a major health menace for humans [29]. Like any other arbovirus, TBEV is associated with complex interactions between vectors, reservoir hosts, and favourable environment for successful viral transmission. Hard ticks are considered to be the vectors as well as the reservoir host for TBEV, and are responsible for transmission of this virus to a variety of animals. Small rodents are the main hosts, and humans are accidental hosts of TBEV. TBEV causes meningitis, meningoencephalitis and also meningoencephalomyelitis in infected patients. The severity of the disease increases with age, and it mainly affects children and adolescents. Long-term sequelae are present for months to years in patients affected with TBEV [29].

22.2.5 La Crosse (LAC) Encephalitis

La Crosse (LAC) encephalitis is also a mosquito-borne arbovirus, also termed bunyavirus [30], that causes inflammation of the brain, i.e., encephalitis. It was first found in La Crosse, Wisconsin in the year 1963 [31].

This disease is endemic in several midwestern and mid-Atlantic states. It mainly occurs in children aged less than 16 years. Symptoms other than encephalitis include fever, headache, vomiting and lethargy. Encephalitis is accompanied by seizures, and in severe cases, coma and paralysis may occur. As no specific treatment for LAC encephalitis is available, supportive drug therapy is used in severe cases [32].

22.2.6 Venezuelan Equine Encephalitis (VEE)

The causative viral pathogen of Venezuelan equine encephalitis (VEE) is Venezuelan equine encephalitis virus (VEEV). Outbreaks commonly involve horses, donkeys, zebras, mules and man. VEEV is a single-stranded positive-sense RNA alphavirus belonging to the *Togaviridae* family [33]. VEEV was first documented in Venezuela, Columbia and Trinidad in 1935 after equine outbreaks [34].

In 1950 it was discovered in humans during an outbreak in Colombia [35]. There are six subtypes of VEEV (I–VI), among which the epizootic strains IAB and IC are responsible for transmission of encephalitis in equines and humans. Equines are the key reservoir hosts and also amplification hosts for epizootic viral strains. Reservoir hosts for enzootic VEE viruses are wild rodents, while birds may be involved in a few cycles. Enzootic VEE viruses often infect opossums, *Didelphis marsupialis*, bats and dogs. The Mexican IE variant causes VEE in equids [36].

Symptoms of VEEV infection vary from flu-like fever to severe encephalitis in infected persons. Patients may develop symptoms such as leukopenia, tachycardia and interstitial pneumonia. Children are likely to develop encephalitis more easily than adults and have a higher rate of morbidity and mortality [36]. Symptoms of more severe cases may include tremors, seizures, changes in behaviour, hemiparesis, hemichorea, cranial nerve palsies, ataxia, myoclonus, confusion, somnolence and eventually coma [33, 37, 38].

22.3 ROLE OF VECTORS IN TRANSMISSION CYCLE OF ENCEPHALITIS

JEV has been isolated from over 30 mosquito species of the genera *Aedes*, *Anopheles*, *Armigeres*, *Culex* and *Mansonia* [39]. These vectors can act as primary or secondary vectors in different geographical areas across the world. However, the presence of JEV does not indicate their capability to be vectors of JEV. These vectors can act as primary or secondary vectors in different geographical areas across the world. Among different vectors of JEV, *Cx. annulirostris* is the primary vector species in Australia; however, *Cx. vishnui*, *Cx. pseudovishnui*, *Cx. gelidus*, *Cx. citiens* and *Cx.*

fuscocephela are among secondary vectors in this region. The main vector of JE in South Asia, Southeast Asia and East Asia is *Cx. tritaeniorhynchus* [39]. In India, *Cx. vishnui*, *Cx. pseudovishnui* and *Cx. tritaeniorhynchus* are considered as primary vectors, while *Ma. indiana*, *Cx. whitmorei*, *Cx. epidesmus* and *Anopheles subpictus* are among the secondary vectors of JE [40]. In China, the main vector is *Cx. tritaeniorhynchus* [41]. Different species of mosquitoes, e.g., *Cx. tritaeniorhynchus*, *Cx. tarsalis*, *Cx. annulirostris*, *Cx. gelidus*, *Cx. fuscocephala*, *Cx. vishnui*, *Cx. sitiens*, *Cx. pseudovishnui*, *Cx. whitmorei*, *Culiseta inornata*, *Cx. pipiens*, *Cx.salinarius*, *Cx. quinquefasciatus*, *Cx. nigripalpus*, *Ae. vexans*, *Ae. albopictus*, *Ae. japonicas* and *Ma. uniformis* are associated with JEV infection in the United States [42].

In the case of EEE, the host–virus–reservoir transmission cycle involves birds, and its predominant vector is *Cs. melanura* mosquitoes, found mainly in freshwater hardwood swamps in America [43]. Other vectors that can transmit the disease are *Coquillettidia perturbans*, *Ae. cinereus* and *Ae. canadensis* [44, 45]. Infected mosquitoes transmit the virus to tree-perching passerine birds and to opportunistic mammals, reptiles or amphibians during blood meal consumption from these species. Dead-end hosts of this virus comprise humans, swine, equids and pheasants. Cross transmission may occur due to changes in environment and global warming.

WEE is one of the mosquito-borne viral infections. The vectors for humans are *Cx. tarsalis*, *Culiseta* and some species of *Aedes* mosquito. Human epidemics lead to outbreaks in mules, horses, pheasants and other birds. Birds serve as amplifying hosts, which makes them reservoirs of the disease [46].

Arthropod vectors of TBEV transmission are primarily the hard ticks. In Europe, the major tick vector is *Ixodes ricinus*; in Russia and Asia, *I. persulcatus* is the principal vector species. In Asia, *Haemaphysalis concinna* also serves as a vector of this virus. Around 22 other tick species have been reported to carry the virus [47].

The principal vector of La Crosse encephalitis is the eastern tree-hole mosquito, *Ae. triseriatus*. Another tree-hole mosquito species, i.e., *Ae. albopictus*, is also an efficient vector of this virus [48]. This zoonotic pathogen is cycled between the daytime-biting tree-hole mosquito, *Ae. triseriatus*, and vertebrate amplifier hosts like chipmunks and tree squirrels in deciduous forest habitats of the Appalachian and midwestern regions and the south east of the United States [49].

VEE virus is usually transmitted by mosquito vectors, although certain types of ticks and mites can transmit the virus as well. The mosquito vector *Culex* (Melanoconion) species is normally responsible for transmission of the enzootic strain of the virus. Other notable vectors are *Ma. titillans, Ochlerotatus taeniorhyn, Psorophora columbiae, Ps. confinnis, Ae. sollicitans,* and *An. aquasalis*, which mainly transmit the epizootic strains of the VEE virus [50].

22.4 EPIDEMIOLOGY AND OUTBREAKS

The epidemiological pattern of JE generally is of two types. Epidemics in northern temperate areas mainly cover countries like Nepal, China, Korea, Japan, Taiwan, northern Vietnam, northern Thailand and northern India, occurring during the summer months, roughly from May to October. In southern tropical areas such as Sri Lanka, southern India, Indonesia, Malaysia, southern Vietnam, southern Thailand and the Philippines, JE tends to be endemic; cases occur sporadically throughout the year, with a peak just after the start of the rainy season during the months of July to September [51–53].

In India, two epidemics generally occur in Karnataka. Severe cases are generally encountered during April to July, and a mild form occurs in September to December [5]. According to WHO, outbreaks of JE occur every 2–15 years. Rainy season is the time when the peak of transmission occurs, as the vector population multiples fast. Agricultural development and rice cultivation with irrigation systems support the spread in new areas.

About 3 billion people live in countries, including China and India, where the JE virus is endemic [6]. However, statistics of JE occurrence should be interpreted with care, because the transmission of JE is exceedingly dynamic in terms of its occurrence in endemic areas and occurs typically in epidemics; thus, the rise and fall in JE cases can be seen globally. Background determinants triggering epidemics mainly include environmental factors and the spillover effects into the human population.

As per the Centers for Disease Control and Prevention (CDC) records, 8 cases of EEEV on an average per year were recorded for the last 15 years in the United States. In 2019, during the months of June to September, 34 cases were recorded. The death rate of patients with EEEV infection was 35% [54].

WEE was reported from the western United States and Canada. Epidemics are related to warm weather and heavy rainfall in endemic areas. The most noteworthy outbreak occurred in 1941, during which 3000 positive WEE human cases were reported. Since 1964, fewer than 700 confirmed cases of WEE have been reported from the United States [46].

VEE was first discovered in South America in horses in the 1930s. It is a highly infectious vector-borne disease, which produces illness in 100% of exposed cases, and is endemic to Central and South America. A VEE outbreak occurred in 1971, when 88 human infections in southern Texas counties, including Cameron, Hidalgo and the Corpus Christi area, were reported [55].

According to CDC reports, annual cases of La Crosse (LAC) encephalitis ranged between 30 and 130 cases per year in the United States [56].

The first confirmed case of TBE was diagnosed in 1993 at Hokkaido Prefecture in Japan. About 10,000–12,000 clinical cases of TBE are reported worldwide each year [57].

22.5 CONTROL STRATEGIES FOR VECTOR-BORNE ENCEPHALITIS

Eradication of vector-borne encephalitis is difficult due to its complex eco-epidemiology, which involves transmission of the virus through natural reservoirs like pigs and birds. Effective preventive measures are the foremost strategy to reduce the JE burden, which will eventually reduce the mortality, morbidity and disability of a large number of people. The strategies include JE vaccination in affected areas, strengthening of surveillance programs and control of vectors.

22.5.1 Vaccination

Elimination of JE is not possible due to its enzootic cycle involving mammals and birds. Hence, vaccination is the most effective strategy and long-term protection to control JE infection. Effective vaccines for JE are accessible, and vaccination in endemic regions significantly decreases the occurrence of JE.

Four types of vaccines are available for JE: inactivated mouse brain-derived vaccines, inactivated Vero cell vaccines, live attenuated vaccines and chimeric vaccines [58]. Different vaccines available against JE so far are listed in Table 22.1.

According to WHO, JE immunization is the most important evidence-based guideline for reducing burden of JE [58].

Though vaccination is available in endemic areas, universal JE vaccination is not planned for various reasons: herd immunity cannot be developed in humans because they are the dead-end host of the virus; piglets acquire antibodies from an immunized mother; immunization coverage is low in rural areas where outbreaks mainly occur; and favourable conditions are favourable for JE transmission in endemic areas.

No human vaccine is available for EEEV; however, a veterinary vaccine is available for use in equids.

TABLE 22.1
List of Vaccines Against Japanese Encephalitis

Type of Vaccine	Strain	Route of Administration	Number of Doses Given	Vaccine Efficacy/ Side Effects	Manufactured/Licensed in Which Country
Inactivated Mouse Brain-Derived Vaccines	Nakayama/Beijing-1 (P1)	Subcutaneous injection	3	Cases of acute disseminated encephalomyelitis associated with the vaccine	Manufactured in Japan (1930s)
Inactivated Vero Cell Vaccines	Beijing-1 (P1)	Intramuscular injection	2		Manufactured in China, recently licensed in South Korea, Nepal, Sri Lanka and India
Live Attenuated Vaccines	SA 14-14-2	Subcutaneous injection	1	Vaccine efficacy is about 100%	Manufactured in China
Chimeric Vaccines	Structural genes (Pre-membrane and E) from SA 14-14-2 strain incorporated into the backbone of attenuated strain of yellow fever (YF) virus YF 17D	Subcutaneous injection	1	Vaccine efficacy is about 100%	Licensed in China, Thailand and India

Source: Kulkarni et al. 2018 [5] and Rustagi et al. 2019 [58]

22.5.2 Vector Control Measures

Vector control measures should be taken against its main mosquito vectors, *Cx. tritaeniorhynchus*, *Cx. vishnui* and *Cx. pseudovishnui*, and other mosquito species that preferably breed in rural agricultural areas in association with rice production and flooding irrigation and can fly up to 5 km, infecting humans with the virus [59]. These vector mosquitoes are zoophilic and considered as primary vectors of JEV [60]. Use of nitrogenous fertilizer and employment of irrigation schemes has a positive effect on the breeding facility of the vector mosquitoes. Thus, monitoring and proper irrigation strategies in rice fields and also cultivation of mosquito larvae-eating fishes in the logged water in ditches and water passages associated with paddy fields may reduce the vector burden by reducing larvae. Use of bio-fertilizers in the paddy field in place of nitrogenous fertilizer may reduce breeding of mosquitoes by creating unfavourable conditions for breeding in rice fields. Control of adult mosquitoes in human habitations and piggeries with ultra-low-volume fogging with malathion is effective against JE vectors [61]. Using novel eco-friendly mosquitocides in breeding areas and controlling adult mosquitoes are also an effective strategy to control JE vector mosquitoes [62–69]. Safety from bite of respective vectors has been reported when mixture of Neem oil and Mustard oil was applied on human body surface [70]. Thus, controlling the vectors can be an effective solution in high-risk JE endemic zones.

In the case of TBE, vaccination is recommended in TBE endemic areas or for travellers to these areas. A booster dose should be given after 3–5 years. Prevention of tick bite is the best protection against TBE infection.

22.5.3 Surveillance

Surveillance of JE in endemic regions is suggested throughout the year to understand the early warning signals of JE outbreaks in a particular area, viral infection in human and animal reservoirs, seasonal variation in JE cases, age groups affected, socio-economic loss measurement and also the rate of transmission in the presence of favourable conditions. Besides, surveillance data is useful in assessment of the effectiveness of vaccination programmes by governments and JE control programmes. Surveillance strategies in the endemic regions may face many challenges: sentinel surveillance sites may often lack laboratory facilities for diagnosis of JE, such that the actual incidence of JE cannot be monitored correctly by the surveillance report.

22.5.4 Behaviour Change Communication

As with malaria, a prevention strategy by effective use of behaviour change communication (BCC) can also be used in vector-borne JE [58]. This strategy includes avoidance of mosquito bites, shifting of pigsties, maintaining good hygiene and keeping the house sanitized. Personal prophylactic measures like application of repellents and using chemical-impregnated mosquito nets, decreasing human contact with reservoir hosts during dusk and dawn, and wearing long-sleeved clothes may be maintained to avoid JE transmission by mosquitoes. Proper immunization with vaccine in childhood is also essential.

22.5.5 Long-Term Therapeutic Measures

JE survivors with disability should have drug therapy, physiotherapy and other rehabilitation measures to get rid of sequelae like seizures, paralysis of body parts, psychiatric problems and difficulties in movement, disorientation, incontinence and even Parkinsonism [59].

22.5.6 Long-Term Preventive Measures

Measures to control mosquitoes, shifting pigsties, environmental sanitation (prevents many other diseases also), avoiding mosquito bites (prevents many other diseases also), and vaccination of the susceptible population (the best of all measures) are useful [61].

22.6 SCENARIO OF VECTOR-BORNE ENCEPHALITIS IN INDIA

The first occurrence of vector-borne JE virus dates back to 1952, while the first case was reported in the year 1955. JEV is widespread in India, and outbreaks have been reported in different states of India [71].

Though JE was already a threat to a large population in India since 2005, outbreaks due to developments in agriculture, mostly irrigated paddy fields, being the breeding place for vector mosquitoes that carry JEV, made it a risky endemic zone of JE [72]. Because of India's increasing JE cases, the National Vector-borne Disease Control Program (NVBDCP) started to monitor the JE scenario around the nation.

The government of India introduced vector control programmes to reduce mosquito load by the application of insecticide, with little impact on the huge burden of the disease.

Children in Andhra Pradesh and other areas were immunized by giving an India-made mouse brain-derived vaccine, but the supply of the vaccine fell short due to India's large number of children across the whole country [73]. On the other hand, this vaccine has a complicated dosing schedule and is also costly. However, another vaccine, CD-JEV, made in China, solved to some extent India's huge requirement for vaccines [74]. It was effective and cost effective. India introduced catch-up campaigns in JE endemic areas; campaigns were run in high-burden and high-priority districts first, followed by routine JE immunization of children. WHO highly recommended India's strategy to control JE in other countries [75].

22.7 SOLUTION TO VECTOR-BORNE ENCEPHALITIS OUTBREAKS

Due to insufficient antiviral treatment against encephalitis, prevention is the first priority to control this brain disease. It is more difficult to control this disease due to high vector abundance, and also, it involves high expense. However, systematic monitoring by surveillance of vector species in endemic areas as well as detection of sero-prevalence of viruses in pigs in particular areas should be done routinely for encephalitis outbreaks in those areas. Rehabilitation for surviving patients with disabilities needs to be included in control programmes. For reducing the paediatric mortality rate in JE endemic zones, vaccination is a better option. Thus, the vaccine deficit in affected areas must be restored. Another method to control the disease is vaccination of horses and pigs. As human infections are linked to the mosquito–swine transmission cycle, pig vaccination or pig slaughter also is an option to eradicate the disease. Vector control always remains a reliable method for the control of any vector-borne disease.

22.8 CONCLUSION

Factors responsible for encephalitis outbreaks are climatic conditions that influence the typical focal distribution of the virus in endemic regions and occurrence of specific vector species in these regions. The other important factor is the virus reservoir host, which amplifies the virus and subsequently infects the mosquitoes, which transmit the viruses to human, cattle and equine hosts.

To fight all types of encephalitis, the most important remedy is to educate people about mosquitoes and the relationship between mosquito bites and incidence of encephalitis. As the disease is maintained by the reservoir host or amplifier host, taking care of their habitats is also an important step regarding eradication of the disease. If we take the necessary steps to stop mosquitoes biting pigs, horses and other domestic hosts, twofold safety measures against encephalitis can be established. As zoophilic female vector mosquitoes of encephalitis imbibe the blood of their hosts for development of their eggs, i.e., to propagate in large numbers, a reduction of mosquito bites to reservoir and amplifier hosts reduces their chances of producing viable eggs. By this method, not only can the mosquito population be controlled to an extent, but also the transmission of the disease to the dead-end host can be diminished. If the reservoir host is not infected by the encephalitis virus, transmission of the virus to humans and equines can be ended, thus providing the second shield of protection against encephalitis for healthy living.

REFERENCES

1. Campbell GL, Hills SL, Fischer M, Jacobson JA, Hoke CH, Hombach JM, Marfin AA, Solomon T, Tsai TF, Tsu VD and Ginsburg AS. Estimated global incidence of Japanese encephalitis: a systematic review. *Bulletin of the World Health Organization*. 2011;89:766–74.
2. Parida M, Dash PK and Tripathi NK. Japanese encephalitis outbreak, India, 2005. *Emerging Infectious Diseases*. 2006 Sep;12(9):1427.
3. Halstead SB and Thomas SJ. New Japanese encephalitis vaccines: alternatives to production in mouse brain. *Expert Review of Vaccines*. 2011 Mar 1;10(3):355–64.
4. Lindenbach BD and Rice CM. Flaviviridae: the viruses and their replication, in *Field's Virology*, Knipe DM and Howley PM, Eds., vol. 1, pp. 991–1041, Lippincott Williams & Wilkins, Philadelphia, PA, USA, 4th edition, 2001.
5. Kulkarni R, Sapkal GN, Kaushal H, and Mourya DT. Suppl-2, M8: Japanese encephalitis: a brief review on Indian perspectives. *The Open Virology Journal*. 2018;12:121.
6. Solomon T. Control of Japanese encephalitis—within our grasp? *New England Journal of Medicine*. 2006 Aug 31;355(9):869–71.

7. Endy TP, and Nisalak AN. Japanese encephalitis virus: ecology and epidemiology. 2002; 267:11–48. In: Mackenzie JS, Barrett ADT and Deubel V., Eds., *Japanese Encephalitis and West Nile Viruses*. Current Topics in Microbiology and Immunology. Springer, Berlin, Heidelberg. https://doi.org/10.1007/978-3-642-59403-8_2.
8. Van den Hurk AF, Ritchie SA, Johansen CA, Mackenzie JS, and Smith GA. Domestic pigs and Japanese encephalitis virus infection, Australia. *Emerging Infectious Diseases*. 2008 Nov;14(11):1736.
9. Solomon T and Vaughn DW. Pathogenesis and clinical features of Japanese encephalitis and West Nile virus infections. 2002; 267:171–94. In: Mackenzie JS, Barrett ADT and Deubel V., Eds., *Japanese Encephalitis and West Nile Viruses*. Current Topics in Microbiology and Immunology. Springer, Berlin, Heidelberg. https://doi.org/10.1007/978-3-642-59403-8_9.
10. Rojanosuphot S and Tsai TF. Regional workshop on control strategies for Japanese encephalitis. *Southeast Asia Journal of Tropical Medicine and Public Health*. 1995;26(S3):1–59.
11. Igarashi A, Tanaka M, Morita K, Takasu T, Ahmed A, Ahmed A, Akram DS and Waqar MA. Detection of West Nile and Japanese encephalitis viral genome sequences in cerebrospinal fluid from acute encephalitis cases in Karachi, Pakistan. *Microbiology and Immunology*. 1994;38(10):827–30.
12. Shimizu T, Kawakami Y, Fukuhara S and Matumoto M. Experimental stillbirth in pregnant swine infected with Japanese encephalitis virus. *The Japanese Journal of Experimental Medicine*. 1954 Dec;24(6):363–75.
13. Yamada M, Nakamura K, Yoshii M and Kaku Y. Nonsuppurative encephalitis in piglets after experimental inoculation of Japanese encephalitis flavivirus isolated from pigs. *Veterinary Pathology*. 2004 Jan;41(1):62–7.
14. Straw, BE. 2006. *Diseases of Swine*. 9th Ed, Blackwell Publishing, Ames, Iowa, pp. 359–364.
15. www.oie.int/fileadmin/Home/eng/Animal_Health_in_the_World/docs/pdf/Disease_cards/JAPANESE_ENCEPHALITIS.pdf
16. Mansfield KL, Hernández-Triana LM, Banyard AC, Fooks AR and Johnson N. Japanese encephalitis virus infection, diagnosis and control in domestic animals. *Veterinary Microbiology*. 2017 Mar 1;201:85–92.
17. Morens DM, Folkers GK and Fauci AS. Eastern equine encephalitis virus—another emergent arbovirus in the United States. *New England Journal of Medicine*. 2019 Nov 21;381(21):1989–92.
18. Halstead SB and Jacobson J. Japanese encephalitis. *Advances in Virus Research*. 2003 Jan 1;61:103–38.
19. Kuno G, Mackenzie JS, Junglen S, Hubálek Z, Plyusnin A and Gubler DJ. Vertebrate reservoirs of arboviruses: myth, synonym of amplifier, or reality? *Viruses*. 2017 Jul;9(7):185.
20. Longbottom J, Browne AJ, Pigott DM, Sinka ME, Golding N, Hay SI, Moyes CL and Shearer FM. Mapping the spatial distribution of the Japanese encephalitis vector, *Culex tritaeniorhynchus* Giles, 1901 (Diptera: Culicidae) within areas of Japanese encephalitis risk. *Parasites & Vectors*. 2017 Dec;10(1):1–2.
21. Curren EJ, Lehman J, Kolsin J, et al. West Nile virus and other nationally notifiable arboviral diseases—United States, 2017. *MMWR Morbidity and Mortality Weekly Report*. 2018;67:1137–42.
22. Shah KJ and Cherabuddi K. Case of eastern equine encephalitis presenting in winter. *Case Reports*. 2016 May 10;2016:bcr2016215270.
23. Garlick J, Lee TJ, Shepherd P, Linam WM, Pastula DM, Weinstein S and Schexnayder SM. Locally acquired eastern equine encephalitis virus disease, Arkansas, USA. *Emerging Infectious Diseases*. 2016 Dec;22(12):2216.
24. https://en.wikipedia.org/wiki/Western_equine_encephalitis_virus. Accessed on 20 February 2021.
25. Simon LV, Sandhu DS, Goyal A and Kruse B. Encephalitis, Japanese. StatPearls [Internet]. 2020 Feb 10.
26. Arrigo NC, Adams AP, Watts DM, Newman PC and Weaver SC. Cotton rats and house sparrows as hosts for North and South American strains of eastern equine encephalitis virus. *Emerging Infectious Diseases*. 2010;16(9):1373–80.
27. www.ncbi.nlm.nih.gov/books/NBK470228/. Accessed on 23 February 2021.
28. www.cfsph.iastate.edu/Factsheets/pdfs/easter_wester_venezuelan_equine_encephalomyelitis.pdf
29. Kaiser R. Tick-borne encephalitis. *Infectious Disease Clinics of North America*. 2008 Sep 1;22(3):561–75.
30. Centers for Disease Control and Prevention (CDC). Possible congenital infection with La Crosse encephalitis virus—West Virginia, 2006–2007. *MMWR. Morbidity and Mortality Weekly Report*. January 2009;58(1):4–7.

31. Virginia Department of Health. La Crosse Encephalitis. 2013 Mar 29. www.vdh.virginia.gov/Epidemiology/DEE/Vectorborne/factsheets/lacrosse.htm. Accessed on 23 February 2021.
32. Centers for Disease Control and Prevention (CDC). La Crosse Encephalitis. 2009 Oct 16. www.cdc.gov/lac/gen/qa.html. Accessed on 21 February 2021.
33. Weaver SC, Ferro C, Barrera R, Boshell J and Navarro JC. Venezuelan equine encephalitis. *Annual Reviews in Entomology*. 2004 Jan;49(1):141–74.
34. Lundberg L, Carey B and Kehn-Hall K. Venezuelan equine encephalitis virus capsid—the clever caper. *Viruses*. 2017 Oct;9(10):279.
35. Aguilar PV, Estrada-Franco JG, Navarro-Lopez R, Ferro C, Haddow AD and Weaver SC. Endemic Venezuelan equine encephalitis in the Americas: hidden under the dengue umbrella. *Future Virology*. 2011;6(6):721–40.
36. Kumar B, Manuja A, Gulati BR, Virmani N and Tripathi BN. Suppl-2, M5: Zoonotic viral diseases of equines and their impact on human and animal health. *The Open Virology Journal*. 2018;12:80.
37. Go YY, Balasuriya UB and Lee CK. Zoonotic encephalitides caused by arboviruses: transmission and epidemiology of alphaviruses and flaviviruses. *Clinical and Experimental Vaccine Research*. 2014 Jan;3(1):58.
38. Sharma A and Knollmann-Ritschel B. Current understanding of the molecular basis of Venezuelan equine encephalitis virus pathogenesis and vaccine development. *Viruses*. 2019 Feb;11(2):164.
39. Pearce JC, Learoyd TP, Langendorf BJ and Logan JG. Japanese encephalitis: the vectors, ecology and potential for expansion. *Journal of Travel Medicine*. 2018 May;25(Suppl_1):S16–26.
40. Chakravarty SK, Sarkar JK, Chakravarty MS, Mukerjee KK, Das BC and Hati AK. The first epidemic of Japanese encephalitis in India-virological studies. *Indian Journal of Medical Research* 1975;63:77–88.
41. Tao Z, Liu G, Wang M *et al.* Molecular epidemiology of Japanese encephalitis virus in mosquitoes during an outbreak in China, 2013. *Science Reports*. 2014;4:4908. https://doi.org/10.1038/srep04908
42. Oliveira ARS, Cohnstaedt LW and Cernicchiaro N. Japanese encephalitis virus: placing disease vectors in the epidemiologic triad. *Annals of the Entomological Society of America*. November 2018;111(6):295–303.
43. Lindsey NP, Staples JE and Fischer M. Eastern equine encephalitis virus in the United States, 2003–2016. *The American Journal of Tropical Medicine and Hygiene*. 2018 May 9;98(5):1472–7.
44. Armstrong PM and Andreadis TG. Eastern equine encephalitis virus in mosquitoes and their role as bridge vectors. *Emerging Infectious Diseases*. 2010 Dec;16(12):1869.
45. Oliver J, Lukacik G, Kokas J, Campbell SR, Kramer LD, Sherwood JA and Howard JJ. Twenty years of surveillance for Eastern equine encephalitis virus in mosquitoes in New York State from 1993 to 2012. *Parasites & Vectors*. 2018 Dec;11(1):1–2.
46. Simon LV, Coffey R and Fischer MA. Western Equine Encephalitis. [Updated 2020 Aug 22]. In *StatPearls* [Internet]. StatPearls Publishing, Treasure Island (FL), 2021 Jan. www.ncbi.nlm.nih.gov/books/NBK470228/
47. Michelitsch A, Wernike K, Klaus C, Dobler G and Beer M. Exploring the reservoir hosts of tick-borne encephalitis virus. *Viruses*. 2019 Jul;11(7):669.
48. Westby KM, Fritzen C, Paulsen D, Poindexter S and Moncayo AC. La Crosse encephalitis virus infection in field-collected *Aedes albopictus*, *Aedes japonicus*, and *Aedes triseriatus* in Tennessee. *Journal of the American Mosquito Control Association*. 2015 Sep;31(3):233–41.
49. https://en.wikipedia.org/wiki/La_Crosse_encephalitis. Accessed on 19 February 2021.
50. www.msdsonline.com/resources/sds-resources/free-safety-data-sheetindex/venezuelan-equine-encephalitis/
51. Tiwari S, Singh RK, Tiwari R and Dhole TN. Japanese encephalitis: a review of the Indian perspective. *The Brazilian Journal of Infectious Diseases*. 2012 Nov 1;16(6):564–73.
52. Erlanger TE, Weiss S, Keiser J, Utzinger J and Wiedenmayer K. Past, present, and future of Japanese encephalitis. *Emerging Infectious Diseases*. 2009 Jan;15(1):1.
53. Koskiniemi M, Korppi M, Mustonen K, Rantala H, Muttilainen M, Herrgård E, Ukkonen P and Vaheri A. Epidemiology of encephalitis in children: a prospective multicentre study. *European Journal of Pediatrics*. 1997 Jun 1;156(7):541–5.
54. Lindsey NP *et al.* Notes from the field: multistate outbreak of Eastern equine encephalitis virus—United States, 2019. *MMWR Morbidity and Mortality Weekly Report.* 2020 Jan 17;69:50. https://doi.org/10.15585/mmwr.mm6902a4

55. Debboun M, Nava MR and Rueda L, eds. 2019. *Mosquitoes, Communities, and Public Health in Texas*. Academic Press. Elsevier.
56. La Crosse Encephalitis: epidemiology and geographic distribution. www.cdc.gov/lac/tech/epi.html. Accessed on 20 January 2021.
57. www.who.int/health-topics/tick-borne-encephalitis#tab=tab_1
58. Rustagi R, Basu S and Garg S. Japanese encephalitis: strategies for prevention and control in India. *Indian Journal of Medical Specialities*. 2019 Jan 1;10(1):12.
59. Rao PN. Japanese encephalitis. *Indian Pediatrics*. 2001 Nov 1;38(11):1252–64.
60. Tuno N, Tsuda Y and Takagi M. How zoophilic Japanese encephalitis vector mosquitoes feed on humans. *Journal of Medical Entomology*. 2017 Jan 1;54(1):8–13.
61. Srivastava VK, Singh A and Thapar BR. Field evaluation of malathion fogging against Japanese encephalitis vector, *Culex tritaeniorhynchus*. *Journal of Vector Borne Diseases*. 2008 Sep 1;45:249–50.
62. Adhikari U and Chandra G. 2012. Laboratory evaluation of ethyl acetate and chloroform: methanol (1: 1 v/v) extract of *Swietenia mahagoni* leaf against Japanese encephalitis vector *Culex vishnui* group. *Asian Pacific Journal of Tropical Disease* 2(6):451–455.
63. Singha S, Adhikari U, Ghosh A and Chandra, G. 2012. Mosquito larvicidal potentiality of *Holoptelea integrifolia* leaf extract against Japanese encephalitis vector, *Culex vishnui* group. *Journal of Mosquito Research* 2(4): 25–31.
64. Bhattacharya K and Chandra G. 2013. Bioactivity of *Acyranthes aspera* (Amaranthaceae) foliage against the Japanese encephalitis vector *Culex vishnui* group. *Journal of Mosquito Research* 3(13): 89–96.
65. Haldar KM, Ghosh P and Chandra G. 2014. Larvicidal, adulticidal, repellency and smoke toxic efficacy of *Ficus krishnae* against *Anopheles stephensi* Liston and *Culex vishnui* group mosquitoes. *Asian Pacific Journal of Tropical Disease* 4:S214–220.
66. Mallick S, Bhattacharya K, Chandra G. 2014. Mosquito larvicidal potentiality of wild turmeric, *Curcuma aromatica* rhizome, extracts against Japanese Encephalitis vector *Culex vishnui* group. *Journal of Mosquito Research* 4(19):1–6.
67. Rawani A, Ray AS, Ghosh A, Sakar M and Chandra G. 2017. Larvicidal activity of phytosteroid compounds from leaf extract of Solanum nigrum against *Culex vishnui* group and *Anopheles subpictus*. *BMC Research Notes* 10(1):1–8.
68. Barik M, Rawani A, Laskar S and Chandra G. 2019. Evaluation of mosquito larvicidal activity of fruit extracts of *Acacia auriculiformis* against the Japanese encephalitis vector *Culex vishnui*. *Natural Product Research* 33(11):1682–1686.
69. Singh A and Chandra G. 2022. Larvicidal, pupicidal, repellence and smoke toxic efficacies of *Nicotiana plumbaginifolia* leaves against *Culex vishnui* mosquito. *International Journal of Tropical Insect Science* 42(2):1187–1195.
70. Chatterjee SN, Banerjee BD and Chandra G. 1996. Effect of Neem (*Azadirachta indica*) oil on the biting population of *Culex vishnui* (complex). *Journal of Basic and Applied Biomedicine* 4:39–41.
71. Dhillon GP and Raina VK. Epidemiology of Japanese encephalitis in context with Indian scenario. *Journal of the Indian Medical Association*. 2008 Oct 1;106(10):660–3.
72. Banerjee A and Chandra G. Role of some factors on the breeding of JE vector *Culex vishnui* group. *The Journal of Communicable Diseases*. 2004 Dec;36(4):260–3.
73. Case study: India's leadership in the fight against Japanese encephalitis. Center for Vaccine Innovation and Access, PATH. www.path.org/programs/center-for-vaccine-innovation-and-access/case-study-indias-leadership-in-the-fight-against-japanese-encephalitis/. Accessed on 24 February 2021.
74. Wijesinghe PR, Abeysinghe MN, Yoksan S, Yao Y, Zhou B, Zhang L, Fleming JA, Marfin AA and Victor JC. Immunogenicity of live attenuated Japanese encephalitis SA 14-14-2 vaccine among Sri Lankan children with previous receipt of inactivated JE vaccine. *Vaccine*. 2016 Nov 21;34(48):5923–8.
75. www.path.org/programs/center-for-vaccine-innovation-and-access/case-study-indias-leadership-in-the-fight-against-japanese-encephalitis/. Accessed on 25 February 2021.

Part IV

Environment and Health Impacts

23 Environmental Contaminants and Male Reproductive Health

Past, Present and Future

Latchoumycandane Calivarathan and Meenu Maniradhan
Molecular Pharmacology and Toxicology Laboratory, Department of Life Sciences, School of Life Sciences, Central University of Tamil Nadu, Neelakudi Campus, Thiruvarur

Pranitha Janardhanan and Premendu Prakash Mathur
Department of Biochemistry and Molecular Biology, School of Life Sciences, Pondicherry University, Pondicherry

CONTENTS

23.1 INTRODUCTION

Exposure to some of the environmental contaminants has been shown to pose a great threat to the reproductive health of the population due to their endocrine-disrupting potential. Endocrine disruptors (EDCs) are naturally occurring or man-made substances that cause adverse health effects in organisms, either directly targeting the normal endocrine physiology or acting as an obesogenic agent, triggering health effects. In the last few decades, the list of molecules considered as endocrine disruptors has been dramatically increasing worldwide due to industrialization. The majority of EDCs are synthetic chemicals, which have been released into the environment by the activities of humans without any prior knowledge of their effects on the ecosystem, animals and human health. Only a few EDCs, like phytoestrogens, are present in nature, and their impact on the

DOI: 10.1201/9781003220824-27

environment is low compared with man-made chemicals. Apart from environmental contaminants with endocrine-disrupting potential, a wide variety of synthetic compounds, including pharmaceutical drugs, pesticides, plastics and plasticizers, are also recognized to possess endocrine-disrupting properties. These chemicals are omnipresent in the environment, and humans are being unknowingly exposed to such chemicals in daily life through different routes such as inhalation, ingestion, dermal absorption, etc. due to their use in industrial and household products, pesticides/herbicides, plastics, detergents, flame retardants and as ingredients of personal care products. The exposure to EDCs from multiple sources greatly increases the chance of obesity, metabolic disorders, organ-related diseases or cancers. EDCs mimic or partially mimic naturally occurring hormones in the body, like estrogens, androgens and thyroid hormones, or they interfere with the way in which natural hormones and their receptors are made or controlled. EDC exposure appears to be more dangerous at specific lifetime periods; especially, an early exposure is supposed to be more dangerous, not only by leading to an immediate onset of specific pathologies but also because it can make the exposed subject more prone to develop several metabolic diseases in the future, even without further significant exposure to EDCs and also due to heritable and epigenetic changes (Goyal, Limesand, and Goyal 2019). The acute or chronic exposure to endocrine disruptors at lower doses during early development may cause adverse health effects at the adult stage due to increased susceptibility of the organism to immaturity in protective mechanisms such as DNA repair and formation of the blood–brain barrier, blood–testis barrier (BTB), etc. Beyond the fetal developmental period, other periods, including childhood and adolescence, also show increased sensitivity to EDCs. EDCs exert their adverse effects through interacting with several membrane and/or cytosolic receptors, including nuclear hormone receptors, steroid hormone receptors, non-nuclear steroid hormone receptors, and orphan receptors (Dallio et al. 2019). These chemical compounds interfere with biological homeostasis not only by acting on steroid biosynthesis and metabolism but also by regulating the release of neurotransmitters to the central and peripheral nervous system (Polyzos et al. 2012, Diamanti-Kandarakis et al. 2009). EDCs have extreme chemical and biological stability and persist in the environment forever, so the continuous exposure to these chemicals results in more adverse effects. In this review, we elaborate an overview on the effect of environmental contaminants on reproductive health in the past, present and future.

23.2 ENVIRONMENTAL CONTAMINANTS AND MALE REPRODUCTIVE HEALTH

Exposure to environmental contaminants has been the major cause for the observed defects in male reproductive health due to reduced sperm quality. Occupational exposure is one of the major contributors to reproductive dysfunctions induced by environmental contaminants. Some of the known chemicals include plasticizers, heavy metals, phthalates, pesticides, solvents and industrial by-products.

23.2.1 Bisphenol A and Male Reproductive Health

Bisphenol A, [2,2-bis (4-hydroxyphenol) propane] (BPA), is one of the most commonly studied EDCs and is synthesized in millions of tons (Witorsch 2002). BPA is used in several consumables, such as CDs, DVDs, water taps, power plugs, safety resins, baby bottles, food and beverage containers, reusable plastics, microwave/ovenware containers, eye lenses, newborn incubators and safety goggles, as epoxy lining in food/beverage cans, as dental fillings, as colorants in thermal papers, in the production of coffee/tea cups and reusable plastic bottles, as part of membranes in hemodialysis and in ultra-water purification units (Covaci et al. 2009, Dodson et al. 2012, Finch 2002, Geens, Goeyens, and Covaci 2011, Liao and Kannan 2011, Poole et al. 2004, Tsai 2006, Van Landuyt et al. 2011). Due to its widespread use, BPA is found to have a wide range of bioaccumulation, including in children, adults, pregnant women, developing fetuses, water bodies, soil and even air, which makes it the most widely occurring substance. Several studies have characterized its toxic effects, and it has been categorized as a reproductive toxic substance under category 3, which is shown to affect human fertility (Manfo et al. 2014, Jenardhanan, Panneerselvam, and Mathur 2016). BPA is also shown to induce cardiovascular problems (Ranciere et al. 2015), obesity (Angle et al. 2013) and diabetes (Mirmira and Evans-Molina 2014); also, it decreases estradiol level and oocyte retrieval in postmenopausal women (Mok-Lin et al. 2010). Irrespective of BPA's presence in different concentrations across various tissues and based on different studies evaluating the toxic effects of BPA, the US Food and Drugs Administration (FDA) has suggested an acceptable human daily intake dose of BPA as 50 mg/kg body weight/day. However, several studies have shown that BPA can exert its toxic effects even at lower levels than the dose designated as safe (Corrales et al. 2015).

With regard to male infertility, BPA can modulate testosterone synthesis, disrupts BTB/apical ectoplasmic specialization, affects meiotic division, causes apoptosis in Sertoli cells, reduces sperm counts, and affects sperm morphology and sperm functions (Peretz et al. 2014). BPA has also been shown to affect enzymes involved in steroidogenesis, antioxidant enzymes, and proteins involved in glucose and insulin metabolism (Meli et al. 2020, D'Cruz et al. 2012). Our previous studies have shown that BPA increases the insulin level in rat testis and substantially decreases the cellular levels of insulin receptor subtype-2 and glucose transporter 8 (D'Cruz, Jubendradass, and Mathur 2012). It also hampers the functions of 17β-hydroxysteroid dehydrogenase and 3β-hydroxysteroid dehydrogenase as well as down-regulating the expression of Steroidogenic Acute Regulatory Protein (StAR) protein and decreasing testosterone synthesis (D'Cruz et al. 2012). Specifically, it deregulates N-cadherin, a component of ectoplasmic specialization; occludin, a component of the occludin-ZO-1 adhesion complex; and Connexin Cx43, a component of gap junctions. By affecting the steady-state level of Cx43, BPA deregulates the spatial localization of cadherin, occludin and other proteins and disrupts the BTB dynamics in testes (Li et al. 2009, Salian, Doshi, and Vanage 2009). In addition, BPA alters the localization of actin regulatory proteins such as Arp3 and Eps8, which affects actin microfilament bundling and nucleation of branched actin filaments, and this in turn changes the BTB restructuring necessary for transport of preleptotene spermatocytes towards the apical compartment (Xiao et al. 2014). BPA can actively target the CaM-CaMKII-ERK1/2 network involved in mediating the Sertoli cell apoptosis; specifically, it increased the concentration of CaM, activated CaM-CaMKII, and increased the mitochondrial concentration of calcium, leading to apoptosis of Sertoli cells (Qian et al. 2014). Subsequently, BPA is shown to activate ERK signaling in Sertoli cells, which increases Sertoli cell number, thereby deregulating the ratio of germ cells that are nursed by Sertoli cells, and finally resulting in reduced sperm count (Ge et al. 2014). Widespread occurrence of BPA, along with its enormous bioaccumulation and its weak estrogenic activity, has paved the way for several detrimental activities that ultimately affect energy metabolism, insulin signaling, decreased testosterone level and induced apoptosis in Sertoli cells.

23.2.2 Heavy Metals and Male Reproductive Health

Heavy metals range as the second most important class of reproductive toxicants, and some of the prominent metals that affect male reproduction include cadmium, mercury, lead, chromium, arsenic, aluminum and nickel. Cadmium (Cd) is released into the environment through mining, waste incineration, battery manufacturing and predominantly by cigarette smoking, and causes significant disruption to the BTB. Cd can exert its negative effects on the BTB by activating the Transforming Growth Factor-β and p38 MAP kinase, which results in down-regulation of tight junction proteins such as occludin (Lui et al. 2003). Cd is also known to down-regulate the expression of two phosphorylated forms of FAK, p-FAK-Tyr397 and p-FAK-Tyr57, both of which have been shown to impart negative regulation of the Sertoli cell's tight junction permeability (Siu, Wong, et al. 2009). Cd acts as an EDC by mimicking natural estrogen and significantly alters the circulating levels of testosterone, follicle stimulating hormone (FSH) and luteinizing hormone (LH) as well as activating androgen receptor expression while down-regulating the expression of LH hormone receptor. Mercury is another heavy metal that exerts a negative effect on male reproductive functions, primarily by inducing oxidative stress that results in decreased sperm count and altered sperm morphology and also causes testicular degeneration (Rignell-Hydbom et al. 2007). Exposure to hexavalent chromium is also proven to exert deleterious effects on male fertility, and the observed effects include decreased lactate dehydrogenase levels, increased FSH levels, altered sperm morphology and decreased sperm concentration (Lafuente et al. 2004). Chromium delocalizes the gap junction protein connexin 43 from the membrane to the cytosol and aids in destabilizing the BTB (Siu, Mruk, et al. 2009).

23.2.3 Nanoparticles and Male Reproductive Health

The use of nanoparticles for various biological applications has increased exponentially in recent years. The size of the nanoparticles ranges between 1 and 100 nm, and they are widely used as transporters for several drugs, for example, mesoporous silica nanoparticles are being used for transporting cancer drugs such as paclitaxel and doxorubicin. Nanoparticles containing metal oxide nanostructures such as zinc oxide (ZnO) are being used as catalysts in electronics, clothing, paints, coatings, cosmetic products and as biosensors in medical devices. In due course, exposure to nanoparticles also cause several undesirable effects, including reproductive toxicity; particularly, exposure to ZnO nanoparticles has been shown to increase reactive oxygen species and mitochondrial damage and also alter the morphology of seminiferous tubules and spermatozoa (Han et al. 2016). Nanoparticles of titanium dioxide (TiO_2) are another example of metal nanoparticles that are used in various consumer products, and exposure is known to disrupt spermeation, alter sperm morphology and reduce sperm count (Meena, Kajal, and R 2015). Several nanoparticles are under the scanner for evaluating their reproductive toxicity.

23.2.4 Phthalates and Male Reproductive Health

Phthalates, like BPA, are categorized as estrogen analogues and are formed by the addition of diesters to benzenedicarboxylic acid. Phthalates are further divided into low- and high-molecular-weight phthalates and translational phthalates. Phthalates find their use in personal care products, plasticizers, solvents, toys, consumable products and the production of papers (Manfo, Nantia, and Mathur 2014). Two well-studied phthalates, diethylhexyl phthalate and its metabolite mono-2-ethylhexyl phthalate, are known to exert intense toxic effects on reproductive functions. Primarily, they act as EDCs, whereby they down-regulate the expression of enzymes such as StAR and cytochrome P450, which ultimately decreases testosterone production (Veeramachaneni and Klinefelter 2014). Butylbenzyl phthalate, another common phthalate, is found to induce collapse of vimentin filaments in Sertoli cells, which leads to disruption of the Sertoli cell–germ cell interface and Sertoli cell apoptosis (Alam and Kurohmaru 2016). Monobutyl phthalate, on the other hand, is shown to delocalize the junction proteins occludin, ZO-1 and β-catenin, resulting in disruption of the BTB (de Freitas et al. 2016).

23.2.5 Pesticides and Male Reproductive Health

Several studies have shown that exposure to natural or synthetic pesticides or metabolites of pesticides affects the male reproductive function. Some pesticides act as EDCs and have been shown to alter the Sertoli cell functions and sperm morphology as well as decreasing the sperm count. 1,1-Dichloro-2,2 bis (p-chlorophenyl) ethylene (p,p′-DDE), the major metabolite of 2,2-bis

(4-chlorophenyl)-1,1,1-trichloroethane (DDT), an organic pollutant still being used in some countries as a pesticide, is shown to exert testicular apoptosis by activating the Fas/FasL pathway (Shi et al. 2010). Fenvalerate, a synthetic pyrethroid insecticide, on the other hand, induces testicular apoptosis and decreases the serum and testicular testosterone levels by actively down-regulating genes such as $P450_{17\alpha}$ and $P450_{scc}$, which are involved in testosterone biosynthesis (Zhang et al. 2009). DDT and DDE have also been shown to induce Sertoli cell apoptosis and decrease the total sperm count (Song et al. 2011). 2,4-Dichlorophenoxyacetic acid, a herbicide, also exerts its reproductive toxicity by decreasing glucose intake by hampering the metabolism in Sertoli cells and also affects lactate intake and metabolism, which altogether impairs spermatogenesis (Amer and Aly 2001). Methoxychlor is an organochlorine insecticide, acts as an EDC and is known to induce oxidative stress in the male reproductive organs by enhancing reactive oxygen species and by suppressing the expression of antioxidant enzymes (Latchoumycandane and Mathur 2002a, b). Lindane, which is also known as γ-hexachlorocyclohexane and benzene hexachloride, is commonly used in shampoo to kill lice and nits and to treat dry scalp conditions. Prolonged usage of such products increases the exposure to lindane, which exerts effects similar to DDT, whereby it increases the oxidative stress in testes, induces testicular apoptosis and decreases testosterone production (Saradha, Vaithinathan, and Mathur 2009). Several other pesticides, like methymol, propoxur, carbofuran, carbaryl, dibromochloropropane, benomyl, atrazine and mancozeb, all exert negative regulation on the endocrine system and induce germ cell apoptosis (Mehrpour et al. 2014).

23.2.6 Solvents and Male Reproductive Health

Recent research activities have turned their concentration to evaluating the ill effects of oil and natural gas extraction on the environment and human health, since the hydraulic fracturing and drilling process can substantially contaminate ground water and surface water, particularly by the toxic substances in waste water released into the environment. Also, the toxic substances that are in oil spills can evaporate into the air, which also becomes a source of exposure (Colborn et al. 2014, Kassotis et al. 2016). Nevertheless, acidic chemicals used in acidizing techniques for the extraction of oil and natural gas are found in the waste water (Vengosh et al. 2014), and since an effective recycling and treatment procedure is not yet developed, the potential of these toxic substances to affect living flora and fauna is unimaginably high (Balise et al. 2016). Recent reports state that more than a thousand highly toxic chemicals are found in the waste water, and of these, about 43% are reproductive toxicants and 40% show developmental toxicity (Elliott et al. 2017). Reports confirm that these toxic substances are present in the range of micrograms to grams, and the typical human exposure will be in the range of 3–30 μg/kg body weight/day, with some of these substances antagonizing the activity of hormones at a concentration of 1.0 μM (Kassotis et al. 2015). Exposure to benzene, toluene, ethyl benzene, and xylene and formaldehyde in the waste water from the oil and gas extraction process is linked with low sperm count, reduced sperm motility, abnormal sperm morphology and semen viscosity (Webb et al. 2014), indicating that the chemicals released into the environment by the extraction of unconventional natural gas can be the source of highly toxic chemicals that can have an intense effect on male fertility.

23.3 LIFESTYLE CHANGES AND MALE REPRODUCTIVE HEALTH

Recently, several studies have shown that changes in lifestyle contribute to many diseases, including reproductive dysfunctions. Personal habits, ranging from the food we take in, clothing style, the medications being taken, consuming alcohol and smoking to the sedentary workstyle are the deciding factors of our reproductive health.

23.3.1 Obesity and Male Reproductive Health

The prevalent belief is that with balanced diet and regular exercise, individuals can avoid being obese, which can avoid all health issues, including reproductive defects in both qualitative and quantitative aspects. Contrarily, the foods, drinks and drugs that the individual consumes, along with his personal habits of performing regular exercise and his choice of clothing, combined with the time spent by the individual driving, sitting in a sedentary job, and even the usage of a mobile phone, is proven to affect an individual's sperm quantity and quality. Both males and females with normal body mass index are observed to have a normal fertility rate, implying that persons who are overweight and underweight are more prone to fertility disorders (Hassan and Killick 2003). It has been shown that obesity in men correlates well with decreased sperm concentration, reduced sperm quality and damaged DNA in sperm (Martini et al. 2010). The molecular level of the protein Inhibin B substantially decreases, and deregulation of Inhibin B is directly correlated with decreased Sertoli cell number, resulting in reduced spermatogenesis (Winters et al. 2006). Obese men have a higher concentration of the hormone leptin, and animal models have shown that higher concentrations of leptin reduce male fertility (Ghanayem et al. 2010). The most promising and direct explanation comes from a report that elucidates the presence of decreased testosterone and increased estradiol ratio in obese men (Hammoud et al. 2008).

The regular intake of soy, red meat and potatoes is also known to affect male fertility. Potatoes, being rich in starches with a high glycemic index and insulinemic response, can directly induce oxidative stress, which is associated with poor semen quality (Salas-Huetos, Bulló, and Salas-Salvadó 2017). Soy, on the other hand, contains phytoestrogens, while red meat or any kind of animal meat is rich in xenoestrogens, both of which are found to have deleterious effects on male fertility. Meat accompanied with dairy products and bakery products contains large amounts of saturated and trans fatty acids, resulting in poor semen quality. Soy foods that are enriched in phytoestrogens mimic the action of estrogens; thus, their intake can profoundly affect the male endocrine system and fertility (Salas-Huetos, Bulló, and Salas-Salvadó 2017).

23.3.2 Alcohol and Male Reproductive Health

As to the effect of alcohol consumption, many studies arrived at a uniform conclusion that drinking alcohol can severely affect male fertility. Testicular atrophy, decreased libido and decreased sperm count are some of the commonly observed negative effects of alcohol consumption. Alcohol consumption impedes the function of gonadotropin-releasing hormone (GnRH) and hampers the cleavage of pre-pro-GnRH to functional GnRH. Also, it affects Sertoli cells, lowers the levels of LH, FSH and testosterone, and induces spermatogenic arrest and Sertoli cell-only syndrome. People who are categorized as heavy and moderate drinkers are found with oligozoospermic and tetrazoospermic sperms, indicating the ill effects of alcohol consumption (Salas-Huetos, Bulló, and Salas-Salvadó 2017).

23.3.3 Smoking and Male Reproductive Health

Smoking is a major determinant of male factor infertility and has harmful effects at all stages of sperm production, including sperm count, sperm concentration and sperm morphology, and also causes erectile dysfunction. Nicotine and trans-3′-hydroxycotinine levels have been shown to affect sperm motility. Nicotine affects sperm viability, promotes spermatozoa apoptosis and induces DNA damage, and also cause Leydig cell apoptosis, resulting in reduced androgen production. Sertoli cell vacuolation, a decrease in the differential germ cell population in the spermatogenic cycle, and degenerative changes in seminiferous epithelium are observed following exposure to cigarette smoke (Dai, Wang, and Qiao 2015). Cadmium is another prominent toxicant present in cigarette smoke, and its toxic effects are well characterized and discussed. Taken together with alcohol, the deleterious effects of cigarette smoking increase exponentially.

23.3.4 Drugs and Male Reproductive Health

As well as cigarette and alcohol, drugs can also adversely affect reproductive health, either directly or through their metabolites. The US Center for Drug Evaluation and Research, which is a part of the FDA, has instructed that animal studies conducted for drug development should include the assessment of reproductive toxicity. The common drugs that are used in the treatment of disorders related to sexual function are commonly found to have unintended side effects that affect male fertility (Ding et al. 2015, Samplaski and Nangia 2015). A few of them include the drugs that fall under the class of 5-alpha-reductase inhibitors, alpha blockers and phosphodiesterase inhibitors. Currently there are two inhibitors available in the market, finasteride and dutasteride, both of which are used to inhibit the action of 5-alpha-reductase, which is involved in converting testosterone into dihydrotestosterone. These inhibitors are primarily used to treat benign prostatic hyperplasia and prostate cancer. Persons consuming this drug are more prone to erectile dysfunction, lack of libido and ejaculatory dysfunction. Besides, these drugs are found to have marginal effects on sperm count, concentration and morphology (Amory et al. 2007). The α-blockers are the next class of drugs that have been shown to impair male fertility; these drugs were first marketed to treat a medical condition called benign prostatic hypertrophy, a condition characterized by the overgrowth of the prostate gland, which is usually observed in aged males. A few examples of α-blockers include Terazosin, Doxazosin, Alfuzoin, Tamsulosin and Silodosin. The negative effect of these drugs is primarily ejaculatory dysfunction. In addition to this, healthy men who have been treated with α-blockers show poor sperm parameters (Furukawa et al. 1995, Hisasue et al. 2006, Gacci et al. 2014). Several other classes of drugs that are used for other purposes also seem to have negative effects on male fertility (Ding et al. 2015); a few of them include the drugs that fall under the class of opioids, which are used as analgesics. Substantial decrease in sperm motility and seminal vesicle secretions and abnormal sperm morphology have been reported following methadone use (Ragni et al. 1985). In addition, several anti-neoplastic agents and hormone antagonists also impose deleterious effects on male fertility; for instance, the drug Busulfan is used for treating chronic myelogenous leukemia, which has been shown to cause damage to spermatozoa and testicular tissue, and causes azoospermia and testicular atrophy (Jackson, Fox, and Craig 1959). Danazol is a synthetic steroid used in the treatment of hereditary angioedema; prolonged use of this drug has been shown to decrease spermatogenesis, cause abnormalities in semen volume, viscosity and sperm count, and affect sperm motility (Ulstein et al. 1975).

23.3.5 Temperature and Mobile Phone Radiations and Male Reproductive Health

The effect of exposure to heat and radiations from mobile phones on reproductive health are critical topics that are under discussion to find out whether they affect male fertility. There is some evidence that intratesticular temperature below the core body temperature is usually required to maintain normal spermatogenesis (Kim, Park, and Rhee 2013). Increase in scrotal temperature has negative effects on sperm production. Continued sitting in the workplace for several hours, driving, using a laptop for too long and wearing restrictive clothing may contribute towards an increase in scrotal temperature (Li et al. 2011, Pacey 2010). At present, it is believed that the effect of cell phone usage or the effect of radiofrequency electromagnetic radiation is detrimental towards spermatogenesis by directly affecting Leydig cells, seminiferous tubules and developing spermatozoa. Studies also elucidate that frequent use of mobile phones has a negative impact on sperm motility and sperm morphology and an increase in reactive oxygen species production (Merhi 2012, La Vignera et al. 2012). However, still more studies are warranted to conclude the effect of cell phone radiations on male reproductive health.

23.4 CONCLUSIONS

Male reproductive health depends on preserving harmonized responses of a network of endocrine signals that function primarily to ensure successful reproduction. However, exposure to environmental contaminants possessing endocrine-disrupting properties and the present sedentary lifestyles pose a great threat to male reproductive health. Several synthetic and natural chemicals are ubiquitously present in our environment, and humans are exposed to such chemicals in day-to-day life in different ways. This exposure leads to various metabolic disorders, such as obesity and insulin resistance, that also contribute indirectly to the decline in reproductive health. To conclude, as well as exposure to EDCs, personal habits, including the food we take in, clothing style, medications and a sedentary lifestyle, are deciding our reproductive health.

REFERENCES

Alam, M. S., and M. Kurohmaru. 2016. "Butylbenzyl phthalate induces spermatogenic cell apoptosis in prepubertal rats." *Tissue Cell* 48 (1):35–42. doi: 10.1016/j.tice.2015.12.001.

Amer, S. M., and F. A. Aly. 2001. "Genotoxic effect of 2,4-dichlorophenoxy acetic acid and its metabolite 2,4-dichlorophenol in mouse." *Mutat Res* 494 (1–2):1–12.

Amory, J. K., C. Wang, R. S. Swerdloff, B. D. Anawalt, A. M. Matsumoto, W. J. Bremner, S. E. Walker, L. J. Haberer, and R. V. Clark. 2007. "The effect of 5α-reductase inhibition with dutasteride and finasteride on semen parameters and serum hormones in healthy men." *J Clin Endocrinol Metab* 92:1659–65. doi: 10.1210/jc.2006-2203.

Angle, B. M., R. P. Do, D. Ponzi, R. W. Stahlhut, B. E. Drury, S. C. Nagel, W. V. Welshons, C. L. Besch-Williford, P. Palanza, S. Parmigiani, F. S. vom Saal, and J. A. Taylor. 2013. "Metabolic disruption in male mice due to fetal exposure to low but not high doses of bisphenol A (BPA): evidence for effects on body weight, food intake, adipocytes, leptin, adiponectin, insulin and glucose regulation." *Reprod Toxicol* 42:256–68. doi: 10.1016/j.reprotox.2013.07.017.

Balise, V. D., C. X. Meng, J. N. Cornelius-Green, C. D. Kassotis, R. Kennedy, and S. C. Nagel. 2016. "Systematic review of the association between oil and natural gas extraction processes and human reproduction." *Fertil Steril* 106 (4):795–819. doi: 10.1016/j.fertnstert.2016.07.1099.

Colborn, T., K. Schultz, L. Herrick, and C. Kwiatkowski. 2014. "An exploratory study of air quality near natural gas operations." *Hum Ecol Risk Assess* 20 (1):86–105. doi: 10.1080/10807039.2012.749447.

Corrales, J., L. A. Kristofco, W. B. Steele, B. S. Yates, C. S. Breed, E. S. Williams, and B. W. Brooks. 2015. "Global assessment of Bisphenol A in the environment: review and analysis of its occurrence and bioaccumulation." *Dose Response* 13 (3):1559325815598308. doi: 10.1177/1559325815598308.

Covaci, A., S. Voorspoels, M. A. Abdallah, T. Geens, S. Harrad, and R. J. Law. 2009. "Analytical and environmental aspects of the flame retardant tetrabromobisphenol-A and its derivatives." *J Chromatogr A* 1216 (3):346–63. doi: 10.1016/j.chroma.2008.08.035.

D'Cruz, S. C., R. Jubendradass, M. Jayakanthan, S. J. Rani, and P. P. Mathur. 2012. "Bisphenol A impairs insulin signaling and glucose homeostasis and decreases steroidogenesis in rat testis: an in vivo and in silico study." *Food Chem Toxicol* 50 (3–4):1124–33. doi: 10.1016/j.fct.2011.11.041.

D'Cruz, S. C., R. Jubendradass, and P. P. Mathur. 2012. "Bisphenol A induces oxidative stress and decreases levels of insulin receptor substrate 2 and glucose transporter 8 in rat testis." *Reprod Sci* 19 (2):163–72. doi: 10.1177/1933719111415547.

Dai, J-B, Z-X Wang, and Z-D Qiao. 2015. "The hazardous effects of tobacco smoking on male fertility." *Asian J Androl* 17:954–60. doi: 10.4103/1008-682X.150847.

Dallio, M., N. Diano, M. Masarone, A. G. Gravina, V. Patane, M. Romeo, R. Di Sarno, S. Errico, C. Nicolucci, L. Abenavoli, E. Scarpellini, L. Boccuto, M. Persico, C. Loguercio, and A. Federico. 2019. "Chemical effect of Bisphenol A on non-alcoholic fatty liver disease." *Int J Environ Res Public Health* 16 (17):3134. doi: 10.3390/ijerph16173134.

de Freitas, A. T., M. A. Ribeiro, C. F. Pinho, A. R. Peixoto, R. F. Domeniconi, and W. R. Scarano. 2016. "Regulatory and junctional proteins of the blood-testis barrier in human Sertoli cells are modified by monobutyl phthalate (MBP) and bisphenol A (BPA) exposure." *Toxicol In Vitro* 34:1–7. doi: 10.1016/j.tiv.2016.02.017.

Diamanti-Kandarakis, E., J. P. Bourguignon, L. C. Giudice, R. Hauser, G. S. Prins, A. M. Soto, R. T. Zoeller, and A. C. Gore. 2009. "Endocrine-disrupting chemicals: an endocrine society scientific statement." *Endocr Rev* 30 (4):293–342. doi: 10.1210/er.2009-0002.

Ding, J., X. Shang, Z. Zhang, H. Jing, J. Shao, Q. Fei, H. Li, and E. R. Rayburn. 2015. "FDA-approved medications that impair human spermatogenesis." *Oncotarget* 8:6–9. doi: 10.18632/oncotarget.12956.

Dodson, R. E., M. Nishioka, L. J. Standley, L. J. Perovich, J. G. Brody, and R. A. Rudel. 2012. "Endocrine disruptors and asthma-associated chemicals in consumer products." *Environ Health Perspect* 120 (7):935–43. doi: 10.1289/ehp.1104052.

Elliott, E. G., A. S. Ettinger, B. P. Leaderer, M. B. Bracken, and N. C. Deziel. 2017. "A systematic evaluation of chemicals in hydraulic-fracturing fluids and wastewater for reproductive and developmental toxicity." *J Expo Sci Environ Epidemiol* 27 (1):90–9. doi: 10.1038/jes.2015.81.

Finch, C A. 2002. "Industrial polymers handbook: products, processes, applications Volume 1, polymerization processes: Synthetic polymers Volume 2, Synthetic polymers (continued) Volume 3, Synthetic polymers (continued); biopolymers and their derivatives Volume 4, Biopolymers and their derivatives." *Polym Int* 51 (3):264–5. doi: 10.1002/pi.840.

Furukawa, K., D. J. Rosario, D. J. Smith, C. R. Chapple, T. Uchiyama, and R. Chess-Williams. 1995. "Alpha 1A-adrenoceptor-mediated contractile responses of the human vas deferens." *Br J Pharmacol* 116:1605–10.

Gacci, M., V. Ficarra, A. Sebastianelli, G. Corona, S. Serni, S. F. Shariat, M. Maggi, F. Zattoni, M. Carini, and G. Novara. 2014. "Impact of medical treatments for male lower urinary tract symptoms due to benign prostatic hyperplasia on ejaculatory function: a systematic review and meta-analysis." *Sex Med* 11:1554–66. doi: 10.1111/jsm.12525.

Ge, L. C., Z. J. Chen, H. Y. Liu, K. S. Zhang, H. Liu, H. B. Huang, G. Zhang, C. K. Wong, J. P. Giesy, J. Du, and H. S. Wang. 2014. "Involvement of activating ERK1/2 through G protein coupled receptor 30 and estrogen receptor alpha/beta in low doses of bisphenol A promoting growth of Sertoli TM4 cells." *Toxicol Lett* 226 (1):81–9. doi: 10.1016/j.toxlet.2014.01.035.

Geens, T., L. Goeyens, and A. Covaci. 2011. "Are potential sources for human exposure to bisphenol-A overlooked?" *Int J Hyg Environ Health* 214 (5):339–47. doi: 10.1016/j.ijheh.2011.04.005.

Ghanayem, B. I., R. Bai, G. E. Kissling, G. Travlos, and U. Hoffler. 2010. "Diet-induced obesity in male mice is associated with reduced fertility and potentiation of acrylamide-induced reproductive toxicity." *Biology of Reproduction* 82:96–104. doi: 10.1095/biolreprod.109.078915.

Goyal, D., S. W. Limesand, and R. Goyal. 2019. "Epigenetic responses and the developmental origins of health and disease." *J Endocrinol* 242 (1):T105–T119. doi: 10.1530/JOE-19-0009.

Hammoud, A.O., N. Wilde, M. Gibson, et al. (2008) "Male obesity and alteration in sperm parameters." *Fertil Steril* 90:2222–5.

Han, Z., Q. Yan, W. Ge, Z. G. Liu, S. Gurunathan, M. De Felici, W. Shen, and X. F. Zhang. 2016. "Cytotoxic effects of ZnO nanoparticles on mouse testicular cells." *Int J Nanomedicine* 11:5187–203. doi: 10.2147/IJN.S111447.

Hassan, M. A. M., and S. R. Killick. 2003. "Effect of male age on fertility: evidence for the decline in male fertility with increasing age." *Fertil Steril* 79:1520–7. doi: 10.1016/S0015-0282(03)00366-2.

Hisasue, S. I., R. Furuya, N. Itoh, K. Kobayashi, S. Furuya, and T. Tsukamoto. 2006. "Ejaculatory disorder caused by alpha-1 adrenoceptor antagonists is not retrograde ejaculation but a loss of seminal emission." *Int. J. Urol* 13:1311–16. doi: 10.1111/j.1442-2042.2006.01535.x.

Jackson, H., B. W. Fox, and A. W. Craig. 1959. "The effect of alkylating agents on male rat fertility." *Br J Pharmacol Chemother* 14 (2):149–57. doi: 10.1111/j.1476-5381.1959.tb01375.x.

Jenardhanan, P., M. Panneerselvam, and P. P. Mathur. 2016. "Effect of environmental contaminants on spermatogenesis." *Semin Cell Dev Biol* 59:126–40. doi: 10.1016/j.semcdb.2016.03.024.

Kassotis, C. D., K. C. Klemp, D. C. Vu, C. H. Lin, C. X. Meng, C. L. Besch-Williford, L. Pinatti, R. T. Zoeller, E. Z. Drobnis, V. D. Balise, C. J. Isiguzo, M. A. Williams, D. E. Tillitt, and S. C. Nagel. 2015. "Endocrine-disrupting activity of hydraulic fracturing chemicals and adverse health outcomes after prenatal exposure in male mice." *Endocrinology* 156 (12):4458–73. doi: 10.1210/en.2015-1375.

Kassotis, C. D., D. E. Tillitt, C. H. Lin, J. A. McElroy, and S. C. Nagel. 2016. "Endocrine-disrupting chemicals and oil and natural gas operations: potential environmental contamination and recommendations to assess complex environmental mixtures." *Environ Health Perspect* 124 (3):256–64. doi: 10.1289/ehp.1409535.

Kim, B., K. Park, and K. Rhee. 2013. "Heat stress response of male germ cells." *Cell Mol Life Sci* 70:2623–36. doi: 10.1007/s00018-012-1165-4.

La Vignera, S., R. A. Condorelli, E. Vicari, R. D'Agata, and A. E. Calogero. 2012. "Effects of the exposure to mobile phones on male reproduction: a review of the literature." *J Androl* 33:350–6. doi: 10.2164/jandrol.111.014373.

Lafuente, A., A. Gonzalez-Carracedo, A. Romero, P. Cano, and A. I. Esquifino. 2004. "Cadmium exposure differentially modifies the circadian patterns of norepinephrine at the median eminence and plasma LH, FSH and testosterone levels." *Toxicol Lett* 146 (2):175–82.

Latchoumycandane, C., and P. P. Mathur. 2002a. "Effect of methoxychlor on the antioxidant system in mitochondrial and microsome-rich fractions of rat testis." *Toxicology* 176 (1–2):67–75. doi: 10.1016/s0300-483x(02)00138-5.

Latchoumycandane, C., and P. P. Mathur. 2002b. "Induction of oxidative stress in the rat testis after short-term exposure to the organochlorine pesticide methoxychlor." *Arch Toxicol* 76 (12):692–8. doi: 10.1007/s00204-002-0388-9.

Li, M. W., D. D. Mruk, W. M. Lee, and C. Y. Cheng. 2009. "Disruption of the blood-testis barrier integrity by bisphenol A in vitro: is this a suitable model for studying blood-testis barrier dynamics?" *Int J Biochem Cell Biol* 41 (11):2302–14. doi: 10.1016/j.biocel.2009.05.016.

Li, W., Z.Q. Wu, J. Zhao, et al. 2011. "Transient protection from heat-stress induced apoptotic stimulation by metastasis-associated protein 1 in pachytene spermatocytes." *PloS One* 6:e26013.

Liao, C., and K. Kannan. 2011. "Widespread occurrence of bisphenol A in paper and paper products: implications for human exposure." *Environ Sci Technol* 45 (21):9372–9. doi: 10.1021/es202507f.

Lui, W. Y., C. H. Wong, D. D. Mruk, and C. Y. Cheng. 2003. "TGF-beta3 regulates the blood-testis barrier dynamics via the p38 mitogen activated protein (MAP) kinase pathway: an in vivo study." *Endocrinology* 144 (4):1139–42. doi: 10.1210/en.2002-0211.

Manfo, F. P., R. Jubendradass, E. A. Nantia, P. F. Moundipa, and P. P. Mathur. 2014. "Adverse effects of bisphenol A on male reproductive function." *Rev Environ Contam Toxicol* 228:57–82. doi: 10.1007/978-3-319-01619-1_3.

Manfo, F. P., E. A. Nantia, and P. P. Mathur. 2014. "Effect of environmental contaminants on mammalian testis." *Curr Mol Pharmacol* 7 (2):119–35.

Martini, A. C., A. Tissera, D. Estofán, R. I. Molina, A. Mangeaud, M. Fiol De Cuneo, and R. D. Ruiz. 2010. "Overweight and seminal quality: a study of 794 patients." *Fertil Steril* 94:1739–43. doi: 10.1016/j.fertnstert.2009.11.017.

Meena, R., K. Kajal, and P. R. 2015. "Cytotoxic and genotoxic effects of titanium dioxide nanoparticles in testicular cells of male wistar rat." *Appl Biochem Biotechnol* 175 (2):825–40. doi: 10.1007/s12010-014-1299-y.

Mehrpour, O., P. Karrari, N. Zamani , A. M. Tsatsakis, and M. Abdollahi. 2014. "Occupational exposure to pesticides and consequences on male semen and fertility: a review." *Toxicol Lett* 230 (2):146–56. doi: 10.1016/j.toxlet.2014.01.029.

Meli, R., A. Monnolo, C. Annunziata, C. Pirozzi, and M. C. Ferrante. 2020. "Oxidative stress and BPA toxicity: an antioxidant approach for male and female reproductive dysfunction." *Antioxidants (Basel)* 9 (5):405. doi: 10.3390/antiox9050405.

Merhi, Z.O. 2012. Challenging cell phone impact on reproduction: a review. *J Assist Reprod Genet* 29:293–7.

Mirmira, P., and C. Evans-Molina. 2014. "Bisphenol A, obesity, and type 2 diabetes mellitus: genuine concern or unnecessary preoccupation?" *Transl Res* 164 (*1*):13–21. doi: 10.1016/j.trsl.2014.03.003.

Mok-Lin, E., S. Ehrlich, P. L. Williams, J. Petrozza, D. L. Wright, A. M. Calafat, X. Ye, and R. Hauser. 2010. "Urinary bisphenol A concentrations and ovarian response among women undergoing IVF." *Int J Androl* 33 (*2*):385–93. doi: 10.1111/j.1365-2605.2009.01014.x.

Pacey, A. 2010. "Environmental and lifestyle factors associated with sperm DNA damage." *Hum Fertil (Camb.)* 13:189–93. doi: 10.3109/14647273.2010.531883.

Peretz, J., L. Vrooman, W. A. Ricke, P. A. Hunt, S. Ehrlich, R. Hauser, V. Padmanabhan, H. S. Taylor, S. H. Swan, C. A. VandeVoort, and J. A. Flaws. 2014. "Bisphenol A and reproductive health: update of experimental and human evidence, 2007–2013." *Environ Health Perspect* 122 (*8*):775–86. doi: 10.1289/ehp.1307728.

Polyzos, S. A., J. Kountouras, G. Deretzi, C. Zavos, and C. S. Mantzoros. 2012. "The emerging role of endocrine disruptors in pathogenesis of insulin resistance: a concept implicating nonalcoholic fatty liver disease." *Curr Mol Med* 12 (1):68–82. doi: 10.2174/156652412798376161.

Poole, A., P. van Herwijnen, H. Weideli, M. C. Thomas, G. Ransbotyn, and C. Vance. 2004. "Review of the toxicology, human exposure and safety assessment for bisphenol A diglycidylether (BADGE)." *Food Addit Contam* 21 (9):905–19. doi: 10.1080/02652030400007294.

Qian, W., J. Zhu, C. Mao, J. Liu, Y. Wang, Q. Wang, Y. Liu, R. Gao, H. Xiao, and J. Wang. 2014. "Involvement of CaM-CaMKII-ERK in bisphenol A-induced Sertoli cell apoptosis." *Toxicology* 324:27–34. doi: 10.1016/j.tox.2014.06.001.

Ragni, G., L. De Lauretis, V. Gambaro, R. Di Pietro, O. Bestetti, F. Recalcati, and C. Papetti. 1985. "Semen evaluation in heroin and methadone addicts." *Acta Eur Fertil* 16 (4):245–9.

Ranciere, F., J. G. Lyons, V. H. Loh, J. Botton, T. Galloway, T. Wang, J. E. Shaw, and D. J. Magliano. 2015. "Bisphenol A and the risk of cardiometabolic disorders: a systematic review with meta-analysis of the epidemiological evidence." *Environ Health* 14:46. doi: 10.1186/s12940-015-0036-5.

Rignell-Hydbom, A., A. Axmon, T. Lundh, B. A. Jönsson, T. Tiido, and M. Spano. 2007. "Dietary exposure to methyl mercury and PCB and the associations with semen parameters among Swedish fishermen." *Environ Health* 6 (1):14. doi: 10.1186/1476-069x-6-14.

Salas-Huetos, A., M. Bulló, and J. Salas-Salvadó. 2017. "Dietary patterns, foods and nutrients in male fertility parameters and fecundability: a systematic review of observational studies." *Hum Reprod Update* 23:1–19. doi: 10.1093/humupd/dmx006.

Salian, S., T. Doshi, and G. Vanage. 2009. "Neonatal exposure of male rats to Bisphenol A impairs fertility and expression of Sertoli cell junctional proteins in the testis." *Toxicology* 265 (1–2):56–67. doi: 10.1016/j.tox.2009.09.012.

Samplaski, M. K., and A. K. Nangia. 2015. "Adverse effects of common medications on male fertility." *Nat. Rev. Urol.* 12:401–13. doi: 10.1038/nrurol.2015.145.

Saradha, B., S. Vaithinathan, and P. P. Mathur. 2009. "Lindane induces testicular apoptosis in adult Wistar rats through the involvement of Fas-FasL and mitochondria-dependent pathways." *Toxicology* 255 (3):131–9. doi: 10.1016/j.tox.2008.10.016.

Shi, Y. Q., Y. P. Wang, Y. Song, H. W. Li, C. J. Liu, Z. G. Wu, and K. D. Yang. 2010. "p,p′-DDE induces testicular apoptosis in prepubertal rats via the Fas/FasL pathway." *Toxicol Lett* 193 (1):79–85. doi: 10.1016/j.toxlet.2009.12.008.

Siu, E. R., D. D. Mruk, C. S. Porto, and C. Y. Cheng. 2009. "Cadmium-induced testicular injury." *Toxicol Appl Pharmacol* 238 (3):240–9. doi: 10.1016/j.taap.2009.01.028.

Siu, E. R., E. W. Wong, D. D. Mruk, K. L. Sze, C. S. Porto, and C. Y. Cheng. 2009. "An occludin-focal adhesion kinase protein complex at the blood-testis barrier: a study using the cadmium model." *Endocrinology* 150 (7):3336–44. doi: 10.1210/en.2008-1741.

Song, Y., Y. Shi, H. Yu, Y. Hu, Y. Wang, and K. Yang. 2011. "p,p′-Dichlorodiphenoxydichloroethylene induced apoptosis of Sertoli cells through oxidative stress-mediated p38 MAPK and mitochondrial pathway." *Toxicol Lett* 202 (1):55–60. doi: 10.1016/j.toxlet.2011.01.020.

Tsai, W. T. 2006. "Human health risk on environmental exposure to Bisphenol-A: a review." *J Environ Sci Health C Environ Carcinog Ecotoxicol Rev* 24 (2):225–55. doi: 10.1080/10590500600936482.

Ulstein, M., N. Netto, J. Leonard, and C. A. Paulsen. 1975. "Changes in sperm morphology in normal men treated with danazol and testosterone." *Contraception* 12:437–44. doi: 10.1016/0010-7824(75)90079-7.

Van Landuyt, K. L., T. Nawrot, B. Geebelen, J. De Munck, J. Snauwaert, K. Yoshihara, H. Scheers, L. Godderis, P. Hoet, and B. Van Meerbeek. 2011. "How much do resin-based dental materials release? A meta-analytical approach." *Dent Mater* 27 (8):723–47. doi: 10.1016/j.dental.2011.05.001.

Veeramachaneni, D. N., and G. R. Klinefelter. 2014. "Phthalate-induced pathology in the foetal testis involves more than decreased testosterone production." *Reproduction* 147 (4):435–42. doi: 10.1530/REP-13-0441.

Vengosh, A., R. B. Jackson, N. Warner, T. H. Darrah, and A. Kondash. 2014. "A critical review of the risks to water resources from unconventional shale gas development and hydraulic fracturing in the United States." *Environ Sci Technol* 48 (15):8334–48. doi: 10.1021/es405118y.

Webb, E., S. Bushkin-Bedient, A. Cheng, C. D. Kassotis, V. Balise, and S. C. Nagel. 2014. "Developmental and reproductive effects of chemicals associated with unconventional oil and natural gas operations." *Rev Environ Health* 29 (4):307–18. doi: 10.1515/reveh-2014-0057.

Winters, S. J., C. Wang, E. Abdelrahaman, V. Hadeed, M. A. Dyky, and A. Brufsky. 2006. "Inhibin-B levels in healthy young adult men and prepubertal boys: is obesity the cause for the contemporary decline in sperm count because of fewer Sertoli cells?" *Journal of Andrology* 27:560–4. doi: 10.2164/jandrol.05193.

Witorsch, R. J. 2002. "Endocrine disruptors: can biological effects and environmental risks be predicted?" *Regul Toxicol Pharmacol* 36 (1):118–30.

Xiao, X., D. D. Mruk, E. I. Tang, C. K. Wong, W. M. Lee, C. M. John, P. J. Turek, B. Silvestrini, and C. Y. Cheng. 2014. "Environmental toxicants perturb human Sertoli cell adhesive function via changes in F-actin organization mediated by actin regulatory proteins." *Hum Reprod* 29 (6):1279–91. doi: 10.1093/humrep/deu011.

Zhang, H., H. Wang, Y. L. Ji, H. Ning, T. Yu, C. Zhang, Y. Zhang, X. F. Zhao, Q. Wang, P. Liu, X. H. Meng, and D. X. Xu. 2009. "Lactational fenvalerate exposure permanently impairs testicular development and spermatogenesis in mice." *Toxicol Lett* 191 (1):47–56. doi: 10.1016/j.toxlet.2009.08.007.

24 Bioaerosol Health Effects from Molecular to Global Scales

Sneha Gautam
Karunya Institute of Technology and Sciences Coimbatore, Tamil Nadu,

Alok Sagar Gautam
Department of Physics, Hemvati Nandan Bahuguna Garhwal University, Uttarakhand

Anjani R.K. Gollakota and Chi-Min Shu
National Yunlin University of Science and Technology, Douliou, Yunlin County, Taiwan

CONTENTS

24.1 INTRODUCTION

According to the World Health Organization (WHO), 3 million deaths per year are caused by strokes, heart disease, chronic respiratory disease, and lung cancer due to ambient air pollution (WHO, 2018). Moreover, around 91% of low- and middle-income countries (i.e., western Pacific regions and Southeast Asian countries) experience the burden of outdoor air pollution. Air pollution occurs due to natural and anthropogenic activities observed at local, regional, and global scales. Substances (particles/gaseous) present in the atmosphere in a concentration which directly or indirectly harms air quality or affects human health are called air pollutants. Higher concentrations of pollutants threaten

DOI: 10.1201/9781003220824-28

human health based on several diseases (i.e., cardiovascular, respiratory, allergic, and lung cancer) (Humbal et al. 2019; Shiraiwa et al., 2017). The fine fraction of particulate matter (PM) determines its potential for causing adverse effects on human health (Gautam and Trivedi 2020; Shiraiwa et al., 2017). Many studies have been conducted to understand the interrelationship between the different particles (PM_{10}, $PM_{2.5}$, PM_1, and lower) and health issues (Shiraiwa et al., 2017).

Fine particles with a lower diameter ($PM_{2.5}$) infiltrate deep into the inner parts of the lungs and cause several respiratory diseases (Shiraiwa et al., 2017). Epidemiological studies describe the connection between atmospheric pollutants and adverse human health effects. In this view, they estimated that approximately 3.3 million humans died because of air pollution in 2010. This scenario has been updated recently to about 4.3 million per year (Shiraiwa et al., 2017). The major risk factors for the burden of diseases worldwide were thought to be indoor and ambient air pollution by pollutants (i.e., aerosols and bioaerosols).

Studies on the effect of bioaerosols on human health are interdisciplinary, highly versatile, and complex, and address a wide range of length and time scales. The health impacts of bioaerosols can be addressed on a short-term and a long-term scale based on regional and global studies (Shiraiwa et al., 2017). Epidemiological and toxicological studies indicate inflammatory, respiratory, and pathogenic effects of bioaerosol exposure and suggest links between bioaerosols and sick building syndromes (Burge, 1990). The assessment of exposure is dependent on the diversity of biotic matter present in the atmosphere (Gautam and Luc 2022; Humbal et al., 2018), which may vary in toxicity.

24.2 REVIEW OF LITERATURE

24.2.1 Epidemiological Literature

According to the US Environmental Protection Agency (EPA), epidemiological studies show a connection between ambient PM concentrations and health effects (U.S. Environmental Protection Agency, 1997). Epidemiological studies emphasize short-term and long-term chronic and acute health outcomes, respectively, depending on the duration of exposure. A few epidemiological strategies (i.e., time-series analyses, ecological studies, semi-ecological studies, prospective cohort studies, and case–control and crossover studies) are generally used to measure fine particles in applications regarding bioaerosol measurement and its impact on human health. Ecological studies are based on responses at the community level with exposure indices and covariates (i.e., annual mortality rates, average yearly bioaerosol concentration, percentage of population greater than 65 years of age). The data are calculated among different communities, which may not reflect individual-level associations between human health outcomes and exposure or the relationship between health effects and exposure.

Moreover, time-series studies are based on changes in information between health outcomes and exposure (i.e., the number of non-accidental deaths in the city and air pollution concentration daily). Prospective cohort studies use different results from an individual, including exposure, health status, and duration (individual), covariates, or risk factors observed over time. Similarly, case–control studies help determine the outcomes of exposure based on cases (a group known to have the outcomes) and the control (a group known to be free of the outcomes). These studies help investigate the exposures with the help of outcomes. Moreover, possible advantages of these studies are that they (I) can find outcomes quickly; (II) can often be undertaken with minimal funding; (III) are efficient for rare diseases; and (IV) study multiple exposures and generally require few study subjects. Intervention studies determine the possible interrelation between ambient air pollution variables (i.e., bioaerosols) and their effect on human health.

Earlier studies focused on short-term and acute effects (i.e., bronchitis and pneumonia; eye, nose, and throat irritation; allergic reactions, asthma, and emphysema, etc.) compared with the daily basis change in the air quality and its effect on humans. It has been suggested that particles with smaller diameter are more hazardous than particles with larger diameter because they directly penetrate deeply into the lungs and are deposited there (Shiraiwa et al., 2017). Many cohort studies have been done in Europe, North America, and Asian countries to understand $PM_{2.5}$ and related diseases (Bentayeb et al., 2015). Some relations have been determined based on these studies, such as the interrelation between global mortality risk and long-term exposure to fine particulate matter mass (≤2.5 μm; $PM_{2.5}$) and also the relationship between long-term exposure to $PM_{2.5}$ and the chance of mortality in terms of cardiovascular disease (CVD). In this view, it can be reported that long-term effects due to $PM_{2.5}$ are responsible for CVD and total non-accidental mortality globally (Thurston et al., 2016). With the reporting of these cohort studies, one can estimate that the consequences of $PM_{2.5}$ are greater than those of PM_{10}, and similarly, it is also observed that bioaerosols can be associated with $PM_{2.5}$. Therefore, cohort studies need to be carried out on the possible effect of bioaerosols on human health, particularly in developing countries.

24.2.2 Assessing Health Effects Based on Epidemiological Studies

With the help of epidemiological studies, the relative risk for diseases caused by air pollution is estimated by integrating the exposure–response function, which is part of the assessment of global BOD (burden of diseases) (WHO, 2018). Earlier, cohort studies were based on identifying IERs for PM, but no such studies were carried out to identify IERs for bioaerosols (Shiraiwa et al., 2017). To estimate the relative risk for a disease caused by air pollution exposure to biotic compounds, an integrated exposure-response function (IER) can be used as it is already used for PM (WHO, 2018). Previous cohort studies were based on the relationship between mortality and long-term exposure to PM from different locations of Europe and North America; studies can be carried out on the basis of the relationship between mortality and long-term exposure to bioaerosols in other places, including the western part of Asia. The annual mean concentration of $PM_{2.5}$ for study areas like North America and Europe is typically 30 μgm^{-3}, and there have been minimal studies from other regions, including most developing countries, on high concentration levels of bioaerosols. Biotic materials include various constituents, like toxic compounds, chemicals, and different living and dead microorganisms (Humbal et al., 2018). Previously, studies were focused only on human health due to PM concentration from anthropogenic sources; with the help of this study, the scope can be widened so that the effect of bioaerosols on human health can be determined. Moreover, identifying the different biotic components, which generally play a critical role from the perspective of health risk assessment, is a widely multidisciplinary area and a new concept. In addition, a few studies have found that short- and long-term exposure to particular constituents such as sulfate, organic carbon, nitrate, and elemental carbon is associated with mortality.

A significant part of atmospheric carbonaceous aerosols consists of biotic matter such as fungi, bacteria, pollens, viruses, and their fragments. Studies based on pollen grains, fungal spores, and plant residues show that they make up 11–15% of PM_{10}. The contribution of organic carbon (OC) mass is 24–33% in an agricultural town in California (Chow et al., 2015). This makes it difficult to assess the health effects of individual constituents due to the high correlation between their respective components. A few studies have also reported that the chemical composition of inhaled pollution (i.e., transition metals, organic compounds, endotoxins, etc.) directly affects the cardiovascular risks to a human (Limaye et al., 2018). Furthermore, Adhikari et al. (2005) reported that 10–15% of mortality is due to non-accidental causes, while a 10 $\mu g/m^3$ increase in $PM_{2.5}$ causes a 30% increase in mortality.

24.2.3 Estimation of Bioaerosol Characteristics for Epidemiological Studies

Air quality monitoring and source–exposure analyses have been used to set up a monitoring station in this context, and the supporting literature (e.g., epidemiological research) has been used to support the findings. Particle sizes with varied chemical contents were once thought to be solely responsible for effects on human health (Shiraiwa et al., 2017). Mass studies also benefit from physical and statistical separation, including size distribution, chemical specificity, temporal duration of exposure, and source apportionment (Shiraiwa et al., 2017). Several investigations have been done to explore the trace makeup of the PM components and their sources (Shiraiwa et al., 2017). Traces of PM were regularly found in the monitoring premises by using X-ray fluorescence to detect PM and thermal and optical analyses to estimate the concentrations of organic and elemental carbon. Furthermore, Sillanpaa et al. (2008) conducted in vitro experiments to identify the biotic matter/fragments linked to PM. Different countries have established the gravimetric mass of PM in accordance with their regional or local regulatory restrictions (Table 24.1). A previous study (Boldo et al. 2006) in 23 European cities found that an increase in $PM_{2.5}$ levels could lead to a decrease in life expectancy and an increase in the number of deaths. Epidemiological surveillance indicators were defined during the first phase of APHEIS 1, which ran from 1999 to 2000. They came up with a set of recommendations for how air pollution affects the health of European citizens.

Moreover, PM_{10} and black smoke (BS) were selected as indicators that reflect the presence and severity of PM in urban Europe, where it has the greatest effect on public health, during the second phase of APHEIS 2 (2000–2001). Data from APHEIS 3's third phase (2002–2003) and its

TABLE 24.1
Standards for PM in Selected Countries

Country	PM Size	Concentration in Ambient Air (μg m^{-3})	Time-Weighted Average	Reference
India	PM_{10}	100	24 h	CPCB (2009)
	$PM_{2.5}$	60		
	PM_{10}	60	Annual	
	$PM_{2.5}$	40		
WHO	PM_{10}	50	24 h	WHO (2014)
	$PM_{2.5}$	25		
	PM_{10}	20	Annual	
	$PM_{2.5}$	10		
US	PM_{10}	150	24 h	NAAQS (2011)
	$PM_{2.5}$	35		
	PM_{10}	–	Annual	
	$PM_{2.5}$	12		
European Union	PM_{10}	50	24 h	EU (2005)
	$PM_{2.5}$	–		
	PM_{10}	40	Annual	
	$PM_{2.5}$	25		
Canada	PM_{10}	–	24 h	CAAQS (2013)
	$PM_{2.5}$	28		
	PM_{10}	–	Annual	
	$PM_{2.5}$	10		
Australia	PM_{10}	50	24 h	Air NEPM (2002)
	$PM_{2.5}$	08		
	PM_{10}	–	Annual	
	$PM_{2.5}$	25		
South Africa	PM_{10}	75	24 h	SAAQIS (2009)
	$PM_{2.5}$	65		
	PM_{10}	40	Annual	
	$PM_{2.5}$	25		

Health Impact Assessment (HIA) were reported. They concluded that $PM_{2.5}$ was connected with an increased mortality rate, and they looked into the specific and total causes of this mortality rate increase. Unidentified or hazardous elements can be discovered using land-use regression models and source apportionment methods (Boldo et al., 2006). The exposure limits are determined by the dose, as well as the duration and level of exposure, at which adverse effects no longer appear (NOVEL). The risk assessment of bioaerosol exposure on outdoor and indoor air quality frequently suffers from fundamental methodological issues.

Furthermore, exposure limits from various studies have not been specified or developed, since the NOVEL or LOAEL for bioaerosols has not yet been identified (Walser et al., 2015). Only a few countries have taken this step, with Germany being the best example, according to this view. It focuses on the concentration of indicator characteristics that occur in lower concentrations and are therefore difficult to measure (Walser et al., 2015). An all-encompassing strategy is important because of the absence of studies on environmental or health-related issues (Shiraiwa et al., 2017). Combining precise measurement with low-cost sensors, monitoring mobile platforms, and personal monitors may necessitate the creation of many micro-environments. It's possible to use less well-known concepts by combining several ideas into a single one (i.e., combing micro-balanced mass measurements with optical size categorization and electrostatics to construct from the resulting data set). Surveys based on approaches and central monitoring that are implemented in a biased way can reduce information if they combine traditional/novel exposure quantification by measuring health, activity, and critical micro-environment components instead (Shiraiwa et al., 2017). While satellite-derived $PM_{2.5}$ has certain limitations in terms of spatial resolution, computer models and remote sensing models are being used to refine the precision concentration field using data gathered from the ground. However, thanks to space–time regression models, information on $PM_{2.5}$ levels can now be reported with a high degree of accuracy and precision.

Bioaerosols can cause allergies and respiratory illnesses in many people, making them an extra health hazard (Walser et al., 2015). As a result of environmental epidemiology investigations, exposure to bioaerosols can be correlated with environmental risk factors. Risk assessment for bioaerosol exposure in indoor and outdoor air is hampered by a lack of core methodological concerns. Bioaerosol concentration must be determined in order to calculate or define an acceptable exposure limit. Depending on the type of exposure, duration of exposure, weather conditions, etc., the concentration may vary (Walser et al., 2015).

24.2.4 Bioaerosol Health Effects Based on Modeling Perspective

According to prior epidemiological research, there is a direct link between the concentration of air pollutants and the health effects they may have, both in the short term and over the long run (Shiraiwa et al., 2017). In order to better understand how pollutants in the air are linked to health problems, modeling is an essential tool (Van Leuken et al., 2015). The features of bioaerosols vary depending on the source and how they can be effectively controlled and mitigated based on their distinct qualities. The properties and concentration of bioaerosols necessitate diverse methods for collecting, identifying, and quantifying them. Bioaerosol concentrations tend to fluctuate widely due to the fact that their sources do not always produce at the same rate. In order to determine the health risks associated with exposure to any air pollution, the method uses health risk assessment. In both outdoor and indoor situations, bioaerosol exposure risk assessment suffers from a number of fundamental methodological issues. Assessment of exposure to causative agents is critical based on related health outcomes. Bioaerosol concentration in occupied rooms or spaces is thought to be a function of a person's personal exposure to it (Van Leuken et al., 2015). Studies on personal exposure (Frohlich-Nowoisky et al., 2016) indicated that respiratory exposure plays an essential role in the transfer of bioaerosol to the human lungs connected with the airborne bioaerosol (i.e., allergies, pollen, microbes, etc.). Respiratory conditions, pulmonary inflammation, allergies, and other health problems are all linked to long-term exposure to bioaerosol (Humbal et al., 2018). Depending on the concentration of $PM_{2.5}$, the concentration of bioaerosol may be affected by the amount of biotic matter that can adhere and create bioaerosol. As a result, the WHO has established regulations for the annual concentration of $PM_{2.5}$ in the air. Air pollution concentrations and health effects must be modeled in a different way. The exact concentration and its composition must be identified and calculated in terms of health outcomes (Van Leuken et al., 2015). Estimating bioaerosol concentrations at the high geographical and temporal resolution of exposure required for assessing potential health concerns is made possible through the use of dispersion models (Williams et al., 2018). Models of air dispersion are commonly used to examine the influence of air pollutants (i.e., bioaerosols) on urban and rural air quality and to address concerns about public health risk associated with exposure from different sources of exposure (i.e., livestock, sources related to biosafety agents, etc.). Additionally, following research has included the addition of turbulence (i.e., fluid dynamics) as well as the basic idea of a Gaussian dispersion equation for the computation of concentrations at a local level (30 km) in a three-dimensional frame (Van Leuken et al., 2015). Models are developed using computational fluid dynamics (CFD) because of the computer's capabilities (i.e., building, trees, etc.). Among the many uses of air pollution measurement and modeling are (1) to determine important physical processes related to atmospheric dispersion and pathogen transport; (2) to focus on the simulation of airborne transmission of pathogenic bioaerosols; and (3) to provide parameterization regarding emission as well as methods for converting concentration to infection probabilities (i.e., concerning quantification microbial risk assessment). Following prior epidemiological cohort research, Anenberg et al. (2010) hypothesized that the mortality rate for a certain community due to exposure to $PM_{2.5}$ is linked to the population's relative risk as well as the harmful effects on humans that this exposure has. Studies on pollutants (such as $PM_{2.5}$) with concentration, characteristics (individual and area-level), and their related adverse health implications are scarce in the field. A third of the world burden is attributed to the developing Asian countries according to this assessment. The relative risk of disease from exposure to $PM_{2.5}$ is estimated using the IER. With the help of epidemiological studies on various sources of pollution (such as outdoor pollution, interior pollution, and house air pollution), IER grows (Hänninen et al., 2014). European countries were assessed by the Environmental Burden of Disease project, which estimated the environmental burden of disease (EBD) attributable to environmental exposure related to European conditions based on their annual health impacts, their environmental risk (i.e., formaldehyde and noise), as well as data availability and applicability for the assessment of environmental exposure. According to this research, $PM_{2.5}$ is responsible for 68% of the total projected EBD. For risk management in public health and resource allocation, EBD assessment studies can help identify vulnerable populations (Hänninen et al., 2014).

24.2.5 Bioaerosols

Bioaerosols are a combination of particulate matter and biotic matter, which may consist of pathogenic and non-pathogenic compounds (i.e., bacteria, fungi, viruses, high-molecular-weight (HMW) allergens, pollen, etc. Bioaerosols play a significant role in distributing infections, allergies, and carcinogenic compounds from the atmosphere to affect human health. They are relevant for the transfer of organisms, the geographic shift of biomass, and genetic exchange between habitats. As single cells or attached to dust particles, microorganisms can survive in bioaerosols, with a size range from micrometer to sub-micrometer (Samake et al., 2017). According to a few studies, coarse-sized particles with diameters more than 1 micrometer (micron) account for 30% to 80% of airborne particles in rainforests (Frohlich-Nowoisky et al., 2016). At Amazonian locations, the percentage of total PM mass in the form of bioaerosols was found to be between 47% and 80%; a few studies reported this value to be between 15% and 25% of total PM mass (Samake et al., 2017). The number and mass concentration of primary bioaerosol over the vegetated area has been

TABLE 24.2
Modeling Perspective on Bioaerosol Health Effects

Author	Equation	Parameter	Remark
Anenberg et al. (2010)	$DM = Yo \times AF \times POP$	AF = Attributable fraction of mortality YO = Baseline mortality rate (BMR) POP = Population within the age range of interest	Defining IER function based on epidemiological cohort studies. Similarly applied by Global Burden of Diseases.
	$AF = (RR - 1) \div RR$	AF = Attributable fraction of mortality RR = Relative risk	
Ezzati et al. (2004)	$RR = [(X+1) / (Xo+1)]^{B}$	X = Concentration of $PM_{2.5}$ XO = Threshold where pollutant is considered as not being harmful	
Burnett et al. (2014)	$RR = 1 + \alpha\left[1 - \exp^{[-\gamma(Xo - Xo)^{\delta}]}\right]$	α, ɣ, δ = unknown parameters	Non-linear regression method based on a collection of epidemiological studies.
Cohen et al. (2005)	$AB = PAF \times B$	PAF = Population attributable fraction AB = Attributable mortality B = Burden of disease.	Defined as a proportional reduction in disease or death that would occur if exposure to risk factors were reduced to zero.
Cole and MacMahon (1971)	$PAF = P(RR - 1) / P(RR - 1) + 1$	RR = Relative risk PAF = Population attributable fraction	
Hänninen et al. (2014)	$EBD = PAF \times BD$	PAF = Population attributable fraction BD = Environmental burden of disease D = Burden of disease	Studies aimed to provide harmonized EBD assessment for countries. For exposure–outcome pairs with a relative risk estimate.
Rockhill et al. (1998)	$PAF = [p \times (RR - 1)] / p(RR - 1) + 1$	p = Proportion of population exposed RR = Relative risk at level of exposure	Used to estimate PAF for exposure–outcome pairs without relative risk estimate available.
	$AC = E \times UR \times P$	AC = Attributable cases E = Exposure level UR = Unit Risk P = Size of exposed population	
	$PAF = AC / I$	AC = Attributable cases I = Total incidence PAF = population attributable fraction	
	$EBD = AC \times DW \times L$	EBD = Environmental burden of diseases L = Average number of years lived with disability for morbidity effect or years of life lost for mortality AC = Attributable cases DW = Disability weight characterizing the severity of diseases, ranging from 0 to 1	
Limaye et al., 2018	$Y = Yo \times e^{-\beta \times PM}$	Yo = baseline mortality rates Y = number of fatalities in a time period β = Relative risk factor	Where PM in μg m^{-3}. Used to estimate the relative risk with exposure to $PM_{2.5}$ and its long-term health effect.
	$RR_{LOGLINEAR}(PM) = \exp^{(\beta \times PM)}$	β = Relative risk factor	With this model, the annual baseline death rate with change as a function of $PM_{2.5}$ exposure and its related relative risk.
	$R = \log(RR_1 - RR_2 + 1) / \Delta PM$	RR = Point estimates of relative risk for specific causes of death for scenarios considered in pairs	It is used for calculating the incremental risk of $PM_{2.5}$ exposure.

estimated to be 104 m^{-3} and 1 g m^{-3}, respectively, by some research (Bauer et al., 2002). Total worldwide bioaerosol mass emission ranges from 10 to 1000 Tg a^{-1}, although this range is not consistent (Frohlich-Nowoisky et al., 2016). There are three main categories of disorders linked to bioaerosol exposure: respiratory diseases (such as asthma), cancer, and infections (Douwes et al., 2003). Aerosol particles that originate in the atmosphere are known as primary biological aerosols (PBA) (Frohlich-Nowoisky et al., 2016). Pathogenic biotic matter is diffused in air, travels with the wind direction, and spreads disease from local to global scales. Bioaerosol concentrations in indoor/outdoor environments ranged from 101 to 105 microorganisms per cubic meter of air, whereas those in rainforests or polluted environments were reported to be 105 to 107 microorganisms per cubic meter of air (Samake et al., 2017). According to a few studies, a single children's classroom contains between 3 and 57 million fungal cells per person per hour, with an emission rate of 0.8 to 35 million bacterium cells per person per hour (Hospodsky et al., 2015). Particles emitted by human bodies and re-suspended from the surface (i.e., indoors and outdoors) are related by two broad routes (Hospodsky et al., 2015). A few studies reported that inhalation of high concentrations of bacterial cells or fungal spores is directly associated with inflammation-related health risk (i.e., chronic obstructive pulmonary disease (COPD), asthma, immunological reactions, etc.) (Samake et al., 2017). As mentioned previously regarding how particle size matters when assessing the health effect due to exposure to particulates, small particles are more dangerous than larger particles, because small particles can easily penetrate and be deposited in various segments of the body parts (i.e., lungs and circulatory system). This causes different adverse health effects that may create complications ranging from damage to a single organ damage to the entire organ system and the body (i.e., infectious diseases, respiratory diseases, cancer, etc.) (Humbal et al., 2018). Exposure to viruses, bacteria, fungi, protozoa, and helminths causes infectious diseases. Several diseases are caused by exposure to fungal spores (i.e., aspergillosis, histoplasmosis, blastomycosis, etc.) (Douwes et al., 2003). Respiratory disease associated with dust can range from acute to chronic health effects. Toxins, inflammatory chemicals, or allergens can produce airway inflammation, which causes these respiratory symptoms. Allergies and non-allergic effects on the airways can be separated into categories. Viability and vitality are among the few bioaerosol assessments that are critical in the biological, ecological and pathological fields of transformation and adaptation, where viability plays a key role. When looking for bioaerosol in the atmosphere, scientists typically look for it in the planetary boundary layer (PBL). The transfer of bioaerosols is critical to the study of the spread of microorganisms, their evolution, and the development of new species/communities. In the context of atmospheric chemistry and physics, bioaerosols play a vital role in cloud droplets, ice crystals, precipitation, and the hydrological cycle. Toxic chemicals are also transported via bioaerosols (such as cyanobacterial neurotoxins and endotoxins, bacterial cell walls, and fungal secondary metabolites). Acute and long-term health effects are caused by toxins (metabolites of fungi) in this way, whereas endotoxins are responsible for significant inflammation that adversely affects human health. Toxins from cyanobacteria, such as neurotoxins and hepatotoxins, have a wide range of harmful effects on animals, both wild and domestic, and humans. A bioaerosol is a mixture of microorganisms from many sources, each of which represents unique environmental conditions. Consequently, the selection of sample methods for a bioaerosol should be based on the features and composition of the material. Samples of bioaerosol can be taken in numerous ways, (e.g., using cyclones, impactors, filters, spore traps, electrostatic or thermal precipitation, etc.) (Table 24.3).

24.2.5.1 Components of Bioaerosols

As mentioned earlier, bioaerosols consist of living microorganisms and dead cells with their components (i.e., fungi, bacteria, endotoxins, allergens, etc.). They can be commonly found in outdoor as well in indoor air.

TABLE 24.3
Sampling Techniques for Bioaerosols

Author	Techniques	Key Observations
Frohlich-Nowoisky et al., 2016	Filters Impactor Impinger Cyclone Samplers	Used for sampling and measurement of bioaerosol.
Baron and Wrilleke, 2001	Immunological Methods Light and Electron Microscopy Optical Spectroscopy Mass Spectroscopy	Used for cultivation and staining of bioaerosol. Also used to trace the chemicals present in bioaerosol.
Despres et al., 2012	Fluorescence Detection Spectroscopy Microscopy X-ray Microscopy Spectroscopy Single Particle Mass Spectroscopy	Used to analyze the presence of ribonucleic acids (DNA/RNA).
	Cultivation Techniques	Used to investigate the concentration of microorganisms present in samples as well as in the environment
Boreson et al., 2004	DNA- and RNA- based methods	Provide information about the entire spectrum of atmospheric microbial diversity.
	Traditional Sanger sequencing approach	Provides sequences that are long enough to identify individual genera or species by comparison with sequences available in online databases.
Whon et al., 2012	Metagenomic	Enable comprehensive determination of diversity and metabolic potential of organism present in aerosol sample. Also characterization of airborne viral diversity and dynamics.

24.2.5.1.1 Fungi and Bacteria

Mycotoxins are present in a wide variety of organic materials, including food waste and trash. Allergic reactions, enzymatic proteins, and poisons are all possible outcomes of this process. There is a strong correlation between weather conditions and the growth of bacteria and fungi, according to much research (Kang et al., 2015). As a result of this, a few studies have indicated a significant seasonal variation in the bioaerosol level (Thorne et al., 1992). Indoor and outdoor locations may have differing concentrations of the substances. According to McGill et al. (2015), bacterial and fungal concentrations in six contemporary bedrooms were between 3–59 and 19–607 and 3–59 colony-forming units (CFU)/m^3 (indoors). A wastewater treatment plant in Tehran, Iran, recorded an average bacterial bioaerosol concentration of 1016 CFU in the winter and 1973 CFU in the summer (Thorne et al., 1992).

24.2.5.1.2 Endotoxins

Gram-negative bacterial cell walls contain a large substance called endotoxin, which is made up of lipid and polysaccharide components (Armstrong et al., 2013). The toxic effects are caused by diverse combinations of the lipid component, o specific polysaccharide side chains (o-antigen), and core polysaccharide chain (lipid A) (Tirsoaga et al., 2007). Endotoxins are constantly inhaled because they can be attached to dust particles that can be sucked in easily. Toxic syndrome and occupational lung illness are exacerbated by the presence of this chemical (Park et al., 2015). Lung function declines as a result of endotoxin compound exposure (Kharitonov and Sjöbring, 2007). Endotoxin exposure can produce a wide range of health problems (such as fever and shivering, leukocytosis in the blood, arthralgia, dyspnea, etc.) (Thilsing et al., 2015).

24.2.5.1.3 *β-Glucans*

These are generally found in the cell walls of various microorganisms. Dong and Yao (2010) reported that the concentration of (1,3)-β-D-glucans in Beijing, China in 2010 ranged from 0.02 to 1.2 ng m^{-3}. Yang et al. (2013) conducted a study on concentrated animal feeding (18 commercial) operations and found concentrations of about 2.4 to 338 ng m^{-3} in Illinois, USA.

24.2.5.1.4 *Mycotoxins*

These are secondary biomolecules that may be produced by several species of fungi and molds. The classification of mycotoxins mainly depends on the chemical structures with their reactive functional groups. Due to exposure to mycotoxins, humans are affected by weakening immune systems, allergies or irritation, and even death. The effect of mycotoxins depends on many parameters, such as exposure period, concentration level, and the conditions of exposure of individual humans based on their age, health, and sex, generally via ingestion, inhalation of spores, and dermal contact.

24.2.5.1.5 *Allergens*

When a bioaerosol is ingested by humans on a local, regional, or global scale, the bioaerosol has a harmful influence on human health. The air is filled with pathogens. Many long-distance journeys can lead to illness (Gollakota et al. 2021). According to a number of studies, fungi can lead to a variety of health issues (i.e., Aspergillus, Cryptococcus, Pneumocystis spp., etc.). Infectious disease and a death rate of up to 95% can also result (Brown et al., 2012). Health care facilities, such as hospitals and clinics, have long-term issues with infection control due to a wide range of airborne diseases. The type of environment has an important influence on infection management despite concerns about health. There are a number of factors that affect the concentration of bioaerosol in the air. Much research has focused on the emission of particles while some human activities (i.e., inhalation, breathing, coughing, swallowing, motions, and re-suspension of dust) are taking place. Consequently (Adems et al., 2015), the inhalation and deposition of bioaerosol in many sections of the body (i.e., lungs, respiratory system, etc.) causes an allergic or poisonous response in people or animals; the inhalation depends on multiple aspects (i.e., airway morphology, breathing characteristics, and particle qualities) (Hussain et al., 2011). If the particle is not living, it can still have detrimental impacts on human health (i.e., endotoxins, exotoxins, mycotoxins, etc.). A number of studies have also found that respiratory and allergic disorders pose a significant problem in several nations (D'Amato et al., 2007). Pollen allergies impact around 40% of the population in industrialized areas and have become a global problem (Shairaiwa et al., 2012a). Also, 30% of people were afflicted by exposure to one or more fungal allergens (Esch et al., 2001). Allergen release from pollen and spore pollination is influenced by a wide range of conditions (i.e., anthropogenic air pollution, thunderstorms, humidity, and so on) (Behrendt and Becker, 2001). One of the most widely used methods for assessing the link between exposure, health outcomes, and risk assessment is disease forecasting, which has been utilized in some research (Aylor, 2003). In addition, the bioaerosol might elicit allergic or harmful reactions in people and animals when inhaled into specific respiratory system locations. Numerous parameters are involved in the deposition of bioaerosols (i.e., particle properties, airway morphology, breathing characteristics, etc.). As an example, sedimentation and impaction mechanisms can be used to gather particles larger than 0.25 micrometers in diameter; diffusion is used to handle particles smaller than 0.25 micrometers.

24.2.6 Potential Health Effects of Bioaerosols

24.2.6.1 Infectious Diseases

Infectious disease is caused by exposure to biological agents by direct and indirect contact (Kim et al., 2017). A total of 75% of the emerging and re-emerging pathogens that are responsible for infectious diseases in animals may be transmitted from animals to humans (zoonotic pathogens). Hill et al. (1998) reported that among all 565 US members of the American Association of Zoo Veterinarians, about 30.2% were infected due to zoonotic infection (Samadi et al., 2013). A few studies have summarized infectious diseases among veterinarians, a population that becomes infected due to the risk of several zoonotic pathogens: Q fever was found in 13.5% of veterinarians in Japan, 12.9% in Sweden, 7.5% in Turkey, 9.5% in Australia, 22% in the USA, and 36% in Slovakia (Samadi et al., 2013).

24.2.6.2 Respiratory Disease

Acute, mild, self-limiting, or severe persistent respiratory problems are the most common bioaerosol side effects. Non-allergic and allergic respiratory indications are primarily separated into two categories based on the inflammatory mechanism. Most people have heard of allergic rhinitis and asthma, both of which can be brought on by inhaling an allergen, such as inorganic dust. Exposure to organic dust has also been associated with COPD. Only a little research has examined the effects of exposure to farm animals on veterinarians' chronic cough, chronic phlegm production, wheezing symptoms, and disease outcomes. Bioaerosol exposure resulted in bronchial hyper-responsiveness and non/atopic asthma in 422 Canadian children (Humbal et al., 2018). Pollen was found to reduce lung function and promote inflammation, according to Baldacci et al. (2015).

24.2.6.3 Communicable Diseases

Airborne viruses and bacteria are the likely culprits behind contagious sickness (Humbal et al., 2018). A few studies have shown that farmers are exposed to germs from veterinary practices, livestock farms, and trapping animals. Birds, particularly pigeons, may be the source of the greatest levels of the bacterium *Chlamydophila psittaci*, according to a few studies (Ling et al., 2015). According to one study, goat farm producers and staff are more vulnerable to infection by microbes (Commandeur et al., 2014).

24.2.7 Reactive Oxygen Species from $PM_{2.5}$

Many effects are strongly dependent on particle surface area, composition, and respective size if its annual average atmospheric concentration is not regulated. Considering PM properties and mass concentration, many communities are working on finding exposure metrics. Health effects like the capacity to transport or form reactive oxygen species (ROS) within lung cells are responsible for oxidative stress and airway inflammation. With the help of antioxidant depletion monitoring, it reveals a cellular assay and biochemical properties of PM in terms of oxidative potential (OP) (Samake et al., 2017). With so little information about the physical and chemical properties of PM, they have been investigated with the help of primary biological airborne particles (PBAP). A few studies mention the fractions of bioaerosols, which shows their negative effect on the climate and humans. The OP of PM is generally related to chemical fraction. For example, overall, 30% of the OP collected in Fresco during the summer was in the form of unidentified components of sub-micron particles of PM (Samake et al., 2017). Along with this, a few studies have provided an initial assessment of the contribution of bioaerosols to the OP of ambient PM.

24.3 CONCLUSIONS

Over the last few years, updated results have been reported from the global to the molecular scale on understanding bioaerosols and their health effects, as mentioned in the previous sections. However, research is still required to know the points based on the complex nature of bioaerosol concerning healthy relationships. Moreover, comprehensive studies are required to add therapies, identify sources, and control bioaerosol pollution. The results observed from the critical analysis indicate that temporal and spatial tasks need to be integrated and synthesized. These results will be an excellent platform for scientists and researchers working on public health planning, policy-making, and regulations to improve air quality. The following points on bioaerosols and their health-related issues are prioritized to develop the evaluation of health effects due to degradation of air quality: (1) mortality due to precise biotic components and their sources could be elucidated by epidemiological studies; (2) cohort studies based on epidemiology are required, especially in Asian and African countries, where studies related to exposure to bioaerosols have not been introduced; (3) developments in the quantification of bioaerosol concentration and their chemical properties concerning source and size are greatly needed.

REFERENCES

Air NEPM, The National Environment Protection Measure for Ambient Air Quality. (2002). National Standards for Criteria Air Pollutants in Australia (www.environment.gov.au/topics/environment-protection/air-quality/air-quality-standards) (last accessed 31 January 2015).

Adhikari A., Reponen T., Grinshpun S.A., Martuzevicius D., and LeMasters G. (2005). Correlation of ambient inhalable bioaerosol with particulate matter and ozone: a two-year study. *Journal of Environmental Pollution* 140:16–28.

Anenberg S.C., Horowitz L.W., Tong D.Q., and West J.J. (2010). An estimate of the global burden of anthropogenic ozone and fine particulate matter on premature human mortality using atmospheric modelling. *Journal of Environmental Health Perspective* 118(9):1189–1195.

Armstrong M.T., Rickles F.R., and Armstrong P.B. (2013). Capture of lipopolysaccharide (endotoxin) by the blood clot: a comparative study. *PLoS One* 8(11):e80192.

Aylor D.E. (2003). Spread of plant diseases on a continental scale: role of aerial dispersal of pathogens. *Ecology* 84(8): 1989–1997.

Baldacci S., Maio S., Cerrai S., Sarno G., Baïz N., Simoni M., Annesi-Maesano I., and Vieg I.G. (2015). Allergy and asthma: effects of the exposure to particulate matter and biological allergens. *Respiratory Medicine* 109(9):1089–1104.

Baron P.A., and Wrilleke K. (2001). Aerosol measurement: principles, techniques and application. *Journal of Aerosol Medicine* 6(4):317–319.

Bauer H., Fuerhacker M., Zibuschka F., Schmid H., and Puxbaum H. (2002). Bacteria and fungi in aerosols generated by two different types of wastewater treatment plants. *Water Research* 36(16):3965–3970.

Behrendt H., and Becker M.W. (2001). Localization, release and bioavailability of pollen allergens: the influence of environmental factors. *Current Opinion in Immunology* 13(6):709–715.

Bentayeb M., Wagner V., Stempfelet M., Zins M., Goldberg M., Pascal M., Larrieu S., Beaudeau P., Cassadou S., Eilstein D., Filleul L., Tertre L.A., Medina S., Pascal L., Prouvost H., Quénel P., Zeghnoun A., and Lefranc A. (2015). Association between long-term exposure to air pollution and mortality in France: a 25-year follow-up study. *Journal of Environment International* 85: 5–14.

Boldo E., Medina S., LeTertre., Hurley F., Mucke G-H., Ballester F., Aguilera I., and Eilstein D. on behalf of the Apheis group. (2006). Apheis: health impact assessment of long-term exposure to PM2.5 in 23 European cities. *European Journal of Epidemiology* 21(6): 449–458.

Boreson J., Dillner M.A., and Peccia J. (2004). Correlating bio aerosol load with PM2.5 and PM10CF concentrations: a comparison between natural desert and urban-fringe aerosols. *Atmospheric Environment* 38(35): 6029–6041.

Brown G.D., Denning D.W., Gow N.A., Levitz S.M., Netea M.G., and White T.C. (2012). Hidden killer: human fungal infections. *Science Translational Medicine* 4(165):1–9.

Burge H. (1990). Bioaerosol: prevalence and health effects in the indoor environment. *Journal of Allergy and Clinical immunology* 86(5):687–701.

Burnett T.R., Pope III A.C., Ezzati M., Olives C., Lim S.S., Mehta S., Shin H.H., Singh G., Hubbell B., Brauer M., Anderson R.H., Smith R.K., Balmes R.J., Bruce G.N., Kan H., Laden F., Prüss-Ustün A., Turner C.M., Gapstur M.S., Diver R.W., and Cohen A. (2014). An integrated risk function for estimating the global burden of disease attributable to ambient fine particulate matter exposure. *Journal of Environment Health Perspective* 122(4):397–403.

CAAQS, Canadian Ambient Air Quality Standard (2013). Canadian Council of Ministers of the Environment (www.ccme.ca/en/current_priorities/air/caaqs.html) (last accessed 1 February 2015).

Chow J.C., Lowenthal D.H., Chen L.W.A., et al. (2015). Mass reconstruction methods for PM2.5: a review. *Air Quality, Atmosphere and Health* 8:243–263 (https://doi.org/10.1007/s11869-015-0338-3)

Cohen J.A., Anderson R.H., Ostra B., Pandey K.D., Krzyzanowski M., Künzli N., Gutschmidt K., Pope A., Romieu I., Samet M.J., and Smith K. (2005). The global burden of disease due to outdoor air pollution. *Journal of Toxicology and Environmental Health, Part A*, 68:1–7.

Cole P., and MacMahon B. (1971). Attributable risk percent in case-control studies. *British Journal of Preventive and Social Medicine*, 25:242–244.

Commandeur, M., Jeurissen, L., van der Hoek, W., Roest, H.J., and Hermans, T.C. (2014). Spatial relationships in the Q fever outbreaks 2007–2010 in the Netherlands. *International Journal of Environmental Health Research* 24(2):137–157.

CPCB (Central Pollution Control Board) (2009) National Ambient Air Quality Standards (NAAQS), Gazette Notification, New Delhi

D'Amato G., Cecchi L., Bonini S., Nunes C., Annesi-Maesano I., Behrendt H., Liccardi G., Popov T., and Cauwenberge V.P. (2007). Allergenic pollen & pollen allergy in Europe. *European Journal of Allergy and Clinical Immunology* 62(9):976–990.

Despres R.V., Huffman A.J., Burrows M.S., Hoose C., Safatov S.A., Buryak G., Frohlich-Nowoisky J., Elbert W., Andreae O.M., Poschl U. and Jaenicke R. (2012). Primary biological aerosol particles in the atmosphere: a review. *Journal Tellus B: Chemical and Physical Meteorology* 64(1):1–58.

Douwes J., Thorne P., Pearce N., Heederik D. et al. (2003). Bio aerosol health effects and exposure assessment: progress and prospects. *The Annals of Occupational Hygiene* 47(3):187–200.

Esch R.E., Hartsell C.J., Crenshaw R., and Jacobson R.S. (2001). Common allergenic pollens, fungi, animals, and arthropods. *Allergy and Immunology* 21(2–3):261–292.

EU (2005). Air Quality standards, European Commission (http://ec.europa.eu/environment/air/quality/standa rds.htm) (last accessed 31 January 2015).

Ezzati M., Lopez A.D., Rodgers A., Vander Hoorn S., and Murray C.J. (2004). Selected major risk factor and global and regional burden of disease. *The Lancet* 360:1347–1360.

Frohlich-Nowoisky J., Kampf J.C., Weber B., Huffman A.J., Pohlker C., Andreae O.M., Lang-Yona N., Burrows M.S., Gunthe S.S., Elbert W., Hang S., Hoor P., Thines E., Hoffmann T., Despres R.V., and Posch U. (2016). Bio aerosols in the Earth system: climate, health, and ecosystem interactions. *Journal of Atmospheric Research* 182:346–376.

Gautam S., and Hens L. (2022). Omikron: where do we go in a sustainability context? *Environment Development and Sustainability* (10.1007/s10668-022-02207-8).

Gautam S., and Trivedi U.K. (2020) Global implication of bioaerosol in pandemic. *Environmental Development and Sustainability* 22:3861–3865.

Gollakota A.R.K., Gautam S., Santosh M., Sudan H.A., Gandhi R., Jebadurai V.S., and Shu C.M. (2021). Bioaerosols: characterization, pathways, sampling strategies, and challenges to geo-environment and health. *Gondwana Research* 99:178–203 (10.1016/j.gr.2021.07.003).

Hänninen O., Knol A.B., Jantunen M., Lim T.A., Conrad A., Rappolder M., Carrer P., Fanetti A.C., Kim R., Buekers J., Torfs R., Iavarone I., Classen T., Hornberg C., Mekel O.C.; EBoDE Working Group (2014). Environmental burden of disease in Europe: assessing nine risk factors in six countries. *Environmental Health Perspectives* 122(5):439–446 (doi: 10.1289/ehp.1206154).

Hill D.J., Langley R.L., and Morrow W.M. (1998). Occupational injuries and illnesses reported by zoo veterinarians in the United States. *Journal of Zoo and Wildlife Medicine* 29(4):371–385.

Hospodsky D., Yamamoto N., Nazaroff W.W., Miller D., Gorthala S., and Peccia J. (2015). Characterizing airborne fungal and bacterial concentrations and emission rates in six occupied children's classrooms. *International Journal of Indoor Environment and Health* 25(6):641–652.

Humbal C., Gautam S., and Trivedi U. (2018). A review on recent progress in observations, and health effects of bio aerosols. *Environmental International* 118:189–193.

Humbal C., Gautam S., et al. (2019) Evaluating the colonization and distribution of fungal and bacterial bioaerosol in Rajkot, western India using multi-proxy approach. *Air Quality, Atmosphere and Health* 12(6):693–704.

Hussain M., Madl P., and Khan A. (2011). Lung deposition prediction of airborne particles and the emergence of contemporary diseases. *The Health* 2(2):51–59.

Kang, S.M., Heo, K.J., and Lee, B.U. (2015). Why does rain increase the concentrations of environmental bioaerosols during monsoon? *Aerosol Air Quality Research* 15:2320–2324.

Kharitonov S.A., and Sjöbring U. (2007). Lipopolysaccharide challenge of humans as a model for chronic obstructive lung disease exacerbations. *Contributions to Microbiology* 14: 83–100.

Kim K.H., Kabir E., and Jahan S.A. (2017). Airborne bio aerosols and their impact on human health. *Journal of Environmental Sciences* 67: 23–35.

Limaye S.V., Schöpp W., and Amann M. (2018). Applying integrated exposure-response functions to PM2.5 pollution in India. *International Journal of Environmental Research and Public Health* 16(1): 1–20.

Ling Y., Chen H., Chen X., Yang X., Yang J., Bavoil P.M., and He C. (2015). Epidemiology of *Chlamydia psittaci* infection in racing pigeons and pigeon fanciers in Beijing, China. *Zoonoses Public Health* 62 (5): 401–406.

McGill, G., Moore, J., Sharpe, T., Downey, D., and Oyedele, L. (2015). Airborne bacteria and fungi concentrations in airtight contemporary dwellings. Mackintosh School of Architecture, Mackintosh Environmental Architecture Research Unit (MEARU).

NAAQS, National Ambient Air Quality Standards for Particulate Matter, Environmental Protection Agency (2011). USA (www.epa.gov/air/criteria.html) (last accessed 31 January 2015).

Park D.U., Ryu S.H., Kim S.B., and Yoon C.S. (2011). An assessment of dust, endotoxin, and microorganism exposure during waste collection and sorting. *Journal of the Air & Waste Management Association* 61(4):461–468.

Rockhill B., Newman B., and Weinberg C. (1998). Use and misuse of population attributable fractions. *American Journal of Public Health* 88:15–19.

SAAQIS, South African Air Quality Information System (2009). (www.saaqis.org.za/Downloads.aspx?type=AQ) (last accessed 31 January 2015).

Samadi S., Wilson C.A.M.E., and Moradkhani H. (2013). Uncertainty analysis of statistical downscaling models using Hadley Centre Coupled Model. *Theoretical and Applied Climatology* 114:673–690 (https://doi.org/10.1007/s00704-013-0844-x).

Samake A., Uzu G., Martins F.M.J., Calas A., Vince E., Parat S., and Jaffrezo L.J. (2017). The unexpected role of bioaerosols in the Oxidative Potential of PM. *Science Reports* 7:10978.

Shiraiwa M., Ueda K., Pozzer A., Lammel G., Kampf J.C., Fushimi A., Enami S., Arangio M.A., Fröhlich-Nowoisky J., Fujitani Y., Furuyama A., Lakey J.S.P., Lelieveld J., Lucas K., Morino Y., Pöschl U., Takahama S., Takami A., Tong H., Weber B., Yoshino A., and Sato K. (2017). Aerosol health effects from molecular to global scales. *Environmental Science and Technology* 51(23):13545–13567.

Sillanpaa M., Geller D.M., Phuleria C.H., and Sioutas C. (2008). High collection efficiency electrostatic precipitator for in vitro cell exposure to concentrated ambient particulate matter (PM). *Journal of Aerosol Science* 39(4): 335–347.

Thilsing T., Madsen A.M., Basinas I., Schlünssen V., Tendal K., and Bælum J. (2015). Dust, endotoxin, fungi, and bacteria exposure as determined by work task, season, and type of plant in a flower greenhouse. *The Annals of Occupational Hygiene* 59(2):142–157.

Thorne P.S., Kickhaefer M.S., Whitten P., and Doriham K.J. (1992) Comparison of bioaerosol sampling methods in barns housing swine. *Applied and Environmental Microbiology* 58:2543–2551.

Thurston D.G., Ahn J., Cromar R.K., Shao Y., Reynolds R.H., Jerrett M., Lim C.C., Shanley R., Park Y., and Hayeset B.R. (2016). Ambient particulate matter air pollution exposure and mortality in the NIH-AARP diet and health cohort. *Environmental Health Perspectives* 124(4):484–490.

U.S. Environmental Protection Agency (1997). Office of Research and Development, Air Quality Criteria for Particulate Matter. Washington, DC, US Government Printing Office, EPA/600/P-95/001aF, Vol. III, pp. 12–31 to 12–75.

Van Leuken J.P.G., Swart N.A., Havelaar H.A., Van Pul A., Van der Hoek W., and Heederik D. (2015). Atmospheric dispersion modelling of bioaerosols that are pathogenic to humans and livestock: a review to inform risk assessment studies. *Microbial Risk Analysis* 1:19–39.

Walser S.M., Gerstner D.G., Brenner B., Bünger J., Eikmann T., Janssen B., Kolb S., Kolk A., Nowak D., Raulf M., Sagunski H., Sedlmaier N., Suchenwirth R., Wiesmüller G., Wollin K.M., Tesseraux I., and Herr C.E. (2015). Evaluation of exposure – response relationships for health effects of microbial bio aerosols – a systematic review. *International Journal of Hygiene and Environmental Health* 218(7):577–589.

Whon W.T., Kim S.M., Roh W.S., Shin R.N., Lee W.H., and Bae W.J. (2012). Metagenomic characterization of airborne viral DNA diversity in the near-surface atmosphere. *Journal of Virology* 86(15):8221–8231.

Williams B., Douglas P., Barcelo R.A., Hansell L.A., and Hayesa E. (2018). Estimating *Aspergillus fumigatus* exposure from outdoor composting activities in England between 2005 and 14. *Journal of Waste Management* 84:235–244.

WHO (2014). Ambient (outdoor) air quality and health. Fact sheet N°313 , World Health Organization (www.who.int/mediacentre/factsheets/fs313/en/) (last accessed 1 February 2015).

WHO (2018). Household air pollution and health (www.who.int/en/news-room/fact-sheets/detail/household-air-pollution-and-health) (accessed on 25 February 2021).

25 Pollution Indicator, Opportunistic, and Pathogenic Bacteria Prevalence in Farmed and Wild Fish of Kashmir Himalaya

Rasy Fayaz Choh, Umer Aziz Mir, and Fayaz Ahmad
Department of Zoology, University of Kashmir, Srinagar

Ibraq Khurshid
Department of Zoology, Central University of Kashmir, Ganderbal, Kashmir

CONTENTS

25.1 INTRODUCTION

Fish is a putrefiable source of protein. Freezing does not slow down decay in fish because of chemical changes and lysis activity that occur after harvest (Huss *et al.*, 1974). Microorganisms found in aquatic settings, in addition to toxins introduced throughout post-harvest processing, hasten the decay of fish. When a fish dies, microorganisms on the surface, as well as those in the belly and gills,

DOI: 10.1201/9781003220824-29

begin to use the protein and nutrients in the fish, resulting in a decrease in biological process value (Ames, 1992). Microbial activity causes undesirable changes such as off-flavors, consistency, and appearance (Johnston *et al.*, 1994).

The initial microbial load, temperature (water temperature 18–21°C), and inappropriate handling all influence the rate of bacterial deterioration. As a result, proper storage is critical in maintaining a high standard of safety in processed fish (Chapman, 1992). Small-scale fishermen employ small-engine boats, dugout canoes, and nets or hooks for fishing, all of which affect the microbiological and chemical quality of the fish sold. Hook fishing, for example, holds previously caught fish on the hook, usually ungutted and without ice, while the fisherman attempts to collect a big quantity of fish to trade (Purvis, 2002).

In expansion from these issues, small-scale fishermen experience a lack of investment in landing, processing, and selling locations, which leads to poor sanitation (Mmopelwa, 1992). These issues result in cross-contamination and microbial proliferation, resulting in low-quality fish being offered to shoppers (Mmopelwa, 1992). According to studies, many bacteria are identified from three principal areas of fresh fish: the slime layer, the gills, and the epithelial duct. According to Cahill, the microbial diversity of fresh fish muscle is determined by fishing regions and natural components. Warm-water fish are largely home to mesophilic Gram-negative bacteria, whereas cold-water fish are home to psychrophilic Gram-positive bacteria. These bacterial segregates are classified into two types: native bacteria and post-harvest bacteria. Prior studies have also proven that the presence of indicator microorganisms in fish is a result of fecal pollution, as well as that these bacteria are opportunistically infectious to humans (Cahill, 1990; Da Silva *et al.*, 2002; Tsai *et al.*, 2005; Ferreira and Pinho, 2006; Tzikas *et al.*, 2007).

Fish rotting has far-reaching repercussions that go beyond the loss of protein. Foodborne disease as a result of consuming infected fish has resulted in significant economic losses. Fish and microbe interactions undermine human food safety and quality; this is especially important when the bacteria are pathogenic and/or opportunistic (World Health Organization, 1996).

Due to difficulties like inadequate handling and insufficient and incorrect storage facilities on the streets, consumers may be at risk of contracting foodborne diseases. As a result of these circumstances as well as increased demand for fish, this study investigated the safety and quality of fish sold in markets and by street sellers. This study describes the microbiological quality of retail fish and street-vendor fish caught from the Jhelum River and several Kashmir Valley fish farms.

25.1.1 Materials and Methodology

25.1.1.1 Bacteriological Testing and Examination

The specimens were collected twice in 2 weeks for bacterial investigations of the skin, gills, and digestive tracts of both common carp and *Schizothorax plagiostomus* in several fish farms and the river Jhelum. The tests were carried out in May and June, when the water temperature ranged between 24.6°C and 25.5°C, excellent for growth.

25.1.1.2 Collection of Samples

After weighing the example in grams (g), the standard length, head length, gill length, and buccal depression were measured in centimeters (cm). The samples were transferred into individual examination sacks and hauled in cooler boxes loaded with ice packs from the examining regions to the research facility in less than 2 hours.

Live samples of common carp (*Cyprinus carpio*, one of the most widely grown carps) and *S. plagiostomus* from Jammu & Kashmir's fish farms and the river Jhelum, respectively, were gathered in May using a cast net and identified using various taxonomic keys (Jhingram, 2007 and Tilak, 1987). Local expertise in the sector was also sought for assistance. The body weights of the collected samples were in the range 50 ± 10 g. Clinically, collected samples were examined for clinical symptoms with an unusual center toward lesions on the skin and gills (El Deen *et al.*, 2014). Swabs from the tests were placed onto nutrient agar (NA) medium from three places on each fish: the skin, the gills, and the buccal depression (Austin and Austin, 2012).

25.1.1.3 Assessment of Morphological/Clinical Objective Indications

Exophthalmia, stomach distension, skin blisters, superficial ulcers, hemorrhages, and in rare cases, intramuscular cavities filled with blood-tinged caseous or necrotic material were all signs of bacterial infection in common carp and *S. plagiostomus*. Infected fish were examined thoroughly, and any macroscopic or gross lesions seen were noted. A post-mortem was performed in almost every case within 1 to 2 hours of death.

25.1.1.4 Isolation and Identification of Bacteria

A culture-dependent methodology and the spread plating method were used to separate fish pathogenic microorganisms. The infected zone's surface was swabbed, and the inoculum was disseminated throughout the surface of a nutrient-rich medium, supplement agar media (Spanggaard *et al.*, 2000; Dar *et al.*, 2013; 2015), with incubation at 25–30°C for 2–3 days (Eddy and Jones, 2002; Al-Harbi and Uddin, 2004; 2005). Purified isolates were used as stocks for morphological and biochemical analysis.

25.1.1.5 Sample Preparation

First, a sterile knife was used to remove 10 g of fish sample from the intestines of the fish, which was then weighed using a weighing scale. The sample was then homogenized in a homogenizer with 90 ml of saline water. For a homogeneous suspension, the sample was properly shaken. After that, the sample was sealed in a conical flask. The original sample stock was made from a 10^{-1} dilution of the homogenate solution. Following the International Standards Organization, 10-fold serial dilutions (1:10) for each fish sample up to 10^{-5} were prepared (ISO, 1995).

The Digital Colony Counter was used to count bacterial colonies. The following formula was used to calculate colony formation units (CFU):

$$\text{CFU} = \frac{\text{Number of colonies}}{\text{Volume of sample}} \text{dilution factor}$$

25.1.1.6 Morphological Characterization

The bacterial films of each suspected pure isolate were stained with Gram's stain (Cruickshank *et al.*, 1979) and inspected under a bright field microscope with an oil immersion lens.

25.1.1.7 Biochemical Characterization

Bacteria are classified following their phenotypic characteristics, and there is a glut of data about microbes' phenotypic traits. The emphasis on molecular-based techniques has shifted in the last few years, whereas phenotypic approaches have dropped. Nonetheless, phenotypic data plays an important function in polyphasic research in which different characteristics of an organism's biology are investigated (Vandamme *et al.*, 1996). The bacteria that had been isolated were identified and characterized using different biochemical and physiological test responses.

25.2 RESULTS

25.2.1 Morphometric Characteristics

The morphometric parameters of *Cyprinus carpio* and *S. plagiostomus* are summarized with substantial differences. The weights are given in grams, while the lengths are given in centimeters.

TABLE 25.1
Summary of Morphometric Characteristics

Morphometrics	*Cyprinus carpio*	*S. plagiostomus*
Weight (g)	364.52 ± 6.23	275 ± 5.45
Standard length (cm)	20.05 ± 5.40	25.60 ± 6.25
Head length (cm)	3.24 ± 0.68	2.57 ± 0.40
Gill length (cm)	2.79 ± 0.52	2.20 ± 0.24
Buccal depth (cm)	3.25 ± 0.20	1.45 ± 0.12

TABLE 25.2
Summary of Viable Bacteria Count (cfu/ml)

Fish Species	Body Parts	Range
Cyprinus carpio	Skin	2.0×10^3–2.90×10^4
	Buccal cavity	1.64×10^4–2.99×10^4
	Gill	1.1×10^3–2.56×10^4
Schizothorax plagiostomus	Skin	1.6×10^3–2.78×10^4
	Buccal cavity	1.5×10^3–2.95×10^4
	Gill	1.1×10^3–2.71×10^4

Table 25.1 displays the means of morphometric features with significant differences. The weights of *Cyprinus carpio* and *S. plagiostomus* in the table are substantially different at $P < 0.05$. Furthermore, the skull lengths of these species differed significantly at $P < 0.05$. The standard length and gill length of the three fish samples differed considerably at $P < 0.05$. The buccal depths of *Cyprinus carpio* and *S. plagiostomus* did not differ substantially at $P < 0.05$.

25.2.2 Viable Bacteria Count

Table 25.2 shows the viable bacteria count (cfu/ml) in various fish samples. *Cyprinus carpio* had 2.0×10^3–2.90×10^4 in the skin, 1.64×10^4–2.99×10^4 in the buccal cavity, and 1.1×10^3–2.56×10^4 in the gills. For *S. plagiostomus*, the range in the skin was 1.6×10^3–2.78×10^4, in the buccal cavity 1.5×10^3–2.95×10^4, and 1.1×10^3–2.71×10^4 in the gills.

25.2.3 Prevalence of Pathogens in Fish Samples

Table 25.3 represents the prevalence of the bacterial isolates in *Cyprinus carpio* fish species as follows: (19%) *Staphylococcus* species, (17%) *Shigella* spp, (3%) *Bacillus* spp, (2%) *Proteus vulgaris*, (2%) *Salmonella* species, (2%) *Escherichia coli*, (1%) *Enterobacter aeruginosa*, (1%) *Klebsiella pneumoniae*, (1%) *Enterobacter cloacae*, and (1%) *Pseudomonas aeruginosa.* In *S. plagiostomus*, the species were (25%) *E. coli,* (20%) *Enterobacter aeruginosa*, (15%) *Salmonella* species, (11%) *Pseudomonas aeruginosa*, (10%) *Bacillus* spp, (8%) *Klebsiella pneumoniae*, (5%) *Shigella* spp, (2%) *Staphylococcus* species, (1%) *Enterobacter cloacae*, and (1%) *Proteus vulgaris.*

TABLE 25.3
Prevalence of Pathogens in Fish Samples

Isolates	*Cyprinus carpio*	*S. plagiostomus*
Enterobacter cloacae	1%	1%
Pseudomonas aeruginosa	1%	11%
Escherichia coli	2%	25%
Bacillus spp	3%	10%
Shigella spp.	17%	5%
Enterobacter aerogenes	1%	20%
Klebsiella pneumoniae	1%	8%
Proteus vulgaris	2%	1%
Salmonella species	2%	15%
Staphylococcus species	19%	2%

25.2.4 Percentage Occurrence of Bacterial Isolates

Table 25.4 and Figure 25.1 indicate the occurrence of bacterial flora in these fish samples' gills, buccal cavity, and skin. They suggest that 10 bacterial species, *Enterobacter cloacae*, *Pseudomonas aeruginosa*, *E. coli*, *Enterobacter aeruginosa*, *Klebsiella pneumoniae*, *Proteus vulgaris*, *Bacillus* spp, *Shigella* spp, *Staphylococcus* species, and *Salmonella* species, were found in the gills, skin, and buccal cavity of *Cyprinus carpio* and *S. plagiostomus.* Six of the recovered bacterium species were Gram-negative, while five were Gram-positive.

25.3 DISCUSSION

The morphometric features tested for these fish species were significantly different ($P < 0.05$). At ($P < 0.05$), *Cyprinus carpio* had the highest mean weight (grams), buccal depth (cm), head length (cm), and gill length (cm). The standard length of *S. plagiostomus* was the longest. The skin of *Cyprinus carpio* had the greatest bacterial count range of 2.0×10^3–2.90×10^4, which contradicts the premise of Jara and Chodyniecki (1999) that the skin, gills, and buccal cavity are in direct touch with the environment and any pathogen.

Microflora is an important component of the aquatic ecosystem, and interactions between pathogenic microflora and the aquatic environment result in the transmission of contagious diseases in aquaculture technologies (Noga, 2000; Ikpi and Offem, 2011). The presence of pathogenic bacterial species has resulted in the pollution of aquatic water bodies by animal feces (Abdelhamid *et al.*, 2006). The presence of *E. coli*, *Shigella*, and *Salmonella* in the fish suggests that the lake has been contaminated by fecal matter, which has resulted in environmental contamination (Yagoub, 2009). The incidence of *Staphylococcus aureus* is important for public health, since it has been linked to food poisoning and illness. This effect is caused by its ability to create enzymes and poisons (Umoh and Odoba, 1999).

A total of 10 bacterial species were collected from the gills, skin, and buccal cavity of *Cyprinus carpio* and *S. plagiostomus* in this investigation. Gram-negative *Proteus* spp, *Escherichia coli*, *Klebsiella* spp, *Pseudomonas aeruginosa*, *Shigella* spp, and *Enterobacter aerogenes* were among the isolated bacterial species. This also verified the findings of Ahmed and Naim (2002), who stated that the bacteria isolated from brackish pond water, sediments, and healthy tilapia grown in Saudi Arabia were mostly Gram-negative, supporting the conclusions of this study. Septicemic bacterial

TABLE 25.4
Screening of Biochemical Tests for the Identification of Fish-Associated Bacterial Flora

Tentatively Identified Bacteria	Morphology	Catalase	Oxidase	Carbohydrate Fermentation Test								H_2S Production	Urease Production
				G	F	L	S	IP	MR	VP	CU		
Enterobacter cloacae	–ve	+	–	Nil	Nil	+		+	+	+	+	–	+
Pseudomonas aeruginosa	–ve	–	+	+	+	+	+	–	–	–	+	–	–
Escherichia coli	–ve	–	–	+	+	+	+	+	+	–	+	–	–
Bacillus spp	+ve	+	–	+	+	+	+	–	–	–	–	–	–
Shigella	+ve	+	–	+	+	–	–	+	+	–	–	Nil	Nil
Enterobacter aerogenes	–ve	+	–	Nil	Nil	+	Nil	–	+	+	+	–	–
Klebsiella pneumoniae	–ve	+	+	+	+	–	+	–	–	+	+	–	+
Proteus vulgaris	–ve	+	–	Nil	Nil	–	Nil	+	+	+	–	+	+
Staphylococcus species	+ve	+	+	Nil	Nil	+	Nil	–	–	–	–	–	+
Salmonella species	+ve	+	–	+	–	–	–	–	+	–	–	+	–

Note: +ve = Positive, –ve = Negative, OF = Oxidative Fermentation Test, MR = Methyl Red Test, VP = Voges-Proskauer Test, H_2S = Hydrogen Sulfide Production Test, Nil = Not Present.

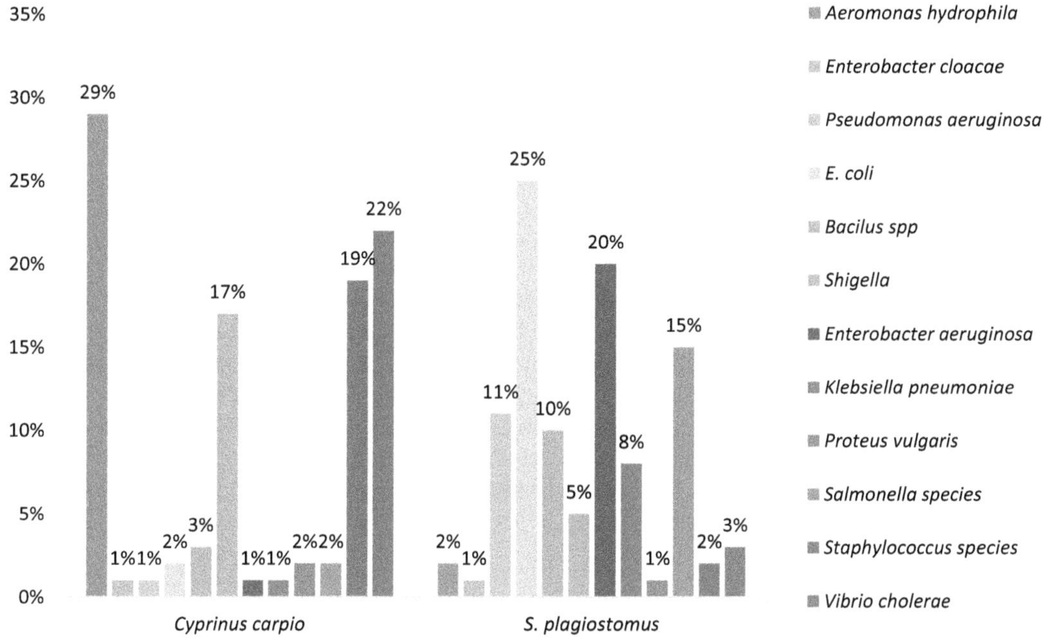

FIGURE 25.1 Percentage distribution of bacterial isolates in both the fish species.

infections by organisms such as *Vibrios*, *Aeromonads*, *Pseudomonads*, *Photobacteria*, *Streptococci*, and *Staphylococci* were found in some marine fish species at various phases of development (Alicia et al., 2005 and Samuelson et al., 2006).

The prevalence of the bacterial isolates in *Cyprinus carpio* fish species was as follows: (19%) *Staphylococcus* species, (17%) *Shigella* spp, (3%) *Bacillus* spp, (2%) *Proteus vulgaris*, (2%) *Salmonella* species, (2%) *E. coli*, (1%) *Enterobacter aerogenes*, (1%) *Klebsiella pneumoniae*, (1%) *Enterobacter cloacae*, and (1%) *Pseudomonas aeruginosa.*

In *S. plagiostomus*, the species were (25%) *E. coli*, (20%) *Enterobacter aerogenes*, (15%) *Salmonella species*, (11%) *Pseudomonas aeruginosa*, (10%) *Bacillus* spp, (8%) *Klebsiella pneumoniae*, (5%) *Shigella* spp, (2%) *Staphylococcus* species, (1%) *Enterobacter cloacae*, and (1%) *Proteus vulgaris.*

The most commonly isolated species were *Vibrio* spp, *Pseudomonas* spp, *Streptococcus faecalis*, *Aeromonas hydrophila*, *Aeromonas sobria*, and *Staphylococcus aureus* according to the findings of Zorrilla et al. (2003) and Moustafa et al (2010), who reported that the primary pathogenic groups isolated from infected gilt-head seabream in Spain were *Vibrio* spp, *Pseudomonas* spp, *Aeromonas* spp, and Gram-positive bacteria. *Aeromonas* spp. are major fish diseases that pose a constant threat to the aquaculture industry and human life. The discussed biochemical investigations are crucial in the hunt for alternative and more effective means of controlling these fish diseases.

Due to ineffective farm management techniques, the aquaculture sectors are prone to numerous diseases, increasing susceptibility to pathogenic infections (El–Sayed, 2006), and the dispersion of mobile *Aeromonas* spp. in the aquatic environment and fish has been recorded (Kaper *et al.*, 1981; Abeyta *et al.*, 1986; Hatha *et al.*, 2005; Hussain *et al.*, 2014).Similar findings were made by Noor El Deen *et al.* (2014), who stated that skin ulcers could be the source of infection caused by *Aeromonas* spp. The findings matched those of Hazen et al. (2012), who stated that the color of a bacterial colony may be used to determine its genus level. *Aeromonas* spp. is an opportunistic pathogen found in freshwater ecosystems worldwide as well as soil, water, and food. This bacterium is capable of causing foodborne and nosocomial illnesses (Cabral, 2010; Gauthier, 2015; Janda and Sharon, 2010).

The existence of these bacteria may raise health concerns among consumers. Pathogenic bacteria, *Salmonella* spp., were found in freshly caught fish in Latvia in the current investigation (Terentjeva *et al.*, 2015). The pathogenicity of *A. hydrophila* for experimentally infected *O. niloticus* fish species may be linked to the generation of toxins and extracellular enzymes by *A. hydrophila* (Saavedra et al., 2004).

The presence of enteric bacteria in fish is a sign of fecal contamination and water pollution. The association of pathogenic bacteria among the fishes suggested that if fishes are prepared in a disorganized way, it could cause food safety hazards for consumers (Akila and Kumaran, 2018). During this study, the fish samples sheltered human disease-causing organisms that spread diseases like food poisoning, diarrhea, typhoid, fever, etc. When microorganisms such as *Pseudomonas*, *Salmonella*, *S. aureus*, and *Shigella* are present in food, they cause predominantly foodborne disorders in the environment and among consumers. Sujatha *et al.* (2011) discovered the same type of human pathogenic bacteria in two edible fish species, *Priacanthus hamrur* and *Megalaspis cordyla*, in Royapuram, Chennai, India, while Akila and Kumaran (2018) also verified the same types of pathogenic species during their study, which are very harmful for the human population.

25.4 CONCLUSIONS

All the fish species obtained from freshwater and hatcheries exhibit satisfactory microbiological quality. The presence of enteric bacteria in fish is a sign of fecal contamination and water pollution. The presence of pathogenic bacteria in the fish species shows that if they are not handled or prepared appropriately, they may provide a food safety risk to customers. In all cases, *Pseudomonas* spp. and *E. coli* were prevalent in fish microbiota. *Pseudomonas* spp. produce rapid degradation of fish and fish products, and their dominance suggests that the fish are susceptible to spoiling process development, while *E. coli* can cause health problems in both aquaculture and the general population.

ACKNOWLEDGEMENTS

The authors would like to express their gratitude to the Head of the Department of Zoology, University of Kashmir, and the Head of the Department of Microbiology, SKIMS Srinagar Kashmir, India for allowing them to use their laboratory facilities to conduct this research. The writers would also like to thank the fishing community on the Jhelum River, as well as the personnel of different hatcheries, for their assistance.

REFERENCES

Abdelhamid, A.M., Gawish, M.M. and Soryal, K.A. (2006). Comparative study between desert cultivated and natural fisheries of mullet fish in Egypt, II-microbiological concern. *Mansoura Univ. J. Agric. Sci.*, 31: 5681–5687.

Abeyta, C., Kaysner, C.A., Wekell, M.M., Sullivan, J.J. and Stelma, G.N. (1986). Recovery of *Aeromonas hydrophila* from oysters implicated in an outbreak of food borne illness. *J. Food Prot.*, 49: 643–646.

Ahmed, H. Al-H. and Naim, U. (2002). Bacterial diversity of tilapia (*Oreochromis niloticus*) cultured in brackish water in Saudi Arabia. *Aquaculture*, 250: 566–572.

Akila, N. and Kumaran, R.S. (2018) Isolation and identification of prevalent bacterial pathogens from an exotic fish *Tilapia zillii* and *Oreochromis mossambicus*. *Int. J. Green Pharm.*, 12(3): S497.

Al-Harbi, A.H. and Uddin, M.N. (2004). Seasonal variation in the intestinal bacterial flora of hybrid tilapia (*Oreochromis niloticus* x *Oreochromis aureus*) cultured in earthern ponds in Saudi Arabia. *Aquaculture*, 229: 37–44.

Al-Harbi, A.H. and Uddin N. (2005). Bacterial diversity of tilapia (*Oreochromis niloticus*) cultured in brackish water in Saudi Arabia. *Aquaculture,* 250: 566–572.

Alicia, E., Toranzo, T., Magarinos, B. and Romalde, J. L. (2005). A review of the main bacterial fish diseases in mariculture systems. *Aquaculture*, 246: 37–61.

Ames, G.R. (1992). *The Kinds and Levels of Post-Harvest Losses in African Inland Fisheries.* Food Agriculture Organization of the United Nations. Rome Italy, CIFA Technical Paper 19.

Austin, B. and Austin, D.A. (2012). *Bacterial Fish Pathogens,* Vol. 481. Dordrecht, The Netherlands: Springer, p. 482.

Cabral, J.P.S. (2010). Water microbiology, bacterial pathogens and water. *Int. J. Environ. Res. Public Health,* 7: 3657–3703.

Cahill, M.M. (1990). Bacterial flora of fishes: a review. *Microb. Ecol.*, 19(1): 21–41.

Chapman, D. (1992). *Water Quality Assessment.* London: Chapman & Hall.

Cruickshank, R., Duguid, J.P., Marmian, B.P. and Swain R.H.A. (1979). *The Practice of Medical Microbiology* (12th edn), Churchill Livingstone, Edinburgh, London. pp. 13: 145–155.

Da Silva, M.V., Pinho, O., Ferreira, I., Plestilová, L. and Gibbs, P.A. (2002). Production of histamine and tyramine by bacteria isolated from Portuguese vacuum-packed cold-smoked fish. *Food Control*, 13(6–7): 457–461.

Dar, G.H., Bandh, S.A., Kamili, A.N., Nazir R. and Bhat R.A. (2013). Comparative analysis of different types of bacterial colonies from the soils of Yusmarg Forest, Kashmir valley India. *Ecol. Balk.*, 5(1): 31–35.

Dar, G.H., Kamili, A.N., Nazir, R., Bandh, S.A., Jan, T.R. and Chishti, M.Z. (2015). Enhanced production of α-amylase by *Penicillium chrysogenum* in liquid culture by modifying the process parameters. *Microb. Pathog.*, 88: 10–15.

Eddy, S.D. and Jones, S.H. (2002). Microbiology of summer flounder *Paralichthys dentatus* fingerling production at a marine fish hatchery. *Aquaculture*, 211(1–4): 9–28.

El Deen, A.N., Dorgham-Sohad, M., Hassan-Azza, H.M. and Hakim, A.S. (2014). Studies on *Aeromonas hydrophila* in cultured *Oreochromis niloticus* at Kafr El Sheikh Governorate, Egypt with reference to histopathological alterations in some vital organs. *World J. Fish Mar. Sci.*, 6(3): 233–240.

El-Sayed, A.F.M. (2006). Stress and diseases. In: *Tilapia Culture* (El-Sayed, A.F.M. ed.), Cambridge: CABI Publishing, pp. 149–151.

Ferreira, I.M. and Pinho, O. (2006). Biogenic amines in Portuguese traditional foods and wines. *J. Food Prot.*, 69(9): 2293–2303.

Gauthier, D.T. (2015) Bacterial zoonoses of fishes: a review and appraisal of evidence for linkages between fish and human infections. *Vet. J.*, 203: 27–35.

Hatha, M., Vivekanandhan, A.A. and Christol, G.J.J. (2005). Antibiotic resistance pattern of motile aeromonads from farm raised fresh water fish. *Int. J. Food Microbiol.*, 98: 131–134.

Hazen, T.C., Fliermans, C.B., Hirsch, R.P., Esch, G.W., Da Cruz, S.P.C., Neto, J.G.M. and De, F.R. (1978). Prevalence and distribution of *Aeromonas hydrophila* in the United States. *Appl. Environ. Microbiol.,* 36: 731–738.

Huss, H.H., Pedersen, A. and Cann, D.C. (1974). The incidence of *Clostridium botulinum* in Danish trout farms: I. Distribution in fish and their environment. *Int. J. Food Sci. Technol.*, 9(4): 445–450.

Hussain, I.A., Jeyasekaran, G., Shakila, R.J., Raj, K.T. and Jeevithan, E. (2014) Detection of hemolytic strains of *Aeromonas hydrophila* and *A. sobria* along with other Aeromonas spp. from fish and fishery products by multiplex PCR. *J. Food Sci. Technol.*, 51(2): 401–407.

Ikpi, G. and Offem, B. (2011). Bacterial infection of mudfish *Clarias gariepinus* (Siluriformes: Clariidae) fingerlings in tropical nursery ponds. *Rev. Biol. Trop.*, 59(2): 751–759.

ISO (1995). Recommendation of the meeting of the subcommittee, International Organization for Standardization, on meat and meat products. ISO/TC-36/Sc-6. The Netherlands. 10–18 November 1995.

Janda, J.M. and Sharon, L.A. (2010) The genus *Aeromonas*: taxonomy, pathogenicity, and infection. *Clin. Microbiol. Rev.*, 23(1): 35–73.

Jara, Z. and Chodyniecki, A. (1999) *Ichthyopathology* (in Polish). Wydawnictwo Akademii Rolniczej we Wroclawiu, Wroclaw. 478 pp.

Johnston, W.A., Nicholson, F.J., Roger, A. and Stroud, G.D. (1994). Freezing and Refrigerated Storage in Fisheries. FAO Fisheries Technical Paper 340.

Kaper, J.B., Lockman, H., Colwell, R.R. and Joseph S.W. (1981). *Aeromonas hydrophila*: ecology and toxigenicity of isolates from an estuary. *J. Appl. Bacteriol.,* 50: 359–377.

Mmopelwa, T.C. (1992). Sun-dried salted fish as a means of reducing post-harvest losses in Botswana, *Proceedings of the FAO Symposium on Post-Harvest Fish Technology, 21-22 October Cairo, Egypt*, CIFA Technical Paper No. 19.

Moustafa, M., Mohamed, L.A., Mohmoud, M.A., Soliman, W.S. and El-gendy, M.Y. (2010). Bacterial infections affecting marine fishes in Egypt. *J. Am. Sci.*, 6: 603–612.

Noga, E.J. (2000). *Fish Disease Diagnosis and Treatment*. Iowa, USA: Iowa State University, p. 321.

Purvis, J. (2002). *Postharvest Fisheries on the Eastern Floodplains*, DEA Research Discussion Paper 51, 1–23, Caprivi Directorate of Environmental Affairs, Ministry of Environment and Tourism, Windhoek, Namibia.

Saavedra, M.J., Guedes-Novais, S., Alves, A., Rema, P., Tacão, M., Correia, A. and Martínez-Murcia, A. (2004). Resistance to β-lactam antibiotics in *Aeromonas hydrophila* isolated from rainbow trout (*Onchorhynchus mykiss*). *Int. Microbiol.*, 7(3): 207–211.

Samuelson, P., Wernérus, H., Svedberg, M. and Ståhl, S. (2000). Staphylococcal surface display of metal-binding polyhistidyl peptides. *Appl. Environ. Microbiol.*, 66(3): 1243–1248.

Spanggaard, B., Huber, I., Nielsen, J., Nielsen, T., Appel, K.F. and Gram L. (2000). The microflora of rainbow trout intestine: a comparison of traditional and molecular identification. *Aquaculture*, 182(1–2): 1–15.

Sujatha, K., Senthilkumar, P., Sangeeta, S. and Gopalakrishnan M.D. (2011). Isolation of human pathogenic bacteria in two edible fishes, *Priacanthus hamrur* and *Megalaspis cordyla* at Royapuram waters of Chennai, India. *Indian J. Sci. Technol.*, 4: 539–541.

Terentjeva, M., Eizenberga, I., Novoslavskij, A., Strazdiņa, V., Valciņa, O., Ošmjana, J. and Bērziņš, A. (2015) Bacterial microflora of freshwater fish originated from Usmas Lake in Latvia. *J. Microbiol. Biotechnol. Food Sci.*, 4: 74–77.

Tilak, R. (1987). *The Fauna of India and the Adjacent Countries: Pisces. Teleostomi. Sub-family, Schizothoracinae*. Zoological Survey of India, New Delhi.

Tsai, Y.H., Chang, S.C., Kung, H.F., Wei, C.I. and Hwang D.F. (2005). Histamine production by *Enterobacter aerogenes* in sailfish and milkfish at various storage temperatures. *J. Food Prot.*, 68(8): 1690–1695.

Tzikas, Z., Ambrosiadis, I., Soultos, N. and Georgakis S. (2007). Seasonal size distribution, condition status and muscle yield of Mediterranean horse mackerel *Trachurus mediterraneus* from the North Aegean Sea, Greece. *Fish. Sci.*, 73(2): 453–462.

Umoh, U.J. and Odoba, M.B. (1999). Safety and quality evaluation of street foods sold in Zaria, Nigeria. *Food Control,* 10: 9–14.

Vandamme, P., Pot, B., Gillis, M., De Vos, P., Kersters, K. and Swings, J. (1996). Polyphasic taxonomy, a consensus approach to bacterial systematics. *Microbiol. Rev.,* 60: 407–438.

World Health Organization. 1996. *Essential Safety Requirements for Street-Vended Foods* (No. WHO/FNU/FOS/96.7 Rev1). World Health Organization, Geneva.

Yagoub, S.O. (2009). Isolation of *Enterobacteriaceae* and *Pseudomonas spp.* from raw fish sold in fish market in Khartoum state. *J. Bacteriol. Res.*, 1(7): 085–088.

Zorrilla, I., Arijo, S., Chabrillon, M., Diaz, P., Martinez-Manzanares, E., Balebona, M.C. and Morinigo, M.A. (2003). Vibrio species isolated from diseased farmed sole, *Solea senegalensis* (Kaup), and evaluation of the potential virulence role of their extracellular products. *J. Fish Dis.*, 26(2): 103–108.

26 Using Monogenoids (Phylum Platyhelminthes) as Indicators of Aquatic Ecosystem Health

Amit Tripathi
Department of Zoology, University of Lucknow, Uttar Pradesh

CONTENTS

26.1 INTRODUCTION

A parasitic relationship occurs when one organism (the parasite) lives on or in the body of another organism (the host), causing harm and possibly death to that organism. Environmental biology, on the other hand, investigates how organisms, species and communities respond to, and are impacted by, the surrounding environment. Environmental parasitology – a combination of parasitological and environmental factors – is a relatively new but well-established discipline (see Goater et al. 2013), and encompasses studies on parasites as (bio)indicators of environmental health (Sures and Nachev 2015).

26.2 PARASITISM

Parasitism is the most successful lifestyle on Earth. Parasites are estimated to account for more than half of global species diversity (Dobson et al. 2008), making them an essential component of all ecosystems, including aquatic environments. As a result, fish parasitology has grown in popularity and importance, with a number of research groups at various universities and institutes across the world practising it. Helminths, the largest and most important group of parasites that infect fish, are represented by five well-defined taxonomic units: Monogenoidea, Trematoda, Cestoda, Nematoda and Acanthocephala. Williams and Jones (1994) estimated that there could be up to 30000 helminth parasites in all known fish species. Monogenoidea includes largely ectoparasitic species that infect the gills and/or skin of fish, as well as a few endoparasitic species that live in the urinary bladder and the rectum of cold-blooded vertebrates (Figure 26.1). None of them infect birds, but one (*Oculotrema hippopotami*) infects the eyes of the hippopotamus, a mammal.

DOI: 10.1201/9781003220824-30

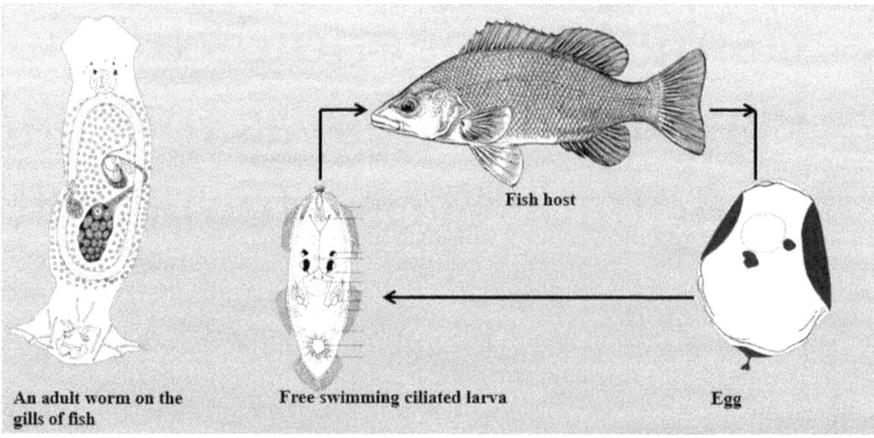

FIGURE 26.1 Direct (host-to-host) lifecycle of parasitic monogenoids. All life stages are in direct contact with and under the influence of aquatic surroundings.

26.3 BIOINDICATORS

The health of aquatic ecosystems has traditionally been assessed using physical and chemical assays, including temperature, dissolved oxygen, salinity, pH and nutrients. In recent years, however, scientists have begun to employ fish parasites instead (reviewed by Khan and Thulin 1991; Lafferty 1997; Galli et al. 2001; Marcogliese 2004; Sures 2004; Palm and Ruckert 2009; Sures et al. 2017). The parasitic organisms used to measure the environmental impacts have come to be known as bioindicators. Although there is no single parsimonious definition, a bioindicator might be defined as a species or a group of species that can respond to specific changes in environmental conditions induced by either anthropogenic or natural stressors. Effects of environmental changes can be detected by observing a significant change in the parasite's physiology, chemical composition, behaviour, abundance and occurrence, or diversity indices (Ruckert et al. 2009).A good bioindicator species must respond to environmental changes in a manner that is predictable and easy to measure and interpret. Bioindicators have a number of advantages over the conventional physicochemical assays, the most important of which is that they include a temporal component, allowing the integration of current, past or future environmental conditions (see Holt and Miller 2010). They are also a low-cost tool for evaluating and monitoring environmental quality.

26.4 MONOGENOIDS AS BIOINDICATORS

Monogenoids exhibit several hallmark characteristics that distinguish them as good bioindicators. They are the most abundant parasites in the aquatic environment (Ivona 2004); they are ectoparasitic, meaning they are in direct contact with and influenced by the surrounding aquatic environment throughout their life; they have short and direct life cycles, allowing them to react immediately to changes in environmental factors (see Sures 2001); they are sensitive to environmental disturbance or stress (see later); they are easy to find and collect, because every known fish species harbours at least one monogenoid species (see Whittington 1988); their infection dynamics can be analysed using ecological principles (see, for example, Bush et al. 1997); they affect host biology and population by inducing mortality through pathogenic effects (see, for example, Johnsen and Jensen 1991); and they are highly diverse and well-documented taxonomically, with around 7000 recognised species (see Gibson et al. 2014).

26.4.1 Environmental Response

Monogenoids were probably the first helminth parasite group to be employed as a bioindicator in environmental parasitology. Over the last 25 years or so, they have proven to be useful experimental bioindicators, since they may respond to two major water quality variables: physicochemical parameters and pollutants, particularly metals. They usually respond in one of three ways: by changing their prevalence and intensity, their presence or absence, and their morphology and/or physiology.

26.4.1.1 Physicochemical Parameters

Numerous studies have explored the linkages between physicochemical parameters of the aquatic habitat and the seasonal variations in the prevalence and intensity of monogenoids, often with predictable results. Temperature can influence the dactylogyrid monogenoids by influencing both parasite reproduction and development as well as host fish immunity (Koskivaara et al. 1992; Mo 1997). As a result, the prevalence of *Dactylogyrus* spp. on *Labeo* spp. increased as the temperature rose in South Africa's Vaal Dam (Crafford et al. 2014) and Limpopo River systems (Mbokane et al. 2015). Similarly, *Dactylogyrus* spp. were shown to be more prevalent and abundant on *Barbus martorelli* during rainy seasons, which was linked to parasite mortality being higher in dry months due to higher water temperature (BilongBilong and Tombi 2005). Additionally, Bayoumy et al. (2008) observed a significantly positive correlation between temperature and monogenoids in five fish species from the Egyptian Red Sea: *Morone labrax*, *Carnagoides bayed*, *Parupeneus forsskali*, *Epinephelus microdon* and *Rhabdosargus sarba*. The seasonal population dynamics of *Quadriacanthus kobinensis* from *Clarias fuscus* in Pearl River, China, were nearly consistent with these observations, with the parasite having the highest prevalence and mean intensity in the summer and the autumn, respectively (Hu and Li 2016). Based on this substantial body of work, the temperature appears to be a determining factor in the growth of monogenoids. However, *Diplectanum aequans*, a diplectanid monogenoid from the sea bass *Dicentrarchus labrax* in Corsica, France, which showed high rates of infestation during the coldest part of the year, did not support such a generalization (Antonelli et al. 2016).

In general, the peak intensities of gyrodactylid monogenoids are also observed at low temperatures (Bauer 1959; Jansen and Bakke 1991). Cone and Cusack (1988), for example, observed the highest numbers of *Gyrodactylus colemanensis* and *G. salmonis* when the water temperature was below 8°C in Nova Scotia, Canada. *Gyrodactylus shulmani*, on the other hand, followed a distinctly different pattern, reproducing best at a temperature of 20°C (Zitnan and Hanzelova 1982). Higher water temperature, on the other hand, has been shown to have a statistically significant negative correlation with the morphometrical variables of haptoral armaments (Ergens and Gelnar 1985; Mo 1991; Geets et al. 1999) and the male copulatory organ (Zolovs et al. 2012) of gyrodactylids. Dmitrieva and Dimitrov (2002) performed a discriminant analysis to examine haptoral measurements of four *Gyrodactylus* species from the Black Sea and concluded that unfavourable environmental conditions shortened the time of embryogenesis in gyrodactylids, increasing reproduction rate but decreasing the size of haptoral parts. Many additional investigations have shown a decrease in the size of the sclerotized body parts of *Gyrodactylus* spp. in conjunction with an increase in water temperature, and vice versa (Malmberg 1970; Ergens 1981; Ergens and Gelnar 1985; Appleby 1996).

Tripathi (unpublished study) found that the mean intensity of *Dactylogyroides tripathii* (Tripathi 1959; Gusev, 1976) was positively correlated with temperature but negatively correlated with dissolved oxygen in India. In the example of *Sciadicleithrum guanduensis* from *Geophagus brasiliensis* in southern Brazil, Lacerda et al. (2018) also observed a similar trend of an inverse relationship between dissolved oxygen and parasite abundance. Meanwhile, Mashaly et al. (2020) uncovered a completely opposite pattern in the Nile Delta, Egypt, namely, a highly significant positive correlation of dissolved oxygen with the mean intensity and abundance of *Quadriacanthus clariadis* on *Clarias gariepinus*. Recently, Cavalcanti et al. (2020) investigated the responses of five monogenoid species of the Nile tilapia *Oreochromis niloticus* (*Cichlidogyrus tilapiae*, *Cichlidogyrus thurstonae*, *Cichlidogyrus sclerosus*, *Cichlidogyrus halli* and *Scutogyrus longicornis*) to variations in the temperature, dissolved oxygen concentration, pH, electrical conductivity and organic matter of a pond ecosystem in Brazil. Their abundances were found to have a significant correlation with changes in the evaluated physicochemical water parameters, with the exception of *C. halli*. Dissolved oxygen, in particular, was found to be positively correlated with the abundance of *C. tilapiae*, whereas it was negatively correlated with the abundances of *C. sclerosus, C. thurstonae* and *S. longicornis*. Soleng and Bakke (1997) found a negative relationship between increased salinity and *Gyrodactylus salaris* survival on Atlantic salmon *Salmo salar*.

Tripathi et al. (2019) investigated the prevalence and intensity of *Cornudiscoides agarwali* on *Mystus bleekeri* in an unpolluted study site in India. These researchers deduced that monogenoids' population dynamics are most likely to thrive under optimal water quality parameters based on their observation that *C. agarwali* reproduced all year. However, laboratory-based experiments to determine *C. agarwali*'s tolerance to pollutants are still lacking.

26.4.1.2 Aquatic Pollutants

Experimental studies have shown that after exposure to a pollutant, particularly heavy metals, the number of monogenoids increases dramatically. This is largely due to the fact that pollutants may impair hosts' immune systems, making them more susceptible to the parasite infections (see McDowell et al. 1999).The absence of *Gyrodactylus unicipula* from the gills of plaice *Pleuronectes platessa* farmed at the Hunterston power station in Scotland, for example, was attributed to chlorination of the cooling water (MacKenzie et al. 1976). In another study, Skinner (1982) examined fishes (*Gerres cinereus*, *Lutjanus griseus* and *Strongylura timucu*) from Florida that were heavily infested with monogenoids (*Neodiplectanum wenningeri*, *Ancyrocephalus* sp. and *Ancyrocephalus parvus*, respectively). He linked these results to the presence of high levels of ammonia, trace metals and pesticides in the aquatic environment, which presumably altered the physicality and physiology of fish, reducing their resistance to parasite infestation. Koskivaara and Valtonen (1992) discovered a higher prevalence of a number of monogenoids (*Dactylogyrus similis*, *D. fallax*, *Gyrodactylus gasterostei*, *G. carassii* and *G. vimbi*) from the roach *Rutilus* in Lake Vatia in central Finland, which was polluted with pulp and paper effluents, when compared with uncontaminated reference lakes. These authors attributed the results to the fish's compromised immune system, which they believed was triggered by the pulp and paper mill effluents. Interestingly, in the same study, *Dactylogyrus* sp. and *Gyrodactylus prostae* were found to be unaffected by pollutants, whether present or absent. In contrast to the previous example, *Paradiplozoon homoion* and *Dactylogyrus* spp. on roach from Sweden's Baltic coast have shown a decrease in infections associated with pulp and paper mill pollution (reviewed in Khan and Thulin 1991).

Polluted sediments are known to allow *Cichlidogyrus sclerosus* infection to persist in Nile tilapia *Oreochromis niloticus* by inducing immunosuppression and subsequent histological damage (Sanchez-Ramirez et al. 2007). Bayoumy et al. (2008) observed a highly significant positive correlation between the prevalence of *Haliotrematoides* sp. and *Neothoracocotyle* sp. and selenium, but a negative correlation between these two species and lead. In a quantitative meta-analysis of the effects of pollution on aquatic parasites, Blanar et al. (2009) also found strong, significant negative effects among monogenoids in response to water contaminants, notably metal pollution. In another study, Gilbert and Avenant-Oldewage (2016) compared the infection variables of *Paradiplozoon ichthyoxanthon* at two sites along the Vaal River system, South Africa and discovered that *P. ichthyoxanthon* was present in the Vaal Dam site but absent in the Vaal River site, despite the presence of the host fish at both sites. The absence of *P. ichthyoxanthon* was attributed to the poorer environmental conditions (higher trace elements concentration and conductivity) at the Vaal River site when compared with the Vaal Dam. El-Naggar et al. (2017) investigated the effects of heavy metals (cadmium, chromium, cobalt, copper, iron, lead, manganese, nickel and zinc) on the number of monogenoid species from *Clarias gariepinus* in the Nile Delta, Egypt. According to these researchers, there was a moderate negative correlation between *Macrogyrodactylus clarii* and cadmium, iron and nickel. They also discovered that *Gyrodactylus* spp. and *Macrogyrodactylus congolensis* had a positive correlation with cobalt and zinc, but a negative correlation with chromium. Mashaly et al. (2020) also studied the impact of heavy metals (cadmium, lead, copper, iron and zinc) on the infestation levels of monogenoids on *Clarias gariepinus* in Nile Delta, Egypt. They found that *Quadriacanthus kearni* was positively related to iron but not to the other heavy metals. Cunha et al. (2021) evaluated the role of *Calceostomella herzbergii*, a parasite from the marine catfish *Sciades herzbergii*, as a bioindicator of environmental quality in Amazonian estuarine ecosystems, and found that the parasite was sensitive to declining water quality, particularly high turbidity.

Pollutants have also been found to influence the morphogenesis of monogenoids. Rodríguez-Gonzalez et al. (2020), for example, discovered fluctuating asymmetry in the size and shape of sclerotized structures of *Haliotrematoides* spp., likely as a result of chemical contamination along the Yucatan Peninsula's continental shelf. The contamination, according to these authors, may have disrupted the developmental stability of parasites. In response to toxic substances of polluted waters, haptors of diplozoid monogenoids have also been shown to develop structural defects such as reduced attachment organs and asymmetry in the arrangement and number of valves (Khan and Thulin 1991; Kuperman 1992; MacKenzie et al. 1995; Dušek et al. 1998; Dzika 2002).

There have been fewer studies on whether or not the monogenoids can accumulate metals. According to Gao and Nie (2000), *Ancryocephalus mogurndae* accumulated more lead than its host fish *Siniperca chuatsi*. Similarly, Shinn et al. (1995) discovered elements in the sclerotized hard parts of *Gyrodactylus* haptor, including vanadium, silicon, sodium and chlorine.

26.5 LIMITATIONS AND FUTURE DIRECTIONS

Fish parasites are an integral component of aquatic biodiversity and are directly affected by abiotic environmental changes. Monogenoids have several characteristics that a fish parasite must have in order to qualify as a good indicator species. While the use of monogenoids, in conjunction with abiotic components, has been shown to be useful in understanding the health status of ecosystems, they have not been included in aquatic environment and health management practises to date. Every bioindicator assessment programme will inevitably have a trade-off between sensitivity to environmental changes/disturbances on the one hand and response reliability on the other. A significant caveat in using monogenoids as bioindicators is that their environmental responses can vary depending on the parasite species, host fish species or even the locality under investigation, making it difficult to evaluate the underlying environmental causes of observed changes. Table 26.1 summarizes the main conflicting responses of monogenoid species to environmental variations. Using monogenoids as a bioindicator will remain a scientific discourse rather than a part of aquatic environment and health management practice until a satisfactory explanation for these variances in parasite response is offered. Until that time comes, individual monogenoid species that have already been demonstrated to be reliable indicators can be used as model taxa for environmental monitoring in aquatic ecosystems. An elaborate assessment protocol would also be required to fully comprehend and examine the relationship between parasite abundance and aquatic ecosystem health. It is also necessary to establish an accurate and up-to-date database of monogenoid species with enough discrimination to be effective as bioindicators.

ACKNOWLEDGEMENTS

A part of this study was supported by the project grant from the Science and Engineering Research Board (SERB), Government of India (SERB–EMR/2017/003232).

TABLE 26.1
Summary of the Major Conflicting Responses of Monogenoid Species to Environmental Variables

Parasite	Host Fish	Habitat	Correlation	Reference
		Environmental variable: Temperature		
Dactylogyrus spp.	*Labeo capensis*, *L. umbratus*	South Africa	Positive	Crafford et al. (2014)
Diplectanum aequans	*Dicentrarchus labrax*	Corsica, France	Negative	Antonelli et al. (2016)
Gyrodactylus salmonis	Salmonids	Canada	Positive	Cone and Cusack (1988)
Gyrodactylus shulmani	Carps	Czechoslovakia	Negative	Zitnan and Hanzelova (1982)
		Environmental variable: Oxygen		
Sciadicleithrum guanduensis	*Geophagus brasiliensis*	Southern Brazil	Negative	Lacerda et al. (2018)
Quadriacanthus clariadis	*Clarias gariepinus*	Egypt	Positive	Mashaly et al. (2020)
Cichlidogyrus tilapiae	*Oreochromis niloticus*	Brazil	Positive	Cavalcanti et. al (2020)
Cichlidogyrus thurstonae	*Oreochromis niloticus*	Brazil	Negative	Cavalcanti et. al (2020)
Environmental variable: Heavy metal (cadmium, iron, nickel)				
Macrogyrodactylus clarii	*Clarias gariepinus*	Egypt	Negative	El-Naggar et al. (2017)
Quadriacanthus spp.	*Clarias gariepinus*	Egypt	Non-significant (cadmium and nickel) Positive (iron)	Mashaly et al. (2020)
		Environmental variable: Pulp and paper mill effluents		
Dactylogyrus species	*Rutilus*	Central Finland	Positive	Koskivaara and Valtonen (1992)
Dactylogyrus spp.	*Rutilus*	Sweden	Negative	Khan and Thulin (1991)

REFERENCES

Antonelli L, Foata J, Quilichini Y, and Marchand B (2016). Influence of season and site location on European cultured sea bass parasites in Corsican fish farms using indicator species analysis (IndVal). *Parasitol Res* 115:561–568. doi:10.1007/s00436-015- 4772-9.

Appleby C (1996) Variability of the haptoral hard parts of *Gyrodactylus callariatis* Malmberg, 1957 (Monogenea, Gyrodactylidae) from Atlantic cod *Gadus morhua* L. in the Oslo Fjord, Norway. *Syst Parasitol* 33:199–207. doi:10.1007/BF01531201.

Bauer ON (1959). The influence of environmental factors on reproduction of fish parasites. *Voprosy Ecologii* 3:132–141.

Bayoumy EM, Osman HAM, El-Bana LF, and Hassanain MA (2008). Monogenean parasites as bioindicators for heavy metal status in some Egyptian Red Sea fishes. *Global Vet* 2:117–122.

BilongBilong CF, and Tombi J (2005). Temporal structure of a component community gill parasites of *Barbus martorelli* (Roman) (freshwater Cyprinidae) in the Centre province, Cameroon. *Cam J Biochem Sci* 13:9–18.

Blanar CA, Munkittrick KR, Houlahan J, MacLatchy DL, and Marcogliese DJ (2009). Pollution and parasitism in aquatic animals: a meta-analysis of effect size. *Aquat Toxicol* 93:18–28. doi:10.1016/j.aquatox.2009.03.002.

Bush AO, Lafferty KD, Lotz JM, and Shostak AW (1997). Parasitology meets ecology on its own terms: Margolis et al. revisited. *Int J Parasitol* 83:575–583. doi:10.2307/3284227.

Cavalcanti LD, Gouveia EJ, Leal FC, Figueir CSM, Rojas SS, and Russo MR (2020). Responses of monogenean species to variations in abiotic parameters in tilapiculture. *J Helminthol* 94:e186. doi:10.1017/S0022149X20000711.

Cone DK, and Cusack R (1988). A study of *Gyrodactylus colemanensis* Mizelle and Kritsky, 1967 and *Gyrodactylus salmonis* (Yin and Sproston, 1948) (Monogenea) parasitizing captive salmonids in Nova Scotia. *Can J Zool* 66:409–415. doi:10.1139/z88-058.

Crafford D, Luus-Powell W, and Avenant-Oldewage A (2014). Monogenean parasites from fishes of the Vaal Dam, Gauteng Province, South Africa I. Winter survey versus summer survey comparison from *Labeo capensis* (Smith 1841) and *Labeo umbratus* (Smith, 1841) hosts. *Acta Parasitol* 59:17–24. doi:10.2478/s11686-014-0205-7.

Cunha KN, Domingues MV, Cunha LDS, and Nunes ZMP (2021). Parasitic monogenoideans of *Sciades herzbergii* as bioindicators of environmental quality in amazonian estuarine ecosystems. *Braz J Vet Parasitol* 30:e024220. doi:10.1590/ S1984-29612021013.

Dmitrieva EV, and Dimitrov G (2002). Variability in the taxonomic characters of Black Sea gyrodactylids (Monogenea). *Syst Parasitol* 51:199–206. doi:10.1023/A:1014594614921.

Dobson AP et al. (2008). Homage to Linnaeus: how many parasites? How many hosts? *Proc Natl Acad Sci USA* 105:11482–11489. doi:10.1073/pnas.0803232105.

Dušek L, Gelnar M, and Sebelova S (1998). Biodiversity of parasites in a freshwater environment with respect to pollution: metazoan parasites of chub (*Leuciscus cephalus* L.) as a model for statistical evaluation. *Int J Parasitol* 28:1555–1571. doi:10.1016/S0020-7519(98)00130-1.

Dzika E (2002). Deformation of the attachment organ in Diplozoidae (Palombi, 1949) (Monogenea). *Wiad Parazytol* 48:69–77.

El-Naggar AM, Mashaly MI, Hagras AM, and Alshafei HA (2017). Monogenean microfauna of the Nile catfish, *Clarias gariepinus* as biomonitors of environmental degradation in aquatic ecosystems at the Nile Delta Egypt. *J Environ Sci Toxicol Food Technol* 8:45–62. doi:10.9790/2402-1108014562.

Ergens R (1981). Variability of hard parts of haptor in *Gyrodactylus truttae* Glaser, 1974 (Gyrodactylidae: Monogenea). *Folia Parasitol* 28:37–42.

Ergens R, and Gelnar M (1985). Experimental verification of the effect of temperature on the size of hard parts of opisthaptor of *Gyrodactylus katharineri* Malmberg, 1964 (Monogenea). *Folia Parasitol* 32:377–380.

Galli P, Crosa G, Mariniello L, Ortis M, and D'Amelio S (2001). Water quality as a determinant of fish parasite communities. *Hydrobiologia* 452:173–179. doi:10.1023/A:1011958422446.

Gao Q, and Nie P (2000). Lead content in the monogenean *Ancryocephalus mogurndae* and in different organs of its host, the mandarin fish, *Siniperca chuatsi*. *China Environ Sci* 20:233–236.

Gibson D et al. (2014). Fauna Europaea: Helminths (Animal Parasitic). *Biodivers Data J* 2:e1060. doi:10.3897/BDJ.2.e1060.

Gilbert BM, and Avenant-Oldewage A (2016). Effects of altered water quality and trace elements on the infection variables of *Paradiplozoon ichthyoxanthon* (Monogenea: Diplozoidae) from two sites in the Vaal River system, South Africa. *Acta Parasitol* 61:52–62. doi:10.1515/ap-2016-0005.

Geets A, Appleby C, and Ollevier F (1999). Host-dependent and seasonal variation in opisthaptoral hard parts of *Gyrodactylus cf. arcuatus* from three *Pomatoschistus* spp. and *C. arcuatus* from *Gasterosteus aculeatus*: a multivariate approach. *Parasitology* 119:27–40. doi:10.1017/S0031182099004400.

Goater TM, Goater CP, and Esch GW (2013). *Parasitism: The Diversity and Ecology of Animal Parasites*. Cambridge, UK: Cambridge University Press.

Gusev AV (1976). Freshwater Indian Monogenoidea. Principles of systematics, analysis of the world faunas and their evolution. *Ind. Jour. Helminth. (N.S.)* 25–26:1–241.

Holt EA, and Miller SW (2010). Bioindicators: using organisms to measure environmental impacts. *Nature Education Knowledge* 3:8. Available from: www.nature.com/scitable/knowledge/library/bioindicators-using-organisms-to-measure-environmental-impacts-16821310/ (accessed 02 July 2021).

Hu JJ, and Li HY (2016). Seasonal population dynamics of the monogeneans *Quadriacanthus kobinensis* parasitising *Clarias fuscus* in Pearl River, China. *Helminthologia* 53:105–108. doi:10.1515/helmin-2015-0070.

Ivona M (2004). Monogenean parasites in Adriatic cage-reared fish. *Acta Adriat* 45:65–73.

Jansen PA, and Bakke TA (1991). Temperature-dependent reproduction and survival of *Gyrodactylus salaris* Malmberg, 1957 (Platyhelminthes: Monogenea) on Atlantic salmon (*Salmo salar* L.). *Parasitology* 102:105–112. doi:10.1017/S0031182000060406.

Johnsen BO, and Jensen A (1991). The *Gyrodactylus* story in Norway. *Aquaculture* 98:289–302. doi:10.1016/0044-8486(91)90393-L.

Khan RA, and Thulin J (1991). Influence of pollution on parasites of aquatic animals. *Adv Parasitol* 30:201–238. doi.org/10.1016/S0065-308X(08)60309-7.

Koskivaara M, and Valtonen ET (1992). *Dactylogyrus* (Monogenea) communities on the gills of roach in three lakes in Central Finland. *Parasitology* 104: 263–272. doi:10.1017/S0031182000061709.

Koskivaara M, Valtonen ET, and Vuori KM (1992). Microhabitat distribution and coexistence of *Dactylogyrus* species (Monogenea) on the gills of roach. *Parasitology* 104:273–281. doi:10.1017/S0031182000061710.

Kuperman BI (1992). Fish parasites as bioindicators of the pollution of bodies of water. *Parazitologiia* 26:479–482.

Lacerda A, Roumbedakis K, Bereta Junior JGS, Nuner APO, Petrucio MM, and Martins ML (2018). Fish parasites as indicators of organic pollution in southern Brazil. *J Helminthol* 92:322–331. doi:10.1017/S0022149X17000414.

Lafferty KD (1997). Environmental parasitology: what can parasites tell us about human impacts on the environment? *Parasitol Today* 13:251–255. doi.org/10.1016/S0169-4758(97)01072-7.

McDowell JE, Lancaster BA, Leavitt DF, Rantamaki P, and Ripley B (1999). The effects of lipophilic organic contaminants on reproductive physiology and disease processes in marine bivalve molluscs. *Limnol Oceanogr* 44:903–909.

MacKenzie K, McVicar AH, and Waddell IF (1976). Some parasites of plaice *Pleuronectes platessa* L. in three different farm environments. *Scott Fish Res Rep* 4:1–14.

MacKenzie K, Williams HH, Williams B, McVicar AH, and Siddall R (1995). Parasites as indicators of water quality and the potential use of helminth transmission in marine pollution studies. *Adv Parasitol* 35:85–144. doi:10.1016/S0065-308X(08)60070-6.

Malmberg G (1970). The excretory systems and the marginal hooks as a basis for the systematics of *Gyrodactylus* (Trematoda: Monogenea). *Arkiv for Zoologi Serie* 2, 23:1–235.

Marcogliese DJ (2004). Parasites: small players with crucial roles in the ecological theatre. *Eco Health* 1:151–164. doi:10.1007/s10393-004-0028-3.

Mashaly MI, Allam HE, and El-Naggar MM (2020). Impacts of physicochemical and heavy metal parameters on infestation level of the monogeneans, *Quadriacanthus* spp. infesting Nile catfish, *Clarias gariepinus* of different water localities in Nile Delta, Egypt. *J Parasit Dis* 44:579–589. doi:10.1007/s12639-020-01231-7.

Mbokane EM, Matla MM, Theron J, and Luus-Powell WJ (2015). Seasonal dynamics and occurrences of three *Dactylogyrus* species on the gills of three cyprinids at Nwanedi–Luphephe dams in Limpopo province, South Africa. *Afr Zool* 50:119–125. doi.10.1080/15627020.2015.1021175.

Mo TA (1991). Seasonal variations of opisthaptoral hard parts of *Gyrodactylus salaris* Malmberg, 1957 (Monogenea: Gyrodactylidae) on parr of Atlantic salmon *Salmo salar* (L) in the River Batnfjordselva, Norway. *Syst Parasitol* 19:231–240. doi:10.1007/BF00011889.

Mo TA (1997). Seasonal occurrence of *Gyrodactylus derjavini* (Monogenea) on brown trout, *Salmo trutta*, and Atlantic salmon, *S. salar*, in the Sandvikselva River, Norway. *J Parasitol* 83:1025–1029.

Palm HW, and Ruckert S (2009). A new approach to visualize ecosystem health by using parasites. *Parasitol Res* 105:539–553. doi:10.1007/s00436-009-1423-z.

Rodríguez-Gonzalez A et al. (2020). Fluctuating asymmetry of sclerotized structures of Haliotrematoides spp. (Monogenea: Dactylogyridae) as bioindicators of aquatic contamination. *Ecol Indic* 117: 106548. doi:10.1016/j.ecolind.2020.106548.

Ruckert S, Hagen W, Yuniar AT, and Palm HW (2009). Metazoan fish parasites of Segara Anakan Lagoon, Indonesia, and their potential use as biological indicators. *Reg Environ Change* 9:315–328. doi:10.1007/s10113-008-0076-2.

Sanchez-Ramirez C et al. (2007). *Cichlidogyrus sclerosus* (Monogenea: Ancyrocephalinae) and its host the Nile tilapia (*Oreochromis niloticus*) as bioindicators of chemical pollution. *J Parasitol* 93:1097–1106. doi:10.1645/GE-1162R.1.

Shinn AP, Gibson DI, and Sommerville C (1995). A study of the composition of the sclerites of *Gyrodactylus* Nordmann, 1832 (Monogenea) using X-ray elemental analysis. *Int J Parasitol* 25:797–805. doi:10.1016/0020-7519(95)00008-P.

Skinner RH (1982). The interrelation of water quality, gill parasites and gill pathology of some fishes from South Biscayne Bay, Florida. *Fish Bull* 80:269–280.

Soleng A, and Bakke TA (1997). Salinity tolerance of *Gyrodactylus salaris* (Platyhelminthes, Monogenea): laboratory studies. *Can J Fish Aquat Sci* 54:1837–4185. doi:10.1139/f97-089.

Sures B (2001). The use of fish parasites as bioindicators of heavy metals in aquatic ecosystems: a review. *Aquat Ecol* 35:245–255. doi:10.1023/A:1011422310314.

Sures B (2004). Environmental parasitology: relevancy of parasites in monitoring environmental pollution. *Trends Parasitol* 20:170–177. doi.10.1016/j.pt.2004.01.014.

Sures B, and Nachev M (2015). Environmental Parasitology. In: Mehlhorn H. (eds) *Encyclopedia of Parasitology*. Berlin, Heidelberg: Springer. doi:10.1007/978-3-642-27769-6_1076-2.

Sures B, Nachev M, Selbach C, and Marcogliese DJ (2017). Parasite responses to pollution: what we know and where we go in 'environmental parasitology'. *Parasit Vectors* 10:65. doi:10.1186/s13071-017-2001-3.

Tripathi A, Wangchu L, and Trivedi AK (2019). Prevalence and intensity of *Cornudiscoides agarwali* (Monogenoidea) on the gills of Day's mystus (*Mystus bleekeri*) in relation to some ecological and biological factors from Arunachal Pradesh, India. *Helminthologia* 56:141–150. doi:10.2478/helm-2019-0005.

Tripathi YR (1959). Monogenetic trematodes from fishes of India. *Ind Jour Helminth* 9:1–49.

Whittington ID (1988). Diversity 'down under': monogeneans in the Antipodes (Australia) with a prediction of monogenean biodiversity worldwide. *Int J Parasitol* 28:1481–1493. doi:10.1016/S0020-7519(98)00064-2.

Williams HH, and Jones A (1994). *Parasitic Worms of Fish*. London, UK: Taylor and Francis.

Zitnan R, and Hanzelova V (1982). Seasonal dynamics of the invasion cycle of *Gyrodactylus shulmani* Ling Mo-En, 1962 (Monogenea). *Folia Vet* 26:183–194.

Zolovs M, Ozuna A, and Kirjusina M (2012). Seasonal variation of attachment apparatus and copulatory organ morphometric variables of *Dactylogyrus crucifer* Wagener, 1857 (Monogenea: Dactylogyridae) on the gills of roach (*Rutilus* L.) in Latvian water bodies. *Acta Biol Univ Daugav* 12:191–198.

27 Haematological Parameters and Cell Morphology of *Cyprinus carpio var. communis* and *specularis* from Dal Lake Kashmir, Himalayan Region

Imtiaz Ahmed, Zubair Ahmad Sheikh, Aamina Hussan, and Kousar Jain
Department of Zoology, University of Kashmir, Srinagar, Kashmir

Francesco Fazio
Department of Veterinary Sciences, University of Messina, Messina, Italy

Rana Samad
Department of Zoology, Kirori Mal College, University of Delhi, Delhi

CONTENTS

DOI: 10.1201/9781003220824-31

27.1 INTRODUCTION

For centuries, fish has been recognized as an excellent food source for humans, and it is preferred as a perfect diet not only because of its excellent taste and high digestibility but also because of the unsaturated fatty acids, essential amino acids and minerals that are essential for the formation of functional and structural proteins in humans (Kumar, 1992; Paul et al., 2016; Ahmed & Sheikh, 2017; Joshi et al., 2017; 2018; Linhartova et al., 2018; Ahmed et al., 2022). The common carp, *Cyprinus carpio* of the Cyprinidae family, is one of the most resilient freshwater fish species, surviving in both cold and warm water environments and withstanding a wide range of environmental variations. It prefers big bodies of water with slow-moving or stagnant water and soft bottom sediments. Plankton and detritus found at the bottom of the water body are used by the common carp. It removes undesirable aquatic vegetation from water bodies by uprooting plants while hunting for food in a culture pond with a high stocking density and intense competition for food (Das & Subla, 1964; Fotedar & Qadri, 1974; Jhingran, 1991; Talwar & Jhingran, 1991; Bozkurt et al., 2017). Common carp farming is widely practised in many tropical and subtropical regions of the world, and it is one of the most widely farmed freshwater fish species. Many investigations on the basic physiology and haematology of common carp species have been undertaken (Adedeji et al., 2000; Svobodova et al., 2008; Bastami et al., 2009; Kumar et al., 2010; Coroian et al., 2015; Babahydari et al., 2018; Lutnicka et al., 2019; Kesbic et al., 2020), although there is limited information available on the haematological examination of two common carp species, *C. carpio communis* and *C. carpio specularius*, that live in Kashmir's climate. However, full research on the shape of blood cells in two kinds of common carp species is yet to be published.

The fish population is dwindling day by day due to a variety of factors. To determine the true cause of fish population decline from natural resources, it is critical to monitor the health status of fish before any clinical symptoms appear. This requires analysing the fish's haematological status, which has been used as an initial marker for assessing fish health and is also important for determining the physiological capacity of fish (Affonso, 2001; Wells et al., 2005; Pradhan et al., 2014; Sharma et al., 2017; Ahmed et al., 2019; Ahmed & Sheikh, 2019; Fazio, 2019; Sheikh & Ahmed, 2019; Sidiq & Ahmed, 2020; Jan & Ahmed, 2021). However, one of the challenges in determining the health of a fish population has been the scarcity of credible references on typical fish values. Many fish physiologists have turned their attention to studies of haematology and associated parameters in order to achieve this goal, which could prove to be a useful diagnostic tool for assessing the health of fish (Pradhan et al., 2014; Farshad et al., 2015; Suleiman et al., 2016; Sharma et al., 2017; Zhao et al., 2018; Sheikh & Ahmed, 2019; Ahmed & Sheikh, 2020; Sheikh & Ahmed, 2020; Jan et al., 2021).

Blood parameters have been studied to discover the systematic links between different species (Filho et al., 1992; Sheikh & Ahmed, 2016; Ahmed et al., 2020; Chen et al., 2021). Blood component values are known to vary due to genetic and physiological factors. Inter-specific influences between species and intra-specific factors within species may cause genetic diversity. A variety of factors influence changes in haematological parameters, including aquatic biotope, fish species, age, sexual maturity and health state (Radu et al., 2009; Pradhan et al., 2014; Ejraei et al., 2015; Fazio et al., 2016; Sharma et al., 2017; Ahmed et al., 2019, 2020; Jan et al., 2021). Blood, which makes up about 1.3–7.0% of a fish's total body weight accompanied by haematopoietic organs, is one of the most active components, contributing to metabolic processes by maintaining gas exchange between the organism and the environment. As a result, blood parameters are increasingly being used as markers of a fish's physiological state or sub-lethal stress in response to endogenous or exogenous alterations (Belanger et al., 2001; Ahmed et al., 2020; Jan & Ahmed, 2021). Furthermore, the metabolic content of fish blood is normally influenced by its environment (Bullis, 1993; Ahmed et al., 2020; Jan et al., 2022).

Fish live in close proximity to their surroundings and are therefore susceptible to physical and chemical changes that may be reflected in their blood components, thus making them ideal for biomonitoring investigations. The blood parameters of *C. carpio communis* and *C. carpio specularis* are of great interest because these species have huge commercial significance due to their low cost, ease of availability in fresh forms, and nutritional value. Blood parameters serve as trustworthy markers of fish health, which is a well-established notion (Satheeshkumar et al., 2011; Fazio et al., 2018; Fazio, 2019). As a result, discovered changes linked with haematological parameters could be used to diagnose diseases and guide the implementation of treatment or preventive measures, both of which are critical in fish farming and the fish business (Roberts, 1981; Sheikh & Ahmed, 2019; 2020).

The following frame of work was used to capture two types of common carp fish from Dal Lake for haematological examination.

27.2 STUDY AREA AND FISH SAMPLING

In the current study, examination of live specimens of *Cyprinus carpio communis* and *Cyprinus carpio specular* collected from Dal Lake was carried out. Three sites, i.e., Telbal, Brein and Gagribal, were selected for the fish collection (Figure 27.1). Within 1 hour of capture, all specimens (fish) were brought to the experimental station in open tanks containing 20 l of water from the same water body. The water tank was artificially oxygenated during transport from the capture site to the experimental station. To compensate for the impacts of capture, handling and transport, the fish were acclimated overnight in a well-aerated 70-l plastic tank connected to a continuous flow-through system with a water volume of 1–1.5 l/min. Fish were transported to a 20-l plastic tank after acclimatization and anaesthetized (MS-222 at 0.3 g/l of water) before blood sampling.

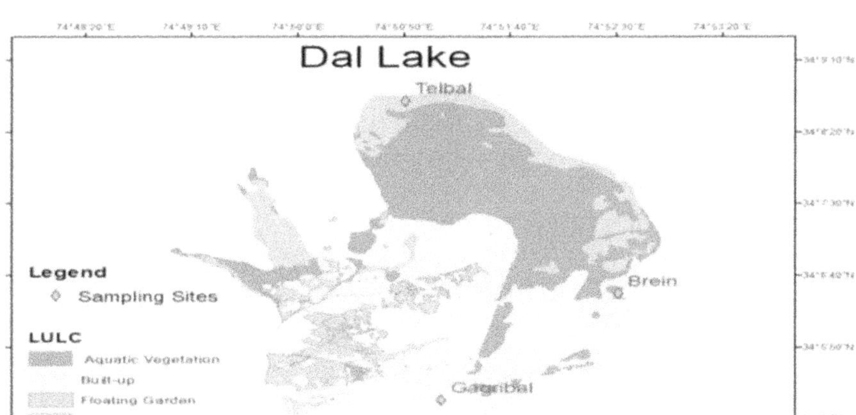

FIGURE 27.1 Map showing the various sites selected from Dal Lake.

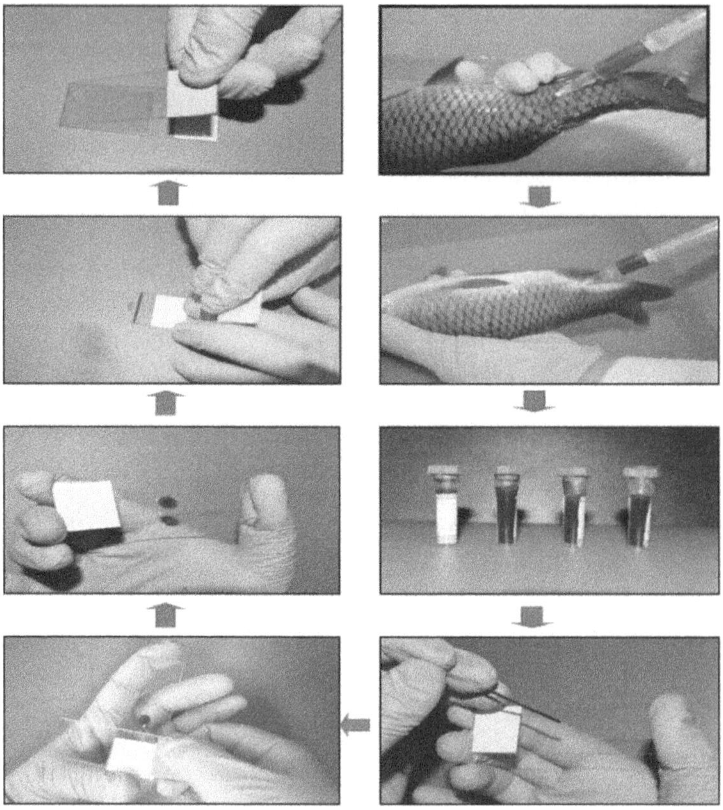

FIGURE 27.2 Haematological analysis of *Cyprinus carpio* collected from Dal Lake.

Haematological analysis was performed on blood samples obtained by venipuncture of the caudal vein (Figure 27.2). Following collection of blood samples, all fish were weighed and measured using an electronic balance (Shimadzu UX320G) and a digital Vernier calliper, respectively.

27.3 METHODS OF HAEMATOLOGICAL ANALYSIS

27.3.1 Estimation of Haemoglobin (Hb)

Drabkin's (1946) approach was used to assess blood haemoglobin concentration. Twenty microlitres of blood was combined with 5 ml of Drabkin solution (Loba chemie, India) and allowed to stand for at least 15 minutes. The solution's absorbance was then measured at 540 nm using a spectrophotometer (Genesys 10S UV-VIS).

Hb content was calculated using the following formula:

$$\text{Hb}(\text{g}/\text{dl}) = \frac{\text{Absorbance of test sample}}{\text{Absorbance of standard}} \times \text{concentration of standard}$$

27.3.2 Total Red Blood Cell (RBC) and White Blood Cell (WBC) Count

A blood sample of 20 µl was obtained with the use of a micro-pipette (Finnpipette, Finland) and diluted with Natt–Herrick's diluent (1952) for RBC count (1:200). The blood cells were counted using a light microscope after the diluted sample was placed in a Neubauer enhanced haemocytometer (Marienfeld-Superior, Lauda-Konigshofen, Germany) (Magnus-MLM, India).

The following formula was used to calculate the RBC count:

$$\text{RBC count} = \frac{\text{Total red blood cells counted} \times \text{dilution factor} \times \text{depth of chamber}}{\text{Area counted}}$$

Where dilution factor is 1 in 200, depth is 1/10 mm and area counted = 80/400 = 1/5 sq.mm.

$$\text{RBC count} = \frac{\text{Number of cells counted} \times 200 \times 10}{1/5}$$

$$\text{RBC} / \text{mm}^3 = \text{number of cells counted} \times 10{,}000$$

The white blood cell count (WBC) was performed using the same technique as for the RBC count, and was calculated using the following formula:

$$\text{WBC count} = \frac{\text{Total white blood cells counted} \times \text{blood dilution} \times \text{chamber depth}}{\text{Number of chambers counted}}$$

$$\text{WBC/mm}^3 = \frac{\text{Total white blood cells} \times 200 \times 10}{4}$$

27.3.3 Haematocrit (Hct%)

The haematocrit (Hct%) was calculated using blood sedimentation. Heparinized blood was collected in a micro-haematocrit capillary (Na-heparinized) and spun at 12,000 rpm for 5 minutes in a micro-haematocrit centrifuge (RM-12C, REMI, India) to acquire haematocrit values. The haematocrit readings were calculated using a haematocrit reader, and the Hct values were expressed as a percentage of the total haematocrit (Zaragoza et al., 2008).

27.3.4 Erythrocyte Indices (MCH, MCHC and MCV)

According to Dacie & Lewis (1991), erythrocyte parameters such as mean corpuscular haemoglobin (MCH), mean corpuscular haemoglobin concentration (MCHC) and mean corpuscular volume (MCV) were calculated.

27.3.4.1 Mean Corpuscular Haemoglobin (MCH)

MCH is the amount of haemoglobin of the average RBC.

$$\text{MCH} = \text{Hb (g dl}^{-1}) \times 10 \text{ micrograms/RBC (millions/microlitre)}$$

27.3.4.2 Mean Corpuscular Haemoglobin Concentration (MCHC)

MCHC in terms of g% for 100 ml erythrocytes was computed as:

$$\text{MCHC} = \text{Hb (g dl}^{-1}) \times 100 \text{ ml/Hct (\%)}$$

27.3.4.3 Mean Corpuscular Volume (MCV)

MCV represents the average volume of RBCs and is calculated as:

$$MCV = Hct\ (\%) \times 10\ \text{cubic microns/RBC (millions/microlitre)}$$

27.4 BLOOD SMEAR PREPARATION AND LIGHT MICROSCOPIC STUDY

A small amount of blood was used to make blood smears on glass slides right away. After air drying, the slides were fixed in 100% methanol for 5 minutes before being stained with Wright–Giemsa solution for 10 minutes. Per sample, one to two blood smears were produced, and photographs were taken using light microscopy. An ocular micrometer was used to measure the size of blood cells. Cells were identified based on their shape and cell structure, which had previously been recorded in fish leukocyte research (Rowley, 1990; Tang et al., 2015). A method published by Zhang et al. (2011) was used to count multiple types of leukocytes at the same time. The erythrocyte length (EL) and erythrocyte width (EW), nucleus length (NL) and nucleus width (NW), thrombocytes, heterophils, eosinophils, lymphocytes and monocytes were measured in dry smears of each slide using an Olympus ocular micrometer at a magnification of ×1000. The sizes of erythrocytes and nuclei were determined using the conventional formulas ($EL \times EW \times \pi/ 4$) and ($NL \times NW \times / 4$) (Metin et al., 2008).

27.5 BLOOD VALUES IN *CYPRINUS CARPIO*

Haematological parameters of two common carp fish species, *C. carpio communis* and *C. carpio specularis*, along with other fish species are presented in Table 27.1. The two healthy fish species that live in Dal Lake revealed significant differences in haematological parameters in the current study. Higher haemoglobin (Hb), red blood cell (RBC) count and Hct (%) value were found in *C. carpio communis*, while *C. carpio specularis* showed lower values of these parameters. In comparison to mammals, fishes have low haemoglobin content. The Hb content of two carp species was estimated to be in the range of 5–10 g/dl, whereas the haematocrit percentage of common carp in this study was reported to be in the range of 24.45–41.87%, indicating that these fish species are very active, since species with Hct content of more than 20 to 45% are considered active fish according to the literature. Fish with a haematocrit of more than 45% are frequently dehydrated, especially when serum osmolality, sodium chloride and total protein levels are high. Fish with haematocrit readings of less than 20% are termed anaemic: a condition characterized by erythrocytes with condensed nuclei, erythroplastids and erythrocyte fragmentation that is really connected with splenic evacuation of senescent erythrocytes from the peripheral circulation (Ellis, 1984; Campbell, 2015). The haematocrit can be artificially raised by splenic constriction and RBC enlargement after the sample is taken (Heath, 1995; Hrubec & Smith, 2010). Haematocrit values differ within and between fish species and are influenced by a variety of factors. The variation in haematocrit values is primarily related to fish activity, with less active fish having a lower haematocrit than energetic or fast active fish species (Rowley, 1988, Wilhelm Filho et al., 1992). The levels of haemoglobin and haematocrit in fish may change over time. Seasonal alterations in water temperature, dissolved oxygen content, photoperiod, feeding regime and the reproductive cycle of fish species are also linked to fluctuations in these factors (Ram Bhasker & Srinivasa Rao, 1989; Groff & Zinkl, 1999; Pradhan et al., 2014; Campbell, 2015; Sharma et al., 2017; Ahmed & Sheikh, 2019; Ahmed et al., 2020). In fish, erythrocytes (RBC) are the most abundant blood cells (98–99%) (Fange, 1994). The erythrocyte number of *C. carpio* was reported to be in the range of $1.5–1.93 \times 10^6/mm^3$ in this case. The quantity of erythrocytes, however, varies with species. According to Soldatov (2005), certain species have a very low erythrocyte count ($0.5–1.5 \times 10^6/mm^3$) yet an extraordinarily high RBC count ($3.0–4.2 \times 10^6/mm^3$). These differences were mainly explained by locomotor activity, stress and the

TABLE 27.1
Comparison of Hematological Parameters of *Cyprinus Carpio* and Various Fish Species

Species	No.	Hb (g/dl)	RBC ($\times 10^6/mm^3$)	PCV (%)	MCH (pg)	MCHC (g/dl)	MCV (fl)	Reference
Cyprinus carpio	65	5.0–10.0	1.5–1.93	24.5–41.87	40.2–75.6	16.84–35.65	132.23–305.0	Present study
***Cyprinus carpio* (koi)**	30	6.3–7.6	1.69–1.91	29.7–33.8	37.7–42.7	20.4–22.9	166–190	Tripathi et al. (2004)
Tilapia	40	7.0–9.8	1.91–2.83	27–37	28.3–42.3	22–29	115–183	Hrubec et al. (2000)
Colossoma brachypomum	29	------	1.68	25	------	-------	------	Tocidlowski et al. (1997)
M. saxatilis* × *M. chrysops	50	8–12	3.66–4.96	23–47	19.6–26.4	22–30	81–106	Hrubec et al. (1996)
Oncorhynchus aquabonita	12	-------	1.19–1.20	44–52	-------	-------	-----	Hunn et al. (1992)
Pleuronectes americanus	20	2.8–6.4	1.8–2.5	21–28	15–26	-------	101–126	Mahoney & McNulty (1992)
Sarotherodon melanotheron	40	7.3–9.0	1.37–1.69	31–34	54	24–26	203–228	LeaMaster et al. (1990)
Chanos	283	5–15	1.70–4.00	22–48	20.9–47.2	11–38	133–302	Ram Bhasker & Srinivasa Rao (1989)
Salmo salar	20	8.9–10.4	0.85–1.1	44–49	94–106	19.4–21.7	441–553	Sandnes et al. (1988)
Morone saxatilis	18	6.2–10.9	2.0–4.2	34–55	31.3	20.5	155	Miller et al. (1983)
Oncorhynchus mykiss	122	1.5–7.7	0.77–1.67	21–44	14.4–70.0	5.6–24.4	192–420	Miller et al. (1983)
Carassius auratus	60	9.7–10.6	1.6–1.8	38–40	63–66	26	241–245	Burton & Murray (1979)
Pleuronectes americanus	20	4.2–6.0	1.7–2.6	17–26	25–33.	-------	90–126	Bridges et al. (1976)
Ictalurus punctatus	35	-------	2.44	40	-------	-----	-----	Grizzl & Roger (1976)
Oncorhynchus mykiss	200	5.3–9.5	------	24–43	---------	21.9–24.1	------	Wedemeyer & Nelson (1975)
Salvelinus fontinalis	74	8.0	1.34	36.6	-------	-------	------	Houston & Dewilde (1972)

temperature of the environment (Hrubec et al., 1996; Roberts & Ellis, 2001; Shahjahan et al., 2018). Fish that are more active and have higher oxygen demands have smaller RBC and consequently lower MCV values.

In fish, the leukocyte count is variable but is generally reported within the range from 30,000 to 100,000 cells/μl. The leukocyte count of two common carp species was recorded in the range from 2.6 to 4.8 × 10^3 mm^{-3} (Table 27.2). The leukocyte count of fish provides reliable information about the health of fish, and any change, i.e., decline in the lymphocyte number, is usually regarded as a stress response (Chen et al., 2002). The value of MCV ranges from 150 to 350 fl in fish species. In our case, the MCV values of common carp fluctuated from 132.23 to 305 fl, while MCH values also varied between 40.2 to 75.0 pg. Similarly, the MCH values also varied significantly between the fish species and were reported within the range from 30 to 100 pg; such differences may occur due to difference in size of circulating RBCs. The MCHC of two common carp species was also noted in the range from 16.84 to 37.65 g/dl. In general, the MCHC values are recorded in the range from 18% to 30% in fishes. The emerging RBCs continue to increase in size and haemoglobin concentration with time, resulting in higher erythrocyte indices for older cells. Moreover, there may be a variety of developing stages present at any period, resulting in a wide range of these indices. The water temperature impacts the RBC indices by its influence on oxygen requirement and metabolism. Smaller fish exposed to higher water temperatures had higher Hb and Hct concentrations than larger fish (Groff & Zinkl, 1999; Svetina et al., 2002; Hrubec & Smith, 2010; Campbell, 2015; Shahjahan et al., 2018).

27.6 BLOOD CELL MORPHOLOGY

Morphometrical analysis of blood cells perform a significant role in fish pathology, because it offers numerical objectification of the most suitable alterations unavailable for visual estimation, and thus has clinical and research applications that are becoming more abundant, particularly in cytological and histopathological studies (Kumar, 2016; Sheikh & Ahmed, 2020; Chen et al., 2021). Essential tissues in fish blood cells are highly susceptible to alterations taking place in physiological status and stimuli from the fish's environment (Li & Wang, 1995; Fang et al., 2014; Sharma et al., 2017; Shahjahan et al., 2018; Sheikh & Ahmed, 2019; Ahmed et al., 2020; Jan et al., 2021). Blood cell characteristics could reflect fish health and the environment in which they thrive (Zhou et al., 2003), besides being useful in determining the systematic associations between certain fish species (Pavlidis et al., 2007). In response to environmental circumstances, fish blood characteristics vary, and hence, a difference in haematological parameters might behave as a biomarker of sub-lethal environment stress.

The morphometric characterizations of blood cells of common carp species were measured, and the results are presented in Table 27.3 along with the information reported on other teleost fish species. Among the blood cells, erythrocytes (Figure 27.3A) are considered as the chief blood cells in fish, comprising about 98–99% of the total blood cell population (Fange, 1994; Witeska, 2013). Mature fish erythrocytes are similar in form to the erythrocytes of aves and reptiles, although they are larger (Witeska, 2013; Campbell, 2015). Normal mature fish erythrocytes are generally oval to ellipsoidal in shape and possess a nucleus. Their size is variable, with a long diameter of about 8.8–17.1 μm and a short diameter of 6.9–12.9 μm (Witeska, 2013). The nucleus is located in the centre of adult fish erythrocytes, which are biconvex in shape with swelling in the centre, while others comprise biconcave flattened erythrocytes (Hibiya, 1985). The cytoplasm is normally homogeneous; however, pale-coloured regions or vacuoles may be present in varying amounts. The nuclei of fish erythrocytes can be quite massive, taking up a quarter or more of the cell's volume. When compared with mature erythrocytes, immature erythrocytes appear spherical and have less cytoplasm (Figure 27.3B). In comparison to mature forms, the nucleus of immature erythrocytes is largely smaller in size and round to oval in shape, with light purple-stained cytoplasm. However,

TABLE 27.2
Comparison of White Blood Cell, Differential Leukocyte and Thrombocyte Counts of Various Fish Species

Species	No.	WBC (/µl)	Lymphocytes (/µl)	Neutrophils (/µl)	Monocytes (/µl)	Eosinophils (/µl)	Thrombocytes (/µl)	Reference
Cyprinus carpio	65	26, 000–48,000	68–75%	15–23%	3.47–6.78%	1.31–2.31%	46.78–54.76%	Present study
Ictalurus punctatus	40	27,460–41,523	5,380–11,581	2,949–7,459	12,529–22,74	-------	58,802–99,569	Tavares-Dias & Moraes (2007)
***Cyprinus carpio* (koi)**	30	19,900–28,100	14,700–23,500	1,570–3,900	460–960	------	-----	Tripathi et al. (2004)
Tilapia	40	21,600–154,700	6,800–136,400	600–9,900	400–4,300	35–1,600	25,100–85,200	Hrubec et al. (2000)
Colossoma brachypomum	22	33,500	21,028	3,183	1,242	209	------	Tocidlowski et al. (1997)
M. saxatilis* × *M. chrysops	50	32,600–115,100	22,500–115,100	400–3,500	1,500–7,500	0–400	30,700–74,100	Hrubec et al. (1996)
Oncorhynchus aquabonita	5	21,000	18,799	1,582	588	------	135,000–310,000	Hunn et al. (1992)
Chanos	195	17,500–92,500	51–68%	3–7%	4–9%	------	12–32%	Bhaskar & Rao (1989)
Carassius auratus	55	10,100–14,700	9,540–13,660	------	--------	-------	30,000–46,100	Burton & Murray (1979)
Pleuronectes americanus	20	88,000–282,000	38,700–154,540	2,470–26,630	-----	-------	36,480–115,500	Bridges et al. (1976)
Ictalurus punctatus	35	164,000	89,900	5,200	500	0	68,400	Grizzle & Rogers (1976)

TABLE 27.3
Cell Dimensions in Microns of Some Teleost Fishes

Species	Erythrocytes	Thrombocytes	Lymphocytes	Monocytes	Neutrophils	Eosinophils	Basophils	Reference
Cyprinus carpio	8.6–14.4 × 4.1–5.49	7.4–5.3 × 3.4–6.4	6.71–10.10 × 6.4–8.8	9.2–12.4 × 6.8–9.7	7.36–9.8 × 5.6–7.2	6.8–9.8 × 5.5–6.4	6.3–9.6 × 5.7–8.0	Present study
Gymnocypris eckloni	14.82 × 10.01 14.88 × 10.02	8.28 × 5.41 8.37 × 5.37	10.49 × 8.57 10.15 × 8.94	12.50 × 10.72 12.32 × 10.55	12.93 × 11.61 12.85 × 11.59	13.35 × 12.32 13.29 × 12.63	---	Tang et al. (2015)
Cichlasoma dimerus	9.4–10 × 6.2–7.3	8.0–9 × 2.1–2.7	3.4–4.7	4.0–7.1	5–9.5	4.8–9.5	---	Vazquez & Guerrero (2007)
Acipenser sinensis	17.9 × 12.6	10.8 × 4.5	12.8 × 11.2	18.1 × 17.7	16.5 × 14.8	15.9 × 13.5	---	Zexia et al. (2007)
Oreochromis hybrid	12.9 × 7	4.7–5.5	4.6–5 × 5.7–6.4	9.4–10.7	6.38±1.04	5.77–6.7	---	Hrubec et al. (2000)
Carassius auratus	7–9 × 12–14	4.7–5.6	7.4–8.4	7.0–17	10.2–12.1	7.4–8.4	--	Groff & Zinkl (1999)
Cyprinus carpio	10.2 × 13.4	4.6 × 7.7	6.6–11.8	10.0–16	10.0–15	13.8	---	Groff & Zinkl (1999)

the number of immature erythrocytes was extremely low as compared with the mature forms. Lay & Baldwin (1999) reported that the erythrocyte dimensions and aerobic swimming ability in teleost fish show an inverse relationship. They found that in the case of smaller cells, the presence of a greater surface area to volume ratio results in a shorter diffusion distance and hence, permits better transfer of oxygen.

The leukocyte count also provides consistent information about the health status of fish. It generally varies between species, and is therefore very difficult to identify the type of cells. These can be recognized by using a comparative method and exclusion process on characteristic smears of blood processed with Wright–Giemsa stain. An assessment of the cell ultrastructure and function of differential leukocyte cells, including lymphocytes, monocytes, thrombocytes, neutrophils, eosinophils and basophils, was undertaken in fish species in order to provide information about the physiological and pathological status of fish living in natural conditions.

27.6.1 Thrombocytes

Fish blood generally clots in response to injury, although the efficiency and rate of coagulation are variable. Thrombocytes are basically involved in coagulation and clot formation in fish species. Fish thrombocytes, being nucleated, are smaller in size than erythrocytes, and by the stage of attaining maturity or level of reactivity, their shape varies and may be rounded, ovoid, spindle shaped or elongated (Figure 27.3C). Immature thrombocytes are of round shape in some fish species; elongated and oval-shaped forms tend to be non-reactive mature thrombocytes, while spindle-shaped thrombocyte cells seem to be reactive forms. Thrombocyte cells were found singly or in clusters of about 2–15 cells with a large nucleus bearing chromatin clusters. The cytoplasm was light purple in colour when stained with Wright–Giemsa. The occurrence of thrombocytes could easily be confused with small mature lymphocytes, though on comparison, lymphocytes consisted of somewhat more profuse, mildly basophilic cytoplasm.

27.6.2 Lymphocytes

Lymphocytes are found to be the most plentiful cells in the peripheral blood of fish, and repeatedly, about 50% of differential lymphocytes occur in healthy fish. The morphological and functional characteristics of fish lymphocytes coincide with the lymphocytes seen in blood smears of aves and mammals; however, their number is higher in fish than in mammals (Blaxhall & Daisley, 1973; Campbell, 2015). In fish, the majority of lymphocytes are small to medium sized, and these constitute the major quantity of the lymphocyte population, although large-sized lymphocytes are also found intermittently in fish blood. Both large- and small-sized lymphocytes were irregularly spherical, oval or round in shape and possessed a purple-stained large nucleus (Figure 27.3D). Small and large lymphocytes were classified on the basis of diameter and relative amount of cytoplasm. The occurrence of undifferentiated cells is believed to signify small lymphocytes, which tend to circulate until antigenic stimulation induces a change in them. Smaller lymphocytes enclosed a nucleus, which approximately occupied the whole cell, whereas the larger lymphocytes possessed a greater amount of cytoplasm. It has been noted that occasionally, lymphocytes containing blue cytoplasm were extended into pseudopodia on their surface.

27.6.3 Monocytes

Monocytes, the largest leukocytes, are infrequently observed in blood smears. Monocytic macrophages were round to oval in shape, distinguished by the presence of a prominent, kidney-shaped, eccentrically placed nucleus with blue-grey agranulocytic cytoplasm (Figure 27.3E). The morphology of fish monocytes was analogous to those found in reptiles, birds and mammals.

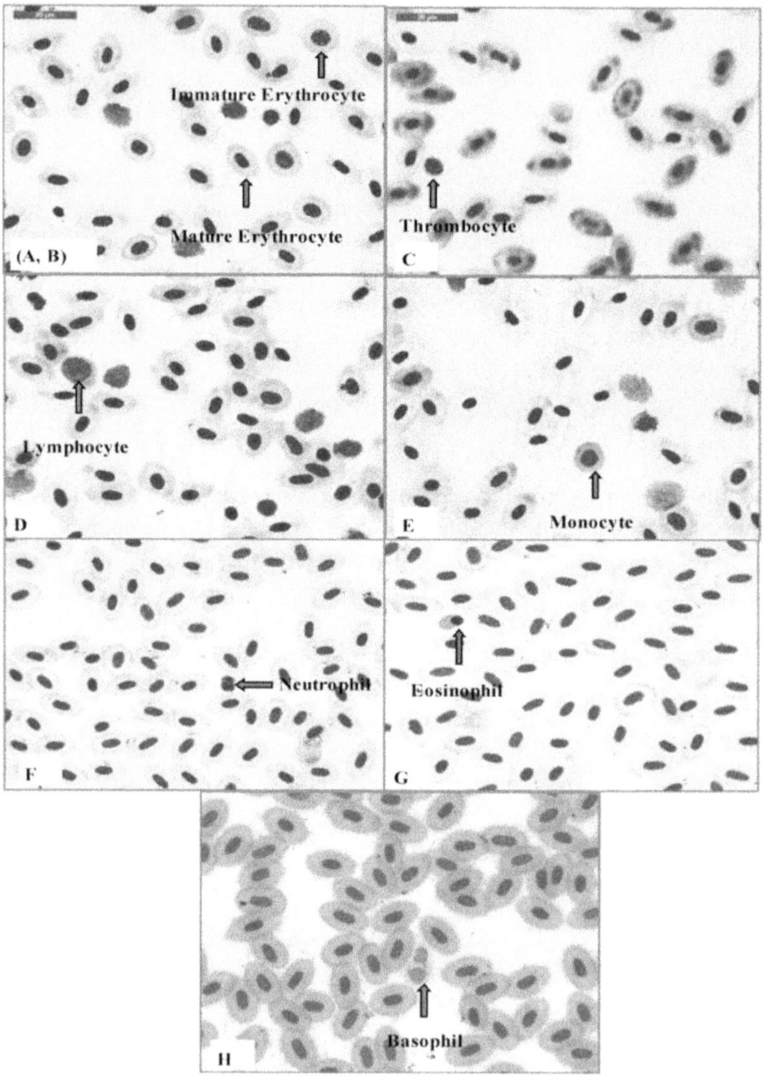

FIGURE 27.3 Pictorial images of different blood cells of *Cyprinus carpio* spp. (A, B) Mature and immature erythrocytes; (C) thrombocytes; (D) lymphocytes; (E) monocytes; (F) neutrophils; (G) eosinophils; (H) basophils.

Monocytes in fish are larger mononuclear leukocytes with rich agranular cytoplasm in blood smears stained with Wright–Giemsa stain.

26.6.4 Neutrophils

The term 'neutrophil' is repeatedly used to label the dominant granulocytes of teleost fish, and their shapes differ with respect to the fish species. The neutrophils of Osteichthyes appear to be irregular or oval in outline. The nucleus of mature erythrocytes varies in shape: band, oval, round, elongated or segmented (Figure 27.3F). Mature neutrophils' cytoplasm contains small pink or purple granules, and immature neutrophils have light purple and pink granules when stained with Wright–Giemsa stain. Instead of neutrophils, these cells are generally regarded as heterophils, though these cells share their cyto-chemical properties with neutrophils of other fishes.

27.6.5 Eosinophils

The presence of eosinophils in fish blood is very confusing, and they are rarely reported in blood smears of fish. These cells are low in number and occur in the peripheral blood as round or oval cells having a band or several segmented nuclei (Figure 27.3G). Medium-sized granulocytes could be distinguished from heterophils and neutrophils by the occurrence of many eosinophilic cytoplasmic granules, which are round to rod-like in shape when stained with Wright–Giemsa stain. The eosinophil has an eccentrically located nucleus, which is notched and partially lobed. The eosinophilic cytoplasm was compactly filled with enormous acidophilic granules of various sizes, larger than in heterophils, which stained pink with Giemsa stain. Eosinophils play a vital role in killing parasites and hence, aid in phagocytosis. A few reports have indicated that fish eosinophils are associated with inflammatory responses, although they have restricted phagocytic capability (Rowley, 1988; Roberts & Ellis, 2001; Fang et al., 2014; Shahjahan et al., 2018; Sheikh & Ahmed, 2020; Chen et al., 2021).

27.6.6 Basophils

In the case of bony fishes, basophils occur rarely in blood peripheral films, and their presence has been reported in only a few species of fishes (Saunders, 1966; Ellis, 1977; Fang et al., 2014; Sheikh & Ahmed, 2020) and usually in lower numbers. Basophil cell morphology was observed with Wright–Giemsa-stained smears. They were found in peripheral blood and were recognized as round cells with basophilic cytoplasmic granules, similar to mast cells. The nucleus, which is big and bilobed, is positioned eccentrically in the cytoplasm (Figure 27.3H). Different staining methods also affect the basophilic cell, which can appear as a fragmented cell or partially dissolved cytoplasmic granules in the stained blood film.

ACKNOWLEDGEMENTS

The authors are grateful to the Head, Department of Zoology, University of Kashmir, Hazratbal, Srinagar, India for providing the laboratory facilities and gratefully acknowledge the support of various funding agencies like the Department of Biotechnology (DBT), Government of India, New Delhi and Ministry of Environment, Forest & Climate Change (MoEF & CC) under its program 'National Mission on Himalayan Studies (NMHS)' for their financial support to the Fish Nutrition Research Laboratory by running various project schemes. The authors also gratefully acknowledge UGC, New Delhi for financial support in the form of 'Himalayan Biodiversity–Documentation, Bio-prospection & conservation' under the scheme 'Center with Potential for Excellence in Particular Areas (CPEPA)'.

REFERENCES

Adedeji OB, Taiwo VO, & Agbede SA. (2000). Comparative haematology of five Nigerian freshwater fish species. *Niger Vet J* 21:75–84.

Affonso EG. (2001). Respiratory characteristics of *Hoplosternum littorale* (Siluriformes, Callichthyidae). *Acta Amaz* 31:249–262.

Ahmed I, & Sheikh ZA. (2017). Study on the seasonal variation in the chemical composition, hematological profile, gonado-somatic index and hepato-somatic index of snow trout, *Schizothorax niger* from the freshwater Dal lake, Kashmir. *Am J Food Technol* 12:1–13.

Ahmed I, & Sheikh ZA. (2019). Hematological and serum biochemical parameters of five freshwater snow trout fish species from river Jhelum of Kashmir Himalaya, India. *Comp Clin Pathol* 28: 771–782.

Ahmed I, & Sheikh ZA. (2020). Comparative study of haematological parameters of snow trout *Schizopyge plagiostomus* and *Schizopyge niger* inhabiting two different habitats. *Eur Zool J* 87:12–19.

Ahmed I, Sheikh ZA, Wani GB, & Shah BA. (2019). Sex variation in hematological and serum biochemical parameters of cultured Chinese silver carp *Hypophthalmichthys molitrix*. *Comp Clin Pathol* 28:1761–1767.

Ahmed I, Reshi QM, & Fazio F. (2020). The influence of the endogenous and exogenous factors on hematological parameters in different fish species: a review. *Aquacult Int* 28:869–899.

Ahmed I, Jan K, Fatma S, & Dawood MA. (2022). Muscle proximate composition of various food fish species and their nutritional significance: a review. *J Anim Physiol Anim Nutr* 106:690–719.

Babahydari SB, Dorafshan S, Heyrati FP, Soofiani NM, & Vahabi MR. (2018). The physiological changes, growth performance and whole body composition of common carp, *Cyprinus carpio* fed on diet containing wood betony, *Stachys lavandulifolia* extract. *J Agr Sci Tech* 16:1565–1574.

Bastami KD, Moradlou AH, Zaragabadi AM, Mir SS, & Shakiba MM. (2009). Measurement of some haematological characteristics of the wild carp. *Comp Clin Pathol* 18:321–323.

Belanger JM, Son JH, Laugero KD, Moberg GP, Dorochov SI, Lankford SE, & Cech JJ. (2001). Effects of short-term management stress and ACTH injections on plasma cortisol levels in cultured white sturgeon, *Acipenser transmontanus*. *Aquaculture* 203:165–176.

Bhaskar BR, & Rao KS. (1989). Influence of environmental variables on haematology and compendium of normal haematological ranges of milkfish, *Chanos* (Forskal), in brackishwater culture. *Aquaculture* 83:123–136.

Blaxhall PC, & Daisley KW. (1973). Routine haematological methods for use with fish blood. *J Fish Biol* 5:771–781.

Bozkurt Y, Yavas I, Gul A, Balci BA, & Çetin NC. (2017). Importance of grass carp (*Ctenopharyngodon idella*) for controlling of aquatic vegetation. In Almusaed A, Al-Samaraee SMS (Eds.) *Grasses-Benefits, Diversities and Functional Roles*. Rijeka: IntechOpen, 158 p.

Bridges DW, Cech Jr JJ, & Pedro DN. (1976). Seasonal hematological changes in winter flounder, *Pseudopleuronectes americanus*. *Trans Am Fish Soc* 105:596–600.

Bullis RA. (1993). Clinical pathology of temperate freshwater and estuarine fish. In: Stoskopf MK (Ed.), *Fish Medicine*. Philadelphia: Saunders Company, pp. 232–239.

Burton CB, & Murray SA. (1979). Effects of density on goldfish blood – I Hematology. *Comp Biochem Physiol A* 62:555–558.

Campbell T. (2015). *Exotic Animal Hematology and Cytology*. Iowa State University Press, John Wiley & Sons: USA, pp. 97–114.

Chen H, Liu H, Li R, Lin X, Luo H, Ji S, … & Luo D. (2021). Blood cell identification and hematological analysis during natural sex reversal in rice field eel (*Monopterus albus*). *Aquaculture* 538:736543.

Chen WH, Sun LT, Tsai CL, Song YL, & Chang CF. (2002). Cold-stress induced the modulation of catecholamines, cortisol, immunoglobulin M, and leukocyte phagocytosis in Tilapia. *Gen Comp Endocrinol* 126:90–100.

Coroian CO, Mireşan V, Cocan DI, Vaţu RD, Raducu CM, & Coroian A. (2015). Growth performance of common carp (*Cyprinus carpio* L.) fingerlings fed with various protein levels. *Aquacult Aquarium Conserv Legis* 8:1038–1047.

Dacies S, & Lewis S. (1991). *Practical Hematology*, 7th edn, London: Churchill Livingstone, p. 633.

Das SM, & Subla BA. (1964). The ichthyofauna of Kashmir: part II the speciation of Kashmir fishes with two new records of species. *Ichthyologica* 3:57–62.

Drabkin DL. (1946). Spectrophotometric studies. XIV. The crystallographic and optical properties of the hemoglobin of man in comparison with those of other species. *J Biol Chem* 164:703–723.

Ejraei F, Ghiasi M, & Khara H. (2015). Evaluation of hematological and plasma indices in grass carp, *Ctenopharyngodon idella*, with reference to age, sex, and hormonal treatment. *Fish Aquat Life* 23:163–170.

Ellis AE. (1977). Leucocytes and related cells in the plaice *Pleuronectes platessa*. *J Fish Biol* 8:143–156.

Ellis AE. (1984). Bizarre forms of erythrocytes in a specimen of plaice, *Pleuronectes platessa* L. *J Fish Dis* 7:411–414.

Fang J, Chen K, Cui HM, Peng X, Li T, & Zuo ZC. (2014). Morphological and cytochemical studies of peripheral blood cells of *Schizothorax prenanti*. *Anat Histol Embryol* 43:386–394.

Fange R. (1994). Blood cells, haemopoiesis and lymphomyeloid tissues in fish. *Fish Shellfish Immunol* 4:405–411.

Farshad E, Maryam G, & Hossein K. (2015). Evaluation of hematological and plasma indices in grass carp, Ctenopharyngodon idella, with reference to age, sex, and hormonal treatment. *Arch Pol Fish* 23:163–170.

Fazio F. (2019). Fish hematology analysis as an important tool of aquaculture: a review. *Aquaculture* 500:237–242.

Fazio F, Marafioti S, Sanfilippo M, Casella S, & Piccione G. (2016). Assessment of immune blood cells and serum protein levels in *Mugil cephalus* (Linnaeus, 1758), *Sparus aurata* (Linnaeus, 1758) and *Dicentrarchus labrax* (Linnaeus, 1758) collected from the Tyrrhenian sea coast (Italy). *Cah Biol Mar* 57:235–240.

Fazio F, Ferrantelli V, Piccione G, Saoca C, Levanti M, & Mucciardi M. (2018). Biochemical and hematological parameters in European sea bass (*Dicentrarchus labrax* Linnaeus, 1758) and gilthead sea bream (*Sparus aurata* Linnaeus, 1758) in relation to temperature. *Vet Arh* 88:397–411.

Filho DW, Elbe GJ, Cancer G, Caprario FX, Dafne AL, & Ohira M. (1992). Comparative hematology in marine fish. *Comp Biochem Physiol* 102:311–321.

Fotedar DN, & Qadri MY. (1974). Fish and fisheries of Kashmir and the impact of carp, *Cyprinus carpio*, on the endemic fishes. *J Sci* 2:79–89.

Grizzle JM, & Rogers WA. (1976). Anatomy and histology of the channel catfish (*Ictalurus punctatus*). Food and Agriculture Organization of the United Nations, Auburn, AL, Auburn University, Agricultural Experiment Station, pp. 86–90.

Groff JM, & Zinkl JG. (1999). Hematology and clinical chemistry of Cyprinid fish. Common carp and goldfish. *Vet Clin North Am Exot Anim Pract* 2:741–746.

Heath AG. (1995). *Water Pollution and Fish Physiology*, 2nd ed. Boca Raton: CRC Press.

Hibiya T. (1985). An atlas of fish histology, normal and pathological features. In: Hibiya T (Ed.), *An Atlas of Fish Histology, Normal and Pathological Features*. Tokyo: Kodansha Ltd. 147 pp.

Houston AH, & Dewilde MA. (1972). Some observations upon the relationship of microhaematocrit values to haemoglobin concentrations and erythrocyte numbers in the carp *Cyprinus carpio* L. and brook trout *Salvelinus fontinalis* (Mitchill). *J Fish Biol* 4:109–115.

Hrubec TC, & Smith SA. (2010). Hematology of fishes. In: Weiss, DJ & KJ Wardrop (Eds.), *Schalm's Veterinary Hematology*. 6th edn, Ames, Iowa: Wiley-Blackwell, pp. 994–1003.

Hrubec TC, Smith SA, Robertson JL, Feldman B, Veit HP, Libey G, & Tinker MK. (1996). Comparison of hematologic reference intervals between culture system and type of hybrid striped bass. *Am J Vet Res* 57:618–623.

Hrubec TC, Cardinale JL, & Smith SA. (2000). Hematology and plasma chemistry reference intervals for cultured tilapia (*Oreochromis hybrid*). *Vet Clin Pathol* 29:7–12.

Hunn JB, Wiedmeyer RH, Greer IE, & Grady AW. (1992). Communications: blood chemistry of laboratory-reared golden trout. *J Aquat Anim Health* 4:218–222.

Jan K, & Ahmed I. (2021). The influence of sex and season on some hematological and biochemical parameters of snow trout *Schizothorax labiatus* in the Indian Himalayan Region. *Fish Sci* 87:39–54.

Jan K, Ahmed I, & Dar NA. (2021). Haematological and serum biochemical reference values of snow trout, *Schizothorax labiatus* habiting in river Sindh of Indian Himalayan region. *J Fish Biol* 98:1289–1302.

Jan K, Ahmed I, Sheikh ZA, & Fazio F. (2022). Impact of three anticoagulants and their storage time on hematological parameters of snow trout, *Schizothorax labiatus*, habiting in river Sindh of Indian Himalayan region. *Comp Clin Pathol* 31:1–9.

Jhingran VG. (1991). *Fish and Fisheries of India.* 3rd edn, New Delhi (India): Hindustan Publication Corporation, p. 727.

Joshi V, Akhtar MS, Sharma P, Kushwaha SS, Baruah DJ, Ciji A, Pande V, & Sarma D. (2017). Himalayan fish manifest higher potential of quality nutrients for human health. *J Aquat Food Prod Technol* 26:843–855.

Joshi V, Akhtar MS, Sharma P, Baruah D, Alexander C, Das P, & Sharma D. (2018). Protein and amino acid composition of Indian Himalayan snow trout and their dietary significance. *Proc Natl Acad Sci India B-Biol Sci* 88:1471–1476.

Kesbic OS, Parrino V, Acar Ü, Yilmaz S, Paro GL, & Fazio F. (2020). Effects of Monterey cypress (*Cupressus macrocarpa* Hartw) leaf essential oil as a dietary supplement on growth performance and haematological and biochemical parameters of common carp (*Cyprinus carpio* L.). *Ann Anim Sci* 20:1411–1426.

Kumar MV. (2016). Morphometric studies of blood cells in *Cyprinus carpio*, *Ctenopharyngodan idella* and *Hypophthalmichthys molitrix* cultured fish in west Godavari region of Andhra Pradesh. *Int J Fish Aquat Stud* 4:489–493.

Kumar V, Makkar HP, Amselgruber W, & Becker K. (2010). Physiological, haematological and histopathological responses in common carp (*Cyprinus carpio* L.) fingerlings fed with differently detoxified *Jatropha curcas* kernel meal. *Food Chem Toxicol* 48:2063–72.

Lay PA, & Baldwin J. (1999). What determines the size of teleost erythrocytes? Correlation with oxygen transport and nuclear volume. *Fish Physiol Biochem* 20:31–35.

LeaMaster BR, Brock JA, Fujioka RS, & Nakamura RM. (1990). Hematologic and blood chemistry values for *Sarotherodon melanotheron* and a red hybrid tilapia in freshwater and seawater. *Comp Biochem Physiol A: Physiol* 97:525–529.

Li YN, & Wang YP. (1995). Advances in research of fish immunology. *Zool Res* 16:83–94.

Linhartova Z, Krejsa J, Zajic T, Masilko J, Sampels S, & Mraz J. (2018). Proximate and fatty acid composition of 13 important freshwater fish species in central Europe. *Aquacult Int* 26:695–711.

Lutnicka H, Bojarski B, Witeska M, Tombarkiewicz B, & Formicki G. (2019). Exposure to herbicide linuron results in alterations in hematological profile and stress biomarkers of common carp (*Cyprinus carpio*). *Ecotoxicology* 28:69–75.

Mahoney JB, & McNulty JK. (1992). Disease-associated blood changes and normal seasonal hematological variation in winter flounder in the Hudson-Raritan Estuary. *Trans Am Fish Soc* 121:261–268.

Metin K, Bashimoghlu Koca Y, Kargin KF, Koca S, & Turkozan O. (2008). Blood cell morphology and plasma biochemistry of captive *Mauremys caspica* (Gmelin, 1774) and *Mauremys rivulata* (Valenciennes, 1833). *Acta Vet Brno* 77:163–174.

Miller WR, Hendricks AC, & Cairns J. (1983). Normal ranges for diagnostically important hematological and blood chemistry characteristics of rainbow trout (*Salmo gairdneri*). *Can J Fish Aquat Sci* 40:420–425.

Natt MP, & Herrick CA. (1952). A new blood diluent for counting erythrocyte and leucocytes of chicken. *Poult Sci* 31:735–738.

Paul BN, Chanda S, Sridhar N, Saha GS, & Giri SS. (2016). Fatty acid, amino acid and vitamin composition of Indian catfish, magur (*Clarias batrachus*) and singhi (*Heteropneustes fossilis*). *SAARC J Agric* 14:189–199.

Pavlidis M, Futter WC, Katharios P, & Divanach P. (2007). Blood cell profile of six Mediterranean mariculture fish species. *J Appl Ichthyol* 23:70–73.

Pradhan SC, Patra AK, & Pal A. (2014). Hematological and plasma chemistry of Indian major carp, *Labeo rohita* (Hamilton, 1822). *J Appl Ichthyol* 30:48–54.

Radu D, Oprea L, Bucur C, Costache M, & Oprea D. (2009). Characteristics of haematological parameters for carp culture and Koi (*Cyprinus carpio* Linneaus, 1758) reared in an intensive system. *Bull UASVM J Anim Sci Biotechnol* 66:1–2.

Ram Bhaskar BR, & Srinivasa Rao KS. (1989). Influence of environmental variables on hematology, and compendium of normal hematological ranges of milk fish, *Chanos* (Forskal) in brackish culture. *Aquaculture* 83:123–136.

Roberts RJ. (1981). *Pathology of the Fish*. Madrid: Mundi-Press, p. 336.

Roberts RJ, & Ellis AE. (2001). The anatomy and physiology of teleosts. In Roberts RJ (Ed.), *Fish Pathology,* 3rd edn, Philadelphia, USA: W. B. Saunders, pp. 12–54.Rowley AF. (1988). Fish. In: Rowley A, & Ratcliffe N (Eds), *Vertebrate Blood Cells.* Cambridge: Cambridge University Press.

Rowley AF. (1990). Collection, separation and identification of fish leucocytes. In Stolen JS, Fletcher TC, Anderson DP, Roberson BS, Van Muiswinkel WB (Eds.) *Techniques in Fish Immunology.* SOS Publications, 3: 113–36.

Sandnes K, Lie O, & Waagbo R. (1988). Normal ranges of same blood chemistry parameters in adult farmed Atlantic salmon (*Salmo salar*). *J Fish Biol* 32:129–136.

Satheeshkumar P, Ananthan G, Senthil Kumar D, & Jagadeesan L. (2011). Hematology and biochemical parameters of different feeding behaviours of teleost fishes from the Vellar estuary, India. *Comp Clin Pathol* 20:127–134.

Saunders DC. (1966). Differential blood cell counts of 121 species of marine fishes of Puerto Rico. *Trans Am Microsc Soc* 85:427–449.

Shahjahan MD, Helaluddin MD, Bain V, & Haque MM. (2018). Increased water temperature altered hematobiochemical parameters and structure of peripheral erythrocytes in striped catfish, *Pangasianodon hypophthalmus*. *Fish Physiol Biochem* 44:1309–1318.

Sharma NK, Akhtar MS, Pandey NN, Singh R, & Singh AK. (2017). Sex specific seasonal variation in hematological and serum biochemical indices of *Barilius bendelisis* from Central Himalaya, India. *Proc Natl Acad Sci India Sect B: Biol Sci* 87:1185–1197.

Sheikh ZA, & Ahmed I. (2016). Seasonal changes in hematological parameters of snow trout *Schizothorax plagiostomus* (Heckel 1838). *Int J Fauna Biol Stud* 3:33–38.

Sheikh ZA, & Ahmed I. (2019). Impact of environmental changes on plasma biochemistry and hematological parameters of Himalayan snow trout, *Schizothorax plagiostomus*. *Comp Clin Pathol* 28:793–804.

Sheikh ZA, & Ahmed I. (2020). Effect of sex on hematology, morphology and blood cell characteristics of *Schizothorax niger*. *Comp Clin Pathol* 29:1069–1078.

Sidiq M, & Ahmed I. (2020). Comparative study of hematological profile of three forage fish species habiting in Dal Lake of Kashmir Himalaya, India. *Comp Clin Pathol* 29:913–920.

Soldatov AA. (2005). Peculiarities of organization and functioning of the fish red blood system. *J Evol Biochem Physiol* 41:272–281.

Suleiman B, Adamu S, & Abdullahi SA. (2016). Influence of sex on haematological and biochemical profiles of pond-reared, *Heterobranchus longifilis* (Valenciennes, 1840). *Nigerian J Fish Aquacult* 4:28–33.

Svetina AZ, Matasin A, Tofant M, Vucemilo M, & Fijan N. (2002). Haematology and some blood chemical parameters of young carp till the age of three years. *Acta Vet Hung* 50:459–467.

Svobodova Z, Kroupova H, Modra H, Flajshans M, Randak T, Savina LV, & Gela D. (2008). Haematological profile of common carp spawners of various breeds. *J Appl Ichthyol* 24:55–59.

Talwar PK, & Jhingran AG. (1991). Inland Fishes of India and Adjacent Countries. Vols I & II. Technical paper, pp. 235–239.

Tang Y, Peng X, Fang J, Cui HM, Zuo ZC, & Deng JL. (2015). Characterization of hematological parameters and blood cells of cultured *Gymnocypris eckloni* Herzenstein, 1891. *J Appl Ichthyol* 31:931–936.

Tavares-Dias M, & Moraes FRD. (2007). Leukocyte and thrombocyte reference values for channel catfish (*Ictalurus punctatus* Raf) with an assessment of morphologic, cytochemical, and ultrastructural features. *Vet Clin Pathol* 36:49–54.

Tocidlowski ME, Lewbart GA, & Stoskopf MK. (1997). Hematologic study of red pacu (*Colossoma brachypomum*). *Vet Clin Pathol* 26:119–125.

Tripathi NK, Latimer KS, & Burnley VV. (2004). Hematological reference intervals for koi (*Cyprinus carpio*), including blood cell morphology, cytochemistry and ultrastructure. *Vet Clin Pathol* 33:74–83.

Vazquez GR, & Guerrero GA. (2007). Characterization of blood cells and haematological parameters in *Cichlasoma dimerus* (Teleostei, Perciformes). *Tissue Cell* 39:151–160.

Wedemeyer GA, & Nelson NC. (1975). Statistical methods for estimating normal blood chemistry ranges and variances in rainbow trout (*Salmo gairdneri*), Shasta strain. *J Fish Res Board Can* 32: 551–554.

Wells RMG, Baldwins J, Seymour RS, Christian K, & Brittain T. (2005). Red blood cell function and haematology in two tropical freshwater fishes from Australia. *Comp Biochem Physiol A* 141: 87–93.

Wilhelm Filho D, Eble GJ, Kassner G, Caprario FX, Dafre AL, & Ohira M. (1992). Comparative hematology in marine fish. *Comp Biochem Physiol A: Physiol* 102:311–321.

Witeska M. (2013). Erythrocytes in teleost fishes: a review. *Zool Ecol* 23: 275–281.

Zaragoza ODR, Rodriguez MH, & Buckle Ramirez LF. (2008). Thermal stress effect on tilapia *Oreochromis mossambicus* (Pisces: Cichlidae) blood parameters. *Mar Freshw Behav Physiol* 41:79–89.

Zexia G, Weimin W, Yi Y, Abbas K, Dapeng L, Guiwei Z, & Diana JS. (2007). Morphological studies of peripheral blood cells of the Chinese sturgeon, *Acipenser sinensis*. *Fish Physiol Biochem* 33:213–222.

Zhang HJ, Xie CX, Liu HP, & Yang XF. (2011). Blood cells of a sisorid catfish *Glyptosternum maculatum* (Siluriformes: Sisoridae), in Tibetan plateau. *Fish Physiol Biochem* 37:169–176.

Zhao H, Panase P, Zhang Z, Yao P, Zhang Y, & Suwannapoom C. (2018). Hematological and plasm biochemical values for *Rhinogobio ventralis* in the Yangtze River, China. *Comp Clin Pathol* 27:741–745.

Zhou YC, Xing YN, & Feng QY. (2003). Research advance in the haemocytes of fishes. *Nat Sci J Hainan Univ* 21:171–176.

28 Use of Nematode Community as Bioindicator in Responses to Wastewater Irrigation in Agro-Ecosystems Near the Yamuna in Haryana, India

Shikha Ahalavat
Department of Zoology & Aquaculture, Haryana Agriculture University, Hisar, Haryana

Nibedita Borgohin
Department of Nematology, Assam Agricultural University, Jorhat, Assam

Puneet Kumar
Department of Zoology, Aligarh Muslim University, Aligarh, Uttar Pradesh

Ashok Kumar Chaubey
Department of Zoology, Chaudhary Charan Singh University, Meerut

CONTENTS

28.1 INTRODUCTION

The growing problem of water scarcity has a significant negative influence on economic development, human livelihoods and environmental quality throughout the planet. Growing global demand for water use, with regional changes in precipitation in many regions, has resulted in increasing

DOI: 10.1201/9781003220824-32

long-term irrigation of agricultural soils with wastewater. Rapid urbanization and industrialization release enormous volumes of wastewater, which is increasingly utilized as a valuable resource for irrigation in urban and peri-urban agriculture. The utilization of treated wastewater for irrigation is one of the most readily available alternative water sources when natural resources are scarce. This water reuse in agriculture is increasing in many places around the world (Hamilton *et al.*, 2007; Qadir *et al.*, 2010), and therefore, the implications for the sustainability of agro-ecosystems deserve careful attention.

Wastewater is any water that has been adversely affected in quality by anthropogenic influence. It comprises liquid waste discharged by domestic residences, commercial properties, industry, and/or agriculture and may encompass a good range of potential contaminants and concentrations. These waters are an important source of irrigation in rural economies like India, where the costs of underground bore wells are very high and difficult to manage for small and marginal farmers. It drives significant economic activity, supports countless livelihoods, particularly those of poor farmers, and substantially changes the water quality of natural water bodies (Makoni *et al.*, 2016). Wastewater may contain various heavy metals, including Zn, Cu, Pb, Mn, Ni, Cr, Cd, etc., depending upon the sort of activities it is related to. Continuous irrigation of agricultural land with sewage and industrial wastewater may cause heavy metal accumulation within the soil and vegetables (Sharma *et al.*, 2007; Marshall *et al.*, 2007). Heavy metals are generally not removed even after the treatment of wastewater at sewage treatment plants and thus, cause a risk of heavy metal contamination of the soil, consequently affecting organic production (Fytianos *et al.*, 2001). Heavy metal pollutants influence the food availability and competitive interactions among species of soil biota and eventually affect the biotic communities in an indirect way (Korthals *et al.*, 1996).

Identifying the effects of soil pollution on soil biodiversity is essential in order to better understand the impact of human activity on ecosystem functioning (Kandeler *et al.*, 1996, Nahmani and Lavelle, 2002; Boyd, 2010), allowing the development of effective management strategies for polluted soils (Rodríguez Martín *et al.*, 2014) and ensuring more sustainable use of this important resource. Several methods have been proposed to assess the environmental quality of soil using chemical and/or biological approaches (Cortet *et al.*, 1999; Edwards, 2002; Caldwell, 2005). Of the biological approaches, nematode communities are relevant bioindicators for assessing soil disturbance in terrestrial systems (Bongers and Bongers, 1998; Bongers and Ferris, 1999; Neher, 2001), in particular in soils polluted by heavy metals (Bongers *et al.*, 2001; Georgieva *et al.*, 2002).

In soil ecosystems, nematodes play an important role in nutrient recycling by feeding plant tissues and microorganisms and thereby liberating minerals for easy absorption by plant roots. They are a successful group of animals placed at a rather low level of taxonomic hierarchy in the animal kingdom. They occupy any niche that provides an available source of organic matter in marine, freshwater or terrestrial environments. They are the most abundant metazoan on earth, and of every five animals, four are nematodes (Gunapala *et al.*, 1998). They are not only numerically abundant but they also show diversity in terms of species. To date, 26,000 species have been identified (Hugot *et al.*, 2001), which may be regarded as the tip of the iceberg considering the estimated number of 10 million species or more. The five major trophic groups of nematodes have been defined as bacterial feeders, fungal feeders, plant feeders, omnivores and predators (Yeates *et al.*, 1993). Plant-parasitic nematodes are considered as primary consumers, and they affect food web resources through direct herbivory (Ferris and Bongers, 2006). Bacterivore and fungivore nematodes graze on decomposer microbes, bacteria and fungi, and thus significantly contribute to nutrient mineralization (Ingham *et al.*, 1985; Ferris *et al.*, 1996; Ferris and Matute, 2003). Bacterivore nematodes also promote rhizosphere colonization of beneficial bacteria (Kimpinski and Sturz, 1996; Knox *et al.*, 2003). Predatory nematodes regulate the food cycle by preying on other nematodes and invertebrates within the soil (Zimmerman and Cranshaw, 1990; Grewal *et al.*, 2005). Nematode taxa are again classified into five groups according to a colonizer-persister (cp) scale based on their lifecycle characteristics and sensitivity to perturbation (Bongers, 1990): from cp1 (opportunistic feeders with a brief generation time and a high reproduction rate) to cp5 (persisters with an extended lifespan, a low reproduction rate and greater sensitivity to soil disturbance). A combination of the cp scale and feeding habits is used to define functional guilds as proposed by Bongers and Bongers (1998). Indices based on these characteristics can be used to analyse nematode community structure (Bongers, 1990; Ferris, 2010) and any changes due to environmental disturbance (Pen-Mouratov *et al.*, 2004; Šalamún *et al.*, 2012), and hence, nematodes can act as bioindicators of soil health and may respond differently to degradation of environmental quality and perturbations (Gupta and Yeates, 1997; Neher, 2001). They have also been extensively used as models for developmental biology, biochemical, nutritional and ageing studies (Dougherty *et al.*, 1959; Nicholas *et al.*, 1959; Vanfleteren, 1973) due to their qualities, i.e., small size, simple multicellular organization, transparent body, ease in cultivation, high fecundity, short generation time and short life span.

Nematodes, as one component of the soil ecosystem, interact with biotic and abiotic factors and adapt themselves to their environment even if the environment threatens to change. If they are unable to do so, they become extinct in due course. So, they are considered as important biological indicators due to their tremendous diversity and participation in many functions at different levels within the soil food cycle. In addition to their diversity, nematodes could also be useful indicators because their populations are relatively stable in response to changes in soil moisture and temperature, but they do respond to land management changes. Changes in nematode populations tend to reflect changes in soil microenvironments. They may be the most useful group for community indicator analysis because more information exists on their taxonomy and feeding roles (Gupta and Yeates, 1997) than is available for any other mesofauna.

Nematodes, being an important component of the underground food web, are influenced by the accumulation of various heavy metals in the soil. Since soil nematodes live in the capillary water, they are assumed to be exposed to the contaminant concentration in the solution, which offers good perspectives for assessing the effects of contaminants in relation to their bioavailability in the soil (Houx and Aben, 1993; Liang *et al.*, 2007). Because nematodes are highly dependent on soil properties, anthropogenic changes in soil systems can strongly affect their community composition (Griffiths *et al.*, 2003), and nematode assemblages have received much attention as indicators of heavy metal pollution in soils. Overall, nematodes are extremely responsive to different types of pressures, including environmental disturbances at spatial and temporal scales, making them an efficient tool for assessing the environmental status of ecosystems (Sroczyńska *et al.*, 2021).

Within the last few decades, nematodes have gained increasing attention in freshwater ecotoxicology studies, and their use in single-species toxicity tests (Traunspurger, 1997; Höss and Williams, 2009; Haegerbaeumer et al., 2016; Schenk *et al.*, 2020), field studies (Beier and Traunspurger, 2001; Burton *et al.*, 2001; Heininger *et al.*, 2007) and model ecosystems (Brinke *et al.*, 2010; Brinke *et al.*, 2011; Faupel *et al.*, 2011; Haegerbaeumer *et al.*, 2016) is now well established. Their ubiquitous occurrence, ecological relevance and universal applicability for variously complex ecotoxicological tools make nematodes excellent bioindicators (Wilson and Kakouli-Duarte, 2009; Sroczyńska *et al.*, 2021). Several studies have shown that nematode communities responded to, e.g., ploughing, crop rotation and water management (Kimpinski and Sturz, 1996) and also to different organic and inorganic pollutants such as polycyclic aromatic hydrocarbons (PAH) or heavy metals (Chen *et al.*, 2009; Ahalavat and Chaubey, 2017; Martinez *et al.*, 2018; Alasmary *et al.*, 2020). But largely, the consequences of urban wastewater irrigation in agro-ecosystems near urban settlements have not been studied in India and elsewhere; therefore, the present work has been initiated.

The present study aimed to review the community structure of the soil-inhabiting nematodes related to crop fields near the Yamuna River area in Haryana to assess the role of nematodes as indicators of soil condition. These fields have been irrigated with wastewaters for about 30 years, and vegetables are grown in these fields and supplied to the local marketplace for consumption. The wastewater comprises both industrial wastewater from factories and domestic sewage. However, it's noteworthy that there is no infallible separation of commercial effluents from domestic wastes, and therefore, the two get mixed because (a) small-scale industries are house based and (b) proper compartmentalization of sewerages has not gained popularity due to lack of strict laws and enforcement procedures (Ahalavat and Chaubey, 2017).

28.2 MATERIALS AND METHODS

Soil samples from agriculture fields near the Yamuna River area in Faridabad, Haryana were collected. These fields have been irrigated with wastewaters and freshwaters for about 30 years, and vegetables like eggplant (*Solanum melongena*), okra (*Abelmoschus esculentus*), mustard (*Brassica juncea*), tomato (*Solanum lycopersicum*), cauliflower (*Brassica oleracea*), carrot (*Daucus carota*), chili (*Capsicum* spp.), cabbage (*Brassica oleracea var. capitata*), potato (*Solanum tuberosum*), coriander (*Coriandrum sativum*), etc. are grown in these fields and supplied to the local marketplace for consumption. From each field, soil samples were collected from a depth of 0–10 cm using a hand spade. Samples were tagged, stored in sealed plastic bags and brought to the laboratory for further processing.

Nematodes were extracted from 100 cc. of fresh weight of soil using Cobb's (1918) sieving and decantation and modified Baerman's funnel techniques. All the nematodes from each extracted sample were counted and identified to genus level. Trophic groups were allocated consistently with Yeates *et al.* (1993), and c-p groups were assigned after Bongers (1990). Qualitative analysis of the soil samples was done at the soil testing laboratory, Indian Agricultural Research Institute (IARI), New Delhi. Nematode diversity was described by using Shannon's diversity index calculated at the genus level (H′). The maturity index (MI) was calculated to estimate the relative state of the two ecosystems studied. Trophic diversity was calculated by the trophic diversity index (TDI) (Heip *et al.*, 1988).

The nematodes were classified into functional groups to calculate four indices: enrichment (EI), basal (BI), structure (SI) and channel index (CI) (Ferris *et al.*, 2001). The EI contains fast-growing, bacteria- and fungi-feeding nematodes with a colonizer-persister (c-p) value of 1 or 2 (Yeates and Bongers, 1999). The BI contains bacteria-feeding and fungi-feeding nematodes with a c-p value of 2. The SI measures the slow-growing and reproducing predatory and omnivore nematodes with c-p values of 3, 4 and 5. The CI is a comparison of the size of the fungal with the bacterial feeding communities. The CI assesses the primary decomposition pathway of soil, a value of 100 being completely fungal and a value of 0 being completely bacterial.

To calculate the varied nematode indices, all genera were assigned weights for EI, BI and SI according to their classification into functional groups (Ferris *et al.*, 2001). For instance, bacteria-feeding nematodes with a c-p value of 1 or 2 received a weight of 0.8 for BI only, while fungi-feeding nematodes with a c-p value of 2 received a weight of 0.8 for EI and BI. Bacteria-feeding nematodes with a c-p value of 3 received a weight of 3.2 for EI only. Plant-parasitic nematodes were omitted from the indices to eliminate host-status effects of the crop sequences. After assigning weights, the sum-products (called E, B and S) were calculated from the assigned weights and numbers of individuals in all genera. SI and EI were calculated to determine the relative stability of the ecosystem studied. All indices were calculated using MS Excel. Differences with $p < 0.05$ were considered significant and $p < 0.01$ as highly significant.

Detailed description of the formulae used are given here:

$$\text{Shannon's diversity (H')} = -\Sigma\,(p_i \ln p_i)$$

28.2.1 Maturity Index (MI)

$$\mathrm{MI} = \sum_{i-1}^{n} V_i . f_i$$

Where
Vi = c-p value of the ith taxon.
f(i) = the frequency of that taxon in a sample.
MI is calculated as the weighted mean of the individual c-p value.

28.2.2 Plant Parasitic Index (PPI)

$$PPI = \sum PP_i X_i / \sum X_i$$

Where
PPi = PP value assigned to taxon i according to Bongers (1990).
Xi = abundance of taxon i in the sample.

Enrichment Index (EI)

$$\mathrm{EI} = (\mathrm{E/E+B}) \times 100$$

Structure Index (SI)

$$\mathrm{SI} = (\mathrm{S/S+B}) \times 100$$

Basal Index (BI)

$$\mathrm{BI} = (\mathrm{B/B+E+S}) \times 100$$

Channel index (CI)

$$\mathrm{CI} = (\mathrm{E_f/E_b} + \mathrm{E_f}) \times 100$$

Where
E, B and S are sum products of assigned weights and number of individuals of all genera.
E_f is E of fungi-feeding nematodes and E_b is E of bacteria-feeding nematodes.

Trophic Diversity index (TDI)

$$\mathrm{TDI} = 1 / \sum \mathrm{pi}^2$$

Where
pi^2 is the proportional contribution of the ith trophic group.

28.3 RESULTS

The soil chemical properties at wastewater- and freshwater-irrigated sites are listed in Table 28.1. It was observed that the concentrations of all three heavy metals were high in wastewater-irrigated fields.

TABLE 28.1
Ecological Indices and Other Qualitative Parameters for Assessing Nematode Community Dynamics

S. No.	Indices/Parameters	Values	
		Freshwater-Irrigated Field	Wastewater-Irrigated Field
1	**Total Organic Carbon (g/kg)**	15.10 ± 1.12 (13.87–15.76)	17.14 ± 2.34 (15.12–17.74)
2	**Total Zinc (mg/kg)**	411.3 ± 14.1 (256.4–445.8)	999.9 ± 5.7 (899.2–1299.4)
3	**Total Lead (mg/kg)**	44.6 ± 8.2 (39.9–46.7)	121.4 ± 9.8 (100.8–123.6)
4	**Total Copper (mg/kg)**	142 ± 10.14 (129.8–146.9)	324.7 ± 12.6 (234.4–359.0)
5	**Shannon's Diversity Index(H′)**	1.55 ± 0.58 (1.4–1.90)	1.85 ± 0.67 (1.56–2.43)
6	**Trophic Diversity Index (TDI)**	1.24 ± 0.33 (1.2–1.33)	1.34 ± 0.45 (1.26–1.39)
7	**Plant Parasitic Index (PPI)**	2.76 ± 0.19 (2.27–3.12)	2.60 ± 0.22 (2.42–3.22)
8	**Maturity Index (MI)**	1.66 ± 0.25 (1.33–1.91)	1.31 ± 0.22 (1.21–1.65)
9	**Channel Index (CI)**	16.56 ± 14.20 (14.45–18.24)	12.50 ± 9.91 (11.12–15.34)
10	**Structure Index (SI)**	40.44 ± 9.81 (34.76–41.43)	32.24 ± 12.62 (27.8–35.14)
11	**Enrichment Index (EI)**	71.11 ± 10.12 (66.78–77.20)	65.67 ± 11.35 (62.44–67.23)

28.4 NEMATODE DIVERSITY

During the present course of study, a total of 35 genera belonging to 8 orders and 21 families were recorded. In terms of abundance, order Tylenchida was most abundant, while in terms of the number of genera, order Rhabditida was most frequent.

A total of 33 genera belonging to 8 orders and 19 families were recorded from the soil samples collected from crop fields irrigated with wastewater (Table 28.2). The number of genera varied from 6 to 19 per sample, while in terms of abundance, the number varied from 222 to 1224 individuals per 100 cc of soil. In terms of the number of genera (Figure 28.1A), the order Rhabditida was most frequent (43%) with 14 genera under 2 families, followed by Tylenchida (27%) with 9 genera under 7 families, Dorylaimida (9%) with 3 genera under 3 families, Aphelenchida (6%) and Araeolaimida (6%) each with 2 genera under 2 families, while Monhystrida (3%), Triplonchida (3%) and Enoplida (3%) were represented by 1 genus each. In terms of trophic diversity, the bacterivores (43%) constituted the most dominant group (Figure 28.2A), followed by herbivores (32%), predators (11%), omnivores (7%) and fungivores (7%). In the case of wastewater-irrigated crop fields, *Acrobeles* was the most dominant genus among bacterivores, while *Basiria*, *Aphelenchoides*, *Eudorylaimus* and *Tobrilus* were the most dominant genera among herbivores, fungivores, omnivores and predators, respectively.

In the case of freshwater-irrigated crop fields, a total of 31 genera belonging to 7 orders and 19 families were recorded (Table 28.2). The number of genera varied from 5 to 16 per sample, while in terms of abundance, the number varied from 190 to 1004 individuals per 100 cc of soil. In terms of the number of genera (Figure 28.1B), the order Rhabditida was most frequent (45%) with 14 genera under 3 families, followed by Tylenchida (26%) with 8 genera under 6 families, Dorylaimida (11%) with 3 genera under 3 families, and Aphelenchida (6%) and Enoplida (6%) each with 2 genera under 2 families, while Monhystrida (3%) and Araeolaimida (3%) were represented by 1 genus each. In terms of trophic diversity, the bacterivores (48%) constituted the most dominant group (Figure 28.2B), followed by herbivores (30%), predators (10%), omnivores (6%) and fungivores (6%). *Acrobeles* was the most dominant genus among bacterivores, while *Rotylenchulus*, *Aphelenchoides*, *Eudorylaimus* and *Mononchoides* were the most dominant genera among herbivores, fungivores, omnivores and predators, respectively.

TABLE 28.2
Soil-Inhabiting Nematodes Population Structure

S.No.	Genera	c-p Value	Order	Family	N	Wastewater	N	Freshwater
			Bacterivores					
1	*Bursilla*	1	Rhabditida	Rhabditidae	7	3.44 ± 7.70	3	3.12 ± 9.7
2	*Mesorhabditis*	1	Rhabditida	Rhabditidae	15	5.11 ± 7.35	6	5.23 ± 6.12
3	*Metarhabditis*	1	Rhabditida	Rhabditidae	9	8.43 ± 5.50	3	2.23 ± 2.34
4	*Rhabditis*	1	Rhabditida	Rhabditidae	1	0.23 ± 0.41	2	1.3 ± 1.7
5	*Cuticularia*	1	Rhabditida	Rhabditidae	3	2.13 ± 2.20	6	6.11 ± 2.55
6	*Pelodera*	1	Rhabditida	Rhabditidae	2	5.42 ± 3.11	3	2.9 ± 10.7
7	*Teratorhabditis*	1	Rhabditida	Rhabditidae	4	4.98 ± 2.76	3	2.8 ± 3.0
8	*Acrobeles*	2	Rhabditida	Cephalobidae	27	26.7 ± 17.15	23	23.2 ± 12.8
9	*Acrobeloides*	2	Rhabditida	Cephalobidae	19	17.23 ± 9.98	16	12.9 ± 6.7
10	*Eucephalobus*	2	Rhabditida	Cephalobidae	12	10.10 ± 8.7	5	5.5 ± 2.1
11	*Pseudacrobeles*	2	Rhabditida	Cephalobidae	3	1.28 ± 6.24	0	0 ± 0
12	*Zeldia*	2	Rhabditida	Cephalobidae	4	3.24 ± 9.27	1	1.2 ± 0.2
13	*Teratocephalus*	2	Rhabditida	Cephalobidae	3	2.68 ± 5.98	2	1.2 ± 0.4
14	*Cervidellus*	2	Rhabditida	Cephalobidae	6	5.12 ± 3.47	2	1.9 ± 8.4
15	*Rhabdolaimus*	2	Araeolaimida	Rhabdolaimidae	2	1.74 ± 4.88	0	0 ± 0
16	*Chiloplectus*	2	Araeolaimida	Plectidae	2	1.80 ± 3.43	2	2.2 ± 7.5
17	*Prismatolaimus*	3	Monhysterida	Prismatolaimidae	10	9.70 ± 6.26	4	3.3 ± 5.4
			Fungivores					
18	*Aphelenchoides*	2	Aphelenchida	Aphelenchoididae	28	26.8 ± 23.60	21	19.6 ± 5.8
19	*Aphelenchus*	2	Aphelenchida	Aphelenchidae	17	16.3 ± 9.35	11	11.8 ± 4.7
			Omnivores					
20	*Eudorylaimus*	4	Dorylaimida	Qudsianematidae	7	5.50 ± 6.98	6	4.45 ± 9.66
21	*Mesodorylaimus*	4	Dorylaimida	Dorylaimidae	4	4.50 ± 12.4	4	4.6 ± 2.22
			Herbivores					
22	*Xiphinema*	5	Dorylaimida	Longidoridae	2	1.50 ± 9.40	3	2.9 ± 1.32
23	*Pratylenchus*	3	Tylenchida	Pratylenchidae	12	10.5 ± 10.9	12	12.6 ± 2.8
24	*Psilenchus*	2	Tylenchida	Tylenchidae	4	3.60 ± 8.65	3	1.6 ± 8.2
25	*Basiria*	2	Tylenchida	Tylenchidae	21	24.9 ± 9.22	9	8.6 ± 2.14
26	*Helicotylenchus*	3	Tylenchida	Hoplolaimidae	19	23.2 ± 12.44	17	15.6 ± 4.55
27	*Hemicriconemoides*	3	Tylenchida	Criconematidae	3	2.40 ± 7.22	0	0 ± 0
28	*Hoplolaimus*	3	Tylenchida	Hoplolaimidae	12	12.1 ± 4.11	16	16.4 ± 6.5
29	*Meloidogyne*	3	Tylenchida	Heteroderidae	4	3.90 ± 1.77	9	8.5 ± 7.1
30	*Rotylenchulus*	3	Tylenchida	Rotylenchulidae	17	20.9 ± 17.45	22	21.6 ± 14.8
31	*Tylenchorhynchus*	3	Tylenchida	Dolichodoridae	12	10.45 ± 9.10	14	15.1 ± 10.4
32	*Trichodorus*	4	Triplonchida	Trichodoridae	2	1.8 ± 1.9	0	0 ± 0
			Predators					
33	*Mononchoides*	1	Rhabditida	Neodiplogastridae	0	0 ± 0	3	2.2 ± 0.4
34	*Tobrilus*	3	Enoplida	Tobrilidae	4	4.23 ± 11.22	2	1.2 ± 0.2
35	*Tripylina*	3	Enoplida	Tripylidae	0	0 ± 0	2	1.1 ± 0.8

Note: Mean abundance per 100 cc soil ± SD; N = 30.

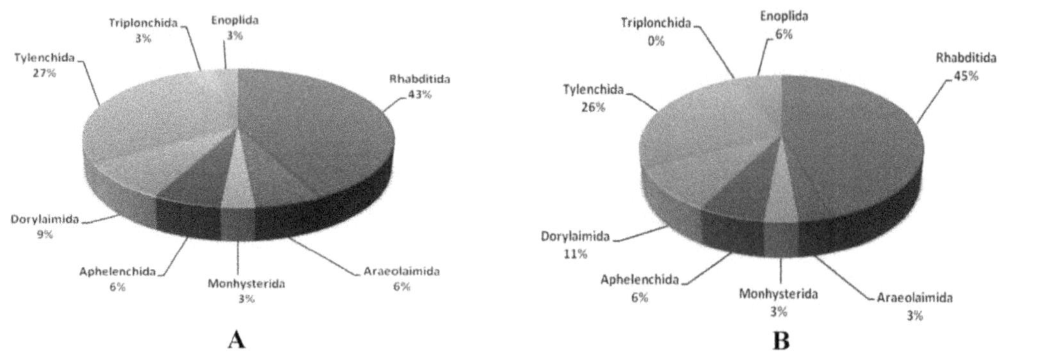

FIGURE 28.1 Genera of nematodes in different orders in wastewater- (A) and freshwater-irrigated fields (B).

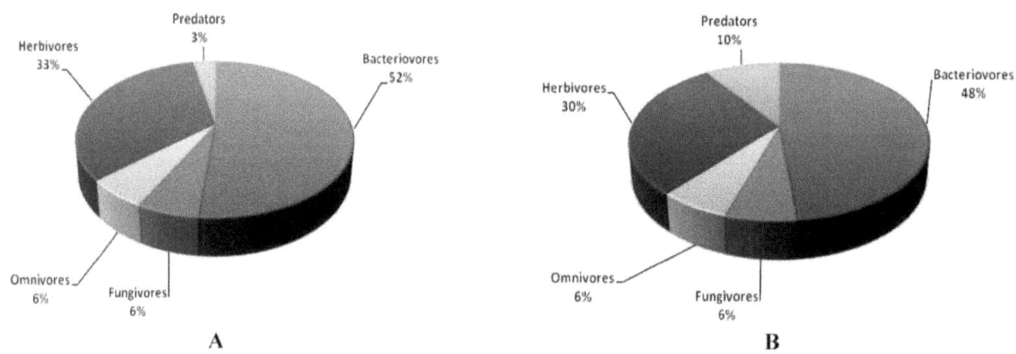

FIGURE 28.2 Genera of nematodes in different trophic groups in wastewater- (A) and freshwater-irrigated fields (B).

Overall, *Acrobeles* was the most abundant among all the nematode genera. Its highest abundance was reported from wastewater-irrigated fields where the concentration of heavy metals was high. *Mononchoides* and *Tripylina*, which are low cp value predators, were absent in wastewater-irrigated fields where the concentration of the heavy metals was high. The presence or absence of these nematode genera at a particular site seems to be influenced by heavy metal concentration.

28.5 NEMATODE COMMUNITY ANALYSIS

Diversity and maturity indices were calculated to assess the diversity of nematodes in freshwater- and wastewater-irrigated fields (Table 28.1). In the present study, total numbers of nematodes have a significantly positive correlation with heavy metals. The MI was calculated to assess the maturity of the agro-ecosystem. The lower values of MI in the present study indicated a disturbed environment due to agricultural practices and heavy metal contamination. The values of MI were higher in freshwater-irrigated fields than in wastewater-irrigated fields. The PPI is an extremely good indicator of plant-parasitic nematode resources. Larger PPI values indicate low levels of disturbance to the community, while smaller PPI values indicate high levels of disturbance. During this study, lower values of PPI in wastewater-irrigated fields indicated a disturbed environment due to heavy metal pollution.

The SI was calculated to assess the structure of the ecosystem. During this study, lower values of SI in wastewater-irrigated fields indicated the polluted and disturbed status of the ecosystem. The EI gives the status of enrichment within the ecosystems due to contamination. The values of EI in the present study were very high at all the sites, giving the impression of an enriched ecosystem. Higher values of EI were observed in wastewater-irrigated fields. The CI was calculated to assess the decomposition pathway within the ecosystem. Higher values of the CI indicate a fungal-dominated decomposition pathway, while lower values indicate a bacteria-based decomposition pathway. The values for CI in the present work were mostly low in freshwater-irrigated fields and high in wastewater-irrigated fields. Shannon's diversity index (H') reflects the diversity of nematodes in an ecosystem. Higher values of H′ show a highly diverse ecosystem, while low values show the opposite. The values of H′ were higher in wastewater-irrigated fields than in freshwater-irrigated fields.

28.6 DISCUSSION

Nematodes are a diverse and highly sophisticated group within the soil environment. They occupy a central position within the soil food web, occurring at multiple trophic levels (Neher *et al.*, 2005). Therefore, nematodes have the potential to supply useful insights into the structure and function of the soil food web (Neher *et al.*, 2005). Nematodes are often grouped into five major trophic groups: bacterivores, fungivores, plant parasites, predators and omnivores (Yeates *et al.*, 1993). Plant-parasitic nematodes are considered as primary consumers, and they affect food web resources through direct herbivory (Ferris and Bongers, 2006; Xiao *et al.*, 2021). Bacterivore and fungivore nematodes graze on decomposer microbes, bacteria and fungi, and thus significantly contribute to nutrient mineralization (Ferris *et al.*, 1996; Ferris and Matute, 2003).Bacterivore nematodes also promote rhizosphere colonization of beneficial bacteria (Kimpinski and Sturz, 1996; Knox *et al.*, 2003). Predatory nematodes regulate the food web by preying on other nematodes and invertebrates within the soil (Zimmerman and Cranshaw, 1990; Grewal *et al.*, 2005). The development of the MI based on the colonizer-persister (c-p) values of nematodes has helped in interpreting the biological and trophic status of the soil food web in several habitats (Bongers, 1990). In terrestrial habitats, MI is routinely used as an ecological measure for assessing the status of soil food webs (Nethi *et al.*, 2002; Nahar *et al.*, 2006; Krashevska *et al.*, 2019). Additionally, the PPI (based on the c-p value of plant-parasitic nematodes) and trophic diversity have been used for assessment of the soil food web condition. Ferris *et al.* (2001) developed enrichment (EI), structure (SI) and channel (CI) indices based on the relative weighted abundance of nematode c-p guilds. EI provides an indication of the response of primary decomposers to the available resources within the soil food web, while SI suggests trophic linkages in a food web, as indicated by the presence of higher c-p value nematodes, particularly predatory and omnivores (Ferris *et al.*, 2001). Therefore, EI and SI can be plotted as a graphical representation of the nematode faunal profile depicting the likely condition of the soil food web in a given habitat (Ferris *et al.*, 2001). As CI provides information about the decomposition channels, a high CI (>50%) indicates fungal decomposition channels, whereas a low CI (< 50%) suggests bacterial decomposition channels (Ferris *et al.*, 2001; Krashevska *et al.*, 2019).

The present work aimed to review the effect of wastewater irrigation contaminated by heavy metals on the community structure of the soil-inhabiting nematodes related to agricultural fields near the Yamuna River in Faridabad, Haryana. Agricultural management practices like tillage, fertilization, irrigation and pesticide application cause a disturbance of the soil ecosystem (Bongers *et al.*, 1997) and as a result, affect the soil nematode community structure (Freckman and Ettema, 1993; Yardim and Edwards, 1998; Porazinska *et al.*, 1999). Although the total number of nematodes did not differ significantly between the wastewater- and freshwater- irrigated fields, trophic groups such as bacterivores, herbivores and predators exhibited significant treatment effects during the study period.

Jagtap *et al.* (2010) and Ahalavat and Chaubey (2017) reported the presence of various heavy metals in crop fields irrigated with wastewaters from Pune and Delhi, respectively, which affects the community structure of the soil-inhabiting nematodes. In the present study, the concentration of heavy metals was extremely high as compared with earlier studies. This may be due to dumping 95% of the waste of nearby factories into the Yamuna River and this water being used by farmers to irrigate their crop fields. As a result, heavy metals are accumulating in agro-ecosystems. The lower abundance of Dorylaims in the crop fields (9% and 11%) indicates more disturbed soil, as cropping practices always involve tillage, fertilization, irrigation and pesticide application. The susceptibility of Dorylaims to these activities is also reported by Thomas (1978), Neher (2001) and Torres *et al.* (2006). Therefore, the sensitivity of the Dorylaims is a good indicator of soil disturbance.

During the present study, the total number of nematodes and the presence of heavy metals has indicated a positive correlation. From the results, it was observed that bacterivores and herbivores were most abundant at polluted sites. Similar results were reported by Pen-Mouratov *et al.* (2008), who found that certain trophic groups are more sensitive to heavy metal pollution than the entire nematode population. This is in agreement with the present study, where total abundance was higher at a wastewater-irrigated site with high heavy metal concentration but omnivores, predators and carnivores were lowest at this site. Shannon's diversity index (H′) reflects the diversity of nematodes in an ecosystem. Higher values of H′ show a highly diverse ecosystem, while low values show the opposite. Within the present study, the values of H′ were 1.55 and 1.85 in freshwater- and wastewater-irrigated fields, respectively. This showed a disturbed environment. Ahalavat *et al.* (2018) also reported low values of H′ in heavy metal-treated soils.

By measuring the relative proportions of colonizers and persisters within the soil, the MI assesses the impact of disturbance in any habitat (Bongers, 1990; Bongers and Bongers, 1998; Neher, 1999; Georgieva *et al.*, 2002). A greater abundance of low c-p value nematodes leads to a lower MI value, which suggests a disturbed environment due to heavy metal contamination in a given habitat (Neher, 1999; Porazinska *et al.*, 1999; Yeates and Bongers, 1999). The SI was calculated to assess the structure of the ecosystem. During this study, lower values of SI in wastewater-irrigated fields indicated the polluted and disturbed status of the ecosystem. Ahalavat and Chaubey (2017) also reported low values of MI and SI in heavy metal-treated soils. The PPI is an extremely good indicator of plant-parasitic nematode resources. Larger PPI values indicate low levels of disturbance to the community, and smaller PPI values indicate high levels of disturbance (Bongers, 1990; Bongers and Ferris, 1999). In this study, values of PPI were lower, which clearly showed that the region is highly disturbed. EI indicates the response of primary decomposers or enrichment opportunists like rhabditids to labile sources of organic material (Ferris *et al.*, 2001; Ferris and Matute, 2003). We observed that all the habitats had moderately enriched food webs with EI ranging from 62 to 77%. CI indicated predominant decomposition channels in the soil food web; a high CI (>50%) indicates fungal decomposition channels, whereas a low CI (<50%) suggests bacterial decomposition channels. Low values of CI in the present study correspond to bacterial decomposition channels. Similar effects of heavy metal pollution on fungivore and bacterivore nematodes have been studied by Nagy *et al.* (2004) and Ahalavat *et al.* (2018). The low values of MI and SI as well as diversity indicated a disturbed environment.

28.7 CONCLUSION

The present study confirms that nematodes are very good indicators of soil conditions. Nematode community analysis provides a powerful tool for the diagnosis of the complexity and status of soil health. Indices developed from nematode community analysis indicate the relative proportions of services and functions, but not their magnitude. If biomass is also taken into account, we can have a better understanding of soil health. The information generated through this investigation will be helpful to provide advisory services to farmers and undertake pest risk analysis.

REFERENCES

Ahalavat S and Chaubey AK 2017. Nematode community structure in crop fields near Yamuna in Faridabad, Haryana. *Int J Agric Inven* 2(2): 210–215.

Ahalavat S, Chaubey AK, Hasan W and Kumar P 2018. Effects of wastewater irrigation on nematode community structure in agro-ecosystems near Yamuna in Haryana, India. *Bull Environ Pharm Life Sci* 8(1): 114–120.

Alasmary Z, Todd T, Hettiarachchi GM, Stefanovska T, Pidlisnyuk V, Roozeboom K, Erickson L, Davis L, and Zhukov O. 2020. Effect of soil treatments and amendments on the nematode community under *Miscanthus* growing in a lead contaminated military site. *Agronomy* 10: 1727; doi:10.3390/agronomy10111727.

Beier S and Traunspurger W 2001. The meiofauna community of two small German streams as indicator of pollution. *J Aquat Ecosyst Stress Recovery* 8: 387–405.

Bongers T 1990. The maturity index: an ecological measure of environmental disturbance based on nematode species composition. *Oecologia* 83: 14–19.

Bongers T and Bongers M 1998. Functional diversity of nematodes. *Appl Soil Ecol* 10: 239–251.

Bongers T and Ferris H 1999. Nematode community structure as a bioindicator in environmental monitoring. *Trends Ecol Evol* 14: 224–228.

Bongers T, Ilieva-Makulec K and Ekschmitt K 2001. Acute sensitivity of nematode taxa to $CuSO_4$ and relationships with feeding-type and life-history classification. *Environ Toxicol Chem* 20: 1511–1516.

Bongers T, van der Meulen H and Korthals G 1997. Inverse relationship between the nematode maturity index and plant-parasitic index under enriched nutrient conditions. *Appl Soil Ecol* 6: 195–199.

Boyd RS 2010. Heavy metal pollutants and chemical ecology: exploring new frontiers. *J Chem Ecol* 36: 46–58.

Brinke M, Höss S, Fink G, Ternes TA, Heininger P, and Traunspurger W. 2010. Assessing effects of the pharmaceutical ivermectin on meiobenthic communities using freshwater microcosms. *Aquat Toxicol* 99(2): 126–137.

Brinke M, Ristau K, Bergtold M, Höss S, Claus E, Heininger P, and Traunspurger W. 2011. Using meiofauna to assess pollutants in freshwater sediments: a microcosm study with cadmium. *Environ Toxicol Chem* 30(2): 427–438.

Burton SM, Rundle SD and Jones MB 2001. The relationship between trace metal contamination and stream meiofauna. *Environ Pollut* 111(1): 159–167.

Caldwell B 2005. Enzyme activities as a component of soil biodiversity: a review. *Pedobiologia* 49: 637–644.

Chen G, Qin J, Shi D, Zhang Y and Ji W 2009. Diversity of soil nematodes in area polluted with heavy metals and polycyclic aromatic hydrocarbons (PAHs) in Lanzhou, China. *Environ Manage* 44(1): 163–172.

Cobb NA 1918. Estimating the nema population of the soil. U. S. Department of Agriculture. *Agricultural Technical Circular of US Department of Agriculture* 1: 48.

Cortet J, Gomot DeVauflery A, Poinsot Balaguer N, Gomot L, Texier C and Cluzeau D 1999. The use of invertebrate soil fauna in monitoring pollutant effects. *Eur J Soil Biol* 35: 115–134.

Dougherty EC, Hansen EL, Nicholas WL, Mollet JA and Yarwood EA 1959. Axenic cultivation of *Caenorhabditis briggsae* (Nematoda: Rhabditidae) with supplemented and unsupplemented chemically defined media. *Ann N Y Acad Sci* 77: 176–217.

Edwards CA 2002. Assessing the effects of environmental pollutants on soil organisms, communities, processes and ecosystems. *Eur J Soil Biol* 38: 225–231.

Faupel M, Ristau K and Traunspurger W 2011. Biomass estimation across the benthic community in polluted freshwater sediment—a promising endpoint in microcosm studies? *Ecotoxicol Environ Saf* 74(7): 1942–1950.

Ferris H 2010. Form and function: metabolic footprints of nematodes in the soil food web. *Eur J Soil Biol* 46: 97–104.

Ferris H and Bongers T 2006. Nematode indicators of organic enrichment. *J Nematol* 38: 3–12.

Ferris H and Matute M 2003. Structural and functional succession in the nematode fauna of a soil food web. *Appl Soil Ecol* 23: 93–110.

Ferris H, Bongers T and de Geode RGM 2001. A framework for soil food web diagnostics: extension of the nematode faunal analysis concept. *Appl Soil Ecol* 18: 13–29.

Ferris H, Venette RC and Lau SS 1996. Dynamics of nematode communities in tomatoes grown in conventional and organic farming systems and their impact on soil fertility. *Appl Soil Ecol* 3: 161–175.

Freckman DW and Ettema CH 1993. Assessing nematode communities in agroecosystems of varying human intervention. *Agric Ecosyst Environ* 45: 239–261.

Fytianos K, Katsianis G, Triantafyllou P and Zachariadis G 2001. Accumulation of heavy metals in vegetables grown in an industrial area in relation to soil. *Bull Environ Contam Toxicol* 67: 423–430.

Georgieva SS, McGrath SP, Hooper DJ and Chambers BS 2002. Nematode communities under stress: the long-term effects of heavy metals in soil treated with sewage sludge. *Appl Soil Ecol* 20: 27–42.

Grewal PS, Ehlers RU and Shapiro-Ilan DI (Eds.) 2005. *Nematodes as Biocontrol Agents.* CABI Publishing, CAB International, Oxon, UK, pp. 1–505.

Griffiths BS, Neilson R and Bengough AG 2003. Soil factors determined nematode community composition in a two year pot experiment. *Nematology* 5: 889–897.

Gunapala N, Venette RC, Ferris H and Scow KM 1998. Effects of soil management history on the rate of organic matter decomposition. *Soil Biol Biochem* 30: 1917–1927.

Gupta VVSR and Yeates GW 1997. Soil microfauna as indicators of soil health. In: Pankhurst CE, Doube BM and Gupta VVSR (Eds) *Biological Indicators of Soil Health.* CAB International, Oxon, UK, pp. 201–233.

Haegerbaeumer A, Höss S, Ristau K, Claus E, Möhlenkamp C, Heininger P and Traunspurger W 2016. A comparative approach using ecotoxicological methods from single-species bioassays to model ecosystems. *Environ Toxicol Chem* 35(12): 2987–2997.

Hamilton AJ, Stagnitti F, Kumarage SC and Premier RR 2007. RIRA: a tool for conducting health risk assessments or irrigation of edible crops with recycled water. *Comput Electron Agric* 57 (1): 80–87.

Heininger P, Höss S, Claus E, Pelzer J and Traunspurger W 2007. Nematode communities in contaminated river sediments. *Environ Pollut* 146(1): 64–76.

Heip C, Warwick RM, Carr RM, Herman PMJ, Huys R, Smol N and Van Holsbeke K 1988. Analysis of community attributes of benthic meiofauna of Frierfjord/Langesundfjord. *Mar Ecol Prog Ser* 46: 171–180.

Höss S and Williams PL 2009. Ecotoxicity testing with nematodes. In: Wilson MJ and Kakouli-Duarte T (Eds) *Nematodes as Environmental Indicators.* CABI International, Wallingford, pp. 208–224.

Houx NWH and Aben WJM 1993. Bioavailability of pollutants to soil organisms via the soil solution. *J Sci Total Environ* 33: 387–395.

Hugot JP, Baujard P and Morand S 2001. Biodiversity in helminths and nematodes as a field of study: an overview. *Nematology* 3: 199–208.

Ingham RE, Trofymow JA, Ingham ER and Coleman DC 1985. Interactions of bacteria, fungi, and their nematode grazers: effects on nutrient cycling and plant growth. *Ecol Monogr* 55: 119–140.

Jagtap MN, Kulkarni MV and Puranik PR 2010. Flux of heavy metals in soils irrigated with urban wastewaters. *Am Eurasian J Agric Environ Sci* 8: 487–493.

Kandeler F, Kampichler C and Horak O 1996. Influence of heavy metals on the functional diversity of soil microbial communities. *Biol Fertil Soils* 23: 299–306.

Kimpinski J and Sturz AV 1996. Population growth of a rhabditid nematode on plant growth promoting bacteria from potato tubers and rhizosphere soil. *J Nematol* 28: 682–686.

Knox OGG, Killham K, Mullins CE and Wilson MJ 2003. Nematode-enhanced microbial colonization of the wheat rhizosphere. *FEMS Microbiol Lett* 225: 227–233.

Korthals GW, De Goede RGM, Kammenga JE and Bongers T 1996. The maturity index as an instrument for risk assessment of soil pollution. In: Van Straalen NM and Krvolutsky DA (Eds) *Bioindicator Systems for Soil Pollution.* Kluwer Academic, Netherlands, pp. 85–93.

Krashevska V, Kudrin AA, Widyastuti R and Scheu S 2019. Changes in nematode communities and functional diversity with the conversion of rainforest into rubber and oil palm plantations. *Front Ecol Evol* 7: 487; doi: 10.3389/fevo.2019.00487.

Liang WJ, Zhong S, Hua JF, Cao CY and Jiang Y 2007. Nematode faunal response to grassland degradation in Horquin Sandy Land. *Pedosphere* 17: 611–618.

Makoni FS, Thekisoe OMM and Mbati PA 2016. Urban wastewater for sustainable urban agriculture and water management in developing countries. In: Younos T and Parece T (Eds) *Sustainable Water Management in Urban Environments. The Handbook of Environmental Chemistry, vol.* 47, Springer, Berlin, pp. 265–293.

Marshall FM, Holden J, Ghose C, Chisala B, Kapungwe E, Volk J, Agrawal M, Agrawal R, Sharma RK, and Singh RP 2007. Contaminated Irrigation Water and Food Safety for the Urban and Peri-urban

Poor: Appropriate Measures for Monitoring and Control from Field Research in India and Zambia, Inception Report DFID Enkar R8160, SPRU, University of Sussex.
Martinez JG, Torres MA, dos Santos G and Moens T 2018. Influence of heavy metals on nematode community structure in deteriorated soil by gold mining activities in Sibutad, southern Philippines. *Ecol Indic* 91: 712–721.
Nagy P, Bakonyi G, Bongers T, Kádár I, Fábián M and Kiss I 2004. Effects of microelements on soil nematode assemblages seven years after contaminating an agricultural field. *Sci Total Environ* 320: 131–143.
Nahar MS, Grewal PS, Miller SA, Stinner D, Stinner BR, Kleinhenz MD, Wszelaki A and Doohan D. 2006. Differential effects of raw and composted manure on nematode community, and its indicative value for soil microbial, physical and chemical properties. *Appl Soil Ecol* 34: 140–151.
Nahmani J and Lavelle P 2002. Effects of heavy metal pollution on soil macrofauna in a grassland of northern France. *Eur J Soil Biol* 38: 297–300.
Neher DA 1999. Nematode communities in organically and conventionally managed agricultural soils. *J Nematol* 31: 142–154.
Neher DA 2001. Role of nematodes in soil health and their use as indicators. *J Nematol* 33: 161–168.
Neher DA, Wub J, Barbercheck ME and Anas O 2005. Ecosystem type affects interpretation of soil nematode community measures. *Appl Soil Ecol* 30*:* 47–64.
Nethi S, Grewal PS, de Nardo EA and Stinner BR 2002. Non-target effect of entomopathogenic nematodes on soil nematode community. *Appl Soil Ecol* 39: 735–744.
Nicholas WL, Dougherty EC and Hansen EL 1959. Axenic cultivation of *Caenorhabditis briggsae* (Nematoda: Rhabditidae) with chemically undefined supplements; comparative studies on related species. *Ann N Y Acad Sci* 77: 218–236.
Pen-Mouratov S, He X and Steinberger Y 2004. Spatial distribution and trophic diversity of nematode populations under *Acacia raddiana* along a temperature gradient in the Negev Desert ecosystem. *J Arid Environ* 56: 339–355.
Pen-Mouratov S, Shukurov N and Steinberger Y 2008. Influence of industrial heavy metal pollution on soil free-living nematode population. *Environ Pollut* 152(1): 172–183.
Porazinska DL, Duncan LW, McSorley R and Graham JH 1999. Nematode communities as indicators of status and processes of a soil ecosystem influenced by agricultural management practices. *Appl Soil Ecol* 13: 69–86.
Qadir M, Wichelns D, Raschid-Sally L, McCornick PG, Drechsel P, Bahri A and Minhas PS 2010. The challenges of wastewater irrigation in developing countries. *Agric Water Manag* 97(4): 561–568.
Rodríguez Martín JA, Gutiérrez C, Escuer M, García-González MT, Campos-Herrera R and Águila N 2014. Effect of mine tailing on the spatial variability of soil nematodes from lead pollution in La Union (Spain). *Sci Total Environ* 473–474: 518–529.
Šalamún P, Renčo M, Kucanová E, Brázová T , Papajová I , Miklisová, D and Hanzelová V 2012. Nematodes as bioindicators of soil degradation due to heavy metals. *Ecotoxicology* 21: 2319–2330.
Schenk J, Höss S, Brinke M, Kleinbölting N, Brüchner-Hüttemann H and Traunspurger W 2020. Nematodes as bioindicators of polluted sediments using metabarcoding and microscopic taxonomy. *Environ Int* 143: 105922; doi: 10.1016/j.envint.2020.105922.
Sharma RK, Agarwal M and Marshall F 2007. Heavy metal contamination of soil and vegetables in suburban areas of Varanasi, India. *Ecotoxicol Environ Saf* 66: 258–266.
Sroczyńska K, Chainho P, Vieira S and Adão H 2021. What makes a better indicator? Taxonomic vs functional response of nematodes to estuarine gradient. *Ecol Indic* 121: 107113; doi: 10.1016/j.ecolind.2020.107113.
Thomas SH 1978. Population densities of nematodes under seven tillage regimes. *J Nematol* 10: 24–27.
Torres GRC, Pedrosa EMR, Montenegro AAA, Michereff SJ and Moura RM 2006. Ecological aspects of nematode community associated with *Cucumis melo* crop in Rio Grande do Norte, Brazil. *Nematol Bras* 30(1): 1–9.
Traunspurger W 1997. Bathymetric, seasonal and vertical distribution of feeding-types of nematodes in an oligotrophic lake. *Vie Milieu* 47(1): 1–7.
Vanfleteren JR 1973. Amino acid requirements of the free-living nematode *Caenorhabditis briggsae. Nematologica* 19: 93–99.
Wilson M and Kakouli-Duarte T 2009. *Nematodes as Environmental Indicators*. CABI Publishing, Wallingford.
Xiao H, Wang W, Xia S, Li Z, Gan J and Yang X 2021. Distributional patterns of soil nematodes in relation to environmental variables in forest ecosystems. *Soil Ecol Lett* 3(2): 115–124.

Yardim EN and Edwards CA 1998. The effects of chemical pest, disease and weed management practices on the trophic structure of nematode populations in tomato agroecosystems. *Appl Soil Ecol* 7: 137–147.
Yeates GW and Bongers T 1999. Nematode diversity in agroecosystems. *Agric Ecosyst Environ* 74: 113–135.
Yeates GW, Bongers T, De Goede RGM, Freckman DW and Georgieva SS 1993. Feeding habits in soil nematode families and genera—an outline for soil ecologists. *J Nematol* 25: 315–331.
Zimmerman RJ and Cranshaw WS 1990. Compatibility of entomogenous nematodes (Rhabditida) in aqueous solutions of pesticides used in turfgrass maintenance. *J Econ Entomol* 83: 97–100.

29 Algae and Macrophytes as Bioindicators of Freshwater Ecosystem

Divya Dubey, Saroj Kumar, and Venkatesh Dutta
Department of Environmental Science, School of Earth and Environmental Sciences, Babasaheb Bhimrao Ambedkar University, Lucknow

CONTENTS

DOI: 10.1201/9781003220824-33

29.1 INTRODUCTION

Freshwater bodies like lakes and rivers are an essential component of the ecosystem and support human sustenance. The healthy condition of aquatic ecosystems is essential for the flourishing diversity of aquatic biota and for its proper functioning. Unfortunately, presently, the actual consumption of natural water resources adds to their deterioration and degradation, leading to the loss of biodiversity and creating imbalance in proper functioning of the aquatic ecosystem (Tórz et al., 2020). Water pollution and various anthropogenic stresses such as land use change in lakes' catchment are a serious concern nowadays, as discharge of polluted water into water bodies degrades natural water quality, causing biological imbalance both quantitatively and qualitatively (Dutta *et al.*, 2020). Increased input of nutrients into the lacustrine ecosystem causes increase in the trophic state, resulting in a major shift and change in aquatic biodiversity of both algae and macrophytes.

29.1.1 Present Scenario of Freshwater Resources

The major factors responsible for degradation of freshwater resources globally are wastewater discharge, urbanization, unhealthy land use change, agricultural activities in the catchment area, solid waste disposal on the littoral area, encroachment activities and habitat fragmentation of large freshwater ecosystems (Fierro *et al.*, 2019), as shown in Figure 29.1. Nutrient enrichment causing eutrophication as a result of human and natural activities, especially due to urban land use changes, is the leading cause of the degradation of freshwater lacustrine ecosystems (Dubey *et al.*, 2021; Bhagowati and Ahamad, 2019). Urbanization, agricultural land use, wastewater discharge, abrupt changes in land use, and encroachment in catchment areas – all contribute to cultural eutrophication. Nutrient enrichment not only causes morphological and physicochemical changes but also severely affects the biological component of the freshwater ecosystem. Major visible effects of nutrient enrichment in the aquatic ecosystem include the proliferation of algal bloom, which results in the degradation, decline and extinction of native and sensitive species of aquatic flora and fauna, including loss of submerged macrophytes, and thereby causing habitat degradation. This adversely affects the normal ecological functioning of the lacustrine ecosystem (Glibert and Burford, 2017).

29.1.2 Freshwater Ecosystem as an Urban Sponge

Freshwater ecosystems, which include rivers and their connected landscape like lakes, ponds and wetlands, are the major component of the hydrological cycle and form an 'urban sponge', which aids in groundwater recharge, purification, utilization and storm water regulation, and helps in ecological restoration (Liang *et al.*, 2020). However, presently, lakes in urban catchments are under highly stressed conditions, as they are severely affected by unplanned development, due to which they are highly susceptible to increasing anthropogenic pressure as well as natural climatic extremes that cause major shifts in the algal and macrophytic communities (Chau, 2005); hence, nowadays, assessment of freshwater ecosystem using bioindicators is highly recommended, as the trophic state and nutrient status of the lake environment are highly susceptible to even small alterations (Lodi *et al.*, 2011).

29.2 TROPHIC STATE OF THE LAKES

The total biomass of biologically living materials existing at a certain place and time is what is referred to as the trophic state in a lake ecosystem. Trophic state is the biological response to nutrient enrichment in the lake ecosystem (Naumann, 1929), but the impact of nutrient enrichment is also influenced by a variety of other factors, including the morphometric and catchment status of the

FIGURE 29.1 Present scenario of freshwater lakes in a catchment stressed by polluted and urban land use. (A) *Trapa* dominance caused by its cultivation in lake. (B) Fragmentation in lakes for fishing and *Trapa* cultivation (C) Dumping of solid waste on the littoral area of the lake. (D) Clothes washing by local washerman along the shoreline. (E) Wastewater discharge into the lake. (F) Excessive growth of invasive macrophytes *Alternaria philoxides* in the lake. (G) Encroachment activities on the littoral area of the lake. (H) Algal blooms and solid waste debris floating on the surface of a nutrient-enriched lake.

lakes, seasons, geographical locations, mixing depth, and grazing. Total phosphorus (TP), chlorophyll a (chl a), and Secchi disc depth are the three factors that have the greatest influence on the trophic status index of lakes, since they are the fundamental determinants of the lakes' primary productivity (Stednick and Hall, 2003). Lake trophic categorization is presently the most established and popular concept for judging and assessing the condition of the lacustrine ecosystem. Lakes are often

TABLE 29.1
Characteristics of the Different Trophic States of Lakes

Major Factors	Oligotrophic	Mesotrophic	Eutrophic
Organic matter	Low level of organic matter	More organic matter	High level of organic matter
Water clarity	Tend to be deep and clear	Moderate clarity	Abundant plant growth, poor clarity
Oxygen concentration	Oxygen-rich bottom supports cold water fish such as trout	Oxygen level in lake bottom is low	Stratified with oxygen-poor bottom, characterized by dead zones where oxygen level falls below 2 mg/l
Phosphorus concentration	Phosphorus is limiting. Phosphorous concentration is in the range of 0–12 mg/l	Phosphorus concentration present in the range of 12–24 mg/l	In eutrophic lakes, phosphorus concentration is very high, i.e., above 24 mg/l, resulting in dominance of algal bloom

TABLE 29.2
Lake Trophic State Classification Based on the Carlson Trophic State Index

TSI	Trophic State	Secchi Depth (SD) in m	Total Phosphorus (TP) in mg/l	Chlorophyll a (chl a)
0–40	Oligotrophic	4–8	0–12	0–2.6
40–50	Mesotrophic	2–4	12–24	2.6–7.3
50–70	Eutrophic	2–0.5	24–96	7.3–56
70–100+	Hypertrophic	up to 0.5	96–384+	56–155+

Source: Carlson, 1977.

divided into three trophic states based on the primary productivity and the level of nutrient enrichment, which are briefly described as follows:

a. Oligotrophic lakes
b. Mesotrophic lakes
c. Eutrophic lakes

Lakes present in different trophic states have different characteristics, which affect the overall functioning of the aquatic ecosystem, as summarized in Table 29.1.

Presently, the most commonly used trophic state classification of lakes is the Tropic State Index (TSI), given by Carlson. Trophic state values ranging from 0 to 100 are assigned to lakes present in different trophic states, that is, from oligotrophic to hypereutrophic. Specific ranges of values are assigned to the major three parameters, Secchi depth (SD), TP and chlorophyll a (chl a), as given in Table 29.2.

29.3 AQUATIC BIOINDICATORS

Bioindicators are taxa or groups of organisms that signal environmental degradation and exhibit major shifts in their diversity and community structure with changes in ambient pressure caused by human activities. The major groups of organisms that have been used as bioindicators of environmental pollution in an aquatic ecosystem include bacteria, fungi, algae, macrophytes and aquatic fauna such as fish, turtles and crabs (Deshmukh, 2017). Algae, being a diverse group of photosynthetic organisms in aquatic ecosystems, are important constituents of biological monitoring programs for evaluating and assessing the water quality status and biological condition of the lake ecosystem. They are appropriate for assessment of various water quality parameters because of their nutrient needs, short life cycle and fast reproduction rate. Algae are important bioindicators of aquatic ecosystems due to their fast responses to alterations in nutritional conditions, in terms of both species composition and densities, in a variety of water conditions. (Dubey and Dutta, 2020).

Algae and macrophytes respond to nutrient enrichment to varying degrees and are taxa specific. Some submerged macrophytes, however, have advantageous adaptations and niche preferences that allow them to coexist alongside dominant species as well as with the changed environment. The spread of invasive macrophytes and algae due to eutrophication and anthropogenic interferences can be detrimental to native macrophytes over the long run, as it leads to their extinction.. Some macrophytes, like *Utricularia* sp., *Triplophyllum dicksonioides* and *Eleocharis interstincta*, can be used as bioindicators of changes in environmental conditions that are affected by anthropogenic activities, such as land use change in the catchment of the lake. Long-term monitoring of macrophytes as a target biological group, along with a phylogenetic and functional approach and the development of an integrity index, have been used for assessment of the integrity of the freshwater ecosystem (Fares *et al.*, 2020). In the humid and shady riparian habitats of the aquatic ecosystem, several species of amphibious ferns are frequently encountered (Paixão et al., 2013). A microclimate with good amount of moisture and shade is favorable for ferns' healthy growth, which is created by the riparian environments (Mackay *et al.*, 2010). These aquatic organisms may therefore be a sign of less altered and heavily shaded environments of the aquatic ecosystem, similarly to shade-tolerant amphibious ferns. These observations reinforce the idea that aquatic macrophytes can serve as excellent bioindicators of altered aquatic conditions. Moreover, amphibious and emergent macrophytic species are important bioindicators of land use change and altered condition (Bleich et al., 2015), and help in representing the environmental succession in freshwater ecosystems. Aquatic species of algae and macrophytes having comparable life forms have varied resource needs in terms of nutrients, light and the availability of minerals(Fares *et al.*, 2020).

29.3.1 Types of Bioindicators in Freshwater Ecosystem

There are two ways of assessing the freshwater lentic ecosystem. Firstly, it can be done by using traditional chemical analysis in the laboratory and secondly, by monitoring the abundance of bioindicator species within the lake. Bioindicators indicate the changes in the lentic environment and help in predicting the natural state or the level of contamination in the ecosystem (Khatri and Tyagi, 2015). Bioindicators are broadly categorized into four categories:

1. *Pollution indicators*: They detect the presence of pollutants in the freshwater ecosystem.
   Examples: Algae, macrophytes, fishes
2. *Environmental indicators*: They monitor changes in the natural environmental conditions and provide information about the state of the environmental pressures on the freshwater ecosystem.
   Examples: Algae, macrophytes, invertebrates, turtles.
3. *Ecological indicators*: Ecological indicators are a subset of environmental indicators. They refer only to ecological processes and assist in monitoring and assessment of the effective measures taken to enhance the ecological process. They indicate the set of environmental conditions and detect changes and their impacts in the natural surroundings of the freshwater ecosystem.
   Examples: Lichens, algae and macrophytes.

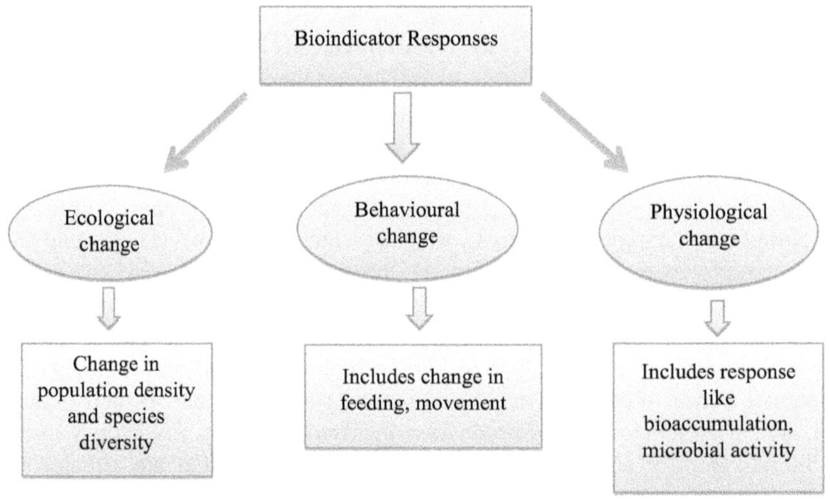

FIGURE 29.2 Various responses of bioindicators in the freshwater aquatic ecosystem.

4. *Biodiversity indicators*: They detect changes in the diversity and the species composition in freshwater ecosystems.
   Examples: Algae, macrophytes and microbial indicators.

Biomonitoring provides additional and accurate information regarding the past trends and present state in environmental behavior They may also detect changes in the aquatic environment resulting from pollutants, heavy metals and nutrients that impact the variety, richness and distribution of aquatic biodiversity (Holt and Miller, 2010). Some of the best examples of bioindicators include plankton, macrophytes, mollusks or other benthic invertebrates, fish, etc. There are some microorganisms that produce stress proteins due to exposure to pollutants. Therefore, a bioindicator can be any organism, ranging from plants to animals, that shows a specific response against pollutants present in the ambient environment (Manickavasagam et al., 2019). Various responses of bioindicators are shown in Figure 29.2.

29.3.2 Advantages of Bioindicators over Conventional Chemical Analysis

There are certain advantages of these bioindicators over performing traditional physicochemical analysis (Manickavasagam et al., 2019). They are summarized as follows:

1. In an aquatic ecosystem, bioindicators such as flora and fauna give clues about the habitat alteration and cumulative effects of chemical pollutants over time.
2. Biological indicators identify the extent of environmental stress on living organisms, which is not possible by physical and chemical analysis.
3. Bioindicator assessment is a more cost-effective method than physicochemical analysis, which entails high-cost instrumentation and chemicals.
4. Bioindicators have the ability to show hazardous effects of pollutants present in the aquatic ecosystem due to tolerance against specific contaminants.
5. Low amounts of chemical substances can be detected in chemical assays, but their potential toxic effects can only be shown by the response of aquatic species inhabiting the aquatic ecosystem.
6. Bioaccumulation and its response within the food web of the aquatic ecosystem is easily studied with the help of bioindicators.

29.3.3 Criteria for Bioindicators of the Freshwater Ecosystem

Criteria set up by various ecologists for bioindicator organisms (Manickavasagam et al., 2019) are listed here:

1. Their distribution must be wide, they must be a part of a trophic system in an ecologically relevant position and should be omnivorous.
2. They must show restricted mobility, must be site specific, with low genetic variability and narrow and specific ecological tolerance.
3. They must have medium to long generation time and show sensitivity towards specific pollutants.
4. The indicator organism should be representative of response action on the other taxa or even the whole aquatic ecosystem.
5. Easy field sampling, sorting, storage, identification, maintain robust condition during handling, should be easily culturable in laboratory conditions, and require low cost and manpower.
6. The indicator organisms should concentrate and accumulate the toxins to measurable levels above those in the surroundings for remote monitoring.

29.3.4 Types of Bioindicators on the Basis of Study Objectives

On the basis of the aim of bioindication, bioindicators are categorized into four types (Gerhardt, 2002; Kumari and Paul, 2020), which are outlined here:

1. *Compliance indicators*: They help in ideal maintenance or restoration goals of the environment or ecosystem. One example is assessment of fish composition, which helps with sustainability of the population or the community as a whole in the aquatic ecosystem.
2. *Diagnostic indicators*: These indicators are measured on the sub-organism or species level for specific biomarkers, for example, biochemical analysis of aquatic flora.
3. *Early warning indicators*: These indicators show the first signs of disturbances in the environment. They reveal signs before most other species are affected, because they respond quickly to the changes happening in the environmental state of the aquatic ecosystem. For example, the alga *Pseudokirchneriella subcapitata* demonstrates potential use as a warning system indicator in the freshwater ecosystem associated with climate change (O'Neill *et al.*, 2019).
4. *Accumulative indicators*: These indicators help to study the effect of various biotic and abiotic stresses that control accumulation processes at different levels of biological organization. These organisms are not normally damaged by stressors present in aquatic environments, as they become resilient to them over time and show their levels in a characteristic way. Examples include lichens, mussels and mosses.

29.4 USE OF BIOINDICATORS IN THE ASSESSMENT OF THE FRESHWATER ECOSYSTEM

The European Union Water Framework Directive (EU-WFD) introduced the most significant environmental regulation to maintain and improve the quality of freshwater bodies (Birk *et al.*, 2013). In order to obtain the status of high ecological condition for the lakes, the EU-WFD recommended using a variety of biological components and hydro-morphology evaluation tools, along with physicochemical properties of water. Bioindicator-based assessments of lake water quality are becoming more and more important internationally because they may serve as the foundation for major decisions by all levels of government engaged in the management of freshwater resources (Poikane *et al.*, 2015). Commonly utilized biological indicators to evaluate the trophic condition of freshwater lakes over geographical and temporal variations include algae, diatoms, macrophytes, benthic macroinvertebrates and fish (Derot *et al.*, 2020). In lacustrine ecosystems, phytoplankton and algae are mostly found in the pelagic zone (Reynolds, 2006). These incredibly diversified primary-producing aquatic flora regulate many environmental biogeochemical cycles as sources of food and energy (Reynolds, 2006). Additionally, they show how different species react to variations in freshwater quality (Toudjani *et al.*, 2018). As a result, algae are one of the most frequently utilized biological elements for evaluating lacustrine ecosystems. The biovolume and community of phytoplankton/algae have been used to construct a number of numerical indicators that evaluate the ecological water quality of lake ecosystems. Important data regarding biotic interactions, community structure, environmental conditions and aquatic ecosystem performance are provided by aquatic bio-assessment studies based on a phytoplankton/ algae matrix (Çelekli and Öztürk, 2014).

Various physical, chemical and biological aspects are closely associated with the environmental conditions of lakes. One method of evaluating the ecological health of lakes is by abiotic assessment of their ecosystems. As this assessment ignores temporal changes in the ecological state of the freshwater ecosystem and is unable to provide comprehensive knowledge about it, the abiotic assessment, however, only provides limited information about the true conditions of the aquatic ecosystems (Çelekli and Kapı, 2019). The assessment of lakes' ecological status using a mix of physicochemical, hydromorphological and biological data has received increasing attention in recent decades (Çelekli *et al.*, 2018). Metrics and biological elements are regarded as essential elements for evaluating the overall ecological health of freshwater ecosystems and assessing the robust responses of aquatic biota to various impacts of disturbances in the aquatic ecosystem, and thus provide a combined view of the ecological quality of the ecosystem (Directive, 2000). Environmental factors such as TP, total nitrogen (TN), pH, water temperature and altitude significantly affect the macrophytic and algal distribution among the lakes (Celekli *et al.*, 2020). With low trophic weights, many algal and diatom taxa are pollution-sensitive species (Çelekli *et al.*, 2019a), which are adversely affected by increasing trophic state of the lake ecosystem. The ecological preferences of each species of algae and macrophytes vary under different climatic and geographic settings, anthropogenic factors and catchment features. In order to categorize the ecological preferences of various species in the aquatic ecosystem, studies on detecting the biological responses of aquatic flora and fauna to varied environmental and anthropogenic situations have significant consequences.

29.5 ALGAE AS BIOINDICATORS IN AQUATIC ECOSYSTEMS

In an aquatic ecosystem, bioindicator organisms can be agents or any species of flora and fauna that define distinct characteristics of the environment they inhabit. Algae are a varied group of organisms found abundantly in aquatic ecosystems and are easy to collect and identify. Algae are known to be good bioindicators of trophic state enrichment and pollution due to their wide spatial and temporal distribution. Many algal species are found throughout the year in the lakes and respond quickly according to the changing environmental and physicochemical conditions in response to additional nutrient loading. Estimation of algal biomass is very common in aquatic ecosystems, as it helps in addressing the status of nutrient enrichment, eutrophication and toxicity in the aquatic ecosystem. High nutrient concentrations can affect recreational activities and the whole ecological balance of the lake ecosystem, as an increase in nutrient concentration results in the dense growth of aquatic vegetation, which is aesthetically unpleasant (Deshmukh, 2017).

In lakes, a significant amount of biological production is carried out by algae, being at the base of the aquatic food chain, as they release considerable amounts of oxygen in the process of photosynthesis. Algae are chlorophyll-containing organisms; they consist of communities that include planktonic, filamentous and benthic algae. They help in maintaining the energy–mineral flux of the aquatic system, which is then passed on to higher trophic levels. Recently, many studies have been conducted in tropical countries on the algal succession with respect to trophic state and seasonal changes in the lake ecosystem. Algal diversity and richness depends upon the physicochemical and morphometric characteristics as well as anthropogenic interferences in the lake ecosystem. As a result, they are employed as indicators to evaluate the lakes' ecological health. Due to their short life cycles and quick rate of reproduction, the algae in the lakes respond quickly to ecological changes and are good indicators of the water quality and trophic conditions of the lakes (Dubey and Dutta, 2020). Under natural circumstances, the presence of algae is influenced by the biotic interconnection, such as mutualism, commensalism and antagonistic relationships among the organisms present in the lake ecosystem, as well as the resistance range in relation to abiotic components, which include temperature, pH, hardness, turbidity and oxygen fixation. Algae are a very delicate flora, and this sensitivity may be seen in their population and community structure as well as rates of photosynthesis in response to various biotic and abiotic stressors. Changes in the variety, abundance and community of phytoplankton, algae and diatom species may be signs of pollution or ecological disruptions in the lacustrine environment. The trophic condition of the lakes may be ascertained using these changes in the algal community as a reference. Anthropogenic activities cause alteration in aquatic processes (Deshmukh, 2018). Various characteristics exhibited by the algal population promote its usage as a bioindicator of temporal and spatial changes in the environment. Benthic as well as pelagic algae are increasingly being used as reliable bioindicators globally (Lange *et al.*, 2016) because they strongly respond to even slight changes in ecological conditions (Dong *et al.*, 2016; Parmar *et al.*, 2016). Three major properties of algae that make them useful in monitoring studies are their sensitivity to changes in water quality, easy sampling and diversity (Hötzel and Croome, 1999). Firstly, they have high sensitivity to a diverse range of pollutants; secondly, their sampling is easy and feasible; and thirdly, most algal species are diverse with well-known autecology (Porter *et al.*, 2008).Algal diversity varies according to the type of polluted water, having different characteristics, which are summarized in Table 29.3.

TABLE 29.3
Algal Species Associated with Different Types of Effluents

Effluent Characteristics	Major Source of Effluents	Algae Generally Found with Industrial Effluents
High acidity	Electroplating industries, heavy metals and high TDS (Total Dissolved Solids)	*Euglena mutabilis*, *Euglena stellata*, *Euglena viridis*, *Ulothrix zonata*
Distillery effluent	Distillery wastewater and spent wash, dark colored with high BOD (Biochemical Oxygen Demand) and COD (Chemical Oxygen Demand) values	*Chlamydobotrys* sp., *Chlorogonium gracilliama*
Industrial effluent	Discharge from pesticide, fertilizers, plasticizers, biphenyl detergents and pharmaceuticals industries	*Chlamydomonas ehrenbergii*, *Cyclotella meneghiniana*, *Navicula pygmea*, *N. subtilissima*, *Nitzschia*
Industrial oil effluent	Crude oil and refinery products and petrochemicals	*Amphora ovalis*, *Diatoma vulgare*, *Gomphonema herculaneum*, *Melosira varians*, *Surirella molleriana*, *S. ulna*, *Navicula radiosa*, *Synedra acus*
Phenolic effluent	Major sources are industrial, agricultural and domestic activities	*Cyelotella kuetzinigii*, *Diatoma vulgare*, *Fragilaria virescens*, *Nitzschia palea*, *Synedra ulna*
Chromium effluent	Various industrial processes such as electroplating, printing, dyeing, tanning and metallurgy	*Euglena acus*, *E. sociabilis*, *Navicula atomus*, *N. cuspidate*, *Nitzschia linearis*, *N. palea*, *E. oxyuris*, *E. viridis*

29.5.1 Temporal Patterns of Algal Community Shift in Freshwater Aquatic Ecosystem

Due to enhanced nutrient cycling and hydrological retention, a rise in TN and chl a concentration simultaneously leads to a high primary production (Meyers, 2003). The increase in algal production is caused by increasing water levels and prolonged euphotic zone, which also causes a decline in the carbon–nitrogen ratio (C/N) in the lake ecosystem. Planktonic *Cyclotella ocellata* is an indicator of a nutrient-poor and turbulent shallow freshwater system (Soylu *et al.*, 2007). This algae species generally preys on other subdominant-dominant species throughout temporal community succession with a continual rise in water level, demonstrating its great sensitivity to changes in the temperature regime in the lake environment (Padisák *et al.*, 2009). In the summer season, thermal stratification along with the raised water levels favors the growth of *Cyclotella*, a fast-growing, small, planktonic diatom (Smol *et al.*, 2005). In tropical countries, regional climate has changed significantly over the past few decades as a result of global warming, resulting in greater and longer thermal stability (Wilhelm and Adrian, 2008). Additionally, there was a high correlation between the abundance of small *Cyclotella* and the average air temperature in the lake ecosystem's water and atmosphere (Wang *et al.*, 2019). The damage and degradation of littoral and shoreline zones due to anthropogenic pressure has probably played a major role in changes in diatom and algal assemblages, thereby leading to a steady decrease in the number and diversity of benthic algae and diatoms (Wang *et al.*, 2019). The species *Aulacoseira granulata* is categorized by competitive advantage and high sinking rates in turbulent waters. *A. minutissima*, the subdominant benthic diatom species, is known to grow favorably not only in the littoral zones, but it can also flourish over a long depth gradient, that is, up to around 24 meters in frequent water level fluctuation in oligotrophic conditions (Cantonati *et al.*, 2009; Laird *et al.*, 2011). Hydro-morphological alteration triggers algae community shift, whereas rise in depth of the water significantly decreases diatom diversity and community heterogeneity. Eutrophication and climate change show an increasing interaction among the hydrological variables that are responsible for anatomical and elemental changes in diatoms, phytoplankton and algae at the community level across the hydrological stages of the lake ecosystem. Lake impoundment not only considerably regulates algal shift and carbon cycling over time but also accelerates the environmental role of climate forcing and nutrient enrichment with rise in intensity of impoundment (Wang *et al.*, 2019).

29.5.2 Seasonal Succession of Phytoplankton/Algae

The seasonal succession and assemblages of algae are important in determining the dynamics of the lake ecosystem. Ambient air temperature, pH, water level and ratio of TN to TP are four important factors determining the rise and drop phase of cyanobacteria bloom and diversity in lakes (Zhou *et al.*, 2019). Relations between natural and anthropogenic drivers can strongly affect the changes in algal diversity and community structure, thereby causing a predictable and periodic algal succession such as the transition from cyanobacteria bloom to Chlorophyceae bloom at a yearly scale (Reynolds, 2006; Thomas *et al.*, 2018). The seasonal algal assemblages and succession are considered as an indicator of changes happening at different spatiotemporal scales in the lake ecosystem (Yang *et al.*, 2017). Widely establishing the mutualist, antagonistic, co-existing and competition behavior among different algal blooms in eutrophic lakes and their responses to changing environmental conditions is a major management task for freshwater ecologists, limnologists, policy makers and water resource managers (Reynolds and Irish, 1997; Zhu *et al.*, 2013).

The limited primary production in the lake might affect the algal succession. Cyanobacteria show tougher capability than Chlorophyceae in adapting to periods of low water level, as cyanobacteria are characteristically associated with water column stability and high temperature (Yang *et al.*, 2016; Zhou *et al.*, 2019). Different literature that used algae as bioindicators in the assessment of the freshwater ecosystem is provided in Table 29.4.

TABLE 29.4
Major Findings from Various Studies That Used Algae as Bioindicators of Aquatic Ecosystem

Water Quality and Trophic State Characteristics	Algae/Bioindicator Taxa Reported	References
Nutrient-enriched eutrophic water	*Microcystis aeruginosa*, *Chlorella vulgaris*, *Cladophora crispate*	Mukherjee *et al.*, 2010; Anderson *et al.*, 2002; Feng *et al.*, 2019
Organic pollution	*Closterium* sp, *Nitzschia* sp, *Chlamydomonas* sp, *Anacystis* sp, *Navicula* sp, *Synedra* sp, *Cyclotella* sp, *Oscillatoria*, *Scenedesmus*, *Euglena*, *Chlorella*, *Nitzschia* and *Ankistrodesmus*	Palmer, 1969; Yusuf., 2020
Eutrophic river contaminated with organic pollution	*Scenedesmus quadricauda*, *Chlorella vulgaris, Oscillatoria limosa* and *Melosira granulate*. *Oscillatoria* common in stretches that had contaminated water	Kshirsagar, 2013
Phosphate, in the range of 1.1 to 1.9 mg/l, critical element responsible for algal production	Bacillariophyceae (31.6%) followed by Chlorophyceae (28.33%), Cyanophyceae (25%) and Euglenophyceae (15%) were abundant	Sayeswara *et al.*, 2011
Positive correlation between phosphate, nitrate, silicate, oxygen and diverse group of phytoplankton	Dominant group of algae in a eutrophic lake receiving treated wastewater was Chlorophyceae (52%). followed by Bacillariophyceae (30%). Cyanophyceae (14%) and Euglenophyceae at lowest with 4%	Ansari *et al.*, 2015
Nitrate, phosphate and organic matter regulated the growth of Cynophyacea	Several studies indicated major representatives of Bacillariophyceae as: *Cymbella cistula*, *Navicula subtilissima*, *Fragilaria capucina*, *Pinnularia nobilis*, *Eunotia minor* and *Synedra ulna*. Major representatives of Chlorophyceae were *Chlorella* sp, *Chlorococcum* sp, *Comarium* sp, *Cladora* sp, *Closterium* sp, *Cladophora* sp, *Spirogyra* sp, *Volvox* sp and *Ulothrix* sp. Major representatives of Euglenophyceae are *Euglena* and *Phacus*	Mukherjee *et al.*, 2010
Eutrophic lake with transparency of water varying between 0.35 and 0.40 m	A total of 79 species were identified, including approximately 29.87% of Cyanophyceae, 18.18% of Bacillariophyceae, 14.28% of Chlorophyceae and Euglenophyceae, respectively, 6.49% Mediophyceae, 3.89% of Dinophyceae, 2.59% of Conjugatophyceae, Coscinodiscophyceae, Cryptophyceae and Trebouxiophyceae, respectively, 1.29% of Pyramimonadophyceae and Ulvophyceae, respectively. Cyanophyceae are the most common and abundant class	Ndjouondo *et al.*, 2020
Benthic or plankto-benthic diatoms preferred low-saline, alkaline, middle oxygenated clear freshwater with low organic pollution and oligo- to meso-eutrophic state	Four hundred and thirty-eight bioindicator taxa were revealed from 558 known diatom algae	Niyatbekov and Barinova, 2018
Altitude was found to be major environmental variable that controls the non-diatom algal diversity. Water quality (mesotrophic) was characterized as neutral to low temperature, acidic, saturated with oxygen, low salinity	Study revealed 154 species of non-diatom algae from four taxonomic divisions: 73 taxa from Charophyta, mainly represented by the genus Cosmarium (37); Chlorophyta (30), Cyanobacteria (42), and Euglenozoa (9) are also present; 20 taxa belong to Staurastrum (5) and Cosmarium (15)	Şahin *et al.*, 2020

29.6 MACROPHYTES IN THE LAKE ECOSYSTEM

Aquatic macrophytes are macroscopic forms of flora, which include macro algae, ferns, mosses and angiosperms. Macrophytes of freshwater ecosystems play crucial roles in the operation and organization of aquatic ecosystems (Roy et al., 2020). Aquatic macrophytes that use photosynthesis are large enough to be seen with the naked eye. They develop periodically or continually when afloat, submerged or growing through the water surface. Seven plant divisions represent aquatic macrophytes, namely Cyanobacteria, Chlorophyta, Xanthophyta, Rhodophyta, Pteridophyta, Bryophytes and Spermatophyta (Chambers *et al.*, 2007). The major part of the aquatic ecosystem is represented by the submerged macrophytes, which help in shaping the physical, chemical and biological environment (Jeppesen and Søndergaard, 1999). These aquatic macrophytes are helpful in offering a significant quantity of ecological niches and support aquatic food chains and food web, thereby maintaining the appropriate ecological balance (McAbendroth *et al.*, 2005).

Macrophytes are the major component of the detritivorous and herbivorous food chain, providing food to aquatic fauna, invertebrates and migratory birds and supplying organic substances to benthic microorganisms. Their leaves, stems and roots provide a substrate for periphyton and refuge to a variety of aquatic microorganisms, including invertebrates, fish, reptiles and amphibians at various developmental phases (Dvořák, 1996).

Macrophytes contribute considerably to aquatic biodiversity at the level of the ecosystem in lakes (Zeng *et al.*, 2012). Other ecosystem processes in which macrophytes are involved are regulation and stabilization of nutrients, transpiration, bio-mineralization and sedimentation (Carpenter and Lodge, 1986). They help in regulating the nutrient availability in the pegic region (water column), improving the stability of the shoreline area of lakes (Blindow *et al.*, 2014). Land use, geology, water and sediment chemistry are the major factors responsible for the macrophyte assemblages (del Pozo *et al.*, 2011). Community structure, composition and distribution of the macrophytes vary with climatic condition, hydrology and nutrient availability.

Various types of macrophytes, which include free-floating, submerged and emergent, are common life forms in an aquatic ecosystem. They have evolved from many diverse groups and often demonstrate extreme flexibility in morphology and structure in relation to changing environmental conditions. The distribution of macrophytes can be used as a key instrument in the determination of pollution threats, nutrient level, water quality and trophic status (Daymond, 2013). Macrophytes are broadly categorized into four types (Chambers *et al.*, 2007):

a. Free-floating macrophytes: Macrophytes that normally float on or under the water surface, for example *Eichhornia crassipes, Azolla pinnata*
b. Floating-leaved macrophytes: Macrophytes rooted to the lake or stream bottom with leaves that float on the surface of the water in the lakes/river
c. Submerged macrophytes: Macrophytes that grow totally submerged under the water, with roots closely associated with the sediments, for example *Myriophyllum spicatum*, *Vallisinera americana*
d. Emergent macrophytes: Macrophytes that are rooted in littoral soils present in the littoral and shoreline area of the lakes, for example *Phragmites australis*, *Typha latifolia*

Macrophytes are affected by various biotic and abiotic stressors, including water and sediment nutrients, underwater light and water-level fluctuations. The growth, propagation and abundance of aquatic and semi-aquatic macrophytes along with other hydrophilic terrestrial vegetation during different seasons helps the enhancement of biodiversity and influences their distribution pattern (Matthews and Fung, 1987; Mandal *et al.*, 2018).

The distribution of macrophytes is organized in lentic systems in structured zones, which start from the edge, that is, the littoral and shoreline area, with emerged macrophytes, followed by plants with floating leaves present on the open area of the lakes. However, abiotic factors like depth, water temperature, light incidence, input of nutrients and interspecific competition are responsible for heterogeneous distribution of macrophytes (Freitas and Thomaz, 2011). Submerged macrophytes have an antagonistic relationship with algae/phytoplankton; they compete for nutrients and reduce water channel, which hinders fish movement, navigation and irrigation prospects (Roy et al., 2020). A major advantage of submerged macrophytes is that they stabilize sediments, thereby reducing the suspension of sediments in the water column and reducing the turbidity of the lacustrine ecosystem (Bamidele and Nyamali, 2008). The macrophytes serve as indicators for the possible degree of damage in the aquatic ecosystem. They serve as a critical interface between the sediments and the overlying water column that drives the biogeochemical cycle and ecosystem productivity (Carpenter and Lodge, 1986).

29.6.1 Macrophytes and Faunal Diversity

The presence of macrophytes in a natural or controlled manner enhances the richness of the local species (Williams *et al.*, 2008). Macrophytes play a major role in controlling algal growth and diversity, which sustains a high diversity and abundance of aquatic fauna, including invertebrates (Thomaz and da Cunha, 2010). The macrophytes provide habitat to epiphytic algae as well as a variety of associated fauna, which includes macrofauna, zooplankton and fish (Christie *et al.*, 2009). Lakes and wetlands, along with their macrophyte assemblages, provide aesthetic appeal, landscape diversity and recreational opportunities (De *et al.*, 2018). Aquatic macrophytes contribute to the structure of aquatic ecosystems, improve habitat complexity and variety, and offer and sustain physical structure (Thomaz *et al.*, 2010). The presence of macrophyte cover maintains communities, diversity and populations, ensuring support for a greater number of aquatic fauna (Dibble *et al.*, 1996).

29.6.2 Macrophytes as Bioindicators

Aquatic flora are delicate tools for forecasting and finding environmental stresses. The ecological health of a lake can be viewed in terms of the ecosystem services provided by it, which maintain the functional and structural characteristics of the aquatic ecosystem. The diversity, depth, density, richness and types of macrophytes are indicators of the pollution level and trophic state of lakes. They help us by providing important clues about the health of the lacustrine ecosystem (Samiyappan *et al.*, 2019). Aquatic macrophytes present in the shoreline area of lakes possess two fundamental properties that make them useful as limnological indicators: the first is that they react progressively and slowly to changes in nutrient concentration (Melzer, 1999), and secondly, as the shoreline zone or littoral zone may experience patterns of nutrient and pollutant concentrations caused by artificial or natural inflows of waste water by diffuse as well as non-point sources of pollution (Melzer, 1999), this patchiness is reflected by the rooted submerged macrophytes. Macrophytes are important as well as sensitive bioindicators and can be used for long-term assessment of the trophic state of freshwater lakes (Melzer, 1999).

29.6.3 Macrophytes as Invasive Species in Eutrophic Lakes

Eutrophic lake ecosystems are especially susceptible to macrophytic invasions caused by excessive nutrient enrichment that enables the rapid proliferation of invasive species. Although wetlands and shallow water bodies make up less than 6% and 9%, respectively, of earth's surface, the proportion of invasive aquatic macrophytes is greater than 30% (Zedler, 2011). Macrophytes have certain adaptations that enable their rapid growth and proliferation (Santamaría, 2002), which increases their invasive potential. Invasion of aquatic macrophytes affects the habitat heterogeneity, leading to homogenization of the lacustrine ecosystem, and changes the structure of the macrophyte assemblage in aquatic ecosystems (Mack et al., 2000; Michelan *et al.*, 2010). They play a significant role in structuring aquatic habitats, as their invasion of lakes changes the waterscape, thereby affecting other aquatic macrophytes present in the lakes (Engelhardt, 2011) and causing serious ecological and economic damage globally (Pieterse and Murphy, 1990). The most problematic invasive free-floating macrophytic species found in the lentic ecosystem are *Eichhornia crassipes*, *Azolla pinnata*, *Pistia stratiotes* and *Salvinia molesta*, widespread in sub-tropical and tropical regions. Other globally widespread invasive macrophytes are *Potamogeton crispus*, *Myriophyllum spicatum*, *Lythrum salicaria* and *Trapa natans* (Rejmankova, 2011). The aquatic biomasses create problems for the maintenance and leisure use of lakes due to their overgrowth. For this reason, the aquatic biomass has to be regularly uprooted, removed and disposed of, which helps the proper functioning of the lake ecosystem.

29.6.4 Role of Aquatic Macrophytes in the Treatment of Eutrophic Lakes

Eutrophication due to excessive loading of phosphorus (P) and nitrogen (N) due to wastewater discharge from agricultural, domestic and industrial sources damages the ecological quality and functioning of the lake ecosystem (Kantawanichkul *et al.*, 2009) and leads to the excess overgrowth of cyanobacterial blooms and extinction of native biota, thereby decreasing the aquatic biodiversity (Conley *et al.*, 2009). Rejuvenation and restoration of water resources and its whole ecological balance is a task of the utmost importance (Song *et al.*, 2006). Among all the natural or biological treatments for controlling nutrient enrichment in lakes, aquatic vegetation, especially free-floating and submerged macrophytes, has been documented as most effective for the treatment. The setting up of macrophyte stands in a shallow lentic ecosystem can increase nutrient recycling and retention time (Jones *et al.*, 1993). During the vegetative seasons of macrophytes, they act as a sink of nutrients, as they accumulate them in their developing body part tissues (Engel, 1998). Floating macrophytes uptake nutrients mainly by their roots, although in some cases uptake through leaves is also significant. Members of aquatic free-floating duckweeds, namely, *Lemna minor*, *Lemna gibba*, *Azolla pinnata*, *Wolffia arrhiza and Pistia stratiotes*, have shown a potential role in the treatment of polluted freshwater resources (Sutton and Ornes, 1975). Many aquatic macrophytes are possible scavengers for the remediation of heavy metals and are used in systems of wastewater treatment, i.e., in constructed wetlands (Abida, 2009; Kumar *et al.*, 2020).

29.6.5 Succession in Macrophytic Community with Increasing Trophic State

Macrophytes respond to increasing trophic state by increased maximum colonization depth (MCD) and changes in species diversity and richness, as they are mainly affected by the TP concentration. Increase or decrease in temperature, pH, water clarity and TP and nitrogen concentration are important factors determining changes in the macrophytic community (Yang *et al.*, 2020).

There is a general shift of aquatic macrophytes from macroalgae found in hypertrophic conditions, like *Enteromorpha* species and *Cladophora* species, through tall angiosperms, which are found in eutrophic conditions. A shift from *Potamogeton pectinatus*, *Myriophyllum spicatum* and short angiosperms found in mesotrophic conditions like *Littorella uniflora* and *Eleocharis acicularis* to characean macrophytes found in oligotrophic conditions, *Chara globularis* and *Nitellopsis obtuse*, and mosses like *Fontinalis antipyretica* was reported to occur as the lake's trophic state decreased from hypertrophic to oligotrophic (Jeppesen *et al.*, 2000). Research showed that in the macrophytic case, recovery was very slow, and that the decrease in species distribution, macrophytes' colonizing depth and species richness that had occurred due to pollution and eutrophication had still not been completely reversed 20–30 years after the reduction of phosphorus inputs (Dudley *et al.*, 2011). During the restoration phase in an aquatic ecosystem, partial recovery of the submerged vegetation occurs in terms of species richness, with the majority of the smaller, native and sensitive species failing to re-appear (Yang *et al*, 2020). The substitution of free-floating and submerged species of macrophytes by emergent macrophytic communities indicates macrophytic community succession (Gołdyn, 2010). Thus, aquatic macrophytes are considered as bioindicators of the ecological quality of lake ecosystems (Kassaye *et al.*, 2016).

Macrophyte composition changes due to various human pressures, as they are susceptible to eutrophication, acidification, fluctuations in water level, shoreline modification, recreation and navigation, fish culture and biological invasions (Poikane et al., 2018). Many European countries, like Austria, Denmark, Germany, Ireland, Poland and the United Kingdom, have been included in the ecological assessment by using macrophytes. They have a sedentary nature and slow growth rate; this is the reason why they serve as long-term bioindicators with high spatial resolution, which is useful for perceiving eutrophication and other adverse impacts on the littoral area of the lakes (Pall and Moser, 2009).

29.7 ANTAGONISTIC BEHAVIOR BETWEEN ALGAE AND MACROPHYTES IN LAKE ECOSYSTEMS

Excessive algal bloom due to eutrophication is a major problem currently. In the present scenario, an increase in the geographical distribution and frequency of outbreak of cyanobacterial blooms has been reported worldwide, mainly due to global warming, which is the major driving force for the growth and proliferation of cyanobacteria (Paerl and Paul, 2012). Aquatic macrophytes in the freshwater ecosystem are an optimistic source of bioactive compounds, which have potential uses and applications in controlling the growth of algal and cyanobacterial bloom due to the antagonistic relationship between them. Macrophytes produce allelopathic compounds, which effectively control or inhibit the algal growth and cyanobacterial blooms, helping to promote the restoration and management of lake ecosystems (Tazart *et al.*, 2020).Several studies have shown that the allelopathic compounds released by macrophytes tend to sustain the clear-water states in shallow and eutrophic lakes, which inhibit the growth of epiphyte and algal biomass (Wang *et al.*, 2015b).

Several macrophyte-produced compounds, including terpenoids, asarone and polyphenols like ellagic, pyrogallic, gallic and catechin as well as fatty acids like hexadecanoic acid, α-linolenic acid and stearic acid, have been shown to significantly slow down the growth of algae and cyanobacterial blooms (Hang, 2017). Diverse allelochemical compounds derived from a variety of aquatic macrophytes confirmed dose-dependent trends in both physiological parameters, including levels of lipid peroxidation, protein and chlorophyll contents, as well as antioxidant parameters such as superoxide dismutase (SOD), catalase (CAT) and peroxidase (POD) (Wang et al., 2015a; Li *et al.*, 2016).

Pyrogallic acid extracted from *Myriophyllum spicatum* causes an increase in malondialdehyde (MDA) levels and inhibits protein synthesis, which are frequently used as important indicators in assessing the mechanism of apoptosis at cellular level and growth conditions of cells (Tang *et al.*, 2015a; Xie *et al.*, 2019). The antioxidant activities of enzymes play an important and significant role in the balancing of free radicals and the metabolism of organisms (Tazart *et al.*, 2020). CAT, POD and SOD activities were considerably increased in cells of the cyanobacterium *Microcystis aeruginosa* under different concentrations of several compounds, such as glufosinate and pyrogallic acid (Lu *et al.*, 2016; Zhang *et al.*, 2017). Different types of extracts, such as crude extract of *Sagittaria trifolia* and aqueous extracts obtained from rice straw *Oryza sativa* (Li *et al.*, 2016; Hua *et al.*, 2018), help in inhibiting the growth of algae. Algicidal allelochemicals produced from macrophytes are mainly terpenoids, phenolic acids, polyphenols and polyether (Tazart et al.,2021). The allelopathic activity of macrophytes mainly depends on the chemical nature of the allelochemical, its production rate, excretion, and the specific or precise toxicological mechanism of each allelochemical product (Mulderij *et al.*, 2007). Three macrophytes, namely *Potamogeton natans*, *Ranunculus aquatilis* and *Nasturtium officinale*, ethyl acetate extract, have a significant or important

TABLE 29.5
Submerged Macrophytes with Allelopathic Activity Against Algal Species

Family	Macrophytic Species	Sensitive Algal Species	References
Hydrocharitaceae	*Stratiotes aloides* L.	*Scenedesmus obliquus* Kütz. *Natural phytoplankton*	Mulderij *et al.*, 2005; Mulderij *et al.*, 2006
	Elodea nuttallii Planch	*Microcystis aeruginosa* Kütz	Wu *et al.*, 2009
	Hydrilla verticillata L.	*Microcystis aeruginosa* Kütz	Wu *et al.*, 2009
	Vallisneria spiralis L.	*Microcystis aeruginosa* Kütz	Wu *et al.*, 2009
Najadaceae	*Najas minor* All.	*Scenedesmus obliquus* Kütz.	He *et al.*, 2008
Ceratophyllaceae	*Ceratophyllum demersum* Linn.	*Oscillatoria limnetica* Lemm.	Jasser, 1995
		Anabaena flos-aquae Breb.	Körner and Nicklisch, 2002
Potamogetonaceae	*Potamogeton malaianus* Miq.	*Microcystis aeruginosa* Kütz.	Wang *et al.*, 2010
		Scenedesmus obliquus Kütz.	He *et al.*, 2008
	Potamogeton maackianus A	*Microcystis aeruginosa* Kütz.	Wang *et al.*, 2010
	Potamogeton pusillus L.	*Microcystis aeruginosa* Kütz.	Takeda *et al.*, 2011
	Potamogeton illinoensis Morong	Phytoplankton	Vanderstukken *et al.*, 2011
	Potamogeton foliosus L.	Cyanobacteria *Chlorella* spp	Hasler and Jones, 1949
Haloragidaceae	*Myriophyllum spicatum* L.	*Microcystis aeruginosa* Kütz.	Nakai *et al.*, 1996
		Pseudokirchneriella subcapitata	Nakai *et al.*, 1999
		Anabaena flos-aquae Breb.	Körner and Nicklisch, 2002
		Phormidium tenue Kütz.	
	Myriophyllum verticillatum L.	*Limnothrix redekei* Goor. *Stephanodiscus minutulus* Kütz.	Hilt *et al.*, 2006
	Myriophyllum aquaticum Vell	*Microcystis aeruginosa* Kütz.	Wu *et al.*, 2008
	Myriophyllum brasiliense Vell	*Microcystis aeruginosa* Kütz. *Anabaena flos-aquae* Breb.	Saito *et al.*, 1989

Source: Gao *et al.*, 2017.

inhibitory effect on the physiological and growth activity of *Microcystis aeruginosa*, which proves that macrophytes are producers of potential allelochemical chemical compounds and are used and suggested as natural substitutes for controlling harmful cyanobacterial bloom. Allelochemicals produced by macrophytes have an inhibitory effect on numerous algal species and can play significant roles in the interaction between macrophytes and algal species (Tazart *et al.*, 2020). A compiled list of macrophytes and their associated sensitive algal species is given in Table 29.5.

29.8 EFFECT OF HYDRO-MORPHOLOGICAL VARIABLES ON AQUATIC ECOSYSTEMS

Hydro-morphological variables in lake ecosystem were found to exert a robust impact on aquatic biota, including diatom, algae and macrophytic communities, with an increasing interaction with climate variables and nutrients over a span of time (Wang *et al.*, 2019). Ecological succession and environmental processes can vary between dammed rivers, free-flowing rivers, regulated lakes, and morphometrically altered and stressed urban aquatic ecosystems (Hall *et al.*, 1999). Hydrological regulation in lakes is known to alter lake hydro-morphology and habitat availability with significant environmental costs (Leira and Cantonati, 2008). An increase in water depth may cause a shift from a mixed to a stratified thermal regime, or improve the stability of the thermocline along with extended stratification (Nowlin *et al.*, 2004). In the catchment area of a lake, anthropogenic activities cause conversion of a shallow lake into a deep system and vice versa, and cause changes in the typology and morphology of the lake basin, which control key limnological processes such as habitat availability and thermocline stability (Wang *et al.*, 2019). Lacustrine basin morphometry causes variation in the habitat quality and quantity, usually showing a loss of benthic producers and littoral environments with a shift towards a deeper lake system (Coops *et al.*, 2003).

In a deep lake, an increase in water level can decrease sunlight penetration into the bottom or deeper area of the lake (Loiselle *et al.*, 2005), with adverse impact on bottom-dwelling aquatic organisms. A continuous water level increase may unfavorably affect the aquatic flora of the littoral area and can result in a shift towards an algae-dominated state due to the decrease in macrophytic population (Kong *et al.*, 2017). The upsurge in water depth beyond a threshold level can support the trophic flow in the pelagic pathway by decreasing resource availability and habitat for bottom-dwelling producers (Devlin *et al.*, 2013) Man-made regulation increases water nutrient retention and its residential time, causing decrease in water transfer to downstream regions (Erwin *et al.*, 2011; Tang *et al.*, 2015b); increased volume of water leads to dilution of soluble nutrients in the aquatic ecosystem (Yang and Yang, 2014). The processes of carbon burial and buildup of organic matter frequently vary considerably depending on lake typology and size, with the rate of burial of organic carbon being larger in shallow lakes of similar pollution status and trophic state (Downing, 2010).

29.8.1 Effects of Water Level Fluctuation on the Lake Ecosystem

The impact of water level fluctuation (WLF) on the biotic component of lakes is frequently limited to shoreline zones, depending on its frequency and amplitude (Evtimova and Donohue, 2016). It also causes variation in community diversity along with prolonged water depth gradient (Aroviita and Hämäläinen, 2008). The moderate range of WLF reported higher species diversity (White *et al.*, 2008), but with amplified WLF, loss of diversity occurs (Aroviita and Hämäläinen, 2008). Various studies have revealed the strong interaction between WLF and lake depth in regulating the assemblages and production of lake biotic components (Evtimova and Donohue, 2016). At an intermediate magnitude of WLF, vertical mixing was found to play a significant role in structuring the algae and macrophytic diversity (Elliott *et al.*, 2001).

A strong variation in shoreline aquatic habitats due to distinct lake regulation can be of vital ecological significance in structuring shoreline communities of the lake ecosystem (Schindler and Scheuerell, 2002). Shoreline area undulation was strongly related to the underwater morphometry of the lakes (Dearing, 1997); steady damage to the aquatic biodiversity of the shoreline zones has been extensively found in lakes that were changed to deeper reservoirs from shallow reservoirs (Hellsten and Riihimäki, 1996).

29.8.2 Effect of Land Use Change on Aquatic Biodiversity

Land use change in the lake's catchment area is considered as one of the most vital drivers that cause changes in aquatic biodiversity. Macrophytes are highly sensitive to fluctuations happening within their physical habitat, and they react at diverse scales to the effects of catchment land use changes (Fares *et al.*, 2020). Land use changes in the lake catchment favor the excessive proliferation of emergent and amphibious species of macrophytes, mostly found in the littoral area, along with some macrophytic species belonging to other life forms that show niche requirements and adaptation favoring the land use change. Changes in macrophyte communities result in proliferation of invasive species, which causes a decrease in aquatic biotic diversity, leading to macrophytic homogenization as changes in land use consequences become abrasive (Dubey *et al.*, 2021). Different features of macrophytic communities, that is, species diversity, richness, life form, composition, presence of invasive species and taxon-specific niche requirements, are to be measured when creating indexes of integrity and when making management decisions concerning the restoration, management and conservation of lake ecosystems (Fares *et al.*, 2020).

29.8.3 Effect of Anthropogenic Activities on the Landscape Change Linked to Aquatic Ecosystems

Globally, freshwater ecosystems have suffered adverse changes due to anthropogenic activities (González-Abraham *et al.*, 2015; Sala *et al.*, 2000). This extreme exploitation due to increase in deforestation activities, reduction in number of lakes and groundwater over-abstraction has directly or indirectly affected the aquatic ecosystem. Deforestation transforms the landscape into a mosaic, which encompasses different land uses (Malhi *et al.*, 2014) and disturbs different components of freshwater ecosystems (Castello and Macedo, 2016). Alterations in landscape cause environmental degradation of both aquatic and terrestrial ecosystems (Dunlap and Jorgenson, 2012), which therefore causes loss of native and sensitive aquatic species, changes in community structure and ecosystem disturbance, affecting global biodiversity (González-Abraham *et al.*, 2015; Fares *et al.*, 2020). Alterations to riparian vegetation structure cause channel siltation along with changes in water quality and chemistry (Castello and Macedo, 2016). Such consequences damage the physical integrity of freshwater ecosystems by causing habitat loss in aquatic ecosystems (Sala *et al.*, 2000; Johnson and Angeler, 2014; Aoki *et al.*, 2017). Due to different and diverse life forms, macrophytes react at different scales to changes in land use effects. The response could be negative, which causes reduction in macrophyte richness, leading to the dominance of a few species, or could be positive (Akasaka *et al.*, 2010), which includes an increase in macrophyte species evenness, leading to biotic homogenization (Elo *et al.*, 2018; Dubey *et al.*, 2021). Therefore, land use change along with nutrient enrichment alters macrophytes' community structure in the lake ecosystem (Elo *et al.*, 2018). Some species become extinct when anthropogenic impacts cause great changes like channel siltation and increased turbidity in water (Sass *et al.*, 2010). The competitive advantages of native species are reduced due to disturbances caused by anthropogenic activities, which add pressure and facilitate invasion (Shea and Chesson, 2002; Michelan *et al.*, 2010). Therefore, macrophytic invasion not only affects the diversity of macrophytes but also causes change in the diversity of other aquatic organisms like algae, fish, insects and periphyton due to the antagonistic and mutualistic relationships they share among them (Lougheed *et al.*, 2008). Umetsu *et al.* (2018) found an excessive growth of emergent macrophytic species on degraded sites, which they attributed to high nutrient content on the sediments of the aquatic ecosystem, while it was reported that the occurrence of submerged macrophytic species richness declined under nutrient-rich conditions (Umetsu *et al.*, 2018). Figure 29.3 shows the relationship between anthropogenic activities and hydro-morphological pressures on biological communities of the lake ecosystem.

29.9 FUTURE PROSPECTS

Freshwater ecosystems have been given great attention globally due to their importance in the sustenance of biodiversity and all life forms on earth. However, deteriorating catchment environments, caused largely by anthropogenic agents and increased exploitation and degradation of water quality, are putting great pressure on their integrity. Globally, they are facing substantial habitat degradation and deterioration in water quality due to various stresses that include bad catchment activities, encroachment, land use change and waste discharge from the nearby locations causing important economic, environmental and social impacts. In recent decades, significant efforts have been made to assess freshwater lakes, taking into account physicochemical, biological and hydromorphological parameters (Çelekli *et al.*, 2018). Pressure–response relations between aquatic biota and various anthropogenic stressors need to be monitored and evaluated (Hering *et al.*, 2010). At

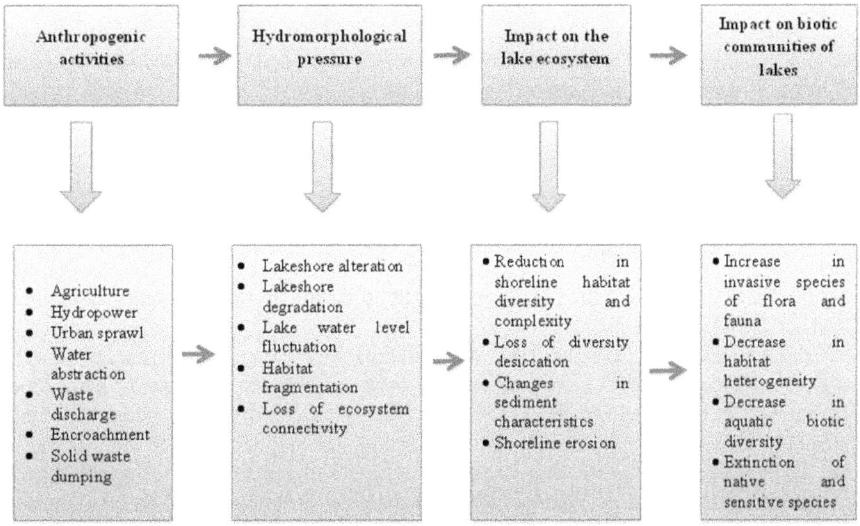

FIGURE 29.3 Schematic representation showing relationship between anthropogenic activities and hydromorphological pressures on biological communities of the lake ecosystem.

present, these relationships have been ignored and have not been tested and documented (Birk *et al.*, 2012). Therefore, the necessary links between management decisions and the environmental status of freshwater resources are unclear in many management plans, thereby creating gaps in management actions (Birk *et al.*, 2013; Hering *et al.*, 2015). To restore and protect every aspect of freshwater resources for healthy aquatic ecosystems, it is very important that all aquatic ecological processes are understood, not just by limnologists or ecologists, but also by water managers, policy makers and citizens (Poikane *et al.*, 2016). Hence, the representation of aquatic biotic communities at different biological classes and trophic statuses (Birk and Willby, 2010), along with a common understanding of aquatic community composition, can be a first step in support of this goal of management of freshwater resources (Palmer *et al.*, 2005).

29.10 CONCLUSION

A significant amount of research effort should be devoted towards the assessment of the ecological condition of freshwater lakes and the role of algae and macrophytes. The next major step towards this challenging goal is to make a diagnosis of agents that cause encroachment, nutrient enrichment and heavy metal pollution, which will help in restoring, managing and conserving the lake ecosystem along with the help of aquatic bioindicator species. Comprehensive understanding of the mutualistic, antagonistic, competitive and co-existence behavior among different macrophytes and algal communities with different trophic states and their responses to ecological or environmental drivers is an important task for freshwater ecologists, limnologists, water resource managers and policy makers. All aspects of freshwater ecosystems, including ecological processes that control the growth of algae and macrophytes, should be considered for the management, restoration and conservation of freshwater ecosystems.

ACKNOWLEDGEMENT

The authors are grateful to the Department of Environment Science, Babasaheb Bhimrao Ambedkar University, Lucknow for providing the necessary facilities for doing this review work.

REFERENCES

Abida B. (2009). Concurrent removal and accumulation of Fe^{2+}, Cd^{2+} and Cu^{2+} from waste water using aquatic macrophytes. *Der Pharma Chemical.*, 1(1), 219–224.

Akasaka M, Takamura N, Mitsuhashi H and Kadono Y. (2010). Effects of land use on aquatic macrophyte diversity and water quality of ponds. *Freshw. Biol.,* 55, 909–922. https://doi.org/10.1111/j.1365-2427.2009.02334.x

Anderson DM, Glibert PM and Burkholder JM. (2002). Harmful algal blooms and eutrophication: nutrient sources, composition, and consequences. *Estuaries*, 25(4), 704–726. https://doi.org/10.1007/BF02804901

Ansari E, Gadhia M and Ujjania NC. (2015). Phytoplankton diversity and water quality assessment of ONGC pond Hazira. *IJRES*, 1(1), 1–5.

Aoki C, Teixeira-Gamarra MC, Gamarra RM, de Medeiros SCH, Pott VJ, Damasceno-Junior GA, Pott A and Scremin-Dias E. (2017). Abiotic factors drive the structure of aquatic plant assemblages in riverine habitats of the Brazilian 'Pantanal.' *Rev. Bras. Bot.*, 40, 405–415. https://doi.org/10.1007/s40415-016-0345-0

Aroviita J and Hämäläinen H. (2008). The impact of water-level regulation on littoral macroinvertebrate assemblages in boreal lakes. In: Wantzen KM, Cantonati M, Rothhaupt KO, Toth LG, Mortl M, Fischer P and Martens K (eds). *Ecological Effects of Water-Level Fluctuations in Lakes*. Springer, Dordrecht, pp. 45–56.

Bamidele JF and Nyamali B. (2008). Ecological studies of the Ossiomo river with reference to the macrophytic vegetation. *Research Journal Botany*, 3(1), 29–34.

Bhagowati B and Ahamad KU. (2019). A review on lake eutrophication dynamics and recent developments in lake modeling. *Ecohydrol. Hydrobiol.,* 19, 155–166.

Birk S and Willby N. (2010). Towards harmonization of ecological quality classification: establishing common grounds in European macrophyte assessment for rivers. *Hydrobiologia*, 652(1), 149–163.

Birk S, Bonne W, Borja A, Brucet S, Courrat A, Poikane S and Hering, D. (2012). Three hundred ways to assess Europe's surface waters: an almost complete overview of biological methods to implement the Water Framework Directive. *Ecol. Indic.*, 18, 31–41.

Birk S, Willby NJ, Kelly MG, Bonne W, Borja A, Poikane S and Van de Bund W. (2013). Intercalibrating classifications of ecological status: Europe's quest for common management objectives for aquatic ecosystems. *Sci. Total Environ.,* 454, 490–499.

Bleich ME, Piedade MTF, Mortati AF and Andre T. (2015). Autochthonous primary production in southern Amazon headwater streams: Novel indicators of altered environmental integrity. *Ecol. Indic.*, 53, 154–161.

Blindow A, Hargeby A and Hilt S. (2014). Facilitation of clear-water conditions in shallow lakes by macrophytes: differences between charophyte and angiosperm dominance. *Hydrobiologia,* 737, 99–110.

Cantonati M, Scola S, Angeli N, Guella G and Frassanito R. (2009). Environmental controls of epilithic diatom depth-distribution in an oligotrophic lake characterized by marked water-level fluctuations. *Eur. J. Phycol.*, 44(1), 15–29.

Carlson RE. (1977). A trophic state index for lakes. *Limnol. Oceanogr.,* 22(2), 361–369.

Carpenter SR and Lodge DM. (1986). Effects of submersed macrophytes on ecosystem processes. *Aquat. Bot.*, 26, 341–370.

Castello L and Macedo MN. (2016). Large-scale degradation of Amazonian freshwater ecosystems. *Glob. Chang. Biol.,* 22, 990–1007. https://doi.org/10.1111/gcb.13173

Çelekli A and Kapı E. (2019). Ecoregion approach in the assessment of aquatic ecosystems in the west of Gaziantep (Turkey): Application of diatom metrics. *Ecol. Indic.*, 103, 373–382.

Çelekli A and Öztürk B. (2014). Determination of ecological status and ecological preferences of phytoplankton using multivariate approach in a Mediterranean reservoir. *Hydrobiologia,* 740, 115–135. https://doi.org/10.1007/s10750-014-1948-8

Çelekli A, Toudjani A, Lekesiz H, Kayhan S and Çetin T. (2018). Biological assessment of water quality: ecological status of three Aegean reservoirs based on phytoplankton metrics. *Turkish J. Water Sci. Manage.,* 2, 76–99.

Çelekli A, Toudjani AA, Gümüş EY, Kayhan S, Lekesiz HÖ and Çetin T. (2019). Determination of trophic weight and indicator values of diatoms in Turkish running waters for water quality assessment. *Turk. J. Botany,* 43, 90–101.
Chambers PA, Lacoul P, Murphy KJ and Thomaz SM (2007). Global diversity of aquatic macrophytes in freshwater. In: Balian EV, Lévêque C, Segers H and Martens K. (eds) *Freshwater Animal Diversity Assessment. Developments in Hydrobiology*. Vol. 198. Springer, Dordrecht, pp. 9–26.
Chau KW. (2005). Characterization of transboundary POP contamination in aquatic ecosystems of Pearl River delta. *Mar. Pollut. Bull.* 51(8–12), 960–965. https://doi. org/10.1016/j.marpolbul.2005.02.028
Christie H, Norderhaug KM and Fredriksen S. (2009). Macrophytes as habitat for fauna. *Mar. Ecol. Prog. Ser.*, 396(9): 221–233.
Conley DJ, Paerl HW, Howarth RW, Boesch DF, Seitzinger SP, Karl EKE and Gene EGE. (2009). Controlling eutrophication: nitrogen and phosphorus. *Science,* 123, 1014–1015.
Coops H, Beklioglu M and Crisman TL. (2003). The role of water-level fluctuations in shallow lake ecosystems–workshop conclusions. *Hydrobiologia*, 506(1–3), 23–27.
De M, Medda S and Dey SR. (2018). Ecological health of wetland ecosystem: an overview. *Int. J. Exp. Res. Rev.,* 17, 20–29.
Dearing JA (1997). Sedimentary indicators of lake-level changes in the humid temperate zone: a critical review. *J. . Paleolimnol.*, 18(1), 1–14.
del Pozo R, Fernandez-Alaez C and Fernandez-Alaez M. (2011). The relative importance of natural and anthropogenic effects on community composition of aquatic macrophytes in Mediterranean ponds. *Mar. Freshw. Res.,* 62, 101–109.
Derot J, Jamoneau A, Teichert N, Rosebery J, Morin S and Laplace-Treyture C. (2020). Response of phytoplankton traits to environmental variables in French lakes: new perspectives for bioindication. *Ecol. Indic.,* 108, 105659.
Deshmukh RN. (2017). An evaluation of algal flora as bioindicator of Futala Lake. *IJRBAT*, V(3): 94–97.
Deshmukh RN. (2018). An estimation of algae in different polluted aquatic ecosystem. *IJRBAT*, III(VI): 162–164.
Devlin SP, Vander Zanden MJ and Vadeboncoeur Y. (2013). Depth-specific variation in carbon isotopes demonstrates resource partitioning among the littoral zoobenthos. *Freshw. Biol.*, 58(11), 2389–2400.
Dibble ED, Killgore KJ and Dick GO. (1996). Measurement of plant architecture in seven aquatic plants. *J. Freshw. Ecol.,* 11, 311–318.
Directive (2000). Directive 2000/60/EC of the European Parliament and of the Council of 23 October 2000 establishing a framework for Community action in the field of water policy. *Off. J. Eur. Union,* 22, 2000.
Dong T, Wychen SV, Nagle N, Pienkos PT and Laurens LML. (2016). Impact of biochemical composition on susceptibility of algal biomass to acid-catalyzed pretreatment for sugar and lipid recovery. *Algal Res.*, 18, 69–77.
Downing JA. (2010). Emerging global role of small lakes and ponds: little things mean a lot. *Limnetica*, 29(1), 0009–24.
Dubey D and Dutta V. (2020). Nutrient enrichment in lake ecosystem and its effects on algae and macrophytes. In: Shukla V and Kumar N (eds) *Environmental Concerns and Sustainable Development.* Springer, Singapore, pp. 81–126.
Dubey D, Kumar S and Dutta V. (2021). Impact of nutrient enrichment on habitat heterogeneity and species richness of aquatic macrophytes: evidence from freshwater tropical lakes of Central Ganga Plain, India. *Int. J. Environ. Sci. Technol.* 19(6), 5529–5546. https://doi.org/10.1007/s13762-021-03438-4
Dudley B, Gunn IDM, Carvalho L, Proctor IO, Hare MT, Murphy KJ and Milligan A. (2011). Changes in aquatic macrophyte communities in Loch Leven: evidence of recovery from eutrophication? *Hydrobiologia*, 681(1), 49–57. doi 10.1007/s10750-011-0924-9
Dunlap RE and Jorgenson AK. (2012). The Wiley-Blackwell encyclopedia of globalization. In: Ritzer G. (ed.) *Environmental Problems*. Blackwell Publishing Ltd, Chichester, UK/Malden, MA, pp. 1–8.
Dutta V, Dubey D and Kumar S. (2020). Cleaning the River Ganga: impact of lockdown on water quality and future implications on river rejuvenation strategies. *Sci. Total Environ.*, 743, 140756.
Dvořák J. (1996). An example of relationships between macrophytes, macroinvertebrates and their food resources in a shallow eutrophic lake. *Hydrobiologia,* 339, 27–36.
Elliott JA, Irish AE and Reynolds CS. (2001). The effects of vertical mixing on a phytoplankton community: a modelling approach to the intermediate disturbance hypothesis. *Freshw. Biol.*, 46(10), 1291–1297.

Elo M, Alahuhta J, Kanninen A, Meissner KK, Seppälä K and Mönkkönen M. (2018). Environmental characteristics and anthropogenic impact jointly modify aquatic macrophyte species diversity. *Front. Plant Sci.,* 9, 1–15. https://doi.org/10.3389/fpls. 2018.01001

Engel S. (1998). The role and interactions of submerged macrophytes in a shallow Wisconsin lake. *Fresh Wat. Ecol.,* 4, 329–341.

Engelhardt KA (2011). Eutrophication aquatic. In: Simberloff D, Rejmánek M (eds) *Encyclopedia of Biological Invasions.* University of California Press, Berkeley, USA, pp. 209–213.

Erwin SO, Schmidt JC and Nelson NC. (2011). Downstream effects of impounding a natural lake: the Snake River downstream from Jackson Lake Dam, Wyoming, USA. *Earth Surf. Process. Landf.,* 36(11), 1421–1434.

Evtimova VV and Donohue I. (2016). Water-level fluctuations regulate the structure and functioning of natural lakes. *Freshw. Biol.,* 61(2), 251–264.

Fares ALB, Calvão LB, Torres NR, Gurgel ESC and Michelan, TS. (2020). Environmental factors affect macrophyte diversity on Amazonian aquatic ecosystems inserted in an anthropogenic landscape. *Ecol. Indic.,* 113, 106231.

Feng C, Jia J, Wang C, Han M, Dong C, Huo B and Liu X. (2019). Phytoplankton and bacterial community structure in two Chinese lakes of different trophic status. *Microorganisms,* 7(12), 621. https://doi.org/10.3390/microorganisms7120621

Fierro P, Valdovinos C, Arismendi I, Díaz G, Ruiz De Gamboa M and Arriagada L. (2019). Assessment of anthropogenic threats to Chilean Mediterranean freshwater ecosystems: literature review and expert opinions. *Environ. Impact Assess. Rev.,* 77, 114–121.

Freitas A and Thomaz SM. (2011). Inorganic carbon storage may limit the development of submersed macrophyte in habitats of the Paraná River Basin. *Acta Limnol. Bras.,* 23, 57–62.

Gao YN, Dong J, Fu QQ, Wang YP, Chen C, Li JH and Zhou, CJ. (2017). Allelopathic effects of submerged macrophytes on phytoplankton. *Allelopathy J.,* 40, 1–22.

Gerhardt A. (2002). Bioindicator species and their use in biomonitoring. *Environ. Monit.,* 1, 77–123.

Glibert P and Burford M (2017). Globally changing nutrient loads and harmful algal blooms: recent advances, new paradigms, and continuing challenges. *Oceanography,* 30, 58–69.

Gołdyn H. (2010). Changes in plant species diversity of aquatic ecosystems in the agricultural landscape in West Poland in the last 30 years. *Biodivers. Conserv.,* 19(1), 61–80.

González-Abraham C, Ezcurra E, Garcillán PP, Ortega-Rubio A, Kolb M and Creel JEB. (2015). The human footprint in Mexico: physical geography and historical legacies. *PLoS ONE,* 10, 1–17. https://doi.org/10.1371/journal.pone.0121203

Hall RI, Leavitt PR, Quinlan R, Dixit AS and Smol JP. (1999). Effects of agriculture, urbanization, and climate on water quality in the northern great plains. *Limnol. Oceanogr.,* 44(3), 739e756.

Hang LIZ. (2017). Short communication. Allelopathic activity of ethyl acetate extracts from typical emergent plants against *Microcystis aeruginosa* KUTZ. *Bangladesh J. Bot.,* 46, 1025–1029.

Hasler AD and Jones E. (1949). Demonstration of the antagonistic action of large aquatic plants on algae and rotifers. *Ecology,* 30, 359–364.

He F, Deng P, Wu XH, Cheng SP, Gao YN and Wu ZB (2008). Allelopathic effects on *Scenedesmus obliquus* Kütz. by two submerged macrophytes *Najas minor* All. and *Potamogeton malaianus* Miq. *Environ. Bull.,* 17, 92–97.

Hellsten S and Riihimäki J. (1996). Effects of lake water level regulation on the dynamics of littoral vegetation in northern Finland. *Hydrobiologia,* 340(1–3), 85–92.

Hering D, Borja A, Carstensen J, Carvalho L, Elliott M, Feld CK and van de Bund W. (2010). The European Water Framework Directive at the age of 10: a critical review of the achievements with recommendations for the future. *Sci. Total Environ.,* 408(19), 4007–4019.

Hering D, Carvalho L, Argillier C, Beklioglu M, Borja A, Cardoso AC and Birk S. (2015). Managing aquatic ecosystems and water resources under multiple stress—an introduction to the MARS project. *Sci. Total Environ.,* 503, 10–21.

Hilt S, Gross EM, Hupfer M, Morscheid H, Mählmann J, Melzer A, Poltz J, Sandrock S, Scharf EM, Schneider S and van de Weyer, K. (2006). Restoration of submerged vegetation in shallow eutrophic lakes—a guideline and state of the art in Germany. *Limnologica,* 36, 155–171.

Holt EA and Miller SW. (2010). Bioindicators: Using organisms to measure. *Nature,* 3, 8–13.

Hötzel G and Croome R. (1999). A phytoplankton methods manual for Australian freshwaters. LWRRDC Occasional Paper 22/99. https://elytus.com/blog/e-waste-and-its-negative-effects-on-the-environment.html

Hua Q, Liu Y-G, Yan Z-L , Zeng G-M , Liu S-B, Wang W-J, Tan X-F, Deng J-Q, Tang X and Wang Q-P. (2018). Allelopathic effect of the rice straw aqueous extract on the growth of *Microcystis aeruginosa. Ecotoxicol. Environ. Saf.*, 148, 953–959. https://doi.org/10.1016/j.ecoenv.2017.11.049

Jasser I. (1995). The influence of macrophytes on a phytoplankton community in experimental conditions. *Hydrobiologia,* 306: 21–32.

Jeppesen E and Søndergaard M. (1999). Lake and catchment management in Denmark. *Hydrobiologia,* 396, 419–432.

Jeppesen E, Peder Jensen J, Søndergaard M, Lauridsen T and Landkildehus F. (2000). Trophic structure, species richness and biodiversity in Danish lakes: changes along a phosphorus gradient. *Freshw. Biol.*, 45(2), 201–218.

Johnson RK and Angeler DG. (2014). Effects of agricultural land use on stream assemblages: taxon-specific responses of alpha and beta diversity. *Ecol. Indic.,* 45, 386–393. https://doi.org/10.1016/j.ecolind.2014.04.028

Jones RI, Shaw PJ and Haan HDE. (1993). Effects of dissolved humic substances on the speciation of iron and phosphate at different pH and ionic strength. *Environ. Sci. Technol.,* 27, 691–698.

Kantawanichkul S, Kladprasert S and Brix H. (2009). Treatment of high-strength wastewater in tropical vertical flow constructed wetlands planted with *Typha angustifolia* and *Cyperus involucratus. Ecol. Eng.*, 35(2), 238–247.

Kassaye YA, Skipperud L, Einset J and Salbu B. (2016). Aquatic macrophytes in Ethiopian Rift Valley lakes; their trace elements concentration and use as pollution indicators. *Aquat. Bot.*, 134, 18–25. https://doi.org/10.1016/j.aquabot.2016.06.004

Khatri N and Tyagi S. (2015). Influences of natural and anthropogenic factors on surface and groundwater quality in rural and urban areas. *Front. Life Sci.*, 8(1), 23–39.

Kong X, He Q, Yang B, He W, Xu F, Janssen AB and Mooij WM. (2017). Hydrological regulation drives regime shifts: evidence from paleolimnology and ecosystem modeling of a large shallow Chinese lake. *Glob. Chang. Biol.*, 23(2), 737–754.

Körner S and Nicklisch A. (2002). Allelopathic growth inhibition of selected phytoplankton species by submerged macrophytes. *J. Phycol.,* 38, 862–871.

Kshirsagar AD. (2013).Use of algae as a bioindicator to determine water quality of River Mula from Pune City, Maharashtra (India). *Univers. J. Environ. Res. Technol.*, 3(1), 79–85.

Kumar S, Pratap B, Dubey D and Dutta V. (2020). Removal of nutrients from domestic wastewater using constructed wetlands: assessment of suitable environmental and operational conditions. *J. Environ. Sustain.,* 3(4), 341–352.

Kumari D and Paul DK. (2020) Assessing the role of bioindicators in freshwater ecosystem. *J. Interdiscipl. Cycle Res.*, XII(IX), 58–74.

Laird KR, Kingsbury MV, Lewis CM and Cumming BF. (2011). Diatom-inferred depth models in 8 Canadian boreal lakes: inferred changes in the benthic: planktonic depth boundary and implications for assessment of past droughts. *Quat. Sci. Rev.,* 30(9–10), 1201–1217.

Lange K, Toownsend CR and Matthaei CD. (2016). A trait based framework for stream alga communities. *Ecol. Evol.,* 6, 23–36.

Leira M and Cantonati M. (2008). Effects of water-level fluctuations on lakes: An annotated bibliography. In: Wantzen KM, Rothhaupt KO, Mörtl M, Cantonati M, Tóth LG and Fischer P (eds). *Ecological Effects of Water-Level Fluctuations in Lakes*. Springer, Dordrecht, pp. 171–184.

Li JY, Zhang P, Zeng G, Cai X, Liu S, Yin Y, Hu X, Hu X and Tan X. (2016). Growth inhibition and oxidative damage of *Microcystis aeruginosa* induced by crude extract of *Sagittaria trifolia* tubers. *J. Environ. Sci. (China),* 43, 40–47. https://doi.org/10.1016/j.jes.2015.08.020

Liang D, Wang Q, Wei N, Tang C, Sun X and Yang Y. (2020). Biological indicators of ecological quality in typical urban river-lake ecosystems: the planktonic rotifer community and its response to environmental factors. *Ecol. Indic.,* 112, 106127.

Lodi S, Vieira LCG, Velho LFM, Bonecker CC, de Carvalho P and Bini LM (2011). Zooplankton community metrics as indicators of eutrophication in urban lakes. *Nat. Conserv.,* 9(1), 87–92. https://doi.org/10.4322/natcon.2011.011

Loiselle SA, Bracchini L, Cózar A, Dattilo AM and Rossi C. (2005). Extensive spatial analysis of the light environment in a subtropical shallow lake, Laguna Iberá, Argentina. *Hydrobiologia*, 534(1–3), 181–191.

Lougheed VL, Mcintosh MD, Parker CA and Stevenson RJ. (2008). Wetland degradation leads to homogenization of the biota at local and landscape scales. *Freshw. Biol.,* 53, 2402–2413. https://doi.org/10.1111/j.1365-2427.2008.02064.x

Lu Z, Zhang Y, Gao Y, Liu B, Sun X, He F, Zhou Q and Wu Z. (2016). Effects of pyrogallic acid on *Microcystis aeruginosa*: oxidative stress related toxicity. *Ecotoxicol. Environ. Saf.*, 132, 413–419. https://doi.org/10.1016/j.ecoenv.2016.06.039

Mack RN, Simberloff D, Lonsdale WM, Evans H, Clout M and Bazzaz FA (2000) Biotic invasions: causes, epidemiology, global consequences, and control. *Ecol. Appl.*, 10, 689–710.

Mackay SJ, James CS and Arthington AH. (2010). Macrophytes as indicators of stream condition in the wet tropics region, Northern Queensland, Australia. *Ecol. Indic.,* 10, 330–340. https://doi.org/10.1016/j.ecolind.2009.06.017

Malhi Y, Gardner TA, Goldsmith GR, Silman MR and Zelazowski P. (2014). Tropical forests in the Anthropocene. *Annu. Rev. Environ. Resour.*, 39, 125–159. https://doi.org/10.1146/annurev-environ-030713-155141

Mandal MH, Siddique G and Roy A. (2018). Threats and opportunities of ecosystem services: a geographical study of Purbasthali oxbow lake. *J. Geogr. Environ. Earth Sci. Int.*, 16(4), 1–24.

Manickavasagam S, Sudhan C and Aanand S. (2019). Bioindicators in aquatic environment and their significance. *J. Aquacult. Trop.*, 34(1/2), 73–79.

Matthews E and Fung I. (1987). Methane emission from natural wetlands: global distribution, area, and environmental characteristics of sources. *Global Biogeochem. Cycles*, 1(1), 61–86.

May L, Olszewska J, Gunn IDM, Meis S and Spears BM. (2020, July). Eutrophication and restoration in temperate lakes. In *IOP Conference Series: Earth and Environmental Science,* 535(1), 012001.

McAbendroth L, Ramsay PM, Foggo A, Rundle SD and Bilton DT. (2005). Does macrophyte fractal complexity drive invertebrate diversity, biomass and body size distributions? *Oikos,* 111, 279–290.

Melzer A. (1999). Aquatic macrophytes as tools for lake management. *Hydrobiologia,* 395/396, 181–190.

Meyers PA. (2003). Applications of organic geochemistry to paleolimnological reconstructions: a summary of examples from the Laurentian Great Lakes. *Org. Geochem.*, 34(2), 261–289.

Michelan TS, Thomaz SM, Mormul RP and Carvalho P. (2010). Effects of an exotic invasive macrophyte (tropical signalgrass) on native plant community composition, species richness and functional diversity. *Freshw. Biol.,* 55, 1315–1326. https://doi.org/10.1111/j.1365-2427.2009.02355.x

Mukherjee B, Nivedita M and Mukherjee D. (2010). Plankton diversity and dynamics in a polluted eutrophic lake, Ranchi. *J. Environ. Biol.*, 31(5 SUPPL.), 827–839.

Mulderij G, Mau B, Van Donk E and Gross EM. (2007) Allelopathic activity of *Stratiotes aloides* on phytoplankton—towards identification of allelopathic substances. *Hydrobiologia,* 584, 89–100. https://doi.org/10.1007/s10750-007-0602-0 Acc

Mulderij G, Mooij WM and Van Donk E. (2005). Allelopathic growth inhibition and colony formation of the green alga *Scenedesmus obliquus* Kütz. by the aquatic macrophyte *Stratiotes aloides* L. *Aquatic Ecol.,* 39, 11–21.

Mulderij G, Smoldes AJP and Van Donk E. (2006). Allelopathic effect of the aquatic macrophyte, *Stratiotes aloides* L., on natural phytoplankton. *Freshw. Biol.,* 51, 554–561.

Nakai S, Hosomi M, Okada M and Murakami A. (1996). Control of algal growth by macrophytes and macrophyte-extracted bioactive compounds. *Water Sci. Technol.,* 34, 227–235.

Nakai S, Inoue Y, Hosomi M and Murakami A. (1999). Growth inhibition of blue-green algae by allelopathic effects of macrophytes. *Water Sci. Technol.,* 39, 47–53.

Naumann E. (1929). The scope and chief problems of regional limnology. *Int. Rev. Hydrobiol.*, 22(1), 423–444.

Ndjouondo GP, Ache NT, Mbah GE and Muyang RF (2020). Phytoplankton characterization and the state of eutrophication of Mefou Lake (Yaounde, Cameroon). *Int. J. Botany Stud.*, 5(4), 139–146.

Niyatbekov T and Barinova S. (2018). Bioindication of aquatic habitats with diatom algae in the Pamir Mountains, Tajikistan. *MOJ Ecol. Environ. Sci.*, 3, 117–120.

Nowlin WH, Davies JM, Nordin RN and Mazumder A. (2004). Effects of water level fluctuation and short-term climate variation on thermal and stratification regimes of a British Columbia reservoir and lake. *Lake Reserv. Manag.*, 20(2), 91–109.

O'Neill EA, Rowan NJ and Fogarty AM. (2019). Novel use of the alga *Pseudokirchneriella subcapitata*, as an early-warning indicator to identify climate change ambiguity in aquatic environments using freshwater finfish farming as a case study. *Sci. Total Environ.,* 692, 209–218.

Padisák J, Crossetti LO and Naselli-Flores L. (2009). Use and misuse in the application of the phytoplankton functional classification: a critical review with updates. *Hydrobiologia*, 621(1), 1–19.

Paerl HW and Paul VJ. (2012). Climate change: links to global expansion of harmful cyanobacteria. *Water Res.,* 46, 1349–1363. https://doi.org/10.1016/j.watres.2011.08.002

Paixão EC, da Costa de Noronha J, Nunes da Cunha C and Arruda R. (2013). More than light: distance-dependent variation on riparian fern community in Southern Amazonia. *Braz. J. Bot.*, 36, 25–30. https://doi.org/10.1007/s40415-013-0003-8

Pall K and Moser V. (2009). Austrian index macrophytes (AIM-Module 1) for lakes: a water framework directive compliant assessment system for lakes using aquatic macrophytes. *Hydrobiologia*, 633(1), 83–104.

Palmer CM (1969). Composite rating of algae tolerating organic pollution. *J. Phycol.*, 5, 78–82.

Palmer MA, Bernhardt ES, Allan JD, Lake PS, Alexander G, Brooks S and Sudduth E. (2005). Standards for ecologically successful river restoration. *J. Appl. Ecol.*, 42(2), 208–217.

Parmar TK, Deepak R. and Agrawal YK. (2016). Bioindicators: the natural indicator of environmental pollution. *Front. Life Sci.,* 9(2), 110–118.

Pieterse AH and Murphy K. (1990). *Aquatic Weeds. The Ecology and Management of Nuisance Aquatic Vegetation*. Oxford Science Publications, New York, pp. 3–16. ISBN 01-985-41813.

Poikane S, Portielje R, Denys L, Elferts D, Kelly M, Kolada A and van den Berg MS. (2018). Macrophyte assessment in European lakes: Diverse approaches but convergent views of 'good' ecological status. *Ecol. Indic*., 94, 185–197.

Poikane S, Birk S, Böhmer J, Carvalho L, De Hoyos C, Gassner H, Hellsten S, Kelly M, Lyche SA, Olin M, Pall K, Phillips G, Portielje R, Ritterbusch D, Sandin L, Schartau AK, Solimini AG, Van Den Berg M, Wolfram G and Van De Bund W. (2015). A hitchhiker's guide to European lake ecological assessment and intercalibration. *Ecol. Indic.,* 52, 533–544.

Poikane S, Kelly M and Cantonati M. (2016). Benthic algal assessment of ecological status in European lakes and rivers: challenges and opportunities. *Sci. Total Environ.,* 568, 603–613.

Porter SD, Mueller DK, Spahr NE, Munn MD and Dubrovsky NM. (2008). Efficacy of algal metrics for assessing nutrient and organic enrichment in flowing waters, *Freshw Biol*, 53, 1036–1054. doi:10.1111/j.1365-2427.2007.01951.x

Rejmankova E. (2011). The role of macrophytes in wetland ecosystems. *J Ecol. Environ*., 34(4), 333–345.

Reynolds CS. (2006). *The Ecology of Phytoplankton*. Cambridge University Press, Cambridge. https://doi.org/10.1017/CBO9780511542145

Reynolds CS and Irish AE. (1997). Modelling phytoplankton dynamics in lakes and reservoirs: the problem of in-situ growth rates. *Hydrobiologia*, 349(1–3), 5–17.

Roy K, Vrba J, Kaushik SJ and Mraz J. (2020). Feed-based common carp farming and eutrophication: is there a reason for concern? *Rev. Aquacult*., 12(3), 1736–1758.

Şahin B, Akar B and Barinova S. (2020). Non-diatom algae of the high mountain protected lakes in the Artabel Lakes Nature Park, Gümüşhane, Turkey. *Bot. Pac.*, 9(2), 47–59.

Saito K, Matsumoto M, Sekine T, Murakoshi I, Morisaki N and Iwasaki S. (1989). Inhibitory substances from *Myriophyllum brasiliense* Vell. on growth of blue-green algae. *J. Nat. Prod.,* 52, 1221–1226.

Sala III OE, Stuart Chapin F, Armesto JJ, Berlow E, Bloomfield J, Dirzo R, HuberSanwald E, Huenneke LF, Jackson RB, Kinzig A, Leemans R, Lodge DM, Mooney HA, Oesterheld M, Pof NL, Sykes MT, Walker BH, Walker M and Wall DH. (2000). Global biodiversity scenarios for the year 2100. *Science* 287, 1770–1774. https://doi.org/10.1126/science.287.5459.1770

Samiyappan M, Sudhan C, Bharathi and Aanand, S. (2019). Bioindicators in aquatic environment and their significance. *J. Aqua. Trop*., 34(1), 73–79.

Santamaría L. (2002). Why are most aquatic plants widely distributed? Dispersal, clonal growth and small-scale heterogeneity in a stressful environment. *Acta Oecol.*, 23(3), 137–154.

Sass LL, Bozek MA, Hauxwell JA, Wagner K and Knight S. (2010). Response of aquatic macrophytes to human land use perturbations in the watersheds of Wisconsin lakes, U.S.A. *Aquat. Bot.,* 93, 1–8. https://doi.org/10.1016/j.aquabot.2010.02.001

Sayeswara HA, Goudar M, Anand and Manjunatha R. (2011). Water quality evaluation and phytoplankton diversity of Hosahalli pond, Shivamogga, Karnataka (India), *Int. J. Chem. Sci*., 9(2), 805–815.

Schindler DE and Scheuerell MD. (2002). Habitat coupling in lake ecosystems. *Oikos*, 98(2), 177–189.

Shea K and Chesson P. (2002). Community ecology theory as a framework for biological invasions. *Ecol. Evol.*, 17, 170–176. https://doi.org/10.1212/WNL.0000000000001675

Smol JP, Wolfe AP, Birks HJB, Douglas MS, Jones VJ, Korhola A and Weckström J. (2005). Climate-driven regime shifts in the biological communities of arctic lakes. *Proc. Natl Acad. Sci.*, 102(12), 4397–4402.

Song C, Cao X, Li J, Li Q, Chen G and Zhou Y. (2006). Contributions of phosphatase and microbial activity to internal phosphorus loading and their relation to lake eutrophication. *Sci. China Ser. D. Earth Sci.*, 49(1), 102–113.

Soylu EN, Maraşlioglu F and Gönülol A. (2007). Phytoplankton seasonality of a shallow turbid lake. *Algol. Stud.*, 123(1), 95–110.

Stednick JD and Hall EB. (2003). Applicability of trophic status indicators to Colorado plains reservoirs. Completion Report (Colorado Water Resources Research Institute); no. 195.

Sutton DL and Ornes WH. (1975). Phosphorus removal from static sewage effluent using duckweed. *J. Environ. Qual.*, 4, 367–370.

Takeda F, Nakano K, Nishimura O, Shimada Y, Fukuro S, Tanaka H, Hayashi N and Inamori Y. (2011). Allelopathic potential of *Potamogeton pusillus* L. community against *Microcystis aeruginosa* Kütz. *J. Water Environ. Technol.*, 9, 21–28.

Tang X, Wu M, Yang W and Scholz M. (2015a). Impact of simulated water level regulation on sediment nutrient release. *Water Air Soil Pollut.*, 226(8), 248.

Tang Y, Tian J, Li S, Xue C, Xue Z, Yin D and Yu S. (2015b) Combined effects of graphene oxide and Cd on the photosynthetic capacity and survival of *Microcystis aeruginosa. Sci. Total Environ.*, 532, 154–161. https://doi.org/10.1016/j.scitotenv.2015.05.081

Tazart Z, Caldeira AT, Douma M, Salvador C and Loudiki M. (2021). Inhibitory effect and mechanism of three macrophytes extract on *Microcystis aeruginosa* growth and physiology. *Water Environ. J.*, 35(2), 580–592.

Tazart Z, Douma M, Caldeira AT, Tebaa L, Mouhri K and Loudiki M. (2020). Highlighting of the antialgal activity of organic extracts of Moroccan macrophytes: Potential use in cyanobacteria blooms control. *Environ. Sci. Poll. Res.*, 27(16), 19630–19637.

Thomas MK, Fontana S, Reyes M, Kehoe M and Pomati F. (2018). The predictability of a lake phytoplankton community, over time-scales of hours to years. *Ecol. Lett.*, 21(5), 619–628.

Thomaz SM and da Cunha ER. (2010). The role of macrophytes in habitat structuring in aquatic ecosystems: methods of measurement, causes and consequences on animal assemblages' composition and biodiversity. *Acta Limnol. Bras.*, 22(2), 218–236.

Tórz A, Bonisławska M, Rybczyk A, Nędzarek A and Tański A. (2020). Susceptibility to degradation, the causes of degradation, and trophic state of three lakes in north-west Poland. *Water*, 12(6), 1635.

Toudjani AA, Çelekli A, Gümüş EY, Kayhan S, Lekesiz HÖ and Çetin T. (2018). Assessment of ecological status using phytoplankton indices and multivariate analyses in the western Mediterranean Basin. *Fundam. Appl. Limnol.*, 191, 155–167.

Umetsu CA, Aguiar FC, Ferreira MT, Cancian LF and Camargo AFM. (2018). Addressing bioassessment of tropical rivers using macrophytes: the case of Itanhaém Basin, São Paulo, Brazil. *Aquat. Bot.*, 150, 53–63. https://doi.org/10.1016/j.aquabot. 2018.07.004

Vanderstukken M, Mazzeo N, van Colen W, Declerck SAJ and Muylaert K. (2011). Biological control of phytoplankton by the subtropical submerged macrophytes *Egeria densa* Planch. and *Potamogeton illinoensis* Morong: a mesocosm study. *Freshw. Biol.*, 56, 1837–1849.

Wang Q, Wang L, Han R, Yang L, Zhou Q and Huang X. (2015a). Effects of bisphenol A on antioxidant system in soybean seedling roots. *Environ. Toxicol. Chem.*, 34(5), 1127–1133.

Wang H, Liang F and Zhang L. (2015b). Composition and anti-cyanobacterial activity of essential oils from six different submerged macrophytes. *Polish J. Environ. Stud.*, 24, 333–338. https://doi.org/10.15244/pjoes/26383

Wang HQ, Cheng SP, Zhang SH, He F, Liang W, Zhang LP, Hu CY, Ge FJ and Wu ZB. (2010). Chemical composition in aqueous extracts of *Potamogeton malaianus* Miq. and *Potamogeton maackianus* A. and their allelopathic effects on *Microcystis aeruginosa* Kütz. *Polish J. Environ. Stud.*, 19, 213–218.

Wang J, Chen G, Kang W, Hu K and Wang L. (2019). Impoundment intensity determines temporal patterns of hydrological fluctuation, carbon cycling and algal succession in a dammed lake of Southwest China. *Water Res.*, 148, 162–175.

White MS, Xenopoulos MA, Hogsden K, Metcalfe RA and Dillon PJ. (2008). Natural lake level fluctuation and associated concordance with water quality and aquatic communities within small lakes of the Laurentian Great Lakes region. In: Wantzen KM, Rothhaupt KO, Mörtl M, Cantonati M, Tóth LG and Fischer P (eds) *Ecological Effects of Water-Level Fluctuations in Lakes.* Springer, Dordrecht, pp. 21–31.

Wilhelm S and Adrian R. (2008). Impact of summer warming on the thermal characteristics of a polymictic lake and consequences for oxygen, nutrients and phytoplankton. *Freshw. Biol.*, 53(2), 226–237.

Williams P, Withfield M and Biggs J. (2008). How can we make new ponds biodiverse? A case study monitored over 7 years. *Hydrobiologia,* 597, 137–148.

Wu C, Chang XX, Dong HJ, Li DF and Liu JY. (2008). Allelopathic inhibitory effect of *Myriophyllum aquaticum* (Vell) Verdc. on *Microcystis aeruginosa* Kütz. and its physiological mechanism. *Acta Ecologica Sinica,* 28, 2595–2603.

Wu ZB, Gao YN, Wang J, Liu BY, Zhou QH and Zhang YY. (2009). Allelopathic effects of phenolic compounds present in submerged macrophytes on *Microcystis aeruginosa* Kütz. *Allelopathy J.,* 23, 403–410.

Xie J, Zhao L, Liu K and Liu W. (2019) Enantiomeric environmental behavior, oxidative stress and toxin release of harmful cyanobacteria *Microcystis aeruginosa* in response to napropamide and acetochlor. *Environ. Pollut.,* 246, 728–733. https://doi.org/10.1016/j.envpol.2018.12.056

Yang J, Lv H, Liu L, Yu X and Chen H. (2016). Decline in water level boosts cyanobacteria dominance in subtropical reservoirs. *Sci. Total Environ.,* 557, 445–452.

Yang JR, Lv H, Isabwe A, Liu L, Yu X, Chen H and Yang J. (2017). Disturbance-induced phytoplankton regime shifts and recovery of cyanobacteria dominance in two subtropical reservoirs. *Water Res.,* 120, 52–63.

Yang W and Yang Z. (2014). Effects of long-term environmental flow releases on the restoration and preservation of Baiyangdian Lake, a regulated Chinese freshwater lake. *Hydrobiologia*, 730(1), 79–91.

Yang W, Yan J, Wang Y, Zhang BT and Wang H. (2020). Seasonal variation of aquatic macrophytes and its relationship with environmental factors in Baiyangdian Lake, China. *Sci. Tot. Environ.*, 708, 135112.

Yusuf ZH. (2020). Phytoplankton as bioindicators of water quality in Nasarawa reservoir, Katsina State Nigeria. *Acta Limnol. Bras.*, 32, e4.

Zedler JB (2011). Wetlands. In: Simberloff D and Rejmanek M (eds). *Encyclopedia of Biological Invasions.* University of California Press, Berkeley & Los Angeles, pp. 609–612.

Zeng J, Bian Y, Xing P and Wu QL. (2012). Macrophyte species drive the variation of bacterioplankton community composition in a shallow freshwater lake. *Appl. Environ. Microbiol.,* 78(1), 177–184.

Zhang Q, Song Q, Wang C, Zhou C, Lu C and Zhao M. (2017). Effects of glufosinate on the growth of and microcystin production by *Microcystis aeruginosa* at environmentally relevant concentrations. *Sci. Total Environ.,* 575, 513–518. https://doi.org/10.1016/j.scitotenv.2016.09.011

Zhou B, Shang M, Zhang S, Feng L, Liu X, Wu L and Shan K. (2019). Remote examination of the seasonal succession of phytoplankton assemblages from time-varying trends. *J. Environ. Manage.*, 246, 687–694.

Zhu K, Bi Y and Hu Z. (2013). Responses of phytoplankton functional groups to the hydrologic regime in the Daning River, a tributary of Three Gorges Reservoir, China. *Sci. Total Environ.,* 450, 169–177.

30 Parasitic Diversity Strategies under the Influence of Pollutants

Reshu, Rashmi Srivastava, Kamal Jaiswal and Neeshma Jaiswal
Department of Zoology, Babasaheb Bhimrao Ambedkar University, Lucknow

Anita Yadav and Neerja Kapoor
Department of Zoology, CMP PG Constituent College, University of Allahabad, Prayagra

Sandeep K. Malhotra
Department of Zoology, University of Allahabad, Prayagraj, India

CONTENTS

30.1 INTRODUCTION

During the last few decades, a lot of research has been conducted to procure environmental information by exploring the parasites prevailing in the ecological community. The extensive work done in this field has led to the development of the field of environmental parasitology, which deals with the consequential environmental regulatory influence on parasites (Goater et al., 2013; Kleinertz et al., 2014; Quiazon, 2015; Morales-Serna et al., 2019). In general, environmental parasitology consists of two different approaches. The first, an ecological approach, is more focused on how environmental health is reflected by the parasites in the ecosystem, whereas the second is the medical approach, which helps us to understand the occurrence and diagnosis of infective parasitic stages that lead to human infection. In the context of the ecologically based approach of environmental

DOI: 10.1201/9781003220824-34

parasitology, parasites express great sensitivity to alterations in the ecosystem (Sures and Nachev, 2015; Adewole et al., 2019; Santoro et al., 2020). At present, parasites are the focus of attention to reveal information about the environment and act as an indicator of environmental health.

The diversity of life comprises an equivalent numerical strength of parasites compared with other species, or more than half of their numbers. The former have a significant ecological function by inducing a debilitating effect on prey, which could not so easily be adapted to by ecosystems (Holt, 2010). The impact of diminishing diversity of sexually dimorphic traits could thus have resulted due to the eventual transformation by the adaptation of parasitic elements to asexual reproduction. An invaluable opportunity is created by parasitic organisms for the exchange of genetic material between species.

A large variety of roundworms, tapeworms, digeneans and monogenoideans, leeches, parasitic crustaceans such as isopods and copepods (Justine, 2010; Justine et al., 2010; 2012a; 2012b), and myxosporidia and microsporidia comprise the structured parasitic fauna of coral reef fish. All of these have contributed to the high biodiversity of the latter. Around 30 species of parasites have been encountered in the coral reef fishes (Justine et al., 2012a; 2010; Poulin and Morand, 2004). The possibility of enhanced intricacies in the interactions of the host–parasitic network emerged because of the high biodiversity of coral reef fishes. Monogeneans infect freshwater fishes and are one of the major groups of parasites that often cause diseases worldwide (Koskivaara 1992; Yoshinaga et al., 2009; Cunha et al., 2021). Helminth parasites are important not only for causing fish diseases but also because they are an essential and integral component of global biodiversity (MacKenzie et al., 1995; Poulin and Morand, 2004; Vidal-Martínez et al., 2006; Al-Hasawi 2019).

Time and again, parasitic helminths are important in the maintenance of the balance of the ecosystem by regulating the abundance or density of the populations of the hosts, which affects trophic chains and structured animal communities (Poulin and Morand, 2004; Pietrock and Marcogliese, 2003). Simultaneously, the diverse influence of quicker changes in the ambient environment, particularly in the aquatic environment, is frequently reflected in the community organization of these parasites, and these alterations could then manifest in their hosts (Lafferty, 1997; Lafferty and Kuris, 2005; Nachev and Sures, 2016). The significant growth diversification of the biodiversity of parasites in multifaceted dimensions is indeed noticeable, maybe due to a major emphasis on the advancements of the infrastructure of fish farms (Thatcher, 2006), as stressed in the recently proposed Pradhan Mantri Matsya Sampada Yojana, 2020 of Hon'ble Prime Minister of India, or their role may be to indicate alterations in the environment under the influence of climate change or other related factors, which contribute to the conservation and maintenance of biodiversity as a whole (Gómez et al., 2011; Palm, 2011). The biodiversity variations in the parasites parasitizing fish of Brazilian basins have been the subject of much analysis in recent years (Eiras et al., 2010).

Many types of research confirm that parasites are of extraordinary significance in the aquatic ecosystem due to their diverse habitats. Because of their wide abundance and distribution, a few specialists began to concentrate on the utilization of parasites as a marker of ecological quality and biodiversity. As parasites showed a diversity of vigorous responses towards anthropogenic pollution, they could be treated as environmental indicators (Parmar et al., 2016). Many recent studies have analyzed how the distribution and abundance of parasites are affected by different types of pollution. These days, there is an expanding number of research works focusing on toxin accumulation in parasites (Gunkel, 1994).

In specific situations, pollution can adversely influence parasites; for instance, i) if a pollutant reduces the life expectancy of infected hosts, it will accordingly hinder the reproduction of parasites; ii) if parasites are more susceptible towards specific contamination as compared with their hosts; and iii) if contamination results in the extinction of necessary hosts. Nonetheless, pollution can likewise increase parasite prevalence and intensity if host defense mechanisms are adversely influenced, which thus expands host susceptibility, or by prompting ecological conditions that enhance the

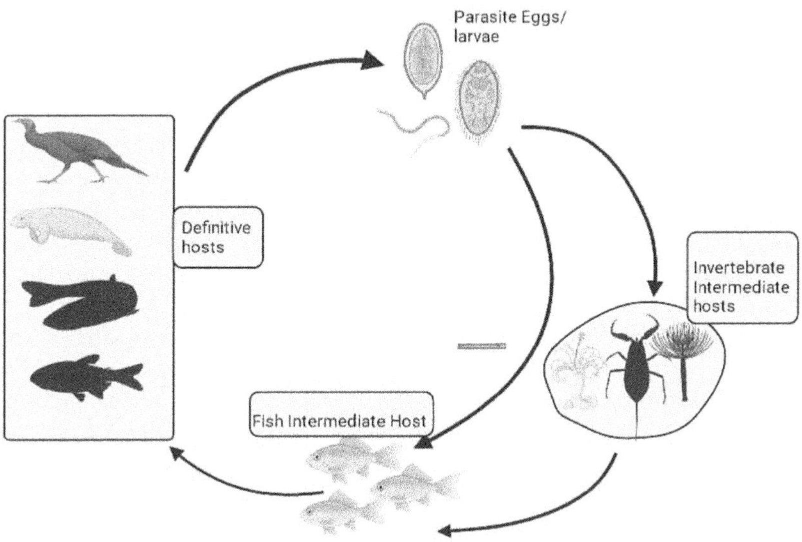

FIGURE 30.1 A general life cycle of nematode parasite in definitive (carnivorous fish, mammals, aves) and intermediate (fish and invertebrates) hosts.

population densities of suitable intermediate or final hosts. The impacts of pollution may change between different species of parasites at different developmental stages (Sures, 2004).

Cestodes, coccidia, digeneans, cercariae and miracidia, nematode larval stages, etc., found in soil or water are referred to as free-living larval stages, which are more sensitive to pollutants than the adult stages inhabiting their definitive hosts (Malhotra et al., 2002; Sures, 2008a and b). The life cycle of a nematode parasite is represented in Figure 30.1.

Just like other organisms, parasites are influenced by environmental conditions and react to them either in a general or in a particular way. The reason behind the vacillations in parasite population dynamics is not yet surely known, but it seems that changes in ecological conditions (e.g., pollution or other anthropogenic effects) play a critical role (Sures, 2004).

30.2 DIVERSITY OF PARASITES ON FISH IN THE CHALLENGED ENVIRONMENT

The physical, chemical, biological, and ecological state of the environment can be recognized by the existence of some organisms. Many studies focus on the alteration in the structure and diversity of different parasites of fish hosts because of the possible use of parasites as a marker of ecosystem health (Kennedy, 1997; MacKenzie et al., 1995, Dušek et al., 1998; Morales-Serna et al., 2019; Santoro et al., 2020). Studies using monogenoidean parasites under the influence of pollutants have been conducted on the gills of fish (Dušek et al., 1998; Bagge and Valtonen, 1996; Sures, 2008a and b).

In ecotoxicology, parasites that act as effective indicators reveal information about the environmental health and well-being of biological systems on earth, which in turn regulate the community biodiversity and population size. The effective indication can be analyzed by direct application of toxic elements to cercariae and miracidia, which are referred to as the free-living parasitic stages of trematodes (Pietrock and Marcogliese, 2003).

Effect indications can be analyzed by treating the parasitic larval stages with synthetic compounds and then investigating the resulting viability, longevity, and infectivity of these larval stages. In contrast, with automatic indication methodology utilizing daphnids, algae, or fish, cercarial test frameworks have all the earmarks of being less encouraging. Now and then, these methodologies can be utilized to assess the impacts of pollution on the transmission of parasites (Sures et al., 2017). Ecological elements can impact the population of parasites directly or indirectly. On the off chance that the intermediate or final host is adversely influenced, the effect is seen first at the level of a free-living organism and consequently on the parasitic level. Fish ecto-parasites, which are in direct contact with the surrounding environmental conditions, respond more rapidly and delicately to stresses created by the environment, such as chemical contamination and eutrophication, which results in decreased species wealth and inconsistent distribution of parasitic abundance in the host (Dušek et al., 1998). However, in the case of endo-parasites with free-living stages, similar impacts were seen in the case of ecto-parasites.

In the context of environmental pollution, changes in the parasite population have been interpreted, but it is arduous to hypothesize these discoveries. To date, only a couple of parasite population combinations reflect unsurprising changes, despite the impressive effort that has been devoted to connecting levels of environmental pollution with parasitism (Vidal-Martínez et al., 2010).

Environmental pollution affects the parasite population directly as well as indirectly by influencing intermediate and final hosts (Sures et al., 2017). As compared with the established indication procedure using a free-living organism, it is preferable to examine parasite community structure, as it shows the effect of the environmental variable at different trophic levels. For instance, the proportion of species richness of heteroxenous (SH) to monoxenous parasites (SM) may be used as an indicator of pollution. The proportion of heteroxenous and monoxenous parasitic species found in or on fishes is higher in unpolluted habitats such as marine habitat as compared with fishes that were tested from polluted habitats. So, we can conclude that heteroxenous parasites are less sensitive to pollution than monoxenous parasites (Perez-del Olmo et al., 2007).

30.3 INTERACTIONS OF POLLUTANTS AND PARASITES ON HOSTS

Although parasites can be profitable by decreasing toxins amassed in the tissues of the host, they can also negatively influence the physiology of the host. Sometimes, parasites can also function as an agent of stress that disrupts the normal physiology of the host organism. Under polluted environmental conditions, these stressors show a synergistic, antagonistic, or neutral impact on host physiology. So, the host's pollutant mechanism is affected by the parasites in many ways. Parasites can influence the metabolism of pollutants in the host body in several ways depending on need. An organism that is exposed to ecological toxins detoxifies chemicals in various ways depending upon the type of pollutant. Poisonous substances are converted into non-poisonous substances, which can be further excreted out of the body or stored in a place where they are not harmful. Parasites inhabiting a host in a polluted environment reflect an alteration in detoxification measures, which is much more devastating than the stressors alone.

30.3.1 TOXICANTS AND STRESSORS

Due to the unavoidable presence of toxic substances several investigation methods have been used for the evaluation of the impacts of pollutants on aquatic organisms. This has prompted the utilization of biological markers, which mark the existence of pollutants in the environment and reflect their effect on living beings. Biomarkers are generally biological parameters based on changes in behavior, physiology, biochemistry, genomic structure, and expression, and are used as an effective warning sign of disturbance at the individual level (Preti et al., 2020). Biomarkers can be categorized into less specific markers and contaminant-specific markers (Morrill et al., 2014). Contaminant-specific markers and less specific markers include metallothioneins as markers for metals and hormones, respectively. The most normally utilized biomarkers, in any case, are sensitive not only to anthropogenic contaminations but also to natural stressors, particularly parasites. Although biomarkers are applied under complex field conditions, information about their possible modulation by normally occurring parasites is still limited (Sures and Nachev, 2015). An increase in anthropogenic activities leads to an increase in the concentration of pesticides, heavy metals, polychlorinated biphenyls (PCBs), polybrominated diphenyl ether (PBDE), persistent organic pollutants (POPs), and polycyclic aromatic hydrocarbons (PAHs) in the environment, which results in a devastating effect on the environmental constituents (Vidal-Martínez et al., 2003; 2006). So, there is an urgent need to recognize these environmental issues and formulate a sustainable method for the proper disposal of contaminants and pollutants (Sciortino and Ravikumar, 1999).

30.3.2 Role of Bioindicators

We must comprehend and forestall the results of noxious environmental degradation procedures and poisonous exposures. One system to obtain this data is the utilization of bioindicators (Vidal-Martínez and Wunderlich, 2017; Debraj Biswal and Chatterjee, 2020). These are the species that perceive the environmental effects, since they react to those modifications occurring in the environment, which is further reflected by making changes in their numbers, physiology, or chemical organization. Such elements could well result in physiological or molecular changes to attain an altered population size. On the other hand, elemental accumulation also results by attracting substances from the surrounding environment proficiently without inducing unwarranted unfavorable impacts (Sures, 2003). Parasites are one of the most important and integral elements in aquatic ecosystems, in which they drive fundamental ecological processes, e.g., by contributing to a system's biodiversity, productivity, and food web structure or ecosystem engineering.

30.4 USE OF SENSITIVITY OF PARASITES TO THE AMBIENT CHEMICAL ENVIRONMENT

The dynamics of certain populations of helminth parasites are the result of available chemical substances in the ambient aquatic environment (Sures, 2003; Vidal-Martínez et al., 2010; Al-Hasawi, 2019). These organisms are capable of accumulating certain chemical substances in their body at much higher concentrations than those found in host tissue as well as surrounding environmental conditions, and this occurs frequently in diverse habitats. In this way, they can give significant data about the chemical conditions of their natural surroundings and convey information about the poisonous substances occurring in that area. Therefore, compared with chemical analyses, such parasites are more advantageous as organisms that reflect the degree of ecological presence of toxic substances in the environment (Sures, 2003). In unspoiled areas such as the Antarctic, anthropogenic pollution and the availability of toxic substances occurring even at the very lowest concentrations can be detected by using such sensitive helminths (Teimoori et al., 2014; Keke et al., 2020). Several investigations have been done on different groups of parasites concerning their own ability to concentrate pollutants in their body. Among different groups of endoparasites, cestodes, as well as acanthocephalans, ended up being profoundly viable in the uptake and amassing of metals (Sures, 2003; Sures, 2008a and b; Malek et al., 2007). It has been found that the concentrations of different harmful metals in acanthocephalans increase many times as compared with those found in the free-living helminths that show sensitivity to chemical substances in the aquatic environment (Mehdi and Mahdi, 2015; Najm and Fakhar; 2015; Sures and Nachev; 2015).

30.4.1 Acanthocephalans – The Dark Horse

Among different groups of intestinal parasites, acanthocephalans are the most promising parasites found in fishes. Four different acanthocephalan species have been recovered from host eels from the river Rhine. Acanthocephalans are abundantly found in an aquatic habitat, but they are rarely considered by scientists for environmental impact studies (Mehana et al., 2020). Parasitization by acanthocephalans is mostly seen in fishes, but many other different vertebrates, such as mammals, also act as the definitive host. Some of the parasites are reported to enhance the fertility rate in human beings. The role of parasites in the study of biological interactions is well known, as most of them are indications of negative interaction. Apart from these facts, parasites are an important pillar of strength regarding the maintenance of the stability of the entire ecosystem. In *Pomphorhynchus laevis*, concentrations of cadmium and lead were, respectively, 400 and 2,700 times higher than those found in the host tissue and 2,700 and 11,000 times higher than those found in the water (Sures et al., 1994a; Sures and Taraschewski, 1995). *Paratenuisentis ambiguus* showed a minimum metal-amassing capacity (Sures et al., 1994b). This is often similar with cestodes inhabiting the intestine of freshwater fish (Sures et al., 1997a and b; Riggs et al., 1987; Tenora et al., 1997). Almost all examined *Acanthocephala* were found to carry a conspicuously higher concentration of metals than that found in the tissue of the host (Sures and Siddall, 1999).

30.4.2 Affinity of Acanthocephalans Towards Heavy Metals

In contrast, *Anguillicola crassus*, a swim bladder nematode, shows much less or miniscule affinity for metals in aquatic bodies (Sures et al., 1997a and b; Sures et al., 1999; Zimmermann et al., 1999). As compared with adult acanthocephalans, the larval stages dwelling in their crustacean intermediate host reflected a lower potential for metal affinity (Brown and Pascoe, 1989; Sures and Taraschewski, 1995; Siddall and des Clers, 1994). For example, the adult stage of *Acanthocephalus lucii* showed 180 times more cadmium and 30 times more lead than the larval stages (Brown and Pascoe, 1989). Many investigations revealed that acanthocephalans also tended to bio-concentrate metal within them from the aquatic environment, and levels were remarkably high as compared with those in *Dreissena polymorpha* (Sures, Taraschewski and Rokicki, 1997; Jankovská et al., 2012a and b).

30.5 PARASITE RESPONSE TOWARDS POLLUTION

Various parasitic taxa have been recognized as helpful sentinels for chemical contamination. Studies on the affinity for metals in various parasites are shown in Tables 30.1, 30.2, 30.3, 30.4, and 30.5.

Parasites more frequently amass chemicals (metals) at much higher concentration levels as compared with their free-living counterparts (Sures, 2003; 2004; 2008a, b; Quiazon, 2015; Nachev and Sures, 2016). Using conventional analytical techniques, the bio-concentrated pollutants in parasites can be detected and quantified. Parasites are used to examine the behavior of chemicals in the environment and to determine to what degree they are available for uptake by the biota (Preti et al., 2020). The cestodes, with around 30 distinct species from various hosts and habitats (limnetic, marine, terrestrial), represent the largest group, trailed by nematodes, acanthocephalans, and digeneans, which have exhibited varied diversity due to variations in the ambient chemical environment. The highest concentrations of metals have been reported in acanthocephalans and cestodes (Tables 30.1, 30.2). The cadmium concentration present in the acanthocephalan parasite *Pomphorhynchus laevis* is 2,700 times higher than that present in the host muscle tissue (Sures et al., 1994a; Sures and Taraschewski, 1995). In the same way, cadmium concentration in cestodes is 1,175 times higher than in the host tissues (Table 30.2). PCBs occurred in cestodes at much higher concentrations than those of the hosts (Oluoch-Otiego et al., 2016). Elevated levels of different elements were also reported for nematodes, which, however, mainly accumulate essential elements rather than toxic ones (Table 30.3).

TABLE 30.1
Summarized Bibliography Exhibiting Affinity of Acanthocephala Infected Fish to Different Metals

Habitat	Parasite Taxa	Host	Host Tissue	Element	Reference
Terrestrial	*Moniliformis*	*Rattus rattus* "urban rat"	l, k, m	Cd, Pb, Cr	Torres et al., 2011; Teimoori et al., 2014
Limnetic	*Acanthocephalus anguillae*	*Perca fluviatilis* *Squalius cephalus*	l i	Ag, Cd, Cu, Fe, Mn, Pb, Zn	Filipović Marijić et al., 2014; Filipović Marijić et al., 2013
Limnetic	*Acanthocephalus lucii*	*Perca fluviatilis*	m, l, go, k, br	Pb, As, Cd, Cr, Cu, Hg, Mn, Ni, Pb, Zn	Jankovská et al., 2011; Brázová et al., 2012; 2015; Jankovská et al., 2012b
Limnetic	*Acanthogyrus* sp.	*Oreochromis niloticus*	m, i, l	Pb	Paller et al., 2016
Limnetic	*Pomphorhynchus laevis*	*Barbus barbus, Perca fluviatilis, Squalius cephalus*	m, i, l	As, Cd, Co, Cu, Fe, Mn, Mo, Ni, Pb, Sn, V, Zn, Se	Nachev et al., 2010, 2013; Filipović Marijić et al., 2014, 2015)

Source: Modified from Sures et al. 2017.

l = liver; m = muscle; i = intestine; go = gonad; br = brain; Cd = cadmium; Pb = lead; Cr = chromium; Ag = silver; Cu = copper; Fe = iron; Mn = manganese; Zn = zinc; As = arsenic; Hg = mercury; Ni = nickel; Co = cobalt; Mo = molybdenum; Sn = tin; V = vanadium; Se = selenium.

TABLE 30.2
Summarized Bibliography Exhibiting Affinity of Tapeworms That Parasitize Fish for a Variety of Metals

Habitat	Parasite Taxa	Host	Host Tissue	Element	Reference
Terrestrial	*Gallegoides arfaai*	*Apodemus sylvaticus*	m, l, k	Cd, Pb	Torres et al., 2004
Terrestrial	*Hymenolepis diminuta*	*Meriones libycus,* *Rattus norvegicus*	m, l, k i, bo	Cd, Cr Pb	Teimoori et al., 2014; Al-Quraishy et al., 2014; Čadková et al., 2014
Terrestrial	*Mesocestoides* spp	*Vulpes*	l, k	Cu, Cr, Mn, Ni, Pb, Zn	Jankovská et al., 2010
Terrestrial	*Moniezia expansa*	*Ovis aries*	m, l, k	Pb	Jankovská et al., 2010
Terrestrial	*Moniezia expansa*	*Ovis aries*	m, k	Cd	Jankovská et al., 2010
Terrestrial	*Mosgovoyia ctenoides*	*Oryctolagus cuniculus*	i, l, k	As, Cd, Pb, Hg	Eira et al., 2005
Terrestrial	*Paranoplocephala dentata*	*Clethrionomy s glareolus, Microtus agrestris*	l, k	Cd, Cr, Cu, Mn, Ni, Pb, Zn	Jankovská et al., 2009
Terrestrial	*Raillietina micracantha*	*Columba livia*	m, l, k	As, Cd, Cr, Cu, Hg, Mn, Pb, Se, Zn	Torres et al., 2010
Terrestrial	*Rodentolepis microstoma*	*Mus domesticus*	m, l, k	Cd, Pb	Torres et al., 2011
Terrestrial	*Skrjabinotaenia lobata*	*Apodemus sylvaticus*	m, l, k	Cd, Pb	Torres et al., 2006
Terrestrial	*Taenia taenaeiformis*	*"Urban rat"*	m, l, k	Cd, Cr	Teimoori et al., 2014

(*continued*)

TABLE 30.2 (Continued)
Summarized Bibliography Exhibiting Affinity of Tapeworms That Parasitize Fish for a Variety of Metals

Habitat	Parasite Taxa	Host	Host Tissue	Element	Reference
Terrestrial	*Tetrabothrius bassani*	*Morus bassanus*	m, l, k	As, Cd, Cr, Cu, Hg, Pb, Se	Mandes et al., 2013
Marine	*Anthobothrium* sp.	*Carcharhinus dussumieri*	m, i, l, go	Cd, Pb	Malek et al., 2007
Marine	*Clestobothrium crassiceps*	*Merluccius*	m, l, k	As, Hg, Se	Torres et al., 2015
Marine	*Gyrocotyle plana*	*Callorhinchus capensis*	m, i, l, k, go	Al, As, Cd, Co	Morris et al., 2007
Marine	*Lacistorhynchus dollfusi*	*Citharichthys sordidus*	m, i, l	As, Cd, Cr, Cu, Fe, Hg, K, Pb	Courtney-Hogue, 2016
Marine	*Paraorigmatobothrium* sp.	*Carcharhinus dussumieri*	m, i, l, go	Cd, Pb	Malek et al., 2007
Marine	*Polypocephalus* sp.	*Himantura cf. gerrardi*	m, i	Cd, Pb	Golestaninasab et al., 2014
Marine	*Rhinebothrium* sp.	*Himantura cf. gerrardi* *Glaucostegus granulatus*	m, i	Cd, Pb	Golestaninasab et al., 2014
Marine	*Tatragonocephalum* sp.	*Himantura cf. gerrardi*	m, i	Cd, Pb	Golestaninasab et al., 2014
Limnetic	*Bathybothrium rectangulum*	*Barbus barbus*	M	Cd, Cr, Ni, Pb	Baruš et al., 2012
Limnetic	*Bothriocephalus acheilognathi*	*Labeobarbus kimberleyensis*	m, l, sc	As, Cd, Co, Cr, Cu, Fe, Hg, Li, Mn, Mo, Ni, Pb	Retief et al., 2006
Limnetic	*Caryophyllaeus laticeps*	*Chondrostoma nasus*	m, i, l, gi	Cd, Cu, Pb, Zn	Jirsa et al., 2008
Limnetics	*Ligula intestinalis*	*Rastreneobola argentea, Tinca tinca* *Tinca Tinca* *Abramis brama, Blicca bjoerkna, Rutilus*	Whole fish m, l, go m, l, gi	Cd, Cr, Cu, Pb, Fe, Mn, Zn, Al, Ni, Sr	Oyoo-Okoth et al., 2010; Tekin-Özan et al., 2005; Tekin-Özan et al., 2008; Morris et al., 2016
Limnetic	*Proteocephalus macrocephalus*	*Anguilla anguilla*	m, l, k	As, Cd, Cr, Cu, Hg, Ni, Pb, Pd, Pt, Zn	Eira et al., 2009
Limnetic	*Proteocephalus percae*	*Perca fluviatilis*	m, l, k, br	As, Cd, Cr, Cu, Hg, Mn, Ni, Pb, Zn	Brázová et al., 2012; 2015
Limnetic	*Senga parva*	*Channa micropeltes*	m, i, l, k	Cd, Cu, Mn, Pb, Zn	Yen Nhi et al., 2013

Source: Modified from Sures et al. 2017.

l = liver; m = muscle; i = intestine; go = gonad; gi = gill; br = brain; Cd = cadmium; Pb = lead; Cr = chromium; Ag = silver; Cu = copper; Fe = iron; Mn = manganese; Zn = zinc; As = arsenic; Hg = mercury; Ni = nickel; Co = cobalt; Mo = molybdenum; Sn = tin; V = vanadium; Se = selenium; Al = aluminum; Pd = palladium; Pt = platinum; Sr = strontium; Co = cobalt; Li = lithium.

TABLE 30.3
Summarized Bibliography Exhibiting the Affinity of Nematoda That Parasitize Fish for a Variety of Metals

Habitat	Parasite Taxa	Host	Host Tissue	Element	Reference
Terrestrial	*Contracaecum* spp.	*Phalacrocorax auratus*	m	Hg	Robinson et al., 2010
Terrestrial	*Toxascaris leonine*	*Vulpes*	l, k	Cu, Cr, Mn, Ni, Pb, Zn	Jankovská et al., 2010
Terrestrial	*Brevimulticaecum tenuicolle*, *Dujardinascaris waltoni*, *Eustrongylides* sp., *Goezia* sp., *Ortleppascaris antipini*, *Terranova lanceolata*	*Alligator mississippiensis*	L	As, Cd, Cu, Fe, Pb, Se, Zn	Tellez and Merchant, 2015
Marine	*Anisakis* sp.	*Dicentrarchus labrax*	m, l, gi	Cd, Cu, Fe, Mn, Ni, Pb, Zn	Morsy et al., 2012
Marine	*Ascaris* sp.	*Liza vaigiensis*	m, i	As, Cd, Fe, Hg, Pb, Zn	Azmat et al., 2008
Marine	*Dichelyne minutus*	*Chasar bathybius*	i, l	Cu, Zn	Amini et al., 2013
Marine	*Echinocephalus* sp.	*Liza vaigiensis*	m, i	As, Cd, Fe, Hg, Pb, Zn	Azmat et al., 2008
Marine	*Hysterothylacium* sp.	*Trichiurus lepturus*	m, i, l, go	Cd, Pb	Khaleghzadeh-Ahangar et al., 2011
Marine	*Hysterothylacium aduncum*	*Pagellus erythrinus*, *Sparus aurata*, *Solea solea*	m, i, l, gi, k	Cd, Cr, Cu, Fe, Hg, Mg, Mn, Pb, Zn	Dural et al., 2010; 2011; Abdel-Ghaffar et al., 2014
Marine	*Hysterothylacium reliquens*	*Nemipterus peronii*	m, l, k	Al, As, Cd, Cr, Cu, Fe, Hg, Mn, Ni, Pb	Mazhar et al., 2014
Marine	*Paraphilometroides nemipteri*	*Nemipterus peronii*	m, l, k	Al, As, Cd, Cr, Cu, Fe, Hg, Mn, Ni, Pb, Se, Sr, Zn	Mazhar et al., 2014
Marine	*Proleptus obtusus*	*Rhinobatos annulatus*, *Rhinobatos blochii*	m, i, l, k, go	Al, As, Cd, Co, Cr, Cu, Mn, Ni, Pb	Morris et al., 2016
Limnetic	*Aguillicola crassus*	*Anguilla anguilla*	m, l, k	As, Cd, Cr, Cu, Hg, Ni, Pb, Pd, Pt, Zn	Eira et al., 2009; Genc et al., 2008
Limnetic		*Acestrorhynchus lacustris*	m, l	Al, As, Ba, Cd, Cu, Cr, Fe, Mg, Mn, Ni, Pb, Ti, Zn	Leite et al., 2017
Limnetic	*Eustrongylides* sp.	*Barbus barbus*	m, i, l	As, Cd, Co, Cu, Fe, Mn, Pb, Se, Sn, V, Zn	Nachev et al., 2013

(continued)

TABLE 30.3 (Continued)
Summarized Bibliography Exhibiting the Affinity of Nematoda That Parasitize Fish for a Variety of Metals

Habitat	Parasite Taxa	Host	Host Tissue	Element	Reference
Limnetic	*Philometra ovate*	*Gobio gobio*	m	Cd, Cr, Cu, Pb, Ni, Zn	Baruš et al., 2007
Limnetic	*Procamallanus* spp.	*Synodontis clarias*	i	Cd, Fe, Mn, Pb, Zn	Akinsanya and Kuton, 2016

Source: Modified from Sures et al. 2017.

l = liver; m = muscle; i = intestine; go = gonad; gi = gill; br = brain; Cd = cadmium; Pb = lead; Cr = chromium; Ag = silver; Cu = copper; Fe = iron; Mn = manganese; Zn = zinc; As = arsenic; Hg = mercury; Ni = nickel; Co = cobalt; Mo = molybdenum; Sn = tin; V = vanadium; Se = selenium; Al = aluminum; Pd = palladium; Pt = platinum; Sr = strontium; Co = cobalt; Li = lithium; Ti = titanium.

TABLE 30.4
Summarized Bibliography Exhibiting the Affinity of Digenea That Parasitize Fish for a Variety of Metals

Habitat	Parasite Taxa	Host	Host Tissue	Element	Reference
Terrestrial	*Drepanocephalus spathans*	*Phalacrocorax auritus*	m	Hg	Robinson et al., 2010
Terrestrial	*Fasciola gigantica*	*Buffaloes*	m, l	Cd, Cr, Cu, Pb, Zn	Lotfy et al., 2013
Terrestrial	*Fasciola hepatica*	*Buffaloes*	m, l	Cd, Cr, Cu, Pb, Zn	Lotfy et al., 2013
Terrestrial	*Acanthostomum pavidum, Archaeodiplostoma acetabulata, Protocaecum coronarium, Pseudocrocodilicola georgiana, P. americana, Timoniella loosi*	*Alligator mississippiensis*	l	As, Cd, Cu, Fe, Pb, Se, Zn	Tellez and Merchant, 2015
Marine	*Neoapocreadium chabaudi*	*Balistes capriscus*	m, l, k	Se, Hg	Torres et al., 2014
Marine	*Robphildollfusium fractum*	*Sarpa salpa*	m, l, k	Se, Hg	Torres et al., 2014
Limnetic	*Siphodera* spp.	*Chrysichthys nigrodigitatus*	I	Cd, Fe, Mn, Pb, Zn	Akinsanya and Kuton, 2016

Source: Modified from Sures et al. 2017.

l = liver; m = muscle; i = intestine; go = gonad; gi = gill; br = brain; Cd = cadmium; Pb = lead; Cr = chromium; Ag = silver; Cu = copper; Fe = iron; Mn = manganese; Zn = zinc; As = arsenic; Hg = mercury; Ni = nickel; Co = cobalt; Mo = molybdenum; Sn = tin; V = vanadium; Se = selenium; Al = aluminum; Pd = palladium; Pt = platinum; Sr = strontium; Co = cobalt; Li = lithium.

TABLE 30.5
Summarized Bibliography Exhibiting the Affinity of Monogenea That Parasitize Fish for a Variety of Metals

Habitat	Parasite Taxa	Host	Host Tissue	Element	Reference
Terrestrial	*Sabekia mississippiensis, Pentastomida*	*Alligator mississippiensis*	l	As, Cd, Cu, Fe, Pb, Se, Zn	Tellez and Merchant, 2015
Limnetic	*Ancyrocephalus mogurndae*	*Siniperca chuatsi*	m, l, k, gi	Pb	Qian and Pin, 2000

Source: Modified from Sures et al. 2017.

l = liver; m = muscle; i = intestine; go = gonad; gi = gill; br = brain; Cd = cadmium; Pb = lead; Cr = chromium; Ag = silver; Cu = copper; Fe = iron; Mn = manganese; Zn = zinc; As = arsenic; Hg = mercury; Ni = nickel; Co = cobalt; Mo = molybdenum; Sn = tin; V = vanadium; Se = selenium; Al = aluminum; Pd = palladium; Pt = platinum; Sr = strontium; Co = cobalt; Li = lithium.

30.6 CONCLUSIONS

The present review sets out the concept that parasites are organisms of environmental importance and their response towards destructive anthropogenic activities in the environment unravels the rate of environmental degradation. In addition, it is an attempt to reflect the effect of parasites on host physiology. The literature cited and the inferences from these citations would assist in spreading awareness amongst non-parasitologists about environmental health through the routine use of parasites as imperative tools of biodiversity. The study so far would strengthen not only the parasitologist but also the environmentalist regarding the new findings, if the established findings are utilized in the formulation of strategies helpful in implementing laws that encompass the elements in the aquatic environment and the respective forms representing biodiversity. The mysteries of environmental health could thus be cracked by an effective utilization of diverse species characteristics during definitive biological interactions.

REFERENCES

Abdel-Ghaffar F, Abdel-Gaber R, Bashtar A-R, Morsy K, Mehlhorn H, Al Quraishy S, et al. (2014). *Hysterothylacium aduncum* (Nematoda, Anisakidae) with a new host record from the common sole *Soleasolea* (Soleidae) and its role as a biological indicator of pollution. *Parasitol Res* 114:513–522.

Adewole SO, Odeyemi DF, Fatunwase OP, Christopher VN, Omoyeni TE and Dada A (2019). Parasites as bioindicator for health status and environmental quality of freshwater fish species in Ekiti State, Nigeria. *J Biomed Eng Med Imaging* 6(2):01–07.

Akinsanya B and Kuton MP (2016). Bioaccumulation of heavy metals and parasitic fauna in *Synodontis clarias* (Linnaeus, 1758) and *Chrysichthys nigrodigitatus* (Lacepede, 1803) from Lekki Lagoon, Lagos, Nigeria. *Asian Pacific J Trop Dis* 6:615–621.

Al-Hasawi ZM (2019). Environmental Parasitology: intestinal helminth parasites of the siganid fish *Siganus rivulatus* as bioindicators for trace metal pollution in the Red Sea. *Parasite (Paris, France)* 26:12.

Al-Quraishy S, Gewik MM, Abdel-Baki AA. (2014) The intestinal cestode *Hymenolepis diminuta* as a lead sink for its rat host in the industrial areas of Riyadh, Saudi Arabia. *Saudi J Biol Sci.* 21(4):387-90.

Amini Z, Pazooki J, Abtahi B and Shokri MR (2013). Bioaccumulation of Zn and Cu in *Chasar bathybius* (Gobiidae) tissue and its nematode parasite *Dichelyne minutus*, southeast of the Caspian Sea. *Iran Indian J Mar Sci* 42:196–200.

Azmat R, Fayyaz S, Kazi N, Mahmood SJ and Uddin F (2008). Natural bioremediation of heavy metals through nematode parasite of fish. *Biotechnology* 7:139–143.

Bagge AM and Valtonen ET (1996) Experimental study on the influence of paper and pulp mill effluent on the gill parasite communities of roach (*Rutilus rutilus*). *Parasitology* 112:499–508.

Baruš V, Jarkovský J and Prokeš M (2007). *Philometra ovata* (Nematoda: Philometroidea): a potential sentinel species of heavy metal accumulation. *Parasitol Res* 100:929–933.

Baruš V, Šimková A, Prokeš M, Peňáz M, Vetešník L. (2012) Heavy metals in two host-parasite systems: tapeworm vs. fish. *Acta Vet. Brno.* 81:313-7.

Brázová T, Hanzelová V, Miklisová D, Šalamún P and Vidal-Martínez VM (2015). Host-parasite relationships as determinants of heavy metal concentrations in perch (*Perca fluviatilis*) and its intestinal parasite infection. *Ecotox Environ Saf* 122:551–556.

Brázová T, Torres J, Eira C, Hanzelová V, Miklisová D and Šalamún P (2012). Perch and its parasites as heavy metal biomonitors in a freshwater environment: the case study of the Ružín water reservoir. *Slovak Sens* 12:3068–3081.

Brown AF and Pascoe D (1989). Parasitism and host sensitivity to cadmium: an acanthocephalan infection of the freshwater amphipod *Gammarus pulex*. *J Appl Ecol* 26:473–487.

Courtney-Hogue C. (2016) Heavy metal accumulation in *Lacistorhynchus dollfusi* (Trypanorhyncha: Lacistorhynchidae) infecting *Citharichthys sordidus* (Pleuronectiformes: Bothidae) from Santa Monica Bay, Southern California. *Parasitology*. 143(6):794-9.

Cunha KN, Domingues MV, Cunha LDDS and Nunes ZMP (2021). Parasitic monogenoideans of *Sciades herzbergii* as bioindicators of environmental quality in amazonian estuarines ecosystems. *Rev Bras de Parasitol Vet* 30(1):e024220.

Debraj Biswal D and Chatterjee S (2020). Fish parasites as biological indicators: a systematic review. *Biosc. Biotech. Res. Comm.* 13(4):1743–1755.

Dural M, Genc E, Sangun MK and Güner Ö (2011). Accumulation of some heavy metals in *Hysterothylacium aduncum* (Nematoda) and its host sea bream, *Sparus aurata* (Sparidae) from north-eastern Mediterranean Sea (Iskenderun Bay). *Environ Monit Assess* 174:147–155.

Dural M, Genc E, Yemenicioğlu S and Kemal Sangun M (2010). Accumulation of some heavy metals seasonally in *Hysterothylaciumaduncum* (Nematoda) and its host red sea Bream, *Pagellus erythrinus* (Sparidae) from Gulf of Iskenderun (north-eastern Mediterranean). *Bull Environ ContamToxicol* 84:125–131.

Dušek L, Gelnar M and Šebelová Š (1998). Biodiversity of parasites in a freshwater environment with respect to pollution: metazoan parasites of chub (*Leuciscus cephalus* L.) as a model for statistical evaluation. *Int J Parasitol* 28:1555–1571.

Eira C, Torres J, Miquel J, Vaqueiro J, Soares AMVM and Vingada J (2009). Trace element concentrations in *Proteocephalus macrocephalus* (Cestoda) and *Anguillicola crassus* (Nematoda) in comparison to their fish host, *Anguilla* in Ria de Aveiro. *Portugal Sci Total Environ* 407:991–998.

Eiras JC, Takemoto RM, Pavanelli GC and Adriano EA (2011). About the biodiversity of parasites of freshwater fish from Brazil. *Bull. Eur. Ass. Fish Pathol.* 31:161–168.

Filipović Marijić V, VardićSmrzlić I and Raspor B (2013). Effect of acanthocephalan infection on metal, total protein and metallothionein concentrations in European chub from a Sava River section with low metal contamination. *Sci Total Environ* 463–464:772–780.

Filipović Marijić V, VardićSmrzlić I and Raspor B (2014). Does fish reproduction and metabolic activity influence metal levels in fish intestinal parasites, acanthocephalans, during fish spawning and post-spawning period? *Chemosphere* 112:449–555.

Genc E, Sangun MK, Dural M, Can MF and Altunhan C (2008). Element concentrations in the swimbladder parasite *Anguillicola crassus* (Nematoda) and its host the European eel, *Anguilla anguilla* from Asi River (Hatay-Turkey). *Environ Monit Assess* 141:59–65.

Goater TM, Goater CP and Esch GW (2013). *Parasitism: The Diversity and Ecology of Animal Parasites*. Cambridge: Cambridge University Press. p. 497.

Golestaninasab M, Malek M, Roohi A, Karbassi AR, Amoozadeh E, Rashidinejad R, Ghayoumi R, Sures B. (2014) A survey on bioconcentration capacities of some marine parasitic and free-living organisms in the Gulf of Oman. *Ecological indicators*. 37:99-104.

Gómez A, Nichols E and Perkins S (2011). Parasite conservation, conservation medicine, and ecosystem health. In: Aguirre AA, Daszak P and Ostfeld RS (Eds.), *Conservation Medicine: Applied Cases of Ecological Health*. New York: Oxford University Press. pp. 67–81.

Gunkel G (1994). *Bioindikation in aquatischen Ökosystemen*. Fischer Verlag, Jena, Stuttgart. ISBN 3-334-60535-3. pp. 106–107.

Holt RD (2010). IJEE Soapbox. *Israel J. Ecol. Evol.* 56(3):239–250.

Jankovská I, Miholová D, Langrová I, Bejček V, Vadlejch J, Kolihová D, Šulc M. (2009) Influence of parasitism on the use of small terrestrial rodents in environmental pollution monitoring. *Environ Poll.* 157(8-9):2584-6.

Jankovská I, Kolihová D, Miholová D, Lukešová D, Romočuský S, Válek P, *et al.* (2012a). Effect of *Acanthocephalus lucii* infection on total mercury concentrations in muscle and gonads of fish host (*Perca fluviatilis*). *Bull Environ Contam Toxicol* 88:967–970.

Jankovská I, Miholová D, Bejček V, Vadlejch J, Šulc M, Száková J *et al.* (2010). Influence of parasitism on trace element contents in tissues of Red Fox (*Vulpes vulpes*) and its parasites *Mesocestoides spp.* (Cestoda) and *Toxascaris leonina* (Nematoda). *Arch Environ Contam Toxicol* 58:469–477.

Jankovská I, Miholová D, Lukešová D, Kalous L, Válek P, Romočuský Š, *et al.* (2012b). Concentrations of Zn, Mn, Cu and Cd in different tissues of perch (*Percafluviatilis*) and in perch intestinal parasite (*Acanthocephalus lucii*) from the stream near Prague (Czech Republic). *Environ Res* 112:83–85.

Jankovská I, Miholová D, Petrtýl M, Romočuský Š, Kalous L, Vadlejch J *et al.* (2011). Intestinal parasite *Acanthocephalus lucii* (Acanthocephala) from European perch (*Percafluviatilis*) as a bioindicator for lead pollution in the stream "Jevanskýpotok" near Prague, Czech Republic. *Bull Environ Contam Toxicol* 86:342–346.

Jirsa F, Leodolter-Dvorak M, Krachler R and Frank C (2008). Heavy metals in the nase, *Chondrostoma nasus* (L. 1758), and its intestinal parasite *Caryophyllaeus laticeps* (Pallas 1781) from Austrian rivers: Bioindicative aspects. *Arch Environ Contam Toxicol* 55:619–626.

Justine J-L (2010). Parasites of coral reef fish: how much do we know? With a bibliography of fish parasites in New Caledonia. *Belgian J Zool* 140(Suppl.):155–190.

Justine J-L, Beveridge I, Boxshall GA, Bray RA, Miller TL, Moravec F, Trilles JP and Whittington ID (2012a). An annotated list of fish parasites (Isopoda, Copepoda, Monogenea, Digenea, Cestodaand Nematoda) collected from Snappers and Bream (Lutianidae, Nemipteridae, Caesionidae) in New Caledonia confirms high parasite biodiversity on coral reef fish. *Aquat Biosyst.* 8(1):22.

Justine J-L, Beveridge I, Boxshall GA, Bray RA, Moravec F, Trilles JP and Whittington ID (2012b). An annotated list of fish parasites (Copepoda, Monogenea, Digenea, Cestoda and Nematoda) collected in groupers (Serranidae, Epinephelinae) in New Caledonia emphasizes parasite diversity in coral reef fish. *Folia Parasitol (Praha)* 57(4):237–262.

Justine J-L, Beveridge I, Boxshall GA, Bray RA, Moravec F and Whittington ID (2010). An annotated list of fish parasites (Copepoda, Monogenea, Digenea, Cestoda and Nematoda) collected from Emperors and Emperor Bream (Lethrinidae) in New Caledonia further highlights parasite biodiversity on coral reef fish. *Zootaxa* 2691:1–40.

Keke U, Mgbemena A, Arimoro FO and Omalu I (2020). Biomonitoring of effects and accumulations of heavy metals insults using some helminth parasites of fish as bio-Indicators in an Afrotropical stream. *Frontiers in Environmental Science* 8:Article 576080.

Kennedy CR (1997). Freshwater fish parasites and environmental quality: an overview and caution. *Parassitologia* 39:249–254.

Khaleghzadeh-Ahangar H, Malek M and McKenzie K (2011). The parasitic nematodes *Hysterothylacium sp.* type MB larvae as bioindicators of lead and cadmium: a comparative study of parasite and host tissues. *Parasitology* 138:1400–1405.

Kleinertz, S, Damriyasa IM, Wagen W, Theisen S and Palm HW (2014). An environmental assessment of the parasite fauna of the reef-associated grouper *Epinephelus areolatus* from Indonesian waters. *J Helminthol* 88(1):50–63.

Koskivaara M (1992). Environmental factors affecting monogeneans parasitic on freshwater fishes. *Parasitol Today* 8:339–342.

Lafferty KD (1997). Environmental parasitology: what can parasites tell us about human impacts on the environment? *Parasitol Today* 13:251–255.

Lafferty KD and Kuris AM (2005). Parasitism and environmental disturbances. *Parasitism and Ecosystems*, Oxford, online edition, Oxford Academic, 113–123. Regents of the University of California.

Leite LAR, Pedro NHO, Azevedo RK, Kinoshita A, Gennari RF, Watanabe S, Abdallah VD (2017). Contracaecum sp. parasitizing *Acestrorhynchus lacustris* as a bioindicator for metal pollution in the Batalha River, southeast Brazil. *Sci Total Environ* 575:836–840.

Lotfy WM, Ezz AM and Hassan AAM (2013). Bioaccumulation of some heavy metals in the liver flukes *Fasciola hepatica* and *F. gigantica*. *Iran J Parasitol* 8:552–558.

MacKenzie K, Williams HH, Williams B, McVicar AH and Siddall R (1995). Parasites as indicators of water quality and the potential use of helminth transmission in marine pollution studies. *Adv Parasitol* 35:85–144.

Malek M, Haseli M, Mobedi I, Ganjali MR and Mackenzie K (2007). Parasites as heavy metal bioindicators in the shark *Carcharhinus dussumieri* from the Persian Gulf. *Parasitology* 134(7):1053–1056.

Malhotra SK, Malhotra A, Ansari Z and Chatterji A (2002). Role of nematodes as bioindicators in marine and freshwater habitats. *Curr Sci* 82(5):505–507.

Mazhar R, Shazili NA and Harrison FS (2014). Comparative study of the metal accumulation in *Hysterothylacium reliquens* (nematode) and *Paraphilometroides nemipteri* (nematode) as compared with their doubly infected host, *Nemipterus peronii* (notched threadfin bream). *Parasitol Res* 113:3737–3743.

Mehana EE, Khafaga AF, Elblehi SS, Abd El-Hack ME, Naiel M, Bin-Jumah M, Othman SI and Allam AA (2020). Biomonitoring of heavy metal pollution using acanthocephalans parasite in ecosystem: an updated overview. *Animals: An Open Access Journal from MDPI* 10(5):811.

Mehdi N and Mahdi F (2015). Helminthic parasites as heavy metal bioindicators in aquatic ecosystems. *Med Lab J* 9:26–32.

Morales-Serna FN, Rodríguez-Santiago MA, Gelabert R and Flores-Morales LN (2019). Parasites of fish *Poecilia velifera* and their potential as bioindicators of wetland restoration progress. *Helgol Mar Res.* 73:1–8.

Morrill A, Provencher JF and Forbes MR (2014). Testing for dual impacts of contaminants and parasites on hosts: the importance of skew. *Environ Rev* 22(4):445–456.

Morris T, Avenant-Oldewage A, Lamberth S and Reed C (2016). Shark parasites as bioindicators of metals in two South African embayments. *Mar Pollut Bull* 104:221–228.

Morsy K, Bashtar A-R, Abdel-Ghaffar F, Mehlhorn H, Quraishy SA, El-Mahdi M, et al. (2012). First record of anisakid juveniles (Nematoda) in the European seabass *Dicentrarchus labrax* (family: Moronidae), and their role as bio-indicators of heavy metal pollution. *Parasitol Res* 110:1131–1138.

Nachev M and Sures B (2016). Environmental parasitology: parasites as accumulation bioindicators in the marine environment. *J Sea Res* 113:45–50.

Nachev M, Schertzinger G and Sures B (2013). Comparison of the metal accumulation capacity between the acanthocephalan *Pomphorhynchus laevis* and larval nematodes of the genus *Eustrongylides sp.* infecting barbel (*Barbus barbus*). *Parasit Vectors* 6:21.Najm M and Fakhar M (2015). Helminthic parasites as heavy metal bioindicators in aquatic ecosystems. *Med Lab J* 9:26–32.

Oluoch-Otiego J, Oyoo-Okoth E, Kiptoo KKG, Chemoiwa EJ, Ngugi CC, Simiyu G, et al. (2016). PCBs in fish and their cestode parasites in Lake Victoria. *Environ Monit Assess* 188:483.

Oyoo-Okoth E, Wim A, Osano O, Kraak MH, Ngure V, Makwali J, et al. (2010). Use of the fish endoparasite *Ligula intestinalis* (L., 1758) in an intermediate cyprinid host (*Rastreneobola argentea*) for biomonitoring heavy metal contamination in Lake Victoria, Kenya. *Lakes Reserv Res Manag* 15:63–73.

Paller VG, Resurreccion DJ, de la Cruz CP, Bandal MZ Jr. (2016) Acanthocephalan Parasites (Acanthogyrus sp.) of Nile Tilapia (Oreochromis niloticus) as Biosink of Lead (Pb) Contamination in a Philippine Freshwater Lake. Bull *Environ Contam Toxicol*. 96(6):810-5.

Palm HW (2011). Fish parasites as biological indicators in a changing world: can we monitor environmental impact and climate change?. In: Mehlhorn H (Eds.), *Progress in Parasitology. Parasitology Research Monographs*, vol 2. Springer, Berlin, Heidelberg. pp. 223–250.

Parmar TK, Rawtani D and Agrawal Y (2016). Bioindicators: The natural indicator of environmental pollution. *Front. Life Sci.* 9:110–118.

Perez-del Olmo A, Raga JA, Kostadinova A and Fernandez M (2007). Parasite communities in Boopsboops (L.) (Sparidae) after the Prestige oil-spill: detectable alterations. *Mar Poll Bull* 54:266–276.

Pietrock M and Marcogliese DJ (2003). Free-living endohelminth stages: at the mercy of environmental conditions. *Trends Parasitol* 19:293–299.

Poulin R and Morand S (2004). *Parasite Biodiversity*. Washington: Smithsonian Books. p. 216.

Preti A, MacKenzie K, Spivey KA, Noble LR, Jones CS, Appy RG, et al. (2020). Spiral valve parasites of blue and common thresher sharks as indicators of shark feeding behaviour and ecology. *J Fish Biol* 97:354–361.

Qian G and Pin N (2000). Lead content in the monogenean, *Ancyrocephalus mogurndae* and in different organs of its host, the mandarin fish, *Siniperca chuatsi*. *China Environ Sci* 20:233–236.

Quiazon KMA (2015). Updates on aquatic parasites in fisheries: implications to food safety, food security and environmental protection. *J Coast Zone Manag* 18:396.

Retief N-R, Avenant-Oldewage A and du Preez H (2006). The use of cestode parasites from the largemouth yellowfish, *Labeobarbus kimberleyensis* (Gilchrist and Thompson, 1913) in the Vaal Dam, South Africa as indicators of heavy metal bioaccumulation. *Phys Chem Earth* 31:840–847.

Riggs MR, Lemly AD and Esch GW (1987). The growth, biomass and fecundity of *Bothriocephalus acheilognathi* in a North Carolina cooling reservoir. *J Parasitol* 73:893–900.

Robinson SA, Forbes MR and Hebert CE (2010). Mercury in parasitic nematodes and trematodes and their double-crested cormorant hosts: bioaccumulation in the face of sequestration by nematodes. *Sci Total Environ* 408:5439–5444.

Santoro M, Iaccarino D and Bellisario B (2020). Host biological factors and geographic locality influence predictors of parasite communities in sympatric sparid fishes off the southern Italian coast. *Sci Rep* 10:13283.

Sciortino JA and Ravikumar R (1999). *Fishery Harbour Manual on the Prevention of Pollution*. Bay of Bengal Programme, Madras, India, p. 215. FAO/BOBP/MAG/22.

Siddall R and des Clers S (1994). Effect of sewage sludge on the miracidium and cercaria of *Zoogonoides viviparus* (Trematoda: Digenea). *Helminthologia* 31:143–153.

Sures B (2003). Accumulation of heavy metals by intestinal helminths in fish: an overview and perspective. *Parasitology* 126:S53–S60.

Sures B (2004). Environmental parasitology: relevancy of parasites in monitoring environmental pollution. *Trends Parasitol* 20:170–177.

Sures B (2008a). Environmental parasitology. interactions between parasites and pollutants in the aquatic environment. *Parasite* 15:434–438.

Sures B (2008b). Host-parasite interactions in polluted environments. *J Fish Biol* 73:2133–2142.

Sures B and Nachev M (2015). Environmental parasitology. In Mehlhorn H (ed.) *Encyclopedia of Parasitology*. Berlin, Heidelberg: Springer-Verlag. pp. 1–4.

Sures B and Siddall R (1999). *Pomphorhynchus laevis*: the intestinal acanthocephalan as a lead sink for its fish host, chub (*Leuciscus cephalus*). *Exp Parasitol* 93:66–72.

Sures B and Taraschewski H (1995). Cadmium concentrations in two adult acanthocephalans, *Pomphorhynchus laevis* and *Acanthocephalus lucii,* as compared with their fish hosts and cadmium and lead levels in larvae of *A. lucii* as compared with their crustacean host. *Parasitol Res* 81:494–497.

Sures B, Taraschewski H and Jackwerth E (1994a). Lead accumulation in *Pomphorhynchus laevis* and its host. *J Parasitol* 80:355–357.

Sures B, Taraschewski H and Jackwerth E (1994b). Lead content of *Paratenuisentis ambiguus* (Acanthocephala), *Anguillicola crassus* (Nematodes) and their host *Anguilla anguilla*. *Dis Aquat Org* 19:105–107.

Sures B, Taraschewski H and Rokicki J (1997a). Lead and cadmium content of two cestodes *Monobothrium wageneri*, and *Bothriocephalus scorpii*, and their fish hosts. *Parasitol Res* 83:618–623.

Sures B, Taraschewski H and Rydlo M (1997b). Intestinal fish parasites as heavy metal bioindicators: a comparison between *Acanthocephalus lucii* (Palaeacanthocephala) and the zebra mussel, *Dreissena polymorpha*. *Bull Environ Contam Toxicol* 59:14–21.

Sures B, Steiner W, Rydlo M and Taraschewski H (1999). Concentrations of 17 elements in the zebra mussel (*Dreissena polymorpha*), in different tissues of perch (*Perca fluviatilis*), and in perch intestinal parasites (*Acanthocephalus lucii*) from the subalpin lake Mondsee (Austria). *Environ Toxicol Chem* 18:2574–2579.

Sures B, Nachev M, Selbach C and Marcogliese DJ (2017). Parasite responses to pollution: what we know and where we go in "Environmental Parasitology." *Parasites Vectors* 10:65.

Teimoori S, Sabour Yaraghi A, Makki MS, Shahbazi F, Nazmara S, Rokni MB, et al. (2014). Heavy metal bioabsorption capacity of intestinal helminths in urban rats. *Iran J Public Health* 43:310–315.

Tekin-Özan S and Kır I (2005). Comparative study on the accumulation of heavy metals in different organs of tench (*Tincatinca* L. 1758) and plerocercoids of its endoparasite *Ligula intestinalis*. *Parasitol Res* 97:156–159.

Tekin-Özan S and Kır I (2008). Concentrations of some heavy metals in tench (*Tinca tinca* L., 1758), its endoparasite (*Ligula intestinalis* L., 1758), sediment and water in Beyşehir Lake, Turkey. *Pol J Environ Stud* 17:597–603.

Tellez M and Merchant M (2015). Biomonitoring heavy metal pollution using an aquatic apex predator, the American alligator, and its parasites. *PLoS One* 10:e0142522.

Tenora F, Krácmar S, Baruš V and Dvořáček J (1997). Some anorganic substances in plerocercoids of *Ligula intestinalis* (Pseudophyllidea). *Acta Univ Agric Silvic Mendelianae Brun* 45:23–30.

Thatcher VE (2006). *Amazon Fish Parasites*. 2th ed. Pensoft Publishers, Sofia.

Torres J, de Lapuente J, Eira C, Nadal J. (2004) Cadmium and lead concentrations in Gallegoides arfaai (Cestoda: Anoplocephalidae) and Apodemus sylvaticus (Rodentia: Muridae) from Spain. *Parasitol Res.* 94(6):468–70.

Torres J, Peig J, Eira C, Borrás M. (2006) Cadmium and lead concentrations in Skrjabinotaenia lobata (Cestoda: Catenotaeniidae) and in its host, Apodemus sylvaticus (Rodentia: Muridae) in the urban dumping site of Garraf (Spain). *Environ Pollut.* 143(1):4–8.

Torres J, Foronda P, Eira C, Miquel J, Feliu C. (2010) Trace element concentrations in Raillietina micracantha in comparison to its definitive host, the feral pigeon Columba livia in Santa Cruz de Tenerife (Canary Archipelago, Spain). Arch *Environ Contam Toxicol.* 58(1):176–82.

Torres J, Eira C, Miquel J, Foronda P and Feliu C (2011). Cadmium and lead concentrations in *Moniliformis moniliformis* (Acanthocephala) and *Rodentolepis microstoma* (Cestoda), and in their definitive hosts, *Rattus* and *Mus domesticus* in El Hierro (Canary Archipelago, Spain). *Acta Parasitol* 56:320–324.

Torres J, Kacem H, Eira C, Neifar L and Miquel J (2014). Total mercury and selenium concentrations in *Sarpa salpa* and *Balistes capriscus* and in their respective digenean endoparasites *Robphildollfusium fractum* and *Neoapocreadium chabaudi* from Tunisia. *Acta Parasitol* 59:580–585.

Torres J, Eira C, Miquel J, Ferrer-Maza D, Delgado E, Casadevall M. (2015) Effect of Intestinal Tapeworm *Clestobothrium crassiceps* on Concentrations of Toxic Elements and Selenium in European Hake *Merluccius merluccius* from the Gulf of Lion (Northwestern Mediterranean Sea). *J Agric Food Chem.* 63(42):9349-56.

Vidal-Martínez V and Wunderlich A (2017). Parasites as bioindicators of environmental degradation in Latin America: a meta-analysis. *J Helminthol* 91(2):165–173.

Vidal-Martínez VM, Aguirre-Macedo ML, Del Rio Rodríguez R, Gold-Bouchot G and Rendón-von Caballero-Pinzón PI (2003). Potential interactions between metazoan parasites of the mayan catfish *Ariopsis assimilis* and chemical pollution in Chetumal Bay, Mexico. *J Helminthol* 77:173–184.

Vidal-Martínez VM, Aguirre-Macedo ML, Del Rio-Rodríguez R, Gold-Bouchot G, Rendón-von Osten J, Miranda-Rosas GA (2006) The pink shrimp *Farfantepenaeus duorarum*, its symbionts and helminths as bioindicators of chemical pollution in Campeche Sound, Mexico. *J Helminthol*, 80(2):159–174.

Vidal-Martínez VM, Pech D, Sures B, Purucker T and Poulin R (2010). Can parasites really reveal environmental impact? *Trends Parasitol* 26:44–51.

Yen Nhi TT, MohdShazili NA and Shaharom-Harrison F (2013). Use of cestodes as indicator of heavy-metal pollution. *Exp Parasitol* 133:75–79.

Yoshinaga T, Tsutsumi N, Hall KA and Ogawa K (2009). Origin of the diclidophorid monogenean, *Neoheterobothrium hirame* Ogawa, 1999, the causative agent of anemia in olive flounder *Paralichthys olivaceus*. *Fish Sci* 75(5):1167–1176.

Zimmermann S, Sures B and Taraschewski H (1999). Experimental studies on lead accumulation in the eel specific endoparasites *Anguillicola crassus* (Nematoda) and *Paratenuisentisambiguus* (Acanthocephala) as compared with their host, *Anguilla*. *Arch Environ Contam Toxicol* 37:190–195.

31 Snow Cover Variability as Climate Change Indicator

A Case Study from Higher Himalaya in Garhwal, Uttarakhand

Anshuman Misra
Department of Geology
Kumaun University, Nainital

CONTENTS

DOI: 10.1201/9781003220824-35

31.1 INTRODUCTION

Himalaya has the largest accumulation of snow and ice outside the two poles (Haeberli et al., 1988; Owen et al., 2002). This vast accumulation is also known as the "water tower of Asia" (Immerzeel et al., 2010). It plays a vital role in regional climate and provides a stable, perennial water source to downstream regions (Thayyen and Gergan, 2010).

Snow, ice and glaciers are very important constituents of the hydrological cycle and play a vital role in the availability of water resources. In the Himalayan region, accumulation of snow cover is due to precipitation (snow events), while depletion takes place with rising temperature during the summer months (ablation period). Depending on the altitudinal location of the catchment, including geomorphic controls and weather conditions, snow cover becomes depleted either partially or totally. Both accumulation and depletion (ablation) are controlled by morpho-dynamic weather and changing climatic conditions/weather conditions. Interpretation of Moderate Resolution Imaging Spectroradiometer (MODIS) data of western Himalaya depicts maximum snow cover in February and March and minimum in August and September (Figure 31.1).

Seasonal snow cover (~30–40%) is one of the important natural resources in assessing the availability of water in Himalayan rivers, which largely contributes to the water resources during summer months. Very little field information on snow/glaciers is collected/available in this region (Singh and Jain, 2002). Therefore, remote sensing has offered a viable tool to obtain snow cover information in Himalayan basins.

Snow cover has a dynamic nature that is largely controlled by latitude and altitude. In Himalaya, the snow cover is dominated by altitude. Snow cover plays an important role in climate control and Earth's albedo (Nolin and Stroeve, 1997). It also affects the hydrological cycle, particularly in regions where water is supplied by melting snow and ice (Barnett et al., 2005).

These catchments are studied to map the extent of snow cover at different levels. Progression and regression of glaciers, accompanied by downward and upward movement as well as increase

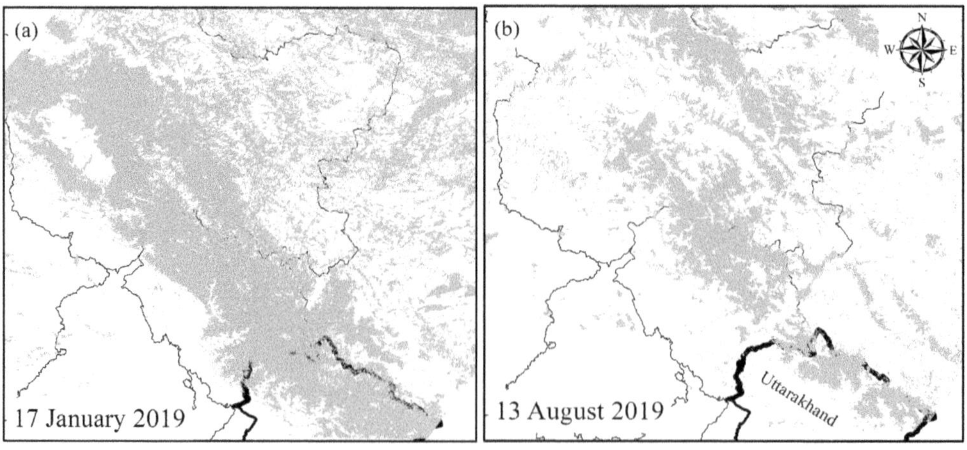

FIGURE 31.1 Snow cover map generated from MODIS data during 2 February showing maximum (a) and minimum on 12 August of same year 2016 (b).

and decrease in snow cover area, are very important proxies to establish climate change. The study highlights the accumulation of snow and its depletion pattern, its variability and topographical controls (elevation, aspect and slope), and their natural relationship. As the study progressed, auxiliary problems related to geomorphological controls, such as elevation, aspect and slope, on snow accumulation and depletion have further motivated and enhanced the research activity. Mapping of snow cover in shadow regions by ratioing multispectral band information has been quite successful. The results of the study are relevant to larger issues such as climate variability and change, management of disasters such as flash floods, avalanches, rockfall and landslides, and estimation and management of meltwater.

The availability of meteorological data (temperature, relative humidity, solar radiation) in digital format from automatic weather stations (AWSs) prompted their integration with remote sensing data sets. This exercise in integrating data sets was expected to bring out matches and mismatches between them and confirm the results of snow cover accumulation and depletion.

Mapping of snow cover and its seasonal and long-term variability has emerged as a very important and reliable parameter to assess climate change over vast regions. Horizontal as well as vertical movement of snow cover extent has been directly related to long-term rise and fall in temperature. Upward movement as well as reduction in snow cover extent indicates rising temperature, while downward movement and increase in extent suggests fall in temperature.

31.2 EARLIER RESEARCH ON THE SUBJECT

The first authentic account of the snow line came from a Swiss expedition to Everest. This expedition had two geologists, Dr Arnold Heim and Dr Augusto Gansser. Although they did not succeed in reaching the highest peak, their monumental work published as "Central Himalaya: Geological Observations of the Swiss Expedition, (1939)" (Heim and Gansser, 1975) has formed the framework of future studies in Himalayan geology.

According to the broad climatic classification described by Köppen-Geiger, Dokriani and Chorabari glacier catchments and neighbouring areas lie in the subtropical highland variety (Cwb) class. Subtropical regions between 30° and 40° latitudes and high-elevation mountainous areas of the Himalayan region are included in this class. Apart from latitude, the higher altitudes (3000–6600 m, indicate drier weather during the lower-sun "winter" season (Sarkar et al., 2016). Monthly average temperatures remain below 22°C in summer and up to −15°C during winter months. The average temperature in the catchments is 5°C and rainfall about 1300 mm (Dobhal et al., 2008; Kesarwani et al., 2015) during the summer.

The weather and climate in this region are controlled by Indian Summer Monsoon (ISM) and Western Disturbances (WDs), and their interaction results in local weather conditions. ISM arrives in this region by the end of June and remains active up to the end of September. The peak of monsoon rains is recorded during July and August. Most of the rainfall is related to low-elevation cloud circulations approaching from the SE. On the contrary, WD approaches from the NW, associated with high-level and low-temperature wind circulations. Most of the snowfall is due to these source disturbances. Some of these are of short duration (hours), while others are of longer duration, resulting in different amounts of snowfall. Smaller ones cause less snowfall, while larger and slow-moving ones cause very heavy snowfall. Extreme conditions of snowfall are recorded when slow-moving disturbances are interconnected in series. Lowering of temperature does not only take place in this region but also affects the entire subcontinent during the winter months. Furthermore, although the source disturbances are very frequent in the winter months, some of them do approach and result in snowfall both before and after the winter months, even as late as May. Occasionally, both ISM and WD systems obstruct each other's movement, resulting in devastating rains and floods (Kedarnath tragedy).

31.3 REMOTE SENSING AND SNOW COVER MAPPING

Many studies were conducted to map snow cover by remote sensing methods from the 1970s onwards. The Defence Meteorological Satellite Program (DMSP) during the mid- and late1970s shows that snow has a higher absorption at 1.49–1.92 μm, whereas at the same wavelength, clouds also have high reflectance. But in the visible spectrum, both have high reflection due to the minimum absorption (Valovcin, 1976). A detailed study by O'Brien and Munis (1975) showed that snow has a high reflectance in red and lowest reflectance at 1.5–2.0 m. Later, during the late 1980s, there was advancement in snow and cloud detection methods based on visible and infrared ratios to map snow cover on regional (Dozier, 1989; Rosenthal and Dozier, 1996) and global scales (Hall et al., 1995; 2002; Riggs et al., 2017). Satellite data was used by several workers in Himalayan regions, such as glacier mass balance studies (Kulkarni, 1992; Berthier et al., 2007), glacier inventory (Sangewar and Shukla, 2009), climate change detection (Vohra et al., 1981) and snow mapping with snowmelt runoff (Gupta et al., 1982; Kulkarni and Bahuguna, 2001; Singh and Jain, 2003).

Since the early 1980s and up till the present, the Geological Survey of India and Wadia Institute Himalayan Geology have carried out several cryospheric studies along the Himalayan length. The majority of them are field-based observations, providing point-based information on snow depth, snow density, depletion rate, water equivalent, and chemical and physical properties of snow. During this period, only a few studies using remote sensing were done by aerial and/or satellite data. Tangri (1999) carried out a study (1986–1989) on spectral response snow cover in Alaknanda and Dhauliganga river basins by Landsat and IRS IA LISS II data sets. It was observed that during the accumulation period, the snow covered area (SCA) ranges between 74.9% and 93%, whereas in the ablation period, the SCA comes down to 41–55% in the study area. An attempt was also made to classify snow/ice on the basis of spectral response and variable moisture content, and it was found that 43% of the total area is covered by dry snow. Space Application Centre (SAC, 2016) Ahmedabad conducted a detailed study on Himalayan glaciers, which involved multiple aspects of the cryosphere. In this study, the analysis of 6 main basins and 33 sub-basins was taken into consideration. Overall, there is 1–7% increase in snow cover in the Indus basin, whereas in the Ganga basin, there is a decline in snow cover. In sub-basins like Bhagirathi and Alaknanda, there was higher variability in the accumulation period, ranging from 20% to 80%, as compared with the ablation period. This is due to uneven precipitation and simultaneous melting. Throughout Himalaya, eastern and central parts showed 30–40% less snow cover in the accumulation period as compared with western Himalaya from 2004 to 2014.

31.4 BRIEF REVIEW OF RESEARCH ON METEOROLOGICAL DATA

Over 30% of the Earth's land surface is seasonally covered by snow, and 10% of it is permanently covered by glaciers (Dozier, 1989). Spatial and temporal variation in snow cover extent provides a useful indication of climate change (Simpson et al., 1998). Water in its frozen state accounts for more than 80% of the total freshwater on Earth and makes the largest contribution to rivers and groundwater. Hoinkes et al. (1967) estimated that snow accounts for about 5% of the total precipitation at the surface of the Earth. Snow and ice cover mapping and monitoring are important for management and planning of water resources, which also helps in delineating the major climatic trends and long-term forecasting (Kargel et al., 2005). Snow is an important input parameter in the models of the global energy balance and water cycle (Wang and Li, 2003). Foster et al. (1982) showed seasonal correspondences between snow cover and temperature in North America and Eurasia. A better relationship was found in Eurasia; if temperatures of autumn and winter are considered, however, when the average winter snow cover is correlated with average winter temperature, it was found that the relationship was better for North America than Eurasia. Rohrer (1989) found that the snow-rain threshold values are also governed by other factors, i.e., type of the meteorological station recordings (i.e., ventilated or non-ventilated) and seasonal and temporal resolution of air temperature values, such as hourly, daily or long-duration measurements. Later, Rohrer et al. (2013) suggested that the transition from snow to rain takes place at threshold temperature, between 0°C and 3°C.

31.5 STUDY AREA

The selection of study areas is based on criteria such as topography, accessibility, similar climatic conditions and most importantly, availability of remote sensing and meteorological data sets. In this region, Dokriani is considered as a benchmark glacier, belonging to Bhagirathi basin, and has been monitored since 1992. Many studies comprising meteorology, mass balance, morphology, paleo-climate, remote sensing and other aspects of glaciology have been described (Baral and Gupta, 1997; Singh et al., 2000; 2021; Dobhal et al., 2004; Gupta et al., 2005; Thayyen et al., 2005b; Dobhal and Mehta, 2010; Kesarwani et al., 2014; Pratap et al., 2015; Shukla et al., 2018). The other Chorabari glacier catchment is selected as it has also been monitored for quite some time (since 2009). Climatic, meteorological and geomorphological details have been described (Chaujar, 2009; Karakoti et al., 2012; Dobhal et al., 2013; Kesarwani et al., 2014; Misra et al., 2020). The major difference is orientation of the catchments. Dokriani glacier catchment is oriented in the east-west direction, whereas Chorabari glacier catchment has a north-south orientation. Furthermore, the selection of two catchments has provided scope for comparative study of snow cover accumulation and depletion. Six AWSs, with three in each catchment, provided the temperature data utilized in this study.

31.6 PHYSIOGRAPHICAL SETUP

Dokriani glacier catchment (DGC) is a part of the Din-Gad basin of Bhagirathi River and later joins the mighty Ganga river system. This basin is surrounded by snow-clad ranges with elevation ranging from ~1800 to 6600 m, oriented in the E-W direction, and has an area of ~77.8 km^2 (Figure 31.2). The catchment has an area of 15.2 km^2 and occupies an elevation range between 3799 (discharge site) and 6600 m. Dokriani glacier is a valley-type glacier, formed by coalescence of two cirque glaciers, one on the northern slope of Draupadi Ka Danda (5716 m) and the other on the western slope of Janoli peak (6632 m). The ablation zone of the glacier is covered by thick debris (~20% of total) and has highly elevated lateral moraines.

Chorabari glacier catchment (CGC) is occupied by a medium-sized, compound valley-type glacier and covers an area of 6.6 km^2 in Mandakini Basin. The river Mandakini originates from Chorabari glacier and joins Alaknanda, also a tributary of the Ganga river system (Figure 31.2). Chorabari glacier has an accumulation zone below Bhartkhunta (6578 m) and Kedarnath (6940 m) peaks and flows from north to south, between 6400 and 3895 m, with an average slope of 20°. This glacier has a small accumulation zone comprising three tributary glaciers. The ablation area is broad, has a gentle slope and is covered by thick debris. The debris-covered area accounts for 53% of total glacier area (Dobhal et al., 2013). Apart from Chorabari glacier, Mandakini river is also fed by Companion glacier from the eastern side, upstream of Kedarnath. This 4.7 km long glacier has ~7.6 km^2 area, its elevation range lies between 3810 and 4250 m, and the contribution of snow mass is mostly from two prominent avalanche sites.

The central points on DGC and CGC are located at latitude 30°51′40″, longitude 78°50′ and latitude 30°46′34″, longitude 79°03′15″, respectively. Glacial inventory by Geological Survey of India (GSI) assigned numbers IN 5O131 02 017 and IN 5O132 02 003 to DGC and CGC, respectively (Sangewar and Shukla, 2009).

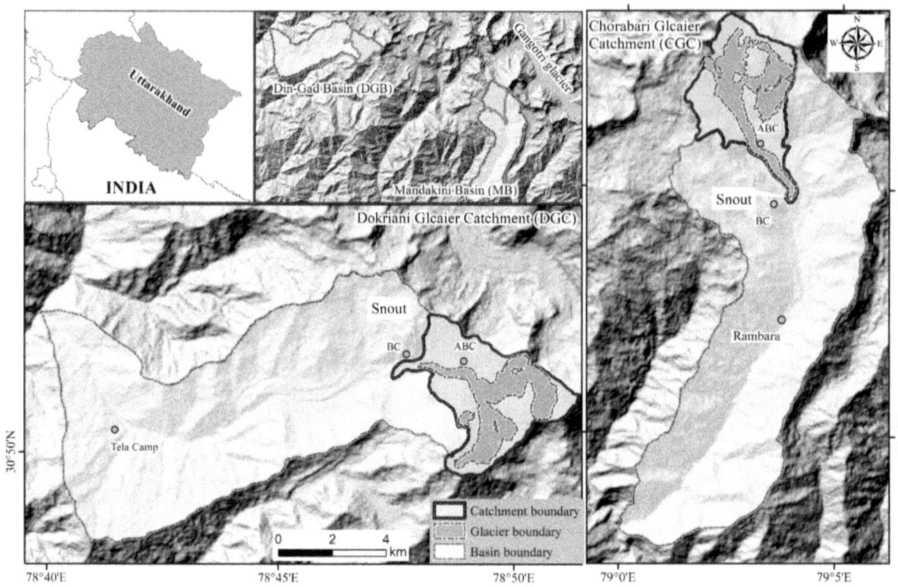

FIGURE 31.2 Location of the study area marked on satellite image (19 September 2016), depicting location of Dokriani glacier catchment along with Din-Gad (left) basin and Chorabari glacier catchments in Mandakini basin (right) showing network of AWSs, satellite images with marked glacier catchments, basin boundaries with location of AWSs and discharge site.

31.7 AUTOMATIC WEATHER STATIONS

A set of three AWSs were installed at different altitudes in both the basins. In Din-Gad basin, the first AWS is at Tela camp (TC), 2540 m, the second at base camp (BC), 3763 m, and the topmost at an elevation of 4364 m. In Mandakini basin, the AWSs were nearly equidistant from each other. Rambara (RB) was located at an elevation of 2760 m and was washed away during flash floods in June 2013 (Kedarnath tragedy). The second, at the snout, was also damaged. The third one is located at the Advance Base Camp (ABC), elevation of 4270 m on left lateral moraine (Figure 31.2). In Din-Gad basin comprising DGC, all the AWSs remained functional and are providing data. The consistency of the data has been fairly good from all the remaining AWSs during the study period.

31.8 MATERIAL AND METHODS

Suitable images that were available at no cost, of good quality and with no cloud cover in the area of interest were selected. Approximately 350 images were perused, and 123 have been used in this study. Taking these parameters into account, the satellite data that fulfilled these criteria were Landsat and Sentinel 2. Landsat satellite series (TM 5 and Operational Land Imager (OLI) 8) data sets for the periods from 2009 to 2011 and from 2013 to 2016 were downloaded from the United States Geological Survey website (USGS, http://earthexplorer.usgs.gov). This was highly dependent on the availability of cloud-free data during the study period. These sets have a spatial resolution of 30 m with orthorectified data type L1T.

Other satellite data sets from Sentinel MSI (https://remotepixel.ca) were also utilized to fill the gaps of the year 2016. This data has a spatial resolution of 10 m in blue, green, red and near-infrared (NIR) and 20 m resolution in short-wave infrared (SWIR) bands. These bands were resampled to match 30 m spatial resolution. Furthermore, the precise catchment boundary of the DGC and CGC catchments was demarcated from Resourcesat-2 images (Oct 2013), since it has a higher resolution of 5.8 m. Both of the catchments extend to the discharge site, so that the data can be directly used for hydrological modelling. Visual checks were conducted to select cloud-free images and to obtain a high level of accuracy.

31.9 PROCESSING OF REMOTE SENSING DATA

For obtaining reflectance and temperature, Landsat DN value data is converted into top of the atmosphere (TOA) reflectance, and at-satellite brightness temperatures are determined by the equation proposed by USGS (https://landsat.usgs.gov/using-usgs-landsat-8-product).

The equation for obtaining TOA reflectance is

$$\rho\lambda = \frac{(M_\rho Q_{cal} + A_\rho)}{\left(\sin\left(\theta_{SE}\right)\right)} \tag{31.1}$$

And, at-satellite brightness temperature (T) in Kelvin is

$$T = \frac{K_2}{\text{In}\left(\frac{K_1}{L_\lambda} + 1\right)} \tag{31.2}$$

where

$\rho\lambda'$ = TOA planetary reflectance (without correction for solar angle)
$M\rho$ = Band-specific multiplicative rescaling factor is from the metadata (reflectance mult band x, where x is the band number)
$A\rho$ = Band-specific additive rescaling factor from the metadata (reflectance is added to band number),
Qcal = Quantized and calibrated standard product pixel values (DN),
$\rho\lambda$ = TOA planetary reflectance
θ_{SE} = Local sun elevation angle
T = At-satellite brightness temperature (K)
$L\lambda$ = TOA spectral radiance (Watts/ (m^2 * srad * μm))
K1 = Band-specific thermal conversion constant from the metadata (K_1 constant band x, where x is the thermal band number),
K2 = Band-specific thermal conversion constant from the metadata (K_2 constant band x, where x is the thermal band number).

Sentinel 2 MSI data are already converted to reflectance values, are devoid of thermal bands and therefore, required no further processing. The spectral ranges and spatial resolution of Landsat sensors OLI and TM are listed along with the equivalent bands of the Sentinel-2A MSI in Table 31.1.

31.10 SNOW IDENTIFICATION AND MAP GENERATION

A common statistical method is normalization, where the difference between two real numbers that range between 0 and infinity is brought down to a notional scale of [0, 1]. This method (Normalized Difference Snow Index, NDSI) is well established worldwide and is the most common technique in the identification of snow, non-snow and cloud cover because of its unique pixel values (Hall et al., 1995; Hall and Riggs, 2007). In this study, two bands belonging to the visible and another NIR or SWIR range are taken. Visible snow has very high reflectance as compared with the SWIR and NIR regions. Therefore, to generate snow maps, the most effective and standard method of extracting snow from non-snow-covered areas and understanding the spatio-temporal pattern,

TABLE 31.1
Spectral Band Ranges of the Optical Sensors (Landsat OLI and Sentinel 2A) That Were Used for Glacier Mapping

	Band Number		
	Landsat series		Sentinel 2
Bands	OLI	TM	MSI
Blue	2	1	2
Green	3	2	3
Red	4	3	4
NIR	5	4	8
$SWIR_1$	6	6	11
$SWIR_2$	7	7	12
PAN	8		
Thermal	10		

NDSI, proposed by Dozier (1989) and modified by Hall et al. (1995), is applied to the reflectance image. The threshold value of 0.4 has proven to be most suitable for most of the images, but in a few images, the values come down to 0.37 also. This range is most suitable and corresponds very well with visual interpretation.

$$\text{NDSI} = \frac{(\text{Green reflectance} - \text{SWIR reflectance})}{(Green\ reflectance + SWIR\ reflectance)}$$

The other method of snow identification is the ratio method suggested by Rott and Markl, 1989; Bronge and Bronge, 1999, in which red/short-wave infrared (SWIR) is used. This method has proven its importance in the winter months, when a large area is engulfed in shadow, and NDSI identifies the entire shadow region as snow cover. Values ranging between 1.5 and 2, recommended by Paul et al. (2004), overcome this problem and give perceptible distinction.

To understand the relation between snow cover and topographical features (elevation, aspect and slope), thematic maps were generated using ASTER Global Digital Elevation Model (ASTER GDEM) Version 2, having 30 m resolution (http://gdem.ersdac.jspacesystems.or.jp/). To understand the SCA variability, DEM is classified into six elevation zones with 500 m interval: zone 1: >6000 m, zone 2: 6000–5500 m, zone 3: 5500–5000 m, zone 4: 5000–4500 m, zone 5: 4500–4000 m and zone 6: <4000 m for both the catchments. In both catchments, the 5000–5500 m elevation zone covers the larger part of the area.

The study area is divided into eight aspect classes: north, northeast, east, southeast, south, southwest, west and northwest. The major difference is that the DGC is E-W oriented (Figure 31.2), and the maximum area lies in the west aspect class (22.1%), followed successively by southwest (20.8%), northwest (19.1%), north (14.6%), south (8.8%), northeast (7.8%), east (4%) and southeast (2.8%). On the other hand, the CGC has N-S orientation, and the southwest aspect class covers the maximum area (21%), while south (19%) and southeast (12%) cover the major part of the catchment, and the north, northeast, east, northwest and west aspect classes cover only 2%, 5%, 8%, 1% and 7% of the area, respectively.

Both these catchments are also classified into seven slope categories (<10°, 10–20°, 20–30°, 30–40°, 40–50°, 50–60° and >60°), and the calculated area percentages in each category are 6.8%, 17.8%, 20.1%, 21.7%, 21.1%, 10.5% and 2.2% in DGC and 8.36%, 16.06%, 18.18%, 19.61%, 19.11%, 13.85% and 4.82% in CGC, respectively. Category 30–40° covers the maximum area, followed by a decrease on either side. Individually, DGC has steeper slopes than CGC.

All AWSs are of identical design and instrumental components and are mounted on a 10 m long metal tower. They are equipped with many sensors to record near-surface air temperature (°C), relative humidity (%), radiation flux (Wm^{-2}) and other parameters. A temperature probe (HMP45C212) mounted 2 m above the ground surface provides near-surface air temperatures, maximum (T_{max}), average (T_{avg}) and minimum (T_{min}), required for calculating lapse rate. The air temperature data for CGC from all the three stations is available from 2012 to June 2013. Data gaps of RB were filled by developing linear equations. These equations were also utilized to generate the data for RB even after the disaster from June 2013 to December 2014, and filled the gap between BC and ABC stations. Relationships between air temperatures of between stations were established statistically. To generate a linear correlation between air temperature (T_{max}, T_{avg} and T_{min}) and weather stations, the Pearson correlation coefficient (PCC) was used as a statistical tool.

31.11 RELATIONSHIP BETWEEN TEMPERATURE GRADIENT AND ELEVATION

Temperature plays a very important role in the melting of snow and ice, which varies from 0 to 5°C. It is also responsible for the phase transition from snow/glacier ice to water (Higuchi et al., 1982; Rohrer, 1989; Singh et al., 2008; MacDonald et al., 2012). The relationship between mean air temperature and elevation of the AWSs is analysed on annual, monthly and daily scales to establish the temperature lapse rate (TLR) in the catchment at 100 m resolution. Marshall et al. (2007) calculated the rate of temperature change with altitude at about 6.5°C km^{-1}; this is also referred to as the adiabatic or environmental lapse rate (ELR) or the free-air lapse rate. With the help of sensor specifications and physical barriers in a mountainous catchment, TLR is calculated by applying equation (31.3) as suggested by Blandford et al.)2008).

$$TLR\left(\frac{dT}{dZ}\right)=\frac{T_1-T_2}{Z_2-Z_1} \qquad (31.3)$$

Where, T_1 and T_2 are the air temperatures between two stations, while Z_1 and Z_2 are their elevations. Monthly lapse rate is obtained by averaging daily temperature and annual lapse rate from January to December. To bring out seasonal variability, four categories, namely, winter (December, January and February), pre-monsoon (March, April and May), monsoon (June, July, August and September) and post-monsoon (October and November) are made.

In the Himalayan region, studies on phase transition between snow and water were conducted by Charbonneau *et al.* (1981), Bocchiola *et al.*, 2010 and Senese *et al.*, 2014. They suggested that the threshold values are based on three conditions:

Melting of snow starts if T_{avg} ≥0°C or ≥2°C,
Melting of snow stops if T_{avg} ≤0°C or ≤2°C and
If the precipitation is a mixture of snow and rain, T_{avg} <2°C and >0°C.

Therefore, to estimate the area of melting within a basin, isotherms 0, 1 and 2°C correspond to different elevation zones.

Altitude-related temperature and isotherms (0°, 1° and 2°) are computed by using the catchment lapse rate from solid to liquid. The point temperature at different elevations is estimated by extrapolation/interpolation, elevation lapse rate and temperature from the base station (Dunn and Colohan, 1999). This method requires at least two meteorological stations in a catchment. In DGC and CGC, elevation-dependent temperature shows a linear relation; the elevation of the isotherm is determined by interpolation/extrapolation using equation (31.4).

$$\delta T\left(\frac{T_i}{T_b}\right)=\delta\left(H_b-H\right) \qquad (31.4)$$

Where δT is the calculated TLR (in °C/100 m), T_i represents isotherms 0, 1 and 2°C air temperature with respect to the index station (from where extrapolation/interpolation is carried out) T_b. H is elevation of the point where temperature is to be considered as 0, 1 and 2°C, H_b is elevation of the index station (m). The elevation represents the isotherms (0, 1, and 2°C) for each month. In the case of DGC, the index station is BC, near the glacial snout, while in CGC, the highest station ABC is considered as the index station. This difference is because of availability of consistent data in all the three stations in DGC from 2012 to 2016, whereas in CGC, the data is inconsistent due to loss and damage of two AWSs during the Kedarnath floods. Therefore, only ABC station is considered for reference, and it corresponds with isotherms 0, 1 and 2°C.

The altitude corresponding to the isotherms is also considered as the temporary or transient snowline for that particular period (month/season). The isotherms derived by equation (31.4) are co-registered on DEM (0, 1 and 2°C) in DGC of Din-Gad basin (2012–2016) and CGC of Mandakini basin (2012–2014) during the study period.

31.11.1 Temperature Gradient of Din-Gad basin (DGB) and Mandakini basin (MB)

To comprehend the melting behaviour of snow, isotherms on a time scale, in particular on an hourly, daily and monthly basis, are integrated with the snow cover extent derived from remote sensing data. Three elevation zones corresponding to 0, 1 and 2ºC isotherms are identified by using BC as index station in DGB and ABC in MB. The complete methodology is described in the flowchart (Figure 31.3).

The observations of TLR between TC–BC and BC–ABC are found not to be suitable for an overall elevation range in DGB because of different ground conditions between stations. Furthermore, the large spatial distance rendered them inapplicable. Therefore, for accuracy, estimated isotherms based on the average lapse rate between TC–BC and BC–ABC were utilized, and isotherms were calculated using T_{avg} on a monthly, daily and hourly scale.

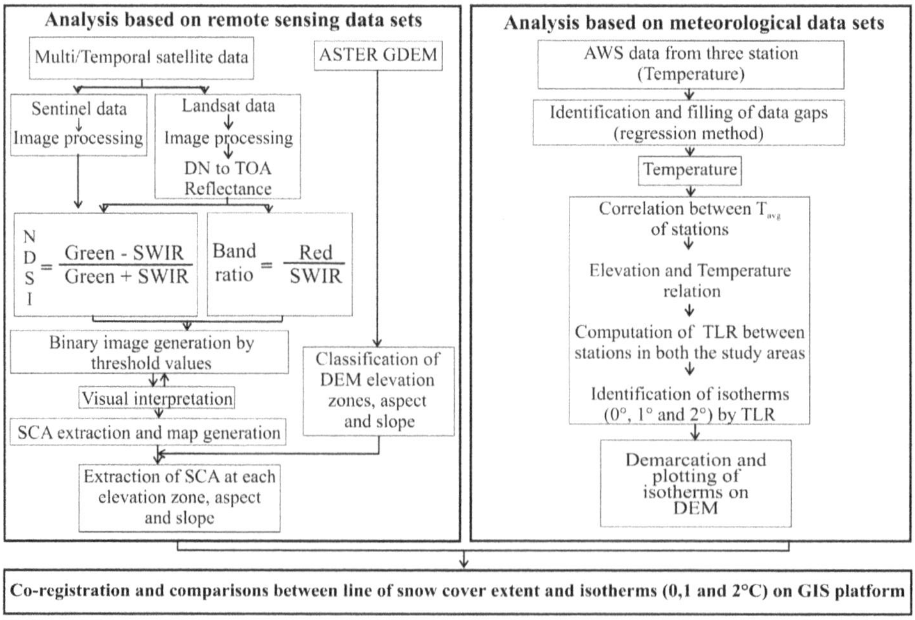

FIGURE 31.3 Complete flow chart showing remote sensing processes to acquire snow cover and meteorological processes to get isotherms and co-registration of both together.

31.12 RESULTS

Various criteria for identifying snow and non-snow-covered area were established. Images from different satellite series were taken into consideration for distinguishing snow-covered areas. The area is quite rugged and required special efforts to demarcate snow-covered areas during both ablation and accumulation periods. Different approaches such as single and multiple band combinations and ratios were attempted. This was possible due to the reflectance and absorption properties of various bands, which can identify subtle differences during the metamorphism of snow to ice. Both Landsat and Sentinel satellite data sets were found to complement each other in this exercise.

Fresh snow has high reflectance in the visible part of spectrum. Reduction in reflectance due to scattering is caused by the air–ice interface as well as time-related metamorphism along with increases in grain size. The albedo of fresh snow is nearly 90%, and this reduces to as low as 30% in the case of dirty or metamorphosed snow. Snowpack is also affected by the presence of liquid water with a consequent reduction in albedo. From a total of ~350 images, 123 were found suitable for the analysis, and 173 snow cover maps were prepared and described.

31.13 YEARLY SNOW COVER MAPPING, ASSESSMENT AND VARIABILITY FROM 2009 TO 2011 AND 2013 TO 2016

The cloud-free Landsat digital data of DGC from 15 March and CGC from 4 February 2009 was used in this study (Table 31.2). During February and March, the snow melting in both the catchments is observed on the southern and western aspects. With the progression of the pre-monsoon months (April and May), SCA gradually reduces and moves upward. Minimum SCA is observed in June; however, it increases in July due to snow events and further reduces by the end of monsoon season (September) (Table 31.2). In post-monsoon months, SCA increases due to snow events to attain a maximum in November.

Gradual reduction along the western and southern slopes during December is marked by lack of snow events. In January 2010, melting along the south and western aspects is observed and is followed by an increase in SCA (<95%) during February and April in DGC as well as CGC (Table 31.2). In DGC, SCA is reduced in June, and the lowest is observed on 24 July (49%). In the post-monsoon months, SCA increases (70% to 85%) and again reduces to 65% in December. In CGC, however, minimum SCA is observed in November and December (60%) along the southern aspect, whereas in monsoon months snow cover is 70% following frequent snow events.

From January to May 2011, 91% to 99% SCA in DGC and 85% to 95% in CGC is recorded because of dominant snow events. High SCA (<88%) continued even during April and May (Table 31.2). Throughout the monsoon months, extraction of SCA could not be done due to cloud cover. In DGC, minimum SCA (44%) is observed in the latter part of September, however, minimum (51%) in CGC is seen in mid-September. In CGC, 15% gain is recorded in the latter part of the October image; after that, no Landsat 5 TM data of this region is available.

In a major shift, the National Aeronautics and Space Administration (NASA) stopped gathering data from the Landsat 5 TM scanning system, which was decommissioned in 2013. Replacing this, Landsat 8 Operational Land Imager (OLI) and Thermal Infrared Sensor (TIRS) data were made available at no cost from April 2013. This system has the added advantages of having a high radiometric resolution, accuracy on a ground scale and high image acquisition rate (Wulder et al., 2019). The first images of the Landsat 8 OLI covering <90% SCA were available from 6 April 2013 for both the catchments. A gradual reduction in SCA is observed during pre-monsoon and monsoon months in DGC, and minimum snow cover is observed on 1 August (35%; Table 31.2). Late monsoon month (September) witnessed an increase in snow cover due to snow events. In post-monsoon and winter months, SCA shows an alternate decrease and increase due to melting and snow events. In CGC, minimum SCA is observed on 30 November, which continues till mid-December (>45%) because of no significant snow event.

TABLE 31.2
Details of Landsat and Sentinel (Cloud-Free) Images Used in This Study. Total SCA % on Different Dates of 2009–2016, Dokriani and Chorabari Glacier Catchments

No.	Date	Scene ID	Dokriani	Total SCA (%)	Chorabari	Total SCA (%)
1.	04-Feb-09	LT51450392009035KHC00			✓	83.4
2.	15-Mar-09	LT51460392009074KHC00	✓	85.9		
3.	24-Mar-09	LT51450392009083KHC00			✓	96.8
4.	18-May-09	LT51460392009138KHC00	✓	84.0	✓	75.1
5.	03-Jun-09	LT51460392009154KHC00	✓	80.9	✓	75.7
6.	19-Jun-09	LT51460392009170KHC00	✓	74.6	✓	70.7
7.	28-Jun-09	LT51450392009179KHC00			✓	59.5
8.	05-Jul-09	LT51460392009186KHC00	✓	79.5	✓	66.2
9.	07-Sep-09	LT51460392009250KHC00	✓	72.2		
10.	16-Sep-09	LT51450392009259KHC00			✓	77.5
11.	23-Sep-09	LT51460392009266KHC00	✓	69.4		
12.	09-Oct-09	LT51460392009282KHC00	✓	91.6	✓	74.8
13.	25-Oct-09	LT51460392009298KHC00	✓	75.7	✓	71.5
14.	03-Nov-09	LT51450392009307KHC00			✓	61.4
15.	10-Nov-09	LT51460392009314KHC00	✓	85.7		
16.	19-Nov-09	LT51450392009323KHC00			✓	87.9
17.	26-Nov-09	LT51460392009330KHC00	✓	78.2	✓	79.0
18.	05-Dec-09	LT51450392009339KHC00			✓	68.4
19.	22-Jan-10	LT51450392010022KHC00			✓	82.2
20.	29-Jan-10	LT51460392010029KHC00	✓	87.2		
21.	14-Feb-10	LT51460392010045KHC00	✓	93.9	✓	88.9
22.	03-Apr-10	LT51460392010093KHC00	✓	93.1	✓	97.7
23.	06-Jun-10	LT51460392010157KHC00	✓	75.5		
24.	22-Jun-10	LT51460392010173KHC00	✓	74.8	✓	71.0
25.	08-Jul-10	LT51460392010189KHC00	✓	70.9	✓	75.1
26.	24-Jul-10	LT51460392010205KHC00	✓	49.8		
27.	26-Sep-10	LT51460392010269KHC00	✓	78.2		
28.	12-Oct-10	LT51460392010285KHC00	✓	69.5	✓	72.8
29.	28-Oct-10	LT51460392010301KHC00	✓	85	✓	82.6
30.	13-Nov-10	LT51460392010317KHC00	✓	70.1	✓	63.4
31.	22-Nov-10	LT51450392010326KHC00			✓	73.1
32.	29-Nov-10	LT51460392010333KHC00	✓	68.7	✓	60.4
33.	08-Dec-10	LT51450392010342KHC00			✓	57.7
34.	15-Dec-10	LT51460392010349KHC00	✓	66.8		83.0
35.	16-Jan-11	LT51460392011016KHC00	✓	91.4	✓	89.7
36.	25-Jan-11	LT51450392011025KHC00			✓	88.1
37.	01-Feb-11	LT51460392011032KHC00	✓	94.4	✓	85.7
38.	17-Feb-11	LT51460392011048KHC00	✓	98.9		
39.	14-Mar-11	LT51450392011073KHC00			✓	94.3
40.	21-Mar-11	LT51460392011080KHC00			✓	96.3
41.	22-Apr-11	LT51450392011121KHC00	✓	98.8	✓	95.9
42.	08-May-11	LT51460392011128KHC00	✓	96.4		
43.	17-May-11	LT51450392011137KHC00			✓	81.9
44.	13-Sep-11	LT51460392011256KHC00	✓	62.5	✓	51.9
45.	29-Sep-11	LT51460392011272KHC01	✓	44.4		
46.	24-Oct-11	LT51450392011297KHC00			✓	66.4
47.	6-April-13	LO08 L1TP 146039 20130406 20170505 01 T1	✓	94.3	✓	93.2

TABLE 31.2 (Continued)
Details of Landsat and Sentinel (Cloud-Free) Images Used in This Study. Total SCA % on Different Dates of 2009–2016, Dokriani and Chorabari Glacier Catchments

No.	Date	Scene ID	Dokriani	Total SCA (%)	Chorabari	Total SCA (%)
48.	11-Apr-13	LC81460382013101LGN01	✓	97.5		
49.	13-May-13	LC81460392013133LGN01	✓	99.3		
50.	22-May-13	LC81450392013142LGN01			✓	80.4
51.	29-May-13	LC81460392013149LGN00	✓	80.6		
52.	30-June-13	LC81460392013181LGN00	✓	63.4		
53.	01-Aug-13	LC81460392013261LGN00	✓	34.8		
54.	18-Sep-13	LC81460392013261LGN00	✓	53.2	✓	51.2
55.	04-Oct-13	LC81460392013277LGN00	✓	54.7		
56.	20-Oct-13	LC81460392013293LGN00	✓	67.7		
57.	29-Oct-13	LC81450392013302LGN00			✓	42.0
58.	21-Nov-13	LC81460392013325LGN00	✓	71.3	✓	64.5
59.	30-Nov-13	LC81450392013334LGN00			✓	39.2
60.	07-Dec-13	LC81460392013341LGN00	✓	65.1	✓	41.8
61.	16-Dec-13	LC81450392013350LGN00			✓	46.7
62.	23-Dec-13	LC81460392013357LGN00	✓	70.2		
63.	01-Jan-14	LC81450392014001LGN00			✓	83.0
64.	24-Jan-14	LC81460392014024LGN00	✓	85.8	✓	87.1
65.	09-Feb-14	LC81460392014040LGN00	✓	93.5	✓	87.5
66.	18-Feb-14	LC81450392014049LGN00			✓	91.3
67.	25-Feb-14	LC81460392014056LGN00	✓	94.4	✓	95.4
68.	06-Mar-14	LC81450392014065LGN00			✓	97.3
69.	13-Mar-14	LC81460392014072LGN00	✓	96.0	✓	95.8
70.	29-Mar-14	LC81460392014088LGN00	✓	99.3		
71.	14-Apr-14	LC81460392014104LGN00			✓	96.8
72.	16-May-14	LC81460392014136LGN00	✓	98.1	✓	95.2
73.	20-Aug-14	LC81460392014232LGN00	✓	32.4		
74.	21-Sep-14	LC81460392014264LGN00	✓	50.5		
75.	16-Oct-14	LC81450392014289LGN00			✓	88.0
76.	23-Oct-14	LC81460392014296LGN00	✓	82.1	✓	67.0
77.	17-Nov-14	LC81450392014321LGN00			✓	56.9
78.	24-Nov-14	LC81460392014328LGN00	✓	69.5	✓	53.4
79.	03-Dec-14	LC81450392014337LGN00			✓	47.1
80.	10-Dec-14	LC81460392014344LGN00	✓	61.9	✓	41.1
81.	26-Dec-14	LC81460392014360LGN00	✓	93.8	✓	88.8
82.	11-Jan-15	LC81460392015011LGN00	✓	80.0	✓	77.3
83.	20-Jan-15	LC81450392015020LGN00			✓	84.1
84.	27-Jan-15	LC81460392015027LGN00	✓	94.8		
85.	05-Feb-15	LC81450392015036LGN00			✓	91.3
86.	12-Feb-15	LC81460392015043LGN00	✓	94.2	✓	91.6
87.	21-Feb-15	LC81450392015052LGN00			✓	95.3
88.	28-Feb-15	LC81460392015059LGN00	✓	99.3		
89.	10-Apr-15	LC81450392015100LGN00			✓	94.9
90.	17-Apr-15	LC81460392015107LGN00	✓	97.3		
91.	03-May-15	LC81460392015123LGN00	✓	95.1		
92.	19-May-15	LC81460392015139LGN00	✓	88.1	✓	83.4
93.	28-May-15	LC81450392015148LGN00			✓	79.7

(continued)

TABLE 31.2 (Continued)
Details of Landsat and Sentinel (Cloud-Free) Images Used in This Study. Total SCA % on Different Dates of 2009–2016, Dokriani and Chorabari Glacier Catchments

No.	Date	Scene ID	Dokriani	Total SCA (%)	Chorabari	Total SCA (%)
94.	08-Sep-15	LC81460392015251LGN00	✓	51.3	✓	51.2
95.	24-Sep-15	LC81460392015267LGN00	✓	64.6	✓	60.3
96.	03-Oct-15	LC81450392015276LGN00			✓	45.2
97.	10-Oct-15	LC81460392015283LGN00	✓	52.2		
98.	19-Oct-15	LC81450392015292LGN00			✓	51.5
99.	11-Nov-15	LC81460392015315LGN00	✓	93.3		
100.	20-Nov-15	LC81450392015324LGN00			✓	44.3
101.	06-Dec-15	LC81450392015340LGN00			✓	46.7
102.	13-Dec-15	LC81460392015347LGN00	✓	91.9	✓	90.0
103.	29-Dec-15	LC81460392015363LGN00	✓	72.0		
104.	23-Jan-16	LC81450392016023LGN00			✓	66.17
105.	30-Jan-16	LC81460392016030LGN00	✓	97.6		
106.	8-Feb-16	LC81450392016039LGN00			✓	95.89
107.	15-Feb-16	LC81460392016046LGN00	✓	91.7		
108.	18-Mar-16	LC81460392016078LGN00	✓	99.4	✓	98.02
109.	23-Mar-16	S2AMSIL1CTLMT20160323T090835T44 RKV	✓	98.6		
110.	03-Apr-16	LC81460392016094LGN00	✓	88.8	✓	85.64
111.	12-Apr-16	LC81450392016103LGN00			✓	93.88
112.	28-Apr-16	LC81450392016119LGN00			✓	87.13
113.	02-May-16	S2AMSIL1CTLSGS20160502T104003T4 4RKV	✓	86.6		
114.	24-Jul-16	LC81460392016206LGN00			✓	58.19
115.	19-Sep-16	S2AMSIL1CTLSGS20160919T104034T4 4RKV	✓	43.4	✓	46.97
116.	09-Oct-16	S2AMSIL1CTLSGS20161009T103950T4 4RKV	✓	53.5		
117.	12-Oct-16	LC81460392016286LGN00	✓	53.9		
118.	28-Oct-16	LC81460392016302LGN00	✓	47.8	✓	52.73
119.	06-Nov-16	LC08_L1TP_145039_20161106_20170318_ 01_T1			✓	51.22
120.	13-Nov-16	LC81460392016318LGN00	✓	50.2	✓	42.07
121.	22-Nov-16	LC81450392016327LGN00			✓	47.59
122.	29-Nov-16	LC81460392016334LGN00	✓	49.4		
123.	08-Dec-16	S2AMSIL1C20161208T052212N0204R062 T44RLV			✓	43.71

From January to May 2014, the snow cover is <86% in DGC. Late monsoon months (August and September) have minimum SCA; fortunately, it was possible to get a cloud-free image of August, where the lowest snow extent is observed (32%) during the entire study period, 2009–2016. Particularly in November and December, the melting of SCA (69% and 61%) is conspicuously seen along both southern and western slopes (-*92). In CGC, maximum SCA is observed on 6 March (97%), whereas the least is observed on 10 December (41%). Generally, melting is along southern slope even at higher elevations in winter months. However, shadow regions and northern slope in both the catchments continue to have snow cover even at lower elevations.

In DGC, the SCA is <80% (Table 31.2) from January to May in 2015 due to frequent snow events. Maximum is attained on 28 February (~ 99%), and minimum is observed in early September (51%). Fluctuations are noticed from late September to December due to melting and few snow events. A similar pattern is observed in CGC, where maximum SCA is identified on 21 February (~95%). During monsoon months, SCA oscillated between 50% and 60% and later between 40% and 50% in post-monsoon and early winter months. Minimum SCA (44%) is observed in November due to a lack of snow events. In mid-December, the catchment is replenished by fresh snow cover (90%).

Maximum SCA, <90% in DGC, remained from January to March 2016. This was followed by a gradual reduction in April and May (88% and 86%, respectively) (Table 31.2). The minimum SCA is observed on 19 September (43%) with a marginal increase in early October (50%). Slight fluctuation (~3%) continued until the end of November. Depletion is seen along the southern aspect (66%) in CGC during January. Maximum (98%) is observed by the end of March and minimum (42%) in mid-November. During monsoon months, it remained between 60 to 45% and fluctuated during post-monsoon and winter months.

The study has demonstrated that the melting pattern is largely controlled by the temperature. Rising temperatures cause continuous melting, and snow cover in shadowed regions escapes melting for a longer time. Accumulation of snow cover is mainly due to solid precipitation in the form of snow and is dependent on the strength and duration of snow events. However, melting progresses gradually and is controlled by rise in temperature. Furthermore, snow events also lower the temperatures and retard the melting process.

In both the catchments, the melting pattern shows similarities in summer and winter months. In summer months, the extent of SCA is mainly controlled by the rise in temperature. However, in winter months, melting is conspicuously seen in southern aspects (S, SE, SW). This is prominently observed in both catchments; however, it is more dominant in southerly-oriented CGC. In Dokriani, during monsoon months (June, July, August and September), the snow cover keeps decreasing and remains at only 36% and 34% in 2013 and 2014, respectively, in August. In September 2016, minimum snow cover (~31%) was observed in DGC. The post-monsoon (October and November) period witnesses a few snow events due to western disturbances that enhance the SCA. Even early winter months are marked by decrease in snow cover due to direct solar radiation and lack of snow events. However, a few patches of snow cover were observed in shadowed regions.

31.14 DOKRIANI GLACIER CATCHMENT (DGC)

31.14.1 Elevation

Elevation zones are derived from DEM generated by ASTER GDEM with snow cover input from Landsat and Sentinel data sets. SCA was plotted with each elevation zone on all available dates throughout the study period. In 2009, all the zones >5500 m are parallel to each other and show no apparent change. Minor changes are seen in the 5000–5500 m zone during pre- and post-monsoon months. Zone 4500–5000 m shows maximum variation. In zone 4000–4500 m, SCA starts to reduce in May, becomes devoid of snow in June and remains snow free till October. Zone <4000 m is the first to lose snow and remain without snow until the beginning of winter (Figure 31.4a).In 2010, zones >5500 m are parallel to each other, and no apparent change is seen. Reduction in SCA is observed within the 5000–5500 m zone and steep depletion in the 4500–5000 m zone in July, whereas the 4000–4500 m zone becomes snow free in June. In 2011, a slight reduction in snow is seen in mid-January within zone 5000–6000 m, which is replenished in February. In zone 4000–4500 m, depletion is seen in May, and the zone becomes snow free by September. Zone 4500–5000 has a maximum variation with SCA <5% in September. After a gap of one and a half years, Landsat 8 data was available for free downloading from April 2013. During this year, zones <4500 m became snow free by the end of July, whereas zones showing maximum variation (4500–5000 m) became snow free in August, and accumulation started from September. Reduced snow cover is seen in zone

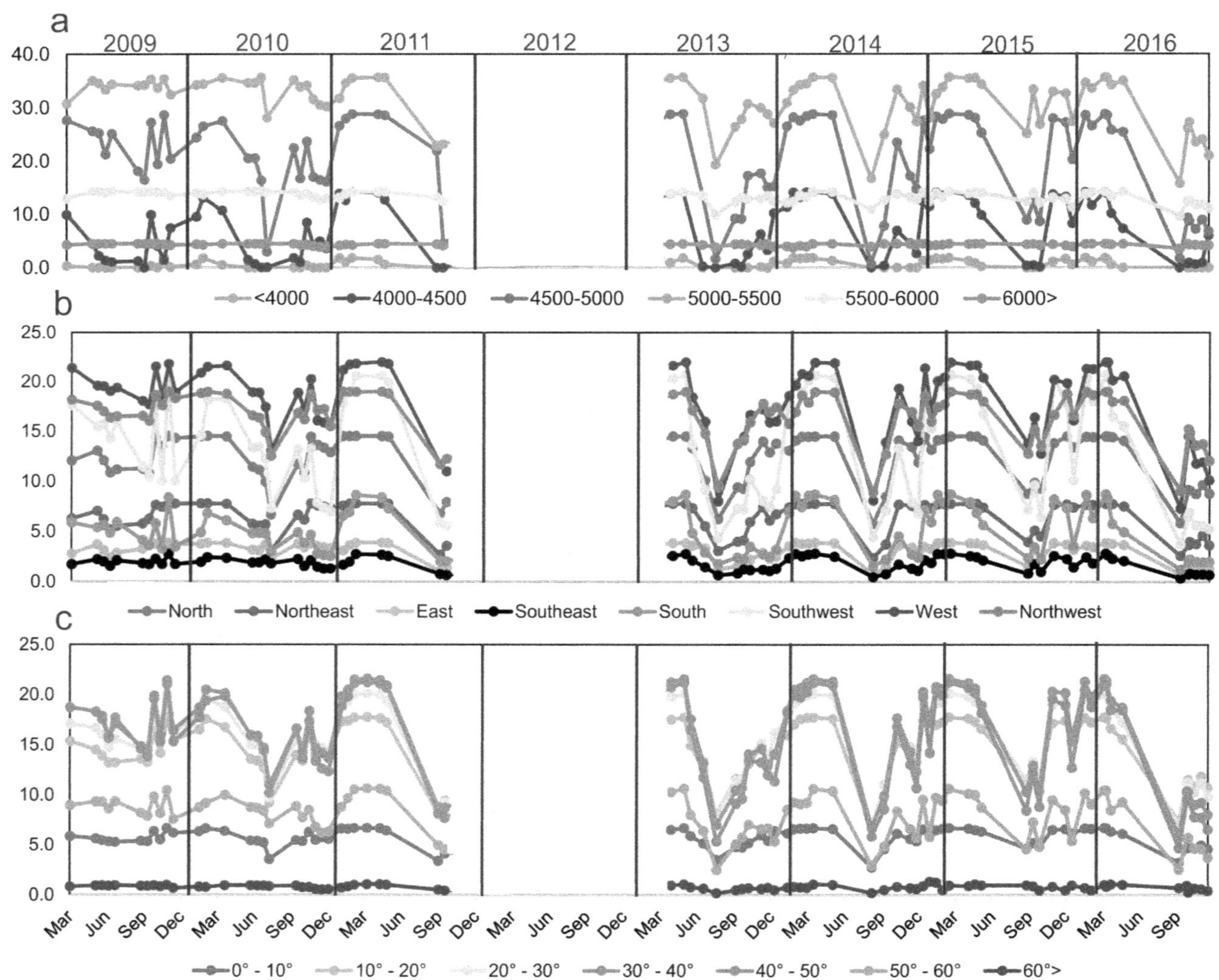

FIGURE 31.4 (a) Lines depicting % of snow-covered area (x-axis) at different elevation zones in the DGC during the study period (y-axis). (b) Lines depicting % of snow-covered area at different aspect classes in the DGC during the study period. (c) Lines depicting % of snow-covered area at different slope categories in the DGC during the study period.

5000–5500 m in August, whereas the next zone, 4500–5000 m, becomes devoid of snow. Continuing upwards, zone 5500–6000 m shows a slight dip in August or else runs parallel to the >6000 m zone. In 2014, all the zones run parallel from January to May. After that, zone 5000–5500 m shows reduction, and other lower zones become snow cover free. Maximum variation, however, continues in 4500–5000 m. October is marked by snow events, replenishing snow cover in all zones. A similar pattern is seen in 2015, where maximum variation is observed in zone 4500–5000 m, and minimum SCA is observed in September. Reduction in SCA is conspicuous in three zones, 4000–4500 m, 4500–5000 m and 5000–5500 m, in early January from the higher SCA of December 2014. In 2016, SCA starts to reduce early in April, and the minimum is observed in mid-September. Zones below 5500 m show reduction till mid-September. Above 4500 m, an increase in SCA is seen, whereas in zones lying below this, SCA is negligible till December. From the description of various zones from the study period, it can be concluded that maximum variability of SCA is seen in zone 4500–5000 m during the study period (2009–2016), while the lowest variability is seen above 5500 m. The results show that almost the entire area remains covered by snow during March due to preceding snow events. In May, melting starts below 4500 m, and the area becomes snow free by the end of June, coinciding with the peak period of the summer season. The elevation zone between 4500 and 5000 m witnesses gradual melting during May, which continues in June.

Within the study area, the elevation zone between 5000 and 5500 m covered the maximum area (36%) of the DGC throughout the year. This zone sustained 16.7% of the total snow-covered area even in peak monsoon season on 20 August 2014. Furthermore, it has been observed that post-monsoon depletion of snow reaches as high as ~6000 m, and, above 6000 m, snow cover is consistent, with variability of less than 0.2%, whereas within the 5500–6000 m zone, 1.6% variation in snow cover is observed during the entire study period.

31.14.2 Aspect

Other controlling factors of snow cover depletion are aspect and slope (Jain et al., 2009; López-Moreno et al., 2013; Shea et al., 2015). During this study, the catchment is divided into eight aspect classes, i.e., N, NE, E, SE, S, SW, W and NW, and the results are described on a yearly basis.

In 2009, snow depletion in DGC is first seen from May on S, SW and W aspects. Reduction in snow cover is observed on all aspects from June to September. During October, gain in snow cover begins due to snow events. This is followed by major fluctuations on S, SW and W aspects from mid-October to the end of November. In 2010, depletion is seen between late January and mid-February on S, SW and W aspects. Reduction is seen in June, and the lowest SCA is observed in July. Gain in snow cover starts in late September, and fluctuations are common on S, SW, W and NW aspects during October and November. The year 2011 is marked by maximum SCA from February to May and the lowest during mid-September. April 2013 showed maximum SCA; a prominent reduction started from May on S and SW aspects, and the minimum was in August. All aspects gained from September to November, with some variation on S and SW in December. Fluctuations are observed between January and March 2014, on S and SW aspects, maximum in April and minimum in August. From September, accumulation starts, with prominent fluctuations on S and SW aspects from October to December. In 2015, the trend is similar, with slight variations during January and November. The year 2016 witnesses fluctuations on S and SW aspects during February and March, whereas the least is seen in September, followed by a slight gain in October. Overall results of the entire period revealed that August and September show high depletion in annual SCA on southern (14–100%), southwestern (18–99%) and western (33–99%) aspects. These aspects also witness high fluctuation during post-monsoon and winter months. North and northwest aspects show SCA variability of 40–100% and 47–100%, respectively. A rapid depletion rate is observed on slopes facing southwest, south and southeast in summer and winter months (Figure 31.4b).

31.14.3 Slope

Apart from elevation and aspect, slope also plays a significant role in the melting of snow. During the present study, the catchment is divided into seven slope categories (<10°, 10–20°, 20–30°, 30–40°, 40–50°, 50–60° and >60°). Calculated area percentages are 6.8%, 17.8%, 20.1%, 21.7%, 21.1%, 10.5% and 2.2%, respectively. Maximum variability is seen in the zones between 30° and 50° from May to December because of solar radiation. From February to April, all the slope categories show parallelism. Zones from 10° to 30° and 50–60° show moderate variation, and 0–10° and >60° show minimum variation. Minimum variability is observed below 20° in the months of January and February, and maximum in 40–50°, followed by 30–40° and 50–60° (Figure 31.4c).

31.15 CHORABARI GLACIER CATCHMENT (CGC)

31.15.1 Elevation

Maximum SCA in total area is observed up to late March 2009; after that, depletion is seen in zone <4500 m. Zone 4500–5000 m shows depletion in mid-May and reduces to a minimum in July (59%). Depletion can also be seen even in zone 5000–5500 m. Zones above 5500 m are parallel during most of the ablation months but show depletion during the winter season, whereas zones below 5500 m show gain in SCA (Figure 31.5a). During 2010, the zone below 4500 m is devoid of SCA in June and July, whereas zones between 4500–5000 m and 5000–5500 m show slight reduction and later gain. Maximum fluctuation is seen in zone 4000–4500 m. Zones above 4000 m show conspicuous fluctuations in post-monsoon and winter months. Year 2011 shows a similar pattern to 2010 during the ablation period, while zones below 4500 m remain snow free and start gaining snow in October. In 2014, fluctuation is seen in all the zones in post-monsoon and winter months. In January and February, zones above 5500 m show reduction in SCA, after which all the zones run nearly parallel to each other till May. From June to mid-October, no cloud-free images are available to observe snow cover change. In November and December, melting is seen in all the zones, and more above zone 5500 m. During 2015, SCA depletion is observed above 4500 m in January, and February to April is marked by maximum SCA. Later, depletion is seen below 5500 m, and zones below 4500 m are snow free by September. Due to snow events in mid-September, all zones above 4500 m gain snow, while fluctuations are seen in November and December. In 2016, reduction is seen in January, while in February to April, gain is observed. Zones below 4000 m become snow free in April, followed by zone 4000–5000 m in July. No change is seen in zones above 5500 m till late September. Due to lack of snow events, lower SCA is observed in all zones in November and December. The results of the present study show that depletion starts in May and continues till the end of ablation season (September) 2015 (Figure 31.5a). Non-availability of cloud-free images during the months of June to September has restricted the estimation of SCA for the years 2011 and 2013–2016. Results indicate large inter-annual variability from a minimum of 53% in 2013 to a maximum of 83% in 2011.

31.15.2 Aspect

A different depletion pattern is seen in CGC in comparison to DGC. Reduction in SCA starts in February 2009 on S, SW and SE, reaches a maximum in March, and becomes gradual on S, SW, SE and E aspect classes. The lowest snow cover is seen in July on all aspects. Fluctuations from September to December are seen on S, SW and SE aspects, and minimum variation on N and W aspects is noticed. During 2010, maximum SCA is seen in April, whereas the lowest is observed during December, with many fluctuations in November and December. During 2011, maximum SCA is seen in March and April, minimum in September and a gain from October. Availability

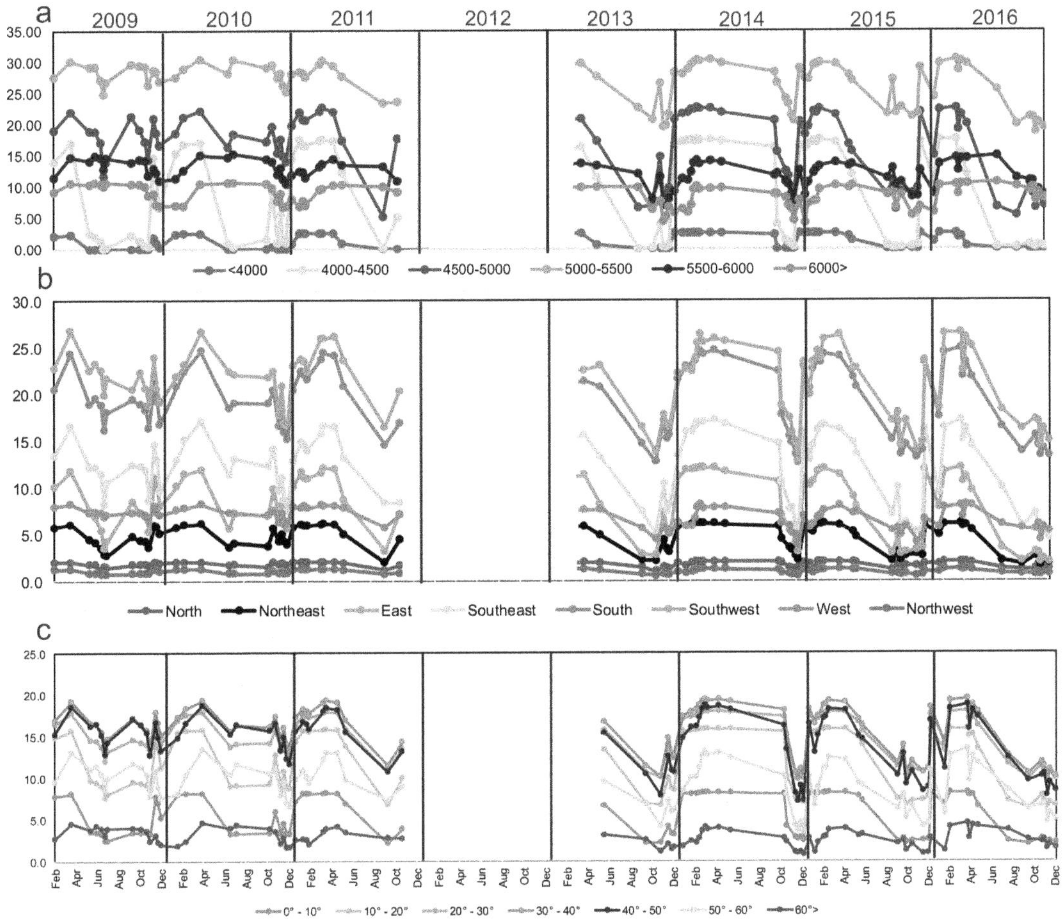

FIGURE 31.5 (a) Lines depicting % of snow-covered area (x-axis) at different elevation zones in the CGC during the study period (y-axis). (b) Lines depicting % of snow-covered area at different aspect classes in the DGC during the study period. (c) Lines depicting % of snow-covered area at different slope catagories in the DGC during the study period

of Landsat data started from April 2013. Lowest SCA is observed in October, with fluctuations in November. In 2014, maximum SCA is observed from March to May. No cloud-free data was available from June to September. Melting occurs on all aspects in October, whereas the lowest SCA is seen in early December. Snow events increased snow cover during mid-December. Nearly the same trend is observed in 2015, with depletion from September to December. In 2016, January shows reduction in SCA on all aspects, maximum from February to April, followed by gradual reduction till December (Figure 31.5b). All the aspects show nearly similar trends and parallelism with each other during the entire study period. Depletion during the monsoon months and fluctuations in winter are seen on all aspect classes.

31.15.3 Slope

The slope-wise distribution of snow cover shows that the zones 30°–40°, 20°–30°, 40°–50° and 10°–20° have an SCA of 16%, 15%, 14% and 12%, respectively, as a percentage of the total basin area, followed by zones 50°–60°, 0°–10°, 50°–60° and >60°, having an SCA of 9%, 5%, 3% and 0%, respectively. The annual SCA for each slope for each year indicates that the year 2013 had the lowest SCA. The coefficient of variation (CV) indicates that the zones >60° and 0°–10° have the highest variability, followed by zones 50°–60° and 10°–20°, while zone 20°–30° depicts the least variability (Figure 31.5c). The zones 50°–60° and >60° have high variability, as whatever snow accumulates on these slopes comes down to the valley floor directly. Similarly, the zones 0°–10° and 10°–20° have large variability due to the fact that these slopes are usually exposed to solar radiation and experience rapid melting.

31.16 RELATIONSHIP BETWEEN SNOW COVER AND METEOROLOGICAL PARAMETERS

This section presents the results obtained by the co-registration of snow cover and meteorological parameters such as temperature, relative humidity and solar radiation. These parameters are studied individually as well as in conjunction with the snow cover area. Very strong correlation exists between elevation and TLR in both the catchments (DGC and CGC).

31.17 SURFACE TEMPERATURE VARIABILITY IN DGC

Acquired air temperature data as minimum (T_{min}), average (T_{avg}) and maximum (T_{max}) from all the AWS is correlated (R^2 = 0.97–0.99) on a month-wise time scale. As described earlier, AWSs are located at different elevations in DGB, encompassing DGC. Consistent data from all the stations are analysed, and large intra-seasonal variability is found both in pre-monsoon (March, April, May) and post-monsoon (October, November) months, while insignificant changes are seen during the monsoon months (June, July, August, September). T_{avg} rapidly decreases between monsoon and post-monsoon months, indicating a strong inter-seasonal variability that marks the beginning of the winter season. The lowest variability with a minimum temperature difference (~ 0.5°C (TC), 0.2°C (BC) and 0.6°C (ABC)) is observed between post-monsoon and pre-monsoon months. A clear-cut thermal difference has emerged between monsoon and winter months. This has enabled the seasons to be thermally distinguished.

Based on the annual cycle of monthly average temperature, all the stations displayed large variation, sub-zero in most of the months at ABC, 0°C and above in all months (except T_{min} in January) at TC. The highest temperature at all sites is recorded in July (T_{avg}: 16.6°C (TC), 10.2°C (BC), 6.8°C (ABC)), followed by August. During July, T_{min} and T_{max} between TC and ABC were observed to be 4.5–14.1°C and 10.3–20.3°C, respectively. The lowest monthly T_{avg}, −8.9 to −1.1°C between TC and ABC, is seen in January, where the monthly T_{min} and T_{max} between TC and ABC varied from −1.1 to −12.2°C and from 7.9 to −3.8°C, respectively. January 2012 is recorded as the coldest month. During monsoon (June, July, August, September), the mean monthly temperature ranged between 12.9 and 20.6°C at TC, from 6.0 to 14.0°C at BC and from 3.0 to 9.2°C at ABC; in winter (December, January, February), from 0.1 to 9.3°C at TC, from −8.1 to 3.2°C at BC, and from −11.0 to −2.2°C at ABC (Singh et al., 2021).

The relationship between the annual cycle (4.5 year average) of temperature (maximum, average and minimum) and 0°C during difference months is highlighted in the DGC.

31.18 VARIABILITY IN TLR OF DGC

TLR has emerged as an important parameter to determine the phase of precipitation (liquid or solid) at a particular altitude. In addition to this, the snowmelt and runoff models also require temperature input. Time series data of air temperatures (T_{avg}) from all AWSs is utilized to find the

TABLE 31.3
Annual Cycle of Month-Wise Derived Temperature Lapse Rate During Study Period (2011–2016) in DGB

	TLR (°C km^{-1})	
	TC–BC	BC–ABC
Months	Average	Average
Jan	6	6.5
Feb	6.1	7.1
Mar	6.4	7.2
Apr	6.4	7
May	6.8	7.5
Jun	6.3	7.3
Jul	5.2	5.9
Aug	5.3	5.7
Sep	5.6	6.3
Oct	6.3	6.4
Nov	5.5	6.3
Dec	5.4	5.9
Mean	**6.6**	**6.6**

relationship with elevations between TC and ABC. Average TLR in lower altitudes (i.e., TC–BC) is lower (~5.8°C km^{-1}) in comparison with higher areas (i.e., BC–ABC) (~6.7°C km^{-1}) (Table 31.3). Reasonable deviation from standard environmental lapse rate (SELR) at a different elevation (6.5°C km^{-1}) is not promising for extrapolation. However, a dense network of AWSs in mountainous region is required for better data (Shea et al., 2015; Yadav et al., 2019).

The annual cyclicity of average TLR is shown in Table 31.3. July and August experienced the lowest TLR (~5.2°C km^{-1} (TC–BC), 5.8°C km^{-1} (BC–ABC)). This represents intense melting over Dokriani glacier under warm and humid conditions. TLR was steepest during the pre-monsoon and gentle during the monsoon (corresponding with the warm, rainy and humid season). On a seasonal basis, TLR was found in the range of 6.4–5.2°C km^{-1}. Thayyen and Gergan (2010) and Thayyen and Dimri (2014) concluded that the monsoon lowers the TLR by reducing the temperature range along the elevations due to heavy rainfall and cloud cover. In pre-monsoon season, TLR varied from 7.5 to 6.3°C km^{-1} in the DGB (Table 31.3). This phenomenon could have been due to incoming short-wave solar radiation being enhanced as compared with the outgoing long-wave radiation (Blandford et al., 2008). This establishes a different radiative heating effect along higher elevations.

Post-monsoon and winter months show almost similar and stable inter-seasonal variability in TLR (Table 31.3) that ranges from 4.6 to 6.4°C km^{-1} for TC–BC and from 5.5 to 7.1°C km^{-1} for BC–ABC. However, higher values of standard deviation (SD > 1.0) display a strong intra-seasonal variability during October and November.

31.19 SURFACE TEMPERATURE VARIABILITY IN CGC

The mean monthly values of air temperature are computed also in the Mandakini basin, which encompasses CGC, from the combined time series for the years 2012–2014. The mean air temperatures at RC, BC and ABC demonstrate that July has a maximum (11°C) while January experiences a minimum (–4.6°C) temperature; a similar trend is recorded in DGB. Mean air temperature in all AWSs follows the same annual cycle with an average of 3.8°C. Mean maximum and mean minimum temperature ranges between 9°C and 0.8°C. Monthly statistics (minimum, mean and maximum) are presented for each month by box and whisker plots. From the box plots, it is obvious that all the three stations experience relatively low variability during the ISM (June–September), whereas high variability is seen during winter (November–April). Box and whisker plots show monthly distribution and variability in air temperatures (T_{avg}) during the study period 2012–2014 at three AWSs (RC, BC and ABC.

31.20 VARIABILITY IN TLR OF CGC

Completed daily time series for each station were further utilized to compute the monthly, seasonal and annual TLR for T_{min}, T_{avg} and T_{max} (Table 31.4). The results show that the annual mean of temperature (T_{min}, T_{avg} and T_{max}) decreases linearly with increasing elevation. The overall trend for the vertical temperature gradient for the whole basin is linear (Table 31.4). To understand the actual temperature variability, results assessed with this method are compared with the standard vertical mean temperature gradient or TLR. The annual mean lapse rates for T_{min}, T_{avg} and T_{max} were 6.2, 6.1 and 5.2°C km^{-1}, respectively (Table 31.4). The annual mean TLRs are very close to each other and apparently lower than the generally used value of 6.5°C km^{-1} in mountain ranges (Wang et al., 2017). The TLR for each month varied to a different extent throughout the year. It is observed that mean monthly TLR for three stations varies from 5.1 to 7.1°C km^{-1}, 5.2 to 6.7°C km^{-1} and 4.0 to 6.5°C km^{-1} for T_{min}, T_{avg} and T_{max}, respectively. The seasonal cycle for the study period (2012–2014) shows that the TLR for T_{max} is lower than the TLR for T_{min}. The mean values of TLR for the snowmelt season (June–September) were computed at 5.3, 5.5 and 4.4°C km^{-1} for T_{min}, T_{avg} and T_{max}, respectively. This may be because of large variations in the range of minimum temperature and less variation in the range of maximum temperature (Misra et al., 2020).

TABLE 31.4
Annual Cycle of Month-Wise Derived Temperature Lapse Rate During Study Period (2012–2014) in CGC

Months	Mean T_{max} (°C)			Mean T_{min} (°C)			Mean T_{avg} (°C)		
(2012–2014)	RC	BC	ABC	RC	BC	ABC	RC	BC	ABC
January		1.7			–7.9			–4.6	
February		3.0			–7.4			–3.9	
March		7.5			–3.6			0.1	
April		9.8			–0.8			2.7	
May		12.7			2.9			6.4	
June		13.9			6.7			9.5	
July		14.0			8.5			10.7	
August		13.5			8.1			10.3	
September		12.6			6.3			8.7	
October		9.4			2.8			5.4	
November		5.9			–1.7			1.2	
December		4.0			–4.2			–1.0	
Mean		9.0			0.8			3.8	

31.21 SNOW COVER EXTENT AND CORRELATION WITH ISOTHERMS

The nature of precipitation (i.e., snow and rain) is generally a function of temperature. The transition between snow and rain depends on a temperature threshold between 0 and 5°C (Higuchi et al., 1982; MacDonald et al., 2012; Martinec and Rango, 1986). In the Himalaya, studies suggested that the phase change from snow to glacier melt may be between 0 and 2°C. Based on these threshold values, three conditions were visualized. First, melting of snow starts if T_{avg} ≥0°C or ≥2°C; second, melting of snow stops if T_{avg} ≤0°C or ≤2°C; and thirdly, intermixing of rain and snow occurs if T_{avg} <2°C and >0°C (Charbonneau et al., 1981; Bocchiola et al., 2010; Senese et al., 2014).

31.22 CORRELATION OF SNOW COVER EXTENT WITH ISOTHERMS IN THE DGC

Altitude isotherms (ZT = i, i = 0, 1 and 2°C) were calculated at hourly and monthly time scales. The seasonal variations in isotherms are given in Table 31.5. The monsoon season is characterized by higher values of isotherms than the winter season. Monthly isotherms show almost a sinusoidal pattern during the entire study period, with the lowest in 2012 (~2000 m) and highest in 2015 (~5600 m). The highest value, ranging from 5000 to 5500 m, was reached in the month of July, followed by August (2012–2016) Table 31.5. The area of the DGB above these isotherms may be considered as snow-covered throughout the year. During the end of the ablation months (i.e., August and September), the 0°C isotherm descends ~5100–5200 m, suggesting a close proximity to the equilibrium-line altitude (ELA) of the Dokriani glacier (~ 5056 m). Field-based studies by Dobhal et al. (2008) also reported nearly similar results during the period from 1992

TABLE 31.5
Calculated Seasonal Isotherm (0, 1, and 2°C) at Different Altitudes Obtained by Daily Near-Surface Temperature Lapse Rate

Years	Winter (DJF)	Pre-monsoon (MAM)	Monsoon (JJAS)	Post-monsoon (ON)
		Elevation of 0°C isotherm (m asl)		
2012	3037	4015	5234	3824
2013	3264	4181	5271	4183
2014	3182	3993	5257	4101
2015	3262	4067	5237	4251
2016	3360	4237	5401	4352
		Elevation of 1°C isotherm (m asl)		
2012	2877	3879	5074	3679
2013	3117	4046	5111	4034
2014	3029	3855	5094	3932
2015	3086	3915	5066	4076
2016	3183	4089	5222	4182
		Elevation of 2°C isotherm (m asl)		
2012	2716	3743	4915	3534
2013	2969	3911	4952	3885
2014	2877	3717	4932	3762
2015	2911	3763	4896	3901
2016	3005	3940	5043	4011

to 2000. Meteorologically, fluctuations of isotherms are attributed to shallow TLR, high relative humidity and low incoming solar radiation. This study also finds that the 0°C isotherm matches the snow cover extent better than other isotherms. In April, the area experiences both ablation as well as accumulation. However, in May, accumulation by snow events diminishes, and ablation dominates till September. In monsoon months, liquid precipitation accelerates melting and further reduces the SCA. The snow cover extent, demarcated by visual analysis of remotely sensed data, lies in the isotherm range (ZT = i) during the ablation months (May to September; Figure 31.6). However, survival of snow patches in certain areas is due to local topography and solar radiation.

The present study reveals that the average monthly isotherm range (ZT = i) does not coincide with snow cover extent during winter months (Figure 31.7). Hourly isotherms (11:00AM) on the same day give better coherence with snow cover extent at higher altitudes. Meteorological observations such as steeper TLRs and strong intra-seasonal variability during winter months also confirm the variance between isotherms and snow cover extent. This suggests that isotherms are not the controlling factor in the accumulation of snow. However, in ablation months, particularly the 0°C isotherm matches the snow cover extent and is characterized by shallow lapse rate between TC–BC as well as BC–ABC. This indicates that isotherms can be suitably utilized to delineate a transition line below which melting takes place. During ablation months, the isotherms and snow cover extent ascend, while in winter months isotherms gradually descend, but the extent of snow cover does not always follow this trend. It is thus concluded that depletion of snow is largely controlled by rise in temperature, whereas accumulation is mainly due to snow events.

31.23 CORRELATION OF SNOW COVER EXTENT WITH ISOTHERMS IN CGC

In order to ascertain the contributing area for melt-runoff generation within a glacierized basin, the elevation zones corresponding to 0, 1 or 2°C air temperature need to be determined. Considering high seasonal variability in the SCA and TLR, three different values of temperature thresholds or critical temperatures (0, 1 and 2°C) have been used to construct the elevations responsible for snowmelt contribution. Utilizing the monthly TLR computed for the study basin, the elevation corresponding to 0, 1 and 2°C air temperature was constructed for each month with reference to the highest station (ABC, 4270 m). This elevation may be considered as the temporary or transient snowline for that month/season. Isotherm altitudes were calculated by using equation 31.3. Based on the isotherms constructed, the ELA for Chorabari glacier lies in the elevation range of 5204, 5025 and 4847 m for 0, 1 and 2°C isotherms, respectively, at the end of September, corroborating that precipitation occurs as rain or rain-on-snow up to a maximum elevation of 5000–5500 m during the Indian summer monsoon. The proximity of ELA with isotherms can be clearly delineated in DGC and CGC (Misra et al., 2020). Also, the variability in snow cover depletion depicts little or no melting above the elevation zone of 5000–5500 m. During the months of May and October, a few events of solid precipitation are observed, which result in bringing the transient snowline down to lower elevations. These events are related to the Westerlies or local recycling of moisture in the region (Kumar et al., 2018). During peak winter season, the snowline is critically restricted to elevations of 2600–2900 m. Further, it is well observed that during the early ablation period (May–June) there is significant reduction in SCA, which is reflected as a rise in meltwater discharge in the months of May and June as reported by Kumar et al. (2016). This indicates that the availability of meltwater discharge for various socio-economic applications is highly dependent on the seasonal distribution of snow cover in the region.

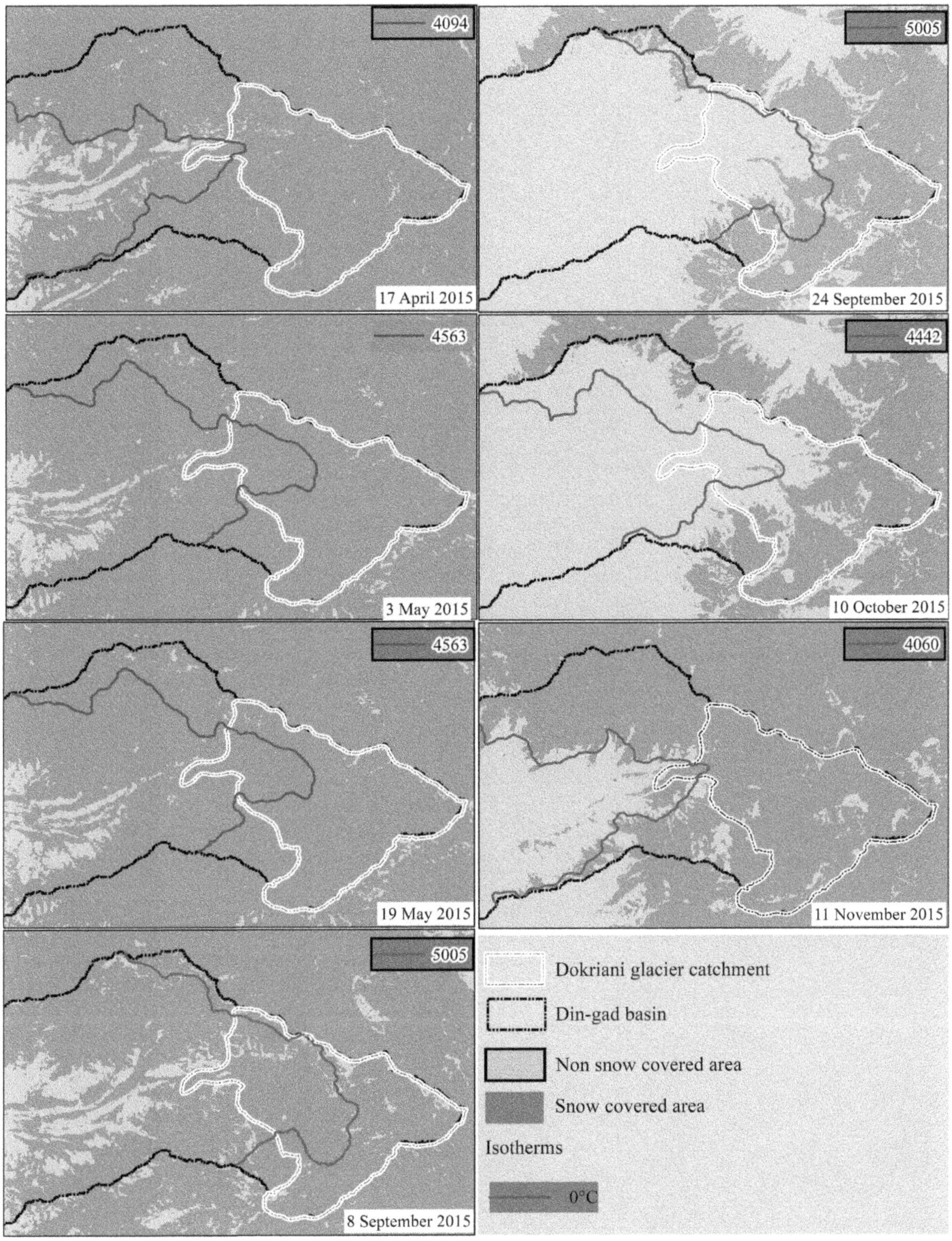

FIGURE 31.6 NDSI classified Landsat images depicting depletion of snow-covered area during ablation period for the year 2015. Monthly isotherms: green 0°C, yellow 1°C, and red 2°C show good correlation with snow cover in blue.

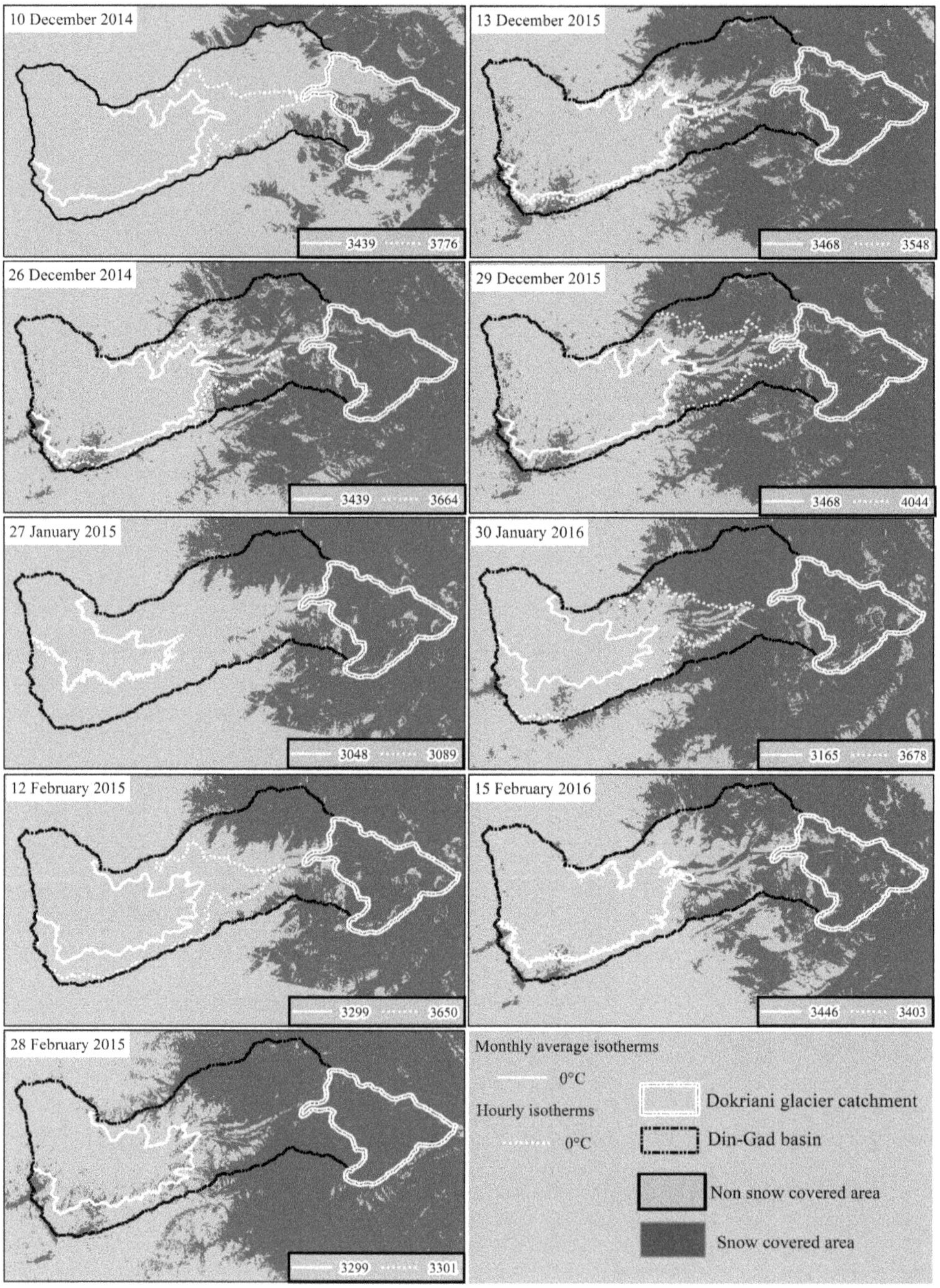

FIGURE 31.7 NDSI classified Landsat images depicting depletion of snow cover during winter months of 2014–2015 and 2015–2016. Monthly isotherms are presented by solid (green 0°C, yellow 1°C, and red 2°C) and hourly isotherms by dotted lines.

31.24 DISCUSSION

Most of the earlier studies covered large areas and used MODIS (500 m resolution) data (Negi et al., 2010; Rittger et al., 2013; Sharma et al., 2014; Kour et al., 2016). This data set is quite useful and provides comprehensive results/study at the regional scale (other places can be added if it is felt to be important). For well-defined glacierized catchments such as Dokriani and Chorabari, Landsat (30 m resolution) data is more suitable and provided details of snow cover at different elevation zone, aspect and slope categories, even with temporal resolution of 16 days.

Reflectance based NDSI is useful for snow cover mapping (Dozier, 1989; Hall et al., 1998). They applied this technique on both Landsat as well as MODIS data and concluded that visual input is also required for textural details to distinguish snow from other entities. NDSI gives better results in low-relief areas; however, in high-relief areas such as the Himalayan region, there are large areas covered by shadows, particularly in the winter months. This method classifies the entire shadowed area as snow cover. In order to overcome this difficulty, band ratio techniques were applied during this study. Various permutations and combinations were tried, and it was found that the Red/SWIR and NIR/SWIR ratios are very useful in distinguishing snow- covered areas from non-snow-covered areas in shadowed regions. Rott and Markl (1989), Bronge and Bronge (1999) and Paul et al. (2004) also utilized a ratioing technique, successfully identified blue ice facies and described the influence of thin cloud and cloud shadows.

A similar annual pattern of snow cover is observed in both the glacier catchments during the study period (2009–2016). Melting of snow starts at the beginning of April, and minimum snow cover is observed in August and September. These findings correspond with studies conducted in Indus and Ganga basins, western and central Himalaya (Kulkarni et al., 2010) and by Kour et al. (2016) in Chenab basin. Aspect-controlled depletion of snow in higher reaches during winter months is a unique phenomenon observed during this study. South-facing aspects (S, SW and SE) are affected by melting in both the catchments. However, N-S-oriented CGC has more prominent melting than E-W-oriented DGC. During this study, it has been observed that the melting pattern is latitude dependent and becomes successively delayed as we move towards polar regions. Shea et al. (2015) reported the melting of snow in mid-May in Tizinafu watershed of Kunlun mountain ranges. Hall et al. (2012) found that in Wind river valley, Wyoming, the melting started during the end of May. These regions are 6° and 12° towards the polar region, respectively. This has led to the inference that with every 6° increase in latitude, there is an approximately 15 days' delay in snow melting. By and large, during September and October, a few snow events cover the area, and depletion is clearly seen below 5500 m. The CV lies between 2.0 and 0.0 in DGC and between 2.3 and 0.0 in CGC in all elevation zones. This is also seen in Tizinafu watershed, where the range starts from 0.85 to 0 above 3000 (Shea et al., 2015).

In DGC, the southern (S, SE, SW) aspects show more snow cover depletion and a higher CV than the northern (N, NE, NW) aspects. This may be due to direct solar radiation and temperature difference on the southern aspect, as suggested by Tang and Fang (2006) and López-Moreno et al. (2013). At lower elevations, depletion of snow is faster, and the same has also been observed by Jain et al. (2008) in the Indus basin, western Himalaya.

In CGC, the eastern aspect has the highest variability in comparison to other aspects. The orientation of the eastern aspect is towards the solar radiation for maximum duration. Maximum SCA is observed on the southern aspects (SE, S and SW) of the catchment, and corresponding results have also been observed in the Chenab basin, western Himalaya, by Kour et al. (2016). As far as the effect of slope is concerned, SCA is found to be minimal on gentle (0–10°) slopes compared with other categories.

Apart from topographical factors, SCA variability is strongly influenced by local meteorological conditions. Earlier studies (Mir et al., 2015) showed that the areal extent of snow cover is also sensitive to near-surface temperature (NST). The decline in snowfall is attributed to a general increase in temperature, which shifts precipitation from solid to liquid (Thayyen et al., 2005b; Dimri and Mohanty, 2007). Similar observations have been made in other mountainous regions in the world (Bednorz, 2004; Hosaka et al., 2005; Gurung et al., 2011; Maskey et al., 2011; Barman and Bhattacharjya, 2015). Singh et al. (2014) suggested that consistently high temperature results in an early start of the ablation period in the Himalayan basins. Islam and Rao (2015) demonstrated that periods of high rainfall and relative humidity also coincide with the periods of low temperature and increased SCA variability. In addition to this, Shafiq et al. (2018) observed that the period of decreased temperature corresponds to an increase in SCA.

Meteorological data collected by AWSs for glaciological studies has earlier been utilized by Oerlemans (2000), Pellicciotti et al. (2008), Azam et al. (2014) and Kumar et al. (2016). These AWSs have significantly reduced errors and temporal delays between measurements in remote sites and record digital data in all weather conditions.

High linearity has emerged by plotting T_{avg} with T_{min} and T_{max} from AWSs at successively higher elevations. Based on this linear relation, the extrapolation/interpolation method can be utilized in areas lacking high-resolution continuous temperature data. Furthermore, selection of lapse rate is also a very important factor in rugged terrains. Different variability is estimated in other regions of the Himalaya by using SELR (6.5°C km^{-1}) or single value of lapse rate to obtain gridded temperature data. These values could be erroneous due to variability along different elevations (Kattel et al., 2013; 2015; Li et al., 2015; Huang et al., 2017; Thayyen and Dimri, 2018). The TLR values are found to be more advantageous and consistently reliable to determine point temperature data.

Similar findings are also reported by several other studies carried out in the Indian sub-continent (Arora et al., 2005; Fowler and Archer, 2006; Bhutiyani et al., 2007), where prominent variation in maximum temperature was seen due to local topography and microclimatic setting.

This observation is consistent with the results by Thayyen and Gergan (2010) in the DGB. Azam et al. (2016) have recently reported similar thermal contrast in Chhota Shigri Glacier, west of the study area. Studies related to prevailing weather conditions in the central Himalaya, carried out in the adjacent Gangotri and Chorabari glaciers (Singh et al., 2005; Thayyen et al., 2005a; Kesarwani et al., 2015), are in broad agreement with the inferences of this study.

In order to ascertain a clear relationship between snow cover and isotherms, statistical correlation (R) is analysed. During ablation months, it stands at ~0.8 and drops to ~0.55 in accumulation months. However, based on hourly analysis, it comes up to ~0.74 during accumulation months. This method has given a general understanding of the relationship between snow cover extent and isotherms. The elevations corresponding to the 0°C isotherm show close proximity between snow-covered areas and non-snow-covered areas in the glacierized catchment. These two data sets are integrated on a geographical information system (GIS) platform to illustrate their mutual relationship, as shown in Figures 31.6 and 31.7. Incorporation of terrain morphology is expected to further reduce the probable gaps between the two. Presently, all the three AWSs are installed at different levels and provide only elevation-related melting behaviour.

31.25 SYNTHESIS AND DISCUSSION

In this study, an attempt is made to develop a methodology for mapping the snow cover and its variability on two glacier catchments in upper Ganga river basin, central Himalaya. Variability in snow cover during different seasons is also brought out by using available multi-date remote sensing data sets. Among all the remote sensing data products, those having high resolution, frequent coverage and availability of free downloading were utilized during a 7-year (2009–2016) study period. This period is considered to be long enough to detail all the phenomena associated with the weather conditions. The study has also demonstrated the combined use of Landsat 5 TM, Landsat 8 OLI and Sentinel 2 MSI, the best-suited data sets in present-day technology.

Accumulation of snow cover is mostly due to snow events associated with high-altitude, low-pressure systems approaching from the NW direction. The severity of snow events controls the accumulation and results in temperature decline, while snow/ice melting/depletion is largely controlled by temperature and effects of elevation, aspect (slope direction) and slope angle. These catchments are sliced into different elevation zones, aspect classes and slope categories. Areas are identified, calculated and presented as maps. Furthermore, individual areas are also statistically analysed, interpreted and represented graphically. The mutual relationship between the two catchments is ascertained, and significant results are highlighted.

Among six identified elevation zones, the maximum area is occupied between 5000 and 5500 m and plays a pivotal role in sustaining snow cover. Minimal effect of depletion of snow is observed above 5500 m asl, while the maximum is in the area below 5000 m asl in both the catchments. Snow cover is also affected by aspect (slope direction). In the northern hemisphere, southern aspects show a faster rate of depletion than northern aspects. The important highlight of the study is that winter depletion of snow along the southern aspects (facing the sun) is not exclusively dependent on elevation and is witnessed even at the elevation of 6000 m. However, during the summer months, depletion is largely controlled by elevation. The amount of prevailing snow cover is also controlled by slope angle. Slopes between 0° and 10° and above 60° show minimum variation in snow cover, whereas the maximum is seen between 30° and 50° in the summer months, and minimum depletion takes place in the slope category 10°–20° in the winter months.

To generate snow cover maps, reflectance-based NDSI has proven its potential in most cases. Wherever NDSI could not classify, band ratioing techniques were explored to overcome problems such as classification of snow and non-snow-covered areas in shadowed regions. These shadowed regions are quite large in the winter months, when the sun angle is low. The band ratio technique applied in this study was based on DN values of NIR and SWIR bands. This increased the contrast and helped in differentiating snow from non-snow-covered areas in shadowed regions.

Furthermore, the present study includes integration between snow cover area and related controlling factors (elevation, slope and aspect) with meteorological data sets. Meteorological data was collected by AWSs located in both the catchments. Since the data was recorded in digital format, it was possible to co-register, integrate and study them together. Meteorological data sets included temperature, relative humidity and solar radiation.

Extrapolation and interpolation techniques are used to construct isotherms from the in situ temperatures of the two basins in order to calculate the TLR. From TLR, isotherms were computed and plotted simultaneously along with snow cover extent on a digital elevation model.

The nature of precipitation (snow and rain) as well as the transition between snow and rain is dependent on the temperature threshold. Earlier studies suggest that this phase change lies between 0 and 2°C. Three isotherms, 0°, 1° and 2°, drawn during this study, show close proximity with snow cover extent and correspond with each other during summer melting. This suggests that isotherms can be utilized as a transition line, below which melting takes place. The isotherms and snow cover extent move upwards during ablation months, while in winter months isotherms gradually descend, but the extent of snow cover does not always follow this trend. This indicates that depletion of snow is largely controlled by rise in temperature, whereas accumulation is mainly due to snow events.

ACKNOWLEDGEMENTS

I wish to profusely thank Prof. R C Sobti, Professor Emeritus and Former Vice-Chancellor, Panjab University for the invitation to contribute this chapter to a prestigious volume on environment studies and climate change to be published by CRC Press. Guidance provided by Prof. Rajiv Upadhya of Kumaun University, Nanital and Dr D P Dhobal Scientist G, of Wadia Institute of Himalayan Geology, Dehradun, during the study period is thankfully acknowledged. Detailed discussions with Dr K S Misra, Distinguished Professor, the University of Petroleum and Energy Studies, Dehradun have motivated me to relate the results of snow cover studies with climate change.

REFERENCES

Arora, M., Goel, N.K., Singh, P., 2005. Evaluation of temperature trends over India. *Hydrological Sciences Journal* 50, 81–93. doi:10.1623/hysj.50.1.81.56330

Azam, M.F., Ramanathan, A., Wagnon, P., Vincent, C., Linda, A., Berthier, E., Sharma, P., Mandal, A., Angchuk, T., Singh, V.B., Pottakkal, J.G., 2016. Meteorological conditions, seasonal and annual mass balances of Chhota Shigri Glacier, western Himalaya, India. *Annals of Glaciology* 57, 328–338. doi:10.3189/2016AoG71A570

Baral, D.J., Gupta, R.P., 1997. Integration of satellite sensor data with DEM for the study of snow cover distribution and depletion pattern. *International Journal of Remote Sensing* 18, 3889–3894. doi:10.1080/014311697216711

Barman, S., Bhattacharjya, R.K., 2015. Change in snow cover area of Brahmaputra river basin and its sensitivity to temperature. *Environmental Systems Research* 4, 16. doi:10.1186/s40068-015-0043-0

Barnett, T.P., Adam, J.C., Lettenmaier, D.P., 2005. Potential impacts of a warming climate on water availability in snow-dominated regions. *Nature* 438, 303–309. doi:10.1038/nature04141

Bednorz, E.W.A., 2004. Snow cover in Eastern Europe in relation to temperature, precipitation and circulation. *International Journal of Climatology* 601, 591–601. doi:10.1002/joc.1014

Berthier, E., Arnaud, Y., Kumar, R., et al., 2007. Remote sensing estimates of glacier mass balances in the Himachal Pradesh (Western Himalaya, India). *Remote Sensing of Environment* 108, 327–338. https://doi.org/10.1016/j.rse.2006.11

Bhutiyani, M.R., Kale, V.S., Pawar, N.J., 2007. Long-term trends in maximum, minimum and mean annual air temperatures across the Northwestern Himalaya during the twentieth century. *Climatic Change* 85, 159–177.

Blandford, T.R., Humes, K.S., Harshburger, B.J., Moore, B.C., Walden, V.P., Ye, H., 2008. Seasonal and synoptic variations in near-surface air temperature lapse rates in a mountainous basin. *Journal of Applied Meteorology and Climatology* 47, 249–261. doi:10.1175/2007JAMC1565.1

Bocchiola, D., Mihalcea, C., Diolaiuti, G., Mosconi, B., Smiraglia, C., Rosso, R., 2010. Flow prediction in high altitude ungauged catchments: A case study in the Italian Alps (Pantano Basin, Adamello Group). *Advances in Water Resources* 33, 1224–1234. doi:10.1016/j.advwatres.2010.06.009

Bronge, L.B., Bronge, C., 1999. Ice and snow-type classification in the Vestfold Hills, East Antarctica, using Landsat-TM data and ground radiometer measurements. *International Journal of Remote Sensing* 20, 225–240. doi:10.1080/014311699213415

Charbonneau, R., Lardeau, J.P., Obled, C., 1981. Problems of modelling a high mountainous drainage basin with predominant snow yields. *Hydrological Sciences Bulletin* 26, 345–361. doi:10.1080/02626668109490899

Chaujar, R.K., 2009. Climate change and its impact on the Himalayan glacier – a case study on the Chorabari glacier, Garhwal Himalaya, India. *Current Science* 96, 703–708.

Dimri, A.P., Mohanty, U.C., 2007. Location-specific prediction of maximum and minimum. *Meteorological Applications* 93, 79–93. doi:10.1002/met

Dobhal, D.P., Mehta, M., 2010. Surface morphology, elevation changes and terminus retreat of Dokriani glacier, Garhwal Himalaya: Implication for climate change. *Himalayan Geology* 31, 71–78.

Dobhal, D.P., Gergan, J.T., Thayyen, R.J., 2004. Recession and morphogeometrical changes of Dokriani glacier (1962–1995) Garhwal Himalaya, India. *Current Science* 86, 692–696.

Dobhal, D.P., Gergan, J.T., Thayyen, R.J., 2008. Mass balance studies of the Dokriani Glacier from 1992 to 2000, Garhwal Himalaya, India. *Bulletin of Glaciological Research* 25, 9–17.

Dobhal, D.P., Mehta, M., Srivastava, D., 2013. Influence of debris cover on terminus retreat and mass changes of Chorabari Glacier, Garhwal region, central Himalaya, India. *Journal of Glaciology* 59, 961–971. doi:10.3189/2013JoG12J180

Dozier, J., 1989. Spectral signature of alpine snow cover from the Landsat thematic mapper. *Remote Sensing of Environment* 28, 9–22. doi:10.1016/0034-4257(89)90101-6

Dunn, S.M., Colohan, R.J.E., 1999. Developing the snow component of a distributed hydrological model: a step-wise approach based on multi-objective analysis. *Journal of Hydrology* 223, 1–16.

Foster, J., Owe, M., Rango, A., 1982. Snow cover and temperature relationships in North America and Eurasia. *Journal of Applied Meteorology and Climatology* 22, 460–469.

Fowler, H.J., Archer, D.R., 2006. Conflicting signals of climatic change in the Upper Indus Basin. *Journal of Climate* 19, 4276–4293.

Gupta, R.P., Duggal, A.J., Rao, S.N., et al., 1982. Snow-cover area vs. snowmelt runoff relation and its dependence on geomorphology — A study from the Beas catchment (Himalayas, India). *Journal of Hydrology* 58, 325–339.

Gupta, R.P., Haritashya, U.K., Singh, P., 2005. Mapping dry/wet snow cover in the Indian Himalayas using IRS multispectral imagery. *Remote Sensing of Environment* 97, 458–469. doi:10.1016/j.rse.2005.05.010

Gurung, D.R., Giriraj, A., Aung, K.S., Shrestha, B., Kulkarni, A. V., 2011. *Snow-cover Mapping and Monitoring in the Hindu Kush-Himalayas*. The International Centre for Integrated Mountain Development (ICIMOD). pp. 1–44. doi:978 92 9115 218 6

Haeberli, W., Bösch, H., Scherler, K., 1988. World Glacier Inventory – Status 1988. doi:10.7265/N5/NSIDC-WGI-2012-02

Hall, D.K., Riggs, G.A., 2007. Accuracy assessment of the MODIS snow products. *Hydrological Processes* 21, 1534–1547. doi:10.1002/hyp.6715

Hall, D.K., Riggs, G.A., Salomonson, V.V., 1995. Development of methods for mapping global snow cover using moderate resolution imaging spectroradiometer data. *Remote Sensing of the Environment* 54, 127–140. doi:10.1016/0034-4257(95)00137-P

Hall, D.K., Foster, J.L., Verbyla, D.L., Klein, A.G., Benson, C.S., 1998. Assessment of snow cover mapping accuracy in a variety of vegetation cover densities in central Alaska. *Remote Sensing of the Environment* 66, 129–137. doi:10.1016/S0034-4257(98)00051-0

Hall, D.K., Riggs, G.A., Salomonson, V.V., et al., 2002. MODIS snow-cover products. *Remote Sensing of the Environment* 83, 181–194.

Hall, D.K., Foster, J.L., DiGirolamo, N.E., Riggs, G.A., 2012. Snow cover, snowmelt timing and stream power in the Wind River Range, Wyoming. *Geomorphology* 137, 87–93.

Heim, A., Gansser, A., 1975. *Central Himalaya: Geological Observations of the Swiss Expedition, 1936*. Vol. 73. Hindustan Publishing Corporation (India).

Higuchi, K., Ageta, Y., Yasunari, T., Inoue, J., 1982. Characteristics of precipitation during the monsoon season in high-mountain areas of the Nepal Himalaya. *International Association of Hydrological Sciences Publications*, 138, 21–30.

Hoinkes, H., Howorka, F., Schneider, W., 1967. Glacier mass budget and mesoscale weather in the Austrian Alps 1964 to 1966. *Proceedings of the International Commission on Snow and Ice* 79, 241–254.

Hosaka, M., Nohara, D., Kitoh, A., 2005. Changes in snow cover and snow water equivalent due to global warming simulated by a 20km-mesh global atmospheric model. *SOLA* 1, 93–96.

Huang, X., Deng, J., Wang, W., Feng, Q., Liang, T., 2017. Impact of climate and elevation on snow cover using integrated remote sensing snow products in Tibetan Plateau. *Remote Sensing of Environment* 190, 274–288. doi:10.1016/j.rse.2016.12.028

Immerzeel, W.W., Beek, V., Ludovicus, P.H., Bierkens, M.F.P., 2010. Climate change will affect the Asian water towers. *Science* 328, 1382–1385.

Islam, Z., Rao, L., 2015. Seasonal and annual rainfall trends: A perspective on climate change in Kashmir Valley, Northwestern Himalaya. *Nature Changes* 2 (2), 1–13.

Jain, S.K., Goswami, A., Saraf, A.K., 2008. Accuracy assessment of MODIS, NOAA and IRS data in snow cover mapping under Himalayan conditions. *International Journal of Remote Sensing* 29, 5863–5878.

Jain, S.K., Goswami, A., Saraf, A.K., 2009. Role of elevation and aspect in snow distribution in Western Himalaya. *Water Resource Management* 23, 71–83.

Karakoti, I., Verma, A., D.P. Dobhal, 2012. Meteorological observations at Chorabari and Dokriani glaciers, Garhwal Himalaya, India. *Journal of Indian Geology Congress* 4, 125–128.

Kargel, J.S., Abrams, M.J., Bishop, M.P., et al., 2005. Multispectral imaging contributions to global land ice measurements from space. *Remote Sensing of the Environment* 99, 187–219.

Kattel, D.B., Yang, K., Tian, L., 2013. Temperature lapse rate in complex mountain terrain on the southern slope of the central Himalayas. *Theoretical and Applied Climatology* 671–682. doi:10.1007/s00704-012-0816-6

Kattel, D.B., Yao, T., Yang, W., Gao, Y., Tian, L., 2015. Comparison of temperature lapse rates from the northern to the southern slopes of the Himalayas. *International Journal of Climatology* 35, 4431–4443. doi:10.1002/joc.4297

Kesarwani, K., Mehta, M., Dobhal, D.P., Durgapal, A., 2014. Effect of wind speed and direction on valley glaciers of Himalaya: A case study of Chorabari Glacier, Central Himalaya, India.

Kesarwani, K., Dobhal, D.P., Durgapal, A., Mehta, M., 2015. High altitude meteorology and cloud cover conditions: A study from the Chorabari glacier catchment, Central Himalaya, India. *Himalayan Geology* 36, 134–142.

Kottek, M., Grieser, J., Beck, C., Rudolf, B., Rubel, F., 2006. World map of the Köppen-Geiger climate classification updated. *Meteorologische Zeitschrift* 15(3), 259–263.

Kour, R., Patel, N., Krishna, A.P., 2016. Effects of terrain attributes on snow-cover dynamics in parts of Chenab basin, western Himalayas. *Hydrological Sciences Journal* 61, 1861–1876. doi:10.1080/02626667.2015.1052815

Kulkarni, A.V., 1992. Mass balance of Himalayan glaciers using AAR and ELA methods. *Journal of Glaciology* 38, 101–104.

Kulkarni, A., Bahuguna, I.M., 2001. Role of satellite images in snow and glacial investigations. *Geological Survey of India* 53, 233–240.

Kulkarni, A.V., Rathore, B.P., Singh, S.K., 2010. Distribution of seasonal snow cover in central and western Himalaya. *Annals of Glaciology* 51, 123–128. doi:10.3189/172756410791386445

Kumar, A., Gokhale, A.A., Shukla, T., Dobhal, D.P., 2016. Hydroclimatic influence on particle size distribution of suspended sediments evacuated from debris-covered Chorabari Glacier, upper Mandakini catchment, central Himalaya. *Geomorphology* 265, 45–67. doi:10.1016/j.geomorph.2016.04.019

Kumar, A., Verma, A., Gokhale, A.A., Bhambri, R., Misra, A., Sundriyal, S., Dobhal, D.P., Kishore, N., 2018. Hydrometeorological assessments and suspended sediment delivery from a central Himalayan glacier in the upper Ganga basin. *International Journal of Sediment Research* 33, 493–509. doi:10.1016/j.ijsrc.2018.03.004

Li, H.Y., He, Y.Q., Hao, X.H., Che, T., Wang, J., Huang, X.D., 2015. Downscaling snow cover fraction data in mountainous regions based on simulated inhomogeneous snow ablation. *Remote Sensing* 7, 8995–9019. doi:10.3390/rs70708995

López-Moreno, J.I., Revuelto, J., Gilaberte, M., et al., 2013. The effect of slope aspect on the response of snowpack to climate warming in the Pyrenees. *Theoretical and Applied Climatology* 117, 207–219.

MacDonald, R.J., Byrne, J.M., Boon, S., Kienzle, S.W., 2012. Modelling the potential impacts of climate change on snowpack in the North Saskatchewan River watershed, Alberta. *Water Resources Management* 26, 3053–3076. doi:10.1007/s11269-012-0016-2

Marshall, S.J., Sharp, M.J., Burgess, D.O., Anslow, F.S., 2007. Near-surface-temperature lapse rates on the Prince of Wales Icefield, Ellesmere Island, Canada: Implications for regional downscaling of temperature. *International Journal of Climatology* 27, 285–398. doi:10.1002/joc

Martinec, J., Rango, A., 1986. Parameter values for snowmelt runoff modelling. *Journal of Hydrology* 84, 197–219.

Maskey, S., Uhlenbrook, S., Ojha, S., 2011. An analysis of snow cover changes in the Himalayan region using MODIS snow products and in-situ temperature data. *Climatic Change* 108, 391–400. doi:10.1007/s10584-011-0181-y

Mir, R.A., Jain, S.K., Saraf, A.K., Goswami, A., 2015. Accuracy assessment and trend analysis of MODIS-derived data on snow-covered areas in the Sutlej basin, Western Himalayas. *International Journal of Remote Sensing* 36, 3837–3858. doi:10.1080/01431161.2015.1070320

Misra, A., Kumar, A., Bhambri, R., Haritashya, U.K., Verma, A., Dobhal, D.P., Gupta, A.K., Gupta, G., Upadhyay, R., 2020. Topographic and climatic influence on seasonal snow cover: Implications for the hydrology of ungauged Himalayan basins, India. *Journal of Hydrology* 585, 124716.

Negi, H.S., Singh, S.K., Kulkarni, A.V., Semwal, B.S., 2010. Field-based spectral reflectance measurements of seasonal snow cover in the Indian Himalaya. *International Journal of Remote Sensing* 31, 2393–2417. doi:10.1080/01431160903002417

Nolin, A.W., Stroeve, J., 1997. The changing albedo of the Greenland ice sheet: Implications for climate modeling. *Annals of Glaciology* 25, 51–57. doi:10.3189/s0260305500013793

O'Brien, H.W., Munis, R.H., 1975. *Red and Near-infrared Spectral Reflectance of Snow* (Vol. 332). Corps of Engineers, US Army, Cold Regions Research and Engineering Laboratory.

Oerlemans, J., 2000. Analysis of a 3 year meteorological record from the ablation zone of Morteratschgletscher, Switzerland: Energy and mass balance. *Journal of Glaciology* 46, 571–579.

Owen, L.A., Finkel, R.C., Caffee, M.W., 2002. A note on the extent of glaciation throughout the Himalaya during the global Last Glacial Maximum. *Quaternary Science Reviews* 21, 147–157. doi:10.1016/S0277-3791(01)00104-4

Paul, F., Huggel, C., Kääb, A., 2004. Combining satellite multispectral image data and a digital elevation model for mapping debris-covered glaciers. *Remote Sensing of the Environment* 89, 510–518. doi:10.1016/j.rse.2003.11.007

Pellicciotti, F., Helbing, J., Rivera, A., Favier, V., Corripio, J., Araos, J., Sicart, J., Carenzo, M., 2008. A study of the energy balance and melt regime on Juncal Norte Glacier, semi-arid Andes of central Chile, using melt models of different complexity. *Hydrological Processes* 22, 3980–3997.

Pratap, B., Dobhal, D.P., Mehta, M., Bhambri, R., 2015. Influence of debris cover and altitude on glacier surface melting: A case study on Dokriani Glacier, central Himalaya, India. *Annals of Glaciology* 56, 9–16. doi:10.3189/2015AoG70A971

Riggs, G.A., Hall, D.K., Román, M.O., 2017. Overview of NASA's MODIS and visible infrared imaging radiometer suite (VIIRS) snow-cover earth system data records. *Earth System Science Data*, 9(2), 765–777.

Rittger, K., Painter, T.H., Dozier, J., 2013. Assessment of methods for mapping snow cover from MODIS. *Advances in Water Resources* 51, 367–380. doi:10.1016/j.advwatres.2012.03.002

Rohrer, M., 1989. Determination of the transition air temperature from snow to rain and intensity of precipitation. In *WMO IASH ETH international workshop on precipitation measurement* (pp. 475–582). Geneva, Switzerland: World Meteorological Organization.

Rohrer, M., Salzmann, N., Stoffel, M., Kulkarni, A.V., 2013. Missing (in-situ) snow cover data hampers climate change and runoff studies in the Greater Himalayas. *Science of the Total Environment* 468–469, S60–S70. doi:10.1016/j.scitotenv.2013.09.056

Rosenthal, W., Dozier, J., 1996. Automated mapping of montane snow cover at subpixel resolution from the Landsat Thematic Mapper. *Water Resources Research* 32(1), 115–130.

Rott, H., Markl, G., 1989. Improved snow and glacier monitoring by Landsat Thematic Mapper. *Proceedings of a workshop on "Earthnet Pilot Project on Landsat Thematic Mapper Applications"*, held at Frascati, Italy in December 1987.

SAC, 2016. *Monitoring Snow and Glaciers of Himalayan Region*, Space Application Centre, ISRO, Ahmedabad, India.

Sangewar, C.V, Shukla, S.P., 2009. *Inventory of the Himalayan Glaciers*, Geological Survey of India Special Publication 1–200.

Sarkar, S., Roy, A.K., Raha, P., 2016. Deterministic approach for susceptibility assessment of shallow debris slide in the Darjeeling Himalayas, India. *Catena* 142, 36–46.

Senese, A., Maugeri, M., Vuillermoz, E., Smiraglia, C., Diolaiuti, G., 2014. Using daily air temperature thresholds to evaluate snow melting occurrence and amount on Alpine glaciers by T-index models: The case study of the Forni Glacier (Italy). *Cryosphere* 8, 1921–1933. doi:10.5194/tc-8-1921-2014

Shafiq, M. ul, Ahmed, P., Bhat, W.A., Islam, Z. ul, Joshi, P.K., 2018. Snow cover area change and its relations with climatic variability in Kashmir Himalayas, India. *Geocarto International* 6049, 1–15. doi:10.1080/10106049.2018.1469675

Sharma, V., Mishra, V.D., Joshi, P.K., 2014. Topographic controls on spatio- temporal snow cover distribution in Northwest Himalaya. *International Journal of Remote Sensing* 35, 3036–3056. doi:10.1080/01431161.2014.894665

Shea, J.M., Wagnon, P., Immerzeel, W.W., Biron, R., Brun, F., Pellicciotti, F., 2015. A comparative high-altitude meteorological analysis from three catchments in the Nepalese Himalaya. *International Journal of Water Resources Development* 31, 174–200. doi:10.1080/07900627.2015.1020417

Shukla, T., Mehta, M., Jaiswal, M.K., Srivastava, P., Dobhal, D.P., Nainwal, H.C., Singh, A.K., 2018. Late Quaternary glaciation history of monsoon-dominated Dingad basin, central Himalaya, India. *Quaternary Science Reviews* 181, 43–64. doi:10.1016/j.quascirev.2017.11.032

Simpson, J.J., Stitt, J.R., Sienko, M., 1998. Improved estimates of the areal extent of snow cover from AVHRR data. *Journal of Hydrology* 204, 1–23.

Singh, P., Jain, S.K., 2003. Modelling of streamflow and its components for a large Himalayan basin with predominant snowmelt yields. *Hydrological Sciences Journal* 48, 257–276.

Singh, J., Misra, A., Dobhal, D.P., Yadav, R.B.S., Upadhyay, R., 2021. Snow cover mapping, topographic controls and integration of meteorological data sets in Din-Gad Basin, Central Himalaya. *Quaternary International* 575–576, 160–177. doi:10.1016/j.quaint.2020.05.030

Singh, P., Jain, S.K., 2002. Snow and glacier melt in the Satluj River at Bhakra Dam in the western Himalayan region. *Hydrological Sciences Journal* 47(1), 93–106. doi:10.1080/02626660209492910

Singh, P., Kumar, N., Ramasastri, K.S., Singh, Y., 2000. Influence of a fine debris layer on the melting of snow and ice on a Himalayan glacier. In *Debris-covered Glaciers: Proceedings of an International Workshop Held at the University of Washington in Seattle, Washington, USA, 13–15 September 2000* (No. 264, p. 63). IAHS.

Singh, P., Haritashya, U.K., Ramasastri, K.S., Kumar, N., 2005. Diurnal variations in discharge and suspended sediment concentration, including runoff-delaying characteristics, of the Gangotri Glacier in the Garhwal Himalayas. *Hydrological Processes* 19, 1445–1457. doi:10.1002/hyp.5583

Singh, P., Haritashya, U.K., Kumar, N., 2008. Modelling and estimation of different components of streamflow for Gangotri Glacier basin, Himalayas. *Hydrological Sciences Journal* 53, 309–322. doi:10.1623/hysj.53.2.309

Singh, S.K., Rathore, B.P., Bahuguna, I.M., Ajai, 2014. Snow cover variability in the Himalayan-Tibetan region. *International Journal of Climatology* 34, 446–452.

Tang, Z., Fang, J., 2006. Temperature variation along the northern and southern slopes of Mt. Taibai, China. *Agricultural and Forest Meteorology* 139, 200–207. doi:10.1002/joc.3697

Tangri, A.K., 1999. Spectral response as a measure of monitoring temporal and spatial variations in snow cover and the prediction of snow-melt phenomena in Alaknanda and Dhauliganga river Watersheds in Chamoli district, Uttar Pradesh. *Geological Survey of India (Special Publication)* 53, 371–378.

Thayyen, R.J., Dimri, A.P., 2014. Factors controlling Slope Environmental Lapse Rate (SELR) of temperature in the monsoon and cold-arid glacio-hydrological regimes of the Himalaya. *The Cryosphere Discussions* 8, 5645–5686.

Thayyen, R.J., Dimri, A.P., 2018. Slope environmental lapse rate (SELR) of temperature in the monsoon regime of the Western Himalaya. *Frontiers in Environmental Science* 6, 1–20. doi:10.3389/fenvs.2018.00042

Thayyen, R.J., Gergan, J.T., 2010. Role of glaciers in watershed hydrology: A preliminary study of a "Himalayan catchment." *Cryosphere* 4, 115–128. doi:10.5194/tc-4-115-2010

Thayyen, R.J., Gergan, J.T., Dobhal, D.P., 2005a. Monsoonal control on glacier discharge and hydrograph characteristics, a case study of Dokriani Glacier, Garhwal Himalaya, India. *Journal of Hydrology* 306, 37–49. doi:10.1016/j.jhydrol.2004.08.034

Thayyen, R.J., Gergan, J.T., Dobhal, D.P., 2005b. Slope lapse rates of temperature in Din Gad (Dokriani Glacier) catchment, Garhwal Himalaya, India. *Bulletin of Glaciological Research* 22, 31–37.

Valovcin, F.R., 1976. *Snow/cloud discrimination* (No. 349). Air Force Geophysics Laboratories, Air Force Systems Command, United States Air Force.

Vohra, C.P., Agarwal, N.K., Srivastava, G.S., 1981. Application of remote sensing techniques in snow and ice hydrology. *Journal of the Indian Society of Photo-Interpretation and Remote Sensing* 9, 49–57.

Wang, J., Li, W., 2003. Comparison of methods of snow cover mapping by analysing the solar spectrum of satellite remote sensing data in China. *International Journal of Remote Sensing* 24, 4129–4136.

Wang, X., Zhu, Y., Chen, Y., et al., 2017. Influences of forest on MODIS snow cover mapping and snow variations in the Amur River basin in Northeast Asia during 2000–2014. *Hydrological Processes* 31, 3225–3241.

Wulder, M.A., Loveland, T.R., Roy, D.P., Crawford, C.J., Masek, J.G., Woodcock, C.E., Allen, R.G., Anderson, M.C., Belward, A.S., Cohen, W.B., Dwyer, J., Erb, A., Gao, F., Griffiths, P., Helder, D., Hermosilla, T., Hipple, J.D., Hostert, P., Hughes, M.J., Huntington, J., Johnson, D.M., Kennedy, R., Kilic, A., Li, Z., Lymburner, L., McCorkel, J., Pahlevan, N., Scambos, T.A., Schaaf, C., Schott, J.R., Sheng, Y., Storey, J., Vermote, E., Vogelmann, J., White, J.C., Wynne, R.H., Zhu, Z., 2019. Current status of Landsat program, science, and applications. *Remote Sensing of the Environment* 225, 127–147. doi:10.1016/j.rse.2019.02.015

Yadav, J.S., Pratap, B., Gupta, A.K., Dobhal, D.P., Yadav, R.B.S., Tiwari, S.K., 2019. Spatio-temporal variability of near-surface air temperature in the Dokriani glacier catchment (DGC), central Himalaya. *Theoretical and Applied Climatology* 136, 1513–1532. doi:10.1007/s00704-018-2544-z

32 Agrofuels Can Beat the Heat for a Sustainable Environment

Amit Kumar
Department of Biotechnology, Keral Verma Faculty of Science, Swami Vivekanand Subharti University, Meerut

R.S. Sengar
Department of Agriculture Biotechnology, Sardar Vallabhbhai Patel University of Agriculture and Technology, Meerut

Rajeev Kumar
Department of Ag. Botany (Genetics and Plant Breeding), C.S.S.S. (P.G.) College, Meerut

CONTENTS

32.1 INTRODUCTION

Bioenergy is renewable energy derived from a biological source. It is a synonym for agrofuel. Agrofuel can be broadly classified as solid, liquid or gas fuel. Agrofuel can be created from any carbon source; however, the most widely recognized by a long shot is photosynthetic plants (Balat, 2007). Agrofuels offer the possibility of producing energy without a net increase of carbon into the atmosphere. They include biomass, biological material used as an agrofuel. Biomass is any organic

DOI: 10.1201/9781003220824-36

material that has stored daylight in the form of chemical energy. As a fuel, it includes wood, wood waste, agriculture straw, manure, cane sugar and plenty of alternative byproducts from a range of agricultural processes (Zarzyycki and Polska, 2007). Therefore, agrofuel is carbon neutral and less likely to increase the atmospheric concentration of greenhouse gases (Sheehan *et al.*, 2003).

Agrofuels are typically produced in one of two ways. One option is to grow sugar-rich crops such as *Saccharum officinarum* (sugar cane), *Beta vulgaris* (sugar beet) and *Sorghum bicolor* (sweet sorghum), and then ferment with yeast to produce ethanol. The second is to create plants with high levels of vegetable oil. for example, oil palm, soybean and Jatropha. When heated, these oils become thinner, allowing them to be burned directly in a diesel engine or chemically prepared for combustion. manufacture agrofuels, for example, biodiesel (Stevens *et al.*, 2004). Wood and its byproducts can also be converted into agrofuels such as wood gas, methanol or ethanol fuel (Richa, 2014).

Agrofuels are the appropriate alternative to reduce the annual oil bill of India, approximately 80,000 crore, 70% of which is met through imports. The current rate of Indian development of agrofuels, particularly biodiesel, is negligible. India's vast and sometimes destructive wastelands were used for biodiesel production, with a modest amount produced – one-tenth of the country's annual oil requirement – because India has around 130 million hectares of wasteland, of which 33 million hectares is available for reclamation through tree planting. Greening these wastelands through agrofuel plantations could completely replace the current use of fossil fuel and provide employment for people.

32.2 TYPES OF AGROFUELS OR BIOFUELS

Agrofuels may be classified into solid (wood and dry dung), liquid (alcohol and biodiesel) and gas (producer and landfill gases) (Figure 32.1).

Agrofuels may be categorized as first-, second-, third- and fourth-generation fuels. The very first generation of agrofuel is manufactured from sugar that has been aged, starch and vegetable oil. Seeds or grains, such as wheat, which produces starch that is fermented into bioethanol, or sunflower seeds, which are crushed to produce vegetable oil that may be used in biodiesel, are examples of seeds or grains that can be used in biodiesel; they are typically used as the primary feedstock for

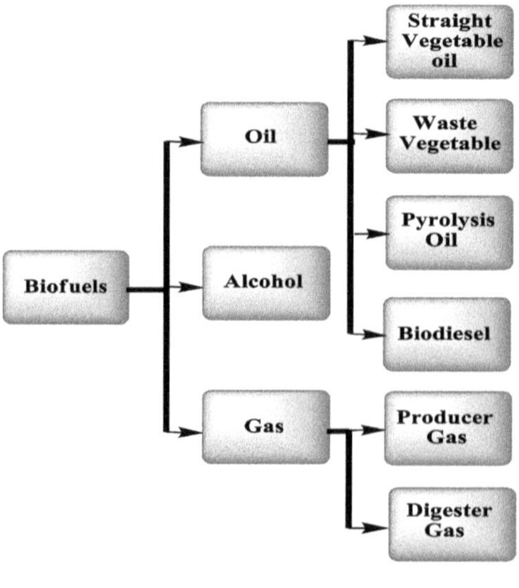

FIGURE 32.1 Types of biofuels.

the production of first-generation agrofuel (Prasad *et al.*, 2009; Kallivroussis *et al.*, 2002). In 2006, the production of first-generation biofuel was about 51 billion liters. India and China contributed 11% to the global ethanol production. Second-generation agrofuels use biomass to power innovation, including cellulosic agrofuels from non-food crops (Schmidt and Dauenhauer, 2007). Various second-generation agrofuels are a work in progress, for example, biohydrogen, wood diesel, biohydrogen diesel, blended alcohols and biomethanol. Third-generation agrofuel is an agrofuel from algae (Ma, 2006). Algae are low-input, high-yield feedstocks that produce agrofuels. They produce 30 times more energy per acre than land. Fourth-generation agrofuel depends on changing vegetable oil and biodiesel into gasoline.

32.2.1 Solid Agrofuels

Solid agrofuels are obtained from wood, grass cuttings, domestic refuse, charcoal and dried manure (Hashem *et al.*, 2007). They are mainly known as primary, secondary and tertiary biomass. Primary biomass resources are produced by photosynthesis. They include perennial short rotation woody crops and herbaceous crops, the seeds of oil crops, and residues resulting from the harvesting of agricultural crops and forest trees. Secondary biomass assets result from the preparation of essential biomass assets, either mechanically (e.g. creation of sawdust plants), artificially (e.g. dark alcohol from pulping measures) or naturally (e.g. fertilizer creation by animals). Tertiary biomass assets are post-consumer deposit streams including animal fats and lubes, utilized vegetable oils, municipal, civil and industrial residues, and construction and demolition debris (Hashem *et al.*, 2007).

32.2.2 Liquid Agrofuels

These may be divided into oils and alcohols. Oils are obtained from straight vegetable oil, waste vegetable oil, pyrolysis oil and biodiesels. For the most part, edible vegetable oil is Although parts are not utilized as fuel, lower-quality oil can be utilized instead. Used vegetable oil is increasingly being converted to biodiesel or cleansed of water and particles before being used as a fuel. To guarantee that the fuel injectors atomize the fuel in the precise pattern for optimum combustion, vegetable oil fuel should be warmed to the viscosity of diesel. Vegetable oil can likewise be utilized in numerous more seasoned diesel engines that don't utilize common rail or unit infusion electronic diesel infusion frameworks.

32.2.2.1 Biodiesel

Biodiesel is the most common agrofuel. It is a non-toxic, biodegradable diesel fuel made from vegetable oils, animal fats, and used or recycled oils and fats (Demirbas, 2008a). Biodiesel could be an excellent renewable fuel for diesel engines. It is derived from vegetable oils that are chemically converted into biodiesel. Biodiesel, when used as a pure fuel, is known as B100. The process is called transesterification (Fukuda, 2001). Its chemical name is unsaturated fat methyl ester. The chemical reaction creates biodiesel and glycerol when oils are mixed with sodium hydroxide and methanol. For every ten parts of biodiesel, one part of glycerol is produced. Biodiesel can be made from animal fats or vegetable oils such as soybean, rapeseed, Jatropha, mahua, mustard, flax, sunflower, palm oil, hemp, and algae (Han *et al.*, 2005). There are a few strategies for performing this transesterification reaction, including the regular batch process (Rashid and Anwar, 2008). A common pathway of biodiesel production is given in Figure 32.2.

32.2.2.2 Potential for Biodiesel in India

For blending with diesel, biodiesel provides a viable solution. Biodiesel based on edible and non-edible oils has already made much headway in the US, major European countries, Australia,

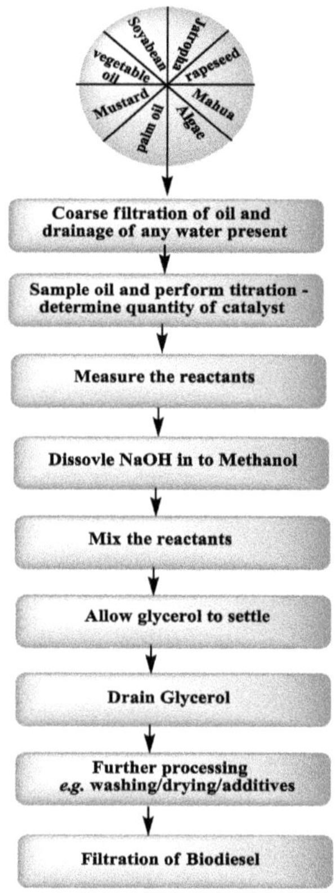

FIGURE 32.2 Major routes and steps of biodiesel formation.

Japan, Malaysia, etc., where regular production plants are already in operation. Major automobile manufacturers have also approved the use of 10–20% biodiesel blended in mineral diesel in their vehicles.

In developed countries like the US, Australia, Germany, France, etc., biodiesel is being extracted from saffola, sunflower, soybean, etc., which are essentially edible in India. But India has vast resources of non-edible/wild seeds (Jatropha seed) from which oil can be derived to develop biodiesel, depending upon the potential of specific seeds in the locality (Berchmans and Hirata, 2008; Reshu C. *et al.*, 2010). Experiments have shown that biodiesel derived from non-edible oil seeds can be used in existing designs of diesel vehicles without substantial modification (Gui *et al.*, 2008). Secondly, there is a shortage of edible oils in India, and as such, our concentration is going to be on development of biodiesel from non-edible oils only. For this reason, the planning commission is also thinking of starting a National Mission on *Jatropha curcas*, which includes large-scale plantation, collection of seeds and setting up of transesterification plants for producing biodiesel.

32.2.2.3 Biodiesel as Fuel

Biodiesel is a fuel made from vegetable oil that runs in any unmodified diesel engine (Demirbas, 2008b). Biodiesel can be made from any vegetable oil, including oils pressed straight from the seed, such as soy, sunflower, canola, coconut and hemp (Patil and Deng, 2009; Saka and Kusdiana, 2001; Han *et al.*, 2005). Biodiesel can also be made from recycled cooking oils from fast food restaurants. Even animal fats like beef tallow and fish oil can be used to make biodiesel fuel.

Biodiesel runs in any unmodified diesel engine. There is no "engine transformation" common to other elective fuels. The diesel engine can run on biodiesel on the grounds that it works on the principle of compression ignition, whereby air is compacted and afterward, fuel is splashed into the ultra-hot, ultra-constrained burning chamber. In contrast to gas engines, which utilize a spark to ignite the fuel/air mixture, diesel engines really use fuel to light hot air. This straightforward interaction permits the diesel engines to run on thick fuels. Since biodiesel is synthetically like diesel fuel, you can empty biodiesel directly into the gas tank of any diesel vehicle. Biodiesel enjoys numerous benefits as a vehicle fuel. Biodiesel has lower outflows, it is made domestically, it doesn't influence engine performance, and biodiesel is created from plants. Since plants are a result of sun-powered energy, biodiesel is "fluid sun oriented fuel".

32.2.2.4 Biodiesel for a Sustainable Environment

Biodiesel is more greasing up than diesel fuel, it extends the life of the motor, and it tends to be used to replace sulfur, a lubricating agent that, when consumed, produces sulfur dioxide – the essential part of acid rain. Rather than sulfur, all diesel fuel sold in France contains 5% biodiesel. It is protected to transport. Biodiesel has a high flash point, or ignition temperature, of about 300°F contrasted with petrol diesel fuel, which has a flash point of 125°F.

Overall biodiesel emissions are lower than gasoline or diesel fuel emissions (with the exception of NO_x). Biodiesel has many emission characteristics when compared with petroleum diesel fuel, such as reduction of carbon dioxide emissions by 100%, reduction of sulfur dioxide emissions by 100%, reduction of soot emissions by 40–60%, reduction of carbon monoxide emissions by 10–50%, reduction of hydrocarbon emissions by 10–50%, reduction of polycyclic aromatic hydrocarbons (PAHs) and specifically, the reduction of the following carcinogenic PAHs.

32.2.2.5 Bio-alcohols

Biologically produced alcohols, such as ethanol and, to a lesser extent, propanol and butanol, are made by bacteria and chemicals fermenting sugars or starches (the least demanding) or cellulose. Biobutanol is a kind of butanol frequently claimed to give an immediate substitution for gasoline, since it tends to be utilized straightforwardly in gasoline. ABE fermentation (butanol and ethanol) produces butanol, and exploratory changes to the cycle demonstrate that butanol is the only fluid component, indicating a potentially high net energy grain. Butanol will produce more energy and is said to be able to be burned "straight" in existing gasoline engines. It is also less damaging and water soluble than ethanol, and might be used as a fuel alternative, dispersed by means of existing frameworks.

Ethanol fuel is the most well-known agrofuel around the world, especially in Brazil. Alcohol fuels are made by fermenting sugars from wheat, sugar beets, sugar stick, and any other sugar or starch that can be used to make alcoholic beverages (e.g. potato and natural product waste and so forth). The ethanol creation techniques utilized are catalyst processing to deliver sugars from stored starches, fermentations of the sugars, refining and drying.

Ethanol can be utilized in petroleum engines as a swap for fuel; at the very least, it can be blended very well with gasoline. Most existing auto petroleum engines can run on bioethanol/petroleum blends of up to 15%. Gasoline containing ethanol has a higher octane rating, implying that the engine may burn hotter and all the more productively. Ethanol is destructive to fuel frameworks, elastic hoses and gaskets, aluminum and burning chambers. Numerous vehicle makers are currently delivering adaptable vehicles that can safely run on any bioethanol–petroleum mixture up to 100% bioethanol,

About 5% of the ethanol produced in the world in 2003 was actually a petroleum product. It is made by catalytic hydration of ethylene with sulfuric acid as the catalyst. It can also be obtained via ethylene or acetylene from calcium carbide, coal, oil, gas and other sources.

32.2.3 Gas Agrofuels

32.2.3.1 Biogas

Biogas is produced through anaerobic absorption of natural material by anaerobes. It could be made from biodegradable waste or by increasing gas output by introducing energy crops into anaerobic digesters. Biogas may be collected from industrial anaerobic digesters and mechanical biological treatment systems because it includes methane.

The other principal type of biogas is wood gas, which is created by gasification of wood or other biomass (Tijmensen *et al.*, 2002). This type of biogas is comprised primarily of nitrogen, hydrogen and carbon monoxide, with trace amounts of methane. Methane within biogas can be concentrated to the same standards as in fossil natural gas; in this case, it is called bio-methane.

Landfill gas is a less environmentally friendly form of biogas, which is created in landfills by the intensive anaerobic processing that happens normally. In the event that it escapes into the air, it is a potent greenhouse gas. Landfill gas is an intensely ozone-depleting substance. It becomes explosive when it escapes from the landfill and blends with oxygen. The lower explosive limit for methane is 5%, and the upper explosive limit is 15% methane.

32.2.3.2 Syngas

Syngas is created by the joint processes of pyrolysis, combustion and gasification (Alavandi and Agrawal, 2008). This gas is generally produced through a Fisher–Tropsch process in which carbon monoxide and hydrogen react to form organic compounds such as methane, propane, ethane, etc., after which synthetic fuel such as diesel and gasoline can be produced (Dry, 2002). Pyrolysis converts agrofuel into carbon monoxide and energy. Help ignition is known to have a limited supply of oxygen. Gasification converts more natural resources into hydrogen and carbon monoxide. Syngas, the resulting gas mixture, is a fuel in and of itself. Syngas could be singed and burned in internal combustion engines.

32.2.3.3 Siloxanes

Biogas contains siloxanes formed from the anaerobic decomposition of materials commonly found in soaps and detergents. During combustion of biogas containing siloxanes, silicon is released and can combine with free oxygen or various other elements in the combustion gas. Deposits are formed, which contain mostly silica or silicates and can also contain calcium, sulfur, zinc and phosphorus.

32.2.3.4 Gobar Gas

In India, biogas produced from the anaerobic digestion of manure in small-scale digestion facilities is called gobar gas (Demirbas, 2006). The digester is an airtight circular pit made of concrete with a pipe connection. The manure is transferred into the pit, usually directly.

32.3 ALGAE AS THE AGROFUEL

Algae can prove to be a cheap and renewable source of energy for internal combustion engines. Tropical countries like India, which have high temperatures and dry weather coupled with strong winds, favor algae production. Algae can provide diesel and petrol fuels by a process of chemical conversion – the method being boiling algae with hydrochloric acid and methanol. The bubbling of carbon dioxide from waste gases is found to increase the yield of algae five times. Algae can be induced to produce more lipids by controlling the supply of nitrogen and silicon, which helps in converting over two-thirds of the algal mass into lipids. For instance, in a pond of 20 meters, about 80% of the converted lipids can produce 3000 liters of fuel each year. Lipids, in turn, produce fuels like diesel and petrol (Chisti, 2007). On boiling with concentrated hydrochloric acid, these lipids are hydrolyzed to fatty acids. Then, they are esterified with methyl alcohol by a process called "transesterification". The non-lipid fraction of the algae also can be used for producing chemicals.

Microalgae are called the most productive biochemical factories in the world (Chisti, 2007). Microalgae can produce up to 30 times more oil per unit of growth area than land plants. Microalgae technology has the potential to produce 150 to 400 barrels of oil per acre per year. They offer an attractive source of lipids for biodiesel production. The lipid content of microalgae can be increased up to 40–60% by manipulating culture conditions (Tsukahara and Sawayama, 2005).

Bio-butanol can be made entirely with solar energy from algae or diatoms by the Centia process. Centia is based on a three-step thermal, catalytic and reforming process that has the potential to turn virtually any lipid compound, e.g., vegetable oils, oils from animal fat and oils from algae, into 1 for 1 replacements for petroleum jet fuel, diesel and gasoline (Thomas *et al.*, 2009). The three steps are hydrolytic conversion, decarboxylation and reforming long-chain alkanes. The octane rating of n-butanol is similar to that of gasoline but lower than that of ethanol and methanol. n-Butanol has a research octane number (RON) of 96 and a mono octane number (MON) of 78, while t-butanol has octane ratings of 105 RON and 89 MON. A fuel with a higher octane rating is less prone to knocking (extremely rapid and spontaneous combustion by compression), and the control system of any modern car engine can take advantage of this by adjusting the ignition timing. This will improve energy efficiency, leading to a better fuel economy than the comparisons of energy content would suggest. Alcohol fuels, including butanol and ethanol, are partially oxidized and therefore need to run at richer mixtures than gasoline.

32.4 BIOHYDROGEN – MICROBIAL FUEL CELLS

Hydrogen has been considered a potential agrofuel because it burns to produce only water and can be stored as a metal hydride, making it both portable and safe for use in automobiles. Hydrogen can be produced by water hydrolysis and biological processes. It can be produced by anaerobic fermentative bacteria such as *Clostridia*. The substrate employed is cellulose, which is hydrolyzed to sugar and then fatty acids by anaerobic bacteria. Some algae also produce hydrogen in a photo-bioreactor and also produce biomass, which can be used as animal feed.

Agrofuel cells for the generation of electrical energy from abundant organic substrates can be organized by different approaches (biohydrogen). One approach includes the use of microorganisms as biological reactors for the fermentation of raw materials to fuel products, e.g., hydrogen, that are delivered into a conventional fuel cell (Ma, 2006). The second approach to utilize microorganisms in the assembly of agrofuel cells includes the *in situ* electrical coupling of metabolites generated in the microbial cells with the electrode, supporting the use of diffusion electron mediators (Mohan *et al.*, 2008). A further methodology to develop agrofuel cells involves the application of redox enzymes for the targeted oxidation and reduction of specific fuel and oxidizer substrates at the electrode, supporting the generation of electrical power output.

The hydrogen-oxidizing bacteria are characterized by their ability to utilize gaseous hydrogen as an electron donor and carbon dioxide as a sole carbon source, i.e., to grow as chemolithoautotrophs. Biotechnologies based on hydrogen-oxidizing bacteria are, in principle, independent of carbon sources derived from fossil fuel or biomass of photosynthetic origin. The capacity for autotrophic growth is furnished by the presence of both hydrogenases for activating molecular hydrogen and enzymes of the ribulose bisphosphate cycle for fixation of carbon dioxide. In several species, at least some of these enzymes are coded for on indigenous mega-plasmids. However, detailed genetic studies have been restricted to the genes for the key enzymes of autotrophic metabolism, hydrogenase and ribulose bisphosphate carboxylase, and to the pathway for pesticide degradation.

Microbial fuel cells (MFCs) provide new opportunities for the sustainable production of energy from biodegradable, reduced compounds. MFCs have operational and functional advantages over the technologies currently used for generating energy from organic matter. MFCs function on different carbohydrates and complex substrates present in wastewaters. In an MFC, substrate is metabolized by bacteria, which transfer the gained electrons to the anode. This can occur either directly through the membrane or via mobile redox shuttles. An MFC converts energy, available in a bio-convertible substrate, directly into electricity. This can be achieved when bacteria switch from the natural electron acceptor, such as oxygen or nitrate, to an insoluble acceptor, such as the MFC anode. This transfer can occur via either membrane-associated components or soluble electron shuttles. The electrons then flow through a resistor to a cathode, at which the electron acceptor is reduced. In contrast to anaerobic digestion, an MFC creates electrical current and an off-gas containing mainly carbon dioxide.

32.5 BENEFITS OF AGROFUEL

As agrofuels, both ethanol and biodiesel, in addition to being renewable and indigenously available, also help in improving the environment. As far as biodiesel extracted from non-edible oils is concerned, it, also enhances the lubricity of diesel and thus improves the efficiency and durability of the engine. Agrofuels are non-toxic, biodegradable and nonflammable with very high flash points. Apart from these benefits, the program on agrofuel will result in employment generation, particularly in rural areas, greening of waste land, drought proofing, energy security for the country and promotion of organic farming. It will also result in utilization of waste and fallow land in addition to the land on agricultural field boundaries, along public roads and beside railway tracks.

The problem with commercialization and widespread cultivation of non-edible oilseeds for production of biodiesel all these years has been the absence of a ready market in the oil sector. The petroleum industry has had a mindset of using only fossil fuels. The oil sector is looking for indigenous sources to reduce its dependence on imported crude oil, and there can be no better source than ethanol and biodiesel in a country where the majority of the population depends on agriculture.

Agrofuel, when blended with conventional fuels, reduces emissions of air pollutant such as sulfur, particulates, carbon monoxide and hydrocarbons. Ethanol and biodiesel are also less of a hazard if they spill or leak, since they are rapidly biodegradable in water. Substituting agrofuels for 1 gallon of gasoline or diesel saves up to 20 pounds of carbon dioxide emissions to the atmosphere, since they are made from carbon "recycled" by plants instead of carbon dug out of the ground in the form of fossil fuels.

Growing perennial energy crops in place of surplus annual crops can reduce soil erosion and compaction, as permanent deep root systems develop and enrich the soil. Perennial crops need less tilling and fewer agrochemical inputs, so they may help to improve the quality of waterways. Their sturdy root systems and more permanent canopies offer a wider variety of habitats for birds and beneficial insects compared with annual row crops. Levels of soil carbon may increase under perennial crops, helping to offset some fossil fuel carbon dioxide emissions. Soil carbon sequestration may even occur under intensively managed annual crops with limited residue removal, such as the harvest of about half the available corn stover. However, the optimal sustainable level of stover removal will depend on many factors, including erosion control, moisture retention and planned tillage reduction, and will be highly specific to local conditions and topography.

32.6 CONCLUSION

Agrofuels are being seriously considered from the multidimensional perspective of depleting fossil fuel resources, environmental health, energy security, economy and new avenues of gainful employment. The coming years will witness greatly enhanced activities and investments in new technologies and infrastructure for cost-effective production and efficient utilization of agrofuels. In the current scenario, biotechnological techniques are also concentrating on the production of energy from renewable resources and biomasses. Starch from corn, potatoes, sugar cane and wheat is already used to produce ethanol as a substitute for gasoline. New crops may be grown specifically for agrofuel production, including native grasses and trees as well as new high-yielding varieties of oil seed crops. In time, these energy crops may be planted in place of millions of acres of surplus arable crops, surpassing even corn stover as an energy resource. Agrofuels are non-toxic, biodegradable and nonflammable with very high flash points. Apart from these benefits, the agrofuel program will result in employment generation, particularly in rural areas, greening of waste land, drought proofing, energy security for the country and promotion of organic farming. In the future, there may even be financial opportunities for farmers through rewards for good stewardship of the land in terms of "carbon credits".

REFERENCES

Alavandi, S. and Agrawal, A. 2008. Experimental study of combustion of hydrogen–syngas/methane fuel mixtures in a porous burner. *International Journal of Hydrogen Energy* 33(4):1407–1415.

Balat, M. 2007. Global bio-fuel processing and production trends. *Energy Exploration & Exploitation* 25:195–218.

Berchmans, H.J. and Hirata, S. 2008. Biodiesel production from crude *Jatropha curcas* L. seed oil with a high content of free fatty acids. *Bioresource Technology* 99:1716–1721.

Chisti, Y. 2007. Biodiesel from microalgae. *Biotechnology Advances* 25: 294–306.

Demirbas, A. 2006. Biogas potential of manure and straw mixtures. *Energy Sources, Part A: Recovery, Utilization and Environmental Effects* 28(1):71–78.

Demirbas, A. 2008a. The importance of bioethanol and biodiesel from biomass. *Energy Sources Part B* 3:177–185.

Demirbas, A. 2008b. Biodiesel: a realistic fuel alternative for diesel engines. *Chemistry and Materials Science.* London: Springer.

Dry, M.E. 2002. The Fischer-Tropsch process: 1950–2000. *Catalysis Today* 71: 227–241.

Fukuda, H. 2001. Biodiesel fuel production by transesterification of oils. *Journal of Bioscience and Bioengineering* 92:405–416.

Gui, M.M., Lee, K.T. and Bhatia, S. 2008. Feasibility of edible oil vs. non-edible oil vs. waste edible oil as biodiesel feedstock. *Energy* 33:1646–1653.

Han, H.W., Cao, W.L. and Zhang, J.C. 2005. Preparation of biodiesel from soybean oil using supercritical methanol and CO_2 as co-solvent. *Process Biochemistry* 40:3148–3151.

Hashem, A., Akasha, R.A., Ghith, A. and Hussein, D.A. 2007. Adsorbent based on agricultural wastes for heavy metal and dye removal: a review. *Energy Education Science and Technology* 19:69–86.

Kallivroussis, L., Natsis, A. and Papadakis, G. 2002. The energy balance of sunflower production for biodiesel in Greece. *Biosystems Engineering* 81:347–354.

Ma, Y.L. 2006. Microbial oils and its research advance. *Chinese Journal of Bioprocess Engineering* 4:7–11.

Mohan, S.V., Babu, V.L. and Sarma, P.N. 2008. Effect of various pretreatment methods on anaerobic mixed microflora to enhance biohydrogen production utilizing dairy wastewater as substrate. *Bioresource Technology* 99:59–67.

Openshaw, K. 2000. A review of *Jatropha curcas*: an oil plant of unfulfilled promise. *Biomass and Bioenergy* 1–15.

Patil, P.D. and Deng, S. 2009. Optimization of biodiesel production from edible and nonedible vegetable oils. *Fuel* 88:1302–1306.

Prasad, K., Maria, S., Anne, B.T., Prawit, K. and Irni, A. 2009. Bioethanol, biohydrogen and biogas production from wheat straw in a biorefinery concept. *Bioresource Technology* 100:2562–2568.

Rashid, U. and Anwar, F. 2008. Production of biodiesel through optimized alkaline catalyzed transesterification of rapeseed oil. *Fuel* 87:265–273.

Reshu, C. Rakesh, S.S. and Kureel, R.S. 2012. Jatropha: promising Global Hope of Biodiesel for next generation. *Biotech Today* 2(2):26–27.

Richa R. 2014. Agrofuel: an alternate energy source. *Biotech Today* 4(2): 14–17.

Saka, S. and Kusdiana, D. 2001. Biodiesel fuel from rapeseed oil as prepared in supercritical methanol. *Fuel* 80:225–231.

Schmidt, L.D. and Dauenhauer, P.J. 2007. Chemical engineering: hybrid routes to agrofuels. *Nature* 447:914–915.

Sheehan, J., Aden, A., Paustian, K., Killian, K., Brenner, J., Walsh, M. and Nelson, R. 2003. Energy and environmental aspects of using corn stover for fuel ethanol. *Journal of Industrial Ecology* 7(3–4):117–146.

Stevens, D.J., Wörgetter, M. and Saddler, J. 2004. Agrofuels for transportation: an examination of policy and technical issues. IEA Bioenergy Task 39, Liquid Agrofuels Final Report 2001–2003, Paris.

Thomas, J.B., Arron, W. and Alexander, N. 2009. Composition-explicit distillation cures for mixtures of gasoline with four carbon alcohols (butanols). *Energy Fuels* 23(4):2295–2306.

Tijmensen, M.J.A., Faaij, A.P.C., Hamelinck, C.N. and van Hardeveld, M.R.M. 2002. Exploration of the possibilities for production of Fischer-Tropsch liquids and power via biomass gasification. *Biomass and Bioenergy* 23:129–152.

Tsukahara, K. and Sawayama, S. 2005. Liquid fuel production using microalgae. *Journal of the Japan Petroleum Institute* 48:251–259.

Zarzyycki A, Polska W. 2007. Bioethanol production from sugar beet – European and Polish perspective. The first TOSSIE workshop on technology improvement opportunities in the European sugar industry, Ferrara, Italy, January 25–26.

33 Holistic Management of Air Pollution Using Modern Technology

Sachin Agarwal and Boyina Rupini
Environmental Studies, SOITS, Indira Gandhi National Open University, New Delhi

Dinesh K. Saxena
Department of Botany, Bareilly College, Bareilly

CONTENTS

DOI: 10.1201/9781003220824-37

33.1 INTRODUCTION

Air pollution is one of the burning problems in the present world, especially in urban areas of developing countries due to the fast growth of population, increase in number of automobiles and industrialization. Vehicular emissions have been regarded as the primary cause and concern of air pollution in the urban areas and account for 60–70% of the pollution found in the vicinity of the urban environment. Urban air pollution is a serious concern, as in 2014, more than half of the world's population lived in cities, and a further growth to 66% by 2050 is expected.

Over the past few decades, there has been a revolution in air monitoring methods worldwide. Air pollution is primarily due to three sources: vehicles, and industrial and domestic sources. Depending upon their origin and sources, the atmospheric particulates may be natural or anthropogenic in nature. The main natural sources are geogenic in origin, caused by non-living things, including sea salt, soil, re-suspended dust, volcanoes, wildfires, lightning and radioactive decay, and some are biogenic, including emission of methane from metabolic activities of cattle, volatile organic compounds (VOCs) from vegetation, etc. Natural sources mainly result in the formation of coarse particulates, whereas fine particulates are generated by anthropogenic sources. Anthropogenic sources are more dominant and concentrated in urban areas, which mainly results in the emission of fine particulate matter. These sources include vehicular emissions, industrial activities, building, road dust and emission from household works. Coal-based power stations, oil-fired furnaces/boilers, stone crushers, hot mix plants, lime kilns, foundry, stationary and portable diesel generators, diesel vehicles (bus and trucks), re-suspension of road dust, dust from huge construction sites and burning of biomass, i.e., tires and tubes, and chlorofluorocarbons (CFCs) are major man-made sources of air pollution. Respirable suspended particulate matter (RSPM) less than 4 microns in diameter, fine particulate matter (FPM) less than 2.5 microns, and particles less than 0.1 microns are termed ultrafine particulate matter. Air pollutants are categorized into two types: primary and secondary. Primary pollutants include carbon compounds (CO, CO_2, CH_4 and VOCs), nitrogen compounds like NO, N_2O and NH_3, sulfur compounds (H_2S and SO_2), halogen compounds and particulate matter. Tropospheric ozone (O_3) is a good example of a secondary pollutant, which is formed during photochemical reactions between automobile and industrial exhausts like VOCs and nitrogen oxides (NO_x) in the presence of sunlight. The combustion of fossil fuels as energy sources releases greenhouse gases, a principal contributor to climate change. The transportation sector is an important contributor, representing about 14% of greenhouse gas emissions globally and twice that amount in the United States (US EPA (Environmental Protection Agency), 2019). In 2017, on-road vehicles in the United States, including light-duty passenger cars and trucks, buses, and commercial and freight trucks, consumed 11.6 million barrels per day (b/d) oil equivalent, which accounted for 80% of all transportation energy use and 31% of all delivered end-use energy in the United States (US Energy Information Administration, 2018). The impact of autonomous vehicles (AVs) on climate change depends on factors such as the impact on vehicle mean time (VMT), the energy usage per vehicle, and the source of energy used. AV use could reduce energy consumption through reduced parking hunting, ridesharing, eco-driving, congestion mitigation, collision avoidance and vehicle/power resizing (US Energy Information Administration, 2018).

Non-exhaust processes involve mechanical abrasion and corrosion, which lead to the generation of particulate matter. The wearing out of tires, brakes and road surface leads to the direct emission of particulate matter into the environment. Other potential sources that may contribute to the particulate matter in the environment include abrasion of automobile parts (wheel bearings, wear out of clutch and engine, corrosion of vehicle components). It has been reported that brake wear is a major source for some metals, while tire wear contributes the least of the non-exhaust sources (Amato et al., 2014). In a report, it was found that Fe, Cu, Zn and Pb are the most abundant metals present in the brake lining material (Bukowiecki et al., 2009).

World Health Organization (WHO) Air Quality Guidelines (AQG) offer guidance on threshold limits for key air pollutants that pose health risks and provide a reference for setting air pollution targets at regional and national levels for better air quality. Air quality guidelines were published by WHO in 1987 and revised in 1997. The 2005 update represents the recent assessment of air pollution health effects, supported by an expert evaluation of the scientific evidence. The set of guidelines have estimated the exposure levels for particulate matter (PM10 and PM2.5), ozone, nitrogen dioxide and sulfur dioxide, as well as a set of interim targets to encourage a progressive improvement in air quality.

33.2 AIR POLLUTION: A GLOBAL ISSUE

Air pollution isn't restricted to a state, country, region or continent; it could have a significant trans-boundary effect on global climate and weather. The effects of acid rain included disappearance of fish from lakes in the Adirondack of North America, corrosion of marble of the Taj Mahal mausoleum situated at Agra, India, the widespread death of forests in the mountains of Europe, and damage to tree growth in the United States and Canada. This has led to the drafting of international agreements to limit the emissions of oxides of sulfur and nitrogen. Depletion of the ozone layer is another global problem caused by air pollution. Ozone can also be categorized as good ozone and bad ozone. At ground level (i.e., in the troposphere), ozone is a pollutant, but at higher altitudes above 12 km, it plays a vital role in blocking ultraviolet radiation (UV) from the sun before it reaches the earth. Exposure to UV radiation has been linked to skin problems such as cancer and other health issues. In 1985, it was found that an oversized "ozone hole" is present above Antarctica every year between August and November. The size of this hole is increased by the presence of CFCs in the atmosphere; common sources of CFCs are aerosol spray cans, refrigerators, industrial solvents and other sources, and the CFCs are transported to Antarctica by atmospheric circulation. It had already been demonstrated in the mid-1970s that CFCs posed a threat to the worldwide ozonosphere, and in 1978 the application of CFCs as propellants in aerosol cans was banned in the United States. Subsequently, their use was restricted in several other countries, and manufacturers were motivated to avoid usage of CFCs. In 1987, representatives from over 45 countries signed the Montreal Protocol, agreeing to impose severe limitations on the generation of CFCs. Another highly significant effect of air pollution is global warming and climate change. With the rise in the fossil fuel consumption, CO_2 levels in the atmosphere have increased steadily since 1900, and the rate continues to be on the rise. It has been predicted that if CO_2 levels are not reduced, average global air temperatures may rise another 4°C (7.2°F) by the end of the 21st century. This may lead to the melting of the polar ice caps, elevation in the level of sea water, and flooding of coastal areas globally. Alterations in precipitation patterns caused by climate change and global warming could cause adverse effects on agriculture and forest ecosystems, and higher temperatures and humidity might increase the incidence of disease in humans and animals in some parts of the earth. Implementation of international agreements on reducing greenhouse gases is required to protect global air quality and to mitigate the consequences of global warming.

It is estimated that by 2050, the global car fleet will result in the emission of greenhouse gases faster than any other sector. The world's countries emit vastly different amounts of heat-trapping gases into the atmosphere. Data compiled by the International Energy Agency estimates that globally, China United States and India (Figure 33.1) are leading emitters of carbon dioxide (CO_2) from the combustion of coal, natural gas, oil and other fuels, including industrial waste and non-renewable municipal waste.

Air pollution is one of the major risk factors for death. In low-income countries, it tops the list. In 2017, it was responsible for an estimated 5 million deaths globally. That means it contributed to 9%, i.e., nearly 1 in 10 deaths. We see that the death rates tend to be highest across Sub-Saharan Africa and South Asia. The burden of air pollution tends to be greater across both low- and middle-income

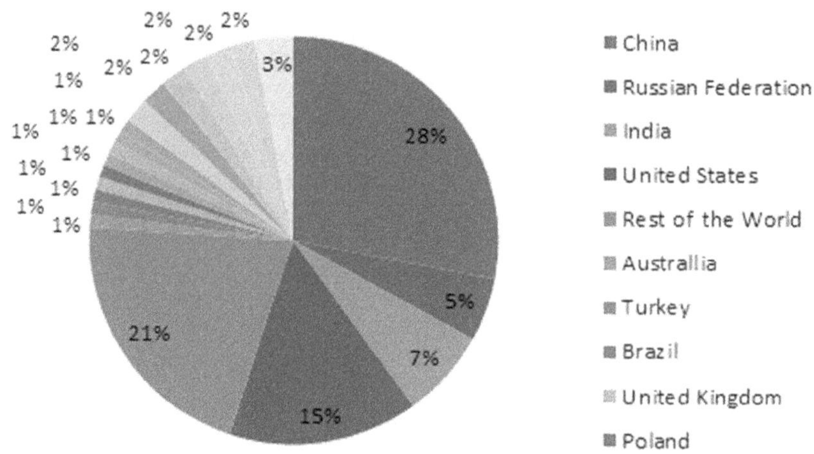

FIGURE 33.1 CO_2 emission data by country, 2018.

(From www.ucsusa.org/.)

countries for two reasons: indoor pollution rates tend to be high in low-income countries as they depend on solid fuels for cooking; and outdoor air pollution tends to increase as countries industrialize and shift from low to middle incomes. Globally, we see that in recent decades the death rate from total air pollution has declined: since 1990, the number of deaths per 100,000 people had nearly halved. But, as we see from the breakdown, this decline has been primarily driven by improvements in indoor air pollution (source: IHME, Global Burden of Disease). WHO reported in 2018 that around 7 million people die each year due to both indoor and outdoor air pollution. The three biggest killers attributable to air pollution are stroke (2.2 million deaths), cardiac disease (2.0 million), and lung disease and cancer (1.7 million deaths). Ambient (outdoor) air pollution accounts for 25% of all deaths and disease from lung cancer, 17% of all deaths and disease from acute lower respiratory infection, 16% of all deaths from stroke, 15% of all deaths and sickness from ischemic cardiovascular disease, and 8% of all deaths and sickness from chronic obstructive pulmonary disease. Air pollution doesn't just kill, however. It also contributes to other illnesses, hampers development and causes neurological disorders. Recent research from the United Nations Children's Fund (UNICEF) shows that inhaling particulate air pollution can damage brain tissue and undermine cognitive development in young children – with lifelong implications. Other studies have linked air pollution to lower intelligence levels, with the average impact equivalent to one lost year of education, and to an increased risk of dementia, with those living closest to major traffic arteries up to 12% more likely to be diagnosed with the condition.

33.3 AIR POLLUTION MANAGEMENT

In order to combat air pollution-related issues, a holistic blueprint is needed that could be efficiently implemented globally. The approach discussed in this chapter consists of four steps (Figure 33.2): (a) monitoring, (b) air dispersion modeling, (c) air pollution management policies and (d) management technologies.

Monitoring of air pollutants will help in planning the mitigation program, which is approached by analysis and interpretation of the air quality data. Next, the data obtained from analysis will be used in modeling. Finally, modeling of the monitored data may play a crucial role in future prediction related to air pollution. On the basis of modeling, we can develop the policies, strategies and technologies to combat challenges of air pollution and air quality-related issues.

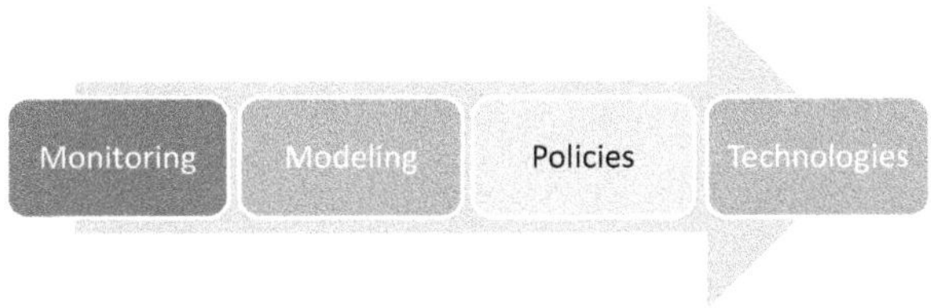

FIGURE 33.2 Holistic approach to air pollution management.

33.3.1 Monitoring of Air Pollutants

Ambient air quality monitoring is required to determine the existing quality of air, to evaluate the effectiveness of the control program and to identify areas in need of restoration and their prioritization. This interpreted data would be very useful as reference material during future studies as well as to observe percentage change. It can be utilized by policy-makers to prepare mitigation programs.

The major objectives for air quality monitoring are as follows:

- (i) Background data
- (ii) Interpreted background data and its use as a reference for the future
- (ii) Status and trend
- (iii) Environmental exposure level determination
- (iv) Scavenging behavior of environment
- (v) Air quality management

Air quality monitoring should be done in areas where a pollution problem exists or is expected, i.e., mainly in industrial areas, urban areas, traffic intersections, etc. One of the objectives of monitoring is to determine status and trends, and the air quality monitoring should be done in metropolitan cities and other urban areas so as to compare their levels and determine trends. Selection of the site is very important, as an incorrect location may result in data that may not meet the objectives of monitoring and will be of limited value. The site should be away from major pollution sources. The distance depends upon the source, its height and its emissions. The station should be at least 25 m away from domestic chimneys, especially if the chimneys are lower than the sampling point; with larger sources, the distance should be greater. The air monitoring site should be away from absorbing surfaces such as absorbent building material. The clearance to be allowed will depend on the absorbing properties of the material for the pollutant in question, but it will normally be at least 1 m. A number of air monitoring methods are available and are summarized in Table 33.1. These efficiently measure the particulate matter, gaseous pollutants and VOCs present in air.

33.3.2 Air Dispersion Modeling

Air pollution modeling is the term used to describe using mathematical theory to understand, or predict, the way pollutants behave in the atmosphere. Similarly, a model can be used to predict alternative situations. It is a mathematical tool used to elaborate the causal relationship between emissions, meteorology, atmospheric concentrations, deposition and other related factors. Dispersion modeling of air plays a vital role in environmental science because of its ability to assess and predict the impact of air quality on human health and provide a reasonable solution to traffic-induced air pollution issues at local as well as global levels. This makes air pollution models indispensable in regulatory, research and forensic applications. The concentrations of substances in the atmosphere are determined by 1) transport, 2) diffusion, 3) chemical transformation and 4) ground deposition. Transport phenomena, characterized by the mean velocity of the fluid, have been measured and

TABLE 33.1
Air Pollutant Monitoring Techniques

Pollutants	Time Weighted Average	Concentration in Ambient Air: Industrial, Residential, Rural and Other Areas	Concentration in Ambient Air: Ecologically Sensitive Area (Noticed by Central Government)	Methods of Measurement
Sulfur Dioxide (SO_2), µg/m^3	Annual[1] 24 Hours[2]	50 80	20 80	- Improved West and Gaeke method - Ultraviolet fluorescence
Nitrogen Dioxide (NO_2), µg/m^3	Annual[1] 24 Hours[2]	40 80	30 80	- Jacob & Hochheiser modified (NaOH-$NaAsO_2$) method - Gas phase chemiluminescence
Particulate Matter (Size less than 10 µm) (PM10), µg/m^3	Annual[1] 24 Hours[2]	60 100	60 100	- Gravimetric - TEOM - Beta attenuation
Particulate Matter (Size less than 2.5 µm) (PM2.5), µg/m^3	Annual[1] 24 Hours[2]	40 60	40 60	- Gravimetric - TEOM - Beta attenuation
Ozone (O_3), µg/m^3	8 Hours[1] 1 Hour[2]	100 180	100 180	- UV photometric - chemiluminescence - Chemical method
Lead (Pb), µg/m^3	Annual[1] 24 Hours[2]	0.50 1.0	0.50 1.0	- AAS/ICP method after sampling on EPM 2000 or equivalent filter paper - ED-XRF using Teflon filter
Carbon monoxide (CO), mg/m^3	8 Hours[1] 1 Hour[2]	02 04	02 04	- NDIR spectroscopy
Ammonia (NH_3), µg/m^3	Annual[1] 24 Hours[2]	100 400	100 400	- Chemiluminescence - Indophenol blue method
Benzene(a) Pyrene (BaP) Particulate phase only, ng/m^3	Annual[1]	05	05	- GC-based continuous analyser - Adsorption and desorption followed by GC analysis
Benzol(a) Pyrene (BaP) Particulate phase only, ng/m^3	Annual[1]	01	01	- Solvent extraction followed by HPLC/GC analysis
Arsenic (As), ng/m^3	Annual[1]	06	06	- AAS/ICP method after sampling on EPM 2000 or equivalent filter paper
Nickel (Ni), ng/m^3	Annual[1]	20	20	- AAS/ICP method after sampling on EPM 2000 or equivalent filter paper

Source: CPCB, 2003.

AAS = atomic absorption spectroscopy; GC = gas chromatography; HPLC = high performance liquid chromatography; ICP = inductively coupled plasma spectroscopy; NDIR = non-dispersive infrared; TEOM = tapered element oscillating microbalance.

1 Annual arithmetic mean of minimum 104 measurements in a year at a particular site taken twice a week 24-hourly at uniform intervals.

2 24-hourly or 8-hourly or 1-hourly monitored values, as applicable, shall be complied with 98% of the time in a year. The remaining 2% of the time, they may exceed the limits, but not on two consecutive days of monitoring.

studied for centuries. For example, the average wind has been studied for sailing purposes. The study of diffusion (turbulent motion) is more recent. Among the first articles that mention turbulence in the atmosphere are those by Taylor (1915; 1921). Parameters considered for the dispersion model were mean height above sea level, geographic information system (GIS) coordinates, wind velocity and direction, ambient air temperature, background pollution, traffic volume, vehicle type and speed. Moreover it provides important information to decision makers for air quality assessment and management. The primary sources of air pollution affecting environmental living quality in urban areas are vehicular exhaust and non-exhaust emissions (Duclaux et al., 2002; Rebolj and Sturm, 1999).

Different air modeling approaches were followed: Gaussian plume, Lagrangian modeling and Eulerian modeling. The Gaussian plume model is the most common air pollution model. It is based on a simple formula that describes the three-dimensional concentration field generated by a point source under stationary meteorological and emission conditions. In Lagrangian modeling, an air parcel (or "puff") is followed along a trajectory and is assumed to keep its identity during its path. In Eulerian modeling, the area under investigation is divided into grid cells in both vertical and horizontal directions. Lagrangian modeling, directed at the description of long-range transport of sulfur, began with studies by Rohde (1972; 1974). Lagrangian modeling is often used to cover longer periods of time, up to years (Eliassen and Saltbones, 1975).

There are several different types of air quality models (Fisher, 1975), all used for differing purposes. The most common models are broadly known as Atmospheric Dispersion Models (ADMs). These models use mathematical assumptions about the way that the atmosphere behaves to assess the impact of emissions. In a Gaussian plume dispersion model, different point source plumes (Figure 33.3) can be observed, such as coning, fanning, looping, lofting and fumigation.

There are a number of air modeling software programs, such as CALINE4, CALPUFF, HYSPLIT and AERMOD, to predict air quality for future reference.

One of the most popular and free access air modeling software programs is the HYSPLIT model. This model (www.ready.noaa.gov/HYSPLIT.php) is a complete system for computing simple air parcel trajectories as well as complex transport, dispersion, chemical transformation and deposition simulations. HYSPLIT continues to be one of the most extensively used atmospheric transport and dispersion models in the atmospheric sciences community. The model calculation method is a hybrid

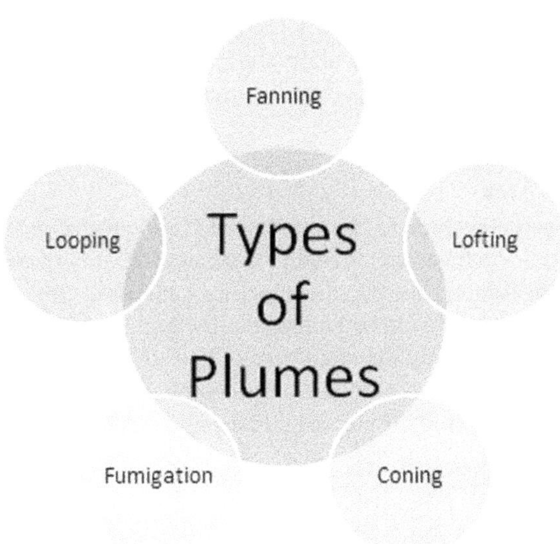

FIGURE 33.3 Different types of point source plumes of Gaussian plume dispersion model.

between the Lagrangian approach, which uses a moving frame of reference as the air parcels move from their initial location, and the Eulerian approach, which uses a fixed three-dimensional grid as a frame of reference. In the model, advection and diffusion calculations are made in a Lagrangian framework following the transport of the air parcel, while pollutant concentrations are calculated on a fixed grid.

Through a joint effort between the National Oceanic and Atmospheric Administration (NOAA) and Australia's Bureau of Meteorology, the model uses advection algorithms, updated stability and dispersion equations, a graphical user interface, and the option to include modules for chemical transformations. HYSPLIT can be run interactively on Air Resources Laboratory's READY (Real-time Environmental Applications and Display sYstem) website, or it can be installed on a PC or LINUX workstation and run using a graphical user interface. The model is designed to support a wide range of simulations related to the atmospheric transport and dispersion of pollutants and hazardous materials to the earth's surface. Some of the applications include tracking and forecasting the release of radioactive material, volcanic ash, wildfire smoke, and pollutants from various stationary and mobile emission sources. Operationally, the model is used by NOAA's National Weather Service through the National Centers for Environmental Prediction and at local Weather Forecast Offices. The HYSPLIT model offers two types of air modeling.

In order to interpret and analyze the air trajectory to study the seasonal and annual trends, HYSPLIT Trajectory model is implemented. NOAA's HYSPLIT model features four different trajectories:

(a) Normal
(b) Matrix
(c) Ensemble
(d) Frequency

The HYSPLIT air dispersion model is free access software to study and analyze the dispersion of air pollutants in the atmosphere. This require certain information such as release type, release quantity, meteorology and starting locations.

33.3.3 Air Pollution Management Policies

The majority of policies for improving air quality focus on discouraging self-owned vehicles as well as investing in infrastructure that promotes alternative transport. The upcoming section describes key air pollution management policies and offers illustrative examples of where they have been put into practice.

33.3.3.1 Low Emission Area

There are certain areas in which vehicular movement needs to be restricted in order to cut down tail-pipe emissions. Such areas are demarcated to ensure that the air quality of cities is in accordance with national emission standards, which have become more stringent over time, particularly in proximity to sensitive areas such as schools, hospitals, and residential and world heritage sites. More than 100 localized air quality measures are implemented across 12 European countries, including England, Italy, Sweden and Holland (Holman et al., 2015). In a study carried out in a few Dutch cities, it is reported that more wide-ranging limitations should be imposed in order to attain improved results for air quality (Boogaard et al., 2012). The use of electric vehicles is encouraged by Oxford City Council in their recently set out plans to accomplish a zero-emission zone starting in 2020 (www.oxford.gov.uk/zez accessed on 6 March 2019). Amongst battery-operated commutator vehicles, Tesla has done tremendous work abroad, while Tata motors has launched battery-operated vehicles after developing indigenous technology.

Another approach to create a positive impact on air quality is to set up congestion charging zones. This may increase the cost of driving, which may encourage the use of public or shared transport instead of private vehicles. However, this may improve air quality very little. A major concern about low emission areas and congestion charging zones is that they do not reduce air pollution; rather, they shift the pollutants from the low emission area to the surrounding areas. By and large, there may not be a significant positive outcome in terms of better air quality (Hawkes, 2015).

33.3.3.2 Use of Public Transport

During the past decade, there has been rapid urbanization in the developing countries; as a result, there is immense population pressure in the cities. The growing interest in road constructions and their impacts on environment and society arises from the fact that globally, more than 25 million kilometers of new roads are expected to be built by 2050, 90% of which will be in developed countries (Laurance et al., 2014).

In developing countries, due to uncomfortable public transport facilities, people are compelled to use their own transport, which eventually results in an increase in emissions. The number of on-road vehicles is increasing at a fast pace, and urban air quality policy discourages the use of self-owned vehicles and promotes less polluting forms of transport, such as cycling and public transport, which could be more appealing. To bring about a behavioral shift towards the use of less polluting means of transport, robust and sustainable infrastructure should be developed, such as a network of dedicated cycle lanes, which could be expensive but may facilitate long-term environmental benefits. This is because the impacts of the schemes depend on the number of participants, the extent to which morning walkers and cyclists are exposed to air pollution, and the population density of the surrounding area, amongst others (www.eunomia.co.uk/report-category/client/sustrans/). In a study carried out in Barcelona City, it was observed that a 40% reduction in car trips caused a decline in PM2.5 concentrations of only 0.64%, and 10.03% in annual casualties related to air pollution (Rojas-Rueda et al., 2012). Serious consideration must be given to limiting the exposure of cyclists to vehicle emissions; otherwise, this may defeat the purpose of the cycling initiative. Safety-related issues, comfortable road infrastructure and dedicated cycling tracks are some factors that influence cycling, including convenience (Mackett, 2003), distance (Scheiner, 2010), safety perception (Bonham and Koth, 2010; Wang et al., 2014) and weather conditions (Sabir, 2011), amongst others. Simpler and less resource-intensive initiatives include minor changes to existing infrastructure, such as ensuring that cyclists are permitted to take their bicycles onto trains and providing safe parking facilities at train stations (Rietveld, 2000). These smaller measures are unlikely to have a measurable impact on air quality individually but may cumulatively assist in creating an environment where cycling is a more appealing option. Comfortable and safe public transport, especially for women and children, will encourage people to adopt eco-friendly transport, which can be further assisted by technology, such as mobile applications for car pools (https://theconversation.com/toxic-emissions-down-but-people-still-dying-from-air-pollution-itstime-for-something-radical-91875). Monetary benefits or discount schemes may be offered to citizens availing themselves of and opting for an ecofriendly transport system, as in San Francisco and in many other countries where employers promote the use of cycling, carpooling and public transport by their staff members through subsidies or pre-tax relaxations on transport costs .A "cycle day" could be organized by employers, encouraging the employees by compulsory usage of cycle or public transport as the only means of transport to reach their work place, just as the UK government runs a "Cycle to Work" Scheme, in which employers can loan employees a bicycle tax free, sometimes in the form of a salary sacrifice arrangement (www.gov.uk/government/publications/cycle-to-work-scheme-implementation-guidance).

33.3.3.3 Speed Management

Apart from highways, it is observed that in towns average vehicular speed is around 40 kph and at some places even 20–30 kph. Unnecessary high speed in town areas not only increases the pollutant levels in the air but also poses a threat of accidents. Lower speed limits serve a dual purpose, which may be an economical way of creating a positive impact on air quality near roads (Porter et al., 2010), and are presently implemented in several European countries to meet air quality standards (D'Elia et al., 2018). Restrictions on vehicular speed also avoid unnecessary acceleration and application of brakes for speed control of the vehicles. The re-suspended particulate matter produced due to the wear and tear of brakes and tires decreases by approximately half the amount of particulate matter from the brakes when stopping from 20 mph instead of 30 mph (www.theguardian.com/environment/2016/sep/11/polluting-effect-wear-brakes-tyres-pollutionwatch). The effects of speed management measures are dependent on local weather conditions and the physical infrastructure surrounding the road. For example, in Amsterdam, a speed limit reduction from approximately 62 mph to 50 mph resulted in decreased particulate matter concentrations but had no effect on NO_x (Dijkema et al., 2008). Speed management has in places been complemented by "eco-driving" campaigns, which educate the public on fuel-efficient forms of driving. An eco-driving program carried out in the Netherlands between 1994 and 2004 is thought to have reduced fuel consumption by between 0.3% and 0.8% (Harmsen et al., 2007).

33.3.3.4 Avoid Burning Solid Fuels

Almost 3 billion people, mostly in low- and middle-income countries, still depend on solid fuels (wood, animal dung, charcoal, crop wastes and coal) burned in inefficient and highly polluting stoves for cooking and heating. In 2016 alone, no fewer than 3.8 million children and adults died prematurely from illnesses caused by such household air pollution, according to estimates by the WHO. Together with widespread use of kerosene stoves, heaters and lamps, these practices also lead to many serious injuries and deaths from scalds, burns and poisoning. These new indoor air quality guidelines for household fuel combustion aim to assist public health policy-makers, as well as specialists working on energy, environmental and other issues, to understand the best approaches for reducing household air pollution, which is a potential environmental health risk around the world today. Open fuel burning during the winter season is very common, which contributes significantly to air pollution. Traffic congestion increases the consumption of petrol and diesel. In India, two Mahatma Gandhi National Rural Employment Guarantee Act (MGNREGA) programs could be run for the cost of the extra fuel India burns due to traffic congestion. In the past few years, significant road and transport infrastructure has been developed in India to avoid jams.

33.3.4 Management Technologies

33.3.4.1 Vehicle Technologies

The major worldwide contribution to carbon emission is from transportation. Particularly, the United States will need to cut emissions from the transportation sector, which accounted for 26% of the country's greenhouse gas emissions in 2014. Of these transportation sector emissions, 61% came from light-duty vehicles, such as passenger cars. China, the United States and India are major CO_2 emitters. Researchers have suggested layering three vehicular technologies to combat the carbon emissions. Autonomous, electric and shared vehicle technology is now supposed to be a solution to this problem.

"Autonomous technology" refers to technology that has the capability to drive a vehicle without active physical control or monitoring by a human operator (State of California 2019). There are six levels of vehicle autonomy, as defined by the Society of Automotive Engineers (SAE): Levels 0–2 are those where the human driver needs to monitor the driving environment, and levels 3–5 are those where an automated driving system monitors the driving environment (also referred to in the US federal policy guidance as highly automated vehicles) (Zmud et al., 2017). A fully autonomous vehicle is a vehicle that has a full-time automated driving system, which undertakes all aspects of driving that would otherwise be undertaken by a human, under all roadway and environmental conditions (Rojas-Rueda, 2017). In 2018, Waymo, the Google subsidiary developing AVs, introduced the first shared AV fleet to the market (Waymo-One) (https://waymo.com/). Recent estimates suggest that by 2020, 5% of car sales will be AVs, representing 2% of the vehicle fleet and 4% of the miles traveled in the United States (Litman, 2019). The same estimates predict that by 2030, AVs will cover 40% of car market sales, representing 20% of the vehicle fleet and 30% of the miles traveled in the United States (Litman, 2019).

Various patterns of AV ownership and use have been suggested; for instance, private AVs imply private vehicle ownership and private use, and shared autonomous vehicles (SAVs) imply shared use with or without vehicle ownership. Variants include car-sharing, personal vehicle sharing, ridesharing, and cab on call (Figure 33.4). Car-sharing is a model of shared transportation in which several people use the same vehicle at a different time without car ownership. Car-sharing may be station based, where the car is picked up and returned to the same location, and free floating, where the car is picked up at one location and left near the user's destination. Personal vehicle sharing is a system in which car owners convert their personal vehicles into shared cars and rent them to others on a short-term basis; this arrangement could be between peers (peer-to-peer) or through shared vehicle ownership (fractional ownership). Ridesharing pools multiple travelers with similar or overlapping paths (origins/destinations) and departure times in the same vehicle (carpooling or vanpooling). Cab on call refer to vehicle sharing with door-to-door services and is classified as ridesourcing or ridesplitting. Ridesourcing is a door-to-door service that uses private vehicles for paid on-demand rides (such as Uber or Lyft). Ridesplitting is a variant of the ridesourcing model, in which passengers with similar or overlapping routes split a fare and ride in a ridesourcing vehicle.

The AV industry and authorities claim that improved traffic safety would be one of the significant beneficial impacts of AV use (General Motors, 2018, NHTSA (National Highway Traffic Safety Administration), 2019). In 2017, 37,133 people were killed in motor vehicle crashes in the United States (including nearly 7,000 pedestrians and cyclists) (Natl. Cent. Stat. Anal. 2018; US Dep. Transp. 2018). Of all serious motor vehicle crashes, 94% involve driver-related factors, such as impaired driving, distraction, and speeding or illegal maneuvers (US Department of Transportation, 2018). Globally, road traffic incidents are one of the leading causes of mortality, with 1.3 million people killed each year (WHO, 2018a), and almost 90% of those road traffic deaths are concentrated in low- and middle-income countries (Luttrell et al., 2015), despite the fact that these countries only

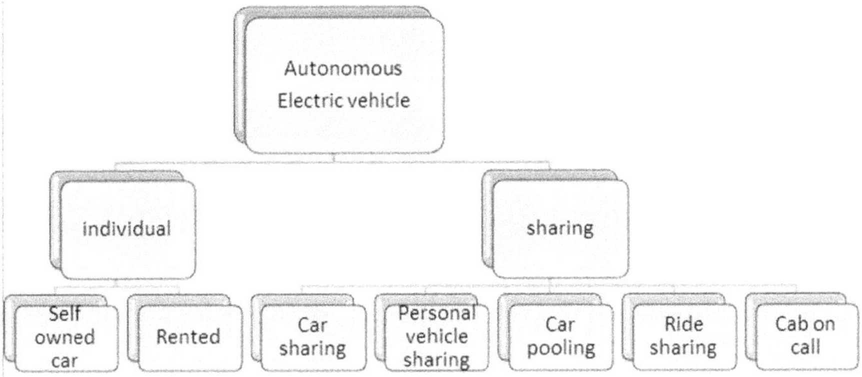

FIGURE 33.4 Autonomous vehicle by type of use and ownership.

(From Annual Review of Public Health 2020.)

have 48% of the world's registered vehicles (WHO, 2018b). Fully automated vehicles could lead to reductions in the number of driver-related crashes (Luttrell et al., 2015). Those authors modeled the expected impacts of AVs on motor vehicle crash injuries and fatalities. They estimated that if 90% of the automobiles in the United States became autonomous, an estimated 25,000 lives could be saved each year, with annual economic savings estimated at more than $200 billion in the United States (Luttrell et al., 2015). Improved road safety related to AV use may lead to a decline in road accidents (Adams and Hobson, 2016). In 2018 in the United States, organ donations from motor vehicle accidents due to excessive speed or poor driving pattern represented 13% of all donations. Transport-related physical activity (walking, cycling, or walking to and from public transport) has been suggested as a strategy for increasing daily levels of physical activity (Dons et al., 2015; Rojas-Rueda et al., 2011; 2012).

If AVs are not fully electric, future higher exposure periods to air pollution may affect AV travelers, and higher air pollution exhaust emissions would then affect the general public. Air pollution from motor vehicles is not limited to exhaust emissions. Other sources include brake and tire wear, road surface wear, and re-suspension of road dust. Together, these may exceed tailpipe emissions, at least with regard to particulate matter (Amato et al., 2014a). Moreover, brake and tire wear particles may have higher oxidative potential than other traffic-related sources, which could translate to worse health impacts (Amato et al., 2014b). Electric vehicles also have been suggested to emit more non-exhaust emissions because they weigh more than nonelectric vehicles. If AV use increases VMTs, even with a shift to electric vehicles, non-exhaust emissions will be an issue for air quality. As in the case of air pollution, if AV use results in increased VMTs, then noise exposure will increase commensurately (Soteropoulos, Berger, Ciari, 2019).

In general, AVs offer an opportunity to promote the transition from fossil fuels to renewable sources of energy if the AVs are implemented as fully electric vehicles together with a supply chain based on renewable energy sources. Electric vehicles will have zero emissions, and there will be a noticeable fall in greenhouse gases worldwide. Apart from the tailpipe emissions, there are numerous benefits of using electric vehicles (Figure 33.5). As the electric vehicle cannot be accelerated beyond a reasonable speed, lower use of brakes will also reduce the chance of non-exhaust emissions. These non-exhaust emissions are generally due to wear and tear of the brakes and tires. This in turn will also reduce abrasion of metal parts of vehicles.

People traveling to a common destination could share the transport, thereby decreasing on-road traffic, which could appreciably reduce the congestion problem. Congestions and traffic jams lead

	Health determinant							
Recommendations	Road safety	Physical activity	Clean energy,energy consumption and climate change	Air Pollution	Noise	Social Interaction	Land use	Social enquiry,autonomy, inclusion,employment, and economy
Favour shared AVs over Private Avs	Yes	Yes	Yes	Yes	Yes	Yes	Yes	yes
Favour Rideshare and Ridesplitting	Yes	Yes	Yes	Yes	Yes	Yes	Yes	yes
Integrate shared-electric AVs in the public transport system	Yes	Yes	Yes	Yes	Yes	Yes	Yes	yes
Integrate shared-electric AVs to promote (not to compete with) active transporation	Yes	Yes	Yes	Yes	Yes	Yes	Yes	yes
Prioritize Shared-Electric AVs on those vulnerable and disadvantaged communities (in all geographical areas),who will benefit more from traffic safety interventions	Yes	Yes	Yes	Yes	Yes	Yes	Yes	yes
Consider marker penetrance of AVs when designing and estimating the road safety impacts	Yes						yes	yes

FIGURE 33.5 Autonomous vehicle (AV) recommendations versus health determinants.

to increased levels of both particulate and gaseous pollutants in the environment. To summarize, if the described vehicular technologies are integrated and implemented successfully across the globe, depending upon the need and application in different countries, better ecology could be established within 5 years. In addition to electric AVs based on renewable energy sources, shared-electric AVs signify the optimal strategy to increase energy efficiency, reducing consumption, especially when integrated into healthy urban and transport environments, and supporting active and public transport.

33.3.4.2 Alternative Fuels or Green Fuels

All the emissions are released once the fuel is burned. Incomplete combustion of fuel could lead to the presence of harmful pollutants in the atmosphere. Fuel is used for indoor, domestic, vehicular, industrial and commercial purposes and applications. Burning of fuel introduces gaseous as well as particulate matter into the environment. Fuel with a low sulfur content (<50 ppm) should be used. With the rapid growth of industries, transport and travel infrastructure worldwide, there is a superfluous need of fuel for both industries and means of transport. To create a balance between environment and vehicular emissions, there is an immediate need to shift towards clean and green fuels. Crude oil contains sulfur as its one of the natural components. A fuel containing more than 0.5% sulfur is termed "sour," whereas fuel with less than 0.5% is termed "sweet." Commonly used fuels such as gasoline and diesel have sulfur content if this is not eliminated during the refining process. Sulfur present in fuels causes a potential threat to the environment. Many engines now burn low-sulfur fuels with less than 0.05% sulfur content. However, some diesel fuels and the heavy oil fuel that very large engines burn may contain significant amounts of sulfur (as much as 3.5% sulfur in the latter case). On the industrial level, a sulfur capture system may also be installed in critical circumstances, which acts as a scrubber. The use of such a system will add to both capital and maintenance costs and affects plant economics. It is only likely to be cost-effective in the very largest reciprocating engine-based power plants (Breeze, 2009). As such, nations adopting soot-free vehicle emission standards that wish to achieve real-world emissions reductions would do well to set a limit of 10 ppm sulfur for diesel fuel imports, and it is planned to move to 10 ppm domestic production in the near future. For domestic purposes, fuels such as biogas, ethanol and liquid petroleum gas, and compressed natural gas are cleaner options that minimize the health risk from household air pollution. Technologies are available to use clean fuels for household cooking, heating and lighting. Improved distribution networks and subsidies are making these cleaner fuels more widely available and used in emerging economies, especially in developing countries.

33.3.4.3 Green Highways

Studies in the past have reported serious health effects due to appreciable amounts of time spent near high-traffic roads with elevated levels of particulate matter, gaseous pollutants and air toxics released by frequent vehicular movements (Health Effects Institute (HEI), 2010). With the mushroom growth of urbanization, vehicular activities have increased across the world in the past decade. Though vehicular exhaust and non-exhaust emission control techniques and programs directly cut down the quantity of air pollutants released, such programs often take a long time to fully implement.

The development of infrastructure such as green highways, green belt and no construction zones could have holistic health benefits, including increased physical activity, lower obesity, improved mental health, overall improved birth outcomes, lower adverse cardiovascular illness and decreased mortality (James et al., 2015; 2016).

Trees and other vegetation have been shown to reduce regional air pollution levels through the interception of airborne particles or through the uptake of gaseous air pollution through leaf stomata on the plant surface (Gallagher et al., 2015). Pollution removal (O_3, PM10, NO_2, SO_2, CO) by urban trees has been estimated across the continental United States using the US Forest Service's i-Tree model. A tall, thick and dense vegetation barrier along highways with a high frequency of vehicular activity could achieve greater reductions in pollutant concentrations. Trees can also act as an obstruction at the local level between the sources of emission and populations. For example, the effectiveness of vegetative barriers at reducing ultrafine particle (UFP) concentration has been shown to be variable (Tong et al., 2015; Hagler et al., 2012; Pataki et al., 2011).

33.3.4.3.1 Vegetation Barrier Height

Fewer studies are available related to varying heights of vegetation barriers. Near-road vegetation barrier studies in non-street canyon settings that measured air pollution reductions behind the vegetation typically had heights ranging from nearly 4 meters or even more (Steffens et al., 2012; Al-Dabbous and Kumar, 2014; Brantley et al., 2014). At these elevations, the barrier will be above the exhaust release height of motor vehicles operating on the adjoining roads, driving the pollutant plume to lob above and over or pass through the vegetation. Heights less than approximately 4 meters may allow the pollutant emissions to proceed unchecked downwind of the low barrier. Trees with large openings under the canopy, particularly ornamental trees, could cause higher downwind concentrations by allowing the plume to pass through (Tong et al. 2015). Thus, the vegetation barrier should be analogous to a solid noise barrier by encumbering the entire plume air flow from the highway. The thickness of the vegetation barrier will provide resistance and residence time for the removal of particulates by impaction or diffusion, as well as lessening turbulence and wind speed, increasing the amount of air flow blocking.

The porosity or density of the vegetation comprising the barrier will determine air movement through the barrier. Also, the lower the porosity (or the higher the density) and the thicker the barrier, the more air flow is forced over the structure. At extremely low porosities, the vegetation will affect pollutant transport and dispersion in a similar manner to a solid noise barrier. At higher porosity, the vegetation can reduce wind speeds, allowing pollutants to stagnate within or behind the vegetation, potentially leading to higher pollutant concentrations. Thus, the vegetation porosity should be high enough that the combination of particle loss within the vegetation and the particle removal mechanisms dominate the lowering wind speed and stagnation effect, leading to reduced concentrations behind the barrier. Since the measurement of porosity and/or density along the lateral plane of air flow from a highway is very difficult, a quantitative technique in the field has not been implemented. Green walls or moderate vegetation in streets with congestion or having a built-up area could be effective in reducing pollutant concentration. Modeling and wind tunnel analyses have used Leaf Area Index (LAI) and Leaf Area Density (LAD) to estimate the porosity/density of vegetation (Lin and Khlystov, 2012; Steffens et al., 2012; Tong et al., 2016; Neft et al., 2016). These studies suggest that thicker and denser vegetation promotes increased pollution reduction, although none provide a quantitative relationship that has been shown to be effective in the field.

33.3.4.3.2 Vegetation Barrier Coverage

Gaps in vegetation barriers, whether from high porosity, missing or dead trees, or space under ornamental trees, can lead to increased pollutant concentrations downwind, sometimes higher than concentrations would be if no barrier were present. These increases can occur because pollutant emissions from the road funnel through the gaps or cause winds to stagnate. This characteristic is important in planning the barrier design as well as maintaining existing or planted roadside vegetation. In order to achieve sufficient coverage, multiple rows and types of vegetation may be most feasible. For example, a barrier could consist of a row of bushy plants and hedges followed by a row of trees to enable a barrier with full coverage from the ground to the top of the canopy at the initial planting, yet achieve higher canopy heights than would be feasible by bushy plants alone. In addition, rows of multiple vegetation types may allow sufficient downwind pollutant removal while the vegetation grows over time after first planting. This approach will ensure sufficient density for pollutant removal at the initial planting, while allowing for increased pollutant removal as the vegetation matures. This process will also limit concerns of promoting plant monocultures.

33.3.4.4 Biomonitoring of Air Pollution

Biomonitoring in simple terms can be understood as the use of organisms and/or biological materials to assess and monitor the ongoing alterations in the existing ecosystem and environment. It is a means to estimate the exposure of a living organism to contamination, a pollutant or a group of pollutants. Different countries could develop, follow or implement their own strategies and programs to perform biomonitoring of air pollution. This makes it obvious that biomonitoring can be applied at different levels, such as national or global scale. Also, it requires further development and integration to enable competent and trustworthy results to be obtained for wider geographical scales. There are numerous benefits of incorporating biological indicators for the purpose of air quality monitoring. Biomonitoring offers the key advantage that it provides a timely alarm for future impacts of air pollutants on biosystems.

Monitoring of air pollution using biological indicators could be cost-efficient when compared with the cost of analytical methods. The type of biological indicator selected could pose limitations on biomonitoring; unexpected alterations in the ecosystem could also affect the cause–effect relationship and make it cumbersome to determine the impact of a pollutant, sampling methods and also the field of study under investigation and monitoring. The most appropriate solution for monitoring of air quality and pollution is the integration of living organisms with chemical analytical methods.

Presently, a number of species are used for biomonitoring of various air pollutants.

Gaseous pollutants in the atmosphere such as oxides of sulfur, oxides of nitrogen, fluorides and ozone were initially examined with the use of lichens, which have been classified as long-term biomonitors of air pollution (Batzias and Siontorou, 2007). In the current scenario, scientists are exploring other plant species for the purpose of biomonitoring gaseous pollutants, for instance, wheat, barley, maize, grass and tobacco (Wolterbeek, 2002). A research study on the monitoring of tropospheric ozone present in the atmosphere with the use of tobacco plants (*Nicotiana tabacum* cv. Bel-W3) was conducted in urban, suburban, rural and traffic-exposed areas (Klumpp et al., 2006). It was found that ozone-induced injury to tobacco plants occurred primarily in suburban and rural areas.

Mosses can also be used for biomonitoring of air polluted with heavy metals and SO_2. Mosses have the ability to show a high capacity for bioaccumulation of heavy metals and retain particulate matter on the plant's surface by absorbing heavy metals in ionized forms (Shakya et al., 2004). Lichens were also used as biomonitors for heavy metals in air particulate matter (Costa et al., 2002), but they found it difficult to establish a quantitative relationship between particulate matter and lichens. Trace elements may be absorbed in the moss from the atmosphere either as soluble chemical species in wet deposition or contained in particles from dry deposition (Glime and Saxena, 1991). Mosses also have several advantages over conventional techniques for atmospheric deposition of metals (Saxena et al., 2007). The high surface to mass ratio of mosses is effective in trapping airborne particles using both native and transplant methods (Kosior et al., 2010). Metals are deposited in moss at very high concentrations, because they pick up nutrients directly from ambient air and can retain them for quite long time without impairment to their physiological processes due to metal binding proteins (Zulem et al., 2018). Therefore, their analysis is a reflection of the atmospheric metal load.

Fluoride present in particulate matter or industrial gases could have a significant impact on plants; this could be due to the direct intake of fluoride present in soil by the root system. Biomonitoring of fluorides due to air pollution by plants and soil helps in predicting the impact of fluoride on a living organism over a specified period of time (Telesiński and Śnioszek, 2009). Examples of other bio-indicators include tulips, crocuses, pines, larches, lichens and even mosses.

33.3.4.5 Fixing of Pollutants by Microalgae

Air pollution is growing day by day, and its threat is becoming a challenge, which we need to overcome. Each individual is affected by this threat in one way or another. Air pollution is primarily due to particulate as well as gaseous pollutants such as SO_2, CO, VOCs, CFCs, etc. or secondary gaseous pollutants like ozone and peroxyacetyl nitrate (PAN). Particulate and gaseous pollutants have different impacts on human health systems, some of which may pose a serious danger to human life. Photosynthesis is a process by plants and other organisms to convert sunlight energy, carbon dioxide and water, to prepare their food to fuel the organism's activities and thus release oxygen as a by-product. In general, plants, algae and cyanobacteria perform photosynthesis, and such organisms are called photoautotrophs. Microalgae are sunlight-driven cell factories that convert carbon dioxide to potential biofuels, foods, feeds and high-value bio-actives.

Photoautotrophic organisms such as algae could absorb gaseous air pollutants. Algae produce 71% of the earth's oxygen by the process of photosynthesis. Algae belong to organisms that are able to absorb oxides of carbon, nitrogen and sulfur, which are potent nutrients for them. Addition of CO_2 stimulates the growth of algae, which could be either marine or freshwater. They have higher photosynthetic efficiencies than terrestrial plants, and therefore, they are more efficient in absorbing carbon. Algae include macroalgae and microalgae. Macroscopic algae are multicellular algae such as seaweeds; on the other hand, microalgae are also known as microscopic algae, and most of them are unicellular (Packer, 2009).

The non-linear rise in the level of CO_2 emissions is predominantly due to fuel combustion from industries and vehicles as well as coal-fired plants. Other constituents of the flue gases may be oxides of nitrogen and sulfur and transition metals, such as: nickel (Ni), vanadium (V) and mercury (Hg) (Packer, 2009). Automobiles, steel plants, cement plants, breweries and fertilizer plants also contribute to the emission of carbon dioxide and nitrogen dioxide. At present, microalgae are one of the emerging technologies to absorb and reduce carbon dioxide emitted by industrial sources like fossil-fueled power plants and fermentation processes (Usui and Ikenouchi, 1997; Benneman, 1997; Braun, 1996). Microalgae can digest carbon dioxide into organic matter, such as carbohydrates, proteins and lipids, that can be converted into useful substances. Algal biomass can also be used for the production of different types of renewable biofuels, such as methane produced by anaerobic digestion of the algal biomass, biodiesel obtained from microalgal oil, bioethanol produced from microalgal carbohydrates, and photobiologically derived biohydrogen (Chisti, 2007).

Among the applications of algal biomass could be its use as an agricultural protein-rich biofertilizer, high-protein animal feed, food supplement, biodegradable polymers/plastics, biosorbents, medicinal applications and in the cosmetic industry.

Currently, the world is facing a potential threat and challenge from climate change caused by increasing emissions of greenhouse gases, predominantly carbon dioxide (Woodward et al., 2009), and other air pollutants released into the atmosphere.

Strategies are being developed to combat the effects of air pollutants, such as limiting the use of fossil fuels and replacing them with renewable sources of energy such as electric, solar, geothermal or nuclear energy. Photosynthesis consists of light reactions and dark reactions. The process can be simplified in this equation:

$$6CO_2 + 12H_2O + \text{energy} \rightarrow C_6H_{12}O_6 + 6O_2 + 6H_2O$$

Greenhouse gases such as CO_2 could be significantly decreased by exhaustive forestation. The increased CO_2 emissions from industries and vehicles will be absorbed by plants and trees, enhancing the process of photosynthesis. This in turn could produce more and more oxygen along with a better crop yield.

33.4 CASE STUDY OF AIR POLLUTION OF BAREILLY (UTTAR PRADESH, INDIA)

Air pollution is a major planetary health risk, with India estimated to have some of the worst levels globally. In India, urban air quality in megacities is deteriorating day by day (Gurjar et al., 2008) due to increasing industrialization and urbanization (Baldasano et al., 2003). Population pressure can't be ruled out for a few reasons. To satisfy the demands of the population explosion, natural resources are being harvested and transported. The rapid urban growth is also associated with the transportation sector and road networks, which supports various vehicular movements on roads (Dubey et al., 2013). Air pollution in India is predominantly caused by three sources: vehicles, and industrial and domestic sources. It was reported that 17% and 28% of total NO_x and particulate matter concentration, respectively, are contributed by vehicular emissions, which was almost equal to the combined sources such as industry, power plants and domestic sectors in Delhi, India (Goyal et al., 2010). In the past few years, there has been a rapid increase in the number of vehicles on the roads of Bareilly city (Figure 33.6).

The following could be the major causes of air pollution in and around the city of Bareilly.

1. Lack of comfortable mass transport compels people to use their own transport; as a result, more combustion of fuel takes place, leading to high levels of pollutants such as SO_2 and NO_x.
2. Traffic congestion due to narrow roads (at times due to encroachment also) makes the situation worse.
3. The high number of vehicles and their faulty parts are responsible for increasing PM2.5 and PM10.
4. Huge construction sites contribute to air pollution, emitting PM10.
5. There is a large number of two- and four-wheelers.
6. A significant source of particulate matter originates from brake, tire and road wear, as well as re-suspension of road dust.
7. Ecofriendly vehicles like bicycles are losing popularity.
8. A large number of old vehicles are still on the roads.

Earlier, Bareilly was reported among the seven most air polluted cities of Uttar Pradesh, where PM10 was almost four times the annual standard of 60 μg/m^3, averaging 22 6μg/m^3 for the years 2015 and 2016 (Greenpeace, 2018).

Two monitoring stations were selected on the basis of traffic frequency and construction sites. A respirable dust sampler was installed at two stations, namely, locations A and B (Figure 33.7), within Bareilly. Station A is at IVRI, Izatnagar, Bareilly, which has low vehicular frequency. Station

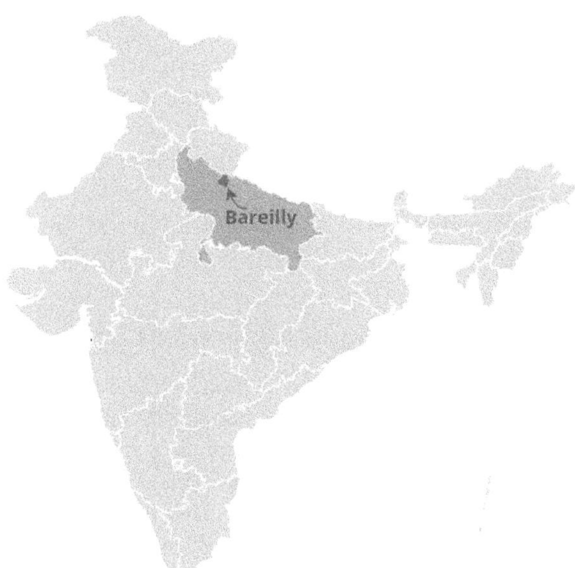

FIGURE 33.6 Map of Bareilly (India).

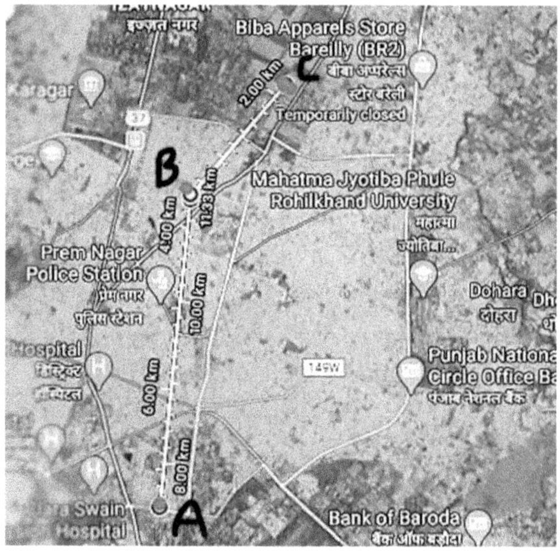

FIGURE 33.7 Satellite image taken from Google Maps of air monitoring stations at Bareilly. Station A: IVRI Station, B: IVRI Bareilly Station, C: DD Puram, Bareilly.

B is situated at Petrol Pump, near highway, civil lines, Bareilly, which has high vehicular frequency and is supposed to be a heavy traffic site. From these stations, the seasonal variation of particulate matter (PM10) was analyzed and interpreted by the author for a period of 2 years, i.e., 2017 and 2018, on an 8-hourly basis as per the guidelines of the Central Pollution Control Board (CPCB), New Delhi (CPCB, 2003). Further monitoring and analysis of air pollutants (PM10, PM2.5, SO_x and NO_2) was carried out during the outbreak of the novel coronavirus in 2020.

33.4.1 Variation of PM10 During Winter Season 2017

PM10 for the winter season was measured for 3 months (December, January and February 2017) from locations A and B on an 8-hourly basis (10pm–6am, 6am–2pm and 2pm–10pm). Both the stations reported PM10 much higher than the specified and permissible National Ambient Air Quality Standards (NAAQS) limit of 80 $\mu g/m^3$ (Figure 33.8(a)–8(c)). During the entire period of study, station A (IVRI, Bareilly) reported PM10 significantly lower than station C. During the monitoring between 10pm–6am, PM10 is more concentrated near the surface, in comparison to the analysis done between 2pm–10pm. Temperature inversion could not be ruled out, and this may cause poor dispersion of air pollutants during night hours in winter. In the month of December, station B reported a similar trend in PM10 irrespective of the season (Figure 33.8(a)–8(c)).

33.4.2 Variation of PM10 During Summer Season 2017

PM10 for the summer season was measured for 3 months (March, April and May 2017) from locations A and B on an 8-hourly basis (10pm–6am, 6am–2pm and 2pm–10pm). In comparison to the winter season, both the stations reported PM10 lower in the summer season (Figure 33.9(a)–9(c)). This could be due to the better dispersion of air pollutants during summer when the average day temperature is high. During the entire period of study during the summer season, station A (IVRI, Bareilly) reported PM10 significantly lower than station B (Petrol Pump) during the month of May 2017 (Figure 33.9(a)–9(c)).

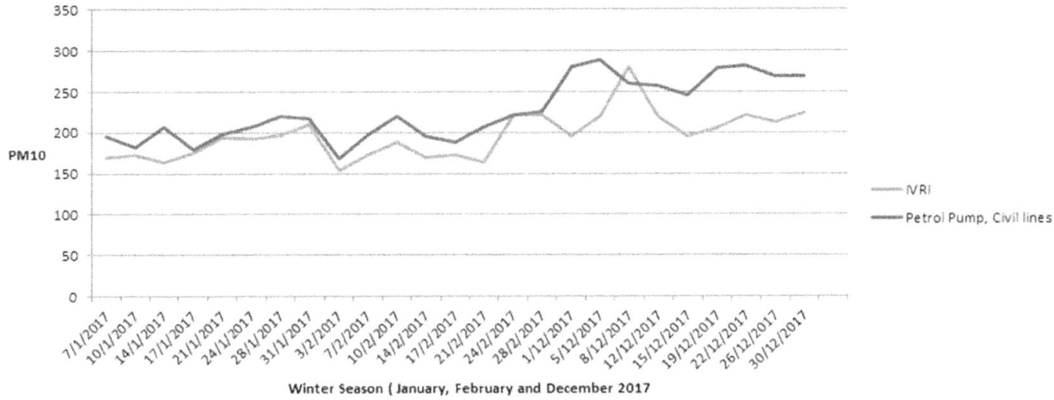

FIGURE 33.8(a) Comparison of PM10 (in μg/m[3]) during winter season at 2pm–10pm from Station A and Station B.

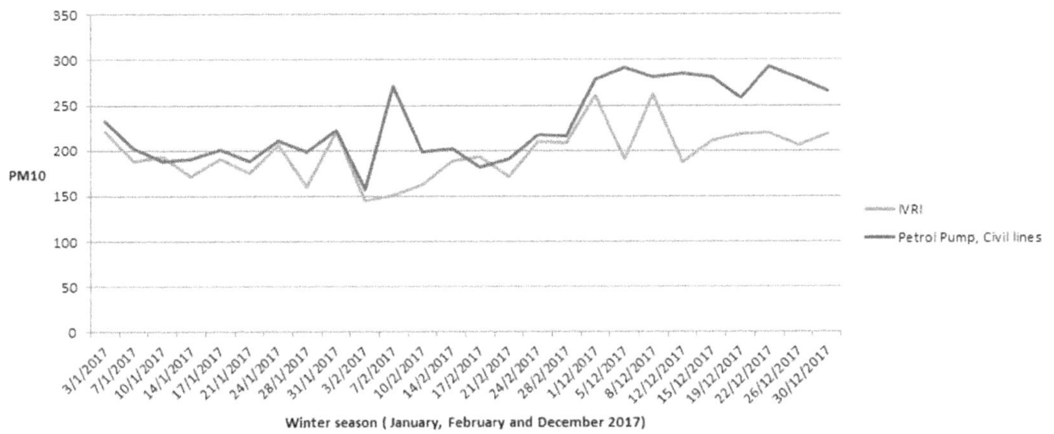

FIGURE 33.8(b) Comparison of PM10 (in μg/m[3]) during winter season at 10pm–6am from Station A and Station B.

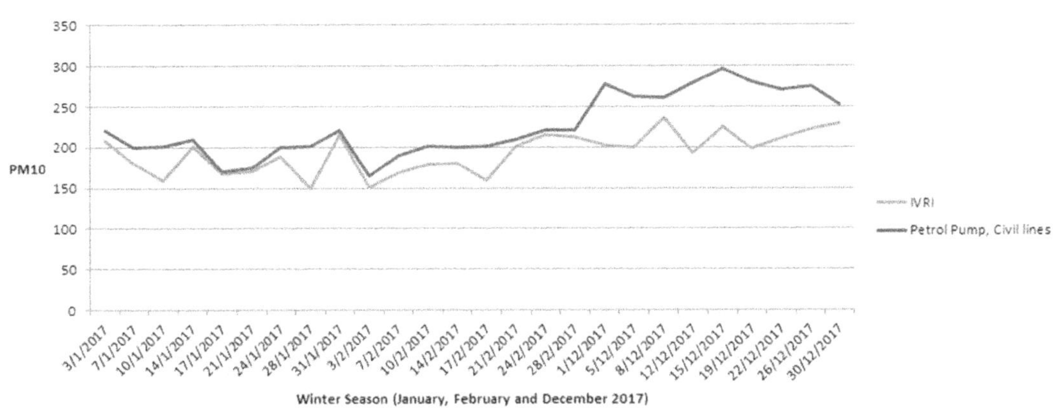

FIGURE 33.8(c) Comparison of PM10 (in μg/m[3]) during winter season at 6am–2pm from Station A and Station B.

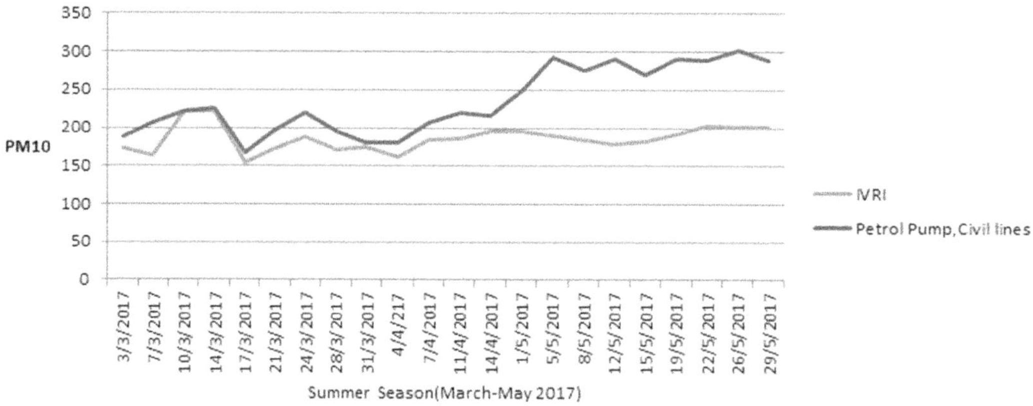

FIGURE 33.9(a) Comparison of PM10 (in μg/m^3) during summer season at 2pm–10pm from Station A and Station B.

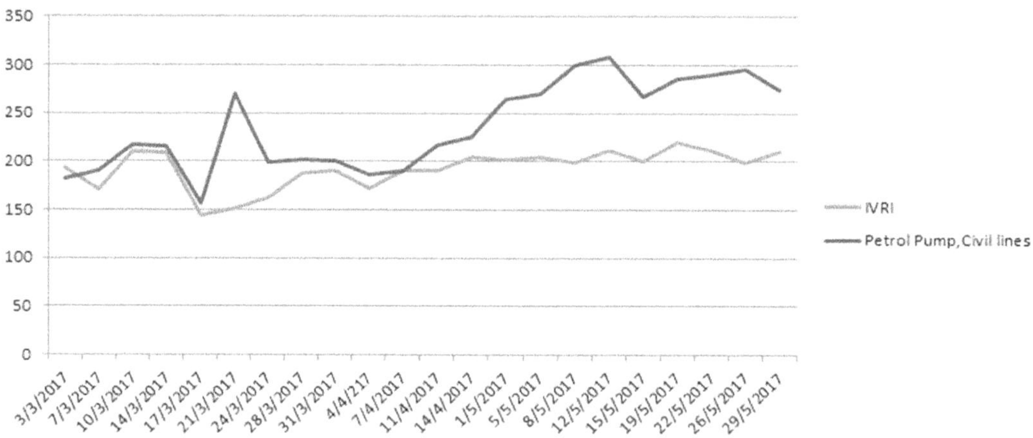

FIGURE 33.9(b) Comparison of PM10 (in μg/m^3) during summer season at 10pm–6am from Station A and Station B.

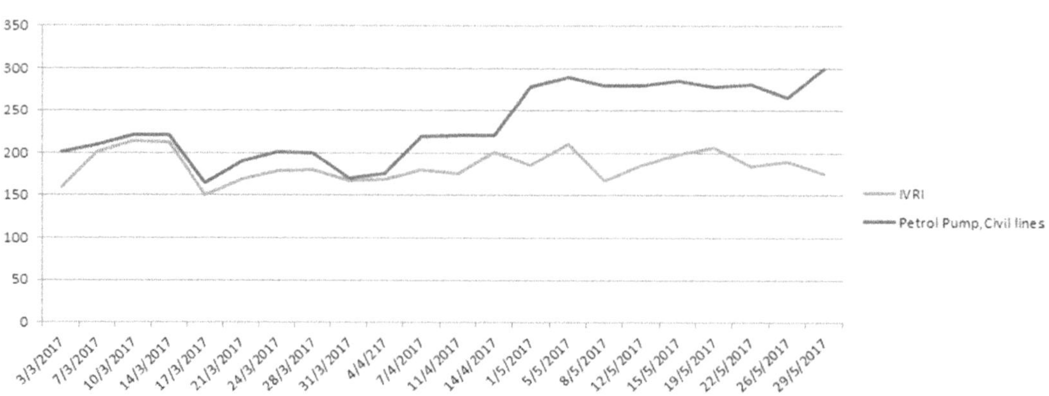

FIGURE 33.9(c) Comparison of PM10 (in μg/m^3) during summer season at 6am–2pm from Station A and Station B.

33.4.3 Variation of PM10 During Monsoon Season 2017

Not much variation is observed in PM10 values during the analysis in monsoon season (June and July 2017) from locations A and B on an 8-hourly basis (10pm–6am, 6am–2pm and 2pm–10pm). In comparison to the other two seasons (winter and summer season), both stations (A and B) reported PM10 much lower during the month of June 2017 (Figure 33.10(a)–10(c)). This is primarily because of the wet deposition of the pollutants due to rain. In the month of July, both station showed a hike in the PM10 during the entire 24-hour analysis, but still the recorded values were found to be lower in comparison to the other two seasons (Figure 33.10(a)–10(c)).

33.4.4 Pre-monsoon Analysis of PM10 During 2018

PM10 for pre-monsoon was measured from February to June 2018 from locations A and B on an 8-hourly basis (10pm–6am, 6am–2pm and 2pm–10pm). Similarly to 2017, both the stations reported PM10 much higher than the specified and permissible NAAQS limit of 80 μg/m³ during pre-monsoon season 2018 (Figure 33.11(a)–11(c)). During the entire course of study, station A (IVRI, Bareilly) reported PM10 significantly lower than station A. During the monitoring between 10pm and 6am,

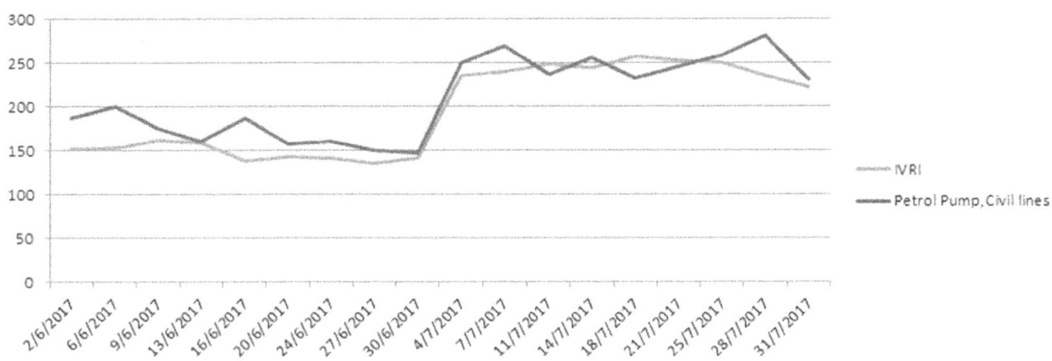

FIGURE 33.10(a) Comparison of PM10 (in μg/m³) during monsoon season at 2pm–10pm from Station A and Station B.

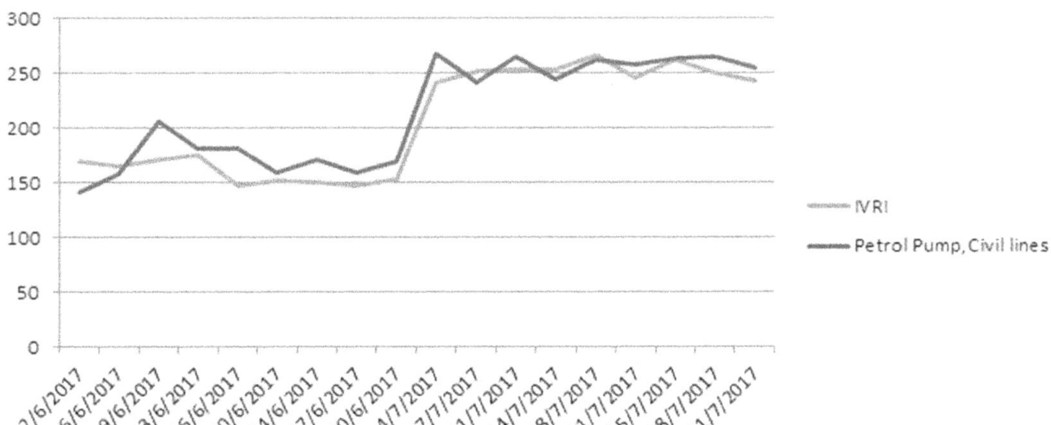

FIGURE 33.10(b) Comparison of PM10 (in μg/m³) during monsoon season at 10pm–6pm from Station A and Station B.

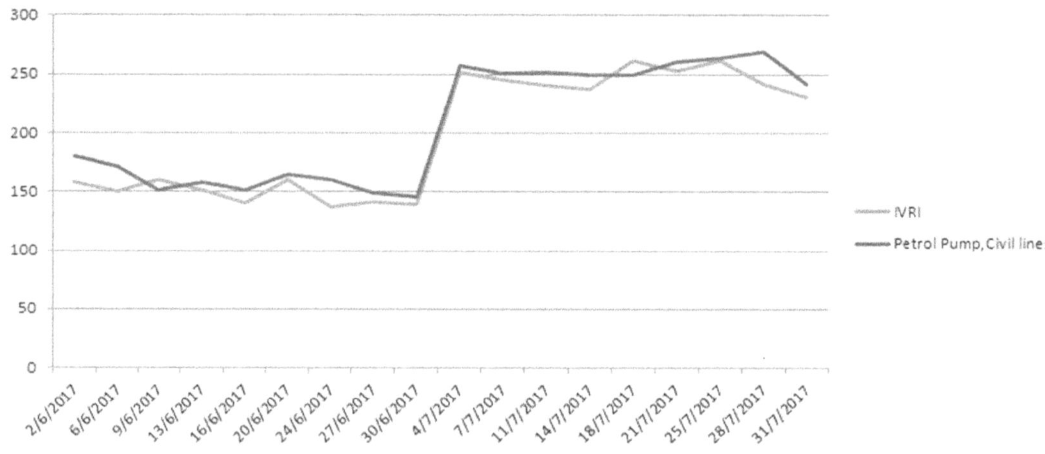

FIGURE 33.10(c) Comparison of PM10 (in μg/m³) during monsoon season at 6am–2pm from Station A and Station B.

PM10 recorded is significantly higher in comparison to the analysis done between 2pm and 10pm. During the month of May, both stations show a dip in the PM10 level (Figure 11(a)–11(c)) during the 24 hours of study. This could possibly be due to the pre-monsoon rainfall in the month of May. The higher value of PM10 at station B (Petrol Pump) is predominantly due to the high traffic activity, as it is located near the state highway.

33.4.5 Post-monsoon Analysis of PM10 During 2018

PM10 for post-monsoon was measured from July to October 2018 from locations A and B on an 8-hourly basis (10pm–6am, 6am–2pm and 2pm–10pm). Both the stations measured lower PM10 values during August and September 2018 but were unable to attain the specified and permissible NAAQS limit of 80 μg/m³ (Figure 12(a)–12(c)). Station A (IVRI) is supposed to be a residential site and was expected to record lower PM10 values. The higher PM10 concentration at station A could be due to the ongoing construction of a flyover just adjacent to IVRI Bareilly. The re-suspended road dust could be a major cause of the elevated PM10 values. The higher value of PM10 at station B (Petrol Pump) is predominantly due to the high traffic activity, as it is located near the state highway.

The seasonal analysis done during the years 2017 and 2018 infers that though station A is considered to be a residential site, it does not have ambient air quality. The situation is worse at location B (Petrol pump). It can be concluded that morning walkers, heart patients and asthmatic patients should avoid outdoor activities during the winter season.

33.4.6 Analysis of Air Pollutants (PM10, PM2.5, SO_2 and NO_2) During Outbreak of COVID-19 in 2020

The following study is focused on the monitoring and analysis of air pollutants (particulate matter and gases) in Bareilly district of Uttar Pradesh during the lockdown period.

During the coronavirus pandemic in India, as a preventive measure, a complete lockdown was imposed all over the country from March 25, 2020, which resulted in a significant reduction in vehicular movement. In the first week of lockdown (March 25, 2020 to March 31, 2020) PM10 and PM2.5 averaged 60 μg/m³ and 47 μg/m³, respectively, which is below the NAAQS average limits of 80 μg/m³ and 60 μg/m³, respectively (Agarwal et al., 2021), whereas the concentrations of gaseous pollutants SO_2 and NO_2 were found to be well below the monthly NAAQS limits of 60 μg/m³,

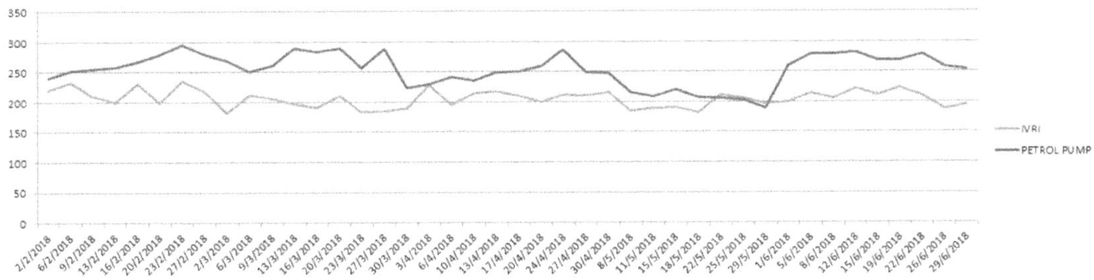

FIGURE 33.11(a) Comparison of PM10 (in μg/m³) during pre-monsoon at 2pm–10pm from Station A and Station B.

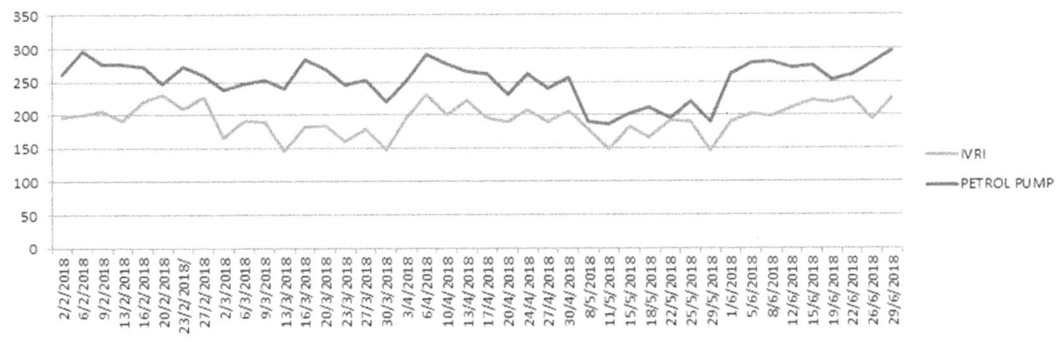

FIGURE 33.11(b) Comparison of PM10 (in μg/m³) during pre-monsoon at 10pm–6am from Station A and Station B.

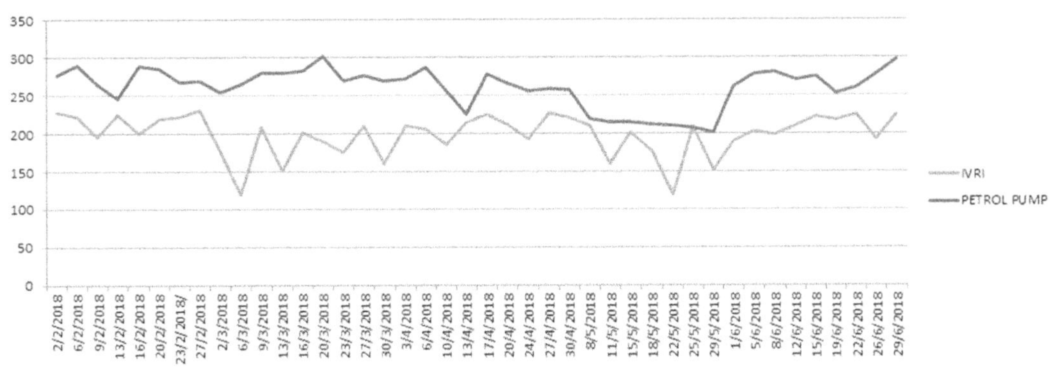

FIGURE 33.11(c) Comparison of PM10 (in μg/m³) during pre-monsoon at 6am–2pm from Station A and Station B.

averaging 21 μg/m³ and 15 μg/m³, respectively. In April 2020, the vehicular movement and construction work were at a minimum, and the levels of air pollutants, PM10, PM2.5, SO_2 and NO_2, were found to be 54 μg/m³, 41 μg/m³, 19 μg/m³ and 14 μg/m³ (Agarwal et al., 2021), respectively, which are the minimum in the 6 months of study from January 2020 to June 2020 and the lowest in comparison to the air quality data of the last 25 years. It can therefore be concluded that restricted vehicular movements and suspended construction work during the lockdown period significantly

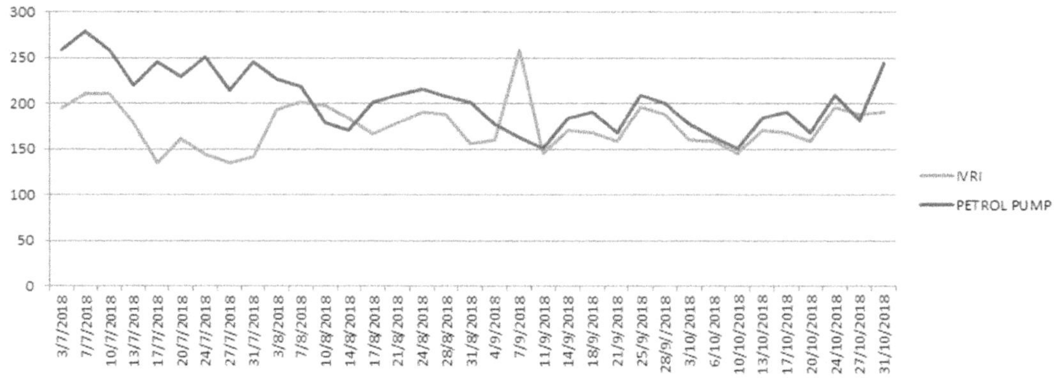

FIGURE 33.12(a) Comparison of PM10 (in μg/m^3) during post-monsoon at 2pm–10pm from Station A and Station B.

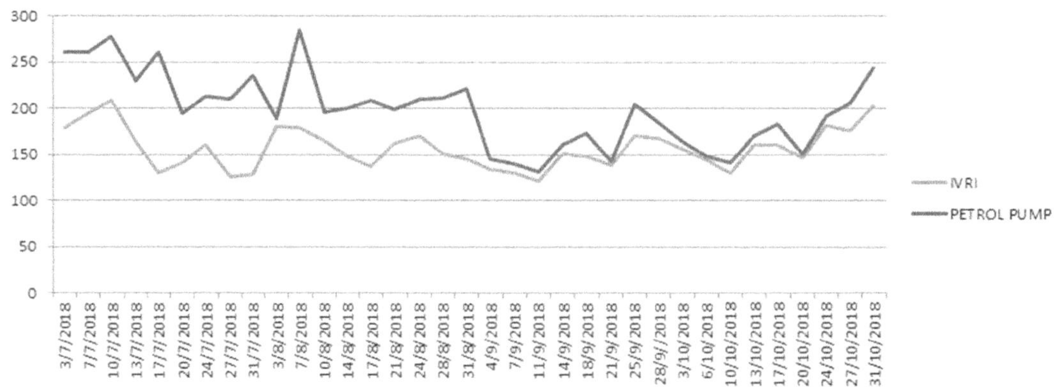

FIGURE 33.12(b) Comparison of PM10 (in μg/m^3) during post-monsoon at 10pm–6am from Station A and Station B.

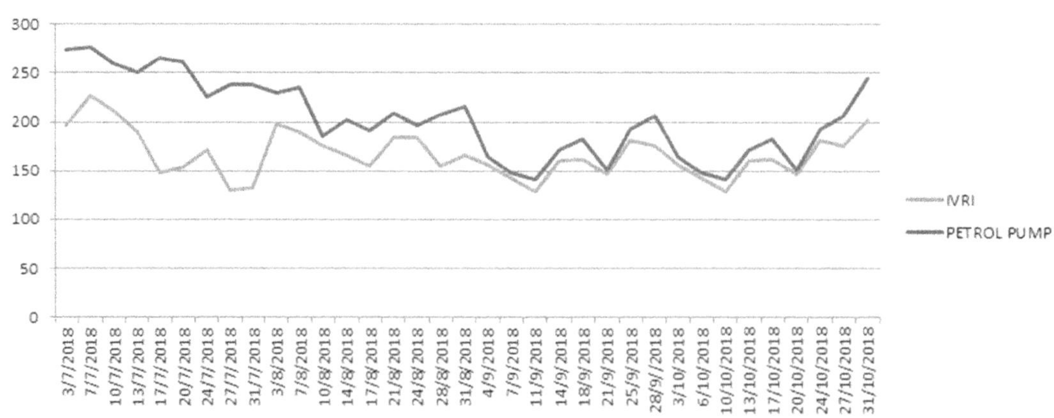

FIGURE 33.12(c) Comparison of PM10 (in μg/m^3) during post-monsoon at 6am–2pm from Station A and Station B.

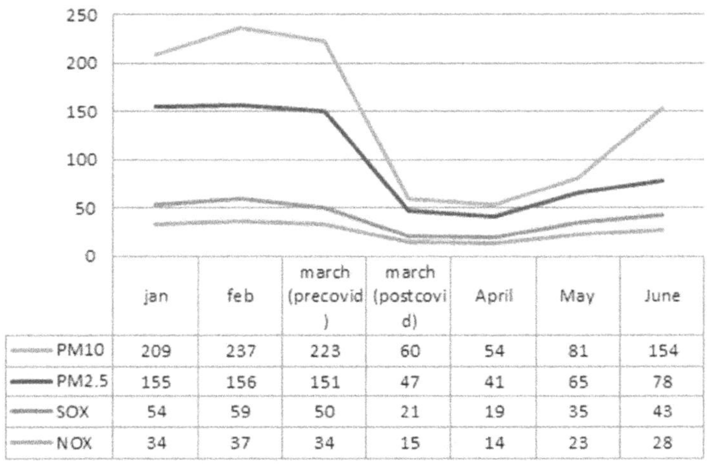

FIGURE 33.13 Pollutant concentrations (in µg/m^3) for a period of 6 months

TABLE 33.2
Monthly Sales of Diesel and Petrol (in Kiloliters) at Three Fuel Pumps in the Vicinity of Study Area: Bareilly During January 2020 to June 2020

Month	Coco Pump Diesel (KL)	Coco Pump Petrol (KL)	DD Puram Diesel (KL)	DD Puram Petrol (KL)	Dhruv Pump Diesel (KL)	Dhruv Pump Petrol (KL)
January 2020	303	349	69	171	66	174
February 2020	283	345	72	192	81	183
March 2020	248	301	51	141	39	129
April 2020	153	139	21	51	21	51
May 2020	203	181	42	102	36	66
June 2020	275	311	75	153	45	123

contributed to the reduction in the level of particulate as well as gaseous pollutants and improved the air quality of Bareilly district.

Three monitoring stations were selected on the basis of traffic frequency and construction sites. A respirable dust sampler with gas analyzer was installed at three stations, namely, A, B and C, within Bareilly city (Figure 33.7). Station A is near IVRI road, Izatnagar, Bareilly, which has low vehicular frequency; station B is situated at Petrol Pump, Civil lines, Bareilly, with maximum vehicular frequency (70–80 vehicles per 5 minutes) and is supposed to be a heavy traffic site. Station C, which is a medium traffic site located in the residential area at DD Puram, Bareilly, has the average vehicular frequency of 50–60 vehicles per 5 minutes. From these stations, sampling of particulate matter (PM10 and PM2.5) was done as per the guidelines of CPCB, New Delhi.

During analysis, there is a sharp decrease in the pollutants measured (Figure 33.13) at all the three stations, and their concentrations of pollutants are found to be within the permissible limits. This could be due to low or no vehicular movement, restricted use of diesel generators and suspended construction work.

A decreasing trend of fuel sales was reported in the month of March 2020 and was lowest in April 2020 during the lockdown period (Table 33.2). This indicates that vehicular emissions contribute significantly to air pollution in Bareilly city. A correlation between a drop in sales of fuel and the decreasing trend of emissions confirms that the major source of air pollution is emissions from vehicles.

33.5 CONCLUSIONS

Modern consumption patterns and energy production contribute significantly to both climate change and air pollution. Instead of tackling these problems separately, there are technological solutions that address both concerns at the same time: for example, switching from fossil fuels to renewable forms of energy cuts down on air pollution emissions (e.g. particulate material, sulfur dioxide and nitrous oxides) while simultaneously reducing emissions of the greenhouse gas carbon dioxide (CO_2) (Bollen et al., 2009).Vehicular emission is one of the major causes of emissions across the world. Switching from fuel-based vehicles to electric vehicles could significantly reduce carbon emissions. Shared autonomous electric vehicles could bring a revolution in improving the air quality. But, with the outbreak of the novel coronavirus pandemic, people's choice has shifted from shared vehicles to private vehicles in order to maintain social distancing. This will in turn increase the number of vehicles on the road, and consequently, this will increase exhaust and non-exhaust emissions. On the other hand, a research case study analyzed by the author(s) during the COVID-19 pandemic at Bareilly (India) shows that during the complete lockdown phase, there was a sharp cut in the air pollutants (Agarwal et al., 2021). Air quality is also influenced by climate policies, which can provide mutual benefits. Climate change mitigation actions can help reduce air pollution, and clean air measures can help reduce greenhouse gas emissions, leading to reductions in global warming. On the other hand, climate change is a leading threat to human health in the present era (Wang and Horton, 2015). Climate change impacts health through multiple pathways, from food access and quality to air pollution to extreme weather (Frumkin and Haines, 2019; WHO, 2018a). Furthermore, air pollution and climate change influence each other through complex interactions in the atmosphere. Direct emissions of air pollutants (e.g., black carbon), or those formed from emissions, such as sulfate and ozone, can also influence this energy balance. Therefore, there is concern that climate change could change the burden of illness and mortality associated with air pollution. However, it is difficult to predict exactly how air quality will be influenced by future climate change and in turn, influence human health. Additional research is thus needed to improve our understanding of climate change's possible impacts on air pollution-related health problems. Of further concern, most studies so far into the impacts of climate change on air quality have focused on developed countries, when in fact, air pollution is likely to be greater in the developing world. As discussed in this chapter. it is evident that air pollution is one of the burning issues in the present world. The approach discussed in this chapter to combat air pollution-related issues could be successful only when the policies and management technologies are efficiently implemented globally by the government as well as private agencies. Industrial, infrastructural and transport development should not compromise the ecology of the world.

REFERENCES

Adams I, Hobson A. 2016. Self-driving cars will make organ shortages even worse. Slate, December 30. https://slate.com/technology/2016/12/self-driving-cars-will-exacerbate-organ shortages.html

Agarwal S, Saxena DK, Rupini B. 2021. Analysis of air pollutants in Covid-19 pandemic lockdown – a case study of Bareilly, UP, India. *Current Research in Green and Sustainable Chemistry*, 4: 100087. https://doi.org/10.1016/j.crgsc.2021.100087

Air Quality Benefits of Active Travel, Final Report for Sustrans. Available online: www.eunomia.co.uk/report-category/client/sustrans/ (accessed on August 31, 2018)

Al-Dabbous AN, Kumar P. 2014. The influence of roadside vegetation barriers on airborne nanoparticles and pedestrians exposure under varying wind conditions. *Atmospheric Environment*, 90: 113–124.

Amato F, Cassee FR, Denier van der Gon HAC, Gehrig R, Gustafsson M, et al. 2014a. Urban air quality: the challenge of traffic non-exhaust emissions. *Journal of Hazardous Materials*, 275: 31–36.

Amato F, Flemming RC, Denier van der Gon H, Gehrig R, Gustafsson M, Hafner W, Harrisson RM, Jozwicka M, Kelly FJ, Moreno T, et al. 2014b. Urban air quality: the challenge of traffic non- exhaust emissions. Available online: www.kau.edu. sa/Files/188/Researches/65862_37290.pdf (accessed on March 2, 2015).

Annu. Rev. Public Health 2020.41:329-345. Downloaded from www.annualreviews.org

Baldasano JM, Valera E, Jimenez P. 2003. Air quality data from large cities. *Science of the Total Environment*, 307: 141–165.

Batzias F, Siontorou CG. 2007. A novel system for environmental monitoring through a cooperative/synergistic scheme between bioindicators and biosensors. *Journal of Environmental Management*, 82: 221–239.

Benneman JR. 1997. CO2 mitigation with microalgae systems. *Energy Conversion and Management*, 38: S475–S479.

Bollen J, van der Zwaan B, Brink C, Eerens H. 2009. Local air pollution and global climate change: A combined cost-benefit analysis. *Resource and Energy Economics*, 31: 161–181.

Bonham J, Koth B. 2010. Universities and the cycling culture. *Transport and Research Part D: Transport and Environment*, 15: 94–102.

Boogaard H, Janssen NAH, Fischer PH, et al. 2012 Impact of low emission zones and local traffic policies on ambient air pollution concentrations. *Science of the Total Environment*, 435–436: 132–140.

Brantley HL, Hagler GSW, Deshmukh PJ, Baldauf RW. 2014. Field assessment of the effects of roadside vegetation on near-road black carbon and particulate matter. *Science of the Total Environment*, 468–469: 120–129.

Braun AR. 1996. Re-use and fixation of CO2 in chemistry, algal biomass and fuel substitutions in the traffic sector. *Energy Conversion and Management*, 37 (6–8): 1229–1234.

Breeze P. 2009. Piston engine based power plant. Chapter 5 in *Power Generation Technologies* (Third Edition). Elsevier.

Bukowiecki N, Gehrig R, Lienemann P, Hill M, et al. 2009. PM10 emission factors of abrasion particles from road traffic (APART). Swiss Association of Road and Transportation Experts (VSS).

Chisti Y. 2007. Biodiesel from microalgae. *Biotechnology Advances*, 25: 294–306.

Costa CJ, Marques AP, Freitas MC, Reis MA, Oliveira OR. 2002. A comparative study for results obtained using biomonitors and PM10 collectors in Sado Estuary. *Environmental Pollution*, 120: 97–106.

D'Elia I, Piersanti A, Briganti G, Cappelletti A, Ciancarella L, Peschi E. 2018. Evaluation of mitigation measures for air quality in Italy in 2020 and 2030. *Atmospheric Pollution Research*, 9: 977–988.

Dijkema M, van der Zee S, Brunekreef B, van Strien R. 2008. Air quality effects of an urban highway speed limit reduction. *Atmospheric Environment*, 42: 9098–9105.

Dons E, Götschi T, Nieuwenhuijsen M, de Nazelle A, Anaya E, et al. 2015. Physical Activity through Sustainable Transport Approaches (PASTA): protocol for a multi-centre, longitudinal study. *BMC Public Health*, 15(1):1126.

Dubey B, Pal AK, Singh G. 2013. Assessment of vehicular pollution in Dhanbad City using CALINE 4 model. *International Journal of Geology, Earth and Environmental Sciences*, 3: 156–164.

Duclaux O. et al. 2002. 3D-air quality model evaluation using the Lidar technique. *Atmospheric Environment*, 36(32): 5081–5095.

Eliassen A, Saltbones J. 1975. Decay and transformation rates of SO_2 as estimated from emission data, trajectories and measured air concentrations. *Atmospheric Environment*, 9: 425.

Fisher BEA. 1975. The long-range transport of sulfur dioxide. *Atmospheric Environment*, 9: 1063.

Frumkin H, Haines A. 2019. Global environmental change and noncommunicable disease risks. *Annual Review of Public Health*, 40:261–282.

Gallagher J, Baldauf R, Fuller CH, Kumar P, Gill LW and McNabola A, 2015. Passive methods for improving air quality in the built environment: a review of porous and solid barriers. *Atmospheric Environment*, 120: 61–70.

General Motors. 2018. 2018 self-driving safety report. Rep., Gen. Motors, Detroit, MI. www.gm.com/content/dam/company/docs/us/en/gmcom/gmsafetyreport.pdf

Glime JM, Saxena DK. 1991. *Uses of Bryophytes*. Today and Tomorrow Printers and Publishers, New Delhi, India.

Goyal, P., Jaiswal, N., Kumar, A., Dadoo, J.K., Dwarakanath, M. 2010. Air quality impact assessment of NOx and PM due to diesel vehicles in Delhi. *Transportation Research Part D*, 15: 298–303.

Gromke C, Jamarkattel N, Ruck B. 2016. Influence of roadside hedgerows on air quality in urban street canyons. *Atmospheric Environment*, 139, 75–86.

Gurjar BR, Butler TM, Lawrence MG, Lelieveld J. 2008. Evaluation of emissions and air quality in megacities. *Atmospheric Environment*, 42: 1593–1606.

Hagler GSW, Lin M-Y, Khlystov A, Baldauf RW, Isakov V, Faircloth J. 2012. Field investigation of roadside vegetative and structural barrier impact on near-road ultrafine particle concentrations under a variety of wind conditions. *Science of the Total Environment*, 419: 7–15.

Harmsen R, van den Hoed R, Harmelink M. 2007, June. Improving the energy efficiency of road transport: the case of eco-driving in the Netherlands. *European Council for an Energy Efficient Economy Conference*, La Colle sur Loup, France: 1703–1709.

Hawkes N. 2015. Air pollution in the UK: The public health problem that won't go away. *BMJ*, 350: h2757.

Health Effects Institute (HEI). 2010. Traffic-related air pollution: a critical review of the literature on emissions, exposure, and health effects. HEI Special Report 17. Health Effects Institute, Boston, MA.

Holman C, Harrison R, Querol X. 2015. Review of the efficacy of low emission zones to improve urban air quality in European cities. *Atmospheric Environment*, 111: 161–169.

James P, Banay RF, Hart JE, Laden F. 2015. A review of the health benefits of greenness. Current Epidemiology Reports, 2:131–142.

James P, Hart JE, Banay RF, Laden F. 2016. Exposure to greenness and mortality in a nationwide prospective cohort study of women. Environmental Health Perspectives. DOI: 10.1289/ehp.1510363

Klumpp A, Ansel W, Klumpp G, Vergne P, Sifakis N, Sanz MJ, Rasmussen S, RoPaulsen H, Ribas A, Penuelas J, Kambezidis H, Garrec JP, Calataguad V. 2006. Ozone pollution and ozone biomonitoring in European cities Part II. Ozone induced plan injury and its relationship with descriptors of ozone pollution. *Atmospheric Environment*, 40: 7437–7448.

Kosior G, Samecka-CymermanKolon A, Kempers AJ. 2010. Bioindication capacity of metal pollution of native and transplanted *Pleurozium schreberi* under various levels of pollution. *Chemosphere*, 81(3): 321–326. DOI: 10.1016/j.chemosphere.2010.07.029

Laurance WF, Clements GR, Sloan S, O'Connell CS, Mueller ND, Goosem M, Venter O, Edwards DP, Phalan B, Balmford A, et al. 2014. A global strategy for road building. *Nature*, 513: 229–232.

Lin MY, Khlystov A, 2012. Investigation of ultrafine particle deposition to vegetation branches in a wind tunnel. *Aerosol Science and Technology*, 46(4): 465–472.

Litman T. 2019. Autonomous vehicle implementation predictions: implications for transport planning. Report, Victoria Transport Policy Institute, Victoria, BC. www.vtpi.org/avip.pdf

Luttrell K, Weaver M, Harris M. 2015. The effect of autonomous vehicles on trauma and health care. *Journal of Trauma and Acute Care Surgery*, 79(4): 678–682.

Mackett R. 2003. Why do people use their cars for short trips? *Transportation*, 30: 329–349.

National Center for Statistics and Analysis. 2018. 2017 fatal motor vehicle crashes: overview. Research Note DOT HS 812 603, National Highway Traffic Safety Administration (NHTSA), Washington, DC. https://crashstats.nhtsa.dot.gov/Api/Public/ ViewPublication/812603

Neft I, Scungio M, Culver N, Singh S. 2016. Simulations of aerosol filtration by vegetation: Validation of existing models with available lab data and application to near-roadway scenario. *Aerosol Science and Technology*, 50(9): 937–946.

NHTSA (National Highway Traffic Safety Administration). 2019. Highly automated or "self-driving" vehicles. Report, NHTSA, Washington, DC. www.nhtsa.gov/sites/nhtsa.dot.gov/files/documents/14269-overview_of_automated_vehicle_technology_042319_v1b.pdf

Office of the Surgeon-General. 2016. Facing addiction in America: the Surgeon General's report on alcohol, drugs, and health. Report, US Department of Health and Human Services, Washington, DC. https://addiction.surgeongeneral.gov/sites/default/files/surgeon-generals-report.pdf

Packer M. 2009. Algal capture of carbon dioxide; biomass generation as a tool for greenhouse gas mitigation with reference to New Zealand energy strategy and policy. *Energy Policy*, 37: 3428–3437.

Pataki DE, Carreiro MM, Cherrier J, Grulke NE, Jennings V, Pincetl S, Pouyat RV, Whitlow TH, Zipperer WC. 2011. Coupling biogeochemical cycles in urban environments: ecosystem services, green solutions, and misconceptions. *Frontiers in Ecology and the Environment*, 9(1): 27–36.

Piritsos SA, Loppi S. 2008. Biomonitoring atmospheric pollution: the challenge of times in environmental policy on air quality. *Environmental Pollution*, 151: 269–271.

Porter C, Kall D, Selin T, Baker R, Kishan S, Preusse D, Stanard A. 2010. Evaluate the Interactions between Transportation-Related Particulate Matter, Ozone, Air Toxics, Climate Change, and Other Air-Pollutant Control Strategies; NCHRP No. 25-25, Task 59. Transportation Research Board, Washington, DC.

Rebolj D, Sturm PJ. 1999. A GIS based component-oriented integrated system for estimation, visualization and analysis of road traffic air pollution. *Environmental Modelling and Software*, 14(6): 531–539.

Rietveld P. 2000. The accessibility of railway stations: The role of the bicycle in The Netherlands. *Transportation Research Part D: Transport and Environment*, 5:71–75.

Rohde H. 1972. A study of the sulfur budget for the atmosphere over northern Europe. *Tellus*, 24:128.

Rohde H. 1974. Some aspects of the use of air trajectories for the computation of large scale dispersion and fallout patterns. *Advances in Geophysics*, 18B: 95.

Rojas-Rueda D. 2017. Autonomous vehicles and mental health. *Journal of Urban Design and Mental Health*, 3(2): 1.

Rojas-Rueda D, de Nazelle A, Tainio M, Nieuwenhuijsen MJ, et al. 2011. The health risks and benefits of cycling in urban environments compared with car use: health impact assessment study. *BMJ*, 343: d4521.

Rojas-Rueda D, de Nazelle A, Teixido O, Nieuwenhuijsen M. 2012. Replacing car trips by increasing bike and public transport in the greater Barcelona metropolitan area: a health impact assessment study. *Environment International*, 49:100–109.

Sabir M. 2011. Weather and Travel Behaviour. PhD Thesis, Department of Spatial Economics, Faculty of Economics and Business Administration, VrijeUniversiteit Amsterdam, Amsterdam, The Netherlands. Available online: https://research.vu.nl/en/publications/weather-and-travel-behaviour (accessed on September 2, 2018).

Saxena DK, Singh S, Srivastava K. 2007. Moss bag technique for monitoring metal precipitation. *Environment Conservation Journal*, 8(2): 16–21.

Scheiner J. 2010. Interrelations between travel mode choice and trip distance: trends in Germany 1976–2002. *Journal of Transport Geography*, 18: 75–84.

Shakya K, Chettri MK, Sawidis T. 2004. Appraisal of some mosses for biomonitoring airborne heavy metals. *Ecoprint. An International Journal of Ecology*, 2(1): 35–49.

Soteropoulos A, Berger M, Ciari F. 2019. Impacts of automated vehicles on travel behaviour and land use: an international review of modelling studies. *Transport Reviews*, 39: 29–49.

State of California. 2019. Key autonomous vehicle definitions. State of California Department of Motor Vehicles. www.dmv.ca.gov/portal/dmv/detail/vr/autonomous/ definitions UN, New York. https://population.un.org/wup/Publications/Files/WUP2018-Report.pdf

Steffens JT, Wang YJ, Zhang KM. 2012. Exploration of effects of a vegetation barrier on particle size distributions in a near-road environment. *Atmospheric Environment*, 50: 120–128.

Taylor GI. 1915. Eddy motion in the atmosphere. *Philosophical Transactions of the Royal Society of London. Series A*, 215: 1.

Taylor GI. 1921. Diffusion by continuous movements. *Proceedings of the London Mathematical Society*, 20: 196.

Telesiński A, Śnioszek M. 2009. Ioindicators of environmental pollution fluo rem. *Bromatology and Toxicological Chemistry*, 4:1148–1154.

Tong Z, Whitlow TH, MacRae PF, Landers AJ, Harada Y, 2015. Quantifying the effect of vegetation on near-road air quality using brief campaigns. *Environmental Pollution*, 201: 141–149.

Tong Z, Baldauf RW, Isakov V, Deshmukh P, Zhang KM. 2016. Roadside vegetation barrier designs to mitigate near-road air pollution impacts. *Science of the Total Environment*, 541: 920–927.

US Department of Transportation. 2018. Automated vehicles 3.0.Preparing for the future of transportation. Report, US Department of Transportation, Washington, DC. www.transportation.gov/sites/dot.gov/files/docs/policyinitiatives/automated-vehicles/320711/preparing-future-transportation-automated-vehicle-30.pdf

US Energy Information Administration. 2018. Autonomous vehicles: uncertainties and energy implications. Report, US Department of Energy, Washington, DC. www.eia.gov/outlooks/aeo/pdf/AV.pdf

US EPA (Environmental Protection Agency). 2019. Global greenhouse gas emissions data. www.epa.gov/ghgemissions/global-greenhouse-gas-emissions-data

Usui N, Ikenouchi M. 1997. The biological CO2 fixation and utilization project by RITE(1) – Highly-effective photobioreactor system. *Energy Conversion and Management*, 38: S487–S492.

Wang H, Horton R. 2015. Tackling climate change: the greatest opportunity for global health. *Lancet*, 386: 1798–1799.

Wang T, Mirza L, Cheung L, Moradi S. 2014. Understanding factors influencing choices of cyclists and potential cyclists: a case study at the University of Auckland. *Road and Transport Research*, 23: 37–51.

WHO (World Health Organization). 2018a. Health and climate change. COP24 Special Report, WHO, Geneva. www.who.int/globalchange/publications/COP24-report-health-climate-change/en/

WHO. 2018b. Global status report on road safety 2018. Report, WHO, Geneva. www.who.int/violence_injury_prevention/road_safety_status/2018/en/

Wolterbeek B. 2002. Biomonitoring of trace element air pollution: principles, possibilities and perspectives. *Environmental Pollution*, 120: 11–21.

Woodward FI, Bardegettz RD, Ravena JA, Hetherington AM. 2009. Biological approaches to global environment review: change, mitigation and remediation. *Current Biology*, 19: 615–623.

Zmud J, Goodin G, Moran M, Kalra N, Thorn E. 2017. *Strategies to Advance Automated and Connected Vehicles*. National Academies Press, Washington, DC.

34 Reducing Air Pollution Can Control Future Pandemics

Amit Krishna De
The Indian Science Congress Association, Kolkata

CONTENTS

34.1 INTRODUCTION

The world is presently facing a pandemic situation. A pandemic is a global outbreak of any particular disease. It is not that this is the first pandemic/endemic of its kind since the beginning of civilization, but it is definitely one of the major pandemics in recent decades.

Since the advancement of human civilization, the level of pollution in the air is increasing, and as a result the environment is changing slowly. A sudden increase in population followed by massive industrialization and urbanization has made an impact on global climatic changes. Increased anthropogenic activities cause changes in the ecology of microorganisms in the environment and induce various types of bacterial and viral diseases. The spread of bacteria and viruses from wild animals because of their displacement for sustainable development slowly infected mankind. The increase in use of animals as food not only infected human beings but also disturbed the biodiversity, leading to major changes in climate. These changes are accompanied by natural disasters and the spread of various types of infectious diseases, as is evident from the number of frequent epidemics and pandemics occurring during the last two centuries throughout the world. Research is concentrated on some particular air pollutants, including particulate matter ($PM_{2.5}$ and PM_{10}) and gaseous pollutants like nitrogen dioxide, sulfur dioxide, carbon monoxide and carbon dioxide to evaluate the toxicity of the air and the spread of pandemic diseases.

34.2 SOME PAST EPIDEMICS/PANDEMICS

In the last century, between 1918 and 1920, more than 500 million people were affected by Spanish flu. The affected area ranged from the South Seas to the North Pole, and as World War I was going on, the flu spread because of unhealthy conditions. Then came the Asian flu pandemic between 1957 and 1958, spreading from China and claiming more than 1 million lives. Singapore, Hong Kong and the USA were the most affected countries. Since 1981, more than 35 million lives have been lost due to human immunodeficiency virus (HIV) causing AIDS, which started from Africa and spread slowly throughout the world. Initially it was considered endemic, but within a few years it turned out to be pandemic.

DOI: 10.1201/9781003220824-38

Between 2000 and 2019, it has been observed that six major pandemic and epidemic outbreaks swept the planet. Asian avian influenza (HPAI) A(H5N1) virus occurs mainly in birds and is especially deadly for poultry. In 1996, the virus was first detected in geese in China, and in humans in 1997. Since then, it has been detected in poultry in more than 50 countries. Since its widespread re-emergence in 2003, rare, sporadic human infections with this virus, associated with severe disease and death, have been reported in Asia, and later in Africa, Europe and the Middle East. It has been observed that flu viruses, especially animal flu viruses, can constantly change. As a result of this change, they acquire the ability to infect people more easily and thus spread among a wide range of people, causing a pandemic.

Severe Acute Respiratory Syndrome (SARS) occurred in China in 2002 and spread between 2002 and 2004. This was caused by a SARS-associated coronavirus. When someone with the disease sneezes, coughs or talks, the droplets from their mouth and nose enter the air, and the virus is transmitted. In 2009, a swine flu pandemic caused by H1N1 originated in Mexico and infected as many as 1.4 billion people across the world within 1 year. Middle East Respiratory Syndrome (MERS) affected around 27 countries in the Middle East, Africa and South Asia between 2012 and 2020. This virus is transmitted between animals and people through direct or indirect contact with infected animals, and around 858 known deaths have been reported since 2012.

Between 2013 and 2016, Ebola virus disease (EVD) or Ebola hemorrhagic fever (EHF), a viral hemorrhagic fever of humans and other primates caused by Ebola viruses, ravaged West Africa, with more than 28,600 cases and 11,325 deaths. The disease spread from Guinea to Liberia and Sierra Leone and finally to Nigeria, Mali, Senegal, the USA and Europe. In 2014, the Zika virus outbreak began in Brazil, spreading across the Caribbean and Latin America and causing an epidemic between 2015 and 2016. The main carrier of Zika virus is the *Aedes* genus of mosquitoes, although it can also be sexually transmitted in humans.

34.3 COVID-19

The COVID-19 endemic, which started in December 2019 in Wuhan City of China, was declared a pandemic by the World Health Organization on March 12, 2020. It is reported to be caused by severe acute respiratory syndrome coronavirus-2 (SARS-CoV-2) (Wang, *et al.*, 2020). The symptoms of COVID-19 infection as reported throughout the world include cough, fever, sore throat, chills, fatigue, nausea, vomiting, diarrhea and breathing difficulty. There is sometimes, in extreme cases, respiratory failure, acute respiratory distress syndrome, cardiac failure and even death. In comparison to the other epidemics and pandemics during this century mentioned earlier, none of them had the widespread impact that the novel coronavirus did.

34.4 PAST EPIDEMICS/PANDEMICS AND AIR POLLUTION

Air pollution is one of the major threats in any epidemic or pandemic. It has been observed in the past that any epidemic or pandemic increased in the cities where air pollution was higher. During the infection with Spanish flu in 1918 and SARS-CoV-1 in 2003, it was observed that air pollution intensified the mortality rate. Pollutants in air increase the virulence of respiratory infections and host susceptibility. In fact, the number of affected patients and the mortality rate were higher in those places with a poor air quality index.

A study related to the 1918 influenza pandemic suggest that air pollution, along with proximity to World War I military bases and baseline city health conditions, contributed to pandemic severity (Clay, *et al.*, 2015). The researchers stated:

> Had coal-fired capacity in above-median cities been reduced to the median level, 3,400–5,860 pandemic-related infant deaths and 15,575–23,686 pandemic-related all-age deaths would have been averted. These results highlight the interaction between air pollution and infectious

disease on health, and suggest that there may be large co-benefits associated with pollution abatement policies.

During the SARS-CoV-1 pandemic, it was reported that patients from areas with high air pollution indices (API) showed a 200% increased relative risk of death compared with people from areas where there is a low API (Cui, *et al.*, 2003).

A major increase is found in nitrogen dioxide and sulfur dioxide, which are generally emitted from the combustion of fossil fuels for transportation (motor vehicle exhaust), industry and electricity generation. These gases cause damage to the pulmonary system and invite microbial infections. They are associated with an increase in influenza infections, asthma, chronic obstructive pulmonary disease (COPD) and increase the risk of cardiac disease (Faustini, *et al.*, 2013). In addition, particulate matter, mainly with a diameter of less than 2.5 μm ($PM_{2.5}$), present in air can reduce phagocytosis of viruses, promote their proliferation, and produce a significant proinflammatory state. The cytokines interleukin (IL)-1, IL-6 and tumor necrosis factor (TNF)-α are released from alveolar macrophages. They also upregulate the angiotensin-converting enzyme 2 (ACE2) receptor and cause acute respiratory distress syndrome (ARDS), particularly in elderly adults (Lin, 2018). Hence, as observed during the past pandemics, there is likely a positive correlation between air pollutants and the infection and mortality rate.

In view of this, the effects of air pollution on the status of the present pandemic of COVID-19, particularly in those areas with a poor air quality index, like China, northern Italy and New York City, were studied (Karan, *et al.*, 2020). It was found that the mortality rate was 12% in Lombardy and Emilia Romagna in northern Italy, which are among the most polluted areas in Europe, as compared with the rate of 4.5% elsewhere across Italy (Conticini *et al.*, 2020). In fact, Johns Hopkins University, Center for Systems Science and Engineering Coronavirus Resource Center, on the basis of death counts due to COVID-19 as collected up to April 22, 2020 from more than 3000 counties of the USA, reported that an increase of 1 μg/m^3 in $PM_{2.5}$ has been shown to increase mortality from COVID-19 by 8%, and in New York state alone by 15% (Wu, *et al.*, 2020).

34.5 COVID-19 PANDEMIC AND AIR POLLUTION

In a recent study, it was reported that long-term exposure to air pollution increases the risk of dying from COVID-19 (Pozzer et al., 2020). In this paper, the researchers write:

> [E]stimates for individual countries show, for example, that air pollution contributed to 29% of coronavirus deaths in the Czech Republic, 27% in China, 26% in Germany, 22% in Switzerland, 21% in Belgium, 19% in The Netherlands, 18% in France, 16% in Sweden, 15% in Italy, 14% in the UK, 12% in Brazil, 11% in Portugal, 8% in the Republic of Ireland, 6% in Israel, 3% in Australia and just 1% in New Zealand.

This study is also supported by another study, which reports that higher rates of COVID-19 infection and mortality may be observed after exposure to short-term and long-term air pollution, especially where $PM_{2.5}$ and nitrogen dioxide levels are high (Ali and Islam, 2020, Liane and Rodrigo, 2020).

The researchers have concluded from their study that when people inhale polluted air, the $PM_{2.5}$ particularly present in such air migrate from the lungs to the blood and blood vessels. This causes inflammation and severe oxidative stress through the production of free radicals, causing damage to the inner lining of arteries and the endothelium. This leads to the narrowing and stiffening of the arteries. NO_2 also causes inflammation in the respiratory system, mediated by cytokines, and causes various types of diseases. The COVID-19 virus causes similar damage to blood vessels by entering the body via the lungs. Moreover, if both long-term exposure to air pollution and infection with the COVID-19 virus come together, then there is an additive adverse effect on the heart and blood vessels, which leads to greater vulnerability to COVID-19. This is why it has been observed in many cases that air pollution and coronavirus infection can lead to heart attacks, heart failure and stroke, particularly in patients who already have heart disease.

Interestingly, during the year 2020, COVID-19 caused major global disruption and brought about several changes in the environment and climate, even though for a short period of duration. It has been observed that many of the major industries reduced their work load or closed down. The number of vehicles on the road fell abruptly. As a result of this, the level of air pollution throughout the world was reduced as compared with 2018/2019. A sudden drop in emissions of greenhouse gases (GHGs) was reported; e.g., a 50% reduction in nitrogen dioxide and carbon monoxide occurred in China (Caine, 2020), a 30–60% reduction in nitrogen dioxide in Barcelona, Madrid, Milan, Rome and Paris (EEA, 2020), a 54.3% decrease in nitrogen dioxide in Sao Paulo, Brazil (Liane and Rodrigo, 2020), and an almost 70% reduction in nitrogen dioxide in Delhi, India (Thiessen, 2020).

A study by the Ozone Monitoring Instrument (OMI) on NASA's Aura satellite, as processed by a team at NASA's Goddard Space Flight Center, Greenbelt, Maryland, indicated that there was a significant reduction in air pollution in the northern USA and other regions of the world, e.g., Canada, China, India, Italy, Brazil, etc., during March 2020 as compared with March 2019 (NASA, 2020). This data is based on the measurement of nitrogen dioxide (NO_2) in the atmosphere, which can also be referred to as an indicator of human activity. It is known that NO_2 causes acid rain with the interaction of oxygen (O_2) and water (H_2O) and several types of respiratory diseases in humans. It is evident from the database during the span from 2005 to 2020 that March 2020 showed the lowest monthly atmospheric nitrogen dioxide levels by about 30% of any March during this period. In fact, during the pandemic-related shutdowns starting from February 2020, human activities were reduced significantly, and the level of nitrogen dioxide decreased by 20% to 30% as compared with the expected pandemic-free environment. In a study on Ontario (Canada), a reduction in the level of NO_2 from 4.5 to 1 ppb was reported (Adams, 2020). This improvement in air quality in many cities was mainly due to the restriction of movement and slowdown of social and economic activities during the period after the spread of COVID-19.

India showed a reduction of around 15% in nitrogen dioxide (NO_2) concentration during the lockdown period, mostly due to the reduction of vehicular traffic. Another pollutant, sulfur dioxide (SO_2), showed similar trends during the months from February 2019 to April 2020 because of reduction in power production, from where sulfur dioxide is mainly emitted. Because of the reduction of nitrogen dioxide and sulfur dioxide, the level of $PM_{2.5}$ was lower, as $PM_{2.5}$ is emitted directly from the combustion of fossil fuels and biomass. In India, during the nationwide lockdown, an overall 46% and 50% reduction in $PM_{2.5}$ and PM_{10}, respectively, was reported (India Environment Portal (IEP), 2020). A study on four major Indian cities indicated that drastic measures such as locking down entire cities, restricting all sorts of transportation, imposing lockdowns, maintaining social distancing, etc. have considerably enhanced the quality of ambient air and water (Somani, *et al.*, 2020).

These reports indicate that during the pandemic situation, air quality in different cities across the world significantly improved, which may assist with the restoration of the ecological system. Hence, it is evident that there is a direct link between endemic or pandemic situations, air pollution, morbidity and mortality. In the past, all types of such endemics have shown enhancement in morbidity and mortality rates, particularly in cities that are more polluted. During COVID-19, because of lockdown, there was a reduction in pollution levels, which may have assisted in controlling the morbidity and mortality rates, particularly in those cities where the pollution level is very high. If such restrictions had not been implemented, the number of cases during COVID-19 might have increased manifold because of the increase in air pollution. However, unfortunately, although restrictions during the lockdown mechanically led to temporary improvements in air quality, as activities return to normal, the level of air pollution will also start to slowly return to its original status until and unless some drastic steps are taken. Moreover, it should be kept in mind that not only air pollution but also other types of pollution, including water, are also important factors in increasing mortality and morbidity during such types of pandemics. But still, out of the different types of pollution, air pollution plays the most important part in spreading the diseases, as it has been seen that coronaviruses affect mainly the respiratory tract (Wang, *et al.*, 2020). It is clearly evident that if necessary, mankind can definitely reduce the level of production of GHGs in the air, when required, by controlling transportation, industry and economic activities.

34.6 CONCLUSION

It has been clearly observed from the infection trends of various type of virus attack (Asian avian influenza, SARS, MERS, COVID-19) during the last century that mankind is prone to mostly flu viruses, especially animal flu viruses. The main reason for this is that these viruses can constantly change, so that they acquire the ability to infect people more easily and thus spread among a wide range of people, causing an epidemic followed by a pandemic. Interestingly, these viruses are transmitted through droplets that enter the air when someone with the disease coughs, sneezes or talks. It is mainly the air that is the carrier, and pollutants in the air increase the susceptibility to this type of infection. As mentioned previously, a study showed that patients affected by SARS (a virus similar to COVID-19) were about 84% more likely to become infected and face death if they lived in a highly polluted area for a longer period of time (Cui, *et al.*, 2003).

In this respect, the Paris Agreement signed in 2016, which calls for "limiting the influence of climate change", stresses as a crucial component the importance of reducing air pollution. The Paris Agreement's goal on long-term temperature control (Article 4.1) is "to keep the rise in global average temperature to well below 2°C above pre-industrial levels", and also to make efforts to reduce emissions and limit the rise to 1.5°C, i.e., to "achieve a balance between anthropogenic emissions by sources and removals by sinks of greenhouse gases (Article 4.1)" in the second half of the 21st century. All countries must take up the agenda of meeting the Paris Agreement's goals through strong commitment to fight against climate change. This can save and protect up to millions of lives per year from mitigation of air pollution alone, particularly in a pandemic situation like COVID-19. Hence, countries must make an effort to establish new systems for pollution monitoring, including satellite imaging, mobilize funding for air pollution mitigation, and build multi-sectoral partnerships for pollution control to improve environmental health. This is the only way to counter such current and future epidemics and pandemics.

REFERENCES

Adams, D.M. 2020. Air pollution in Ontario, Canada during the COVID-19 State of Emergency. *Sci Total Environ* 742: 140516.

Ali, N . and Islam, F. 2020. The effects of air pollution on COVID-19 infection and mortality—a review on recent evidence. *Front Public Health* 8: 580057.

Caine, P. 2020. Environmental impact of COVID-19 lockdowns seen from space. https://news.wttw.com/2020/04/02/environmental-impact-covid-19-lockdowns-seen-space (accessed April 28, 2021).

Clay, K., Lewis, J. and Severnini, E. 2015. *Pollution, Infectious Disease, and Mortality: Evidence from the 1918 Spanish Influenza Pandemic*, IZA Discussion Papers, No. 9399, Institute for the Study of Labor (IZA), Bonn.

Conticini, E. Frediani, B. and Caro, D. 2020. Can atmospheric pollution be considered a co-factor in extremely high level of SARS-CoV-2 lethality in Northern Italy? *Environ Pollut* 261: 114465.

Cui, Y., Zhang, Z-F., Froines, J., *et al.* 2003. Air pollution and case fatality of SARS in the People's Republic of China: an ecologic study. *Environ Health* 2: 15. https://doi.org/10.1186/1476-069x-2-15.

EEA. 2020. *Air Pollution Goes Down as Europe Takes Hard Measures to Combat Corona Virus*. European Environmental Agency (EEA), Copenhagen. www.eea.europa.eu/highlights/air-pollution-goes-down-as.

Faustini, A., Stafoggia, M., Colais, P., *et al.* 2013. Air pollution and multiple acute respiratory outcomes. *Eur Respir J* 42 (2): 304–313.

India Environment Portal (IEP). 2020. Impact of lockdown (25th March to 15th April) on air quality. www.indiaenvironmentportal.org.in/content/467415/impact-oflockdown-25th-march-to-15th-april-on-air-quality/.

Karan, A., Ali, K., Teelucksingh, S. *et al.* 2020. The impact of air pollution on the incidence and mortality of COVID-19. *Glob Health Res Policy* 5: 39.

Lin, C.I., Tsai, C.H., Sun, Y.L., *et al.* 2018. Instillation of particulate matter 2.5 induced acute lung injury and attenuated the injury recovery in ACE2 knockout mice. *Int J Biol Sci* 14 (3): 253–265.

Liane, Y.K.N., and Rodrigo, C.U. 15 August 2020. COVID-19 pandemic: impacts on the air quality during the partial lockdown in São Paulo state, Brazil. *Sci Total Environ* 730: 139087. Published online April 29, 2020. doi: 10.1016/j.scitotenv.2020.139087

NASA. 2020. NASA Satellite Data Show 30 Percent Drop in Air Pollution over Northeast U.S. 9 April 2020. www.nasa.gov/feature/goddard/2020/drop-in-air-pollution-over-northeast

Pozzer, A., Dominici, F., Haines, A., et al. 2020. Regional and global contributions of air. *Cardiovasc Res* 116 (14): 2247–2253.

Somani, M., Srivastava, A.N., Gummadivalli, S.K. and Sharma, A. 2020. Indirect implications of COVID-19 towards sustainable environment: an investigation in Indian context. *Biores Technol Rep* 11: 100491.

Thiessen, T. 2020. How clean air cities could outlast COVID-19 lockdowns. www.forbes.com/sites/tamarathiessen/2020/04/10/h

Wang C., Pan R., Wan X., *et al.* 2020. Immediate psychological responses and associated factors during the initial stage of the 2019 corona virus disease (COVID-19) epidemic among the general population in China. *Int J Environ Res Publ Health* 17: 1729.

Wu, X., Nethery, R.C., Sabath, M.B., *et al.* 2020. Exposure to air pollution and COVID-19 mortality in the United States: a nationwide cross-sectional study. *MedRxiv*. www.medrxiv.org/content, https://doi.org/10.1101/2020.04.05.20054502v2

35 Integration of Constructed Wetland Technology as Decentralized Wastewater Treatment Facility and Water Reuse in Agriculture

Saroj Kumar, Bhanu Pratap, Divya Dubey, and Venkatesh Dutta
Department of Environmental Science, School of Earth and Environmental Sciences, Babasaheb Bhimrao Ambedkar University, Lucknow

CONTENTS

35.1 INTRODUCTION

Water scarcity has emerged as one of the biggest challenges to affect the economy and lives of human beings throughout the globe (Lakho et al., 2020). Freshwater pollution and depletion due to rapid industrialization, population growth and climate change are the main factors that worsen the condition (Kumar et al., 2020a; Magwaza et al., 2020). Not all regions in the world have access to necessary water requirements from streams, rivers and underground reservoirs (Correia et al., 2020). A report provided by the World Resources Institute on water stress highlights that North Africa and the Middle East regions are the major water-stressed parts in the world, facing hot and dry weather (World Resources Institute, 2015). Numerous continents of the world are facing water stress, and the situation may become worse in the future (Jiang et al., 2020). It is estimated that in the European Union, about 11% of the population and 17% of the territory are facing adverse water-inadequate

DOI: 10.1201/9781003220824-39

conditions (Pedro-Monzonís et al., 2015; Lakho et al., 2020). In the Mediterranean zone, more than 20% of the inhabitants survive under continuous water shortage, and several municipalities of Northern Europe depend on treated water as their drinkable water supply throughout the summer (Lakho et al. 2020). The majority of peri-urban and rural areas throughout the world are not linked through the municipal water supply and wastewater collection infrastructure. Creating these distribution and collection networks by a centralized wastewater remediation system is challenging and needs substantial investment (Moreira and Dias, 2020). Traditional wastewater treatment plants are also considered to eliminate organics, nutrients and solids only but are unsuccessful in eliminating the emergent pollutants (Fito and Van Hulle, 2020). Consequently, decentralized wastewater treatment schemes that fulfill the local needs are known as a viable and sustainable wastewater treatment solution at the site, including their reuse for several purposes (Libralato et al., 2012; Lakho et al., 2020; Moreira and Dias, 2020). Reuse of wastewater after suitable treatment is also measured as a possible solution to lessen the water crisis for several activities (Kumar and Dutta, 2019b). Constructed wetlands (CWs) are nature-based wastewater treatment systems designed especially for the treatment of wastewater from different sources. At present, several CWs effectively treat wastewater from diverse sources such as urban and agricultural runoff, municipal waste, storm water, river water, groundwater, and mining and industrial effluents (Dotro et al., 2017; Kumar and Dutta, 2019a). Despite being a low-cost, simple and robust solution to handle wastewater generated in decentralized locations, CWs require to be designed suitably for optimum efficiency. Several designs and configurations of CWs and their integration with other wastewater treatment technologies are available for efficient performance in economic terms (Kumar et al., 2020b). In this review-based study, we have assessed the strengths and limitations of various wastewater treatment methods for a better understanding of appropriate wastewater management in urban settlements. Integration of CWs for decentralized wastewater treatment in urban communities and the further reuse of treated wastewater in urban agricultural activities are also discussed.

35.1.1 CWs

CWs are artificially designed natural ecosystems that are used to remediate the several types of pollutants present in various kinds of wastewater (Dell'Osbel et al., 2020). A CW consists of a basin filled with different types of filtering materials and lodged with vegetation, especially macrophytes. As the applied wastewater comes into the basin, there are mainly two remediation processes that occur: abiotic and biotic (Sudarsan et al., 2015; Kumar and Dutta, 2019a; Choudhary and Kumar, 2020). In the abiotic process, the filter material assists in settling down the solid substances and filtering the wastewater. As the macrophyte roots develop, they become highly effective biological filters (Bhatia and Goyal, 2014). However, in the biotic process, solid substances start to pile up, and the microbial world develops. The available microbes associated with fallen plant litter initiate chemical conversions. Such a mixture helps in the breakdown and transformation of the contaminants existing in the wastewater (DuPoldt et al., 2015). Availability and growth of vegetation offer attachment sites to microorganisms for development and remediation activity. Several microorganisms produce various enzymes that help in the biodegradation of contaminants. Plants after death and decay also provide organic carbon to further enhance microbial metabolism (Kumar et al., 2020b).

35.1.1.1 Classification of CWs

CWs are classified into two major types – subsurface flow and surface flow. According to the flow of wastewater, these are further characterized into vertical flow and horizontal flow. Based on the growth of macrophytes, they are further subcategorized into free-floating, emergent, submerged and rooted-submerged (Kumar and Dutta, 2019a). Recently, another subdivision of CWs has been described as river diversion wetlands. They are mainly surface flow CWs situated nearby or inside a stream or river system (Ahmed et al., 2020). These are differentiated by their site as in-stream and off-stream wetlands.

In-stream CWs are created within the river or stream bed, and all water flows of the channel pass through the CW system. However, off-stream CWs are designed adjoining a stream or river, where only a portion of the total flow passes into the wetland system (Westerhoff et al., 2014). Several crucial benefits of these river diversion CWs are in terms of river/stream water quality enhancement, reduction in flood events, growing connectivity with floodplains, and development of a mixed environment of fauna and flora (Jia et al., 2016). The key benefit associated with the off-stream CWs is that they may be designed and operated to alleviate pollution originating from non-point sources such as agricultural activities, decentralized communities and urban runoff before meeting the water channel. Nevertheless, an off-stream CW system requires storage and flow regulator assemblies to adjust flow and a large area for design and operation, which might require a high initial investment. Moreover, only a portion of the total flow can be managed at once. For in-stream CWs, space is not an issue, because they are designed within the river stream, and the total flow may be applied for the treatment, although it may be tough to control flow, particularly throughout the peak flows of the river (Ahmed et al. 2020).

35.2 EXISTING WASTEWATER TREATMENT TECHNOLOGIES FOR NUTRIENT RECOVERY

The existing wastewater treatment technologies working in the urban areas of developing countries are failing to meet the wastewater discharge standards for surface waterbodies. Therefore, integrated approaches that support the protection of freshwater resources through the recovery of nutrients from municipal wastewater are most desirable. Currently, the expansion of alternate wastewater treatment technologies utilizing the natural processes within the environment is gaining significant interest (Bawiec et al., 2016). To minimize the pressure of enormous population growth, these technologies may be associated with urban agriculture to regulate the long-term impacts on freshwater reserves. Integration of wastewater treatment with urban agriculture needs a change in pattern from optimization of biodegradation, denitrification, nitrification and absorption to elevate the recycling of nitrogen and phosphorus. This development comprises decentralized wastewater management systems and nutrient recovery approaches, which not only accomplish discharge standards but also require lower costs for sanitation (Lakho et al., 2020). Numerous techniques have been established recently, mainly for the elimination and reuse of both liquid and solid portions of organics, removal and recovery of nutrients, and water reuse (Sheets et al., 2015). For that, the decentralized wastewater treatment system (DEWATS), pit latrines, urine diversion toilets (UDDT), and pour and flush toilets are useful to improve the partitioning of liquid and solid portions. Such technologies may also recover various byproducts after processing, such as nutrients, biogas and soil conditioners (Pittman et al., 2011). Additionally, the solid portion may be used to develop valuable waste-based fertilizers. However, the liquid portion can be managed to develop phosphate fertilizers, as struvite, and a nitrogen source, as nitrified urine concentrate (Magwaza et al., 2020). The growth and productivity of several crops enhanced by irrigation with treated wastewater have been described earlier by Aziz et al. (1996) for wheat, Adewoye et al. (2010) for vegetable crops, Alikhasi et al. (2012) for cotton and Musazura et al. (2015) for Swiss chard. It is also evaluated that the continuous irrigation with treated wastewater and the use of such fertilizers can damage environmental quality due to soil salinization, groundwater contamination and heavy metal accumulation, alter soil structure, and adversely affect the microbial population and its activity (Prazeres et al., 2017). Change in soil structure can also negatively affect the growth and development of plants and their productivity (Travis et al., 2010; Prazeres et al., 2014).

According to IPCC (2013), one of the prime necessities for maintaining sustainability in biotic mechanisms to treat municipal wastewater using plants is its capability to reduce the release of nitrous oxides. The generation of nitrous oxides within a biological wastewater treatment system takes place by the activity of ammonia-oxidizing bacteria and by incomplete denitrification. Therefore, design and working parameters must be chosen to alter the variability and magnitude of nitrous oxide emission (Kampschreur et al., 2009).

35.2.1 Mechanisms of Nutrients and Pollutant Removal

Wastewater treatment involves several physicochemical and biological processes synchronously to remove nutrients and other pollutants. These processes comprise microbial degradation, sedimentation, adsorption, nitrification, plant uptake, etc. (Matamoros and Salvadó, 2012; Li et al., 2015). Aquatic plants play a crucial part in wastewater remediation by taking up nutrients and other elements and also by favoring several removal mechanisms (Ong et al., 2010; Ko et al., 2011). The mechanism of nutrient uptake by various plants follows several means, such as phyto-degradation, phyto-filtration, phyto-extraction, phyto-transformation and phyto-concentration (Tangahu et al., 2011). Roots of plants give off exudates that allow the stabilization and immobilization of pollutants through binding. The capacity of particular plants for the removal of wastewater contaminants like heavy metals and nutrients has been studied earlier by several researchers around the world (Vymazal, 2007; Ha et al., 2011; Liu et al., 2013; Kumar et al., 2020a). Removal of nitrogen is mainly attained by denitrification, nitrification, physical settlement and plant uptake. The nitrification and denitrification processes are enabled by the interaction between microorganisms and plant roots. The microbial population breaks down the inorganic nitrogen, generally via a denitrification process, into nitrogen gas, which is ultimately freed up into the ambient environment (Gebeyehu et al., 2018). Several factors are responsible for altering the nitrogen removal mechanism, such as dissolved oxygen, temperature, pH, substrate properties, plant species, availability of chelating agents and root zone activities (Saeed and Sun, 2012; Kumar et al., 2020c). This may be due to the sensitivity of these factors to both microorganisms and plants that affect the degradation potential. The optimum temperatures that favor the denitrification and nitrification process are in the range of 20–25°C and 16.5–32°C, respectively. The optimum pH values required for nitrification, denitrification and ammonification to remove nutrients adequately are in the range of 7–9, 6.5–8.5 and 8–9, respectively (Magwaza et al., 2020).

35.2.2 Removal of Infectious Pathogens

Infectious pathogens and their impact on human health are a major concern associated with wastewater irrigation that limits the implementation of wastewater reuse in agriculture. These pathogens are known to cause several infectious diseases to farmers in both developed and developing nations (Almuktar and Scholz, 2016). CWs have been identified to eliminate pathogens from wastewater together with other contaminants during the treatment process (Ottoson et al., 2005). Several exclusion mechanisms for pathogens have been established, such as filtration by substrate materials, antibiosis and a biological interaction among the microorganisms' population, ultimately reducing their population (Martín et al., 2013). Research carried out by Ndulini et al. (2018) exhibited 92.77% removal of fecal coliforms from domestic wastewater. Furthermore, Ottoson et al. (2005) analyzed treated, untreated and partly treated wastewater samples for microbial indicators such as fecal coliforms, *Clostridium perfringens* and *Escherichia coli*. The study indicated that the removal efficacy of microbes extended from 60% to 87% for treated wastewater through CWs. From these findings, it is estimated that the risk of pathogens must be taken into serious consideration when irrigated crops are used for human consumption directly.

35.3 VARIOUS STAGES OF WASTEWATER TREATMENT IN SEVERAL PROCESSES

Based on the treatment required, the mechanisms are generally categorized into four stages: preliminary, primary, secondary and tertiary treatment (Von Sperling, 2007; Ullah et al., 2020). The degree of treatment essential for these stages relies on the source, characteristics, type and flow of wastewater, and each stage is designated to eliminate different pollutants (Crini and Lichtfouse, 2019). In each treatment stage, wastewater is treated via various unit operations and processes, as

TABLE 35.1
Advantages and Disadvantages of Various Processes Utilized in Wastewater Treatment at Primary Levels

Wastewater Treatment					
Level	**Process**	**Unit Operations**	**Advantages**	**Disadvantages**	**References**
Preliminary (Level-1)	Physical processes	Screens/bar racks	Prevents clogging and damage, less maintenance, improved flow regimes	High backwater, enhanced labor expenses	Laqbaqbi et al., 2019
		Comminutors/ grinders	Reduce odor and deterioration, improve plant aesthetics, reduce screenings, management and disposal	The buildup of rag on diffusers, accumulation of plastics, refusal of bio-solids, extra cost	
		Grit chambers	Avoid abrasion and deposition, ease of operation, constant performance, minimize septic situations	Cost, space, needs power for pumping, volatile organics and odors, does not satisfy discharge standards	
Primary (Level-2)	Physical processes	Flotation systems	Shorter HRT[1], low cost, no chemicals, removal of suspended and floating organic particles	Unable to eliminate nutrients and dissolved or colloidal solids	Li and Yang, 2018
		Primary sedimentation tank	Energy-efficient, easy to operate, settleable solids removal, hydraulic regulation	Inappropriate to remove nutrients and colloids, odor, less efficient for peak loads	Oraeki et al., 2018
		Equalization tank	Dilute toxic waste, remove shocks, homogenize wastewater, easy operation, during maintenance it can hold influent	Large area requirement, odor, require aeration to avoid anaerobic environments	Tchobanoglous et al., 2003; Goel et al., 2005
	Chemical processes	Neutralization tank	Compatible for joint wastewater treatment, advance efficacy of subsequent processes, prevent corrosion, saves aquatic life	Chemical costs, inaccuracy of pH sensors, specific neutralizing agents, incorrect selection of electrodes	Zanil et al., 2014; Ullah et al., 2020

[1]HRT= Hydraulic Retention Time

explained in Tables 35.1, 35.2 and 35.3, respectively. In unit operations, only physical mechanisms are utilized, whereas in unit processes, physical and/or chemical mechanisms can be employed to treat the wastewater. Wastewater generated from a source that may have coarse solids that can harm working equipment requires initial preliminary treatment. In preliminary treatment, screens, grit chambers, comminutors and grinders are used to remove large materials and reduce their size for easy handling and operation of the system. Such preliminary treatment is mostly employed in the treatment of municipal wastewater and industries such as tannery and sugar (Laxmeshwar et al., 2018; Lv et al., 2019). The effluent received from preliminary treatment might contain organic

TABLE 35.2
Advantages and Disadvantages of Various Processes Utilized in Wastewater Treatment at Secondary Levels

Wastewater Treatment					
Level	**Process**	**Unit Operations**	**Advantages**	**Disadvantages**	**References**
Secondary (Level-3)	Aerobic methods	Biological nitrogen elimination	Organics and nutrient exclusion, avoids eutrophication, reduces aquatic toxicity, viable	Unstable pH, inhibits microbial growth, detention time, high O&M cost, sludge bulking	Peng and Zhu, 2006; Akpor and Muchie, 2010
		Sequencing batch reactor	High efficiency, less land requirement, stable, flexible, energy from biogas	Low pathogen removal, expertise, power supply, time for sludge settling	Ghodeif, 2013; Li and Yang, 2018
		Activated sludge process (ASP)	Appropriate to both large and small societies, efficient, less area	Low pathogen removal, skilled workforce, power supply	Wilas et al., 2016
		Aerated lagoons	Major BOD removal, needs less area, less unpleasant odors	Larger area than ASP, energy-intensive, high O&M cost, sludge issues	Wilas et al., 2016; Kamali et al., 2019
		Constructed wetlands	Eliminates almost all pollutants and pathogens, unskilled personnel, low energy, eco-friendly and sustainable	Land intensive, requires native plants, little flexibility in operating conditions, high detention time	Kumar et al., 2019b; 2020a
		Membrane biological reactors	Simple, robust, stable, can be applied for both large and small groups	Energy-intensive, expensive, skilled manpower, fouling, high O&M	Wilas et al., 2016; Li and Yang, 2018
		Waste stabilization ponds/lagoons	Efficient, low O&M cost, less technical manpower	Large area, odor, unstable pH may affect bacterial growth	Wilas et al., 2016; Sanborn et al., 2018
		Trickling filters	Cost-effective, smaller area, easy operation, less sludge	Clogging due to biomass growth, high O&M cost	Bressani-Ribeiro et al., 2018
		Rotating biological contactors	Low maintenance, do not need skilled workers, VOC and ammonium elimination	Require energy, low efficiency, less pathogen removal, reactor clogging	Ghodeif, 2013; Ravi et al., 2015
		Fluidized-bed reactor	Shorter HRT, appropriate for microbial growth, high biomass retention, mass transfer rates	Energy requirements, high O&M cost; needs time for biofilm development	Papirio et al., 2013; Özkaya et al., 2019

Anaerobic processes	Up-flow anaerobic sludge blanket	Cost-effective, efficient, compact design, do not need motorized stirring	Complex management, toxicity can influence microbial growth	Li and Yang, 2018; Tara et al., 2019
	Anaerobic/facultative ponds	Effluent reuse without disinfection, sustainable, low cost	Large area, pH can inhibit bacterial growth	Butler et al., 2017
	Anaerobic filters	Low cost, efficient, smaller area, reliable, easy operation, manage hydraulic shocks	Costly due to powered devices, filter blockage	Wilas et al., 2016;
	Membrane separation anaerobic treatment	Energy production, remove trace organic contaminants, low energy requirement, less sludge	Need expertise, O&M cost, low nutrient removal, increase salinity	Hu et al., 2018; Song et al., 2018
Chemical processes	Coagulation-flocculation	Less settling time, simple, economical, advances biodegradability	Expertise, contact time, chemical requirement, toxic end-products	Uday et al., 2017
	Chemical precipitation	Simple, low capital cost, efficient for inorganics, rapid	Poor settling and contact time, toxic intermediates	Kurniawan et al., 2006
	Chemical oxidation	Simple, rapid, high effluent quality, able to disinfect, eliminates odor and color	Sludge production, chemical use, toxic intermediates, needs a low-cost and stable catalyst	Li and Yang, 2018; Crini and Lichtfouse, 2019
	Electrochemical processes	Stable, efficient, eliminate color, lower area requirement, no waste management issues	High energy requirement, contact time, produce toxic intermediates	Gupta et al., 2012; Li and Yang, 2018
	Ion-exchange process	High performance, no sludge, recovers valuable metals	High O&M costs, needs expertise, needs pre-treatment	Ullah et al., 2020

TABLE 35.3
Advantages and Disadvantages of Various Processes Utilized in Wastewater Treatment at Tertiary/Advanced Levels

Level	Wastewater Treatment Process	Unit Operations	Advantages	Disadvantages	References
Tertiary/ advanced (Level-4)	Physicochemical processes	Adsorption	Effective, lessen organic matter, high performance, cheap, remove nutrients and micropollutants	Costly due to synthetic adsorbents, clogging of active sites, expertise	Li and Yang, 2018
		Membrane filtration	Can be applied for different wastewaters, no use of chemicals, does not require special apparatus	Sensitivity, high membrane costs, requires expertise and backwashing	Sanborn et al., 2018
		Membrane distillation	Water purification, ammonium removal, desalination of seawater and brackish water	Requires expertise, membrane wetting and fouling, high energy consumption	Alkhudhiri and Hilal, 2018
		Distillation	High purity, able to handle high flow rates, treatment of seawater and brackish to potable water	Slow output, energy-intensive, controlled temperature, costly	Gupta et al., 2012; Ullah et al., 2020
		Solvent extraction	Efficient in the elimination of organics, metal ions, chemicals, oil and grease	Compatibility issues, excess use of chemicals	Gupta et al., 2012
		Advanced oxidation processes	High performance, reliable, simple, rapid, mild reaction, cheap, ease of operation	Chemical use, production of toxic end-products, sludge issues, need expertise	Comninellis et al., 2008; Li and Yang, 2018
		Disinfection	Prevents microbial growth and spread of pathogens, no toxic metabolites	Short life, expensive, high sensitivity	Li and Yang, 2018; Sanborn et al., 2018

matter and non-settleable and settleable solids. For this reason, primary treatment is employed to eliminate these organics and solids by gravitational forces (Patel and Vashi, 2015). Several processes, such as primary sedimentation tanks, flotation systems, and equalization and neutralization tanks, are extensively used in primary treatment. Such treatment is required mainly for wastewater originating from the textile (Tara et al., 2019), pharmaceutical (Sanborn et al., 2018), and pulp and paper industries (Liang et al., 2019). Primary treatment is capable of eliminating more than 25–40% biochemical oxygen demand (BOD), 30–40% total coliforms and 50–70% solids from wastewater (Mara and Horan, 2003), However, these removal percentages do not fulfill the discharge criteria (Pepper and Gerba, 2015). To overcome these limitations, continuous research is required to increase their removal percentage by altering their working and/or replacing them with other effective alternative treatment methods. Several modifications have been reported to improve the efficiency, such as the use of coagulants in sedimentation and dissolved air flotation (DAF) systems

(Scholz, 2015; Pepper and Gerba, 2015). The secondary treatment process is employed to eliminate organics, nutrients, metals and solids, etc. by the action of microorganisms and chemicals. The biological treatment utilizes anaerobic or aerobic processes by attached, suspended or combined microbial growth (Tchobanoglous et al., 2003). Various factors are identified that have a significant impact on the removal processes, such as dissolved oxygen, temperature, pH, types of microorganism, hydraulic retention time (HRT) and presence of toxic materials (Rana et al., 2017). The unit processes employed in chemical treatment comprise chemical oxidation, coagulation, flocculation, ion-exchange, chemical precipitation and electrochemical processes (Crini and Lichtfouse, 2019). Chemical approaches are extensively used for the remediation of wastewater originating from the textile (Dotto et al., 2019) and pharmaceutical industries (Zhan et al., 2019). Nutrients and other contaminants, especially micropollutants, existing in secondary wastewater are eliminated mainly by tertiary treatment. The processes employed in this level comprise adsorption, membrane filtration, solvent extraction, distillation, advanced oxidation, membrane distillation and disinfection (Alkhudhiri and Hilal, 2018; Crini and Lichtfouse, 2019; Zhang et al., 2020). Disinfection and advanced oxidation processes are applied to remove inorganic and organic pollutants (Gupta et al., 2012) and to kill pathogens (Laqbaqbi et al., 2019). The factors responsible for altering the efficiency of the tertiary treatment are surface area, temperature, pH, contact time, technical suitability, cost and concentration of pollutants (Zhang et al., 2020). These unit processes can offer ultra-purity, though their application is inadequate in developing nations due to lack of expertise and high cost involvement (Hai et al., 2018).

35.4 INTEGRATION OF CW TECHNOLOGY IN FUTURE SCHEMES OF UPCOMING SMART CITIES

In recent times, cities and towns have been facing more frequent extreme events like urban floods or extended dry periods. Therefore, it is considered that the ecological services that nature offers to human beings, as well as nature-based solutions that minimize the tremendous costs, should not be ignored (Potschin and Haines-Young, 2011). One of the key areas where CWs can play a crucial role and put forward a fresh approach to manage wastewater and its reuse is the management of stormwater together with municipal wastewater. Management of stormwater runoff is a key challenge in modern cities. Moreover, stormwater also contributes to combined sewer overflow, especially in older cities where runoff and municipal wastewater are mixed and passed through identical pipe systems to a centralized wastewater treatment facility (Barbosa et al., 2012). The generated volume of wastewater is often higher as compared with the available treatment capacity, which means that the additional volume of wastewater is discharged into nearby freshwater streams, creating flood events in urban areas (Mayer et al., 2013). This situation also leads to further pollution of surface waterbodies, particularly damaging the receiving ecosystems. During the floods, wastewater from the sewer washout, sanitary items, concrete surfaces and road runoff poses high loads of nutrients, organics, microbiological contaminants and solids (Gasperi et al., 2010; Stefanakis, 2019). Managing these wastewater volumes is likewise vital to maintain the well-being of the community and the ecosystems and to avoid damage to civic infrastructure and private belongings. A conventional wastewater treatment system working in urban areas requires the execution of centralized treatment units. However, the functioning of such technologies is not considered eco-friendly due to their heavy installations made of non-renewable resources, high energy inputs, utilization of chemicals, and high maintenance and operational costs (Stefanakis et al., 2014). Installing traditional wastewater treatment machinery also degrades the quality of the surrounding environment through aesthetic value and market rate. Therefore, an innovative, alternate environmentally friendly mechanism that can fulfill the necessities of wastewater treatment and hygiene with additional environmental benefits is essential. Together with this, creating green areas in the city's environment to partly compensate for the absence of adequate green areas is also required. Nature-based eco-friendly technologies such as CWs seem like a perfect choice that can offer the desired additional values to ecological services and encourage water circulation in the urban settings (Masi et al., 2018). The applications of CWs together with urban wastewater management are to improve air quality and reduce the impacts of urban heat islands. In recent years, the applications have been further extended to include auxiliary benefits such as providing ecological services, development of new wildlife territories, and reducing the emissions of greenhouse gases to enhance the resilience of urban atmospheres (Kumar and Dutta, 2019b; Stefanakis, 2019). Consequently, the integration of CWs with smart cities is strongly connected to sustainability.

CWs may provide several functions to a community when they are integrated with urban wastewater treatment systems (Masi et al., 2018):

- The polishing of secondary treated wastewater for possible reuse.
- Organics and nutrient recovery.
- Treatment of greywater for local reuse and recreational purposes.
- Conservation of rainwater through treatment and storage.
- Removal of low concentrations of persistent organic pollutants.
- Production of biomass and fertilizer from activated sludge through sludge treatment CWs.
- Integration of CWs with anaerobic reactors also produces biogas together with polishing of wastewater.
- Provides several ecological services such as wildlife habitat, flood control, groundwater recharge, fisheries, aquaculture, carbon sequestration, silt capture, aesthetic values to the surroundings, green areas for walking, and parks and permaculture productive areas for organic food production (DuPoldt et al., 2015; Kumar and Dutta, 2019b).
- CWs may also offer animal protection and heat storage (Stefanakis, 2020).
- CWs develop small oases to bring flora and fauna into that area.
- Protects groundwater from pollution and lessens the dependency on other water resources by providing irrigation to urban agriculture, gardens and trees.
- Awareness among the residents about their water usage.
- CWs offers a low-cost resolution to societies in terms of reducing their sewage cost. The construction and operational costs of CWs are 50–90% lower than those of conventional wastewater treatment systems (Stefanakis, 2020).
- They are capable of removing infectious pathogens and contaminants that would usually distress human health and hygiene.

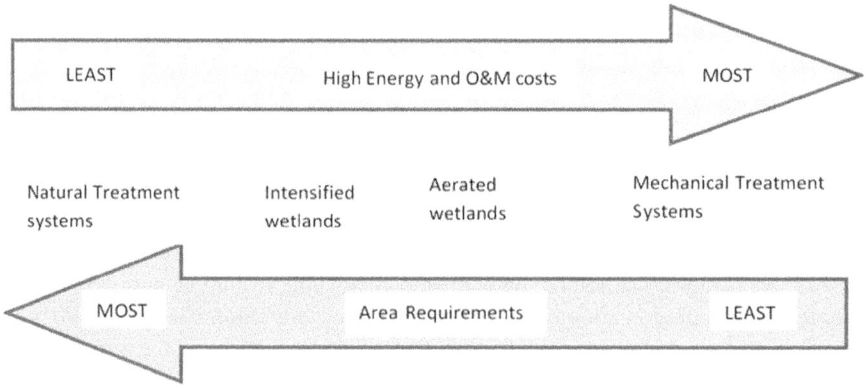

FIGURE 35.1 Conventional wastewater treatment systems versus constructed wetlands.
(Modified from Stefanakis, 2019.)

From these findings, it is considered that the incorporation of CWs as green infrastructure into urban advancement is an excellent opportunity to generate a circular economy and numerous positive influences on the atmosphere (Figure 35.1). However, the integration process for health and economy demands strict collaboration of various professionals such as urban planners, biologists, habitat and biodiversity experts, climatologists, agronomists, sociologists, economists and engineers.

35.5 REUSE OPPORTUNITIES OF TREATED MUNICIPAL WASTEWATER IN URBAN AGRICULTURE

The reuse of municipal wastewater for the irrigation of agricultural crops lessens the dependency on additional fertilizer requirements and available freshwater resources (Jiménez and Asano, 2008). At present, such irrigation is adopted in nearly all arid areas throughout the globe (Edwards and Dudi, 2004). Both untreated and treated wastewater may be utilized in the irrigation of agricultural crops in place of being released into the water streams. Nevertheless, the management of infectious pathogens associated with wastewater irrigation is essential before use to ensure the safety of farmers and food. Since agricultural production utilizes approximately 70% of the total water (Pedrero et al., 2010), the reuse of municipal sewage after treatment for agriculture represents a great choice as an alternative water source (Meda and Cornel, 2010), particularly for nutrient reuse (Norton-Brandão et al., 2013). The reuse of treated municipal wastewater in agriculture is extremely reliant on the characteristics of municipal wastewater, which govern the approaches and level of essential treatment. Besides this, the major problem associated with the reuse of treated water is its effects on the environment and human beings due to the subsistence of pathogens. Treated water quality standards for the irrigation of agricultural crops have been developed by the United States Environmental Protection Agency towards contaminants such as nutrients, heavy metals, chemicals, particles, biodegradable organics and microbes (USEPA, 2012; Alcalde-Sanz and Gawlik, 2017) (Table 35.4). Various studies have reported that the treated water contains plenty of nutrients required for the growth and development of plants (Almuktar et al., 2015; Bonvin et al., 2015). It is observed by Lubello et al. (2004) that the irrigation of plants grown in nurseries by tertiary treated municipal wastewater has no major effects on the environment or human health. Similarly, Rusan et al. (2007) expressed that the growth of cereal crops increased significantly when irrigated with treated wastewater in fields. Several other researchers have claimed that the treated wastewater

TABLE 35.4
Suggested Guidelines for Wastewater Reuse in Agriculture

Crop category	Treatment requirement	BOD	TSS	Turbidity	pH	E. coli	Total coliforms
All food crops whose edible part is not in contact with treated water	Filtration, secondary treatment followed by disinfection	≤10	≤10	≤5	6.0–9.0	≤10	ND
Food crops consumed raw	Secondary treatment and disinfection	≤10	≤10	–	–	≤100	ND
Processed food crops	Secondary treatment and disinfection	≤30	≤30	–	–	≤1,000	≤200
Crops to feed milk- or meat-producing animals	Secondary treatment and disinfection	≤30	≤30	–	–	≤10,000	≤200

BOD = Biochemical Oxygen Demand; ND = Not Detected; TSS = Total Suspended Solids.

Source: Modified from USEPA, 2012; Alcade-Sanz and Gawlik, 2017.

may be reused as nutrients only for potted plants, and proposed that additional fertilizers are also required for sustaining plant growth (Fonseca et al., 2007; Almuktar et al., 2015). It is observed that the growth responses of plants in fields and pots irrigated with treated wastewater varies significantly depending upon wastewater quality and physicochemical characteristics of the growth medium (Barton et al., 2005). This is due to the presence of organic matter, soil microbial activity and nutrient retention capacity, which influence the absorption capacity and nutrient accessibility (Acosta-Martínez et al., 2007). Irrigation of agricultural fields with treated wastewater also enhances the microbial activity within the soil and nutrient recycling (Chen et al., 2008). However, several other studies found that the irrigation of agricultural fields with treated wastewater increases the soil salinity and heavy metal concentration, depending on the characteristics of wastewater (Qian and Mecham, 2005; Prazares et al., 2017). Therefore, an appropriate wastewater treatment system and adaptation of agricultural systems that permits efficient reuse can play a vital role in protection from harmful impacts associated with wastewater irrigation. For this, integration of wastewater remediation together with crop production can diminish the risk of environmental contamination and provide an alternative water source. It is reported that anaerobic digestion used for the remediation of municipal wastewater removes almost all infectious pathogens associated with sludge (Appels et al., 2008). On the field scale, the selection of crops and irrigation methods can also minimize the degree of environmental contamination and protect farmers' health (Allende and Monaghan, 2015). Increasing the time of irrigation and implementation of hygienic practices are suggested to lessen the risks related to the reuse of treated wastewater from infectious pathogens (World Health Organization, 2006; Mateo-Sagasta et al., 2013).

35.6 FUTURE PROSPECTS

The generated volume of wastewater is often higher than the available treatment capacity in many cities and towns. This means that the additional volume of wastewater is discharged into nearby freshwater streams, which can cause contamination of water and flood events in urban areas. Managing these wastewater volumes is likewise vital to maintain the well-being of the community and the ecosystems, and to avoid damage to civic infrastructure and private belongings. The major concern linked with the reuse of treated wastewater in agriculture is its effects on the environment and farmers' health due to the subsistence of infectious pathogens. Management of these pathogens is essential before use to ensure the safety of farmers and food. The risks could be lessened and managed by increasing the time of irrigation and the implementation of hygienic practices. Continuous use of treated water in irrigation also causes severe environmental contamination such as soil salinization, heavy metal accumulation, groundwater pollution, adverse effects on microorganisms' activity and diversity, and alteration of soil structure. The change in soil structure and quality can also affect the growth and expansion of plants, productivity and fruit quality. To address this problem, the selection of appropriate treatment mechanisms and crops is vital.

35.7 CONCLUSION

The existing technologies for wastewater treatment in the urban areas of the majority of developing countries are failing to meet the wastewater discharge standards within surface waterbodies. Therefore, integrated approaches that can protect freshwater resources by the recovery of water and nutrients available in municipal wastewater are gaining significant attention in the field of wastewater management. Integration of CWs as a green infrastructure into urban development for wastewater treatment and also reuse in urban and peri-urban agriculture is an excellent opportunity to generate a circular economy and numerous other environmental benefits. However, the integration process pertaining to public health and the economy demands strict collaboration of various professionals. The use of treated municipal wastewater for the irrigation of agricultural crops could further lessen the dependency on additional fertilizer requirements and available freshwater resources. Applications of CWs together with urban wastewater management are also helpful in reducing the impacts of urban heat islands and improving the water quality of receiving streams. Recently, the applications have been further extended to provide several ecological services, development of new wildlife territories, adding aesthetic value, groundwater recharge, carbon sequestration, aquaculture, flood control and for lessening the emissions of greenhouse gases to enhance the resilience of the urban environment.

ACKNOWLEDGMENT

The authors are thankful to Department of Environmental Science, Babasaheb, Bhimrao Ambedkar University, Lucknow and University Grants Commission, New Delhi for providing laboratory facilities and financial support (Ref. no. 3525/SC/NET-JULY 2016) to the first author throughout this study.

REFERENCES

Acosta-Martínez, V., Cruz, L., Sotomayor-Ramirez, D., & Pérez-Alegría, L. (2007). Enzyme activities as affected by soil properties and land use in a tropical watershed. *Applied Soil Ecology*, 35(1), 35–45.

Adewoye, A. O., Okunade, D. A., & Adekalu, K. O. (2010). Assessing the yields and nutrient uptake of okra *Abelmoschus esculentus* using diluted stabilized wastewater for irrigation in South-Western Nigeria. *Journal of Waste Water Treatment & Analysis* 1, 104.

Ahmed, S. D., Agodzo, S. K., & Adjei, K. A. (2020). Designing River Diversion Constructed Wetland for Water Quality Improvement. In: Devlin, A., Pan, J., & Shah, M. M. (Eds.) *Inland Waters-Dynamics and Ecology*, Intech Open, London, United Kingdom, 55–72.

Akpor, O. B., & Muchie, M. (2010). Bioremediation of polluted wastewater influent: phosphorus and nitrogen removal. *Scientific Research and Essays*, 5(21), 3222–3230.

Alcalde-Sanz, L., & Gawlik, B. M. (2017). Minimum quality requirements for water reuse in agricultural irrigation and aquifer recharge. *Towards a Legal Instrument on Water Reuse at EU Level.* EUR 28962 EN, Publications Office of the European Union, Luxembourg, *JRC Publications Repository*, doi: 10.2760/887727.

Alikhasi, M., Kouchakzadeh, M., & Baniani, E. (2012). The effect of treated municipal wastewater irrigation in non-agricultural soil on cotton plant. *Journal of Agricultural Science and Technology*, 14(6), 1357–1364.

Alkhudhiri, A., & Hilal, N. (2018). Membrane distillation—principles, applications, configurations, design, and implementation. In: Veera, G. G. (Ed.) *Emerging Technologies for Sustainable Desalination Handbook,* Elsevier, Amsterdam, Netherlands, 55–106.

Allende, A., & Monaghan, J. (2015). Irrigation water quality for leafy crops: a perspective of risks and potential solutions. *International Journal of Environmental Research and Public Health*, 12(7), 7457–7477.

Almuktar, S. A. A. A. N., & Scholz, M. (2016). Mineral and biological contamination of soil and *Capsicum annuum* irrigated with recycled domestic wastewater. *Agricultural Water Management*, 167, 95–109.

Almuktar, S. A. A. A. N., Scholz, M., Al-Isawi, R. H. K., & Sani, A. (2015). Recycling of domestic wastewater treated by vertical-flow wetlands for irrigating chilies and sweet peppers. *Agricultural Water Management,* 149, 1–22.

Appels, L., Baeyens, J., Degrève, J., & Dewil, R. (2008). Principles and potential of the anaerobic digestion of waste-activated sludge. *Progress in Energy and Combustion Science*, 34(6), 755–781.

Aziz, O., Inam, A., Samiullah, & Siddiqi, R. H. (1996). Long term effects of irrigation with petrochemical industry wastewater. *Journal of Environmental Science & Health Part A,* 31(10), 2595–2620.

Barbosa, A. E., Fernandes, J. N., & David, L. M. (2012). Key issues for sustainable urban stormwater management. *Water Research*, 46(20), 6787–6798.

Barton, L., Schipper, L. A., Barkle, G. F., McLeod, M., Speir, T. W., Taylor, M. D., ... & Pandey, S. P. (2005). Land application of domestic effluent onto four soil types: plant uptake and nutrient leaching. *Journal of Environmental Quality*, 34(2), 635–643.

Bawiec, A., Pawęska, K., & Pulikowski, K. (2016). Seasonal changes in the reduction of biogenic compounds in wastewater treatment plants based on hydroponic technology. *Journal of Ecological Engineering*, 17(2), 128–134.

Bhatia, M., & Goyal, D. (2014). Analyzing remediation potential of wastewater through wetland plants: a review. *Environmental Progress & Sustainable Energy*, 33(1), 9–27. doi:10.1002/ep.11822.

Bonvin, C., Etter, B., Udert, K. M., Frossard, E., Nanzer, S., Tamburini, F., & Oberson, A. (2015). Plant uptake of phosphorus and nitrogen recycled from synthetic source-separated urine. *Ambio*, 44(2), 217–227.

Bressani-Ribeiro, T., Almeida, P. G. S., Volcke, E. I. P., & Chernicharo, C. A. L. (2018). Trickling filters following anaerobic sewage treatment: state of the art and perspectives. *Environmental Science: Water Research & Technology*, 4(11), 1721–1738.

Butler, E., Hung, Y. T., Al Ahmad, M. S., Yeh, R. Y. L., Liu, R. L. H., & Fu, Y. P. (2017). Oxidation pond for municipal wastewater treatment. *Applied Water Science*, 7(1), 31–51.

Chen, W., Wu, L., Frankenberger, W. T., & Chang, A. C. (2008). Soil enzyme activities of long-term reclaimed wastewater-irrigated soils. *Journal of Environmental Quality*, 37(S5), S-36.

Choudhary, A. K., & Kumar, P. (2020). Constructed wetland: a green technology for wastewater treatment. In: Singh, A., Srivastava, S., Rathore, D., & Pant, D. (Eds.) *Environmental Microbiology and Biotechnology,* Springer, Singapore, 335–363.

Comninellis, C., Kapalka, A., Malato, S., Parsons, S. A., Poulios, I., & Mantzavinos, D. (2008). Advanced oxidation processes for water treatment: advances and trends for R&D. *Journal of Chemical Technology & Biotechnology: International Research in Process, Environmental & Clean Technology*, 83(6), 769–776.

Correia, A., Huynh, Q., Swan, K., Bar-on, I., Tuler, S., Cicelsky, A., … & Lotan, K. (2020). Decentralized Wastewater Treatment in Southern Israel. Doctoral dissertation, *Worcester Polytechnic Institute*.

Crini, G., & Lichtfouse, E. (2019). Advantages and disadvantages of techniques used for wastewater treatment. *Environmental Chemistry Letters*, 17(1), 145–155.

Dell'Osbel, N., Colares, G. S., Oliveira, G. A., Rodrigues, L. R., da Silva, F. P., Rodriguez, A. L., … & Machado, Ê. L. (2020). Hybrid constructed wetlands for the treatment of urban wastewaters: increased nutrient removal and landscape potential. *Ecological Engineering,* 158, 106072.

Dotro, G., Langergraber, G., Molle, P., Nivala, J., Puigagut, J., Stein, O., & Von Sperling, M. (2017). *Treatment Wetlands*. IWA Publishing, London, UK.

Dotto, J., Fagundes-Klen, M. R., Veit, M. T., Palácio, S. M., & Bergamasco, R. (2019). Performance of different coagulants in the coagulation/flocculation process of textile wastewater. *Journal of Cleaner Production,* 208, 656–665.

DuPoldt, C., Edwards, R., Garber, L., Isaacs, B., Lapp, J., Murphy, T., & Harold. (2015). *A Handbook of Constructed Wetlands: A Guide to Creating Wetlands for Agricultural Wastewater, Domestic Wastewater, Coal Mine Drainage, Stormwater in the Mid-Atlantic Region. General Considerations*. US EPA, Washington, DC.

Edwards, M., & Dudi, A. (2004). Role of chlorine and chloramine in corrosion of lead-bearing plumbing materials. *American Water Works Association*, 96(10), 69–81.

Fito, J., & Van Hulle, S. W. (2020). Wastewater reclamation and reuse potentials in agriculture: towards environmental sustainability. *Environment, Development and Sustainability*, 23(3), 2949–2972.

Fonseca, A. F. D., Herpin, U., Paula, A. M. D., Victória, R. L., & Melfi, A. J. (2007). Agricultural use of treated sewage effluents: agronomic and environmental implications and perspectives for Brazil. *Scientia Agricola*, 64(2), 194–209.

Gasperi, J., Gromaire, M. C., Kafi, M., Moilleron, R., & Chebbo, G. (2010). Contributions of wastewater, runoff and sewer deposit erosion to wet weather pollutant loads in combined sewer systems. *Water Research*, 44(20), 5875–5886.

Gebeyehu, A., Shebeshe, N., Kloos, H., & Belay, S. (2018). Suitability of nutrients removal from brewery wastewater using a hydroponic technology with *Typha latifolia*. *BMC Biotechnology*, 18(1), 74.

Ghodeif, K. (2013). Baseline Assessment Study for Wastewater Treatment Plant for Al Gozayyera Village, West Kantara City, Ismailia Governorate, Egypt. Network of Demonstration Activities for Sustainable Integrated Wastewater Treatment and Reuse in the Mediterranean, Cairo, Egypt.

Goel, R. K., Flora, J. R., & Chen, J. P. (2005). Flow equalization and neutralization. In: Wang, L. K., Hung, Y. T., Shammas, N. K. (Eds.) *Physicochemical Treatment Processes. Handbook of Environmental Engineering*. Humana Press, New Jersey, United States. 3, 21–45.

Gupta, V. K., Ali, I., Saleh, T. A., Nayak, A., & Agarwal, S. (2012). Chemical treatment technologies for wastewater recycling—an overview. *RSC Advances*, 2(16), 6380–6388.

Ha, N. T. H., Sakakibara, M., & Sano, S. (2011). Accumulation of indium and other heavy metals by *Eleocharis acicularis*: an option for phytoremediation and phyto mining. *Bioresource Technology*, 102(3), 2228–2234.

Hai, F. I., Yang, S., Asif, M. B., Sencadas, V., Shawkat, S., Sanderson-Smith, M., … & Yamamoto, K. (2018). Carbamazepine as a possible anthropogenic marker in water: occurrences, toxicological effects, regulations and removal by wastewater treatment technologies. *Water*, 10(2), 107.

Hu, Y., Wang, X. C., Ngo, H. H., Sun, Q., & Yang, Y. (2018). Anaerobic dynamic membrane bioreactor (AnDMBR) for wastewater treatment: a review. *Bioresource Technology,* 247, 1107–1118.

IPCC (2013). Climate Change 2013: The Physical Science Basis. In: Stocker, T. F., Qin, D., Plattner, G. K., Tignor, M., Allen, S. K., Boschung, J., Nauels, A., Xia, Y., Bex, V., Midgley, P. M. (Eds.) *Contribution of Working Group I to the Fifth Assessment Report of the Intergovernmental Panel on Climate Change*, vol. 2013 (p. 1535). Cambridge University Press, Cambridge, United Kingdom and New York, NY, USA.

Jia, Z., Tang, S., Luo, W., & Hai, Y. (2016). Water quality improvement through five constructed serial wetland cells and its implications on nonpoint-source pollution control. *Hydrological Sciences Journal*, 61(16), 2946–2956.

Jiang, L., Bao, A., Yuan, Y., Zheng, G., Guo, H., Yu, T., & De Maeyer, P. (2020). The effects of water stress on croplands in the Aral Sea basin. *Journal of Cleaner Production,* 254, 120114.

Jiménez, B., & Asano, T. (Eds.) (2008). *Water Reuse: An International Survey of Current Practice, Issues and Needs*. IWA, London.

Kamali, M., Costa, M. E., Aminabhavi, T. M., & Capela, I. (2019). Sustainability of treatment technologies for industrial biowastes effluents. *Chemical Engineering Journal,* 370, 1511–1521.

Kampschreur, M. J., Temmink, H., Kleerebezem, R., Jetten, M. S., & van Loosdrecht, M. C. (2009). Nitrous oxide emission during wastewater treatment. *Water Research*, 43(17), 4093–4103.

Ko, C. H., Lee, T. M., Chang, F. C., & Liao, S. P. (2011). The correlations between system treatment efficiencies and aboveground emergent macrophyte nutrient removal for the Hsin-Hai Bridge phase II constructed wetland. *Bioresource Technology*, 102(9), 5431–5437.

Kumar, S., & Dutta, V. (2019a). Efficiency of constructed wetland microcosms (CWMs) for the treatment of domestic wastewater using aquatic macrophytes. In: Sobti, R. C., Arora, N. K., & Kothari, R. (Eds.) Environmental Biotechnology: *For Sustainable Future*. Springer, Singapore, 287–307.

Kumar, S., & Dutta, V. (2019b). Constructed wetland microcosms as sustainable technology for domestic wastewater treatment: an overview. *Environmental Science and Pollution Research*, 26(12), 11662–11673.

Kumar, S., Nand, S., Dubey, D., Pratap, B., & Dutta, V. (2020a). Variation in extracellular enzyme activities and their influence on the performance of surface-flow constructed wetland microcosms (CWMs). *Chemosphere,* 251, 126377.

Kumar, S., Pratap, B., Dubey, D., & Dutta, V. (2020b). Microbial communities in constructed wetland microcosms and their role in treatment of domestic wastewater. In: Bharagava, R. (Ed.) *Emerging Eco-friendly Green Technologies for Wastewater Treatment.* Springer, Singapore, 311–327.

Kumar, S., Pratap, B., Dubey, D., & Dutta, V. (2020c). Removal of nutrients from domestic wastewater using constructed wetlands: assessment of suitable environmental and operational conditions. *Environmental Sustainability,* 3(4), 341–352.

Kurniawan, T. A., Chan, G. Y., Lo, W. H., & Babel, S. (2006). Physico–chemical treatment techniques for wastewater laden with heavy metals. *Chemical Engineering Journal*, 118(1–2), 83–98.

Lakho, F. H., Le, H. Q., Mattheeuws, F., Igodt, W., Depuydt, V., Desloover, J., … & Van Hulle, S. W. (2020). Decentralized grey and black water reuse by combining a vertical flow constructed wetland and membrane based potable water system: full scale demonstration. *Journal of Environmental Chemical Engineering,* 9(1), 104688.

Laqbaqbi, M., García-Payo, M. C., Khayet, M., El Kharraz, J., & Chaouch, M. (2019). Application of direct contact membrane distillation for textile wastewater treatment and fouling study. *Separation and Purification Technology,* 209, 815–825.

Laxmeshwar, L. S., Jadhav, M. S., Akki, J. F., Raikar, P., & Raikar, U. S. (2018). Elemental analysis of wastewater effluent using highly sensitive fiber Bragg grating sensor. *Optics & Laser Technology,* 105, 45–51.

Li, J., Cheng, W., Xu, L., Strong, P. J., & Chen, H. (2015). Antibiotic-resistant genes and antibiotic-resistant bacteria in the effluent of urban residential areas, hospitals, and a municipal wastewater treatment plant system. *Environmental Science and Pollution Research*, 22(6), 4587–4596.

Li, Z., & Yang, P. (February 2018). Review on physicochemical, chemical, and biological processes for pharmaceutical wastewater. In *IOP Conference Series: Earth and Environmental Science*, 113, 012185.

Liang, L., Shi, Y. Q., Fang, G., Pan, A., Tian, Q., & Ran, M. (2019). The changes of bamboo pulping waste water residuals in biological treatment process. *Water and Environment Journal*, 33(1), 14–20.

Libralato, G., Ghirardini, A. V., & Avezzù, F. (2012). To centralise or to decentralise: an overview of the most recent trends in wastewater treatment management. *Journal of Environmental Management*, 94(1), 61–68.

Liu, L., Liu, Y. H., Liu, C. X., Wang, Z., Dong, J., Zhu, G. F., & Huang, X. (2013). Potential effect and accumulation of veterinary antibiotics in *Phragmites australis* under hydroponic conditions. *Ecological Engineering,* 53, 138–143.

Lubello, C., Gori, R., Nicese, F. P., & Ferrini, F. (2004). Municipal-treated wastewater reuse for plant nurseries irrigation. *Water Research*, 38(12), 2939–2947.

Lv, X., Dong, Q., Zuo, Z., Liu, Y., Huang, X., & Wu, W. M. (2019). Microplastics in a municipal wastewater treatment plant: fate, dynamic distribution, removal efficiencies, and control strategies. *Journal of Cleaner Production,* 225, 579–586.

Magwaza, S. T., Magwaza, L. S., Odindo, A. O., & Mditshwa, A. (2020). Hydroponic technology as decentralised system for domestic wastewater treatment and vegetable production in urban agriculture: a review. *Science of the Total Environment,* 698, 134154.

Mara, D. & Horan, N.J., 2003. *Handbook of Water and Wastewater Microbiology*. *Elsevier*, London, UK,.

Martín, M., Gargallo, S., Hernández-Crespo, C., & Oliver, N. (2013). Phosphorus and nitrogen removal from tertiary treated urban wastewaters by a vertical flow constructed wetland. *Ecological Engineering,* 61, 34–42.

Masi, F., Rizzo, A., & Regelsberger, M. (2018). The role of constructed wetlands in a new circular economy, resource oriented, and ecosystem services paradigm. *Journal of Environmental Management,* 216, 275–284.

Matamoros, V., & Salvadó, V. (2012). Evaluation of the seasonal performance of a water reclamation pond-constructed wetland system for removing emerging contaminants. *Chemosphere*, 86(2), 111–117.

Mateo-Sagasta, J., Medlicott, K., Qadir, M., Raschid-Sally, L., & Drechsel, P. (2013). *Proceedings of the UN-water Project on the Safe Use of Wastewater in Agriculture*. United Nations University. UN-Water Decade Programme on Capacity Development (UNW-DPC).

Meda, A., & Cornel, P. (2010). Aerated biofilter with seasonally varied operation modes for the production of irrigation water. *Water Science and Technology*, 61(5), 1173–1181.

Moreira, F. D., & Dias, E. H. O. (2020). Constructed wetlands applied in rural sanitation: a review. *Environmental Research*, 190, 110016.

Musazura, W., Odindo, A. O., Bame, I. B., & Tesfamariam, E. H. (2015). Effect of irrigation with anaerobic baffled reactor effluent on Swiss chard (*Beta vulgaris* cicla.) yield, nutrient uptake and leaching. *Journal of Water Reuse and Desalination*, 5(4), 592–609.

Ndulini, S. F., Sithole, G. M., & Mthembu, M. S. (2018). Investigation of nutrients and faecal coliforms removal in wastewater using a hydroponic system. *Physics and Chemistry of the Earth, Parts A/B/C,* 106, 68–72.

Norton-Brandão, D., Scherrenberg, S. M., & van Lier, J. B. (2013). Reclamation of used urban waters for irrigation purposes—a review of treatment technologies. *Journal of Environmental Management,* 122, 85–98.

Ong, S. A., Uchiyama, K., Inadama, D., Ishida, Y., & Yamagiwa, K. (2010). Performance evaluation of laboratory scale up-flow constructed wetlands with different designs and emergent plants. *Bioresource Technology*, 101(19), 7239–7244.

Oraeki, T., Skouteris, G., & Ouki, S. (2018). Optimization of coagulation-flocculation process in the treatment of wastewater from the brick-manufacturing industry. *Water Practice & Technology*, 13(4), 780–793.

Ottoson, J., Norström, A., & Dalhammar, G. (2005). Removal of micro-organisms in a small-scale hydroponics wastewater treatment system. *Letters in Applied Microbiology*, 40(6), 443–447.

Özkaya, B., Kaksonen, A. H., Sahinkaya, E., & Puhakka, J. A. (2019). Fluidized bed bioreactor for multiple environmental engineering solutions. *Water Research,* 150, 452–465.

Papirio, S., Villa-Gomez, D. K., Esposito, G., Pirozzi, F., & Lens, P. N. L. (2013). Acid mine drainage treatment in fluidized-bed bioreactors by sulfate-reducing bacteria: a critical review. *Critical Reviews in Environmental Science and Technology*, 43(23), 2545–2580.

Patel, H., & Vashi, R. T. (2015). *Characterization and Treatment of Textile Wastewater*. Elsevier, Oxford, UK.

Pedrero, F., Kalavrouziotis, I., Alarcón, J. J., Koukoulakis, P., & Asano, T. (2010). Use of treated municipal wastewater in irrigated agriculture—review of some practices in Spain and Greece. *Agricultural Water Management*, 97(9), 1233–1241.

Pedro-Monzonís, M., Solera, A., Ferrer, J., Estrela, T., & Paredes-Arquiola, J. (2015). A review of water scarcity and drought indexes in water resources planning and management. *Journal of Hydrology,* 527, 482–493.

Peng, Y., & Zhu, G. (2006). Biological nitrogen removal with nitrification and denitrification via nitrite pathway. *Applied Microbiology and Biotechnology*, 73(1), 15–26.

Pepper, I. L., & Gerba, C. P. (2015). Environmental sample collection and processing. In: Pepper, I. L., Gerba, C. P., & Gentry, J. T. (Eds.) *Environmental Microbiology*), Academic Press, 157–175.

Pittman, J. K., Dean, A. P., & Osundeko, O. (2011). The potential of sustainable algal biofuel production using wastewater resources. *Bioresource Technology*, 102(1), 17–25.

Potschin, M. B., & Haines-Young, R. H. (2011). Ecosystem services: exploring a geographical perspective. *Progress in Physical Geography,* 35(5), 575–594.

Prazeres, A. R., Albuquerque, A., Luz, S., Jerónimo, E., & Carvalho, F. (2017). Hydroponic system: a promising biotechnology for food production and wastewater treatment. In: Grumezescu, M. A., & Holban, M. A. (Eds.) *Food Biosynthesis.* Academic Press, 317–350.

Prazeres, A. R., Carvalho, F., Rivas, J., Patanita, M., & Dôres, J. (2014). Reuse of pretreated cheese whey wastewater for industrial tomato production (*Lycopersicon esculentum* Mill.). *Agricultural Water Management,* 140, 87–95.

Qian, Y. L., & Mecham, B. (2005). Long-term effects of recycled wastewater irrigation on soil chemical properties on golf course fairways. *Agronomy Journal*, 97(3), 717–721.

Rana, R. S., Singh, P., Kandari, V., Singh, R., Dobhal, R., & Gupta, S. (2017). A review on characterization and bioremediation of pharmaceutical industries' wastewater: an Indian perspective. *Applied Water Science* 7, 1–12.

Ravi, R., Philip, L., & Swaminathan, T. (2015). Modified rotating biological contactor for removal of dichloromethane vapors. *Environmental Technology*, 36(5), 566–572.

Rusan, M. J. M., Hinnawi, S., & Rousan, L. (2007). Long term effect of wastewater irrigation of forage crops on soil and plant quality parameters. *Desalination*, 215(1–3), 143–152.

Saeed, T., & Sun, G. (2012). A review on nitrogen and organics removal mechanisms in subsurface flow constructed wetlands: dependency on environmental parameters, operating conditions and supporting media. *Journal of Environmental Management,* 112, 429–448.

Sanborn, M., McGee, T., Macdonell, M., & Kosch, S. (2018). Apparatus for treating pharmaceutical waste. U.S. Patent No. 10,046,993. U.S. Patent and Trademark Office, Washington, DC, United States.

Scholz, M. (2015). *Wetland Systems to Control Urban Runoff.* Elsevier, Oxford, UK.

Sheets, J. P., Yang, L., Ge, X., Wang, Z., & Li, Y. (2015). Beyond land application: emerging technologies for the treatment and reuse of anaerobically digested agricultural and food waste. *Waste Management,* 44, 94–115.

Song, X., Luo, W., Hai, F. I., Price, W. E., Guo, W., Ngo, H. H., & Nghiem, L. D. (2018). Resource recovery from wastewater by anaerobic membrane bioreactors: opportunities and challenges. *Bioresource Technology,* 270, 669–677.

Stefanakis, A. I. (2019). The role of constructed wetlands as green infrastructure for sustainable urban water management. *Sustainability*, 11(24), 6981.

Stefanakis, A. I. (2020). Constructed wetlands: description and benefits of an eco-tech water treatment system. In: Information Resources Management Association (Eds.) *Waste Management: Concepts, Methodologies, Tools, and Applications*. IGI Global, USA, 503–525.

Stefanakis, A., Akratos, C. S., & Tsihrintzis, V. A. (2014). *Vertical Flow Constructed Wetlands: Eco-engineering Systems for Wastewater and Sludge Treatment*. Elsevier, Oxford, UK.

Sudarsan, J. S., Roy, R. L., Baskar, G., Deeptha, V. T., & Nithiyanantham, S. (2015). Domestic wastewater treatment performance using constructed wetland. *Sustainable Water Resources Management,* 1(2), 89–96.

Tangahu, B. V., Sheikh Abdullah, S. R., Basri, H., Idris, M., Anuar, N., & Mukhlisin, M. (2011). A review on heavy metals (As, Pb, and Hg) uptake by plants through phytoremediation. *International Journal of Chemical Engineering*, 2011, 939161, 1–31.

Tara, N., Arslan, M., Hussain, Z., Iqbal, M., Khan, Q. M., & Afzal, M. (2019). On-site performance of floating treatment wetland macrocosms augmented with dye-degrading bacteria for the remediation of textile industry wastewater. *Journal of Cleaner Production,* 217, 541–548.

Tchobanoglous, G., Burton, F., & Stensel, H. D. (2003). Wastewater engineering: treatment and reuse. *American Water Works Association Journal*, 95(5), 201.

Travis, M. J., Wiel-Shafran, A., Weisbrod, N., Adar, E., & Gross, A. (2010). Greywater reuse for irrigation: effect on soil properties. *Science of the Total Environment*, 408(12), 2501–2508.

Uday, U. S. P., Mahata, N., Sasmal, S., Bandyopadhyay, T. K., Mondal, A., & Bhunia, B. (2017). Dyes contamination in the environment: ecotoxicological effects, health hazards, and biodegradation and bioremediation mechanisms for environmental cleanup. In: Bharagava, R.N. (Ed.) *Environmental Pollutants and their Bioremediation Approaches* (pp. 127–176). CRC Press, London, UK.

Ullah, A., Hussain, S., Ahmad, W., & Jahanzaib, M. (2020). Development of a decision support system for the selection of wastewater treatment technologies. *Science of the Total Environment*, 731, 139158.

USEPA. (2012). *Guidelines for Water Reuse AR-1530 EPA/600/R-12/618.* US Environmental Protection Agency National Environmental Research Centre.

Von Sperling, M. (2007). *Basic Principles of Wastewater Treatment* (p. 210). *IWA Publishing*, London, UK.

Vymazal, J. (2007). Removal of nutrients in various types of constructed wetlands. *Science of the Total Environment*, 380(1–3), 48–65.

Westerhoff, P., Sharif, F., Halden, R., Herckes, P., & Krajmalnik-Brown, R. (2014). *Constructed Wetlands for Treatment of Organic and Engineered Nanomaterial Contaminants of Emerging Concerns.* Water Research Foundation, Denver, CO.

Wilas, J., Draszawka-Bołzan, B., Daniszewski, P., & Cyraniak, E. (2016). Wastewater treatment technologies. *World News of Natural Sciences,* 4, 33–43.

World Health Organization. (2006). *WHO Guidelines for the Safe Use of Wastewater Excreta and Greywater* (Vol. 1). World Health Organization.

World Resources Institute. (August 26, 2015). *Ranking the World's Most Water-Stressed Countries.* Retrieved from www.wri.org/blog/2015/08/ranking-world-s-most-water-stressed-countries-2040.

Zanil, M. F., Norhuda, A. M., Hussain, M. A., & Omar, R. (2014). Hybrid model of pH neutralization for a pilot plant. *Journal of Intelligent & Fuzzy Systems*, 26(2), 551–561.

Zhan, J., Li, Z., Yu, G., Pan, X., Wang, J., Zhu, W., … & Wang, Y. (2019). Enhanced treatment of pharmaceutical wastewater by combining three-dimensional electrochemical process with ozonation to in situ regenerate granular activated carbon particle electrodes. *Separation and Purification Technology,* 208, 12–18.

Zhang, X., Zhang, Y., Ngo, H. H., Guo, W., Wen, H., Zhang, D., … & Qi, L. (2020). Characterization and sulfonamide antibiotics adsorption capacity of spent coffee grounds-based biochar and hydrochar. *Science of the Total Environment,* 716, 137015.

Index

Note: Page numbers in *italics* refer to figures; those in **bold** refer to tables.

B

C

E

G

H

I

J

K

L

M

O

T

For Product Safety Concerns and Information please contact our EU
representative GPSR@taylorandfrancis.com
Taylor & Francis Verlag GmbH, Kaufingerstraße 24, 80331 München, Germany

www.ingramcontent.com/pod-product-compliance
Ingram Content Group UK Ltd.
Pitfield, Milton Keynes, MK11 3LW, UK
UKHW052217190926
475702UK00006B/1054